Society, Culture, and Drinking Patterns Reexamined

Society, Culture, and Drinking Patterns Reexamined

EDITED BY
David J. Pittman
and
Helene Raskin White

Publications Division
Rutgers Center of Alcohol Studies
New Brunswick, New Jersey USA

This book is the first in the Alcohol, Culture, and Social Control Monograph Series, funded in part by the Wine Institute. Editors of the series are David J. Pittman and Helene Raskin White.

ISBN 911290–21–4 (cloth)
ISBN 911290–22–2 (paper)

Library of Congress Catalog Card Number: 91–061680

Printed in the United States of America

To Bennett Barnes
and
Sidney and Madeleine Raskin

Contents

SECTION V: RESPONSIVE MOVEMENTS AND SYSTEMS OF CONTROL

Foreword

WE LIVE in a time of renewed public concern about drinking and the problems of alcohol. The rise of a new temperance movement paralleling the outbreak of a national war on drugs is but one manifestation of this. Thinking people these days seem generally to appreciate that the use of alcoholic beverages is somehow socially and culturally patterned and that pathologies like alcoholism cannot be explained entirely by individualistically phrased psychologies or even genetics. It is one thing, however, to affirm in generalities the influence, even preeminence, of social and cultural factors in shaping drinking behavior and quite another to depict concretely and in depth just what this involves. Yet it is precisely such a firmly grounded understanding that David Pittman and Helene White, the editors of this welcome new book, have sought to bring us. They have done so by carefully choosing and knitting together in a single well-conceived volume much of the best of the heretofore widely scattered social scientific research on the use and abuse of alcoholic beverages.

The occasion for the reexamination noted in the title of Professors Pittman and White's book was provided, as some readers will recognize, by an earlier, formative work simply called *Society, Culture, and Drinking Patterns,* which was first published in 1962. This, too, was intended to give substance to the social and cultural aspects of drinking, but for a different period. It emerged on the crest of an extraordinary wave of pioneering research as social scientists of the 1940s and 1950s opened up, on an unprecedented scale, the systematic investigation of drinking behavior. After an initial creative surge, however, the pace of significant development in research slackened. The original volume settled comfortably into the scientific literature, eventually becoming known as the bible of social science research on drinking patterns.

But there can be no closed canon in science, and since the appearance of that first volume the question among interested scholars has never been whether a new version might someday be in order but rather when the fruits of the field would be sufficiently ripe to justify such an undertaking. In my judgment that time has come and for a variety of reasons, a few of which may be mentioned here.

In the first place, nearly 3 decades have elapsed since 1962 and, notwithstanding slow development at the outset, the accumulation of new and, on the face of it, worthwhile studies is virtually overwhelming.

These investigations and their findings, often lying unnoticed in specialized journals, need not only to be sorted out and evaluated individually but also to be organized and juxtaposed in a coherent and sociologically relevant way if we are to assess their collective meaning and experience their full impact.

In the second place, the lapse of time and accumulation of research have been sufficient by now to enable tests of hypotheses and theories, as well as to allow for the introduction of new research techniques and conceptual schemes quite beyond the bounds of possibility earlier on. Nowhere is this state of affairs more evident than in regard to proposed explanations of the etiology of alcoholism.

In the third place, society itself and the perspectives of social scientists have changed significantly over the last generation or two so as to make salient topics that were largely neglected or altogether absent from the research informing the 1962 publication. This change is plain to see, for example, in the growing interest of researchers in the drinking patterns of women, of the elderly, or of certain minorities, as well as in the temperance movement, which has recently emerged on the social scene in a new form.

Surely considerations such as these lay behind Professors Pittman and White's decision to move ahead with a new volume bearing a title resonant with that of the 1962 book and inviting us through reexamination to discover the newer realities of society, culture, and drinking patterns. It cannot be overemphasized that theirs is, on the one hand, a fundamentally new book and not just a minor revision of the earlier volume. The bulk of the chapters consist of recent studies carefully culled from the scientific literature or, in some instances, specially commissioned to fill major gaps. These are all studies that in one way or another either break new ground or revise or extend previous research in important ways. On the other hand, Professors Pittman and White have thoughtfully preserved features of the earlier volume that weathered well. The focus, for instance, remains as before primarily on sociological and social-psychological research in the United States, while the comparative and anthropological dimension of the original is also preserved and critically developed. With good reason, too, a few chapters and parts of introductory notes from the 1962 volume have been reprinted along with a couple of classics from the older literature of alcohol studies. Most important, the overall framework of the earlier volume has been kept essentially intact. For it is this sociological framework that helps locate the drinking of alcoholic beverages in those larger social structures, cultural configurations, and processes crucial to understanding but so often lost to view.

The upshot is a remarkably comprehensive and stimulating new book about drinking patterns seen from the perspectives of social scientists. It is a work that encourages comparison of the more important studies of the past and present, exhibits major trends in research over the span of three decades, draws together and highlights the main findings of that research, and suggests the lines along which future investigation ought logically to proceed. This is a book of interest to social scientists, but it should be welcomed as well by students, educators, health professionals, public servants, policymakers, and all citizens who want to deepen their understanding of drinking behavior and the problems of alcohol as social and cultural phenomena.

Charles R. Snyder
Carbondale, Illinois
1991

Preface

OVER THREE DECADES ago a number of social science researchers in the field of alcohol studies found themselves associated in the Committee on Alcoholism of the Society for the Study of Social Problems (SSSP). In those days, the study of drinking behavior in general and alcoholism in particular was viewed with some degree of concern, not only by the lay but by the scientific community—the former because Prohibition (1920–1933) was fresh in their memories and the latter because the stigma that was attached to alcoholics had partially spoiled the identity of alcohol research studies. Thus, in 1956–57 the Committee on Alcoholism of the SSSP deemed it desirable and feasible to prepare "a collection of social science studies centering primarily on patterns of drinking alcoholic beverages and more narrowly on alcoholism" (Pittman & Snyder, 1962, p. ix). That effort resulted in the publication of the volume edited by David J. Pittman and Charles R. Snyder, *Society, Culture, and Drinking Patterns* (1962), composed of 35 chapters, of which 12 were previously published and 23 were original works. The book was well received by social scientists and was later referred to as the "social science 'bible' of research and writing on drinking behavior" (Conrad & Schneider, 1980, p. 109).

Both Charles Snyder and David Pittman considered editing a revised edition of the volume during the late 1970s and early 1980s; it was apparent to them and others in the field of alcohol studies that there had been a literal explosion of research investigations on alcohol use and alcoholism since the original volume was published in 1962. The new edition, envisioned by the editors, never came to fruition because of their both being chairpersons of sociology departments and having other commitments. When both had relinquished their respective chairs, they again discussed the idea of updating *Society, Culture, and Drinking Patterns.* Professor Snyder had retired from his professorship at Southern Illinois University at Carbondale and after much thought and consideration decided against becoming intimately involved in this endeavor, although his counsel and advice were invaluable throughout the preparation of this volume. Before his withdrawal, however, Professor Helene Raskin White, of the Rutgers Center of Alcohol Studies, had agreed to become a coeditor of the new edition.

Discussions were conducted with Peter Nathan, then director of the Rutgers Center of Alcohol Studies, concerning a new series of monographs under the rubric "Alcohol, Culture, and Social Control," to be published by the Center's Publications Division. Arrangements were effected by which the editors of this series would be David J. Pittman and Helene Raskin White, with an editorial advisory board composed of Gail Gleason Milgram, Robert J. Pandina, Barbara S. McCrady, and Michael S. Goodstadt of the Rutgers Center of Alcohol Studies. The current volume, *Society, Culture, and Drinking Patterns Reexamined,* is the first in this monographic series. The second volume in the series is *Alcohol: The Development of Sociological Perspectives on Use and Abuse,* which was edited by Paul Roman (1991).

In the original *Society, Culture, and Drinking Patterns,* the aim was "to bring under a single cover a wide selection of the best current social science research on drinking patterns, normal and pathological" (p. ix). To a large extent this is the aim of this volume; however, given the large amount of both theoretical and empirical literature on the social aspects of alcohol use, misuse, and alcoholism, it was impossible to include all of the important and excellent contributions which have been made to this field of research since the publication of the first volume in 1962. In many respects the organization is similar to that of the original book. We begin with an overview of drinking in a comparative perspective. From there we move to an overview of observations on the modern setting, with an emphasis upon drinking patterns and social processes in drinking. Next we present chapters on social structure, subcultures, and drinking patterns, as they are reflected in such crucial divisions as age, gender, sexual orientation, religion, race, and ethnicity. In the next section we focus on the genesis and patterning of alcoholism and alcohol-related problems, as well as their relationship to selected institutional structures that focus on the family, workplace, and the legal system. We conclude with a section on responsive social movements and systems of control of alcohol use and alcohol-related problems. Each section begins with a brief introductory note. In some places these notes will be identical in part to those Pittman and Snyder wrote for the original volume. Of course, there is nothing sacred about this particular organizational framework, but it is our desire that this format will highlight the relevance of studies of drinking patterns to social theory and social structure. In selecting materials for this volume, we have included 5 chapters that appeared in the original *Society, Culture, and Drinking Patterns,* because of their relevance to contemporary studies of alcohol use and misuse. Of the remaining 35 chapters, 23 are original contributions written and edited especially for this book. In this category will be found both critiques and syntheses of studies in particular subareas and reports of new research. The other 12 chapters represent either abridgements, excerpts, or reprintings of previously published material, with varying degrees of modification by the authors or editors.

In our emphasis upon the sociocultural approach to drinking patterns, we have, in effect, not covered other major aspects of the study of the human relationships to alcoholic beverages. These areas have been excluded not only because of the nonencyclopedic nature of this endeavor but because of our aim to emphasize the social aspects of drinking behavior. Therefore, investigations of the physiological and psychological effects of alcohol on humans, the historical aspects of drinking behaviors and efforts to control them, the evaluations of various treatment programs for alcoholics, the economics of the manufacture and distribution and marketing of alcoholic beverages, the techniques of the production of beverage alcohol, and prevention methods including education to reduce alcohol-related harm are, among a number of other subjects, beyond the boundaries of this book. Moreover, some scholars may feel that there are relevant contributions that are not included in this volume; we would not argue this point, but would simply point to the reality imposed by space constraints. This latter fact made it impossible for us to include all of the behavioral scientists in sociology, psychology, psychiatry, and cultural anthropology who have conducted research that has culminated in the publication of excellent papers.

Our book is directed toward two major groups of readers. First, it should be of value to those social scientists—researchers and teachers—who are generally concerned with or who are specializing in the relationship between alcohol and human behavior. In the university setting, the book may prove useful as a text for courses on alcohol, alcoholism, and society as well as serve as supplementary reading for courses in social problems, criminology, social psychiatry, deviant behavior, and social control, and it should be useful to researchers in these fields. Second, the book should be of interest to professionals and lay persons in so-called applied fields directly concerned with the problems of drinking and especially with alcoholism. This group would include psychiatrists, social workers, physicians, makers of public health policy, health educators, members of public interest and consumer groups, and beverage industry leaders. Persons connected with the large number of international, national, state, and local programs on alcoholism may also find this book of interest in their diverse research, educational, and policymaking activities.

As editors, we would like to acknowledge the support of the many individuals and groups who made this book possible. First, this volume was supported in part by a grant from the Wine Institute to the Rutgers Center of Alcohol Studies. The opinions expressed by the editors as well as the authors of the various chapters are their own and not necessarily those of either the Wine Institute or the Rutgers Center of Alcohol Studies. Second, we would like to express our particular thanks to Ruth Krueger-Singleton, Washington University, who has been involved in this endeavor since its inception and who has provided invaluable editorial assistance at all stages of this project. Special thanks are due to Hugh Klein, and William Staudenmeier, former doctoral students at Washington University who provided helpful suggestions. We would also like to thank the members of the Editorial Board, Gail Gleason Milgram, Robert J. Pandina, Barbara S. McCrady, and Michael S. Goodstadt, for their suggestions on the content of the book. Special thanks are also due to Jane Mosley, a student research assistant and a sociology major at Washington University; Adele Tuchler and Elsie Beck Glickert of the Sociology Department of Washington University; and Marian Parra of the Center of Alcohol Studies at Rutgers University for their help in preparing the manuscript for publication. We are appreciative of the support provided by Alex Fundock III, managing editor of the *Journal of Studies on Alcohol,* who provided invaluable aid in bringing this book to publication. Finally, we are extremely appreciative of colleagues who have contributed chapters to this book.

David J. Pittman
Helene Raskin White

References

Conrad, P., & Schneider, J.W. (1980). *Deviance and medicalization from badness to sickness.* St. Louis: C.V. Mosby.

Pittman, D.J., & Snyder, C.R. (Eds.). (1962). *Society, culture, and drinking patterns.* New York: Wiley.

Roman, P. (Ed.). (1991). *Alcohol: The development of sociological perspectives on use and abuse.* New Brunswick, NJ: Rutgers Center of Alcohol Studies.

SECTION I

Drinking: An Anthropological Perspective

Introductory Note

Emphasis in this book has been placed intentionally on studies of drinking patterns in the context of complex modern society, and particularly U.S. society. Yet there are good reasons for beginning with a brief excursion into studies of drinking in preliterate societies, which are the province of anthropologists. The advantages to be gained from such a procedure are several. Familiarity with anthropological investigations of peoples and cultures seemingly remote from our own helps cultivate a kind of detachment from parochial judgments, which is needed in approaching a subject as fraught with emotion as the use of beverage alcohol.

At the same time, the momentary intellectual identification with the ways of other people afforded by these investigations may bring into sharp relief crucial features of our own patterns of behavior and attitudes—features that might otherwise have remained unnoticed. More important, perhaps, is the fact that anthropologists who focus their attention on relatively small homogeneous groups often are able to portray with boldness and clarity the larger social and cultural structures in which drinking patterns are anchored. The understanding that is gained highlights a goal toward which the necessarily fragmentary studies of drinking in our own complex society may strive, even though a firm grasp of the relevance of larger structures continues to elude us. Finally, anthropological inquiry opens up that wider range of fact that must be accounted for in any effort to generalize about drinking behavior, thereby paving the way for comparative sociology.

Important as they may be, these broad considerations yield only the most general criteria for the inclusion of anthropological work and hardly suffice to provide a rationale for the specific selections to follow. After all, other than those concerned with the cultures of the minority of peoples who traditionally lacked beverage alcohol, virtually all ethnographers have had something or other to report on the subject of drinking customs, however incidentally; and in principle we were at liberty to choose from the nearly endless list of their works. Moreover, recent years have witnessed the development of a rapidly growing body of anthropological literature that has deliberately sought to bring into clearer focus the drinking behavior of preliterate and other peoples marginal to Western culture. (Chapter 5, by Dwight Heath, reviews this enormous body of literature and provides citations for other review articles.)

Because of this situation and in the absence of any monistic principle of selection, it is probably as important to indicate what we have not tried to do in introducing the five chapters that follow as it is to provide brief justification for each selection. It has not been our purpose to sample the anthropological literature representatively, either in terms of the varieties of drinking patterns extant or in terms of cultural areas. Nor do we propose to take readers on a global Cook's tour by exposing them to smatterings of drinking customs from each continent. On the contrary, each selection has been made on the grounds of relevance to current theoretical and methodological concerns in the study of drinking behavior.

First we include Donald Horton's (chapter 1) classic study of the functions of alcohol in preliterate societies. This 1943 study marks the beginning of the sociocultural approach to the study of drinking behavior. Because Horton's study had been quoted more often than any other cross-cultural study in the alcohol field, it earned him the nickname "Footnote Horton." Horton postulates that the degree of inebriety is directly related to the degree of anxiety in a culture. He defines three sources of anxiety (or fear) in preliterate societies: (1) the level of subsistence economy, (2) the presence or absence of subsistence hazards, and (3) the degree of acculturation. Horton's work suggests that in preliterate societies (and presumably in modern society) one of alcohol's primary functions is to relieve tension and anxiety. Other sociologists, such as Bales (see chapter 27) have also included stress as a factor in their theories of the etiology of alcoholism. On the other hand, recent examinations of laboratory and clinical studies question the validity of the tension reduction theory, citing evidence indicating that alcohol use actually increases rather than decreases anxiety and stress in alcoholics (see Cappell & Greeley, 1987).

Next, we include a reprint from the original volume: Peter Field's cross-cultural study of drunkenness in preliterate societies, which builds upon and modifies Horton's study. Field's study (chapter 2) aims at generalizing, on the basis of extensive and systematic cross-cultural study, about the factors responsible for the enormous range of variation in drunkenness in preliterate societies. The reader who is mindful of possible loopholes in the cross-cultural method and who has digested material on drinking patterns in particular cultures will discover places where the chain of inference in Field's reasoning appears tenuous or where his interpretations seem opposed to those of other authors. Whatever may prove eventually to be the most satisfactory total explanation, Field's findings seem clearly to indicate that the nature of the social organization is a crucial determinant of the extent of drunkenness in preliterate societies. It seems equally clear that the degree of elaboration of durable and well-defined corporate kin groups is of far greater significance in this respect than the presence, absence, or development of certain other types of social structure.

These findings will surely warm the hearts of those social scientists who have long pressed for a thorough consideration of variables of social organization in exploring ranges of variation in drinking patterns. They also lend support to lines of speculation and research that, broadly speaking, envisage the incidence of alcohol problems in complex society and various sectors thereof as dependent upon the vitality, dissolution, or resurgence of those social structures beyond the individual or nuclear family—be they based upon nominally familistic, ethnic, religious, or other criteria—that may reasonably be supposed to have a potential for organizing and controlling social life equivalent to that of the structures Field sees as capable of inducing sobriety in preliterate societies. In consequence, Field's study will serve for many readers as a natural bridge for a return from the so-called preliterate to a consideration of drinking patterns in modern society.

We have also reprinted from the original volume Dwight Heath's chapter on the Bolivian Camba (chapter 3), in which he offers one of the most thoughtful descriptions of drinking and related behavior that has appeared in the anthropological literature. Anyone who tries to address to this literature rather elementary questions deriving from the study of drinking patterns in Western society immediately runs afoul of the lack of sufficient information to make clear-cut inferences and conclusions possible. Often enough, this is not so much a problem of poverty in descriptive material, although ethnographies vary greatly in this respect, as it is a lack of concern for negative evidence. In considering preliterate societies where drunkenness is prevalent we may, for example, wish to know if something comparable to alcoholism exists, if manifestations of aggression or sexuality in the course of drinking are to be found, or if sequels to drinking such as the hangover or hallucinations are experienced. Of course the simple fact that such phenomena are often not reported may genuinely reflect their absence in certain preliterate groups, but it may also reflect the failure of ethnographers to note their absence. It is characteristic of Heath, however, to assert positively that a variety of phenomena are not to be found in the Camba group even though drunkenness is widespread—an order of facts that may prove quite useful. Beyond this, his tentative interpretation of Camba drinking ritual as an institutionalized mechanism for stimulating interaction and dramatizing collective identification and solidarity in an otherwise vacuous context points to more general functions of drinking.

Dwight Heath has returned to visit and study the Camba several times since his original study and the present. In chapter 4, which was written especially for this book, he describes the economic, political, and environmental changes that have occurred in Camba society. Yet in spite of these major changes in the lives of the Camba, traditional drinking practices have endured, and the "modern" Camba have also avoided drinking problems and alcoholism. Heath feels that the study of the Camba supports sociocultural rather than control models of prevention. The Camba, 30 years ago and still today, are but one cultural example supporting MacAndrew and Edgerton's (1969) famous proposition on drunken comportment: "The way people comport themselves when they are drunk is determined not by alcohol's toxic assault upon the seat of moral judgement, conscience or the like, but by what their society makes of and imparts to them concerning the state of drunkenness" (p. 165).

In the last chapter in this section (chapter 5), Heath reviews the current state of anthropological research on alcohol use. This chapter highlights the issues of concern in anthropological research. In addition to focusing on use patterns in various societies, anthropologists have extended their work to include a diversity of alcohol-related topics, such as drinking rituals, drinking contexts, clinical populations, and prevention approaches. The recent work in this field, which is methodologically more sophisticated than the earlier work, still emphasizes the importance of sociocultural factors in the development of drinking behaviors and problems. Anthropologists have made significant contributions to the study of alcohol use, especially because they not only have concentrated on the negative aspects of use and on alcoholism, but rather have included the study of the positive benefits of use. Research in other disciplines could benefit from such a varied and comprehensive approach.

References

Cappell, H., & Greeley, J. (1987). Alcohol and tension reduction: An update on research and theory. In H.T. Blane & K.E. Leonard (Eds.), *Psychological theories of drinking and alcoholism* (pp. 15–54). New York: Guilford Press.

MacAndrew, C., & Edgerton, R.B. (1969). *Drunken comportment: A social explanation.* Chicago: Aldine.

CHAPTER 1

Alcohol Use in Primitive Societies

Donald Horton

Since the middle of the 19th century anthropologists have been collecting observations on the life and customs of the hundreds of primitive societies now existing in the world or known to have existed recently. A majority of these societies had native alcoholic beverages before their contact with Europeans. Those of the minority who, for one reason or another, had never invented for themselves, or had never borrowed from their neighbors, the techniques of winemaking or of brewing were very much in the market for alcoholic beverages when Europeans made them available. They became eager customers of the traders except where governments intervened and controlled or prohibited the liquor traffic.

Viewed historically, the use of alcoholic beverages is a very ancient custom. We know, judging from the present distribution of brewed beverages, that the brewing of beer was probably discovered almost simultaneously with the discovery of agriculture itself, and it seems to have spread wherever the art of agriculture spread subsequently in the world. We may presume that the use of natural fermentation to produce wine is even more primitive and ancient. Of course, the origin of this technique is lost in the prehistoric period, but the chances are that it could have been an independent invention, that it could have occurred simultaneously in different parts of the world because the process itself is so simple. It seems likely that all that would be required of a primitive people to discover this property of fruits would be suitable containers—pottery perhaps, or, even antedating pottery, the bark container, which would have been sufficiently substantial to contain a fruit mash and, eventually, wine. We know, as a matter of fact, that at the time of their discovery by Europeans, a number of small, primitive tribes in scattered parts of the world, which apparently had never been in contact with people who had the techniques of wine-

This chapter was first published under the title "The Functions of Alcohol in Primitive Societies," in *Alcohol, Science and Society* (New Haven: Quarterly Journal of Studies on Alcohol), 1945, and is reprinted here with minor modifications.

making and brewing, did have primitive, simple wines made from tree sap or berry juice. In any event, we may presume that fermentation came before brewing, and that brewing itself is a very ancient culture trait, probably going back to the very origins of agriculture. Distillation we know to be a relatively recent invention, probably not antedating the Christian era by very many years; and it is generally thought, on the basis of present evidence, that the art of distilling was developed somewhere in India and from there spread throughout the oriental world and thus to the West.

At the time of first European contact with primitive peoples there were a number of areas, however, in which alcoholic beverages were lacking. Among the very simplest hunting peoples and seed gatherers the use of alcoholic beverages had apparently not yet been discovered. The Australian bushmen had none; the peoples of Siberia, in fact, of the whole polar region, northern Siberia and, of course, the Arctic coastal regions of North America, populated by the Eskimos, had no alcohol. Throughout northern and western North America and the southernmost part of South America the hunting people had not yet obtained, invented, or borrowed the techniques for producing alcoholic beverages. But these people were very eager to use alcohol when they had the opportunity; thus, we can assume that their failure to utilize it was a matter of their not having made the invention or not having come in contact with people who had it. The American Indians, particularly, were very enthusiastic over the use of alcoholic beverages, thus creating a considerable problem during the frontier period. I should mention, however, that there is one other area where alcoholic beverages were not used, namely, the island area of the Pacific—specifically Polynesia and Melanesia. Polynesia includes Hawaii and Samoa, and Melanesia includes the Solomon Islands. In these areas alcoholic beverages were not used, presumably because, owing to the isolation of their position, the inhabitants had neither invented the technique of brewing nor received it from the mainland of Asia.

If we review the kinds of beverages now used, or formerly used by the primitive people of the world, we find that the variety is amazing. Among the horticultural and agricultural folk, whether they did gardening or large-scale farming, whether they worked with a hoe or used draft animals and plows, alcoholic beverages were nearly universal. A very great number of natural materials were used in their fermentations and brews—berries, fruits, honey, plant saps of various kinds (especially the sap of the palm tree in its many varieties), juice of the sugarcane, cow's milk, and, in central Asia, mare's milk. It has been reported, too, that cow's milk had been used in Europe. Practically every natural source of sugar has been used to produce some kind of a fermented beverage. There is a tremendous variety of sources of starch for the brewing of beer—a great variety of tubers and cereals. Among the tubers a noteworthy one is the manioc of South America. Manioc beer is almost a universal drink in the jungle region of South America.

Maize, or Indian corn, produced the main beer of the American Indians in North, South, and Central America; rice, millet, and barley were used especially in Asia and its outlying districts.

Of course, distilled spirits are available to a great many people, mainly in the Asiatic area. These are distilled from rice, beer, sugarcane juice, and so forth.

In the face of this worldwide distribution of alcoholic beverages, and their great variety and antiquity, what can we learn from primitive peoples that will be useful to us? Let me say frankly that a new recital of strange and mysterious primitive customs is of no great interest to me. We are looking for information and ideas that will be useful to us in the solution of our own problems.

Survival of Customs

Even our brief review of the antiquity and distribution of alcoholic beverages suggests at least one important generalization, namely, that the antiquity of the use of alcohol, and its nearly universal occurrence as an item of human behavior, tell us that this custom is a very strong one as measured by its power to survive in the face of competing customs. This surmise or generalization is based on the anthropological doctrine, which is, after all, very easily confirmed by common sense, that a custom, a traditional way of thinking and acting, does not survive and spread from its point of origin unless it gives men some satisfaction, unless it solves some human problem. Think back over the long history of the use of alcoholic beverages. Then compare the changes that have occurred during the same period in other aspects of culture, other aspects of primitive customs—the ease with which many forms of behavior that seemed to be an essential part of primitive life have changed and disappeared under the impacts of higher civilization. That should give you an idea of what I mean when I say that alcohol appears to be a strong and successful custom in the face of competition with other customs. Think of how men, faced with the problem of obtaining game, invented the bow and arrow and various kinds of traps. Give them an opportunity to trade the bow and arrow for a firearm, or their native log-trap for a steel trap, and the native inventions disappear.

The primitive man, at the time of the invention of alcoholic beverages, undoubtedly carried his burdens on his back or on a rough sledge dragged along the ground. Somewhere along the way these customs disappeared and were replaced by others that fitted his needs in a more successful way. The beast of burden, the wheelcart, took the place of the old modes of transportation. But the use of alcoholic beverages has continued and spread, and every year penetrates into areas where it did not exist before.

Not only has this custom been successful in terms of sheer survival, but it has been successful in the face of very severe opposition. We know, for instance, that a good many of the higher civilizations of the past have fought against alcoholic beverages and tried to control and prohibit them. We know that in China at various times, in India, in Mesopotamia, and among the Incas and the Aztecs—the high civilizations of the Americas—attempts were made either to prohibit alcoholic beverages entirely or to control their use, and these attempts invariably failed. In other words, the use of alcoholic beverages as a custom prevailed in the face of definite, organized, and consciously directed opposition. At one period the Hindus went probably further than any others in attempting to make the manufacture, transportation, sale, barter, or use of alcoholic beverages a capital offense. But this severe law apparently had as little success in stopping the development or continuation of this custom as any of the others. We have to conclude from this stubborn reality, then, that some important human value is involved here that makes alcohol hard to abolish. This observation, in turn, leads to another generalization, namely, that despite this value, the use of alcoholic beverages is frequently regarded as a dangerous custom, and attempts are made to control or abolish it.

Restriction of Customs

Sometimes the controls or the restraints are of a relatively limited sort. For instance, in a great many primitive societies men are permitted to drink as much as they choose, but women are severely restricted in the amount they may drink. In a few instances, women are prohibited entirely from drinking the alcoholic beverage even though they may be the ones who make it in the course of their preparation of food. In some societies, permission to drink is obtained only after one has reached the age of maturity, however maturity may be defined.

There are also many kinds of limitations with regard to the proper time or the proper circumstances under which one can drink. These limiting conditions do not go nearly so far as the attempt at prohibition in higher civilizations. But it seems quite clear, even from these examples, that there is definitely an ambivalent attitude, an attitude that is contradictory, an attitude that, on the one hand, approves drinking and permits it to exist and has given it a long life, as the life of a custom goes, and, on the other hand, displays anxiety, fear, and suspicion and leads to attempts to restrict the use of alcohol in various ways.

Two Aspects of the Use of Alcohol

These two generalizations suggest that one way that might be useful to us in approaching primitive drinking customs would be to try to get further information on these two aspects of the use of alcohol. First, its value to mankind that has enabled it to persist against serious opposition; and second, the nature of those dangers inherent in its use that have led to opposition. As a matter of fact, this would be a very sensible kind of request to put to anthropology. Many of the primitive societies studied by the anthropologist are so small and so simply organized that they show us, in almost microscopic form, some of the social and psychological interrelations of different aspects of human life. These are aspects that in our own complex civilization are so overlaid with elaborations and details that most of us have trouble comprehending even some of the major ones. This complexity in our own society, in our own culture, has led us inevitably to specialization in scholarship and research, even to a specialization in understanding. One scholar is an economist. We leave it to him to understand those aspects of our civilization concerned with the production and distribution of goods. Another is a legal specialist. His job is to understand the rules and institutions through which we control our social and economic systems. Another is a psychologist. He is concerned with child welfare and development. Another is a sociologist. He may become the authority on the cause and treatment of crime, juvenile delinquency, and marriage problems. Few of us can be authorities in several of these fields simultaneously, and even our experts have trouble in pooling their separate stores of information. Consequently, it is difficult for us to see a social problem in its full relationship to our way of life. We tend to approach any problem from the viewpoint of our special knowledge and perhaps never do succeed in assessing all of its social, economic, psychological, and historical aspects. An anthropological approach that attempts to see in the simple society, in the life habits of simple people like the Indians or the Africans, some of the general patterns of human social behavior, fits in very well with the present approach to the problems of alcohol.

If you ask a member of a primitive or semiprimitive society why he values alcohol, he will probably say it is because his ancestors found it good, or because the gods gave it to his people; or he might say that the gods gave it to his people and it is also a very good food. He might add that it is good for his health, that it makes people fat and jolly, and that he values it because, when he comes home from a hard day's work in the forest or field, he enjoys his drink. But you would be surprised to find that the majority of the primitive people you question say, "Because it's wonderful to get drunk!"

In point of fact, among primitive peoples and those not so primitive, few drink in moderation as we understand it. Few primitive people drink as moderately as most Americans do. While the individual inebriate is conspicuous by his absence from most primitive communities, the whole community, or at least all of the men, will customarily drink to a degree that we consider to be excessive. Observations on the drinking behavior of primitive peoples have shown that whether the primitive man says that he values drinking because his fathers honored it, or whether he says he values it because he likes to get drunk, the fact remains that most primitive men apparently do drink in order to reach intoxication, for many of them celebrate the glories and happiness of intoxication in song and elegy, as well as in the kind of statement they make to ethnographers.

We know that many native beverages do have a food value, and this fact is concordant with the statements of many that their alcoholic beverage is a food, and a valued food. Take, for instance, the fermented mare's milk of central Asia. It certainly has a very considerable food value. So do the beers, which frequently are not strained and are therefore rather thick liquids with a good bit of the original grain in them. We know, then, that there is some truth in the statement that there is a food value in the native beverage. We can understand very well its being valued for that reason. We also know that alcohol alleviates the discomforts of fatigue.

A characteristic of most drinking among non-Europeans is that it goes far beyond the requirements for food, or the reduction of fatigue. We note, for instance, that there are communities in which it is customary, whenever there is enough of the native beverage accumulated, to have a beer-drink that may last for several days, in which the men quickly drink themselves into a state of intoxication that they maintain until they go into a stupor. Whenever they wake up, they drink some more and go into another stupor. There are many places in South America where these drinking feasts last for a week or more. Then there may be a long period of abstinence while new batches are being prepared. Or it may be, among some who are dependent on European liquor, that they can only indulge themselves when the trader comes on his infrequent visits. Then they get as much as they can and drink every drop of it, after which it may be a month or two before they have another drink. But it is quite characteristic that excessive drinking—excessive by our standard—is the normal practice.

Alcohol for Relief of Anxiety

This finding would seem to imply that the appeal to tradition, the religious validation, the food value, are, in a sense, rationalizations. They may have some subsidiary function, but there is something more important than that. Intoxication is the thing. In the state of intoxication, and in what it means to men, apparently lies the answer to the positive value of alcohol. By positive value I mean that this is the thing for which the alcoholic beverage is treasured, honored, preserved as a custom through the ages. The best possible explanation that I can offer you, based on a review of the use of alcohol in a good many primitive societies, is that the value is primarily in its anxiety-reducing function. This is apparently the only explanation that will serve as a key that has universal validity, with which you can begin to understand the use of alcoholic beverages, the customs surrounding their use, the attitudes that people have toward them in any society anywhere in the world, whether it is a highly sophisticated and civilized society or a very simple society of hunters and gatherers.

We do not have to seek very far for a justification of this hypothesis. Alcohol is an effective sedative. Among its sedative effects is that of reducing the activity of those physiological mechanisms that produce anxiety. Anxiety, or fear, a state of tension, is a painful condition. Any person who is anxious, who is in a state of anxiety, is actually faced with a problem. In anthropology we frequently find that it is very useful to try to analyze a custom in terms of problems and the solution of problems. If we say to ourselves, "Now, to what problem could this custom be a solution?—how could it be solving a difficulty?—how could it be resolving some kind of conflict for the primitive society in which we find it?"—then we begin to get further insight into the meaning of the custom that, on the surface, might be quite mysterious to us. I suggest that when you have a state of anxiety existing in any individual, or in all the members of the population, you have a problem, an unresolved situation, a tension, that seeks to be reduced. You have a need that has to be met, even though at times it may be unconscious—the people who have this state of tension may not be aware of it. We know that anxiety is a universal phenomenon because it functions very simply. It is merely the anticipation of danger, as when a person has experienced pain or has been exposed to a danger. Thereafter, whenever he comes into the presence of the same danger, whenever it is suggested to him, he will have a slight rise in tension, will have this anxiety, which warns him that the danger is close and prepares him to evade or avoid or counteract it in some way.

Human life is full of dangers of all kinds. Among primitive peoples there is the danger of external enemies, the enemies who surround the tribe. Remember that in many societies, especially before the coming of the European government, when one man kills another it is incumbent on the family—the father, brothers, and near relatives of the murdered man—to exact vengeance, to kill the killer. Sometimes these feuds become automatic and reciprocal, so that once a feud starts it never ends, because when the killer is killed his family, in turn, avenges his death, and so on.

In many societies there is a constant state of tension because of the operation of this mechanism, because all people except close kinsmen are potential enemies to everyone.

This method of law and order involves a certain anxiety toward everybody, because it means that each individual is a potential victim of somebody else's crime, the potential victim of the thing that his brother or father does, and so on. Of course, there were peculiar customs that were not universal by any means. For example, in many societies in many places in the world, head-hunting was a custom. If a region held a tribe of head-hunters, then all other tribes in the surrounding region had to be constantly on guard against the raids of these head-hunters and thus felt constant anxiety. Such problems are of course not universal; but nearly universal is the fact that being a primitive man means being subject to instability of food supply, to famine, and to the dangers of an undeveloped productive technique. The primitive man is always very close to the margin of existence. There is never a very great supply of food ahead. He is rather helpless in the face of such dangers as, say, a plague of grasshoppers devouring his crops, other insects of various kinds, floods, and droughts. There is no secure water supply; there are only limited facilities for meeting a drought. The primitive man can be devastated by unpredictable seasonal variations in the yield of his crops; or, if he is a hunter, he is faced with the danger that wildlife may suddenly disappear, that the game on which he depends may be decimated by one of the mysterious epidemic diseases that sweep periodically through various species. Then there are the equally mysterious variations in the migratory patterns of animals that frequently cause great distress to hunting peoples—animals that for generations move back and forth each year on a set course and then suddenly desert that route for no apparent reason and follow a new one, hundreds of miles away, leaving the people who had been dependent upon them for a food supply stranded.

Then, consider the constant danger of sickness and epidemics to all people who do not have our medical science. Now, there is a special danger to primitive peoples that has to be mentioned in this regard: the danger of contact with us. Almost universally, contact between native peoples and Europeans or Americans has been an experience that, in some way or other, has destroyed, damaged, or harmed their own way of life, sometimes creating hardships and unhappiness for them merely by producing changes—even, though rarely, changes that objectively are for the best. Characteristically, in this contact between the natives and the Europeans the native suffers in tangible ways. If he depends upon game, the Europeans come in and settle the forest, and the game is driven out. An example is our destruction of the buffalo and, consequently, the life and culture of all the Indians of the plains. Or the Europeans come in and, in various ways, interfere with the established native way of life and thus create the potentialities of hardship and unhappiness. This is one of the most common sources of the anxiety, the tensions, of native peoples.

The degree of anxiety aroused in an individual by such threats as these is proportional in a general way to the nature of the threat, its intensity, and the amount of damage that is potentially involved. But of course the actual danger is a function not only of the objective situation, the objective danger itself, but of the competence of native institutions, of the native way of life, to meet the threat—for even a very great threat can be met readily by a competent society. But regardless of the fact that this anxiety, with which man suffers the world over, is a relative thing, there is always some anxiety present. There is always some source of danger to which men are reacting.

The use of alcoholic beverages to reduce this anxiety is, of course, not the only remedy that the primitive man tries. His reaction, in the first place, is one of practical activity. He tries to overcome the difficulty, to escape from the danger, by doing something sensible. Primitive men are by no means fools. Testimony to this truth is the fact that we ourselves have borrowed and made a part of our culture, even if a minor part, many of the things that they invented. No European goes to the region north of the Arctic Circle without wearing the clothes designed by the Eskimos for this purpose. We have never been able to invent anything better than the snowshoe, which, in its present form, was the invention of the American Indian, or the dog sledge, which was an Eskimo invention. These are practical and sensible and very clever ways of overcoming the dangers that produce anxiety.

Furthermore, primitive man has still other means, and one of them is magic. When the practical effort is not successful in reducing the threat, he takes recourse to something over and above his practical activity. He performs comforting magical ritual, which gives him a sense of control over the great danger. Or, in his primitive religious practices, he sacrifices or offers libations to the spirits, the bad as well as the good, in an effort to ward off the danger he can no longer completely control by his own means. Then, of course, he has all kinds of social responses to dangers. One of them, the most obvious, is that he lives in a closely knit community, where he develops between himself and the other people strong ties of friendship and support. The primitive community is therefore a much more intimate kind of organization than the kind of social organization that we generally rely on. Each of us lives separately in his own house; we come together only on specified occasions. But in the primitive society, people live together in a very close, huddled way, as it seems to us. This close contact and constant daily association with one another gives them strength and support and encouragement in the face of the innumerable dangers over which they have so little control. Of course, they have not only their religious rituals and ceremonies but their purely social entertainments, their dances, and so forth.

Alcohol is only one of the substances capable of fulfilling the anxiety-reducing function. Among the Asiatic peoples, for instance, many prefer opium. It is characteristic of some parts of the Asiatic world that where alcohol is not used, opium is, and where opium is not used, alcohol is. Among the American Indians there are many who seem to prefer certain drugs such as peyote, the fruit of a certain cactus from which they get a trancelike effect with visual hallucinations, color visions, and so forth. It is characteristic of these Indians, too, that those who use peyote do not use alcohol, and vice versa. Others use various other drugs. Jimsonweed, for example, is very common throughout California and Mexico.

Alcohol as a Sedative

Alcohol has a special virtue for primitive peoples as a sedative inasmuch as it has relatively few or no harmful physiological effects and thus there is no necessary interference with productive activities. It does not put the drinker in a semicoma, so to speak, for an indefinite period, and recovery is not accompanied by a long period of illness, as in the case of some drugs. It can be indulged in for an evening, one can obtain a sedation that reduces anxiety, and yet the next day the drinker can perform his daily tasks without any serious hangover effect. Another factor is that the materials for the production of alcohol are universal. All this great variety of fruits and berries and tubers and cereals I have already mentioned makes it possible for virtually every people in the world to produce an alcoholic beverage. The production of drugs, on the other hand, is limited by the fact that they occur only in certain places and that in many places they are simply not available. Furthermore, not only are the materials abundant and the technique for the production of alcohol simple and easily mastered, but the process itself is cheap. It does not take a great deal of labor to produce fairly large quantities of an alcoholic beverage.

Another important consideration, it seems to me, is the fact that alcohol's effects can be enjoyed along with other gratifications. I have never specifically studied the use of opium, but my impression is that to use it, one has to withdraw somewhat from society. It is a kind of private practice, and you cannot very well conceive of smoking or eating it in an ordinary social gathering as you can use alcohol in the course of a normal social event. Alcohol has the advantage of reducing those tensions that arise among the members of a society when they get together. Inasmuch as people always stand in a slightly competitive relationship with one another in regard to their functions in life, the goods they need, and so on, their relationships are always fairly complicated and subject to the possibility of getting out of kilter. A slight tension can and frequently does arise when certain people get together in a social situation. It is not the kind of thing that makes them enemies by any means, but it is there, and the alcoholic beverage, because it can be used in a social situation, can reduce these tensions and thus actually facilitate the social intercourse that is a part of, and the desirable end of, the occasion.

The Need for Alcohol

In summary, then, alcohol appears to have the very important function throughout the world, in all kinds and levels of human social activity, of reducing the inevitable anxieties of human life. We find, in fact, that there is a general tendency for the amount of drinking, as measured by the degree of drunkenness obtained, to be roughly proportional to the strength of the dangers threatening the society. The range of variation is, of course, very great—from societies in which there is only slight intoxication as the usual thing, to others, and many of them, in which what we would describe as debauchery is the common thing. The worst cases, inevitably, are in those societies that are undergoing a process of destruction at the hands of Europeans. There, always, we find the ultimate in the excessive use of alcohol, because there the anxieties are very powerful.

In other words, alcohol solves the problem of anxiety reduction. For every human being who is anxious there is the problem of reducing the anxiety, whether he knows it or not, and eventually something must be done toward reducing it. From that time on, that something will become customary with the individual, will become a habit, because it succeeded in relieving him of some pain, succeeded in releasing him from some tension. In other words, alcohol satisfies a need, the consciously or unconsciously felt need for relief from anxiety; and when this relief is found, a deep, though temporary, gratification ensues. This result makes the use of alcohol a desired end and gives the alcoholic beverage an economic value. Upon this function of anxiety reduction, which may be unconscious, are built other, secondary functions. Of course the food value of many alcoholic beverages used in primitive societies is undoubtedly real, and so we might regard this as an additional function. But then there are secondary functions that can exist only because of the fact that psychologically alcohol has a deep meaning and, therefore, a value.

The general category of its economic functions, its economic value, exists only because it has a psychological value. But once a substance like this has a psychological value, it rapidly develops an exchange value. It can be used, as it is invariably used by primitive societies, in the reciprocal exchanges that are a basic part of their life. I refer to that very primitive form of commerce in which two people do not get together in the market and exchange their goods for money, but one gives to the other a gift he himself does not need, and he expects that sometime later he will receive a gift the other does not need; in this way surpluses are gradually distributed throughout the population. On a more advanced level there is actual barter, where people come together and exchange directly. Still more advanced is the market for exchange with cowry shells, money, and so on. Thus, the native alcoholic beverage becomes an economic value. It can then be used as a medium of trade and can become a part of important economic transactions, a way of disposing of surplus food products—fruit that cannot be stored because there is no refrigeration, other organic matter that would spoil unless immediately brewed. Then, of course, all of these values can be used as an incentive to labor, as the alcoholic beverage frequently is in many African societies when people get together for joint labor, for instance in the building of a house or garden. The pay they receive from the man in whose interest they are working is usually food and drink; and the more drink he gives, the more liberal he is, the higher his prestige. As a matter of fact, having started with a psychological value and having then developed an economic value, the beverage naturally takes on the function of social ostentation, facilitating the display of wealth and the obtaining of prestige, whereby the man who can give the largest feast with the most to drink, who can permit the most people to get the most drunk, becomes the great man of the community. Of course, he has to do other things as well, but this capability is one of his distinguishing attributes.

Naturally, anything that has become of value is pleasing and acceptable to the gods and becomes one of the most valued of sacrifices, the most valued of offerings. Where a native beverage is of old standing, that is, has been a part of the native way of life for a good many generations, it has generally worked its way throughout the whole way of life until it appears in the context of every kind of human activity. It then becomes an integral part of the system of religious observances, and no one would think of drinking without first offering a few drops in the proper way to satisfy, in a spiritual sense, the needs of the spiritual beings presumed to be watching over him. Thus, these various derived values and secondary elaborations of the whole system exist on the basis of the fundamental psychological value of alcohol, which is, as I see it, primarily the reduction of anxiety and fatigue.

The final derived value from all these others is the fact that drinking becomes a necessary part of the system of life. Since it has been so from time immemorial, and has nearly innumerable ramifications throughout the whole system, it becomes one of those things the absence of which would interfere with all other life habits into which it is integrated. Each of these subsidiary functions obviously could be replaced with a substitute. That is, there could be a substitute in exchanges, a substitute as an offering to the gods, a substitute liquid for giving libation. But the one thing for which there can be no substitute is not the beverage, but the alcohol in it. Again, the alcohol seems to be the significant thing; what alcohol does is the crux of the whole matter.

In retrospect, we are now in a position to understand better our first generalization, namely, that the custom of drinking alcoholic beverages has been a highly successful one in that it has survived and has spread in competition with other customs. We can see now that this success is probably due to the universality of anxiety—and of conditions producing anxiety—among semiprimitive peoples, and to the fact that in all of these societies the ability to meet danger is relatively limited, that there are some dangers these people apparently cannot meet in any rational, practical way, and that the anxieties arising from these dangers are temporarily successfully reduced by the use of alcoholic beverages.

Alcohol and Aggression

Now, what about the other generalization, that there are almost universal manifestations of opposition to, or concern about, the use of alcohol? To analyze this problem we must again return to the psychological effects of alcohol.

We have said that alcohol reduces anxiety and that anxiety is a signal of—a response to—danger. But some of the dangers to which men learn to respond with anxiety are dangers within themselves. That is to say, over and above the dangers involved in threats from external enemies or failure of the food supply, there are also dangers that take the form of antisocial impulses within the individual himself. On frequent occasions he may be in a situation in which he has the momentary impulse to harm someone—to be aggressive, to strike a blow, even to kill. Everywhere, to some extent, the aggressive impulse is forbidden and punished when it occurs within the cooperating social group, particularly within the family group, within the clan, within the village, or even within the tribe.

This group—the family, the clan, or the village group—must work together, must have a harmonious system of social relations, must have a minimum of conflicts, in order to survive, for all the people in the group are dependent upon one another for their survival. Impulses of aggression toward the members of one's own group must therefore be inhibited. They must be inhibited during childhood. The child must learn that when he has such impulses they must be stifled. This is a part of the process of making a child into a social human being.

A child starts out as a mere organism without social character, without culture, without habits to fit him for human life. To this child must be imparted all the behavior that will be required of him as an adult member of the particular society into which he was born. As part of this process of becoming a human being capable of carrying on joint social activities that will enable him and the other members of the community to survive in the face of dangers in nature, the child must learn, among other things, to suppress all his constant impulses to respond with aggression, with hostility, to those who temporarily frustrate him. Especially will he be punished for aggression toward his parents, because his relations with them must be maintained in a certain form, otherwise the parents may fail in making the child into a productive member of society. The parents must have absolute authority. Parents who have the wisdom of their culture, inherited in turn from their parents, must have control over the child's life. No hostility on the part of the child may be tolerated for long.

From the very beginning every individual in any society learns to control his aggressive impulses, and eventually he forces them down to the point where he is no longer aware of the fact that these impulses are being generated and being repressed. To be sure, some forms of aggression are permitted him. He may be given an opportunity to release all the stored-up aggressive impulses he has felt toward parents, neighbors, and close associates, against some permitted objects, against enemies in warfare and in head-hunting, and against evildoers and criminals. He may be permitted to take part in their punishment. He is always permitted to exercise his aggression against those who are regarded as public enemies. Or he may take it out against those individuals who turn out merely to be queer, who turn out to be not like others; against these he may express his aggression in the form of ridicule. Or there may be some very special forms of physical contests, sports, games of a warlike, gladiatorial character, and he may release some of his aggression in them. But these various exceptional ways in which aggression may be relieved in a socially accepted form merely emphasize the fact that the taboo on aggression within the society is absolutely universal:

Now, the strength of the aggressive impulses arising out of social experience depends largely on how much frustration of hopes and ambitions occurs within the group. Some societies produce a human character that is highly aggressive, in which you find an aggressive response occurring in all members of the society very frequently in day-to-day situations. This trait is especially present in those societies in which the whole nature of the society is solely competitive, in which people are accustomed to striving to get the best of one another, to be superior to one another. There are some societies in which the highest social achievement is to be able to shame your rival by showing that you can accumulate a greater mass of property than he and then destroying it in front of his eyes. The rival is challenged to accumulate as much property as you were able to get and destroy it. Greatness is manifested by the ability to achieve great things and then to destroy them. In other societies we find that a man's advancement along the road of life and his achievement of distinction and power depend entirely on the good will of certain older people who stand in a particular relationship to him. If these older people choose, they can forever thwart his hopes of attaining the status to which he might aspire. In a situation such as that, a tremendous amount of aggression may be accumulated. On the other hand, there are societies in which there is no ruthlessness toward weaker members, in which there is very little competition, in which there are very slight inequalities in wealth, in which opportunity is a matter of no importance between men, where corporate activities are the order of the day, and where anybody who chooses to be too self-seeking is frowned upon. In such societies there is very little evidence of hidden, repressed aggression.

The expectation of punishment for the aggressive impulse arouses anxiety. The psychological mechanism is fairly simple. Punishment is a danger; whether it is direct corporal punishment or a matter of being excluded from some special occasion that is valued highly, it is a punishment. This punishment, when first perceived, gives rise to anxiety. To escape the anxiety the aggressive impulse is inhibited or repressed. If alcohol has been consumed, this mechanism fails, because it depends upon anxiety. The anxiety is reduced and the aggression is aroused. There is nothing to prevent it, nothing to inhibit it, nothing to repress the aggressive thought, and it then manifests itself in action. In point of fact, aggressive behavior under the influence of intoxication is almost universal. Among primitive societies it ranges from its very mildest form, which is simply the exchange of insults and harsh words, to its extreme form, in which assault and murder occur.

Release of Aggression

As an example of the first case we may take the Lepcha, a people of about 25,000, who live on the slopes of Mt. Kinchinjunga in the Himalayas in northeastern India. These are among the most peaceful people ever reported in the anthropological literature. Their life is almost without conflicts. It does not run to ostentation, display, or the accumulation of wealth. It is a life that involves a year-round series of operations in the fields, growing rice and millet, and religious rituals. These go on year after year in their accustomed way—everybody has enough, and nobody has too much. There are no sharp distinctions of class or wealth. These people, who have a calm and even temperament, have a beer called chi, which is made from millet. They have it in abundance because they have more than enough millet to supply their food needs. They customarily drink until they are pretty well intoxicated, but this intoxication takes the form of increased jollification and loquacity. Some of them show slightly heightened sexual behavior, but nothing serious, and in the end the men gradually go to sleep under the effects of the beverage.

For a contrast, we may take the Indians of eastern Canada. These were a people—they have now been destroyed—who, at the time the Europeans first came here, were living what we would call a marginal existence. They were primarily hunters, with no other source of food than the game they were able to kill, the fish they caught in the rivers and along the Atlantic shore, and the berries and seeds they were able to gather. These people were constantly subject to the dangers of their existence. Frequently game became scarce, fish failed to run up a certain river, and the berry crop proved inadequate. There were many small tribes, and they were constantly at war with one another over the hunting grounds, because there often was insufficient game to support their populations. There was also great competition among the tribes to see who could reach and hold the best hunting grounds, who could surround the largest herd of caribou and fight off all comers. These people were very simple; they had a formal social organization that involved no real government. They had learned for their own good to keep under control all the aggression that arose among them as a result of the frustrating conditions of their natural life. They took out their aggression on their enemies and on their captives. It is interesting, too, that this pattern was accepted by all of the Indians, so that each man knew what to expect if he were captured by another tribe. It was accepted and regarded as a duty to show fortitude and stoicism and not to give one's enemies the gratification of hearing one scream. A man would, if possible, go to his death, perhaps eventually being torn into shreds, without letting out a whimper. This was the accepted way of things, and it had a deep psychological meaning for these people. They had managed, in one way or another, to keep their aggression under control, to give it outlet in warfare and in various ceremonies and rituals.

Their aggression took another form: they practiced sorcery against one another or believed that sorcery was being practiced. This was a kind of projection of their own aggressive impulses. When alcohol was brought into the situation, these people were overcome by it because the aggression that was released was something they could not cope with. They had no institution for handling this effect; they had no police, they had no government, the chief had only nominal power. When a warrior was given his first bottle of gin by a trader, and then became a maniac and went about killing people, who was to stop him, who would organize the people to stop him? They were not accustomed to dealing with this type of situation. When the whole community got gin and all the able-bodied men began to fight, to kill one another as well as their wives and children, there was just no power available to bring the situation under control. Thus, some of these tribes actually were destroyed more by their own behavior as a result of the alcohol than by the direct assault of the Europeans. Of course this situation was complicated by the fact that when the Europeans came, the game became even more scarce, and the Europeans gave firearms to some tribes, which then were able to behave much more arrogantly and successfully in their own interests against others. The whole situation became much more tense.

I have cited these two extreme examples to show the different kinds of effect that one can get with the difference in psychological conditions. We see, next, that in solving the problem of anxiety by the use of alcohol, people may create for themselves a new problem, that of dealing with aggression, with the drunken aggression that follows the reduction of anxiety. In the case of the Lepcha this second problem does not arise, so that for them the use of alcohol to reduce their relatively moderate anxiety does not have the social consequences that would create a new problem. But for many other primitive peoples the new problem emerges very sharply.

A similar situation exists with regard to the sexual impulse. It is restricted to some degree in every society. Each one we have ever heard of, and we know of thousands, has had the basic incest taboo prohibiting, under the severest penalties, sexual relations between parents and children or between siblings. This is an absolutely universal taboo, but in many societies it now extends in various other directions. In one society, the taboo will mean that a man cannot have sexual relations with or marry any girl who is related to his mother's family, or to all the families that are related to his mother's family, or to his father's family, and so forth. In some extreme cases this extension will mean that perhaps two thirds of all the women in his tribe are women whom he may not approach sexually, women whom he may never marry, and thus his choice is severely restricted. In other societies, the incest taboo is very much like our own, applying to members of the immediate family and, sometimes, to first cousins. Few societies are as rigid as ours with regard to sex, not merely with regard to the incest taboo but with regard to sexual activities in general, forbidding sexual activity until the time of marriage and then insisting on absolute monogamy, so that, according to our standards, a man should have sexual relations with only one woman in the course of his life. While there are few primitive societies in which the sexual regulations are as severe as this, few are actually promiscuous, and none absolutely promiscuous. A characteristic in most primitive societies is that a certain amount of sexual activity is permitted before marriage, and in some it starts very early. Children are permitted to have sexual play, which gradually becomes serious sexual activity and interest, without any interference as long as it does not transgress the incest lines. Thus, a little boy might be permitted to play sexually with all the little girls of a certain clan but not with those of his own clan.

There are various forms of marriage among primitive peoples. Some of these look like monogamy superficially, but they are monogamous only in a particular sense. There may be a series of spouses, but the individual must be faithful to the spouse of the moment. However, the marriage is easily dissolved, each seeks a new partner, and then faithfulness to that partner is required. There are societies, also, in which such practices as wife lending are permissible; and there are societies in which polygamy is practiced, where a man of sufficient wealth may have several wives. But in any case, even with all these variations and relative freedoms, in contrast to the rigidity of our own system in this regard, any man attending a social gathering in any society in the world will always be in the presence of some attractive women who are forbidden objects to him. If his sexual impulse toward these women is aroused, it must be stopped, it must at least be inhibited, just as his aggressive impulse toward those members of the social gathering who stand in a competitive or thwarting relationship to him must be stopped. One can very easily see why. A society that permitted these impulses to come out with absolute freedom would soon be riven by antagonisms and jealousies.

We have seen how primitive societies control the sexual impulses of their members. If there is drinking, however, the reduction of the anxiety that inhibits the sexual impulse will, as in the case of aggression, permit the thwarted impulse to come to life in the form of action. With alcohol a new problem arises: the problem of restraining the sexual impulses that may be temporarily set off before drinking goes so far that the general anesthetic effect represses them, as it eventually does. We can say, thus, that using alcohol to solve one problem raises two new ones.

There are various ways of solving these problems of aggression and sex. In many societies the aggression loosed is controlled by the removal of the offender. He is expelled from the drinking situation when he becomes aggressive. In other societies he may be rendered incapable of expressing his aggression. Very frequently—and this is a rather amusing thing—one finds, particularly in some of the very primitive tribes of the South African jungles, that the women, who are forbidden to drink at all or permitted to drink very little, are given the role of policing the drinking bout; and if a man gets aggressive, the women just set on him, tie him up, put him in a hammock, and leave him there until he sobers up. In other instances the women go around the day preceding the party and make sure that all the spears and bows and arrows are gathered up and hidden in some place known only to them, so that nobody can get his hands on a dangerous weapon. Then, when the drinking begins, if anybody gets obstreperous he can fight it out with his fists. They consider that sort of combat to be relatively superficial, with no danger of anybody's being harmed. Removing the offender, or simply removing his weapons, or punishing him afterward (and usually the punishment is only for those who are extremely aggressive) really means that aggression is being given a certain amount of permission: You can be aggressive, provided you do not use weapons and provided it does not go too far. So very frequently these drinking bouts will end up in a brawl in which a lot of people get pushed around but no harm is done. Then the convention not to hold a grudge against the man who assaulted you comes into play, because it is understood that brawling is one of the effects of the particular drink.

Of course, primitive peoples do not know that there is alcohol in their drinks. All they know is that they have certain beverages they regard as god-given, remarkable, and very valuable, which have the effect of producing a certain degree of happiness, and that occasionally this happiness somehow mysteriously becomes transformed into fighting and somebody perhaps gets hurt a little, but that it cannot be helped. That result is seen as an inherent quality of this particular beverage, and thus no grudge is held afterward.

They control the aggression problem by limiting it—not trying to prevent it entirely but giving it a certain permission and always steering it along certain channels. Very frequently the sexual impulse is treated in the same way. It becomes recognized by the society that you can do things in the drinking situation that you may not do at any other time without being punished, and consequently the expression of sexuality, too, becomes a permitted form of activity as long as it is kept within certain limits. Or it may be prohibited entirely, and in that case the problem is to punish the offender so severely that the next time he will not be able to reduce his anxiety to the point where this behavior will emerge. Another method, very nearly universal, is to exclude women from the drinking situation. If they are not permitted to drink, they are not permitted to be present, and this naturally reduces the instigation to sexual activity.

How successful these measures are will depend on local conditions. Absolute punishment for such behavior can only be successful where there is a fairly strong government with police powers. Many primitive societies lack these institutions, and thus their forms of punishment are more like ostracism and other social penalties for bad behavior. In a very few instances, the only solution that the people have found for the problems of aggression and sex is to do without alcohol. This abstinence is quite remarkable, because very few primitive societies have been capable of it. Among the few are some of the Pueblo societies of the southwestern United States—the Hopi and the Zuni. They had their chance at alcohol. When the Spaniards came in, both of these tribes learned to drink distilled liquor and wine. Although their historical record has never been carefully analyzed from this point of view, I know that there are indications that the effect of drinking was to release these tensions, to produce aggression and sexual behavior that was dangerous to them. But in their case they had something special at stake. They had a society that had survived in a desert environment for hundreds of years, and had been able to maintain itself only by demanding a degree of social cohesiveness and solidarity and cooperation remarkable even for a primitive society. These people were so closely bound together, and their whole culture depended so much on this question of cooperation, that they had made the nonaggressive, cooperative man their ideal, and they could not tolerate any deviation from this ideal. For them, it was a sheer matter of survival. Not only that, but they did not have to depend on the slow process of trial and error in applying social controls. They had a very highly cultivated system of governmental and priestly authority that included men of no mean philosophical achievement, men who were real thinkers, who actually perceived the problem. This again is a rare occurrence among primitive peoples—they generally do not have the techniques of thinking that enable them to see their social problems in a clear light. The Hopi and Zuni, however, saw their problems, and therefore passed a decree that thenceforward no member of their community could drink. This rule is breaking down somewhat. It is still forbidden to drink in the pueblos, but the young people who go out into American society—to boarding schools in American cities or to jobs on the railways or in the factories in the surrounding region—do occasionally become drinkers. Usually they give it up when they go back to their homes, because they know that it will not be tolerated there. I mention this last to show that the anxiety that would motivate drinking is present. There is a drive that can be satisfied with alcohol, and in certain situations they will use it, but in their home community it is taboo. Usually, however, the solution of the primitives is to isolate the drinking from the other activities of social cooperation so that drinking and its consequences can happen without doing too much damage to the social structure. When drinking is done in a special ceremonial context—all the cares and work of the day have been finished, the special arrangements made, the weapons put away, and the women excluded—then it is safe to drink to the degree of intoxication that apparently is always the goal of the primitive drinkers.

Social Need and Social Danger

Now we are in a position, it seems to me, to understand why this valued thing—the alcoholic beverage—is used with so much concern, why the stories of its origin, the myths that account for its development, usually imply in some way or other that an evil spirit was involved as well as a good spirit, that it is a gift, but a gift with the sting of the scorpion in it. Now we understand why it is that, very frequently, before drinking begins rituals are performed, why prayers are made to the gods to see that nothing happens, to keep the alcohol demon from seizing people, to keep evil spirits from taking advantage of their drunken condition to enter into them and to create trouble. We understand why, among the Ifugaos of the Philippines, the special rite of tying up the stomach of a newborn child, a practice designed to control the spirits in him, is performed so that when he grows up he will not be able to liberate these spirits when he drinks. The same ritual, or a modification of it, is performed before each drinking ceremony, just before each party. In other words, the social need for relief from pain, tension, fatigue, and anxiety stands in opposition to the social dangers that inhere in the reduction of certain anxieties. To the extent that the conditions of life are frustrating, frightening, we see the development of anxiety and, at the same time, the development of aggressive impulses. When attempts are made to solve this set of complex problems by means of alcohol, new problems arise.

In general, however, our observation of the peoples of the world indicates that almost invariably, with the exception of people like the Hopi, the Palaung of Indochina, and a few others, the need for alcohol, the need for anxiety reduction, overrides the dangers inherent in the release of aggression, and all kinds of compensatory mechanisms are developed to take care of the subsidiary problems that arise. In other words, people will endure the small ills of drunken aggression and sexuality in return for getting rid of tensions borne of anxiety.

Alcohol and Culture

From this concept of the universal function of alcohol, it seems to me, there are some obvious implications that might be useful to us. Here, of course, I step out of my role as an anthropologist reporting what goes on in primitive societies and really venture a few steps as a social philosopher.

First of all, it is quite apparent, as one reviews this field, that drinking behavior is not determined by such qualities as race. We have heavy drinkers and light drinkers, we have fighting drinkers and passive and peaceful drinkers, within every known race. Evidently there can be very little determination here by constitution. While constitutional differences might account for one man's ability to drink more than another before going into a stupor, we know that culture does not follow any constitutional line. The same people racially, genetically, may be divided into different groups with different cultures and thus behave very differently in similar situations. One feels, in reviewing these data, that drinking habits are rather a question of the social conditions that are reflected through the individual and his behavior.

The anthropological view of the relation of individual behavior to society is something like this: In the first place, a man is what he learns to be; he comes into this world knowing nothing, and he has the potentialities for being any kind of a man that he is taught to become. This teaching we understand not as the formal education of the schoolroom. That is only part of it. His whole life experience teaches him. It rewards him for doing certain things and punishes him for doing others. Through this process over the years he eventually becomes the man who does the things that are proper and rewarded in his society. From infancy, those who have preceded him and who are therefore the standard bearers of the society, who have themselves learned from their ancestors what is right, shape him into an individual who is acceptable to his society. Those who guide him through his infancy and make him into a social being impart an unconscious system of habits, attitudes, and opinions that they can change very little as they transmit them to him. They change them somewhat; each generation acts a little differently from the preceding generation. In modern societies, of course, we see this process more strikingly, because in our society change is much more rapid, whereas in primitive societies the trend is a uniform one with only slow variations in time. As we look back through the anthropological records we can see that some societies have gone for hundreds of years changing only very, very slightly from one generation to another, with the result that, over a period of centuries, the changes are still what we would regard as minor. We know, too, that you can take an individual from one culture into another—if he has not gone too far in the development of his personality, in the development of that whole system of habits that marks him as a member of the first society—in infancy or even in childhood, put him in a new society, and make of him an entirely different creature from what he would have become. You can take an Indian, put him in a mission school, and make him psychologically—in terms of all his habits, in everything except his actual physical appearance—into an American; and of course this sort of thing is being done all the time.

This system of standards of behavior, of habits, of norms of conduct, we anthropologists call the culture of a people. Each society has its own culture, its own way of life. Each individual learns to incorporate these traditional habits into his own system of behavior. When danger threatens him in his relations with other people, or when it threatens him in facing the crises of life—illness and death—or when external dangers face him—starvation or enemies—his reactions will be those his culture demands of him. If he has any spontaneous reactions that are not acceptable to the culture, they will usually be stopped, prohibited in some way.

In primitive societies the culture is fairly uniform. The society is little, the population is usually small, and individuals are much more alike in their behavior than in our society. There are, in primitive societies, relatively few differentiated groups, and there is no great differentiation in labor skill. You do not have the teacher on the one hand and the ditch digger on the other. Each man is something of a teacher to his son. Each is able to take what measures are necessary. Each labors in the same industry. Where there is a division of labor, it is usually in certain minor specialties. There will be the priest, the medicine man; there will be the few expert craftsmen who have a peculiar natural talent for making carved wooden objects or dugout canoes, or something like that. But, in general, there will be relatively few differences in culture. There will be no class differences, no great economic differences, and thus no opportunity for the development of divergent personality types. That is why we are accustomed to say in anthropology that in primitive society the personality, the psychology, of individuals is likely to be fairly uniform for all members of the same sex. There will be differences between the sexes, and slight differences due to age, but, in general, all the mature men will be very much alike in their inner psychological development and structure because they have all experienced the same things, they have had the same teachers, they have grown up under the force of the same tradition.

In our society, obviously, things are very different. We have a tremendous diversification of functions, roles, and activities. We have the basic differences of occupations. We have differences in wealth, all kinds of differences of opportunity, differences in national background, and racial differences. All of these make our society infinitely more complex than the primitive society. There is more opportunity for individuals to deviate from the norm because there are conflicting norms. Even a criminal, who in a primitive society would be immediately exposed and punished for his crime, may, in our society, seek out a society of criminals in which it is normal to be abnormal, in which one can flout the standards of society and yet be among people who approve the flouting attempt. Of course, such a thing is unheard of among primitive societies. In our society, if one has sexual abnormalities, it is possible to join a society of perverts in our big cities, consisting of a small group of people who agree that because they are perverted they will allow their practices, they will not take the attitude toward their abnormalities that society at large does. And as long as they act within their protected precincts and do not go so far as to come into conflict with the other members of society, they can manage. In other words, the possibility of deviation is much greater, and the conflicts that give rise to tension within the individual are much greater and much more diverse and can express themselves in more varied kinds of actual deviation from the norm.

In our society it is much more difficult, merely by an examination, to relate the behavior of an individual to his social context. It is much more difficult to study an insane man and state the social conditions that produced his insanity, yet we know that they must exist unless the insanity is due to an organic disturbance. In the case of psychiatric disorders for which there is no organic basis, however, we know of only one possibility: those disorders come out of the individual's experiences with society. But the society is so complex in this individual's experience, it has carried him through so many different competing situations, situations in which there were conflicting standards of behavior and conflicting attitudes, that it becomes a major job of the psychoanalyst to try to reconstruct the life experience of this individual and to see at what points his contacts with society have produced in him the reactions that now express themselves in a symptom that requires psychiatric treatment.

For us, then, this problem of personality in relation to culture is a far more complicated matter than it might be in a primitive society, in which there is much greater uniformity and in which the experiences of all individuals are seen to be very similar if their effects are studied. But some of the larger issues are the same, even though we cannot take, say, the alcohol addict and immediately refer, merely by an examination of the society, to the specific conditions that would give rise to alcohol addiction. We would need a special technique of psychiatry to analyze the character structure in relation to experience, and then experience in relation to mode of life. Nevertheless, it seems to me that we can recognize certain relationships, again speaking of the social philosophy of alcohol addiction. It seems clear to me from all this that certain large factors must be implicated—factors of poverty, of insecurity, of a type of job situation that so often prevails in which employment is uncertain and old age represents the threat of dependency and even starvation, factors of jealous strivings among people for power and security. In all these characteristics of our society, and in the rapid changes going on in our culture between the generations, we can see some of the necessary background for the development of the alcohol problem as we know it.

There are general psychological conditions arising out of the conditions created and transmitted by our culture that it is now the task of the psychiatrist and the social psychologist to analyze for us. If we carry over what we have learned about primitive societies to our own society, we certainly cannot regard our problem as specifically the problem of alcohol—because alcohol is merely the agent. The problem is first seen as the use of alcohol, but from there we are led back directly to the anxiety, tension, unhappiness, and frustration that led to the use of alcohol and make it rewarding to the individual. The reduction of excessive anxiety may be tackled as a problem of individual therapy, and that is the standard way of doing it. But if I were asked to give one overall statement, on the basis of my survey of primitive societies and the reference of this survey to our own society, it would be that the fundamental problem is one of social engineering rather than of individual therapy.

CHAPTER 2

A New Cross-Cultural Study of Drunkenness

Peter B. Field

The cross-cultural method has several important potential advantages for extending our knowledge of drinking behavior. First, this technique examines relations between variables across a number of tribes at once, whereas the typical anthropological report is restricted to just a single tribe. Furthermore, the method deals simultaneously with a great range of modal drinking patterns—from abstention in some tribes to periodic bouts of extreme drunkenness in others. Finally, the tribes that are included are separated from each other and also from Western industrial society, so that relationships emerging across a sample of tribes cannot be attributed to the widespread influence of a small number of cultural heritages. This method is therefore a potentially valuable tool in the development of general explanations of wide scope and applicability concerning drunkenness. These generalizations may in turn advance our knowledge of alcoholism in modern society.

Previous Cross-Cultural Research on Drunkenness

Cross-cultural research of this kind depends upon both the development of valid rating categories for assessing the extent of drunkenness and their successful application to the available anthropological literature

This chapter is reprinted with permission from *Society, Culture, and Drinking Patterns* by D.J. Pittman and C.R. Snyder, Eds. (New York: Wiley, 1962), pp. 48–74. A short version of this chapter was read at the August 1960 meeting of the Society for the Study of Social Problems. This chapter is based on an unpublished doctoral dissertation, "Social and Psychological Correlates of Drunkenness in Primitive Tribes," Harvard University, 1961.

on primitive societies. Next, the anthropological reports for each society must also be scored on other theoretically relevant variables, which may then be related to the index of drunkenness. This method has been employed only once in the past—in the classic cross-cultural research of Horton (1943).

To explain his results, Horton presented an anxiety theory of drunkenness. He proposed that a major factor determining the degree of drunkenness in a society is the level of anxiety or fear among the individual members. Drunkenness, however, can be inhibited, as well as produced, by anxiety. Consequently, according to Horton, the level of drunkenness in a society is a resultant of a complex interaction of anxiety reduction and anxiety induction. He states, for instance, that the "strength of the drinking response in any society tends to vary directly with the level of anxiety in that society." However, because the "drinking of alcohol tends to be accompanied by the release of sexual and aggressive impulses," and because sexual and aggressive impulses are often punished, drinking will be weaker where these responses lead to anxiety about punishment—that is, the "strength of the drinking response tends to vary inversely with the strength of the counteranxiety elicited by painful experiences during and after drinking" (Horton, 1943, p. 230).

Horton's study was one of a series of research reports directly influenced by the efforts of Hull and his collaborators to integrate behavior theory and psychoanalysis. A hallmark of this approach was the explanation of complex sociocultural phenomena in terms of individual drives—especially frustration or anxiety. Before Horton's work, for example, two members of Hull's school of thought reported an inverse relation between the price of cotton and the number of lynchings a year in the Deep South, and explained this finding in terms of frustration and aggression (Hovland & Sears, 1940). In view of theoretical precedents of this kind, it is not surprising that Horton should have interpreted the level of drunkenness in a society as an expression of the level of fear or anxiety in the society. His research is, however, one of the very few reporting an important relation of any kind between the level of anxiety in a society and any aspect of alcohol use. Indeed, it is the only social research cited in a recent review of the most important research evidence that alcohol reduces fear (Conger, 1956).

Not all of Horton's results support his own theory unequivocally, and those that appear to do so vary greatly in quality. The most convincing support comes from his finding that aggressive and sexual responses are often released at drinking bouts, suggesting that there is ordinarily an inhibiting force holding these responses in check. This fearlike force is temporarily weakened at drinking bouts. However, his two major indices of the social anxiety level—that is, an insecure food supply and acculturation by contact with Western civilization—are very indirect and questionable measures of fear. Furthermore, as Lemert (1954, p. 370) has pointed out, Horton failed to "include a well-defined variable of social organization. Correlating such things as the technological level of a society or the amount of its wealth or the death rate of its population with a psychological reaction such as anxiety can only partially reveal the dynamics behind primitive drinking, because these non-sociological facts always take on variable meaning and impact for the members of societies through the media of social organization." Several authors have also made the very important point that a simple anxiety theory of alcoholism and drunkenness is inadequate because it does not explain why other modes of reducing anxiety instead of drinking are not used (e.g., Lemert, 1956). Why does a society develop drunkenness in response to food shortages rather than, say, resort to ritual magic aimed at inducing a reluctant god to supply food? Why, moreover, does a psychoneurotic become an alcoholic instead of developing a different form of anxiety-reducing personality disorder?

Since Horton wrote the above-cited paper, a number of studies of primitive drinking have been reported, but these have not unequivocally supported his position. As a case in point, Lemert (1954) found that the greatest drunkenness among the Indians of the American Northwest occurred when they were enriched by the fur trade; it is hard to understand why this drunkenness might have been associated with fear. For his part, Lemert identified a number of other factors—including cultural conservatism, anomie, and interclan rivalry—affecting drunkenness on the Northwest coast. Other studies of individual tribes, such as Heath's study of the Camba [see chapters 3 and 4, by Heath, in this volume], have suggested social disorganization or social isolation as important causal variables in drunkenness.

Methods and Procedures

The reexamination of Horton's theory, which is an integral part of the present study, was made possible by the development of new cross-cultural scales not available to him. This reexamination began with the assembling of a group of scales on the basis of their theoretical relevance to alcohol problems.[1] These scales were originally developed by many different raters for a wide variety of purposes and vary widely in degree of reliability. Some, such as Horton's drunkenness measure or Murdock's

(1957) social structure ratings, have been scored by only one judge, and consequently their interscorer reliability cannot be directly estimated (although we know from the magnitude of the correlations with other variables that it must be substantial). In certain other instances where reliability can be estimated, the measures are reliable enough to indicate the presence of a non–chance relationship but are too crude for more accurate prediction.

The most important scale in the tables to follow is Horton's measure of "strong vs. moderate and slight degrees of insobriety" (1943, pp. 265–266). Tribes with "strong" insobriety were described by ethnographers as drinking to unconsciousness, drinking for many hours or days, or getting "excessively" drunk. The last category probably indicates extreme functional impairment. Tribes with "slight" or "moderate" insobriety either did not get drunk or, if they did, did not drink for days and did not drink to unconsciousness. This scale measures degree of drunkenness at periodic drinking bouts and not, of course, the number of alcoholics in the society or the death rate from alcoholism.[2]

After the scales were assembled, Horton's 56 tribes were divided into two subsamples of roughly equal size. The first subsample consisted of all of Horton's tribes that had also been rated by Whiting and Child (1953). Preliminary correlations were examined with only this subsample. Although only a few of the Whiting and Child child-rearing ratings showed promising relationships, a group of highly suggestive relationships appeared with the use of Murdock's (1957) social structure ratings. These relationships were checked in the second subsample. As a rule, the preliminary relationships emerged just as clearly in the second sample, and for no important relation did an association reverse or change markedly between the samples. This result strongly indicates that the relations found between drunkenness and social structure are not the result of capitalization upon chance associations. For this reason the subsamples were combined for ease of presentation. Finally, unpublished social structure ratings on six additional tribes were kindly provided by George P. Murdock in a personal communication, and were added to the two subsamples to complete the final sample.

Unless it is otherwise indicated, scales were dichotomized at the median. Associations were tested for statistical significance by chi-square corrected for continuity with one degree of freedom. Tetrachoric correlation coefficients for each association were estimated by the method given in Wert, Neidt, and Ahmann (1954). All probability values reported in this chapter are based on a conservative two-tailed criterion. Any association based on fewer than 56 cases indicates that either an ethnographer or a rater omitted relevant information on one or more of Horton's tribes.

TABLE 1
The Relations of Degree of Drunkenness to Indices of Fear

	χ^2	r_{tet}
Fear of sorcerers	0.00	−.06
Fear of spirits	0.00	.07
Fear of others (sorcerers and spirits combined)	0.00	.06
Fear of ghosts	0.34	.29
Fear of animal spirits	0.91	−.39
Fear of ghosts at funerals[a]	0.00	.08

[a]N = 30; 27 for the other measures.

Results and Discussion

Drunkenness, Fear, and Subsistence Insecurity

The first section of Table 1 shows the relations between Horton's measure of drunkenness and the best available measures of fear (Friendly, 1956; Whiting & Child, 1953). None of these relationships is statistically significant and no clear positive or negative trend appears. The first five measures of fear are drawn from beliefs concerning the causation of illness. Fear of spirits, for example, means that the society attributes illness to malevolent spirits. The last rating is an even more direct measure of fear based on records of direct statements of fear of ghosts at funerals. All of these scales of fear have shown significant relations to other measures such as sex anxiety or severe child rearing (Whiting, 1959; Whiting & Child, 1953), a fact that suggests they are genuine measures of fear.[3] They are all reliable enough to show significant relations to Horton's drunkenness score, if they existed. Horton himself was unable to find a relation between degree of drunkenness and sorcery, but the reason for this result was not clear. The consistent negative evidence in Table 1 suggests that, within the limits of these measures, variation in the level of fear is not related to the extent of drunkenness in primitive tribes.

The data in the first column of Table 2 clearly confirm Horton's finding that tribes with very primitive hunting-and-gathering economies tend to have more drunkenness than tribes with more advanced herding and agricultural economies. The second column in Table 2 shows a very strong relation between Horton's measure of an insecure food supply and the absence of agriculture; indeed, this relation is so strong that it is perhaps justifiable to assume that Horton's "subsistence insecurity" is more of an

TABLE 2
The Relations of Type of Subsistence Economy to Degree of Drunkenness and Insecure Food Supply[a]

Subsistence Economy	Drunkenness		Insecure Food Supply	
	χ^2	r_{tet}	χ^2	r_{tet}
Predominance of herding and agriculture over hunting and fishing[b]	4.14[c]	−.50	1.45	−.41
Agriculture	3.24	−.47	12.32[d]	−.85
Animal husbandry	1.34	−.32	0.02	−.13
Fishing and marine hunting	3.86[c]	.49	2.64	.49
Hunting and gathering	5.71[c]	.57	2.42	.48

[a]For all insecure food supply correlations, $N = 37$; N is either 49 or 50 for the other correlations. The Murdock (1957) scales on economy were divided at "important" versus "present but unimportant."
[b]See Herbert, Child, and Bacon, 1959.
[c]$p < .05$.
[d]$p < .001$.

indication of the lack of a stable agricultural basis for society than of the level of fear.

The implications of this latter, alternative assumption are far-reaching. We know that the social organization of a hunting tribe is quite different in some ways from the social organization of a tribe with an advanced herding and agricultural subsistence economy. Tribes that live by hunting are frequently nomadic, but agriculture requires a settled, stable community. With the development of agriculture and herding, it is also possible to support larger concentrated populations that would starve under a hunting economy (Murdock, 1949, pp. 80–81). When such a population is present in one place over several generations, community divisions will be formed, occupational specialization will be more obvious, and social stratification and centralized political authority will become possible. At the same time, complex corporate-kin structures may emerge with the addition of a unilocal rule of marital residence and the development of unilinear descent. Thus, with the transition from the hunting band to the compact village, formalized social control over the individual's behavior may well become more diffuse and pervasive. Parallel psychological differences may also appear. For instance, Barry, Child, and Bacon (1959) have shown that herding and agricultural tribes expect obedience and responsibility from their children, while hunting and fishing societies demand independence, achievement, and self-assertion instead.

Drunkenness and Social Organization

These results and speculations suggest that something other than fear, perhaps a social variable, explains primitive drunkenness. Perhaps this variable can be detected by an examination of the primitive drinking bout.

Horton has described the sequence of behavior in drinking bouts in this way: first one observes gaiety, laughter, friendliness, lewd jokes, jollity, and loquacity. Later, sexual behavior, verbal quarrels, and fights are often the rule. In the tribes Horton classifies as extremely drunken, this sequence goes on for many hours or days, extreme functional impairment is usual, and the bouts end with most of the participants unconscious.

Although anxiety is certainly being reduced, something more important is happening. Respect and reserve are giving way to friendly equality. Usually inhibited aggressive and sexual impulses are emerging. Behavior is directed at momentary personal pleasure and self-indulgence rather than duty, responsibility, and task performance. There is loosely organized companionship rather than formal execution of institutionalized obligations. In one sense the group is clearly becoming disorganized, since its members eventually withdraw into unconsciousness; but this disorganization is the final outcome of informal release of tension in a companionship setting. In summary, a personal, informal organization has replaced a formal, well-structured organization: alcohol has facilitated informal, personal interaction by temporary removal of social inhibitions; and it has disorganized precisely controlled behavior, making difficult the performance of duties, labor, respect and avoidance, ceremonials and rituals, and other legalized obligations requiring inhibition of impulses.

In tribes with a great deal of this kind of informal, friendly, loosely controlled behavior at drinking bouts, there is a strong possibility that ordinary social relationships are also informal, personally organized, and based on friendly equality. Marriages may be primarily unions of individuals for companionship and mutual support, rather than socially extended alliances designed to fulfill broader economic or social functions, such as inheritance and lineage continuity. Formal alignments of kinsmen with clearly formulated mutual rights and obligations may be weak or absent. Social solidarity may be stronger between individuals of the same generation than between adjacent generations—that is, the drunken tribes may have a wide-ranging kinship system emphasizing friendly solidarity among siblings, cousins, and spouses, rather than hierarchical, lineal solidarity between parents and children. This state of affairs in turn might mean that the drunken tribes place less importance on the conservation and transmission across generations of an elaborate social tradition.

In these tribes the organization of tribesmen to attain common objectives may be weak and informal. That is, in these tribes functions such as ceremonials, military operations, and transmission of property from one generation to another may be carried on by individuals or informal groups rather than assigned to specialists such as priests, a warrior caste, or a land-owning group. In the informally-structured tribes with primitive hunting and gathering economies there may be few differentiated structures serving well-defined social purposes.

Another possibility suggested by this line of reasoning is that there may be a general weakness of extreme power, respect, or reserve differentials between individuals in the extremely drunken tribes. If a society is highly organized, prestige and power will be vested in a few individuals rather than diffused throughout the tribe, as may happen both in political organization and in the family. The leader of a South American band of hunters has little power, but an African monarch wields a great deal. Similarly, in the drunken tribes there should be no great differences between husband and wife in centralized power and in control of the family, and therefore polygyny should not be important. This informal equality may be reflected in the child-rearing practices: we would not expect a great deal of pressure for submission and obedience on the part of children in the drunken tribes.

Form of kin group. Preliminary evidence for the relevance of social organization in explaining drunkenness among primitive peoples is presented in Table 3. The following facts will help to clarify the meaning of this table. Tribes with kin groups organized into units tracing descent through males are called *patrilineal*; those organized into units tracing descent through females are called *matrilineal*; and tribes with both these forms of kin group are referred to as having *double* descent. In contrast, tribes lacking kin groups organized into special corporate units (tribes that, for this reason, are more loosely organized and might be expected to exhibit more drunkenness) are referred to as *bilateral*, which means that kinship is traced through males and females without formal distinction.

Thus, in Table 3, although patrilineal tribes appear to have somewhat less drunkenness than matrilineal tribes, the outstanding fact is the striking association of drunkenness with bilateral descent—in other words, its association with the absence of patrilineal or matrilineal kin groups of a corporate nature. Since, according to Murdock (1949), kin groups based upon patrilineal and matrilineal descent parallel one another almost exactly, it seems permissible for statistical purposes to classify these together with double descent groups in a single category of *unilineal* groups. Viewed in relation to the presence or absence of unilineal kin

TABLE 3
The Relation of Degree of Drunkenness to Type of Kin Group

Degree of Drunkenness	TYPE OF KIN GROUP			
	Bilateral	Matrilineal	Patrilineal	Double
Extreme	19	5	9	1
Moderate or slight	2	3	12	4

groups so defined, degree of drunkenness yields a strong negative correlation of −.67, which is highly significant.[4]

A major distinction between the bilateral kindred, the most widespread and characteristic form of arrangement of kin in a bilateral society, and unilineal kin group is this: a unilineal kin group is a "corporate" group, whereas the bilateral kindred is not. The term *corporate* means that unilineal kin groups are structured groups with "perpetuity through time, collective ownership of property, and unified activity as a legal individual" (Pehrson, 1954, p. 200). An individual and his kindred die, but unilineal kin groups, like tribes, extend indefinitely into the past and the future.

Murdock (1949, p. 61) points out that because the kindred is not a group except from the point of view of one individual and because it has no temporal continuity, it "can rarely act as a collectivity. One kindred cannot, for example, take blood vengeance against another if the two happen to have members in common. Moreover, a kindred cannot hold land or other property." Murdock notes further that a member of a bilateral kindred may be involved in conflicting or incompatible obligations, whereas these conflicts cannot arise under unilineal descent. He also states that unilineal kin groups are discrete social units, automatically defining the role of every participant in a ceremonial activity or every bystander in a dispute (p. 62). By contrast, according to Pehrson (1954), bilateral societies have often been called "amorphous," "unstructured," "loosely organized," or "infinitely complex"—the latter term being synonymous with *fragmented*. He suggests, however, that bilateral tribes are not truly disorganized; they have a flexible, noncorporate organization featuring a network of horizontal ties uniting individuals of the same generation.

Similarly Davenport (1959 p. 569) concludes that the kindred structure "occurs where collective and corporate control is absent or minimal." Davenport also points out that in addition to kindreds, some bilateral tribes have nonunilineal descent groups that have some corporate functions. However, affiliation with them is not as well institutionalized as it is in unilineal societies: an individual's affiliation is usually determined by choice, not by birth as it is in unilineal tribes; he may move from one group to another during his lifetime; and his claim to membership in a group may depend on his decision to perform actions maintaining reciprocal ties with the group, or on performance of some specialized technique for validating membership, such as giving a potlatch.

In a later article Pehrson (1957) points out that the bilateral Lappish band has a flexible, variable structure permitting the individual a wide range of alternative courses of action. The structure of the band, he reports, is a wide-ranging alliance of sibling groups held together—sometimes loosely—by conjugal ties. Devices such as the extension of sibling terminology to cousins increase the solidarity between members of the same generation. Sahlins (1960) points out that in other bilateral tribes friendly relations between tribesmen may be secured at the expense of marital ties, as in Eskimo wife lending; in the most primitive tribes similar friendly relations are secured by the continuous circulation of food and other goods among individuals—a kind of mutual sharing that has also been noted in drinking groups. Finally, Wolf (1955) reports that in open or noncorporate peasant communities social, economic, and political arrangements are based on informal, personal ties between individuals and families.

Because there are several different kinds of unilinear kin groups, the nonbilateral societies in Table 3 were reexamined for evidence of possible differences in drunkenness in each type of group. The association of unilinearity with little drunkenness was most marked for societies segmented into unilineal sibs: only 7 of the 21 societies below the median on drunkenness did not have sibs. Of the 22 societies with this structure, 14 were below the median on drunkenness. (Societies that had more inclusive structures—moieties or phratries—in addition to sibs were excluded from this tabulation and considered below.) According to Murdock (1949, p. 73), "the sib is associated with totemism and ceremonial, acts as a unit in life crisis situations, and regulates marriage and inheritance." In other words, it is associated with the most central symbols of formal solidarity, and regulates the individual's most important links to other individuals. Normally a sib is composed of several lineages and thus is differentiated internally. With sib exogamy, members of one sex marry into other sibs, thus forming collateral ties in the community. These tribes, then, have complex webs of formal kin affiliations spreading temporally and spatially throughout the entire society.

There were four tribes divided into unilineal moieties, or half-tribes, and all were above the median on drunkenness. Lowie (1948, p. 247) points out that such moiety divisions are "fluid" and are not often found in tribes with large populations (p. 245). These indirect relations may well explain this result. On the other hand, there may be a failure to cre-

TABLE 4
The Relation of Degree of Drunkenness to Form of Marital Residence

	FORM OF RESIDENCE				
Degree of Drunkenness	**Bilocal or Neolocal**	**Uxoripatri-local**	**Matri-local**	**Avuncu-local**	**Patri-local**
Extreme	11	5	8	1	9
Moderate or slight	0	0	4	2	15

ate strong solidarity by overinclusion of members, since the solidarity of all kin groups necessitates the exclusion of some tribesmen. In any event, if these four tribes with moiety divisions had been classified empirically with the bilateral groups, the already high correlation between drunkenness and form of kin group would have been still higher ($r = -.78$, $p < .001$). (There were only two tribes with phratries—one drunk and one sober—and thus no generalization is possible here.)

These results indicating the importance of social disorganization in primitive drunkenness also suggest a new explanation of Horton's findings on acculturation. His best predictor of drunkenness was severity of acculturation by contact with Western civilization. He examined the possibility that this correlation might have been caused simply by the introduction of distilled liquors, and concluded that although there was some relationship, it was not an important one. He concluded that anxiety was produced during acculturation and that it caused drunkenness. However, it is not obvious that acculturation always increases anxiety, although it certainly may do so in special instances. Acculturation may in fact reduce anxiety by diminishing supernatural fears, or by providing rational solutions for anxiety-arousing cultural problems. It is very clear, however, that prolonged, intensive contact with Western civilization eventually disorganizes and destroys the social structure of a tribe. It also seems likely that rapid, far-reaching acculturative changes will be facilitated by an originally loose tribal social organization. For these reasons, it seems reasonable to suppose that Horton's relation between drunkenness and acculturation indicates an underlying process of loosening of a traditional social organization, not increased anxiety.

The general conclusion indicated by the findings to this point is: drunkenness in primitive societies is determined less by the level of fear in a society than by the absence of corporate kin groups with stability, permanence, formal structure, and well-defined functions.

Marital residence. Table 4 shows that there is a strong relation between degree of drunkenness and the form of marital residence. *Bilocal* residence means that the newlyweds customarily make a decision at marriage whether to reside near the bridegroom's parents or near the bride's parents; *neolocal* residence means that they reside in a new location. All 11 tribes with bilocal or neolocal residence are above the median on degree of drunkenness. The five tribes with *uxoripatrilocal* residence, in which residence with the bridegroom's family follows a period of residence with the bride's family, are also uniformly above the median of drunkenness. The large number of societies in which the bride leaves her home and lives with the groom near his family, a practice called *patrilocal* residence, includes, on the other hand, most of the tribes below the median on drunkenness. Tribes with *matrilocal* or *avunculocal* residence (in which the newlyweds live in the first case with the wife's family and in the second with the groom's mother's brother) are harder to place. The former seems to show the extreme drunkenness pattern while the latter seems to be much closer to the sober patrilocal pattern.

Since two of the avunculocal tribes have residence shifts partially similar to the uxoripatrilocal tribes, they have been classified with the nonpatrilocal tribes, in accordance with conservative procedure. If, on the other hand, the crucial variable here is immediate residence with a male exercising authority over the groom, and if avunculocal tribes are therefore grouped with patrilocal tribes, the relation between residence and drunkenness will be still stronger ($r = .71$, $\chi^2 = 11.81$, $p < .001$, $N = 55$).

The correlation between nonpatrilocal residence and drunkenness ($r = .64$) is almost as high as the correlation between bilateral descent and drunkenness.[5] There is a correlation of .67 between nonpatrilocal residence and bilateral descent. Although these two predictor variables are importantly related to each other, each predicts some independent portion of the total variance in drunkenness scores. The multiple correlation using these two variables as joint predictors of drunkenness is .72. This finding means that the degree of dependence of drunkenness on form of residence and kin group is so strong that its level can be predicted with moderate accuracy in individual tribes from just these two variables alone. If the tribes with unilineal moieties had been grouped with the bilateral tribes, and the avunculocal with the patrilocal tribes, the multiple correlation would have been considerably higher.

The extreme drunkenness of the neolocal and bilocal tribes appears to reflect the fact that it is difficult for a society to develop extended lineal kinship structures if independent choice of residence is permitted the newlyweds. It is also probable that if the married couple is permitted to choose its residence, it will not be constrained by lineal and corporate rules in many other areas. There seems to be an intimate relationship between drunkenness and personal choice, absence of institutionalized constraints, and isolation of both the nuclear family and the individual from corporate kin structures.

Absence of drunkenness, on the other hand, appears to be associated with male dominance reflected in patrilocal and perhaps avunculocal residence. We know that the father or the paternal grandfather usually wields authority in the household (Murdock, 1949, p. 39) but we also know that he loses some of his power to his wife in matrilocal societies (Simmons, 1937, pp. 507–508; Strodtbeck, 1951). There are several reasons. In husband-wife conflicts, the wife will be supported by her own kinsmen, who are close at hand under matrilocal residence. Furthermore, since the man does not own land or the household in these tribes, his ties to his wife's home are weaker. Moreover, since the mother's brother can take on child-rearing duties, the father will be loosely attached to the nuclear family. He may divide his time between his sister's household and his wife's household, with both his authority and his obligations divided between them. For reasons such as these, Linton (1949) concludes that marriage is particularly unstable in matrilineal societies.

By contrast, according to Linton, in patrilineal societies the woman is integrated into her husband's lineage by a series of marriage-stabilizing devices motivated by the interest of the husband's family in the children. With patrilocal residence, which means that the wife's kinsmen are at a distance and cannot support her, the father's formal and de facto authority grows. With increased centralization of familial power and the formation of hierarchical chains of command in the family, extreme drunkenness diminishes.

The tribes with uxoripatrilocal residence are significantly more drunken than the patrilocal tribes. In these five tribes, as in many of the matrilocal tribes, the bridegroom is required to serve as a hired hand for his wife's family, and must often undergo trials to determine his suitability for the marriage. It is only after this that he acquires full marital rights or, with uxoripatrilocal residence, permission to remove the bride from her own family. One implication of this custom is that the husband's authority over his wife develops only gradually and is not clearly reinforced by external social influences. A second implication follows from the fact that there are at least two shifts of marital residence in these tribes, and that authority over a man is not consistently held by a single person but is successively taken by his father and father-in-law. This system means the attenuation of lineal structures and so, perhaps, more drunkenness.

Although sobriety is obviously associated with a father-centered family, the crucial factors mediating this relationship are not entirely clear. One possibility is that there is a general increase in respect, reserve, and self-control in these tribesmen, because most of their interpersonal relations are organized hierarchically. Another possibility is that with patrilocal residence there is decreased mobility on the part of the heads of the households; therefore there will be a greater opportunity for the formation of formalized social ties in the community and, in turn, strong pressure exerted against interpersonal aggression, informal self-indulgence, and related individualistic behavior prominent at drinking bouts.

Still another possibility is that there may be a general stabilization of society's interpersonal relationships when the authority of the father in any society is legitimized and extended by rules of residence and descent; that is, in societies whose internal power struggles by individuals or groups are minimized in normal interpersonal relationships, prolonged drinking bouts—with their inevitable changing, transient coalitions—and conflicts will also be minimized. Certainly these factors are not mutually exclusive. In fact, each to some extent implies the other.

One further possible stabilizing influence might be role differentiation and specialization within the family; with a highly stable power structure it would often be possible for the father to delegate increased authority to his wife in defined areas such as child rearing and routine household management while he specializes in, for example, ceremonial or political activities. This role differentiation would in turn make husband and wife more interdependent, minimize conflicts over allocation of rights and privileges, make the family a more closely cooperating unit pursuing mutually accepted goals, and therefore suppress highly personal, individualistic drinking behavior. Quite a different result would be expected, on the other hand, in modern urban society, where corporate division of labor means that the average worker is minimally involved in company goals, has relatively little commitment to his co-workers and supervisor, is free after work hours, and can change jobs at will. These conditions have little in common with those just discussed as prevailing in father-centered tribes and may have opposite effects on drunkenness.

To summarize, drunkenness increases markedly if the authority of the man in the household is lessened or diffused and if the nuclear family is less integrated into larger kin structures through bilocal or neolocal residence.

The clan-community. A clan has been defined by Murdock (1949) in terms of three criteria: (1) it is a kin group based on a unilinear rule of descent uniting its central core of members; (2) it has residential unity; and (3) it shows clear evidences of social integration, such as positive group sentiment and recognition of in-marrying spouses as an integral part of the membership. Our data show that as the community approaches an exogamous clan organization, drunkenness decreases. More precisely, there is a significant correlation of −.59 between drunkenness and approach to a clan system in community organization (see Table 5). The development of clan segmentation within a tribe implies both increased internal solidarity and increased separation or cleavage within the community. In other words, an individual is more strongly integrated

TABLE 5
The Relation between Drunkenness and Additional Measures of Social Structure

	χ^2	r_{tet}	*N*
Approach to an exogamous clan-community	6.67[a]	−.59	51
Presence of a bride-price (versus bride-service or no material consideration)	4.64[a]	−.51	52
Nonsororal polygyny (versus monogamy and sororal polygyny)	3.26	−.46	46
Patrilineal extension of exogamy	7.24[b]	−.65	45
Matrilineal extension of exogamy	2.32	−.44	43
Bilateral extension of exogamy	0.00	−.02	46
Degree of political integration	2.88	−.41	55
Degree of social stratification	0.54	−.22	52
Presence of slavery	1.49	−.35	46
Nuclear household (versus extended, mother-child, or polygynous household)	0.56	.25	49

[a]$p < .05$. *Approach to a clan community* compares societies with exogamous clan communities, exogamous communities, and exogamous divisions within the community against societies without reported clans or with demes (communities without localized exogamous units and with a marked tendency toward local endogamy). *Degree of political integration* compares politically independent local communities of less than fifteen hundred average population and societies where family heads acknowledge no higher political authority with minimal states, little states, dependent societies, and societies with peace groups transcending the local community. *Slavery* compares societies with hereditary and nonhereditary slavery with societies with absence or near absence of slavery.
[b]$p < .01$.

within his own clan, but at the same time he is probably more isolated from individuals belonging to other clans. Seen from the outside, clans are exclusive corporations; this clannishness probably militates against the diffuse friendliness and broadly inclusive brotherhood that is an integral part of social drunkenness.

The clan differs from sibs and lineages by excluding as members siblings who have married out of the group. With the clan structure, then, the legal ties of mutual obligation linking siblings are not as close as with sib and lineage structures, and dispersed families are not unified by these sibling relations. Community integration, therefore, may well be weaker with the clan than with the sib structure, although household integration localized around the residentially unified clan should be at least as strong. If the extension of impersonal ties of obligation and control throughout the community helps diminish drunkenness, the negative relationship between the clan and drunkenness should not be as strong as that between drunkenness and siblike structures. This is, in fact, what is found: approach to the clan community correlates −.59 with drunkenness, while sibs and segmentary lineage organizations correlate −.76 with drunkenness.[6]

Approach to a clan-community is correlated .74 with unilineal descent and .60 with patrilocal residence, with both relations significant at the .01 level of confidence. Since these three predictors of drunkenness are all positively intercorrelated to a moderate or strong degree, they appear to be variations on a single theme, such as corporate, formal social organization or traditional, hierarchical social solidarity. Although the intercorrelation matrix was not computed, many of the predictors of drunkenness reported here may well be positively intercorrelated. If they are, both the predictors and the drunkenness score could be conceived as loading highly on a general factor of corporate (formal) versus personal (informal) social organization.

Bride-price and form of marriage. In primitive tribes there are several forms of exchange of services, goods, and gifts at marriage. One form previously discussed is bride-service, which is associated with lessened paternal authority and with extreme drunkenness. However, there is another form of payment for the bride that is more likely to reinforce the husband's authority over his wife, namely, the bride-price.[7] As might be anticipated, there is a moderate but statistically significant relation between the presence of a bride-price and the absence of extreme drunkenness (Table 5). Undoubtedly this relation reflects some of the previous factors shown to be associated with little drunkenness, since the bride-price is highly correlated, for example, with patrilocal residence.

Certain important functions of the bride-price, however, suggest that it may have some independent influence. One such function suggested by Linton (1949) and Gluckman (1950) is the stabilization of marriage.[8] If the husband mistreats his wife, she may return to her parents' home and the husband must forfeit the goods he has paid for his bride. If, on the other hand, the bride is lazy or shrewish, she may be returned to her parents' home, and they must return the bride-price. Payment of a bride-price may also stabilize marriage by vesting in the father legal rights over his wife's children; if the wife returns to her parental home, she must abandon them. Moreover, the bride-price also creates ties between separated lineages. Malinowski (1929) points out that the presents given at marriage by the husband have often been contributed in part by his kinsmen, and are often shared by the bride's relatives and clansmen as well as her parents. He concludes that these transactions bind two groups rather than two individuals. All such legalized, impersonal ties between individuals and groups may be an important factor in diminishing loosely structured, informal behavior in ordinary social relations, including drinking bouts.

With regard to the actual form of marriage, extreme drunkenness appears to be slightly more extensive in societies with monogamy or sororal polygyny than in societies with nonsororal polygyny (Table 5). (This correlation, of borderline significance, reaches the 5% significance level by comparing only societies having more than 20% of the marriages nonsororally polygynous with monogamous societies [$\chi^2 = 3.87$, 1 *df*; $p < .05$, $r = -.66$, $N = 27$].) This correlation is probably an indirect one. Societies with nonsororal polygyny usually have patrilocal residence, whereas sororal polygyny is more frequent and technically more feasible under matrilocal residence. Furthermore, it is doubtful whether nonsororal polygyny itself plays a direct causal part in diminishing drunkenness. In these societies plural marriages are limited to the wealthy. The poorer tribesmen are usually monogamous and would be expected to exhibit the tendency toward more extreme drunkenness evident for other monogamous groups.

Rules of exogamy. Whereas all societies forbid marriage with a small number of close relatives, societies differ in the extent to which rules of exogamy are extended to more distant kinsmen. Patrilineal extension of rules of exogamy, prohibiting marriage between individuals tracing descent through males to a common ancestor, is associated with the presence of patrilineal kin groups and should therefore be negatively correlated with drunkenness. Matrilineal extension of exogamy should also be negatively related to drunkenness for similar reasons, but since there are fewer matrilineal than patrilineal societies in the sample, the relationship should be a good deal weaker. The results in Table 5 confirm this expectation.

The expected relationship between drunkenness and degree of bilateral extension of exogamy, that is, to kinsmen traced through both sexes, is not so obvious. If extended rules of exogamy reflect a stable society, or by themselves create structure in a society, they might be related negatively to drunkenness. If, on the other hand, they indicate the absence of a corporate unilineal group, they might be related positively to drunkenness. The relationship actually found in Table 5 shows no important trend in any direction and is difficult to interpret.

Household type and settlement pattern. The absence in our data of any important relation between household type and drunkenness was unexpected, especially because household type tends to predict a number of important psychological phenomena, including child-rearing practices (Murdock & Whiting, 1951; Whiting, 1959). The correlation between household type and drunkenness is empirically maximized by the testing of nuclear and polygynous households against extended and mother-child households. Even so, it is still an insignificant and unimpressive .36, and this combination of household types is hard to rationalize theoretically. No other combinations of household type yield a significant or nearly significant relation to drunkenness. Nuclear households tested alone against other household types (Table 5) correlate only .25.

There is, however, a significant relation between drunkenness and patterns of settlement, as rated by Murdock (1957).[9] Because it was not possible to order his categories confidently with respect to degree of community organization, a correlation coefficient based upon a dichotomization would not be very meaningful. However, Murdock's two extreme categories—"nomadic bands" versus "complex villages and towns"—show a significant difference in degree of insobriety. All 9 nomadic tribes are characterized by extreme drunkenness, and 13 of 27 tribes with compact villages or towns exhibit this pattern. It seems likely that factors of the sort already mentioned explain these relations, because tribes with very small population wandering over an extensive territory would have little chance of elaborating the kind of social organization apparently conducive to sobriety. There may well be an independent stabilizing influence exerted by the small village, however. The only two bilateral tribes that were relatively sober (Macusi and Taulipang) had a village settlement pattern.

Social stratification and political integration. The final aspects of social organization investigated in this study in relation to degree of drunkenness were extent of social stratification (including the presence of slavery) and degree of political integration into units transcending the local community. Although the elaboration of these aspects of social organization signifies population growth and complexity in social arrangements relative to small nomadic tribes, they are not of the same order in the structuring of social relations as the development of corporate kin groups is. Indeed, stratification and political integration may, in the course of social evolution, proceed at the expense of corporate kin groups and with consequences other than sobriety.

Hence it is not surprising, but nevertheless important, that none of the measures considered in this connection yielded a significant relation with degree of drunkenness. Within the range of complexity of social organization represented by our sample, there was a tendency for political integration and, to a lesser extent, stratification to be negatively correlated with drunkenness, but not significantly so. The lack of very strong relationships in connection with these variables implies that kin group and residence factors may be the important causal elements, in that they do not appear to influence drunkenness indirectly by mutual correlation with stratification or political factors.

TABLE 6
The Relation of Degree of Drunkenness to Selected Child-Rearing Measures[a]

Child-Rearing Measures	N	χ^2	z	r_{tet}
Aggression socialization anxiety	26	1.39	1.97[b]	−.47
Pressure for obedience and responsibility rather than self-reliance and achievement (from 5 to 12 years)[c]	20	3.23	2.76[d]	−.71
Indulgence during childhood (from 5 to 12 years)	25	6.74[d]	1.66	.81
Indulgence of the infant	25	2.06	1.00	.55

[a] χ^2 was computed at 1 *df*; z was estimated by the Mann-Whitney test for rank-ordered data.
[b] $p < .05$.
[c] See Herbert, Child, and Bacon, 1959.
[d] $p < .01$.

Tests of Some Social-Psychological Variables

In a study of this kind it is important to explore variables other than those that obviously reflect social organization. Since this study is reexamining many of the hypotheses of Horton—hypotheses that were essentially psychological in conception—it is necessary to pursue their implications in the light of all available new evidence. Moreover, the social-psychological and clinical literature contains a number of points of view on drunkenness and alcoholism testable by the cross-cultural method. The recent proliferation of cross-cultural indices not only makes tests of alternative hypotheses possible, but also permits a crude comparison of the power of variables of social organization with other kinds of variables in explaining drunkenness in primitive societies.

Aggression anxiety. An important feature of Horton's theory attributes the relative absence of drunkenness to fear concerning the expression of previously punished aggressive impulses. Yet in point of fact, all but one of Whiting and Child's (1953) measures of aggression training or aggression in fantasy failed to relate significantly to Horton's drunkenness score. The single exception is shown in Table 6: there is a barely significant inverse relation between drunkenness and punishment for aggression in childhood, supporting this aspect of Horton's theory. However, the inverse relation between drunkenness and general pressure for responsible performance of duties and obedience is slightly stronger, suggesting that drunkenness is correlated with generalized lack of discipline instead of with permissive aggression training alone.

This interpretation seems to be supported by the other results in Table 6, which show that indulgence during childhood correlates significantly with drunkenness. Childhood indulgence is defined as a function of: protection from environmental discomforts; amount of overt affection shown the child; degree, immediacy, and consistency of drive reduction; constancy of presence and absence of pain inflicted by the nurturant agent; and lowness of degree of socialization demanded (Barry, Bacon, & Child, unpublished). (There were only a few tribes that were exceptions to this relationship—that is, that were both sober and indulgent to their children, or drunken and severe to their children—but these exceptions were quite marked. Since a rank test reflects the degree of exception as well as the presence of an exception, the association drops from high significance with a median division to borderline significance with the use of a rank test.) Indulgence of the infant shows a similar trend to indulgence of the child, but is not statistically reliable.[10]

Wright (1954) has previously shown that societies that punish aggression severely do not thereby eliminate aggressive responses; instead, the object of aggression is changed. In societies with high aggression socialization anxiety, agents and objects of aggression in folktales are likely to be strangers, signifying displacement along a scale of similarity-dissimilarity to the hero and his family. My interpretation of these results is that in the sober societies aggression is displaced toward, and expected from, an out-group, which is a logical corollary of the proposition that little drunkenness is found in societies segmented into a number of internally solidary lineal units. A reasonable general conclusion suggested by these facts is that fear of aggression directed against close relatives at drinking bouts is a more genuine causal influence diminishing drunkenness than is generalized fear of aggression, and that generalized pressure for strict obedience and self-control is still more important.

Wright (1954) has also shown that acts of fantasy aggression are in fact less intense in tribes that allow their children freedom to express aggression—that is, in the drunken tribes. This finding suggests, on the one hand, that the drunken tribes do not express aggression against distant or inappropriate objects; in other words, they are neither unusually conflicted about aggression nor "unconsciously hostile" in the psychiatric sense. On the other hand, the aggression of the drunken tribes will be much more obvious against family members and friends, because the psychological mechanisms enforcing submissiveness and respect are weak. Interpersonal relations seem to be stronger and more lasting in the sober tribes, but only at the cost of repression of aggression. The superficially strong interpersonal ties in the sober tribes are much more ambivalent than the loose, informal relations in the drunken tribes. This tentative formulation of drunken aggression needs further testing, on the other hand, in view of the lack of relation between drunkenness and the indices of projected aggression in Table 1.

Sex anxiety. In his earlier study, Horton reported an association between premarital sex freedom and drunkenness, and in this chapter we have noted an inverse relation between degree of drunkenness and the extension of certain rules of exogamy. There are two major possible interpretations of these findings. Horton's explanation is that they both simply reflect sex anxiety, which is strong in the tribes that have little drunkenness. A more logical possibility, however, is that they both reflect hierarchical control of a variety of marital choices that might disrupt a corporate social organization.

As a further test of the sex anxiety hypothesis, drunkenness was correlated with the five best currently available measures of sex anxiety—measures based on the content of fantasy as well as actual punishment for sex play in childhood. More specifically, the measures used were: (1) sex socialization anxiety, (2) sex explanations of illness, (3) sex avoidance therapy, (4) duration of sex taboo during pregnancy, and (5) duration of sex taboo after childbirth.[11] None of these measures is related significantly to degree of drunkenness, nor is any trend apparent. The most promising nonsignificant negative relationship—with sex socialization anxiety—was not improved by use of a more powerful statistical test. The scale with the best reliability, as determined by its correlations with other measures, is the duration of the taboo on sexual intercourse following childbirth (Ayers, 1954). Although there are very few cases, there does not appear to be a better correlation with this measure than with others. The correlations between sex anxiety and drunkenness prove to be so generally weak that if a larger sample shows a genuine negative relation, it will probably be merely an indirect reflection of other factors. Factors that might produce such an indirect relation include generally restrictive social control, extended rules of exogamy, or incest fears generated in a tightly knit community.

In addition to the above, three more sets of tests bearing upon the relations of drunkenness and sex anxiety were made; each set involved the use of indices suggested by the frequent assertion that a major conflict in the lives of alcoholics centers around latent homosexuality.[12]

The first of these made use of Ford and Beach's (1951) list of societies in which homosexual activities are normal and permitted for some members, frequently a *berdache* (male transvestite) who assumes a female role in life. This scale, however, is probably a much better index of overt than latent homosexuality. When societies permitting homosexual activities on the part of some members are compared with those in which such activities are reportedly absent, rare, or secret, the findings indicate that there is no correlation whatsoever, positive or negative, with degree of drunkenness.

The second set of tests made use, in contrast, of a very indirect measure, namely, Stephens' (1959, 1961) ingenious Guttman scale of taboos on menstruating women. Stephens noted that societies that isolate menstruating women in special huts or forbid them to cook believe that at this time they are dangerous to men, although not to themselves, to other women, or to children. In these societies there is also severe sex training, and the father is the main disciplinarian. These facts suggested to Stephens that menstrual taboos might be an indirect index of the psychoanalytic concept of castration anxiety, which has been claimed to be a factor in the causation of homosexuality. There are, of course, other logical interpretations of this scale; it may quite possibly indicate just sex anxiety, without any necessary implication for the development of homosexuality. However, there is no significant relation between Stephens' scale and the extent of drunkenness in the primitive societies sampled in this study.

Finally, consideration was given to the possibility of a correlation between drunkenness and male initiation rites at puberty, because Whiting and his collaborators (Whiting, Kluckhohn, & Anthony, 1958) have contended that one of the several functions of these rites is the attempted removal of a latent female identification in males. Once again, however, the findings show no significant relation using the ratings in Anthony (1956).

Although none of these latter variables is significantly related to drunkenness, there is a hint of a moderate negative relation with menstrual taboos sufficient to suggest that a significant correlation might be established with a larger sample. Yet if there is a genuine relation of this kind, it may reflect structural factors; periodic isolation of women demands a variety of special and relatively complex social arrangements, including provision for nonmenstruating women to assume the duties of the isolated women, which may be technically difficult for a small band of hunters to maintain. In any event, we are forced in the light of the present evidence to conclude that however important they may be in individual cases of alcoholism, latent or overt homosexuality or sex anxiety are not important causal variables in extreme drunkenness in primitive societies.

It is impossible to rule out an indirect association between drunkenness and one possible kind of surface femininity. The personal, friendly behavior of the drunken tribes seems to have something in common with the traditional female sentimental, expressive role, while the impersonal, self-controlled behavior of the sober tribes shows similarities to the instrumental, command behavior associated with the traditional male role.

At the same time, the emphasis upon diffuse brotherhood in the drunken tribes may produce friendly, companionate behavior among men. This possible surface femininity, however, seems to be quite different from the paranoid fear and hostility that has been claimed to be associated with unconscious homosexual impulses.

Orality and self-injury. Many psychoanalysts and psychiatrists (Fenichel, 1945) have identified oral fixations or frustrations as important causes of inebriety and alcoholism, following Freud's opinions concerning the relation between oral-erogenous sensitivity and drinking (1953, p. 182). Without implying an equation between drunkenness and alcoholism, it seems pertinent in these circumstances to explore the relations between drunkenness and the best available measures of oral fixations and frustrations in primitive societies. Tests of significance and correlations were therefore computed for degree of drunkenness in relation to the following measures taken from Ayres, 1954; Friendly, 1956; and Whiting and Child, 1953: (1) age of weaning; (2) oral socialization anxiety; (3) oral satisfaction potential; (4) oral explanations of illness; (5) oral performance therapies (eating herbs or medicine to cure illness); (6) oral avoidance therapies (spitting, vomiting, adhering to food taboos to cure illness); (7) number of food taboos during pregnancy; (8) feasts during mourning; and (9) food taboos during mourning.

The results indicate no important relation between drunkenness and any of these measures. Nor do the variables with the highest reliability (oral socialization anxiety and oral explanations of illness) appear to show stronger relations than the other variables. Thus, while oral fixations may play a role in particular instances of alcoholism, they do not appear to explain variations in drunkenness among primitive peoples insofar as they are adequately indexed by these measures.

Because psychoanalysts and psychiatrists have often singled out self-destructive motives as important factors in alcoholism,[13] an attempt was made to determine what bearing measurable cultural tendencies toward self-injury have upon drunkenness in primitive societies. For an index of self-injury, reliance was placed on the work of Friendly (1956), who measured a number of variables of ascetic mourning. Friendly found that many of these (including vigils by the mourner, change in appearance, property destruction, purification, etc.) were positively interrelated, and she consequently developed, by summation, an overall scale of ascetic mourning that proved to be positively related to severe child-training practices.

One important variable of ascetic mourning, namely, degree of self-inflicted bodily injury, did not correlate positively with Friendly's other measures and was therefore treated separately in this analysis. The results show, however, that neither this latter measure nor the scale of ascetic mourning are significantly related to drunkenness in the societies considered in this study. Hence, the limited cross-cultural evidence suggests that there is no important general association between drunkenness and self-injury.[14]

A Note on Ethnic Differences in Rates of Alcoholism

Caution is necessary in applying these results on primitive drunkenness directly to national or ethnic variations in alcoholism, because these phenomena have important differences as well as similarities. But it is a fact that some father-centered ethnic groups with a high degree of lineal social solidarity—such as the Jews and the Chinese (Snyder, 1958)—have very low rates of alcoholism and drunkenness, as the conclusions reported in this chapter would predict. These cross-cultural findings also suggest a highly speculative possibility that might be worth checking: that some father-centered Western countries with a highly traditional, lineal social organization (such as Italy) have developed preferences for wine or beer rather than distilled liquors in part *because* of their social structure and that these preferences should discourage extreme drunkenness.

The Irish present a different problem. They are the group most overrepresented among American alcoholics, and yet they seem on the surface to be a traditional father-dominated group. The most prominent current hypothesis concerning Irish drunkenness was proposed by Bales (1946), and was based in part on Horton's anxiety theory. Bales noted instances of anxiety in Irish life and, following Horton, suggested that high anxiety played a causal role in Irish drunkenness. However, some of the anxiety-arousing conditions he mentioned, such as absentee landlordship and penal restrictions on Irish industry, can be explained just as well as indices of the alienation of the Catholic majority from the controlling jural, formal aspects of their own society (the imposed legal, governmental, and economic systems, and the established Protestant church).

One writer on Ireland, in fact, uses the phrase "disorganized social system" (Potter, 1960, p. 46) to describe the wholesale evictions and the general loosening of ties to the land in the 18th and 19th centuries. Furthermore, the authority of the Irish father in the home does not seem to be as impressive as paternal authority in southern Europe. Bales, for example, comments that the Irish-American father "in many cases seems to have dropped into a role of impotence and insignificance." Bales also noted a close mother-son attachment in many of his Irish alcoholics, and Opler and Singer (1956) found a very close, dependent relation of the

Irish son to his mother, who keeps him a "boy and a burden." These authors also call the Irish father "shadowy and evanescent." Factors such as these have been shown to be associated with drunkenness in primitive tribes in the present study. These suggestions underscore the need for a more extensive treatment of Irish drinking patterns in the light of these new cross-cultural findings.

Methodological Check

By way of a methodological check on the findings and conclusions of this chapter, degree of drunkenness was examined in relation to Horton's five culture areas, and a significant relation between these variables was found[15]—the African tribes being the least drunken, the South American the most drunken. This finding raises the possibility that the associations found might be explained on the basis of historical accident and parallel diffusion to contiguous tribes, rather than on the basis of cause and effect.

An unpublished technique to deal with questions of this kind has recently been invented by Roy D'Andrade of the Harvard Laboratory of Human Development and is summarized here with his permission. He reasons that the diffusion hypothesis would be supported if an expected relationship generally failed to occur in each of a pair of neighboring tribes, whereas the causal hypothesis would be supported if diffusion of both the antecedent and consequent had not usually occurred from one to another of a pair of neighboring tribes—between which diffusion might easily take place. He therefore divides the sample of tribes into the geographically closest possible pairs, excluding societies that do not form close pairs, and examines the patterns of relationship within these pairs.

D'Andrade examines the few cases in which the expected relationship fails to occur in each of a pair of geographically contiguous societies—that is, the usual antecedent does not produce the expected consequent in either tribe. Each of these cases he calls a victory for the historical-diffusion hypothesis. He then examines the few cases in which the predicted relation occurs in only one of a pair of geographically contiguous societies—that is, in which only one tribe has both the predicted antecedent and consequent, and the other tribe has neither. Each of these cases is a victory for the causal-functional hypothesis. D'Andrade's method was usable with Horton's sample even though Horton had excluded close pairs of similar tribes, since relatively close and relatively distant tribes could be distinguished.

Application of D'Andrade's technique to the relation between drunkenness and form of kin group, residence, and clan-community organization indicated victories for the causal-functional hypothesis by 5 to 1, 4 to 1, and 3 to 1, respectively. Two of the three apparent victories for the historical-diffusion hypothesis were accounted for by the Hopi-Zuni pair. These tribes have a strong degree of traditional community solidarity, unilinear sibs, and little drunkenness, but matrilocal residence and agamous or deme community organization. This is probably not a genuine case of diffusion. It seems rather to signify two independent examples of the dominant influence of sib solidarity over opposed factors (such as marital weakness) in reducing drunkenness. The available evidence, then, does not support a diffusionist interpretation of the correlations reported in this chapter.

Summary

This chapter has reported an extensive reexamination of Horton's cross-cultural study of the functions of alcohol in primitive tribes. The most important single conclusion was that the degree of drunkenness at periodic communal drinking bouts is related to variables indicating a personal (or informal) rather than a corporate (or formal) organization but is substantially unrelated to the level of anxiety in the society.

Following are some of the variables found to be positively correlated with relative *sobriety* in primitive tribes: (1) corporate kin groups with continuity over time, collective ownership of property, and unified action as a legal individual; (2) patrilocal residence at marriage; (3) approach to a clan-community organization; (4) presence of a bride-price; (5) a village settlement pattern (rather than nomadism). It was suggested that societies with these features are likely to be well organized, to have a high degree of lineal social solidarity, and to have interpersonal relationships structured along hierarchical or respect lines. It was hypothesized that these factors in turn controlled extremely informal, friendly, and loosely structured behavior at drinking bouts. This interpretation was supported by the fact that the sober tribes control aggression severely in their children, whereas the drunken tribes are relatively indulgent with their children and permit disobedience and self-assertion.

No indices of fear were found that correlated significantly with drunkenness, and it was concluded that Horton's measures of an insecure food supply and acculturation indicated a loose social organization rather than fear. Indices suggested by psychoanalytic hypotheses about alcoholism (oral fixations, latent homosexuality, and drives toward self-injury) did not predict primitive drunkenness. Suggestions about Irish alcoholism in the light of these results were made.

Acknowledgments

The author is greatly indebted to John W. M. Whiting and David C. McClelland for research guidance, to George P. Murdock and Herbert Barry for suggestions and unpublished ratings, and to the National Science Foundation and the U.S. Public Health Service for predoctoral fellowship support.

Notes

1. The source of each scale used in the tables is given in the References, following these notes.
2. The scale also excludes solitary drinking and drinking by women. The few societies without available alcohol were excluded, and so were the societies with no satisfactory ethnographic information about drunkenness. Societies similar to a neighboring society already in the sample were excluded. The societies that were included come from all over the world, with a number of representatives in each major culture area. Horton excluded the Moslem societies of North Africa, since they have a religious taboo on alcohol; this was a questionable decision, but it probably affects neither his general conclusions nor mine.
3. A few other measures of anxiety given by Whiting and Child also showed no significant relation to drunkenness, but these either were based on such small numbers or were so unclear, theoretically, that they are not presented here.
4. $\chi^2 = 9.94$, 1 *df*; $p < .01$; $N = 55$. The degree of drunkenness was dichotomized at "extreme" versus "moderate or slight" in accordance with Horton's definitions noted above.
5. For patrilocal versus all other rules of residence, $\chi^2 = 8.92$, 1 *df*; $p < .01$; $N = 55$.
6. This last correlation was derived by grouping societies with lineages, moieties, and phratries together with bilateral societies and comparing them to societies with sibs and segmentary lineage organizations.
7. This is quite different from the familiar European dowry; it is a payment by the bridegroom to his wife's family compensating them for the loss of an economically productive member and giving the groom the right to remove his bride from her parents' home.
8. Gluckman (1950) notes, however, that there is no simple relation between the presence of a bride-price and the frequency of divorce in a society. Ratings on frequency of divorce (Murdock, 1950) are available for only 15 of Horton's tribes. A statistically unreliable correlation of $-.11$ ($\chi^2 = 0.00$) between drunkenness and the divorce rate was obtained. Because of the small number of cases little confidence can be placed in this relationship. The bride-price may well stabilize the jural (contractual-obligatory) aspects of marriage more than the affectional-companionship aspects; for information on the complex issues involved, see Murdock (1950) and Schneider (1953).
9. $\chi^2 = 11.03$, 4 *df*; $p < .05$, $N = 55$.
10. It is probably relevant in this connection that Parker (1959) has found 15 articles reporting a close mother-son attachment in alcoholics; and several authors have reported that youngest children in a family are both relatively indulged and overrepresented among alcoholics (Navratil, 1959).

11. The *N*s for the tests of significance and correlations utilizing these measures were, respectively: 21, 27, 23, 16, and 14. The scales were taken from Ayres (1954) and from Whiting and Child (1953).
12. *N* ranges from 20 to 24 in these three cases. On the theme of alcoholic conflict over latent homosexuality, see for example Parker (1959) and Fenichel (1945).
13. The work of Karl Menninger (1938) is an outstanding example. A similar relation is suggested by anomie theories of suicide.
14. This finding does not preclude the possibility of a positive relation between drunkenness and other forms of guilt. However, the available measures of guilt (Whiting & Child, 1953) were based on too few tribes to permit an estimate of such a relation.
15. $\chi^2 = 9.66$, 4 *df*; $p < .05$, $N = 56$.

References

Anthony, A.S. (1956). A cross-cultural study of factors relating to male initiation rites and genital operations. Unpublished doctoral dissertation, Harvard University.

Ayres, B.C. (1954). A cross-cultural study of factors relating to pregnancy taboos. Unpublished doctoral dissertation, Radcliffe College.

Bales, R.F. (1946). Cultural differences in rates of alcoholism. *Quarterly Journal of Studies on Alcohol*, *6*, 480–499.

Barry, H.,III, Child, I.L., & Bacon, M.K. (1959). Relation of child training to subsistence economy. *American Anthropology*, *61*, 51–63.

Conger, J.J. (1956). Alcoholism: Theory, problem, and challenge, II: Reinforcement theory and the dynamics of alcoholism. *Quarterly Journal of Studies on Alcohol*, *17*, 296–305.

Davenport, W. (1959). Nonunilinear descent and descent groups. *American Anthropologist*, *61*, 557–572.

Fenichel, O. (1945). *The psychoanalytic theory of neurosis*. New York: Norton.

Ford, C.S., & Beach, F.A. (1951). *Patterns of sexual behavior*. New York: Harper and Brothers.

Freud, S. (1953). Three essays on the theory of sexuality. In J. Strachey (Ed. and Trans.), *The Standard edition of the complete psychological works of Sigmund Freud* (Vol. 7). London: Hogarth Press. (Original work published 1923)

Friendly, J.P. (1956). A cross-cultural study of ascetic mourning behavior. Unpublished undergraduate honors thesis, Radcliffe College.

Gluckman, M. (1950). Kinship and marriage among the Lozi of northern Rhodesia and the Zulu of Natal. In A.R. Radcliffe-Brown & D. Forde (Eds.), *African systems of kinship and marriage*. London: Oxford University Press.

Heath, D.B. (1958). Drinking patterns of the Bolivian Camba. *Quarterly Journal of Studies on Alcohol*, *19*, 491–508.

Horton, D. (1943). The functions of alcohol in primitive societies: A cross-cultural study. *Quarterly Journal of Studies on Alcohol*, *4*, 199–320.

Hovland, C.I., & Sears, R.R. (1940). Minor studies of aggression, VI: Correlation of lynchings with economic indices. *Journal of Psychology*, *9*, 301–310.

Lemert, E.M. (1954). Alcohol and the Northwest coast Indians. *University of California Publications in Culture and Society*, *2*, 303–406.

Lemert, E.M. (1956). Alcoholism: Theory, problem, and challenge, III: Alcoholism and the sociocultural situation. *Quarterly Journal of Studies on Alcohol, 17*, 306--317.

Linton, R. (1949). The natural history of the family. In R.N. Anshen (ed.), *The Family: Its function and destiny*. New York: Harper and Brothers.

Lowie, R.H. (1948). *Social organization*. New York: Holt, Rinehart and Winston.

Malinowski, B. (1929). Marriage. In *Encyclopedia Britannica* (14th ed., Vol. *14*). London and New York: Encyclopedia Britannica.

Menninger, K.A. (1938). *Man against himself*. New York: Harcourt, Brace and World.

Murdock, G.P. (1949). *Social structure*. New York: Macmillan.

Murdock, G.P. (1950). Family stability in non-European cultures. *Annals of the American Academy of Political and Social Science, 272*, 195–201.

Murdock, G.P. (1957). World ethnographic sample. *American Anthropologist, 59*, 664–687.

Murdock, G.P., & Whiting, J.W.M. (1951). Cultural determination of parental attitudes: The relationship between the social structure, particularly family structure, and parental behavior. In M.J.E. Senn (Ed.), *Problems of infancy and childhood*. New York: Josiah Macy Foundation.

Navratil, L. (1959). On the etiology of alcoholism. *Quarterly Journal of Studies on Alcohol, 20*, 236–244.

Opler, M.K., & Singer, J.L. (1956). Ethnic differences in behavior and psychopathology: Italian and Irish. *International Journal of Social Psychiatry, 2*, 11–23.

Parker, F.B. (1959). A comparison of the sex temperament of alcoholics and moderate drinkers. *American Sociological Review, 24*, 366–374.

Pehrson, R.N. (1954). Bilateral kin groupings as a structural type: A preliminary statement. *Journal of East Asiatic Studies, 3*, 199–202.

Pehrson, R.N. (1957). The bilateral network of social relations in Könkämä Lapp District (Publication 3, Indiana University Research Center in Anthropology, Folklore, and Linguistics). In *International Journal of American Linguistics, 23*, (Pt. 2).

Potter, G. (1960). *To the golden door: The story of the Irish in Ireland and America*. Boston: Little, Brown.

Sahlins, M.D. (1960). The origin of society. *Scientific American, 203*, pp. 76–87.

Schneider, D.M. (1953). A note on bridewealth and the stability of marriage. *Man, 53*, 55–57.

Simmons, L.W. (1937). Statistical correlations in the science of society. In G.P. Murdock (Ed.), *Studies in the science of society*. New Haven: Yale University Press.

Snyder, C.R. (1958). *Alcohol and the Jews*. Glencoe, IL: Free Press.

Stephens, W.N. (1959). Child rearing and oedipal fears: A cross-cultural study. Unpublished thesis, Harvard University.

Stephens, W.N. (1961). A cross-cultural study of menstrual taboos. *Genetic Psychology Monographs, 64*, 385–416.

Strodtbeck, F.L. (1951). Husband-wife interaction over revealed differences. *American Sociological Review, 16*, 468–473.

Wert, J.E., Neidt, C.O., & Ahmann, J.S. (1954). *Statistical methods in educational and psychological research*. New York: Appleton-Century-Crofts.

Whiting, J.W. (1959). Sorcery, sin and the superego: A cross-cultural study of some mechanisms of social control. In M.R. Jones (Ed.), *Nebraska symposium on motivation*. Lincoln: University of Nebraska Press.

Whiting, J.W., & Child, I.L. (1953). *Child training & personality: A cross-cultural study*. New Haven: Yale University Press.

Whiting, J.W., Kluckhohn, R., & Anthony, A. (1958). The function of male initiation ceremonies at puberty. In E.E. Macoby, T.M. Newcomb, & E.L. Hartly (Eds.), *Readings in social psychology* (3rd ed.). New York: Holt, Rinehart and Winston.

Wolf, E.R. (1955). Types of Latin American peasantry: A preliminary discussion. *American Anthropologist*, *57*, 452–471.

Wright, G.O. (1954). Projection and displacement: A cross-cultural study of folk-tale aggression. *Journal of Abnormal Social Psychology*, *49*, 523–525.

CHAPTER 3

Drinking Patterns of the Bolivian Camba

DWIGHT B. HEATH

An approach to evaluating the role of sociocultural aspects of drinking behavior can perhaps best be made through the comparison of drinking patterns and associated traits in the different cultures of several societies. It is the anthropologist who has the conceptual tools, broad ethnographic knowledge, and opportunities for research of this kind. The most ambitious cross-cultural study in this field was undertaken by Horton (1943) and was based on general ethnographic descriptions of 77 primitive societies. His major conclusion was that "the primary function of alcoholic beverages in all societies is the reduction of anxiety." [See chapter 1, by Horton, in this volume.]

One of the difficulties encountered in Horton's study was the frequent ambiguity or inadequacy of data pertaining to drinking as an aspect of culture. It is simply impossible for the ethnographer in the field to report in equal detail on every phase of the culture he is studying, however "simple," "primitive," or "homogeneous" that culture may appear. It is hardly surprising, therefore, that Horton found few systematic descriptions of prosaic and familiar drinking patterns in the anthropological literature. The orgiastic tribal drinking sprees so vividly depicted in many reports are not typical in worldwide perspective. Yet these uses of alcohol commanded more attention than social drinking in small groups or other less spectacular forms of drinking behavior.

A few investigators have recently made studies specifically concerned with the relationship of drinking to other aspects of culture in primitive societies.[1] The discussion that follows adds to this growing literature a brief characterization of the culture of a relatively homogeneous society, the Camba of eastern Bolivia, together with a description of Camba drink-

Dwight B. Heath is with the Department of Anthropology, Brown University, Providence, Rhode Island. This chapter is reprinted from the original D.J. Pittman and C.R. Snyder (Eds.), *Society, Culture, and Drinking Patterns* (New York: John Wiley & Sons, 1962), pp. 22–36. It was a revised and somewhat shortened version of an article that originally appeared in *Quarterly Journal of Studies on Alcohol, 19*, 491–508.

ing patterns and some tentative interpretations of the relationships between drinking and other cultural institutions. The peculiar relevance of such an analysis of this particular group to the study of drinking behavior in general should become apparent in the course of this discussion.

The Cultural Context

The Camba are a mestizo people. They are descendants of colonial Spaniards and local Indians, and their physical and cultural characteristics evidence both sides of their ancestry. In reality, the Camba constitute what might be called an *emergent society*. They have rejected traditional tribal ways of life, yet have been admitted as a laboring caste only to the periphery of another society dominated by a small group of whites. Even today, Indians are "becoming" Cambas by assuming clothing, learning Spanish, and moving to haciendas as farmhands. Numbering about 80 thousand, the Camba occupy an area of alternating jungle and prairie that stretches north from the city of Santa Cruz in eastern Bolivia. Enormous distances and natural barriers have effectively isolated them from any regular or sustained contact with other centers of population. Hence the Camba constitute an enclave with a slightly modified colonial Spanish way of life, virtually surrounded by nomadic Indian tribes.

A primitive form of agriculture is the primary economic activity in this tropical lowland area. Trees are felled in a small plot of jungle land; the undergrowth is cut and burned; and rice, corn, or manioc is planted for a few years before the soil is exhausted and abandoned. Such an inefficient system is possible only because so small a population occupies a vast area of rich land—some of the larger haciendas are measured in terms of "so many days' ride by horseback" in different directions from the house of the landlord. Even on such estates no more than 20 tenant farmers are regularly employed because only a tiny portion of the land is cultivated at any time.

Although it is at most a few generations since their Indian forebears were lured or bought from the Chiriquano, Guarayú, Siriono, or Tapieté tribes by gifts of steel tools and alcohol, these farmers, who speak an archaic dialect of Spanish and wear clothing of Western cut, regard the neighboring nomadic tribesmen of today as "savages" or "barbarians." Virtually none of the aboriginal religion, folklore, or social organization persists among the Camba; some elements of handicraft and a few native words are all that has been retained of the indigenous cultures.

The Camba may be described as peasants. A few of them are independent farmers, living on isolated homesteads and cultivating small plots in order to grow enough produce to feed their families. The great majority, however, live as tenants on haciendas where they are given food, housing, tools, clothing, and a tiny token wage in exchange for their labor. These haciendas are rarely large-scale commercial enterprises. Enough produce is raised to feed the owner, his family, and his workers, and to provide a small surplus for trade in local village markets. Unlike most peasants, the Camba have no love of the land and are extremely mobile. The few who own small tracts of land live in isolation on their scattered homesteads. But the majority who work as tenants move about frequently "just to try a new place," unless they are bound to a landlord by debt. Others live in isolation as squatters on the uncultivated land on the edges of the large estates, using it as their own until dispossessed. Work patterns are such that cooperative effort is rarely encountered. Even during harvests on haciendas each man works a small area assigned to him, much as the small landholder works alone.

There is little opportunity for accumulation of capital in an economy neither marginal nor overproductive. The individual Camba is not beset either by the stress of competition or by subsistence anxieties. Especially insulated from strain are the tenant farmers, whose few needs are met by their landlords and who accept a secure dependent status unquestioningly.

Geographic mobility is such that membership in neighboring groups is fluid and enduring friendships are rare. Wage work is conceived of as an impersonal business relationship, and there is no special loyalty to the employer or to the hacienda. The only associations that might be considered sodalities are the Farmers' Union, a supposed labor organization that really serves only as an agency for political patronage, and the Veterans of the Chaco War, which dates from the early 1930s, when virtually every able-bodied man in the country was fighting against Paraguay.

Although nominally Roman Catholic, the Camba have little knowledge of Church doctrine. They observe a few rites in a way that bespeaks magical faith rather than profound religious conviction. Fundamentalist Protestant missionaries have made some devoted converts, but their numbers are few. Perhaps the most striking feature of Camba religious life is that not one of the indigenous elements survives.

Just as in the religious and economic spheres social integration is minimal, kinship ties are tenuous, and solidarity is lacking. Common-law marriage is the rule, and such consensual unions are extremely brittle. Often a man will leave a woman after 3 or 4 years and move to another area where he takes another wife. The deserted woman does not return to her family but works for a time until another man comes on the scene. Because the individual couple usually set up housekeeping independently, apart from either spouse's parents, the extended family of three or more generations, so common among primitive and peasant peoples, does not occur. The independence of the nuclear family is stressed, and geographic dispersion results in few cooperative enterprises.

Camba socialization gives little indication of severity, trauma, or discontinuity. Children are valued and well treated. There are no ceremonies marking the birth, but christening takes place as soon as the family comes to town where the church is located, usually within 4 months. Babies are swung in hammocks or cradled in the mother's arms most of the time until they begin to crawl, and then they have almost complete freedom. Because the mother is always near, she can offer sympathy, petting, or the breast whenever an infant cries. Weaning is abrupt only when a newborn baby displaces an older child; otherwise children may nurse occasionally for comfort until the age of 4 or 5. Gradual toilet training is started shortly after the first birthday.

Children over 3 are left very much to their own devices, except that girls are sometimes obliged to care for younger siblings when their mother is occupied with a new baby. This they do ungrudgingly but with childish irresponsibility. Discipline is rarely severe and usually takes the form of a brief scolding; physical punishment is considered brutish. Aggression is rarely shown by adults or children. The family usually are together for meals, for siesta, and in the evening. The small one-room thatch hut is used only for sleeping and as shelter against inclement weather; otherwise, most domestic activities center around the yard. Because of familial isolation there is little opportunity for group play. Children have responsibilities such as gathering firewood, carrying water, running errands, and so forth, but spend much of their time playing alone or with siblings until adolescence when, with no ritual observance, they assume the duties of adults and are ready for marriage.

The Role of Alcohol

The Camba drink a highly concentrated alcoholic beverage on their numerous festive occasions. Both drinking and drunkenness are the norm on these occasions and an integral part of their social ritual. Drunkenness is consciously sought as an end in itself, and consensus supports its value. Aggression and sexual license are conspicuously absent on these occasions when beverage alcohol is used. Moreover, there is no evidence whatsoever of individual instances of dependence upon alcohol comparable to alcoholism or addiction as it is known in the United States. These facts highlight the strategic significance of the Camba for the understanding of alcohol problems.

The Beverage

Despite an abundance of tropical fruits, palms, corn, manioc, and other foods from which neighboring tribes make beers and wines, the Camba do not make or drink home brews. After meals they take strong black coffee and, as a refreshment, an unfermented corn chicha. But their festive beverage is not so innocuous. It is probably the most potent alcoholic drink in customary usage.

Almost the only commercial industry in the area is the distillation of cane alcohol. The soil and climate are well suited to the cultivation of sugar cane. However, neither sugar nor molasses could have been sold for enough profit to cover the cost of the 2-month mule trek over the mountains—until recently, the only route to the nearest market city in the western portion of the country. The margin of profit from alcohol, however, was sufficiently great to justify the expense of transportation. Hence, nine distilleries are now operating. These range from a small homemade contraption that produces about 200 liters a day during the 4-month season to a huge mill with modern French machinery and a daily capacity of 12,000 liters.

The product of these distilleries, called *alcohol,* is shipped throughout the country in 16-liter cans. Chemical analysis indicates that it contains 89% ethyl alcohol. Watered to 30 or 40%, *alcohol* becomes the *aquardiente* so often encountered in the Andean region. Among the Camba, however, it is not diluted. It is a colorless drink with little flavor but has an extremely irritating effect upon the drinker's mouth and throat. The Camba report that this burning does not diminish with habituation, and admit readily that they enjoy neither the taste nor the feel of *alcohol.* What they profess to enjoy is the drunkenness it brings. On a particularly hot day, the acrid fumes which may bring tears to a seasoned drinker's eyes occasion good-natured jokes about the strength of the beverage and derision of the unseemly "tame" drinks of other peoples. A special advantage of drinking the pure *alcohol* frequently cited by informants was that "it kills the [intestinal] parasites" that infest virtually all of the Camba. Although the validity of this "medical" opinion is questionable, the Camba attribute "stomach trouble" and "liver sickness" among other peoples to their use of additives—water or fruit juice—that "spoil" *alcohol.*

In each village some shopkeepers sell *alcohol*, and on the haciendas it can usually be purchased at the commissary. Sometimes a shrewd speculator keeps a few bottles hidden in his hut so that he can resell them at night, when the normal sources of supply are closed, realizing a profit of 20 or 30%. The standard unit of measure is the *botella*, a used beer or soda bottle with a capacity of 0.7 liters. The cost of a bottle of *alcohol* is equivalent to three days' wages of a farm laborer.

The fact that buying *alcohol* keeps men poor must not be construed as an indication of constant drinking or neglect of family obligations. The basic needs of a tenant and his family are met by his landlord, and cash wages represent only a small portion of a man's real earnings. This informal "social security," including care of the ill and aged, stems from an old, strictly feudal system that has only recently undergone slight modification.

Occasions for Drinking

The Camba drink *alcohol* only during fiestas, but these are not infrequent. Every national or religious holiday throughout the entire year, including local patron saints' days, is such an occasion. There are no religious prescriptions or proscriptions about drinking, but the Camba sometimes confide that they are putting one over on the village priest when they have a party. *Alcohol* plays no part in the rites of the church, and drinking during a religious service or procession is frowned upon. However, the rest of each holiday is devoted to carousal, and Carnival, the week preceding Lent, is treated as a week-long occasion for revelry. Peasants who come into town from outlying haciendas sit in small groups in the street and drink, while those at home often have open house for friends and neighbors. Each weekend, too, is an occasion for rest from the week's labors and for drunkenness in an atmosphere of companionship with whoever joins the party in a house or yard.

Rites of passage are also occasions for festive drinking. When the family members return home from a christening, the godparent "sponsors" a drinking party.[2] After his wedding the groom is expected to give a party for all comers. Some men have even given avoidance of this expense as their chief reason for not solemnizing a stable common-law marriage. Regardless of the age and sex of the deceased, a wake is another occasion for drinking, usually at the expense of surviving relatives. It is perhaps noteworthy, in marked contrast to American patterns, that the annual reunion of the Veterans of the Chaco War is a sober occasion when warm camaraderie reigns but drinking is completely absent.

Fiestas may last several days. Most people spend their entire weekends drinking; wakes last a night and a day; national and religious holidays are not only used to the full but almost always are stretched a day longer. The Camba recognize only two possible reasons for bringing a fiesta to a close—lack of *alcohol* and obligation to return to work. The former applies mostly to independent farmers (small landholders and squatters) to whom no source of credit is available; tenants are permitted and even encouraged to take small loans from landlords who consider their extension of credit as cheap assurance of a stable labor force. The obligation to return to work, by contrast, rarely pertains to the independent farmer, who sets his own schedule, unlike the tenant, who is assigned working hours. Most landlords have made a compromise that is effective in getting cooperation from their workers: the day following every holiday is also a day of fiesta, but after that everyone must go back to work. Thus, the workers feel that they are getting an extra holiday on each occasion and return ungrudgingly when the time comes. In like manner, after a weekend people are resigned to the obligation of working through the new week.

Participants

It is probably a valid generalization that all Camba adults, except the few Protestants, drink *alcohol*; and most of them become intoxicated at least twice each month. Despite frequent and gross inebriety, alcoholism, in the sense of addiction, does not occur.

Drinking takes place only within a social context; solitary drinking is inconceivable to the Camba. Observed groups varied in size from 3 to 16, and were as often composed of both men and women as they were of men only. Women do not drink except with men. Kinship does not figure in any consistent manner or significant degree in the composition of such groups even during festivities connected with life crises, reflecting the dispersal of kinsmen and the tenuousness of relationship bonds. Usually unmarried men and women drink only in the company of married couples. *Muchachos* and *muchachas* (boys and girls) stand around but are not invited to join until they are considered *jóvenes* (young people), a status of maturity with criteria no less vague than those of adolescence in our own society. The youngest drinker encountered was 12; the oldest claimed to be 92.

Fiestas never comprise invited guests only—they grow by aggregation of neighbors, passers-by who are hailed from the road, and watchful freeloaders, who come uninvited. For this reason, they show no marked stratification by age, apart from the exclusion of children. In the course of a fiesta a group may expand considerably but would in no event subdivide into smaller units. Social class lines are so sharply drawn between the peasant majority and the landed gentry that these groups never drink together and, in fact, rarely mix in any social situation.

Drinking Ritual

The behavioral patterns associated with drinking are so formalized as to constitute a secular ritual. Members of the group are seated in chairs in an approximate circle in a yard or, occasionally, in a hut. A bottle of *alcohol* and a single water glass rest on a tiny table that forms part of the circle. The "sponsor" of the party pours a glassful (about 300 cc) at the table, turns and walks to stand in front of whomever he wishes, nods, and raises the glass slightly. The person addressed smiles and nods while still seated; the sponsor toasts with "*Salud*" (health) or "*A su salud*" (to your health), drinks half of the glassful in a single quick draught, and hands it to the person he has toasted, who then repeats the toast and finishes the glass in one gulp. While the sponsor returns to his seat, the recipient of the toast goes to the table to refill the glass and to repeat the ritual.

There are no apparent rules concerning whom one may toast, and in this sense toasts proceed in no discernible sequence. A newcomer is likely to receive a barrage of toasts when he first joins a drinking group, and sometimes an attractive girl may be frequently addressed, but there tends to be a fairly equal distribution of toasts over a period of several hours. To decline a toast is unthinkable to the Camba, although as the party wears on and the inflammation of mouth and throat makes drinking increasingly painful, participants resort to a variety of ruses in order to avoid having to swallow an entire glassful of *alcohol* each time. These ruses are quite transparent (such as turning one's head aside and spitting out a fair portion) and are met with the cajoling remonstrance "Drink it all!" Such behavior is not an affront to the toaster, however, and the other members of the group are teasing more than admonishing the deviant.

After the first 3 or 4 hours virtually everyone "cheats" this way, and almost as much *alcohol* is wasted as is consumed. Also, as the fiesta wears on, the rate of toasting decreases markedly: during the 1st hour a single toast is completed in less than 2 minutes; during the 3rd hour it lasts 5 minutes or more. A regular cycle of activity can be discerned, with a party being revived about every 6 hours. When a bottle is emptied, one of the children standing quietly nearby takes it away and brings a replacement from the hut. When the supply is exhausted, members of the group pool their funds to buy more; they send a child to bring it from the nearest seller.

The ritual sequence described above is the only way in which the Camba drink, except at wakes, where a different but equally formalized pattern of behavior is followed. At a wake there is still a single bottle and a single glass, but they are carried by an adolescent girl who is not necessarily related to the deceased. She stops in front of each male mourner and, if he nods, pours a glassful of *alcohol* and hands it to him. He looks around the room nodding slightly and silently to all present, and they return the silent nod. He drinks in a single quick draught and returns the glass to the girl who continues making the rounds. A mourner may decline by simply shaking his head. Female mourners are not offered *alcohol* because "women always have so much sadness [that] they don't need it."

Behavior and Attitudes Associated with Drinking

A fiesta is a time for easy social intercourse. Despite the elaborately patterned behavioral sequence, drinking is not a solemn rite. There is usually a low undercurrent of nondescript small talk—conversations about the weather or crops, a few anecdotes, or simply recounting the events of the past week. Occasionally there is music—a young man may play the guitar and sing love songs, or men may form a band of flute, snare drum, and bass drum. In either case, men ask women to dance and couples shuffle awkwardly around inside the circle while the drinking ritual continues.

During a fiesta the Camba usually begin drinking shortly after breakfast. A stomachful of *masaco* (a pasty mash of manioc and lard) retards the absorption of alcohol into the system. Nevertheless, as a party wears on, the effects of intoxication become apparent. After 2 or 3 hours of fairly voluble and warm social intercourse, people tend to become thick-lipped, and intervals of silence lengthen. By the 4th hour there is little conversation; many people stare dumbly at the ground except when toasted, and a few who may have fallen asleep or lost consciousness are left undisturbed. Once a band or guitarist starts playing, the music is interminable and others take over as individual players lose consciousness. The 6th hour sees a renewed exhilaration as sleepers waken and give the party a second wind. This cycle is repeated every 5, 6, or 7 hours, day and night, until the *alcohol* gives out or the call to work is sounded.

During a lull, a woman may slip away to prepare *masaco* or bread, which is then put in a conspicuous place outside the circle where people may help themselves as they go to urinate in the shadows a few feet away from the group. No matter how long a fiesta may be, no other food or drink is taken, yet nausea was observed only once.

The general pattern of individual behavioral changes in inebriation is fairly standard. There is an appreciable increase in sociability during the early stages. The ordinarily taciturn Camba peasant becomes more voluble; his normal attitude of almost complete indifference to the world around him is replaced by a self-confidence which is shown in positive but not dogmatic expression of opinions. As he imbibes more *alcohol*, the individual becomes befuddled—he thinks more slowly and has trouble voicing his ideas. There follows a sort of retreat inward; he stops talking, slumps in his chair, and may fall asleep. When he revives, the earlier euphoria is restored and he enthusiastically rejoins the party as though he had never withdrawn. Women usually "cheat" by swallowing a smaller portion of each draught, and they rarely lose consciousness. Although no more subject to censure for drunkenness than men, women claim to be more sensitive to irritation of the mouth and throat.

It is perhaps worthwhile to underscore a few ways in which this pattern of drunkenness differs from that of certain other peoples. For instance, among the Camba drinking does not lead to expressions of aggression in verbal or physical form, as is so often the case among other peoples when alcohol overrides normal inhibitions to asocial behavior. Neither is there a heightening of sexual activity: obscene joking and sexual overtures are rarely associated with drinking. Even when drunk, the Camba are not given to maudlin sentimentality, clowning, boasting, or baring of souls. Drunkenness is considered an inevitable consequence of drinking and, far from being despised, is highly valued and even eagerly sought by the men. Repeated questioning in various forms and contexts about their reasons for drinking inevitably evoked the same answer: "*Para emborracharse, no mas*" (just to get drunk). The Camba have no conception of an ideal stage of drunkenness, such as "just feeling good," nor do they differentiate between stages of intoxication such as "high," "tight," and "drunk."

Because a Camba almost never leaves the permissive context of the drinking group while intoxicated, no one has occasion to condemn his behavior. Occasional untoward actions are explained and excused on the basis that "sometimes when a man is drunk he doesn't always know what he's doing." Examples of such readily excused blunders were a man's unwittingly stepping on a sleeping pig, another's failing to button his trousers after urinating, and another's burning himself by accidentally brushing against the globe of a lantern. Such minor misbehaviors are not held against the drinkers, nor do they feel obliged to apologize then or later. There is apparently no guilt associated with drinking, drunkenness, or drunken behavior.

Although it has never been explicitly formulated by the Camba, an implicit general system of reciprocity seems to hold with respect to fiestas. Whenever a peasant feels he can afford it, he buys a few bottles of *alcohol*, and all comers are welcome. Often, in fact, a man will sponsor a fiesta when he obviously cannot afford it. A few unmarried men are notorious freeloaders but are tolerated because they provide creditable performances on the guitar that add to the atmosphere of a fiesta.

Magical properties are not attributed to *alcohol* by the Camba, apart perhaps from its supposed medicinal value as an internal parasiticide. Drinking does not appear to have connotations of proof of masculinity, and individual differences in tolerance to alcohol go unnoticed. However, the fact that the investigator, a *gringo* (foreigner), drank with the Camba was to his credit and was made much of as a sign of true friendship even on his first day in the area. Indeed, he soon came to be called in consequence *muy Camba* (very much a Camba).

Hangovers and hallucinations are unknown among these people, as is addiction to alcohol. A farmhand snaps out of intoxication abruptly when the call to breakfast is sounded after a fiesta. He eats a bowlful of *masaco*, drinks a cup of black coffee, and then performs a normal morning of work. Sometimes on an especially hot day workers' systems are so full of alcohol that fumes from their perspiration make their eyes water, but neither their work nor they themselves show any ill effects. And so they labor quietly, diligently, apart, until the next fiesta.

Tentative Interpretations

An attempt has been made above to describe drinking patterns and their cultural context in sufficient detail that other investigators may analyze and interpret them in terms of their own diverse interests and theoretical orientations. A few tentative interpretations will be offered here, however, to suggest approaches that may be fruitful in understanding the role of alcohol in Camba society. It is hoped that these may also suggest, by analogy, certain functions of alcohol in other areas and throughout history, although this analysis makes no pretense at comprehensiveness or finality.

The Camba data supplement those from other primitive societies indicating that alcoholism is not a function of the alcohol concentration of beverages used or of the quantities imbibed. Although these people drink large amounts of almost pure alcohol, the investigation encountered no instance of addiction, and drinking is completely restricted to specific social situations. Furthermore, the Camba convincingly demonstrate that extensive inebriety does not necessarily result in manifest troubles. An important contrast with most primitive societies that have marginal economies must be noted, however, in that the periodicity of Camba drinking cannot be attributed to irregularity in the supply of beverage. Neither does their high rate of drinking appear to be correlated with obvious threats in the environment. In terms of the sources of anxiety on which Horton (1943) based his ratings for cross-cultural comparison—subsistence, sexuality, and aggression—the Camba are virtually free from fears and it would be a dubious matter to attribute their drinking to any of those areas of insecurity.

If it be assumed that all human behavior is motivated and that the patterns of behavior that persist are those that are in some way rewarding to the individuals who compose the group or society, then any attempt to understand the relationship of one aspect of culture to the broader ongoing sociocultural system must deal with functions. The following is an attempt to suggest, by inference, some of the functions alcohol may serve for the individual Camba, as a start in explaining the role of alcohol in that society.

The outstanding characteristic of Camba drinking is the elaborately patterned sequence of behavior that is the only context in which drinking takes place. The strict conformity and the implicit equality of communion expressed in the use of a single glass seem to make of drinking a socially significant rite. The importance of this factor cannot be fully appreciated without the realization that the rest of Camba culture is virtually lacking in forms of communal expression. The child is early trained to self-sufficiency, which prepares him well for the relative isolation of adult life. Geographically dispersed nuclear families are virtually independent of each other, and kinship ties are tenuous and unstable.

There is also, as we have noted, little sense of identification with neighborhood, church, or voluntary associations. People live alone and work alone, with few opportunities for sociability throughout the week. Fiestas provide occasions for intense interaction, and drinking groups constitute primary reference groups lacking in other phases of Camba life. If it be postulated, in keeping with a universal implicit assumption, that there is some element of value inherent in human association per se, this function of alcohol can be seen to have enormous potential importance for individual adjustment as well as for social cohesion. To be sure, there are activities other than drinking that could fulfill the same function of facilitating social interaction. Comparative ethnographic data suggest that the act of drinking, like eating, may in itself have a peculiar significance as a means of expressing corporate solidarity. Furthermore, there are certain characteristics inherent in beverage alcohol that make it particularly suitable as a stimulus to sociability. Despite the overall depressing effects of alcohol on physical and mental functions, its apparent stimulating effect has long been observed in group situations.

This illusion of stimulation is given by increased volubility and camaraderie that, presumably, result from the lowering of inhibitions and restraints that normally operate to limit the individual's rapport with others. This function would be of especial importance among persons whose experience is particularly individualistic and whose frequent or prolonged isolation results in introversion and a loss of the basic social skills of relating themselves to others in a meaningful way. The effects of alcohol in overcoming such personal reserve are quite rapid; but since alcohol is readily oxidized in the body, the effects are short-lived and its occasional use does not disable people from their normal round of activities. A similar function of alcohol may be cited in the frequent association of drinking with rare and sporadic occasions for social interaction in other highly atomistic societies or subgroups, such as sailors ashore after a voyage, cowboys or lumberjacks "on the town," the farmers of Chichicastenango at market (Bunzel, 1940), or the homeless men of Skid Row (Straus & McCarthy, 1951).

It is interesting in this connection to compare Camba drinking with that of the surrounding tribes to which they are closely related. Although detailed descriptions are not available, it is plain that each of the major groups that contributed to the Camba population had some form of indigenous alcoholic beverage even before the Spaniards arrived and that their drinking takes place now, as then, in occasional large-group sprees or drinking bouts in which intoxication is normal. Some similarities in sociocultural situation are shared by all of these relatively loosely structured groups.

The few exceptions to the normal pattern of drinking behavior that were observed among the Camba must also be considered in the light of this interpretation. It was noted that the only Camba who never drink are the few Protestants. The fact that they are all members of fundamentalist sects (Assembly of God, Baptist, Free Brothers) for whom abstinence is an article of doctrine would not appear to be wholly adequate as an explanation of their deviance from the norms of the larger society. It is important to note that they have, in the church, the very kind of stable primary group lacking for the great majority. During meetings, 3 or 4 nights weekly, each individual plays an active role, and they call each other brother. Between services, members of these churches often play volleyball and interact in other ways that differ markedly from the social isolation of Catholics who, by and large, take no active part in religious or extraceremonial activities together. It appears, in similar manner, that the presence of a prevailing atmosphere of genuine camaraderie stemming from a past of significant shared experiences and a common chauvinistic pride may be sufficient basis to unite the veterans, during their reunion, in a way that allows warm and easy fellowship without dependence on alcohol to overcome initial reserve.

The pattern of drinking at a wake is not the usual excessive drinking to oblivion that is normal for the Camba, and in other societies has often been described as serving to overcome the effects of grief. On the contrary, it is a sedate rite in which the drinker silently pledges to all present, and it has been said that it is precisely those who grieve the most (women) who "have no need" of alcohol in this context. It may be that on such an occasion the impact of grief is itself a unifying force for most mourners, and that those who need something more to bring them out of their introversion find in alcohol a means, symbolic or physiological, of achieving rapport with others in a situation in which social reinforcement is especially important.

This discussion has emphasized the importance of the drinking group as a primary reference group. But this is by implication a relational conception—the focusing of emotion on an in-group implies awareness of an out-group. That such an awareness does prevail among the Camba, despite the ephemeral nature of their associations in drinking, is suggested by their confiding that they are "putting one over" on the priest whenever they have a party, in their ridiculing the alcoholic beverages of other peoples, and in the investigator's becoming "very much a Camba" in large part because he drank with them.[3]

Another striking feature of Camba drinking is the high alcohol concentration of the common beverage. It is difficult to understand why a people should consistently drink so strong a beverage, especially when they themselves frankly admit it is distasteful and even painful to do so. The presence of a constant and plentiful supply, together with ease of acquisition, are not sufficient explanations, especially in the light of the expense involved. It is possible that desire for immediacy or certainty of effect is involved here. Related to this problem is the curious fact that the Camba make no use of the several wild crops plentiful in the environment that could be rendered into beers or wines at no expense and with little effort and that are so used by the very tribes from which they are descended.

It is generally agreed that distillation is a technological refinement unknown throughout the New World in pre-Columbian times. Certainly it was the Spanish colonists who introduced cane alcohol to the nomadic hunting tribes that occupied this area. Alcohol was an important commodity in the 17th-century South American slave trade, and some historians suggest that the Jesuits and Franciscans used it to lure Indians to their missions and to keep them in controlled settlements there.

Still another problem of interpretation is raised by the Camba custom of drinking to the point of gross inebriation. The fact that they consistently cite the quest of drunkenness as their reason for drinking may be no more explanatory than the usual middle-class American cliché that one drinks "just to be sociable." Discrepancy between stated reasons and actual objectives is a commonplace in all phases of human behavior. However, the frequent occurrence of complete intoxication, even to the point of losing consciousness, would be inconsistent with the use of alcohol as a stimulus to general sociability. The Camba do not long retain the manic euphoria that would give grounds for the sociability rationale, but almost invariably drink beyond that euphoria to a stage of increased introversion and subsequent oblivion before regaining the easy rapport that so contrasts with their usual reserve. This loss of euphoria may result simply from faulty judgment—miscalculation of the effects of *alcohol* and subsequent excess. Quite apart from the momentum of social drinking and toasting, and from the progressive effects of *alcohol* in distorting the powers of discrimination, it must be kept in mind that small amounts of so strong a beverage can appreciably change the concentration of alcohol in the organism of relatively small, lightweight people. Consequently, a single toast (comprising about 300 cc of 178-proof alcohol) could easily mean the difference between exhilaration and oblivion.

In summary, it is suggested that alcohol plays a predominantly integrative role in Camba society, where drinking is an elaborately ritualized group activity and alcoholism is unknown. The anxieties often cited as bases for common group drinking are not present, but fiestas constitute virtually the only corporate form of social expression. Drinking parties predominate among rare social activities, and alcohol serves to facilitate rapport between individuals who are normally isolated and introverted.

Notes

1. Bunzel's pioneering work (1940), in which she compared the role of alcoholism in a Guatemalan and a Mexican Indian community, is a model of integration and interpretation. Alcoholism has been treated as a major focus of anxiety also in two Indian communities in Mexico by Viqueira and Palerm (1954), and in two Indian and three white cultures that exist within a small area in the southwestern United States by Geertz (1951). Honigmann and Honigmann (1945) compared in some detail the drinking patterns and cultures of Athapaskan Indians and their white neighbors in a small Canadian trading post community. The function of alcohol in Mohave society was interpreted by Devereux (1948) in the light of psychoanalytic concepts, and the drinking behavior of the Navajo was described by Heath (1952) and that of some Salish tribes by Lemert (1954, 1958). Berreman (1956) related inebriety to social strain among the Aleuts, as did Sayres (1956) in three communities in rural Colombia. Marroquín (1943) briefly described the use of alcoholic beverages by the Quechua in Peru, and Mangin (1957) stressed the predominantly integrative functions of drinking among them. Rodriguez (1945) summarized reasons that another group of the same tribe (in Ecuador) gave for drinking, and Sariola (1956), in characterizing the drinking customs of the Indians of Bolivia, generalized Quechua patterns to other tribes. Platt's (1955) description of traditional alcoholic beverages among the south African Bantu emphasized their socially integrative uses.
2. No special prestige accrues from sponsoring a party, and no one expresses thanks when it is over. In fact, the concept of sponsorship is used here simply as a convention for the more cumbersome phrase *providing the first bottle*. The Camba do not think in such terms, and no one "gives" a party; it is simply held "at so-and-so's place." When the sponsor's supply of *alcohol* is exhausted, other participants readily and without comment pool their funds to buy more.
3. In like manner, Lemert (1958) has noted an implicit conflict between Catholicism and drinking parties among the Salish Indians of northwestern North America.

References

Berreman, G.D. (1956). Drinking patterns of the Aleuts. *Quarterly Journal of Studies on Alcohol, 17*, 503–514.

Bunzel, R. (1940). The role of alcoholism in two Central American cultures. *Psychiatry, 3*, 361–387.

Devereux, G. (1948). The function of alcohol in Mohave society. *Quarterly Journal of Studies on Alcohol, 9*, 207–251.

Geertz, C. (1951). Drought, death, and alcoholism in five Southwestern cultures. Unpublished manuscript, Harvard University Values Study.

Heath, D.B. (1952). Alcohol in a Navajo community. Unpublished Harvard College thesis.

Honigmann, J.J., & Honigmann, I. (1945). Drinking in an Indian-White community. *Quarterly Journal of Studies on Alcohol, 5*, 575–619.

Horton, D. (1943). The functions of alcohol in primitive studies: A cross-cultural study. *Quarterly Journal of Studies on Alcohol, 4*, 199–320.

Lemert, E.M. (1954). Alcohol and the Northwest Coast Indian tribes. *University of California Publications in Culture and Society, 2*, 303–406.

Lemert, E.M. (1958). The use of alcohol in three Salish Indian tribes. *Quarterly Journal of Studies on Alcohol, 19*, 90–107.

Mangin, W. (1957). Drinking among Andean Indians. *Quarterly Journal of Studies on Alcohol, 18*, 55–66.

Marroquín, J. (1943). Alcohol entre los Aborígenes Peruanos. *Crónica Médica, 60*, 226–231.

Platt, B.S. (1955). Some traditional alcoholic beverages and their importance in indigenous African communities. *Proceedings of the Nutrition Society, 14*, 115–124.

Rodriguez Sandoval, L. (1945). Drinking motivations among the Indians of the Ecuadorian Sierra. *Primitive Man, 18*, 39–46.

Sariola, S. (1956). Indianer och alkohol. *Alkoholpolitic, 2*, 39–42.

Sayres, W.C. (1956). Ritual drinking, ethnic status and inebriety in rural Colombia. *Quarterly Journal of Studies on Alcohol, 17*, 53–62.

Straus, R., & McCarthy, R.G. (1951). Nonaddictive pathological drinking patterns of homeless men. *Quarterly Journal of Studies on Alcohol, 12*, 601–611.

Viqueira, C., & Palerm, A. (1954). Alcoholismo, brujería, y homicidio en dos comunidades rurales de México. *América Indígena, 14*, 7–36.

CHAPTER 4

Continuity and Change in Drinking Patterns of the Bolivian Camba

DWIGHT B. HEATH

It was rare, in 1958, that a paper written by a graduate student about a topic he never set out to study should be published in a major international journal (Heath, 1958). It is rare, in 1991, that a primarily descriptive article should be reprinted in a major transdisciplinary anthology. Having revisited the Camba often in the intervening years, which were marked by a variety of significant changes in demography, ecology, economy, and politics, I find this an appropriate occasion for some brief retrospective comments on the original work and for an update of the data and interpretations concerning Camba drinking. This is, therefore, both a critical reevaluation of a naive ethnographic study that assumed unexpected importance and a longitudinal study of changes in drinking patterns in sociocultural context. This unusual case study also suggests some practical implications concerning alcohol use and its outcomes throughout the world.

The Original Paper

Ironically, this "classic" in the genre of ethnographic studies of alcohol use was a serendipitous byproduct of doctoral research that had as its focus land tenure and social organization in a frontier region that was undergoing many changes, at least partly because of a then recently enacted agrarian reform law. [See previous chapter, by Heath, in this volume.] I have elsewhere (Heath, 1991) described how the interest of E.M. Jellinek, Mark Keller, and Charles Snyder prompted me to dredge the data out of my field notes and to work them up in a way that fit with multi-

Dwight B. Heath is a professor of anthropology in the Department of Anthropology at Brown University, Providence, R.I. This chapter was written especially for this book. It is meant to be read after chapter 3.

disciplinary interests in alcohol (with which I had been totally unacquainted while conducting the research). The fact that I was able to write so rich a description and so plausible an interpretation is a telling illustration of the value of the inductive method, and a tribute to the old-fashioned ethnographic aim of learning as much as one can about "everything" when one studies an alien way of life.

Having quickly surveyed the scanty written corpus that was then available concerning alcohol in other cultures, I set out to address each of the issues that had been raised by other authors. After briefly sketching the cultural setting, I described in some detail the beverage (178-proof rum), occasions for drinking (episodic fiestas), participants (virtually all Camba adults), the secular ritual that was about the only context for drinking (formal toasting from a single glass), behavior and attitudes associated with drinking, and some tentative interpretations, which emphasized social organization. After noting comparisons with drinking in other cultures, and the few exceptions to modal norms that had been observed among the Camba, I emphasized that "fiestas provide occasions for intense interaction, and drinking groups constitute primary reference groups which are lacking in other phases of Camba life" and that "alcohol serves to facilitate rapport between individuals who are normally isolated and introverted" (Heath, 1958, pp. 504, 507). In short, my interpretation was unabashedly functionalist, with a focus on social organization. The sociability of formalized drinking seemed to have strong positive value, as (I have since learned) was spelled out earlier, in general sociological terms, by Simmel (1949).

The study became well known, presumably because the Camba represent an extreme in worldwide perspective with respect to the frequency of drunkenness and the potency of their customary beverage, on the one hand, and virtual absence of alcohol-related problems on the other. It was promptly translated (Heath, 1961), and reprinted (Heath, 1963), and the data figured predominantly in the major hologeistic study linking drinking patterns with child training (Keller, 1965), as well as in a volume that was a key foundation for subsequent emphasis on expectancies as they relate to the effects of alcohol (MacAndrew & Edgerton, 1969). Many psychiatrists, psychologists, and sociologists, as well as anthropologists, use Camba drinking patterns as a vivid illustration of behavior that may appear deviant in the view of their audiences but that makes sense in that distinctive sociocultural context.

Changes in Camba Culture

One of the factors that had originally attracted me to the jungle of eastern Bolivia was the prospect of rapid social and economic change. It had long been an isolated enclave of archaic *mestizo* (mixed Hispanic and indigenous) culture, surrounded by a number of Indian tribes that retained many of their native patterns of belief and behavior. The first railroad had just arrived at the state capital, about 60 miles south of where I was working, and a paved highway was being built to connect the region with the much more densely populated highlands. Immigrants were flooding into this frontier area as homesteaders, including Bolivians of a different linguistic and ethnic group, and foreigners from many distant lands. Recent political revolution had unseated the traditional oligarchy, and land reform was an integral part of the new populist government's effort to redistribute wealth and power. In that fluid context, many of the Camba who had been tenant farmers were becoming independent small-scale freeholders while formerly landless peasants were becoming homesteaders. It was a time of rapid and exciting change in many respects (Heath, 1964).

Fortunately, I was able to return to Bolivia at least every couple of years during the first decade after my initial study. As before, I paid special attention to political socialization, ethnohistory, economic development, and other aspects of change that dramatically accompanied the "opening of the frontier" (Heath, Erasmus, & Buechler, 1969). But, as before, I still tried to pay attention to other aspects of culture as well and noted some changes in drinking patterns.

As awareness of land reform spread among the Camba, they were quick to form *sindicatos,* peasant leagues that were the preferred corporate plaintiffs in appeals for the expropriation and reallocation of large estates. The methods and rationale for affiliating with a neighborhood-based sindicato have been explained in considerable detail elsewhere (Heath, 1972), and need not concern us here; the important point is that each sindicato constituted a strong and meaningful reference group for Camba peasants during the 1960s, whereas they had previously had no such primary social affiliation. Political awareness was an integral aspect of the sindicatos, with frequent reminders that it was the newly incumbent Nationalist Revolutionary Movement (MNR) that had given peasants the vote, expropriated major industries, promoted educational reform, and promised land to those who worked it. Grass-roots decision making was enthusiastically embraced, with respect to such issues as local public works, land tenure, participation in political demonstrations, construction of a school, and so forth. This decision making often occurred in the format of a town meeting, with ample discussion until a consensus (often unanimous, rather than simply majority rule) was achieved.

Members of a sindicato had a unity of purpose, a sense of common identity and concern, and ample opportunity to meet with others and to be heard as an equal when their community's affairs were discussed—all of which contrasted markedly with the atomistic pattern of social relationships that had prevailed earlier.

With the progressive sense of social interrelatedness, there was a progressive diminution of drinking parties. They occurred less often, and they tended to be shorter when they did take place. Those individuals who were most active in pursuing the affairs of the sindicato tended often to attend meetings, whether in the community or elsewhere when they went as delegates to regional, state, or national conventions, and they participated proportionately less in traditional drinking parties. Even those who held no offices in the nationwide network of sindicatos—which, not so incidentally, also served as channels for political patronage—developed a sense of class consciousness that had previously been lacking and readily adopted the salutation *compañero* (comrade), which was fostered by the MNR.

It was in reference to that heady period of so-called liberation and markedly increased social interaction that I subsequently wrote, "In many respects, the *sindicato* now serves as a primary reference group for most Camba farmers, in much the same way that the Protestant congregation [did] for its members. The coordinate decline in drinking bouts is striking, and sindical activity has replaced frequent heavy drinking in the lives of many individuals" (Heath, 1965, p. 290). My emphasis on the social structure as a major factor in shaping drinking patterns not only was in keeping with my original interpretation of the earlier very different reality but was also offered as a more plausible alternative to Mandelbaum's (1965) "Culture and Personality Analysis of [Camba] Drinking Patterns." Using my data, he also put considerable weight on the atomistic social organization that had earlier characterized the Camba, but he interjected a vaguely Freudian element by incorrectly positing "fear and distrust of others" among the Camba, as Simmons (1960) had noted among the people of Lunahuaná, Peru.

The late 1960s and 1970s were a period of rapid and significant change, both political and economic, throughout much of Bolivia, but especially in the northern Santa Cruz area, which had long been the home of the Camba (for more details see also Gill, 1987; Stearman, 1985). Economic stabilization, following a long period of runaway inflation, was combined with an exceptionally long period of relative political stability. Local and foreign investors took advantage of that unusual combination, and of governmental subsidies, to create a veritable boom in what had been a sleepy backwater of quasi-feudal subsistence farming.

Modern sugar refineries built with European funds consumed the forest in two ways: vast tracts were cleared for the cultivation of sugar cane, and the enormous furnaces were fueled with tons of firewood each day. Local entrepreneurs who were not out of favor politically were allowed to buy or keep huge farms, despite agrarian reform. A new crop, cotton, flourished on highly mechanized plantations as well, allowing the country in just a few years not only to become self-sufficient but even to export a surplus in exchange for scarce foreign currency. Subsidized credit, tax breaks, rental of machinery at less than cost, and several other advantages allowed large-scale farmers to become wealthy, although few breaks were offered to the small-scale farmers.

As prices rose and they were increasingly involved in the money economy, many of the Camba *campesinos* (peasants) were pressured to sell their land to large landholders. Even when they received more than they had originally paid for the land, they lost the long-term improvements they had made and were obliged either to seek work (in a context in which unskilled labor was abundant and cheap) or to move on and carve a homestead out of the jungle to the north or west. Homesteading itself became less attractive as Kollas (Quechua-speaking Indians from the highlands) and other outsiders (such as Japanese, Okinawans, Italians, and others) competed for those parcels of land that were not subject to seasonal flooding and that were near enough to roads so that one could afford to send produce to market. According to the Camba, even the market was being taken over. Whether the Kollas had more entrepreneurial skill, more willingness to work long hours, better social networks among the flood of immigrants, or a combination of those and other factors, they readily came to dominate the small-scale commerce that provides slow but sure economic mobility for those willing to work at it.

The MNR was ousted in a military coup, and sindicatos were not only declared illegal but actively harassed, their leaders were killed or exiled, and grass-roots participatory democracy was effectively stifled. A combination of flood, drought, pests, and plant diseases ended the cotton boom and curtailed the expansion of sugar cultivation.

By the 1980s, the large-scale operators had turned to cattle ranching or to the production of cocaine paste. Coca had long been grown in high semitropical valleys in the eastern escarpment of the Andes (notably Chapare and Yungas), but it was mostly for domestic consumption by Quechua- and Aymara-speaking Indians whose religious and habitual dietary chewing of the dried leaves involved neither psychoactive "highs" nor addiction in the sense of chemical dependency. A burgeoning world market for cocaine brought international drug traffickers to Bolivia, where coca production could easily be increased and where large quantities of leaves could be converted quickly and easily to relatively valuable paste, for further refinement in Colombia. Common chemicals and simple procedures, using unskilled labor in small and easily concealed facilities, became the basis for an underground regional economy that appears soon to have eclipsed the legal economy of the entire nation. Military and other political leaders not only were aware of these developments but appear to have been actively involved (Henkel, 1987). A worker willing to "stamp" coca leaves, with his feet in chemicals that are not immediately painful, could earn more in a day than he would otherwise in a fortnight. Few of the Camba peasants took any part in the drug trade, but they suffered from the recurrence of galloping inflation and from the utter lawlessness that still dominates in much of the region.

The jungle is now far away—that free and beautiful treasure-trove in which the Camba used to hunt for game and gather wild honey, as well as fruits, nuts, herbal medicines, timbers for their house frames, and palm-leaves for thatched roofs and walls. The rivers no longer yield fish and are increasingly treacherous as erosion from ruthless overexploitation of the land results in silting, erratic changes in course, and flooding, and the waste from production of cocaine paste (kerosene, sulphuric acid, ether) combines with herbicides, insecticides, fertilizers, and other pollutants. Many have lost their land, or better parts of it; some have lost their places in the market and are fearful that no one will take care of them as age and illness take their toll. Some even reminisce nostalgically about the relative sense of security that they had in the prerevolutionary days as tenant farmers in a quasi-feudal system. They resent much that outsiders would interpret as progress or development, longing for the simpler, more tranquil life that they remember—or at least think that they remember.

They no longer live on haciendas, as some did when I first studied there. Those who are still small-scale farmers, whether homesteading on the moving fringe of the frontier or as dogged hold-outs on the plots they got as beneficiaries of the land reform that was underway at that time, feel inundated by noisy, aggressive, dirty, and generally unsympathetic people of alien cultures, who have stolen and ruined the paradise that had been theirs.

Few in numbers, but quietly proud of their Camba heritage, the old men and women get together on local saints' days and on major holidays to drink in a way that looked, in mid-1989, very much like that I first encountered in 1955. The groups are smaller, and the younger people do not attend—they are more likely to be listening to the radio at home or quietly in town, walking about in small same-sex groups, hoping for brief liaisons, or "just being where the lights and action are."

The drinking parties again have the single bottle and single glass, with the same secular ritual of toasting and quick drinking. The beverage is the same—*alcohol,* that insufferably high-proof rum that still makes them wince but that they still agree is more healthful than weaker drinks. The undercurrent of talk tends to be more about the past than about the present, and the pace of toasting has slowed so that although a few people may nod briefly in the wee hours, no one passes out on the ground.

Tentative Interpretations

It would be tempting to assert that my structural-functional interpretation of Camba drinking patterns had again been vindicated. Even after all the changes that have taken place in political, economic, and other realms of Camba culture, they drink in a way that fits well with their newly atomistic social organization, and in a way that is similar to that of other populations for whom the drinking group is important, at least on those rare and sporadic occasions when they enjoy interaction: for example, sailors ashore, cowboys or lumberjacks "on the town," farmers of Chichicastenango at market (Bunzel 1940), or the homeless of Skid Row (Spradley, 1970).

It would be gratifying if I could say more now than I was able to say earlier about women's reactions to men's drinking. Although I originally noted the fact that women drank somewhat less than men, I didn't pursue it to elicit the emotions, values, and attitudes that may have been associated with that difference. Unfortunately, I had no more access to women's thoughts and feelings this time, although I conscientiously tried to fill that gap.

In an age when the interactive role of the ethnographer with the subjects of his study is increasingly recognized as having an important influence on what one perceives, I am struck by how little has changed as well as how much has changed. Quite apart from the vastly different political economy in which Camba life is now embedded, and quite apart from my having laid aside the shield of logical positivism that played so important a role in my early research, we few survivors picked up our conversations—and our drinking—as if I had been away for a weekend, rather than for seven years (with some especially close friends, and fully a quarter of a century with other Camba acquaintances).

The immediate practical implication that makes the Camba case so important in cross-cultural perspective remains unchanged: despite a relatively high rate of consumption of an extraordinarily high-proof beverage, with frequent intoxication, they still cannot conceive of any kind of alcohol-related problems. Furthermore, although I tried, I could discern no indication of the kinds of problems that are so labeled in some cultures, such as spouse- or child-abuse, homicide, suicide, injurious accidents suffered by persons "under the influence," aggression of any sort, job interference, psychological distress (on the part of the drinker or close relatives), social strain in the family, trouble with legal authorities, or even physical damage that differs in any significant degree from that suffered by others in the area who drink less or abstain. In short, the Camba remain a striking exception to the popular rule of thumb that drinking problems are in proportion to alcohol consumption.

The control model of prevention, based on that premise, has been increasingly espoused by policymakers and others throughout the world, calling for increasing restrictions on the availability of alcohol as the best way to lessen alcoholism or a wide range of alcohol-related problems. In light of this case study (among others), the sociocultural model of prevention appears more plausible, stressing that the meanings, values, norms, and expectations associated with drinking have more effect than sheer quantity does in determining how many and what kinds of problems may be associated with alcohol—or whether, as is strikingly the case among the Bolivian Camba, such problems appear not to occur at all.

Acknowledgments

Although I never set out to study drinking, I am grateful to Mark Keller and E.M. Jellinek for having impressed upon me the intrinsic interest and potential importance of using alcohol as a window on various ways of life. It was Clyde Kluckhohn who first sparked my interest in anthropology, and George Murdock and Floyd Lounsbury were exacting but supportive teachers of the research process. Charles Snyder and David Pittman introduced me to alcohol studies as a broadly interdisciplinary enterprise. Throughout all of this, A.M. Cooper has been my best friend and colleague, and the Camba shared their lives with us in important ways.

References

Bunzel, R. (1940). The role of alcoholism in two Central American cultures. *Psychiatry, 3,* 361–387.

Gill, L. (1987). *Peasants, entrepreneurs, and social change: Frontier development in lowland Bolivia.* Boulder, CO: Westview Press.

Heath, D.B. (1958). Drinking patterns of the Bolivian Camba. *Quarterly Journal of Studies on Alcohol, 19,* 491–508.

Heath, D.B. (1961). *Normas de beber del camba boliviano* [Drinking patterns of the Bolivian Camba]. *Revista de la Universidad Autónoma "Gabriel René Moreno," 8 (15),* 112–126.

Heath, D.B. (1963). Drinking patterns of the Bolivian Camba. In D. Pittman & C. Snyder (Eds.), *Society, culture, and drinking patterns.* New York: Wiley.

Heath, D.B. (1964). Camba: A study of land and society in eastern Bolivia. (University Microfilms No. 64–11, 374). (Original work 1959)

Heath, D.B. (1965). Comment. *Current Anthropology, 6,* 289–290.

Heath, D.B. (1972). New patrons for old: Changing patron-client relationships in the Bolivian Yungas. In A. Strickon & S. Greenfield (Eds.), *Structure and process in Latin America: Patronage, clientage, and power systems.* Albuquerque: University of New Mexico Press.

Heath, D.B. (1991). The mutual relevance of anthropological and sociological perspectives in alcohol studies. In P. Roman (Ed.), *Alcohol: The Development of Sociological Perspectives on Use and Abuse.* New Brunswick, NJ: Rutgers Center of Alcohol Studies.

Heath, D.B., Erasmus, C.J., & Buechler, H.C. (1969). *Land reform and social revolution in Bolivia.* New York: Praeger.

Henkel, R. (1987). The Bolivian cocaine industry. In E. Morales (Ed.), *Drugs in Latin America* (Studies in Third World Societies 37). Williamsburg, VA: Department of Anthropology, College of William and Mary.

Keller, M. (Ed.). (1965). *A cross-cultural study of drinking. Quarterly Journal of Studies on Alcohol,* Suppl. 3. New Brunswick, NJ: Rutgers Center of Alcohol Studies.

MacAndrew, C., & Edgerton, R.E. (1969). *Drunken comportment: A social explanation.* Chicago: Aldine.

Mandelbaum, D.C. (1965). Alcohol and culture [with comments by V. Erlich, K. Hasan, D. Heath, J. Honigmann, E. Lemert, & W. Madsen]. *Current Anthropology, 6,* 281–294.

Simmel, G. (1949). The sociology of sociability. *American Journal of Sociology, 55,* 254–261. (Original work published in German, 1911)

Simmons, O. (1960). Ambivalence and the learning of drinking behavior in a Peruvian community. *American Anthropologist, 62,* 1018–1027.

Spradley, J.P. (1970). *You owe yourself a drunk: An ethnography of urban nomads.* Boston: Little, Brown.

Stearman, A.M. (1985). *Cambas and kollas: Migration and development in Santa Cruz, Bolivia.* Orlando, FL: University of Central Florida Press.

CHAPTER 5

Alcohol Studies and Anthropology

DWIGHT B. HEATH

Anthropological data, methods, and concepts have contributed significantly over many years to our understanding of the interactions between alcohol and the human animal. In this chapter I briefly review the major impacts of anthropologists on the broad field of alcohol studies. I emphasize the striking changes that have taken place since the 1970s and discuss current issues that are focusing multidisciplinary attention on such work.

Brief discussion provides an historical context for both the multidisciplinary realm of alcohol studies and the roles of anthropology within it. The specifically anthropological issues within alcohol studies include types of populations studied, links made between alcohol and other factors in cultural context, and the range of research methods used. I note the practical implications of all the foregoing and offer general conclusions, emphasizing the impact of anthropological perspectives.

Historical Context

Although humankind has experimented with innumerable psychoactive drugs, none has been used longer or more widely than ethanol, that form of alcohol produced by the natural process of fermentation and readily concentrated to a high degree by the simple process of distillation. Around this simple chemical have grown up highly complex and diverse patterns of belief and behavior.

Surrounded with paradoxes, being often recognized as both a stimulant and a depressant, a food and a poison, ethanol in its use symbolizes an

Dwight B. Heath is with the Department of Anthropology, Brown University, Providence, R.I. This chapter is adapted, with permission, from "Anthropology and Alcohol Studies: Current Issues," in B.J. Siegel, A.R. Beals, & S.A. Tyler (Eds.), *Annual Review of Anthropology,* Vol. 16 (Palo Alto, Calif.: Annual Reviews, Inc., 1987), pp. 99–120. This chapter is an abridged version of the original article. The reader is referred to the original for greater detail and a more complete list of references.

enormous range of both positively and negatively valued things and feelings. Its rapid and sometimes dramatic effects on mood and behavior can easily be discerned by anyone, and acute observers have long paid attention to diverse aspects of the substance and its effects on various animals including human beings.

The Field of Alcohol Studies

The psychoactive and motor effects of alcohol have been variously noted, interpreted, and evaluated in the folk wisdom, mythology, art, and traditions of many societies for millennia, and a few pioneers began making systematic observations only sporadically just two centuries ago. Academic and scientific studies of alcohol were rare before the 1940s. Investigations of the substance require a combination of different kinds of expertise.

So distinctive is ethanol that since the middle of the 20th century an interdisciplinary cadre of researchers have been engaged in a field that has come to be called alcohol studies, some even proposing that "alcohology" be recognized as an emergent academic discipline. From a modest epidemiological and bibliographic project at Bellevue Hospital and experiments at Yale Laboratory of Applied Physiology emerged the Yale Center of Alcohol Studies, which, having moved to New Jersey, continues to flourish as the Center of Alcohol Studies, Rutgers University. A few other major centers have sprung up around the world, but this is a field in which "the university without walls" is a vital and enduring reality, with collegial liaison and even collaboration commonplace not only among faculty of different institutions but also across national boundaries. Anthropological data and viewpoints have played important roles in this field from the outset and have had an impact out of all proportion to the involvement of professional anthropologists. For these reasons, it seems timely to provide a summary of current anthropological issues in alcohol studies for the benefit of colleagues who have not been actively engaged.

Precursors and Early Anthropological Work

Much of what is anthropological in the history of alcohol studies cannot be attributed directly to the efforts of people in the discipline. Counselors and alcoholics as well as social scientists seem comfortable in discussing the relevance of sociocultural factors, and there has been a flurry of cross-cultural studies by psychologists and others who have little interest in conducting close and sustained field work with any non-Western population. Until the 1960s, anthropological perspectives had not been strongly represented in the growing field of alcohol studies, although ethnographic descriptions of drinking patterns among various world populations proliferated. (Many languages use the word *drink* to denote both the ingestion of any beverage and, also, with a whole set of special connotations, the use of alcohol. My recent exploration [Heath, 1986a] of some of the semantic and symbolic implications of this duality need not be summarized here. In this chapter *drinking* denotes use of alcohol.)

At first it was a few sociologists, using concepts, methods, and data commonly associated with anthropology, who made a lasting impression by adding *social* and *cultural* to the vocabulary of scholars and practitioners who thought and talked about alcohol. The classic illustration of the fact that differences in the cultural context of drinking affect the occurrence of drinking problems is the contrast between Jewish-Americans and Irish-Americans. Bales (1946) pointed out that although virtually all Jews drink some alcoholic beverages at least weekly few Jewish alcoholics had been identified. By contrast, although Irish women and children tended to abstain, a high rate of psychic and other problems had been linked to drinking among Irish men. On the basis of literary and thematic evidence, Bales suggested that learning to drink with one's family at home, in the context of religious ritual, tended to make Jews moderate drinkers. Sexual and economic frustration, combined with a tradition of barroom conviviality among men, made heavy drinking acceptable among the Irish, although drunken comportment often led to guilt. [See chapter 27, by Bales, in this volume.] Snyder's (1958) survey of an urban community added the findings that drinking problems were inversely related to religious orthodoxy among Jews and that drunkenness was a significant ethnic boundary marker, attributed to outsiders. [See also chapter 15, by Glassner, in this volume.]

Horton's (1943) pioneering experiment with large-scale cross-cultural correlation of traits and indexes set the pattern for all of the hologeistic studies that followed, using samples of societies from the Human Relations Area Files. Many people who have no familiarity with ethnographic data and know nothing about the methodological subtleties and problems associated with such research accept as virtually axiomatic his finding that "the primary function of alcoholic beverages in all societies is the reduction of anxiety" [See chapter 1 by Horton, in this volume.] Bacon (1943) articulated "foundations for a sociological study of drinking behavior" that remain valid as a charter for our enterprise more than 40 years later. It was also sociologists, notably Ullman (1958) and Pittman (1967), whose discussions of the role of various kinds of norms in regulating drinking behavior have come to be viewed broadly under the heading *the sociocultural model,* especially in the context of ameliorating alcohol-related problems.

In the early years, however, alcohol studies cast in the traditional anthropological mold had little cumulative impact. The uneven quality of early anthropological studies of alcohol use can be traced to their data collection techniques. Few who went into the field before 1960 had any intention of studying drinking patterns, much less any sophisticated hypotheses to test or practical problems to resolve. In most instances, they paid attention to alcohol for the simple reason that it was important in the lives of the people among whom they were working; and it is a tribute to their insight and thoroughness that they were often able later to provide a meaningful picture on the basis of field notes that were incidental to their major interests. Once having done so, researchers often recognized that alcohol use—like kinship, religion, or sexual division of labor—can provide a useful window on the linkages among many kinds of belief and behavior. Unfortunately, not all who produced such notes invested the time and effort necessary to master the large, diverse, and broadly scattered literature on alcohol in other disciplines and so were not always adept at fitting their observations into a context meaningful to others. Still, around 1970 a few anthropologists, without rejecting or diminishing the major contributions to alcohol studies made by colleagues in other behavioral and social sciences, took various initiatives that went far toward articulating anthropological contributions to our understanding of alcohol.

Recent Anthropological Contributions

I have elsewhere presented several broad overviews of the literature that deals with alcohol from an anthropological perspective (Heath, 1975, 1976, 1986b). Rather than reiterate even the major points made in those articles, I here focus on significant recent anthropological contributions to alcohol studies and on current issues of interest to those not already familiar with the subject.

In 1965 Mandelbaum contributed a milestone synthetic article on alcohol to *Current Anthropology.* He effectively summarized such key issues as the similarities and differences in alcohol use among cultures and culture areas, stability and change, culture-and-personality interpretations, and so forth; this article and the accompanying commentary focused attention on the subject at a time when a large-scale cross-cultural study linked patterns of drinking and drunkenness with child-rearing techniques (Bacon, Barry, & Child, 1965). An outstanding book by MacAndrew and Edgerton (1969) combined ethnographic and ethnohistorical evidence to demonstrate conclusively that drunken comportment, however much it may be affected by biochemical and neuropharmacological factors, is also a product of expectations and culturally shared values.

At the same time, the first major conference to focus on major anthropological aspects of alcohol use was held. The work reported there differed from much other research on the subject. Rather than focus on the deleterious consequences suffered by a few excessive long-term users, the authors dealt matter-of-factly with drinking as a workday activity that generally had beneficial effects. The proceedings of the conference were eventually published in the belated *World Anthropology* series (Everett, Waddell, & Heath, 1976). Establishment in 1971 of the National Institute on Alcohol Abuse and Alcoholism (NIAAA) gave new visibility and funding to alcohol studies. NIAAA sponsored a postdoctoral program that emphasized anthropological research (Heath, 1981a), supported widespread and varied research projects proposed by individuals, and backed conferences.

Also in the mid-1970s, a book by Madsen (1973) combined a sociocultural analysis of Alcoholics Anonymous, a remarkably effective self-help group, with his synthesis of biomedical and behavioral perspectives on the causes of alcoholism in the United States. Multidisciplinary research became an ideal in theory if not in practice, and alcoholism became widely labeled a "biopsychosocial" disease. "The disease concept" of alcoholism has a long history (Levine, 1978) but was popularized by Jellinek (1960), together with the large and influential constituency of recovering alcoholics and practitioners in the alcoholism treatment industry. The concept's popularization served to lessen the stigma that had been associated with alcoholism and to engage physicians and other researchers. However, many people who give lip service to the disease concept nevertheless condemn alcoholics as immoral or weak-willed (Rodin, 1981), and many scientists are beginning to suggest that the disease concept has outlived its usefulness (Galanter, 1984). [See also chapters 21, 22, and 39, by Fingarette, Hore, and Roman & Blum, respectively, in this volume.] In the meantime, general acceptance of the relevance of at least some social and cultural factors was afforded by the so-called public health model. In that paradigm, any disease or illness can be viewed as a resultant of the interplay of vector, host, and environment, and any drug problem as a resultant of the interplay of drug, set, and setting. The presence of an alcoholic beverage is a crucial part of the environment or setting, as are attitudes, expectancies, and other social and cultural factors. Thus, even a multivolume series entitled *The Biology of Alcoholism* (Kissin & Begleiter, 1971–1983) contains chapters by anthropologists. Clearly, sociocultural perspectives have gained credibility and relevance even among influential practitioners of the so-called hard sciences.

The volume of work on the subject has increased markedly since the early 1970s, and the quality of research is significantly higher in many respects (Heath, 1986b). Issues that are of current concern in both anthropology and alcohol studies are discussed below under the following broad rubrics: populations, cultural and other contexts, research methods, and implications for action.

Drinking Populations

A major difference between anthropologists and most others who study alcohol is in the choice of populations. In many fields, the only aspect of alcohol use that is of interest is alcoholism. Although it has been defined in many ways, *alcoholism* generally denotes any injury or combination of injuries (physical, psychic, social, economic, or other) suffered as a result of drinking. Deleterious consequences of drinking that are not seriously debilitating are sometimes called *drinking problems,* and those that are associated with addiction have recently been given the label *alcohol dependence syndrome.* Each of these terms is itself the subject of continuing controversy among specialists, but this is not an appropriate context in which to pursue such fine points. Anthropologists, on the other hand, generally focus on the majority, who drink moderately or with impunity. Thus, whereas work in other fields addressed a wide range of pathologies, often among institutionalized individuals, anthropologists tended to deal with alcohol as used in the normal course of workaday affairs in integral communities.

Another difference is that until recent years most studies of alcohol use by anthropologists concerned tribal or peasant societies, with each case study representing a sort of "natural experiment" in terms of the variability of the human experience. Drinking was examined in its natural context, with no expectation that great revelations would be forthcoming about why people (as a whole) drink or why some people feel compelled to drink in ways that hurt themselves and others. Like ethnographic findings in many other realms of behavior, such data may have been viewed by some nonanthropologists as relatively unbiased but little more than quaint and curious reports of exotic customs, occasionally providing interesting anecdotes, but not the kind of methodologically rigorous or quantitatively impressive data that would shed light on what was viewed as a major problem in health and social welfare. A thoroughly indexed bibliography (Heath & Cooper, 1981) has made this widely dispersed literature more accessible; and Marshall's anthology (1979), although it contains little that is new, may have considerable impact if it used as a sourcebook in college courses.

Prehistoric and historic populations came in for only sporadic analysis until recent years, but there has been so great an expansion in historical studies that the interdisciplinary Alcohol and Temperance History Group was formed, which publishes its own newsletter (most recently called the

Social History of Alcohol Review). Much attention is focused on the temperance movement and Prohibition in the United States (Blocker, 1979), but other kinds of social history, such as links between migration and changes in drinking patterns, are not ignored. Few studies combine historical and anthropological approaches as effectively as Taylor's (1979) analysis of colonial Mexican villages. In a meticulous analysis of court records and other documents, he showed how the widely held view that liquor destroyed the remnants of Aztec society grew out of the Spaniards' cultural prejudice against drunkenness, which was not only acceptable but highly esteemed among the Indians in specific religious contexts. A recent article by Hill (1984) provides an excellent review of ethnohistory and alcohol studies, and some good regional studies are available on Oceania (Marshall, 1982) arctic America (Hamer & Steinbring, 1980), sub-Saharan Africa (Pan, 1975), and the Indians of the United States (Heath, 1983; Leland, 1980).

A recurring problem in alcohol studies is that social groups and categories are referred to vaguely, inconsistently, and often inaccurately. For example, "Jews" may be compared with "Irish," "Italians," "French," or "ascetic Protestants" (Maghbouleh, 1980). Those especially interested in supposed racial differences in susceptibility to the effects of alcohol have sometimes grouped "Indians" with other "Orientals" (e.g., Reed, Kalant, Gibbons, Kapur, & Rankin, 1976). By contrast, the careful work of a team of geneticists (Goedde, Agarwal, & Harada, 1980) even includes detailed genealogies of individuals. Within the United States, there was a flurry of effort in the 1970s to alert clinicians that "minorities" in trouble with alcohol might not respond to the usual modes of treatment or of education for abuse prevention. Soon those populations were identified, sometimes with the use of the broad and vague labels devised in connection with federal antidiscrimination laws ("Hispanic," "Indian," "Asian" or "Pacific Islander," "Black," et al.), sometimes with variants like "Latino," "Asiatic," "Oriental," and so forth. It is little wonder that the cultural sensitivity that had been hoped for among clinicians did not materialize; it is also little wonder that epidemiological and other data collected under such rubrics are virtually meaningless.

Even though such gross categories bear no relation to the reference groups with which people normally identify themselves, one might expect them at least to yield useful statistics at the national level. Unfortunately, they are utterly useless in that respect as well. Such segments of the population are so small that a variety of periodic surveys on nutrition, disease, and other aspects of health consistently conducted on national samples are just as consistently ignored because they are statistically meaningless. This view is not the biased judgment of an ethnographer, but a candid admission of frustration by the epidemiologists and statisticians of the U.S. Public Health Service, who pride themselves on their technical skill at such work (National Institute on Alcohol Abuse and Alcoholism, 1989).

In an attempt to escape the sometimes pejorative association of *minorities,* the term *special populations* was soon applied (NIAAA, 1982) to such categories but often also to women, adolescents (or "youth"), and the elderly (or the "aged"). In one respect such broad social categories made some sense at an early stage of discussion, if only because until then our experimental knowledge about drinking and its effects had been based overwhelmingly on white adult males. A few people seemed suddenly to recognize that other segments of the population might think and act differently. The terms *ethnic groups* and *subcultures* had been used indiscriminately to refer to most of the categories mentioned. Most of what were called cross-cultural studies were comparisons between samples in France, Mexico, Israel, California *[sic],* Zambia, or other large and culturally diverse political entities, a pattern that persists (Babor, 1986).

Anthropologists have made the salutary and distinctive contributions of increasing the specificity of the populations studied and paying closer attention to intrasocietal variation. In fact, the predominant theme in a major compilation of original papers on *The American Experience with Alcohol* (Bennett & Ames, 1985) is variation among populations in both drinking patterns and associated attitudes. For example, whereas most of the literature lumps migrants from Latin America within the United States generically as "Hispanics," "Mexican-Americans" alone are diverse (Gilbert & Cervantes, 1986), with six "lifestyle divisions" just in the Lower Rio Grande Valley (Trotter, 1985). Also, very different patterns occur in a single small city among Guatemalans, Puerto Ricans, and Dominicans (Gordon, 1984). Among the Navaho, the largest Indian tribe in the United States, Topper (1985a) identified five "subcultures." Major differences have been noted, for example, among Jews according to degrees of orthodoxy (Glassner & Berg, 1985), and along generational lines among several enclaves of immigrants from Europe (e.g., Stivers, 1978). With respect to age groups, scholars have vividly described drunken "hell raising" among adolescents who as adults become moderate drinkers (Burns, 1980; Hill, 1978).

One way in which the problems inherent in identifying or delimiting groups, categories, subcultures, and so on, can be overcome is to deal with social networks, and many anthropologists have done so with considerable success while studying drinking and its outcomes (Maida, 1984). Another valuable trend in alcohol studies by anthropologists is the longitudinal focus on families that is being effectively carried out by a few small teams. By dealing with entire families, usually in their own homes, researchers can identify discrepancies between ideal and real norms, observe intergenerational cultural transmission, and analyze the complex dynamics of workaday interaction. Ablon (1976) pioneered in such work, and others are now doing it, with cross-national comparisons under way in Poland, West Germany, and the Scandinavian countries. In the interest of efficiency, most of the families selected for such close scrutiny already have alcohol problems (Wolin, Bennett, Noonan, & Teitelbaum, 1980).

Although anthropologists have not usually focused their attention on individuals who are institutionalized or otherwise impaired, neither do they ignore them. Spradley's (1970) work among what we would now call homeless men helped change their treatment at the hands of police. [See chapter 38, by Rubington, in this volume.] Researchers who are also clinicians or social workers (e.g., Topper, 1985b) have learned from their clients and often bring immediate practical concerns to their interpretations and recommendations. Others look at how therapeutic programs work (Strug, Priyadarsini, & Hyman, 1986; Waddell, 1984), or try to understand how Alcoholics Anonymous helps so many—sometimes in ways all AA members would recognize (e.g., Madsen, 1973), and sometimes in ways distinctive to the local ambience (Jilek-Aall, 1981).

In sum, although the basis and composition of our samples are not always clearly specified, anthropologists have appropriately paid more attention to these criteria in recent years. For that matter, work throughout the field of alcohol studies, including contributions with rigorously controlled experimental conditions and quantitative analysis, is sometimes based on samples that fall short of the scientific ideal.

Cultural and Other Contexts

One of anthropology's greatest values lies in how it reveals human diversity, documenting the broad range of beliefs and behaviors associated with any trait. With relation to alcohol, this diversity ranges from abstinence to chronic excessive drinking and habitual drunkenness, from admiration of its mystical power to deprecation of it as immoral and taboo. Emotions run high on the subject of who may drink what, how much, when, in the company of whom, and so forth. For these reasons, alcohol tends to be linked in at least some way with almost every aspect of culture, and the many roles it plays in most cultures are surprisingly diverse. We also find that the setting for drinking affects how people drink and how they respond to alcohol. Brief mention should be made of where drinking fits in the context of cultural meanings, and to settings as contexts for drinking.

Drinking in the context of meaning. I have provided elsewhere (Heath, 1975, 1986b) a broad overview of how alcohol fits in sociocultural systems around the world, including such factors as nutrition, health, social organization, religion, technology, economics, politics, communication, sex, recreation, aggression, criminality, and social and psychological dynamics. Here I suggest some of the myriad ways alcohol is imbedded in various sociocultural systems. For example, the fundamental idea of ingesting ethanol is anathema to some populations, whereas others embrace it to the point of approving drunken stupor as an adjunct to religion (Kearney, 1970). Beers and wines are often considered general tonics and good food as well as enjoyable adjuncts to eating and sociability, and there is scientific justification for some such beliefs (Baum-Baicker, 1985a, 1985b).

Social factors are evident in contrasting drug use by class and caste within a single community (e.g., Honigmann & Honigmann, 1970), by the sexes (Bacon, 1976), or even at different stages in the life cycle (Hill, 1978). Drinking is often an ethnic boundary marker (Glassner & Berg, 1985) and a symbol of status and prerogatives (Morris, 1979). The sharing of drinks can be a valuable way of building up social credit, especially among those who are poor (Waddell, 1980). Just as religion can prompt some to drink, it can serve as a basis for others to stop drinking; in a community where participation in folk-Catholic rituals is costly, some become Protestants (e.g., Kearney, 1970), and rituals that are indigenous (Jilek, 1981) or syncretic (e.g., Singer & Borrero, 1984) can be therapeutic for some alcoholics. Whereas some deplore drinking as a disruptive modern influence at Navaho ceremonies, others recognize it as providing a context that is often used to facilitate the transmission of traditional religious lore (Topper, 1985a). Details of production and distribution of alcoholic beverages can be linked with sex roles (Bacon, 1976), social stratification (Cawte, 1986), changes in agricultural production (DeWalt, 1979), reciprocal labor exchanges (Kennedy, 1978), as well as hospitality, the manipulation of social networks (Weisner, Weibel-Orlando, & Long, 1984), and so forth. Except where prohibition is cast in religious terms, it is often foiled by illicit production (Bunzel, 1976).

Most people who drink talk both during and about drinking; a variety of ethnosemantic studies concern such talk (e.g., Spradley & Mann, 1975). The relaxing and supposed disinhibiting qualities of alcohol make drinking a common form of recreation (Lex, 1980), sometimes associated with casual sex (e.g., Read, 1980). Despite the linkage that many presume between alcohol and aggression, well-documented studies in many societies show that expectancy plays at least as great a role as biochemistry in that respect (MacAndrew & Edgerton, 1969).

Stress and strain on individuals and on social relations are often cited as the reasons people drink more than they should. A large body of literature, not all of it impressionistic, has grown up around the tensions generated by political domination, acculturation, or marginality and other kinds of anomie. A few exceptions to that pattern are noteworthy (e.g., DeWalt, 1979), and it is important to recognize that drunkenness can itself be a defiant gesture rather than a retreat (e.g., Hill, 1978).

On the basis of large-scale cross-cultural (or hologeistic) studies, a variety of measures of drinking, drunkenness, and drunken comportment have been correlated variously with witchcraft, war, and subsistence anxiety (Horton, 1943); sex roles, child training, and dependence (Bacon, Barry, & Child, 1965); feelings of power (McClelland, Davis, Kalin, & Wanner, 1972); social organization (Field, 1962); drinking problems (Frankel & Whitehead, 1981); and aggression (Schaefer, 1981). The methods and logic of such studies are controversial, but the findings usually command considerable attention from many who otherwise pay little to anthropology.

Settings as contexts for drinking. Just as drinking and its effects are imbedded in other aspects of culture, so are many other aspects of culture imbedded in the act of drinking. Studies that pay attention to the ecology of place and to other conceptual aspects of alcohol use serve as important foils to the assumption, often explicit in other kinds of writing about alcohol, that little matters but how many cubic centimeters of ethanol are ingested for each kilogram of body weight. The anthropological bias toward the observation of normal behavior in natural settings has also resulted in some interesting insights, especially since drinking is, in most cultures, primarily a social act.

Public drinking places can be special in the sense of allowing relative relaxation of racial segregation (Wolcott, 1974), facilitating informal contacts with partners of the same sex (Read, 1980) or opposite sex (Cavan, 1966), or relaxed conviviality with co-workers (LeMasters, 1975) and compatriots (Freund, 1985). Bars, taverns, and related establishments have a long history (Popham, 1978), and some authors have even developed functional typologies in major cities (Cavan, 1966; Thomas, 1978). Even groups that were legally forbidden alcohol until recent years may gravitate to bars, and some of them become important multipurpose clearinghouses for information on jobs, housing, and so forth, for people unaccustomed to life in the city (e.g., Weibel-Orlando, 1982). From such studies have also come interesting insights about the pace of drinking, ethnicity and aggression (Bach & Schaefer, 1979), and class differentiation (LeMasters, 1975). Studies of drinking in private places are fewer but invaluable, especially for what they reveal about how young people learn to drink (Glassner & Berg, 1985) or, sometimes, not to drink (Bennett, Wolin, & Noonan, 1977).

Research Methods

A major change in anthropological studies of alcohol during the past decade or so is that markedly greater attention to research methods has yielded data that are significantly more reliable and are generally recognized as creditable and useful by colleagues in other disciplines. In large part this fact derives from another: that such studies are no longer incidental post facto byproducts of other projects but are often major, long-term efforts well planned to illuminate the subject, with ample attention paid to ways of collecting and assessing appropriate data. This development has taken place because a cadre of anthropologists have paid special attention to the subject, often reading widely in the broadly multidisciplinary literature on alcohol studies and making such work at least one of the foci of their professional activity. Well aware of the shortcomings of early studies, some have paid special attention to developing rigorous and quantitative methods, and most have paid more attention to identifying their samples, spelling out the range of individual variation (rather than relying on modal generalizations, as had earlier been commonplace), and explicitly setting each of their contributions in the context of some theoretically or conceptually relevant problem.

This is not to say that observational and other traditional ethnographic methods have been abandoned. On the contrary, those qualitative approaches to data gathering proved their value when urban ethnographers were discussing PCP ("angel dust") almost a year before it came to the attention of the elaborate Drug Awareness Warning Network, a nationwide agency of the United States government that constantly monitors police and hospital reports about drugs. It is ironic that researchers and policymakers concerned with alcohol and other drugs—like many in education—have become enthusiastic about the strength and potential of observational ethnography at the same time that many anthropologists, for whom such methods were long considered standard, spurn them as "soft" and outdated (Akins & Beschner, 1980; Heath, 1981b).

While others have been adopting traditional anthropological methods of research, anthropologists have also been adopting methods from history, sociology, semiotics, and other fields. Careful attention to documents yields remarkably detailed information on some aspects of drinking in earlier times (Heath, 1982; Hill, 1984). Survey instruments, when prepared and administered in a manner appropriate to the local way of life, can be invaluable as guides to intracultural variation, large and representative samples, and so forth (Bennett & Ames, 1985; Jessor, Graves, Hanson, & Jessor, 1968). They can equally be misleading if they couch questions in ways that are alien or provide only for responses that the subjects consider not appropriate. Too many hypercritical observers seem to view survey and observational approaches as antithetical, whereas seasoned practitioners of both tend to emphasize the value of their complementarity (Heath, 1981b; Room, 1984).

Network analysis is another approach that combines quantitative and qualitative data and that can be fruitfully applied to many aspects of alcohol use and its outcomes. Maida's (1984) review article is an excellent guide to this fresh approach; topics studied include the kinds of social relationships people have (often in the absence of families or corporate groups), kinds of social identity and reference group boundaries, sources of social support, channels of referral to treatment, and much about how treatment modalities themselves work (Singer, 1985).

Multidisciplinary team studies with anthropology as a major component focused their research on alcohol use in relation to other aspects of behavior in the 1960s, both in the field (Jessor, Graves, Hanson, & Jessor, 1968) and in interpretations of cross-cultural correlations (Bacon, Barry, & Child, 1965). Insensitivity to local cultural attitudes in a more recent survey, based on drinking among Eskimos, caused major problems that anthropologists might have avoided (Klausner, Foulks, & Moore, 1979); and much of the interpretation that sociologists and psychologists weave around their quantitative data springs from thematic, attitudinal, anecdotal, and other qualitative data gleaned incidentally from their subjects or from experience (Babor, 1986).

Ethnographers can help geneticists who are attempting to link enzymatic variation with differences in susceptibility to the physiological effects of alcohol. By paying close attention to genealogy, and by helping gain access to hair roots for spectrographic analysis (Goedde, Agarwal, & Harada, 1980), we may gain a better understanding of the still-uncertain links between genetics and alcoholism and of the popular "firewater myth" of "racial" differences (e.g., Leland, 1976).

We have already noted the varying usage of the term *cross-cultural* in alcohol studies. This variation should not obscure the fact that Horton's (1943) method of the large-scale correlation of traits among a sample of cultures—what is now often called by practitioners the hologeistic approach—has been used to test many different hypotheses and associations. A recent review (Heath, 1984) is a convenient summary of this subfield, in which methods of sampling and rating have been made much more rigorous and samples have been expanded over the years.

Longitudinal work with families is one of the more exciting and fruitful approaches shedding new light on cultural transmission and on the integration and disintegration of social relationships (e.g., Ablon, 1985; Wolin, Bennett, Noonan, & Teitelbaum, 1980). Such work requires not only rare skill and patience on the part of investigators but perhaps even rarer qualities of personal warmth and trustworthiness. Participants must open their homes and their lives over long periods, allowing at least some observations to be made when they may not feel their best—observations, for example, of occasional unpleasant encounters they might prefer to keep "backstage."

In a few highly urban and industrial countries, researchers have recently made insightful comments about cultural themes and alcohol based on scrutiny of films produced there (e.g., Cook & Lewington, 1979). Analysis of verbal action plans lends special precision to the description of how people drink and what they think about it (Topper, 1976), just as thematic analysis of stories told by recovering alcoholics provides insight into their cognitive deficiencies and the solace they find in Alcoholics Anonymous (Rodin, 1985). Clearly anthropologists are being innovative and increasingly effective in conducting alcohol research.

Conclusions and Practical Implications

Even so summary a review of current issues in anthropology and alcohol studies reveals that many earlier shortcomings have been overcome. The importance of sociocultural perspectives has been generally accepted on the basis of earlier work that, though not always theoretically or methodologically sophisticated, demonstrated that intellectual and other environmental considerations play crucial roles in interaction with more rigorously proven biomedical factors. A new generation of scholars with specialized knowledge and some methodological innovations are making different kinds of contributions, often in collaboration with colleagues in other disciplines.

A special strength of anthropology continues to be its anomalous role as "the science of leftovers." What this distinctive role means with respect to the study of alcohol is that, unlike many other investigators, we study "moderate" or "normal" drinking—and abstaining—as well as "excessive" or "alcoholic" drinking. [See also chapter 6, by Klein, in this volume.] (After all, even the federal agency that stands to gain most from inflating the estimated prevalence of alcohol-related problems acknowledges that only two thirds of U.S. adults drink at all and only about one tenth of them ever suffer any deleterious consequences.) In this way, we add to knowledge about a significant aspect of culture in industrialized as well as other societies.

The fact that ethnographers so often remark on unproblematic drinking, even in populations that have relatively high per capita consumption of alcohol, prompted Room (1984) to speculate about "problem deflation" and the reasons for it. As an international spokesperson for increasing legislative control of alcohol use, he appeared unimpressed by frequent exceptions to the general rule that problems occur in populations in direct proportion to average per capita consumption of alcohol; he doubted whether people can judge whether drinking causes them problems. Citing my long-outdated (Heath, 1975) self-critical comments, he deplored functional interpretations as biased against the identification of problems. He also noted the antitemperance sentiment of a "wet generation" of ethnographers, said to be reacting against both the ascetic legacy of Prohibition and the risk of being mistaken for missionaries in exotic places. In the same article, however, he warned against the opposite tendency, "problem amplification," which is encouraged by a loose coalition (including alcoholics, treatment and health professionals, etc.) who benefit from making alcohol appear more harmful than it is. In view of the falling rates of cirrhosis and traffic fatalities even before the self-styled New Temperance Movement gained broad publicity with increasing pressures for warning labels, higher taxes, price indexing, advertising restrictions, and other measures to lessen alcohol sales throughout the world, his warning seemed timely. An especially disconcerting aspect of the antialcohol coalition is their readiness to ignore or distort data for political ends (Heath, 1988). [See chapter 40, by Pittman, in this volume.]

Because norms, values, and attitudes play such dominant roles in drinking and associated behavior (Heath, 1980), it is evident that education holds far more promise as a preventive strategy than legislative controls (Frankel & Whitehead, 1981) do. The ineffectiveness of some poorly conceived and executed programs of classroom training is cited by critics to discredit the value of education, with no apparent recognition that entire populations can be informed in other ways. In fact, knowledge of the rapidity with which drinking patterns can change is a special contribution by anthropologists, and such studies belie the idea that we deal only with modal behaviors in static settings. Also on the subject of change, anthropologists have made a major contribution by showing that traditionalists in some populations drink more than those who might be presumed to be suffering from the stress of acculturation (e.g., DeWalt, 1979).

In a later article, Room (1985) presumed that our literal adherence to some culturally specific definition of alcoholism had led anthropologists to overlook problems elsewhere; most, however, disprove such an interpretation by not using technical terms or by stating that few or no drinking problems of any kind could be discerned in specific populations. In some populations, of course, alcohol is often implicated in a variety of physical, social, and economic damage; but policymakers prefer to ignore the clear benefits of moderate drinking (Baum-Baicker, 1985a, 1985b) and seem displeased that we do not endorse their view of a worldwide pandemic of alcohol-related problems.

Even within the United States, it is evident that heavy drinking does not inevitably lead to the supposed progressive disease of alcoholism. Social surveys have consistently shown that 18–24-year-old males drink more than other segments of the population, but that few develop dependence or other enduring problems and most greatly reduce their drinking in later life. A few longitudinal studies demonstrate the fact (Vaillant, 1983), but only field studies reveal how kin and others treat such short-term excess as an integral part of the development cycle in the lives of such young men (Burns, 1980; Hill, 1978; Marshall, 1979). This analysis also illuminates our understanding of normative behaviors that could easily be misinterpreted as deviant.

The natural history approach that is an important aspect of much anthropological work also contributes to our understanding of family dynamics and of alcohol. Observations about the intergenerational transmission of drinking problems (Wolin, Bennett, Noonan, & Teitelbaum, 1980) clearly deal with education, in a nonscholastic sense. Workers in this field have been unusually consistent, clear, and explicit in articulating recommendations that would improve mental health and social relations in such stressed families. Few others have studied how families deal with drinking problems, except in the limited context of therapeutic settings.

As anthropologists become more engaged with individuals who have experienced problems with alcohol, the charge that naive functionalism distorts our view of alcohol use is shown to be irrelevant. Similarly, bringing observational methods to bear has helped to identify strengths and weaknesses in various treatment modalities (Strug & Hyman, 1981). Paying attention to informants' views has let us recognize the therapeutic value of many native and other institutions (e.g., Albaugh & Anderson, 1974), and has revealed how the guidelines for Alcoholics Anonymous can fruitfully be adapted to accommodate the life patterns of diverse populations (Jilek-Aall, 1981; Leland, 1980). The analysis of social networks also has practical relevance, enabling early identification of individuals who may benefit from counseling (Maida, 1984; Strug & Hyman, 1981).

A major portion of the work discussed in this article deals with the benefits of drinking and how these are imbedded in a people's way of life. It is also important to emphasize that anthropology and alcohol studies have been mutually beneficial. The general acceptance among scientists of the relevance of sociocultural factors should help allay misgivings that anthropology has little to offer other than empathetic and literary perspectives on the lives of others or that qualitative approaches to human behavior are outdated and useless.

Acknowledgments

As usual, my friend and colleague A.M. Cooper provided the context and stimulation for writing, having originally developed the bibliographic system that allows broad and continuing coverage. Clyde Kluckhohn sparked my original interest in anthropology, and Mark Keller in alcohol studies.

References

Ablon, J. (1976). Family subculture and behavior in alcoholism: A review of the literature. In B. Kissin & H. Begleiter (Eds.), *The biology of alcoholism* (Vol. 4, pp. 205–242). New York: Plenum.

Ablon, J. (1985). Irish-American Catholics in a West Coast metropolitan area. In L.A. Bennett & G.M. Ames (Eds.), *The American experience with alcohol: Contrasting cultural perspectives* (pp. 395–409). New York: Plenum.

Akins, C., & Beschner, G. (Eds.). (1980). *Ethnography: A research tool for policymakers in the drug and alcohol fields.* Rockville, MD: National Institute Drug Abuse.

Albaugh, B., & Anderson, P. (1974). Peyote in the treatment of alcoholism among American Indians. *American Journal of Psychiatry, 131,* 1247–1250.

Babor, T.F. (Ed.). (1986). *Alcohol and culture: Comparative perspectives from Europe and America (Annals of the New York Academy of Sciences, 427).* New York: New York Academy of Sciences.

Bach, P.J., & Schaefer, J.M. (1979). The tempo of country music and the rate of drinking in bars. *Journal of Studies on Alcohol, 40,* 1058–1064.

Bacon, M.K. (1976). Cross-cultural studies of drinking: Integrated drinking and sex differences in the use of alcoholic beverages. In M. Everett, J.O. Waddell, & D.B. Heath (Eds.), *Cross cultural approaches to the study of alcohol: An interdisciplinary perspective* (pp. 23–33). The Hague: Mouton.

Bacon, M.K., Barry, H. III, & Child, I.L. (Eds.). (1965). A cross-cultural study of drinking. *Quarterly Journal of Studies on Alcohol* (Suppl. 3), 1–114.

Bacon, S. (1943). Sociology and the problems of alcohol: Foundations for a sociological study of drinking behavior. *Quarterly Journal of Studies on Alcohol, 4,* 399–455.

Bales, R.F. (1946). Cultural differences in rates of alcoholism. *Quarterly Journal of Studies on Alcohol, 6,* 480–499.

Baum-Baicker, C. (1985a). The health benefits of moderate alcohol consumption: A review of the literature. *Drug and Alcohol Dependence, 15,* 207–227.

Baum-Baicker, C. (1985b). The psychological benefits of moderate alcohol consumption: A review of the literature. *Drug and Alcohol Dependence, 15,* 305–322.

Bennett, L.A., & Ames, G.M. (Eds.). (1985). *The American experience with alcohol: Contrasting cultural perspectives.* New York: Plenum.

Bennett, L.A., Wolin, S.J., & Noonan, D.L. (1977). Family identity and intergenerational recurrence of alcoholism. *Alcoholism, 13,* 100–108.

Blocker, J.S. (Ed.). (1979). *Alcohol, reform and society: The liquor issue in social context.* Westport, CT: Greenwood.

Bunzel, R. (1976). Chamula and Chichicastenango: A re-examination. In M. Everett, J.O. Waddell, & D.B. Heath (Eds.), *Cross-cultural approaches to the study of alcohol: An interdisciplinary approach* (pp. 21–22). The Hague: Mouton.

Burns, T.F. (1980). Getting rowdy with the boys. *Journal of Drug Issues, 10,* 273–286.

Cavan, S. (1966). *Liquor license: An ethnography of bar behavior.* Chicago: Aldine.

Cawte, J. (1986). Parameters of kava used as a challenge to alcohol. *Australia and New Zealand Journal of Psychiatry, 20,* 70–76.

Cook, J., & Lewington, M. (Eds.). (1979). *Images of alcoholism.* London: Educational Advisory Service of Maudsley Hospital.

DeWalt, B.R. (1979). Drinking behavior, economic status, and adaptive strategies of modernization in a highland Mexican community. *American Ethnologist, 6,* 510–530.

Everett, M., Waddell, J.O., & Heath, D.B. (Eds.). (1976). *Cross-cultural approaches to the study of alcohol: An inter-disciplinary approach.* The Hague: Mouton.

Field, P.B. (1962). A new cross-cultural study of drunkenness. In D.J. Pittman & C.R. Snyder (Eds.), *Society, culture, and drinking patterns* (pp. 48–74). New York: Wiley.

Frankel, B.G., & Whitehead, P.C. (1981). *Drinking and damage: Theoretical advances and implications for prevention* (Monograph 14). New Brunswick, NJ: Rutgers Center of Alcohol Studies.

Freund, P.J. (1985). Polish-American drinking: Continuity and change. In L.A. Bennett & G.M. Ames (Eds.), *The American experience with alcohol: Contrasting cultural perspectives* (pp. 77–92). New York: Plenum.

Galanter, M. (Ed.). (1984). *Recent developments in alcoholism* (Vol. 2). New York: Plenum.

Gilbert, M.J., & Cervantes, R. (1986). Patterns and practices of alcohol use among Mexican Americans: A comprehensive review. *Hispanic Journal of Behavioral Science, 81,* 1–60.

Glassner, B., & Berg, B. (1985). Jewish Americans and alcohol: Processes of avoidance and definition. In L.A. Bennett & G.M. Ames (Eds.), *The American experience with alcohol: Contrasting cultural perspectives* (pp. 93–107). New York: Plenum.

Goedde, H.W., Agarwal, D.P., & Harada, S. (1980). Genetic studies on alcohol metabolizing enzymes: Detection of isozymes in human hair roots. *Enzyme, 25,* 281–286.

Gordon, A.J. (1984). Alcohol use in the perspective of cultural ecology. In M. Galanter (Ed.), *Recent developments in alcoholism* (Vol. 2, pp. 355–375). New York: Plenum.

Hamer, J., & Steinbring, J. (Eds.). (1980). *Alcohol and native peoples of the North.* Lanham: University Press of America.

Heath, D.B. (1975). A critical review of ethnographic studies of alcohol use. In R. Gibbins, Y. Israel, H. Kalant, R. Popham, W. Schmidt, & R. Smart (Eds.), *Research advances in alcohol and drug problems* (Vol. 2, pp. 1–92). New York: Wiley.

Heath, D.B. (1976). Anthropological perspectives on alcohol: An historical review. In M. Everett, J.O. Waddell, & D.B. Heath (Eds.), *Cross-cultural approaches to the study of alcohol: An interdisciplinary approach* (pp. 41–101). The Hague: Mouton.

Heath, D.B. (1980). A critical review of the sociocultural model of alcohol abuse. In T.C. Harford, D.A. Parker, & L. Light (Eds.), *Normative approaches to the prevention of alcohol abuse and alcoholism* (NIAAA Research Monograph 3, pp. 1–18). Rockville, MD: NIAAA.

Heath, D.B. (1981a). Social science research training on alcohol: Profile of a training grant. *Alcohol Health and Research World 5*(4), 48–52.

Heath, D.B. (1981b). Observational studies into alcohol-related problems. In World Health Organization (Ed.), *Community response to alcohol-related problems, Phase I: Final Report.* Arlington, VA: National Technical Information Service.

Heath, D.B. (1982). Historical and cultural factors affecting alcohol consumption in Latin America. In Institute of Medicine (Ed.), *Legislative Approaches to Prevention of Alcohol-Related Problems.* Washington: National Academy Press.

Heath, D.B. (1983). Alcohol use among North American Indians: A cross-cultural survey of patterns and problems. In R.G. Smart, F.B. Glasser, Y. Israel, H. Kalant, R. Popham, & W. Schmidt (Eds.), *Research advances in alcohol and drug problems* (Vol. 7, pp. 343–396). New York: Plenum.

Heath, D.B. (1984). Cross-cultural studies of alcohol use. In M. Galanter (Ed.), *Recent developments in alcoholism* (Vol. 2, pp. 405–415). New York: Plenum.

Heath, D.B. (1986a). Cultural definitions of drinking: Notes toward a semantic approach. *Drinking and Drug Practice Surveyor, 21,* 17–22.

Heath, D.B. (1986b). Drinking and drunkenness in transcultural perspective: Parts I and II. *Transcultural Psychiatric Research Review, 23,* 7–42, 103–126.

Heath, D.B. (1988). Alcohol control policies and drinking patterns: An international game of politics against science. *Journal of Substance Abuse, 1,* 109–115.

Heath, D.B., & Cooper, A.M. (1981). *Alcohol use and world cultures: A comprehensive bibliography of anthropological sources* (Bibliography Series 15). Toronto: Addiction Research Foundation.

Hill, T.W. (1978). Drunken comportment of urban Indians: "Time-out" behavior? *Journal of Anthropological Research, 34,* 442–467.

Hill, T.W. (1984). Ethnohistory and alcohol studies. In M. Galanter (Ed.), *Recent developments in alcoholism* (Vol. 2, pp. 313–337). New York: Plenum.

Honigmann, J.J., & Honigmann, I. (1970). *Arctic townsmen: Ethnic backgrounds and modernization.* Ottawa: St. Paul University, Canadian Research Centre for Anthropology.

Horton, D.J. (1943). The functions of alcohol in primitive societies: A cross-cultural study. *Quarterly Journal of Studies on Alcohol, 4,* 199–230.

Jellinek, E.M. (1960). *The disease concept of alcoholism.* New Brunswick, NJ: Rutgers Center of Alcohol Studies.

Jessor, R., Graves, T.D., Hanson, R.C., & Jessor, S.L. (1968). *Society, personality and deviant behavior: A study of a tri-ethnic community.* New York: Holt, Rinehart and Winston.

Jilek, W.G. (1981). Anomic depression, alcoholism and culture-congenial Indian response. In D.B. Heath, J.O. Waddell, & M.D. Topper (Eds.), *Cultural factors in alcohol research and treatment of drinking problems. Journal of Studies on Alcohol,* Suppl. 9, 159–170.

Jilek-Aall, L. (1981). Acculturation, alcoholism, and Indian-style Alcoholics Anonymous. In D.B. Heath, J.O. Waddell, & M.D. Topper (Eds.), *Cultural factors in alcohol research and treatment of drinking problems. Journal of Studies on Alcohol,* Suppl. 9, 143–158.

Kearney, M. (1970). Drunkenness and religious conversion in a Mexican village. *Quarterly Journal of Studies on Alcohol, 31,* 1332–1352.

Kennedy, J.G. (1978). *Tarahumara of the Sierra Madre: Beer, ecology, and social organization.* Arlington Heights: A.H.M.

Kissin, B., & Begleiter, H. (Eds.). (1971–1983). *The biology of alcoholism* (7 vols.). New York: Plenum.

Klausner, S.Z., Foulks, E.P., & Moore, M.H. (1979). *The Inupiat: Economics and alcohol on the Alaskan North Slope.* Philadelphia: Center for Research on the Acts of Man.

Leland, J. (1976). *Firewater myths: North American Indian drinking and alcohol addiction* (Monograph 11). New Brunswick, NJ: Rutgers Center of Alcohol Studies.

Leland, J.H. (1980). Native American alcohol use: A review of the literature. In P.D. Mail & D.R. McDonald (Eds.), *Tulapai to Tokay: A bibliography of alcohol use and abuse among Native Americans of North America* (pp. 1–56). New Haven: HRAF Press.

LeMasters, E.E. (1975). *Blue collar aristocrats: Life styles at a working class tavern.* Madison: University of Wisconsin Press.

Levine, H.G. (1978). The discovery of addiction: Changing conceptions of habitual drunkenness in America. *Journal of Studies on Alcohol, 39,* 143–174.

Lex, B.W. (Ed.). (1980). The recreational and social uses of dependency-producing drugs in diverse social and cultural contexts. *Journal of Drug Issues, 10*(2), 1–310.

MacAndrew, C., & Edgerton, R.B. (1969). *Drunken comportment: A social explanation.* Chicago: Aldine.

Madsen, W. (1973). *The American alcoholic: The nature-nurture controversy in alcoholic research and therapy.* Springfield, IL: Charles C. Thomas.

Maghbouleh, M.D. (1980). *Psychocultural dimensions of alcoholism, witchcraft, ethnic relations, and asceticism: A comparative study* (3 vols.). Irvine, CA: University of California.

Maida, C.A. (1984). Social-network considerations in the alcohol field. In M. Galanter (Ed.), *Recent developments in alcoholism* (Vol. 2, pp. 339–353). New York: Plenum.

Mandelbaum, D.G. (1965). Alcohol and culture [with comments by V. Erlich, K. Hasan, D. Heath, J. Honigmann, E. Lemert, & W. Madsen]. *Current Anthropology, 6,* 281–294.

Marshall, M. (Ed.). (1979). *Beliefs, behaviors, and alcoholic beverages: A cross-cultural survey.* Ann Arbor: University of Michigan Press.

Marshall, M. (Ed.). (1982). *Through a glass darkly: Beer and modernization in Papua, New Guinea* (Research Monograph 18). Boroko, P.N.G.: Institute for Applied Social Economics.

McClelland, D.C., Davis, W.N., Kalin, R., & Wanner, E. (1972). *The drinking man.* New York: Free Press.

Morris, C. (1979). Maize beer in the economics, politics, and religion of the Incan Empire. In G. Gastineau, W. Darby, & T. Turner (Eds.), *Fermented food beverages in nutrition* (pp. 21–34). New York: Academic Press.

National Institute on Alcohol Abuse and Alcoholism (1982). *Special population issues* (NIAAA Alcohol Health Monograph 4). Rockville, MD: NIAAA.

National Institute on Alcohol Abuse and Alcoholism. (1989). *Alcohol use among U.S. ethnic minorities* (NIAAA Research Monograph 18). Rockville, MD: NIAAA.

Pan, L. (1975). *Alcohol in colonial Africa* (Monograph 22). Helsinki: Finnish Foundation of Alcohol Studies.

Pittman, D.J. (Ed.). (1967). *Alcoholism.* New York: Harper & Row.

Popham, R.E. (1978). The social history of the tavern. In Y. Israel, F.B. Glaser, H. Kalant, R.E. Popham, W. Schmidt, & R.G. Smart (Eds.) *Research advances in alcohol and drug problems* (Vol. 4, pp. 225–302). New York: Wiley.

Read, K.E. (1980). *Other voices: The style of a male homosexual tavern.* Novato, CA: Chandler Sharp.

Reed, T.E., Kalant, H., Gibbins, R.J., Kapur, B.M., & Rankin, J.G. (1976). Alcohol and acetaldehyde metabolism in Caucasians, Chinese and Amerinds. *Canadian Medical Association Journal, 115,* 851–855.

Rodin, M. (1981). Alcoholism as a folk disease: The paradox of beliefs and choice of therapy in an urban American community. *Journal of Studies on Alcohol, 42,* 822–834.

Rodin, M.B. (1985). Getting on the program: A biocultural analysis of Alcoholics Anonymous. In L.A. Bennett & G.M. Ames (Eds.), *The American experience with alcohol: Contrasting cultural perspectives* (pp. 41–58). New York: Plenum.

Room, R. (1984). Alcohol and ethnography: A case of problem deflation? [with comments by M. Agar, J. Backett, L.A. Bennett, S. Casswell, D.B. Heath, J. Leland, J.E. Levy, W. Madsen, M. Marshall, J. Moskalewicz, J.C. Negrete, M.B. Rodin, L. Sackett, M. Sargent, & D. Strug]. *Current Anthropology, 25,* 169–191.

Room, R. (1985). Dependence and society. *British Journal of Addiction, 80,* 133–139.

Schaefer, J.M. (1981). Firewater myths revisited: Review of findings and some new directions. In D.B. Heath, J.O. Waddell, & M.D. Topper (Eds.), *Cultural factors in alcohol research and treatment of drinking problems. Journal of Studies on Alcohol,* Suppl. 9, 99–117.

Singer, M. (1985). Family comes first: An examination of the networks of Skid Row men. *Human Organizations, 44,* 147–153.

Singer, M., & Borrero, M. (1984). Indigenous treatment for alcoholism: The case of Puerto Rican spiritism. *Medical Anthropology, 8,* 246–253.

Snyder, C.R. (1958). *Alcohol and the Jews: A cultural study of drinking and sobriety.* Glencoe, IL: Free Press.

Spradley, J.P. (1970). *You owe yourself a drunk: An ethnography of urban nomads.* Boston: Little, Brown.

Spradley, J.P., & Mann, B.J. (1975). *The cocktail waitress: Woman's work in a man's world.* New York: Wiley.

Stivers, R. (1978). Irish ethnicity and alcohol use. *Medical Anthropology, 2*(4), 121–135.

Strug, D.L., & Hyman, M.M. (1981). Social networks of alcoholics. *Journal of Studies on Alcohol, 42,* 855–884.

Strug, D.L., Priyadarsini, S., & Hyman, M.M. (Eds.). (1986). *Alcohol interventions: Historical and sociocultural approaches.* New York: Haworth.

Taylor, W.B. (1979). *Drinking, homicide, and rebellion in colonial Mexican villages.* Stanford: Stanford University Press.

Thomas, A.E. (1978). Class and sociability among urban workers: A study of the bar as social club. *Medical Anthropology, 2*(4), 9–30.

Topper, M.D. (1976). The cultural approach, verbal action plans, and alcohol research. In M. Everett, J.O. Waddell, & D.B. Heath (Eds.), *Cross-cultural approaches to the study of alcohol: An interdisciplinary approach* (pp. 379–402). The Hague: Mouton.

Topper, M.D. (1985a). Navajo "alcoholism": Drinking, alcohol abuse, and treatment in a changing cultural environment. In L.A. Bennett & G.M. Ames (Eds.), *The American experience with alcohol: contrasting cultural perspectives* (pp. 279–296). New York: Plenum.

Topper, M.D. (1985b). "Syntax error": Alcohol, drugs, data, and anthropologists in clinical care settings. *Alcohol and Drug Study Group Newsletter, 15,* 1–3.

Trotter, R.T. (1985). Mexican-American experience with alcohol: South Texas examples. In L.A. Bennett & G.M. Ames (Eds.), *The American experience with alcohol: Contrasting cultural perspectives.* New York: Plenum.

Ullman, A.D. (1958). Sociocultural backgrounds of alcoholism. *Annals of the American Academy of Political Science, 315,* 48–54.

Vaillant, G. (1983). *The natural history of alcoholism: Causes, patterns, and paths to recovery.* Cambridge, MA: Harvard University Press.

Waddell, J.O. (1980). Drinking as a means of articulating social and cultural values: Papagos in an urban setting. In J.O. Waddell & M.W. Everett (Eds.), *Drinking and behavior among Southwestern Indians: An anthropological perspective* (pp. 1–32). Tucson: University of Arizona Press.

Waddell, J.O. (1984). Alcoholism-treatment-center-based projects. In M. Galanter (Ed.), *Recent developments in alcoholism* (Vol. 2, pp. 397–404). New York: Plenum.

Weibel-Orlando, J.C. (1982). American Indians, urbanization, and alcohol: A developing urban Indian drinking ethos. In National Institute on Alcohol Abuse and Alcoholism (Ed.), *Special population issues* (NIAAA Alcohol Health Monograph 4, pp. 206–227). Rockville, MD: NIAAA.

Weisner, T.S., Weibel-Orlando, J.C., & Long, J. (1984). Serious drinking, white man's drinking, and teetotaling: Drinking levels and styles in an urban Indian population. *Journal of Studies on Alcohol, 45,* 237–250.

Wolcott, H.F. (1974). *African beer gardens of Bulawayo: Integrated drinking in a segregated society* (Monograph 10). New Brunswick, NJ: Rutgers Center of Alcohol Studies.

Wolin, S.J., Bennett, L.A., Noonan, D.L., & Teitelbaum, M. (1980). Disrupted family rituals: A factor in the intergenerational transmission of alcoholism. *Journal of Studies on Alcohol, 41,* 199–214.

SECTION II

Observations on the Modern Setting

A. DRINKING PATTERNS

Introductory Note

From this point, the emphasis of this volume shifts from the drinking patterns of less complex societies to those of more complex ones. Societal complexity has implications both for the social use and the evaluation of alcoholic beverages. Selden Bacon's (1945) essay was an early attempt to relate some aspects of modern life—social stratification, division of labor, institutional interdependence, and individualism, among others—to drinking behavior and attitudes. Other facets of modern industrial societies, such as the bureaucratization of life, the development of instant means of communication, modernity, the changing social ethic, character structure, and modes of social control, all affect drinking behavior and the reactions to it by societal members and organizations.

In this subsection we first present Hugh Klein's (chapter 6) portrayal of cultural determinants of the use of beer, distilled spirits, wine, and wine coolers in U.S. society with emphasis on the nonproblematic use of the beverages. Klein's chapter is based on a 1986 national probability sample of 2,401 U.S. citizens aged 21 and over, 1,069 of whom were deemed "drinkers" on the basis of having drunk at least one alcoholic beverage within the previous seven days. This study investigated, among other things, the respondents' differences in the perceived situational appropriateness of drinking various alcoholic beverages in each of six social occasions and the perceptions of adverse consequences associated with drinking each type of alcoholic beverage. Thus, this empirical study demonstrates the significant role culture plays in determining individuals' attitudes toward drinking, their patterns of alcohol use, and the types of alcoholic beverages they consume. In sum, in 1986 the majority of the alcohol consumed in the United States was beer, followed by distilled spirits, and then wine and wine coolers.

The selection that follows, by the Norwegian scholar Ole-Jörgen Skog (chapter 7), is concerned with drinking and the distribution of alcohol consumption. The investigation of the distribution of consumption was pioneered by Ledermann in the late 1940s, but Skog's chapter is an analysis of the shape of actual alcohol consumption curves. He finds that empirical studies confirm that alcohol distributions are very skewed in both low- and high-consumption countries. However, simple knowledge of the shape of the curve does not offer any theoretical explanation of what significance this distribution has for alcohol-related problems or for heavy drinking in any society. Later in this book (chapter 29) Skog presents a model—based on social network theory—that aids in understanding the distribution of consumption.

We conclude this subsection with Michael Hilton and Walter Clark's "Changes in American Drinking Patterns and Problems, 1967–1984" (chapter 8), which empirically tests a distribution-of-consumption model. Their analyses are based on comparisons between 1967 and 1984 population surveys. The authors report changes in beverage preferences but few differences in drinking patterns. Their findings are equivocal for changes in drinking problems and provide neither strong support for nor a refutation of the single-distribution-of-consumption theory.

More recent data indicate that apparent per capita consumption of absolute alcohol by the drinking age population of 14 years of age or older continues to decline and is at its lowest levels since the early 1960s. This decrease is most graphically reflected in the decrease in sales of distilled spirits, but sales of both wine and beer are slightly lower or stagnant (National Institute on Alcohol Abuse and Alcoholism, 1990). Crucial indices of alcohol-related harm have been declining. Deaths from liver cirrhosis, which were used by Jellinek and others in the period from the 1940s through the 1960s to construct rates of alcoholism, peaked in the United States in 1973 at 15.0 deaths in every 100,000 people. It has steadily decreased since that year to 9.3 deaths in every 100,000 in 1986, the lowest rate since 1955 (National Institute on Alcohol Abuse and Alcoholism). This change indicates that individuals are now seeking treatment for their alcoholism at an earlier stage in the disease process. Statistics from Alcoholics Anonymous (AA) appear to support this hypothesis. Membership in AA has increased from an estimated 612,000 in 1977 to around 1.6 million in 1987 (Hall, 1989)—almost 1 million more U.S. citizens now define themselves as being alcoholic.

Furthermore, the number of U.S. victims of alcohol-related car crashes has been decreasing since the early 1980s, according to the Center for Disease Control (CDC). The proportion of fatal crashes involving alcohol declined during the period 1982–87 from 57% to 51%; the number of deaths was 25,165 in 1982 and 23,630 in 1987. The number of crashes in which one person tested with a blood alcohol level of 0.10% dropped from 20,356 (46%) in 1982 to 18,529 (40%) in 1987 ("CDC Reports Drop," 1989).

Thus, there has been not only a reduction of per capita consumption of alcoholic beverages in the United States but a decrease in measures of alcohol-related harm in the decade of the 1980s. Some possible explanations for this decline are discussed in section V, on responsive movements and social control.

References

Bacon, S. (1945). Alcohol and complex society. In *Alcohol, Science and Society,* New Haven, CT: Journal of Studies on Alcohol.

CDC reports drop in alcohol-related crashes (1989, January/February). *The Big Issue.*

Hall, T. (1989, March 15). A new temperance movement is taking root in America. *New York Times.*

National Institute on Alcohol Abuse and Alcoholism. (1990). *Seventh special report to the U.S. Congress on alcohol and health* (DHHS Publication No. ADM 90–1656). Rockville, MD: Author.

CHAPTER 6

Cultural Determinants of Alcohol Use in the United States

HUGH KLEIN

In recent years, as the field of alcohol studies has experienced considerable growth in popularity and legitimacy as a scholarly discipline, a number of topics that had previously been studied only minimally have been the subjects of much research. Among these topics are fetal alcohol syndrome, drunk driving, controlled drinking for alcoholics, protective effects of moderate drinking (as is often claimed for alcohol's relation to coronary heart disease), minimum purchase age, and, most recently, codependency. All of these areas of study have a common theme: They all focus on the adverse consequences resulting from drinking, rather than on the nonproblematic use of alcoholic beverages.

Indeed, in the last decade, many scholars (e.g., Heath, 1982; White, 1982) have lamented the lack of research on the customary use of alcohol—that is, drinking that occurs without resulting in alcohol-related problems. Since estimates usually place the proportion of problem drinkers and alcoholics in the adult population at 10% to 15% (Haglund & Schuckit, 1982; U.S. Department of Health and Human Services, 1987), the converse is that the large majority of drinkers (by more than a 4:1 margin) drink without significant impairment. Although it is important to study people who misuse alcohol, to focus so much attention on the alcohol abusers is to ignore the fact that most adults are able to drink without incurring problems. Social drinkers' drinking patterns are just as worthy of investigation as those of their impaired counterparts, if for no other reason than that understanding moderate consumption of alcoholic

Hugh Klein is an assistant professor of sociology in the Department of Sociology-Anthropology at Indiana University of Pennsylvania, Indiana, Pa. 15705.
This chapter was written especially for this book. Parts of this chapter are based on two other articles: (1) Klein, H., & Pittman, D.J. (1990a), Social occasions and the perceived appropriateness of consuming different alcoholic beverages, *Journal of Studies on Alcohol, 51,* 59–67; (2) Klein, H., & Pittman, D.J. (1990b), Perceived consequences associated with the use of beer, wine, distilled spirits, and wine coolers, *International Journal of the Addictions, 25,* 471–493.

beverages could provide some insight into how alcohol problems in society at large might be minimized or treated more effectively.

Understanding the cultural factors that influence drinking is, therefore, crucial. The social settings in which people drink, the occasion at hand, the social and psychological cues that induce drinking, and the perceptions to which people adhere regarding alcohol use are all likely to be key determinants of when people drink and how much they consume on any given occasion. The purpose of this chapter is to investigate some of these cultural variables and describe how they shape contemporary American drinking patterns. Differences among beer, distilled spirits, wine, and wine cooler use will be analyzed and discussed in light of the differential impact that culture has on the consumption of these various alcoholic beverages.

Method

Data Set and Sample Description

This study's data were collected in November 1986 by a national research firm for a study of drinking occasions. A national probability sample was obtained via telephone interviewing; random digit dialing was used to obtain access to potential respondents to ensure the inclusion in the sample of people whose telephone numbers were unlisted.

To be included in the sample, respondents had to be at least 21 years of age and had to have consumed beer, distilled spirits, wine, or wine coolers within the previous week. People who had not consumed any alcoholic beverages within the previous 7 days were deemed nondrinkers. The only information available for these individuals is their age and sex. The drinkers (i.e., those people who had consumed at least some alcohol during the past week), on the other hand, were asked a series of questions about their alcohol use.

In the 4,104 telephone contacts that were made, 1,603 people (or 39.1%) refused to be interviewed. Of the 2,501 respondents, 1,432 (or 57.3%) were deemed nondrinkers through the criterion outlined above. Thus, 1,069 interviews were completed with eligible respondents. A brief look at the sample shows that 55.4% were male, 87.8% white, 56.9% had at least some college education, 60.9% were married, and 34.8 was the median age. Further details about the sample can be found in Klein & Pittman (1990a).

Survey Instrument and Measures Used

The survey instrument included separate sections for each of the four types of alcohol, but questions were otherwise identical within each of these sections. Drinking context information was gathered only about the last time that respondents had had a particular beverage within the past week; all respondents were asked about each type of alcohol that they had consumed within the past seven days. Among the topics of study were the respondents' perceptions of how appropriate it would be to consume different alcoholic beverages in a variety of social situations, their perceptions of what types of adverse consequences are associated with the four alcoholic beverage types, their emotional state at the time of drinking, the location of the drinking episode, and the relationship of the alcohol use to a meal. These will each be briefly discussed.

For the perceptions of how appropriate it would be to consume the four alcoholic beverage types in different social settings, respondents were asked whether they felt it would be "very appropriate," "somewhat appropriate," or "not at all appropriate" to drink one particular type of alcohol (which was always a type that they themselves had consumed in the past week) in each of six social situations. The social contexts chosen for study—namely, (1) "at a celebration like a birthday party or wedding," (2) "as a perfect complement to a nice dinner," (3) "in a bar, after work or class," (4) "at a ballgame," (5) "when the party's really rolling," and (6) "after a particularly rough day"—were selected as surrogate measures to represent drinking on integrative social occasions (as in the first two contexts), drinking to facilitate relaxation (as with the third and fourth occasions), drinking as simultaneously integrative and disintegrative (as in the fifth context), and drinking in a disintegrative or anxiety-reductive fashion (as in the last occasion). These notions of integrative, disintegrative, and anxiety-reductive drinking were previously suggested by Spindler (1964).

For the evaluation of their perceptions of the adverse consequences that may result from drinking, respondents were asked which one type of alcoholic beverage—if any—they most associated with each of six adverse consequences of drinking: underage drinking, fighting and rowdy behavior, alcoholism, drunk driving, severe health problems like liver disease, and birth defects. Subjects could volunteer that all four types of alcohol were, to them, equally involved in any or all of these problems. But in no instance was someone allowed to select two or three types of alcohol for any one problem.

The interviewer assessed the emotional state for each type of beverage respondents consumed by telling them, "Next, I'm going to read you some words which may or may not describe how you were feeling when you drank [beverage type] on that [i.e., the last] occasion. For each word I read, please tell me how much that applies to how you were feeling. Does it apply 'very much,' 'somewhat,' or 'not at all'?" The words that were read were asked in a random order to prevent response set bias and consisted of 10 emotions, of which 5 were negative (*sad, tense, bored, irritable, lonely*), 4 were positive (*romantic, stimulated, festive, happy*), and 1 was generally neutral in affect (*calm*).

Information about the location of the drinking episode was gathered on a beverage-by-beverage basis and involved several questions. First, respondents were asked whether they drank at home or away from home. Those who drank at home were then asked whether they drank alone or with someone else.[1] Those who initially indicated that they drank away from home were then asked whether they drank at a friend's home, a hotel or restaurant, a bar, at work or in the office, or at some other location.

The relationship of the alcohol use to a meal was also ascertained during the interview. For each type of beverage they consumed, respondents were asked whether they drank the alcohol before, during, or after a meal, or at a time that was unrelated to a mealtime setting. If they responded in any of the first three ways, subjects were additionally asked the meal to which they were referring.

Throughout these analyses, whenever amounts of alcoholic beverages consumed are discussed, they are based on the ethanol equivalencies for the various types of alcoholic beverages under study, rather than on the number of drinks consumed, per se. Thus, alcohol consumption was calculated by multiplying the number of drinks of a given type of beverage that a respondent had had by the number of ounces of alcohol that drink contained (in the respondent's estimation) by the ethanol equivalency factor. The equivalency factors used—0.045 for beer, 0.129 for wine, 0.411 for distilled spirits, and 0.05 for wine coolers—can be found in Williams, Doernberg, Stinson, and Noble (1986).

Results

Perceptions of Situational Appropriateness

In an examination of the first of the two social contexts in which drinking is being termed integrative (i.e., as a perfect complement to a nice dinner), the data indicated that drinkers tended to associate wine consumption closely with the social context of dinner (see Figure 1). More than three quarters of the respondents surveyed felt that wine makes a perfect complement to a nice dinner—nearly twice the proportion that felt the same way about wine coolers ($p < .0001$), more than twice the proportion that felt that way about distilled spirits ($p < .0001$), and more than quintuple the proportion that felt that way about beer ($p < .0001$).[2] In addition, respondents were significantly less likely to say that it is very appropriate to drink beer as a perfect complement to a nice dinner than they were to say the same thing about wine, distilled spirits, and wine

FIGURE 1
Percentage of Respondents Reporting "Very Appropriate" Consumption, by Beverage Type and Social Occasion

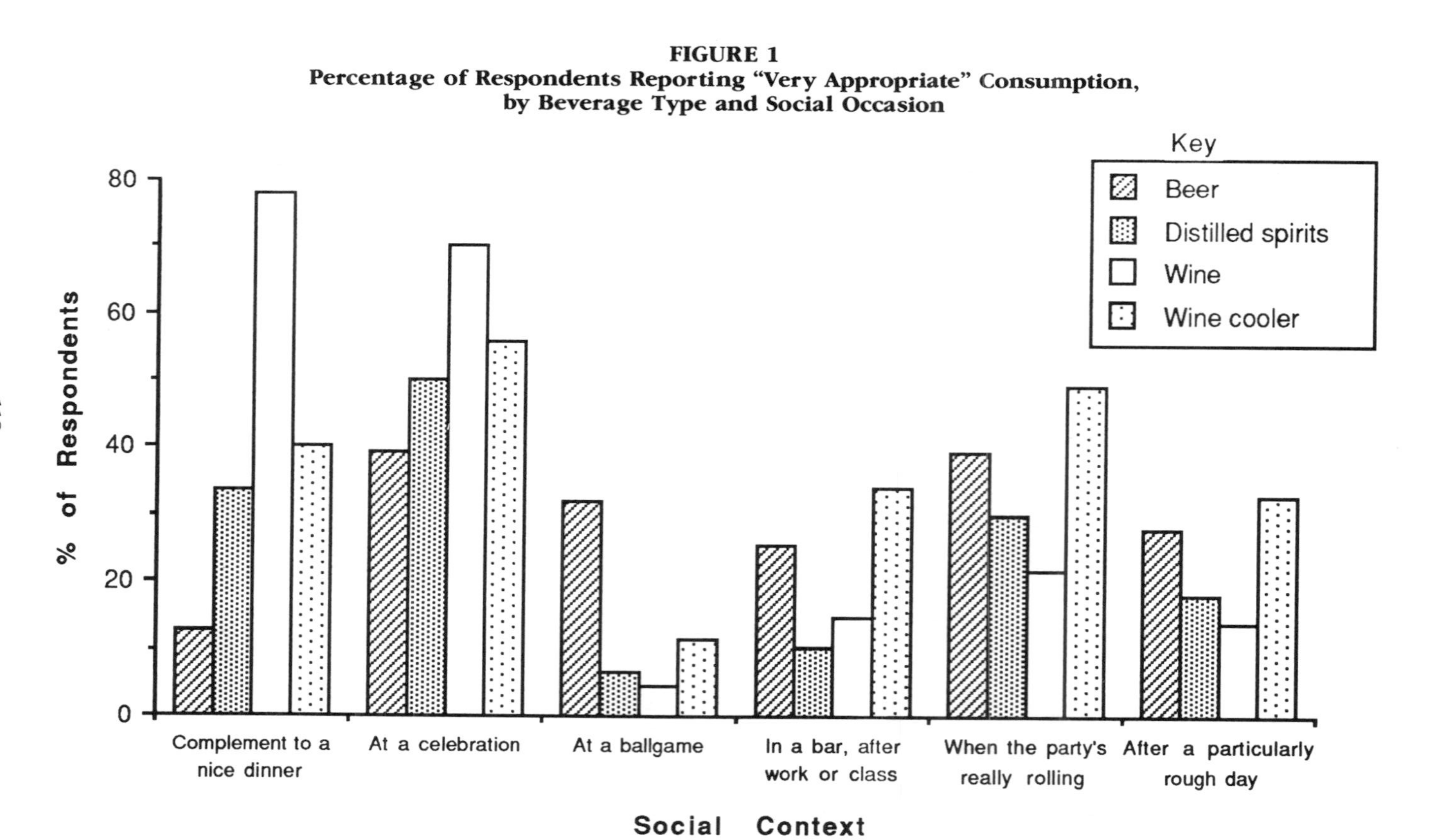

coolers ($p < .0001$ for all comparisons). Sex, age, income, and marital status did not affect these relationships.

The respondents clearly associated alcohol use with drinking at a celebration like a birthday party or wedding. This finding is perhaps best illustrated by the fact that the response "not at all appropriate" did not exceed 13% for any of the types of beverage (see Figure 2). As was true for the social context of drinking as a perfect complement to a nice dinner, people once again made rather sharp differentiations among the four types of beverage. First, and perhaps most interesting, analyses showed that only beer was not deemed by a majority of the respondents to be very appropriate for consumption at a celebration. Meanwhile, more than two thirds of the people surveyed felt that wine was very appropriate for consumption on such an occasion. In fact, subjects were significantly more likely to endorse wine consumption for this social context than to endorse the consumption of any other type of alcoholic beverage ($p < .0001$ for beer and distilled spirits, $p < .004$ for wine coolers). In addition, they were significantly less likely to approve of the consumption of beer at such celebrations than to approve any other type of alcohol ($p < .005$ for distilled spirits, $p < .0001$ for wine, $p < .004$ for wine coolers).

A number of significant differences, based on sex, age, income, and marital status, were found for drinking at a celebration; but age was found to be by far the most influential of these factors. First, respondents aged 65 or older were significantly less likely than their younger counterparts to say that it is very appropriate to drink distilled spirits at a celebration ($p < .0001$ for all age groups). In addition, respondents aged 65 or over were also significantly more likely than their younger counterparts to endorse consumption of wine coolers at a celebration like a birthday party or wedding ($p < .0001$ for all age groups).

The next type of social context to be discussed is drinking for relaxation, with drinking in a bar after work or class and drinking at a ballgame being used as surrogate measures. Respondents were significantly more likely to say that it is very appropriate to drink wine coolers in a bar after work or class than they were to say the same about wine or distilled spirits ($p < .0001$ for both) (see Figure 1). They also felt that it is more appropriate to consume beer in a bar after work or class than distilled spirits ($p < .0001$). At the same time, they were more likely to say that it is not at all appropriate to drink wine in this context than they were to say the same about beer or wine coolers ($p < .0001$ for both). For the most part, the effects of sex, age, income, and marital status did not alter these findings.

When the other relaxation-oriented context (drinking at a ballgame) is examined, only for beer was even a simple majority of the sample saying that it is at all appropriate to drink at a ballgame (see Figure 2). Yet even here it should be noted that nearly one third of the respondents did not

FIGURE 2
Percentage of Respondents Reporting "Not at All Appropriate" Consumption, by Beverage Type and Social Occasion

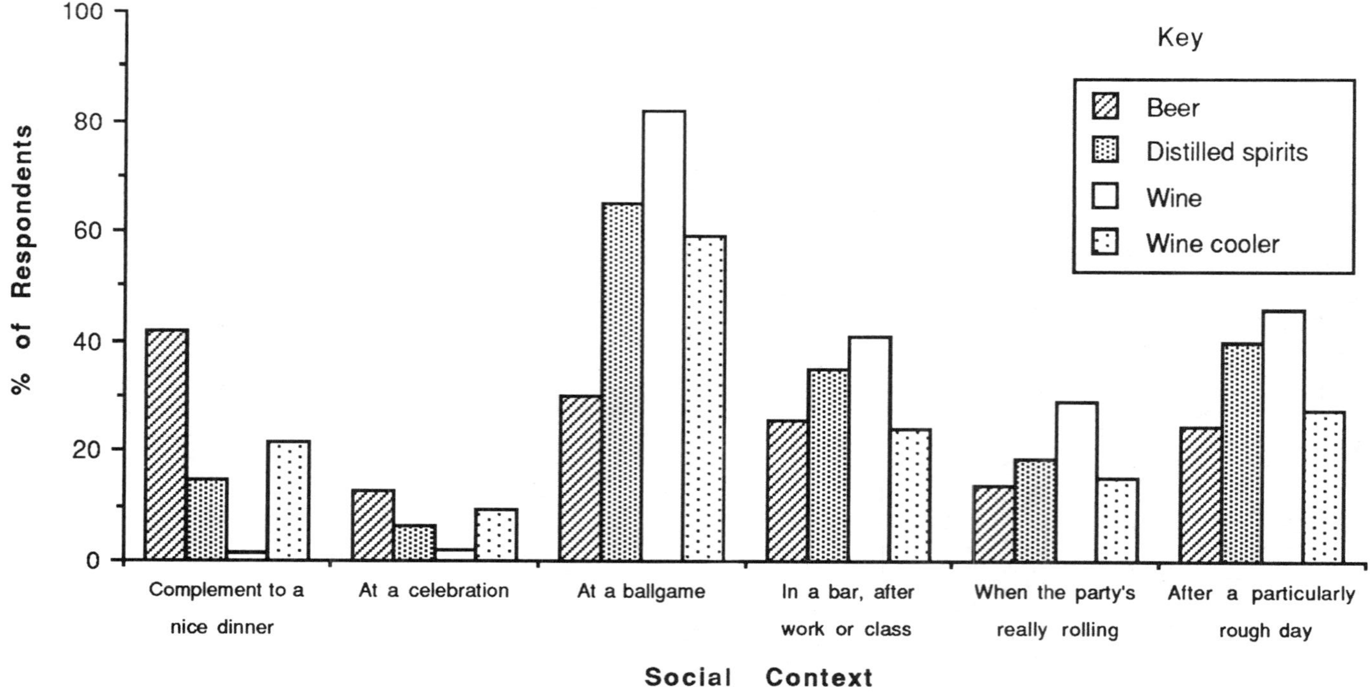

% of Respondents

100
80
60
40
20
0

Key
Beer
Distilled spirits
Wine
Wine cooler

Complement to a nice dinner
At a celebration
At a ballgame
In a bar, after work or class
When the party's really rolling
After a particularly rough day

Social Context

find this practice acceptable. In addition, subjects were significantly more likely to say that it is not at all appropriate to drink wine at a ballgame than they were to say the same about any other type of alcohol ($p < .0001$ for all comparisons).

Although some significant differences were found for sex, education, and income, no consistent patterning of results was noted. Age and marital status were not influential in these relationships.

For drinking "when the party's really rolling," drinkers felt that wine was significantly less appropriate for consumption than beer or wine coolers ($p < .0001$ for both). And wine coolers were thought to be significantly more appropriate for consumption "when the party's really rolling" than distilled spirts were ($p < .0001$). No other significant relationships were found for this occasion, and only one significant difference was found for sex, age, income, or marital status.

Finally, more than for any of the other occasions, respondents were reluctant to endorse drinking alcohol of any kind "after a particularly rough day." Nevertheless, they were significantly less likely to cite wine as being very appropriate for consumption "after a particularly rough day" than they were to cite beer or wine coolers ($p < .0001$ for both) (see Figure 1). Conversely, wine was also significantly more likely than beer and wine coolers ($p < .0001$ for both) to be said to be "not at all appropriate" for consumption in such circumstances (see Figure 2). The latter association was also found for use of distilled spirits compared with beer drinking ($p < .0001$).

Once again, for this occasion no significant differences based on sex, income, or marital status were found; age, however, was a discerning variable. Respondents aged 65 or older were significantly more likely than those who were younger to say that it is very appropriate to drink wine coolers after a particularly rough day ($p < .0001$ for all comparisons).

Perceptions of Adverse Consequences Associated with Drinking

These analyses indicated that no overall pattern of beverage consequence associations can accurately describe all six of the problems (i.e., underage drinking, fighting and rowdy behavior, alcoholism, drunk driving, severe health problems like liver disease, and birth defects) under study. It is therefore necessary to analyze each of these problems separately.

Beer was significantly more closely associated with teenage or underage drinking than any other single type of alcohol was ($p < .0001$ for all three beverages) or all four types of alcohol equally ($p < .0001$), with nearly two thirds of the respondents associating it with this consequence (see Figure 3). Furthermore, subjects were significantly more likely to implicate all four beverages equally than they were to single out wine, dis-

FIGURE 3
Adverse Consequence Associations with Use of Alcoholic Beverage

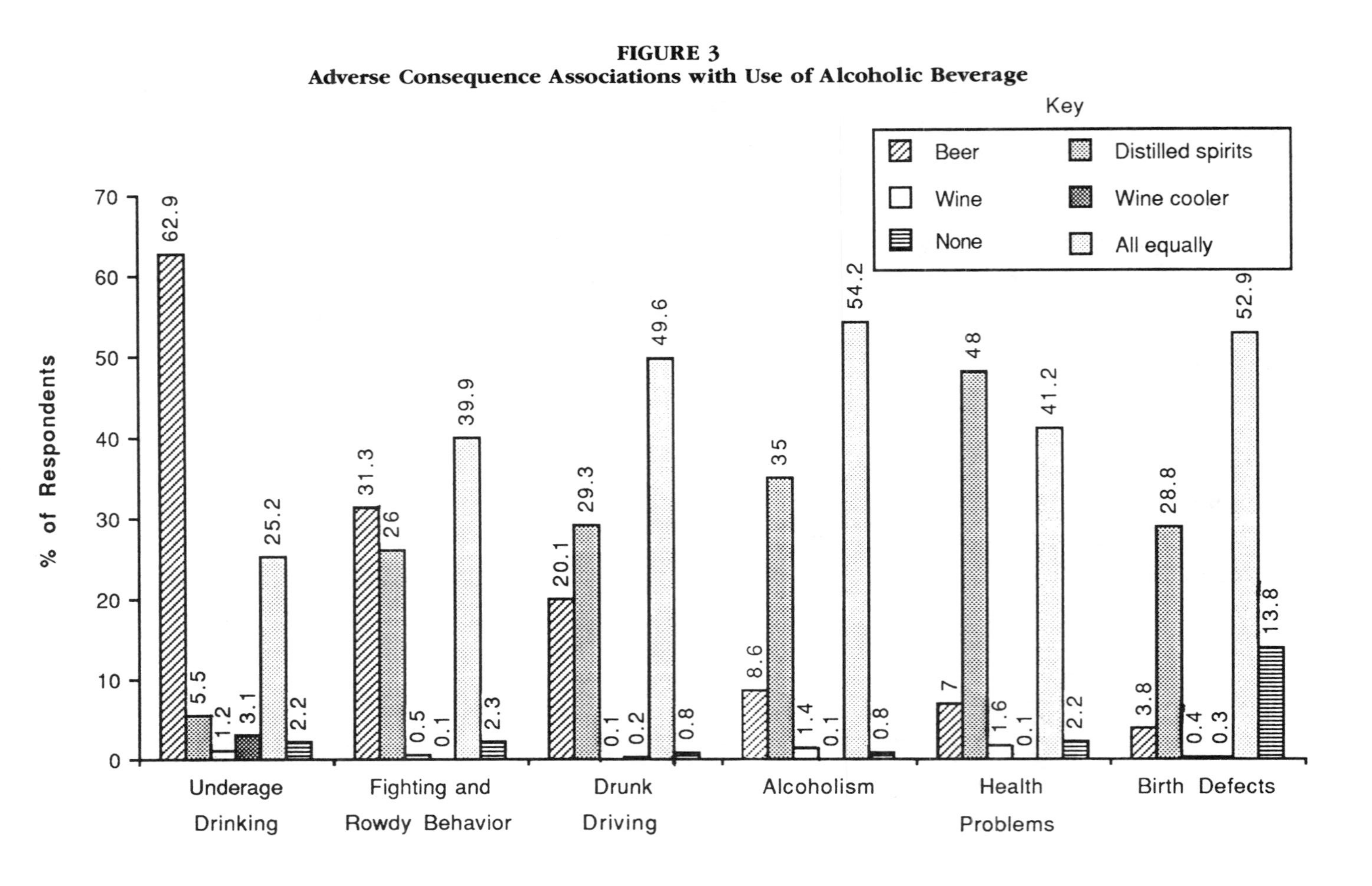

tilled spirits, or wine coolers ($p < .0001$ for all three) as being related to teenage or underage drinking. None of these relationships was mediated by the effects of sex, age, education, income, or marital status.

A very different pattern was observed with fighting and rowdy behavior, though (see Figure 3). In this category, drinkers were significantly more likely to identify all four types of alcohol as equally involved in the occurrence of this problem than they were to single out any one type of alcoholic beverage ($p < .0001$ for all four beverages). In addition, beer and distilled spirits were cited significantly more often for causing fighting and rowdy behavior than wine or wine coolers were ($p < .0001$ for all comparisons).

For these relationships, education was the most important control variable. Respondents who had had some graduate school experience were significantly more likely than their less well educated counterparts to say that beer is most closely related to fighting and rowdy behavior ($p < .001$ for those with less than a high school education, $p < .0001$ for those with a high school education, $p < .003$ for those with some college, $p < .005$ for those with a college degree). At the same time, respondents who had been to graduate school were significantly less likely than those from all other educational levels to implicate distilled spirits for this problem ($p < .005$ for less than high school and for college degree, $p < .0001$ for high school and for some college), whereas those with a high school education were significantly more likely than those from all other educational levels to blame fighting and rowdy behavior on distilled spirits ($p < .05$ for less than high school, $p < .02$ for some college, $p < .004$ for college degree, $p < .0001$ for graduate school).

Significant age and gender effects were also discovered for this variable. Interested readers will find these in Klein and Pittman (1990b).

A pattern fairly similar to that noted for fighting and rowdy behavior was also found for their perceptions of drunk driving. Once again, individuals were significantly more likely to implicate all four types of alcohol equally than they were to choose any one of them alone ($p <. 0001$ for all comparisons). Yet, beer and distilled spirits were still more likely to be blamed for drunk driving than wine or wine coolers were; and individuals were significantly more likely to attribute drunk driving to distilled spirits than to beer ($p < .0001$ for all comparisons).

Differences for these relationships based on sex, age, education, income, and marital status did not form any consistent pattern. Specific findings are provided in Klein & Pittman (1990b).

Alcoholism as a problem was significantly more likely to be associated with all four types of alcohol equally than with any one type by itself ($p < .0001$ for all comparisons). But there was also a strong association in respondents' minds between consumption of distilled spirits and the development of alcoholism (see Figure 3)—more so than was true for beer, wine, or wine coolers ($p < .0001$ for all comparisons). Likewise, beer was significantly more likely to be believed to be associated with alcoholism than wine or wine coolers were ($p < .0001$ for both). Once again, differences based on sex, age, education, income, and marital status did not form any consistent pattern for these relationships.

Severe health problems like liver disease (the fifth adverse consequence of drinking) were associated most closely with use of distilled spirits—significantly more so than any other single type of alcoholic beverage ($p < .0001$ for all comparisons) or all four types equally ($p < .002$). When distilled spirits were not singled out as the culprit, all four beverages were significantly more likely to be equally implicated in the development of severe health problems than beer, wine, or wine coolers alone were ($p < .0001$ for all comparisons).

Sex and age were found to be the most influential variables in these relationships. For example, men were significantly more likely than women to say that use of distilled spirits is most closely associated with severe health problems ($p < .004$). Women, though, were significantly more likely than men to say that all four types of alcohol are equally involved in this problem ($p < .0001$). With respect to age, 21–29-year-olds were significantly more likely than those aged 45 or older to relate liver disease to beer consumption ($p < .0003$ for 45–64-year-olds; $p < .01$ for those aged 65 or older). Furthermore, respondents in their 20s were significantly less likely than their older counterparts to say that all four beverages are equally involved in the development of severe health problems ($p < .003$ for 30–44-year-olds; $p < .007$ for 45–64-year-olds; $p < .03$ for those aged 65 or older). Other significant differences were also obtained for education, marital status, and income (see Klein & Pittman, 1990b).

The findings in reference to birth defects partially resemble those already discussed for drunk driving, alcoholism, and severe health problems (see Figure 3). In essence, respondents were significantly more likely to say that all four types of alcohol are equally involved in the occurrence of this problem ($p < .0001$ for all comparisons). But when one type of alcohol was singled out as being culpable, it was usually distilled spirits ($p < .0001$ for all comparisons).

One very different finding here, though, was the fact that nearly one seventh of the respondents stated the belief that alcohol use is not related to birth defects. Moreover, the nearly one in seven subjects who reported this opinion also constitute a proportion significantly greater than those reporting "no association" for each of the other five problems under study ($p < .0001$ for all comparisons). Clearly, this was an area of considerable misunderstanding and undereducation for these respondents, who tended to be aged 45 or older and male.

Emotional State

In terms of main effects, emotional state at the time of drinking had relatively little impact on the amount of a given type of alcoholic beverage that people consumed. In fact, for six of the emotional states under investigation—namely, happiness, sadness, calmness, tension, festiveness, and irritability—no significant associations were found. Increased beer consumption was found only among respondents who reported feeling lonely at the time of their drinking ($p < .004$), while greater than usual consumption of distilled spirits and wine was noted when drinkers felt stimulated ($p < .04$ for both). More than the other beverage types, consumption of wine coolers was found to be associated with emotional state at the time of drinking. Three affective states—stimulation, boredom, and romanticism—were predictive of increased consumption of wine coolers ($p < .002$, $p < .005$, $p < .0002$, respectively).

The apparent reason for the general lack of association between emotional state and amount of alcohol consumed was the opposing effects of certain control variables. That is, oftentimes it appeared that one group of drinkers (for example, men) drank a certain type of beverage when experiencing a positive affective state, whereas other drinkers (here, women) did not drink when feeling that same way. The results of the sex-, age-, and education-related analyses—and, to a much lesser extent, the findings related to marital status and income, too—support this interpretation.

For instance, with respect to sex, the feelings men and women reported having had when they last drank beer, distilled spirits, or wine were entirely different from one another. Women drank significantly more beer when they felt festive ($p < .0004$), stimulated ($p < .0002$), or romantic ($p < .001$), whereas men drank beer the most heavily when they felt lonely ($p < .002$). For distilled spirits, increased consumption among men was noted in response to their feeling sad ($p < .006$), whereas no emotional state led to significantly greater use of this beverage among women. Feeling festive led to increased wine consumption among women ($p < .03$), whereas no emotional state was related to heavier wine drinking among men. Finally, regarding wine coolers, men reported drinking more when feeling stimulated ($p < .0005$), whereas women drank more when experiencing boredom ($p < .007$). The only relationship between emotional state and alcohol consumption that men and women had in common involved greater consumption of wine coolers, in conjunction with romantic feelings ($p < .001$ for men, $p < .04$ for women).

Age, too, was a highly influential variable in the relationship of consumption to emotional state. First, the data highlighted a strong tendency for the younger people sampled (i.e., those aged 21–29) to show increased consumption of distilled spirits in response to many more emotional states than was true for older respondents. Respondents in their 20s drank significantly more when feeling happy ($p < .05$), stimulated ($p < .004$), festive ($p < .02$), romantic ($p < .0002$), or irritable $p < .008$), whereas only sadness predicted greater consumption of distilled spirits among the 45–64-year-olds ($p < .03$), and only irritability led to heavier drinking among those aged 65 or older ($p < .01$). This trend toward decreased reliance on emotional state as a drinking cue with increasing age was highly significant ($p < .0001$).

The same thing was observed for wine consumption, too ($p < .0001$). The younger respondents drank wine most heavily when feeling stimulated ($p < .04$) or festive ($p < .003$). No significant differences were found for emotional state and wine consumption in any of the groups of respondents aged 30 or older.

With respect to wine coolers, a similar patterning of results was obtained. Once again, the older the respondents, the less their consumption of wine coolers depended upon their emotional state at the time they were drinking ($p < .0001$). Those in their 20s drank wine coolers most heavily in response to feelings of boredom ($p < .005$), stimulation ($p < .03$), and romance ($p < .03$), whereas 30–44-year-olds drank more when feeling stimulated ($p < .005$) and romantic ($p < .0006$). No emotional state predicted greater consumption for respondents over age 45.

Consumption of beer, too, showed a similar trend, but to a lesser extent than noted for the other beverages. As was true for wine, distilled spirits, and wine coolers, the older the respondent, the less tied beer consumption became to emotional state ($p < .0001$). But only one significant age-related relationship was found: loneliness led to increased beer drinking among 30–44-year-olds ($p < .0001$). No other differences were statistically significant.

Educational attainment was the third major variable to have significant effects on the amount of alcohol consumed. But since the sheer number of education-related significant differences obtained was large and formed no distinct pattern, these results will not be reported here (see Klein & Pittman, 1989, for more information).

Location of the Drinking Episode

Respondents who consumed beer, wine, or wine coolers were significantly more likely to drink these beverages at home than away from home ($p < .0005$, $p < .0001$, $p < .01$, respectively), whereas the reverse was true for distilled spirits ($p < .001$). Nevertheless, even for the former types of beverage, a sizable proportion of the drinkers did their drinking away from home, making further analyses of these situations both possible and necessary.

Results indicated that respondents were significantly more likely to consume wine and distilled spirits at a restaurant than at a friend's house ($p < .02$, $p < .0005$, respectively), whereas the opposite was true for wine cooler drinkers ($p < .03$). Respondents were also significantly more likely to have wine and wine coolers at a friend's home than in a bar ($p < .002$, $p < .01$, respectively), whereas the reverse was true for distilled spirits ($p < .03$). In addition, wine drinkers were significantly more likely to drink in a restaurant than in a bar ($p < .0001$)—a finding that was not obtained for the consumption of beer, distilled spirits, or wine coolers.

But where people drink is only half the story here; for location not only influenced which beverage (or beverages) drinkers consumed, but also how much alcohol they actually ingested. Thus, even though drinkers of distilled spirits were more likely to drink away from home than at home, they consumed significantly more distilled spirits when they did their drinking in their own home ($p < .008$). On the other hand, not only were wine drinkers more likely to drink at home, but they ingested more wine there than when they drank away from the home ($p < .03$). And as for drinkers of beer and wine coolers, despite their tendency to drink at home, their amounts of consumption did not vary significantly from one location to another.

Further analyses also revealed that respondents drank significantly more beer in bars than in restaurants ($p < .006$) or at friends' homes ($p < .003$), whereas they did not differ in their amounts of consumption in restaurants and at friends' homes. Drinkers of distilled spirits also consumed significantly more in bars than in restaurants ($p < .007$), but their amount of consumption at friends' homes did not differ significantly from their alcohol use in bars or restaurants. There were no significant differences for drinking wine or wine coolers in bars, restaurants, or friends' homes.

Relationship of Drinking to a Meal

Given the respondents' tendency to differentiate quite sharply in both their attitudes toward and their use of the four beverages, it is not surprising that they also drank different types at different times where meals were concerned. When drinking before a meal, subjects were significantly more likely to drink distilled spirits than any other beverage type ($p < .005$ for wine; $p < .004$ for beer; $p < .05$ for wine coolers). Yet when they drank during a meal, they were more likely to choose wine than any other alcoholic beverage type ($p < .0001$ for distilled spirits and beer; $p < .002$ for wine coolers). Conversely, these individuals were significantly less likely to drink wine than any other alcoholic beverage in a nonmeal situation ($p < .002$ for beer; $p < .009$ for distilled spirits; $p < .01$

for wine coolers). But for drinking after a meal, these respondents preferred beer, especially compared with wine ($p < .02$). There was also a tendency for subjects to select wine coolers in a nonmeal setting rather than during a meal ($p < .05$).

Counterintuitively, although the type of beverage consumed was strongly influenced by the mealtime situation at hand, this variable made very little difference in the amount of alcohol consumed. Respondents who had beer after a meal drank more than those who did so during a meal ($p < .03$). And those who drank distilled spirits in nonmeal settings ingested more than those who did so during a meal ($p < .03$). But no other significant differences were found. The foregoing suggests that the relationship to a mealtime situation helps determine which beverage people will consume but not how much they will imbibe once their drinking is underway.

Discussion

This study demonstrates the salient role that culture plays in determining people's attitudes toward drinking, their patterns of alcohol use, and the types of alcoholic beverages they consume. For without cultural influences, there is no reason to expect that people's perceptions of various types of alcoholic beverages would differ—least of all to the extent noted here. Beer, wine, distilled spirits, and wine coolers all contain ethanol; therefore, differences in perceptions of these beverages are not traceable to differential effects caused by the consumption of these four beverages. It is U.S. culture—"the customs, beliefs, values, knowledge, and skills that guide a people's behavior along shared paths" (Light & Keller, 1979)—that shapes such values. And conversely, it is the different values—"the general ideas people share about what is good or bad, right or wrong, desirable or undesirable" (Light & Keller)—to which Americans adhere regarding alcoholic beverages that best illustrate the power of cultural norms.

Interacting with each other, culture and values emphasize that beer is not the same as wine, which, in turn, is not the same as distilled spirits, which, in turn, are not the same as wine coolers. As a culture, Americans tend to perceive wine and wine coolers to be benign beverages, while viewing beer and distilled spirits as being potentially more harmful. To talk about "alcohol use" is somewhat of a misnomer, then, because to do so is to treat the cultural aspects of the consumption of beer, wine, distilled spirits, and wine coolers as though they were the same. These findings have shown that they are not. Thus, understanding alcohol use in the United States requires separate discussions for the different types of beverage; for each says something quite different about the norms governing alcohol consumption.

For example, in our society wine is clearly considered the beverage of choice for integrative social occasions. Its use is associated with sociability and the enhancement of pleasure (hence its association with the affective state of stimulation) and is almost always moderate in nature. Few, if indeed any, major alcohol-related problems are thought to arise from the consumption of wine. Wine is deemed most appropriate for consumption at home, usually during mealtime—which, it should be noted, is yet another drinking occasion that has been related to moderate alcohol intake (Harford, 1979; Rodin, Morton, & Shimkin, 1982)—and wine drinkers who drink more than average amounts have a strong tendency to regret their "excessive" consumption of this beverage (data not presented). Thus, even in our ambivalent society, attitudes toward the consumption of wine most closely resemble the permissive cultural type ("in which the prevailing attitude is positive toward the use of alcoholic beverages [even though] there may be attitudes . . . which proscribe drunkenness") (Pittman, 1980, p. 8).

But although our cultural perceptions of wine may not necessarily be aligned with our overall societal classification along this schema (i.e., as ambivalent), our attitudes toward beer certainly are. In this study, the use of this type of beverage was neither strongly approved nor strongly condemned in any of the six social contexts investigated in this research. Furthermore, respondents associated certain adverse consequences of drinking (e.g., teenage drinking) rather closely with the use of beer but did not see others (e.g., birth defects) similarly. Additionally, this study has found that beer is typically consumed in one's own home at times unrelated to mealtime settings, especially among drinkers who are feeling lonely. It is much more likely to be drunk by men than by women, by less well educated individuals than by well-educated ones, and by younger than by older ones. That certain types of people (e.g., men, young people) are disproportionately likely to drink beer whereas others in the same society (e.g., women, well-educated people) are underrepresented in the beer-drinking population can be due only to the differing cultural and normative expectations that different segments of the population have about beer. Thus, Americans' attitudes toward beer consumption exemplify the ambivalent culture.

Essentially the same thing can be said for distilled spirits. Respondents in this study seemed to favor drinking distilled spirits during social occasions that were integrative, disfavor their consumption during situations that reduced anxiety, and be rather mixed in their perceptions of the appropriateness of having distilled spirits on occasions that were disintegrative or simultaneously integrative and disintegrative. Add to this the findings for the adverse-consequence associations—namely, that when respondents singled out one alcoholic beverage type as blameworthy, they almost always cited distilled spirits—and the picture becomes quite complicated. The United States has a long-standing antiliquor legacy (Critchlow, 1986; Levine, 1984) that has, over the years, shaped our attitudes toward and use of alcoholic beverages in general and of distilled spirits in particular. Critchlow (p. 753) has pointed out that

> temperance ideology solidified the image of liquor as a powerfully perverting substance that, by dissolving moral restraints over behavior, leads the drinker to behave like a brute. Although American attitudes toward alcohol have changed considerably since, these doctrines continue today in some quarters.

As a result, in this study it appeared as if respondents, on the one hand, wanted to be able to consume distilled spirits moderately but, on the other hand, feared the consequences that are thought to accompany its excessive consumption. Consequently, these conflicting attitudes toward consumption of distilled spirits became manifest in respondents' drinking behaviors. Drinkers of distilled spirits were more likely to drink away from home (typically in a bar) than at home (unlike the case of any other type of alcoholic beverage) usually before a meal (also unlike the case of any other type of alcoholic beverage); yet they drank most heavily when they drank at home at times unrelated to meals. Combined, these attitudes and drinking practices form another example of an ambivalent pattern of alcohol use.

For their part, wine coolers remain more of an enigma in our culture at the present time. Clearly, respondents did not associate adverse consequences with the use of this type of beverage. Yet, at the same time, they did not seem to share any consensus about when it was or was not appropriate to consume wine coolers. Drinkers tended to perceive them as acceptable for integrative social occasions; but they judged them equally acceptable in disintegrative contexts. Wine coolers were usually consumed in the home during nonmeal settings. But otherwise their use was widely dispersed in both the social context of drinking and the characteristics of the drinkers. At the present time, drinkers of wine coolers do not, by any means, constitute a homogeneous group; and consumption of wine coolers is not well patterned in our society.

Since wine coolers are relatively new to the alcoholic beverage market in this country (having first appeared commercially in the early 1980s), perhaps the general lack of normative patterning of their use is due to lack of sufficient time having passed for people to formulate distinctive, crystallized opinions regarding their consumption. Bacon (1976) has suggested that drinking norms (and, presumably, specific types of individuals likely to adhere to these norms) are likely to become more highly developed and more specific as a given culture gains experience with and knowledge about particular alcoholic beverages. If this suggestion has merit, then perhaps with time and more experience with wine coolers, U.S. society will develop norms about the appropriateness of their use. Such a development is now in doubt, however, because wine cooler sales have steadily dropped during the past few years (Rothman, 1988).

To summarize, this study has demonstrated that no one generalization can be made about alcoholic beverages in the aggregate. To understand people's attitudes toward the use of these beverages or their drinking patterns, or both, one must discuss each type of alcohol separately from the others. At the same time, one must also specify the social context in which these beverages are being consumed if any meaningful analysis is to be undertaken. This is necessary because people's attitudes regarding the use of alcoholic beverages vary so much from occasion to occasion.

And just as this research has highlighted the importance of not studying alcoholic beverages as though they constitute a homogeneous category, it has also demonstrated the importance of looking beyond the main effects kinds of findings obtained for "drinkers" to determine more precisely what kind of drinker is being discussed. Repeatedly throughout this study, significant differences based on sex, age, education, income, and marital status were discovered. Although the latter three variables were not found to be as influential as the former two, none of these control variables could rightfully be dismissed as unimportant.

Sex and age in particular were found to exert powerful influence over these findings. As a general rule, women were found to be better informed about the link between alcohol and adverse consequences resulting from drinking than men were. Women, the findings showed, drank different types of alcoholic beverages in different settings than men did, and consumed lesser amounts of alcohol than men did. Men and women also differed quite a bit in terms of the emotional states that apparently catalyzed their drinking. To talk only about "drinkers" would be to overlook these many, interesting, and important differences between the sexes. Therefore, this research suggests that a more meaningful way of investigating drinking data is to look at men drinkers and women drinkers separately.

Likewise, the findings obtained for age also indicate the need to get away from the tendency to treat drinkers of all ages as though they were the same, simply because they consume alcoholic beverages. The differences between young (i.e., people in their 20s) and old (i.e., those aged 65 or older) respondents in this sample are truly striking. Young respondents drank in response to a great number of emotional states, regardless of the type of beverage under consideration; older subjects drank what they wanted whenever they wanted, regardless of their affective condition at the time. Those aged 65 and older differed from their younger counterparts in their perceptions of when it is or is not appropriate to drink certain types of alcohol. And age was also an influential variable in the findings about respondents' perceptions of the relationship between the use of alcoholic beverages and fighting and rowdy behavior, severe health problems like liver disease, and birth defects.

Conclusion

There is nothing inherent in different types of alcoholic beverages (other than, perhaps, their strength and taste) to make people perceive them differently from one another. Physiologically, they all create the same changes in users when consumed in ethanol-equivalent amounts; and if misused, all of them can cause many unwanted, potentially damaging adverse consequences. Essentially, as the educational slogan goes, "Alcohol is alcohol is alcohol."

But this statement is only valid when one is discussing the alcoholic beverages themselves. For when we talk about drinking as a social phenomenon, "alcohol is not alcohol is not alcohol," and "drinkers are not drinkers are not drinkers." Culturally, beer is very different from wine, distilled spirits, and wine coolers (and the same is true for all of the other beverage comparisons). Culturally, male drinkers are very different from female drinkers. Culturally, young drinkers are not the same as older drinkers. Knowing these things is essential; for only by taking into account the influence of culture on drinking behaviors and attitudes toward alcohol use can we understand such complex phenomena. And only by studying cultural attitudes towards alcohol use, alcohol misuse, and people's drinking-related values can we begin to understand why certain cultures or certain people within these cultures develop problems with alcohol.

Acknowledgments

I wish to thank the Winegrowers of California and the Graduate School of Arts and Sciences at Washington University for cosponsoring this research, and the Field Research Corporation for its participation in the collection of the data for this study. Also, William Staudenmeier, Jr., John Helzer, J.L. Fitzgerald, Cynthia Wooldridge, and James Wells have provided many helpful comments about earlier drafts of this work; their suggestions are appreciated. Finally, I also want to thank David Pittman for his significant contribution to this work.

Notes

1. Those who admitted to drinking in the presence of someone else were asked whether this person was a spouse, another adult family member, a friend, a business or professional associate, a child under age 18, or some other person. These data are not, however, reported in this chapter.
2. The statistical significance of these differences was assessed with the use of proportions-difference tests. The formula for calculating this statistic can be found in Bohrnstedt and Knoke (1982).

References

Bacon, M. (1976). Cross-cultural studies of drinking: Integrated drinking and sex differences in the use of alcoholic beverages. In M. Everett, J. Waddell, & D.B. Heath (Eds.), *Cross-cultural approaches to the study of alcohol.* Chicago: Aldine Publishers.

Bohrnstedt, G., & Knoke, D. (1982). *Statistics for social data analysis.* Itaska, IL: F.E. Peacock.

Critchlow, B. (1986). The powers of John Barleycorn: Beliefs about the effects of alcohol on social behavior. *American Psychologist, 41,* 751–764.

Haglund, R.M.J., & Schuckit, M.A. (1982). The epidemiology of alcoholism. In N.J. Estes & M.E. Heinemann (Eds.), *Alcoholism: Development, consequences, and interventions* (2nd ed.). St. Louis: C.V. Mosby.

Harford, T. (1979). Beverage specific drinking contexts. *International Journal of the Addictions, 14,* 197–205.

Heath, D.B. (1982). In other cultures, they also drink. In E.L. Gomberg, H.R. White, & J.A. Carpenter (Eds.). *Alcohol, science, and society revisited.* Ann Arbor, MI: University of Michigan Press.

Klein, H., & Pittman, D.J. (1990a). Social occasions and the perceived appropriateness of consuming different alcoholic beverages. *Journal of Studies on Alcohol, 51,* 59–67.

Klein, H., & Pittman, D.J. (1990b). Perceived consequences associated with the use of beer, wine, distilled spirits, and wine coolers. *International Journal of the Addictions,* 25.

Levine, H. (1984). The alcohol problem in America: From temperance to alcoholism. *British Journal of Addiction, 79,* 109–119.

Light, D., Jr., & Keller, S. (1979). *Sociology* (2nd ed.). New York: Knopf.

Pittman, D.J. (1980). *Primary prevention of alcohol abuse and alcoholism: An evaluation of the control of consumption policy.* St. Louis: Social Science Institute of Washington University.

Rodin, M., Morton, D., & Shimkin, D. (1982). Beverage preference, drinking, and social stress in an urban community. *International Journal of the Addictions, 17,* 315–328.

Rothman, A. (1988, October 4). Wine-cooler industry scrambles to revive once thriving market. *Wall Street Journal, 70,* B6.

Spindler, G.D. (1964). Alcoholism symposium: Editorial preview. *American Anthropologist, 66,* 341.

U.S. Department of Health and Human Services. (1987). *Sixth special report to the U.S. Congress on alcohol and health.* Rockville, MD: National Institute on Alcohol Abuse and Alcoholism.

White, H.R. (1982). Sociological theories of the etiology of alcoholism. In E.L. Gomberg, H.R. White, & J.A. Carpenter (Eds.). *Alcohol, science, and society revisited.* Ann Arbor, MI: University of Michigan Press.

Williams, G., Doernberg, D., Stinson, F., & Noble, J. (1986). State, regional, and national trends in apparent per capita consumption. *Alcohol Health and Research World, 10,* 60–63.

CHAPTER 7

Drinking and the Distribution of Alcohol Consumption

OLE-JÖRGEN SKOG

We are all different and we all behave differently, although the degree to which we are different varies greatly. The main theme of the two chapters written for this volume is to describe and explain individual differences in the consumption of alcoholic beverages, mainly from a sociological and epidemiological point of view. In this chapter the focus is primarily on the epidemiological point of view, whereas in chapter 29 the focus is primarily sociological.

When we talk about alcohol users in every-day language, we typically use constructs like *abstainers, light drinkers, moderate drinkers, heavy drinkers, alcohol abusers, alcoholics,* and the like. Such terms easily give the impression of more or less distinct categories of drinkers. These categories may correspond to distinct objects in the world of experience, or they may be linguistic artifacts, mainly reflecting the inability of everyday language to describe very heterogeneous phenomena and variations along continuous dimensions. One way of resolving this problem is to study more closely individual differences with respect to one of the main dimensions of alcohol use, namely, the quantities consumed over some suitable period of time.

Individual differences in the consumption of alcoholic beverages can be described accurately by the distribution of the population along the consumption scale. A consumption scale is constructed by converting all beverages into absolute alcohol, and for each drinker one obtains a summary measure, say in liters per year. Obviously, the consumption of alcoholic beverages may vary from zero to some physiological maximum. Although the latter may not be exactly the same for all individuals, very few seem

Ole-Jörgen Skog is with the National Institute for Alcohol and Drug Research, Oslo, Norway. This chapter is the first of two chapters prepared especially for this volume. The reader is referred to chapter 29 for a discussion of a social network model of alcohol consumption.

to be able to drink more than about 50 cl (400 g) of pure alcohol a day, or 180 liters a year, for an extended period of time.

The main question to be addressed below is how the population of drinkers is distributed along this continuum. Are there distinct subgroups, corresponding to the above-mentioned categories? Are there any regularities or instances of "lawfulness" in distributions that may lend themselves to quantitative description? How does the distribution vary across drinking cultures, and how does the distribution change when the overall level of consumption in a culture changes? Does the prevalence of heavy drinkers (*abusers, alcoholics,* etc.) vary much across cultures and within cultures over time? What are the explanations for such variations? How stable are individual drinking habits, and are there any typical patterns of individual change?

These and similar questions will be discussed below. First, some theoretical arguments for the existence of regularities in the distribution of alcohol consumption are outlined. These arguments suggest that the distribution may resemble a class of theoretical distributions called the lognormal family. The empirical evidence for this hypothesis is reviewed. In chapter 29, drinking as a social behavior and the social genesis of individual drinking habits are discussed in the context of social network theory.

The Shape of the Distribution

Skewness of the Distribution

A priori, the distribution of the population along the consumption scale could have many different shapes (Figure 1.) One possibility would be a uniform distribution (A), with the same number of persons in each consumption interval. The Gaussian normal distribution (B) is another possible form, which is known to apply to some human attributes. A bimodal shape (C) is also conceivable, with normal drinkers at the lower part of the scale and alcoholics more or less clearly separated at higher levels. A fourth alternative would be a skew distribution (D) without any sharp division between normal and alcoholic drinking.

Some of these possibilities can be disregarded on the basis of common knowledge. Symmetrical distributions, like the uniform and the Gaussian, would not allow anyone to drink more than twice as much as the average, since nobody can drink less than zero. We know that in most cultures in which alcohol is consumed there are a significant number of very heavy drinkers (alcoholics) and their intake is substantially higher than twice the average. This argument strongly suggests that the distribution must be asymmetrical and skew to the right (i.e., positive skewness). Although many possible shapes remain (cf. Johnson & Kotz, 1970, for an overview

FIGURE 1
Hypothetical Distributions of the Population of Drinkers, According to Annual Intake.

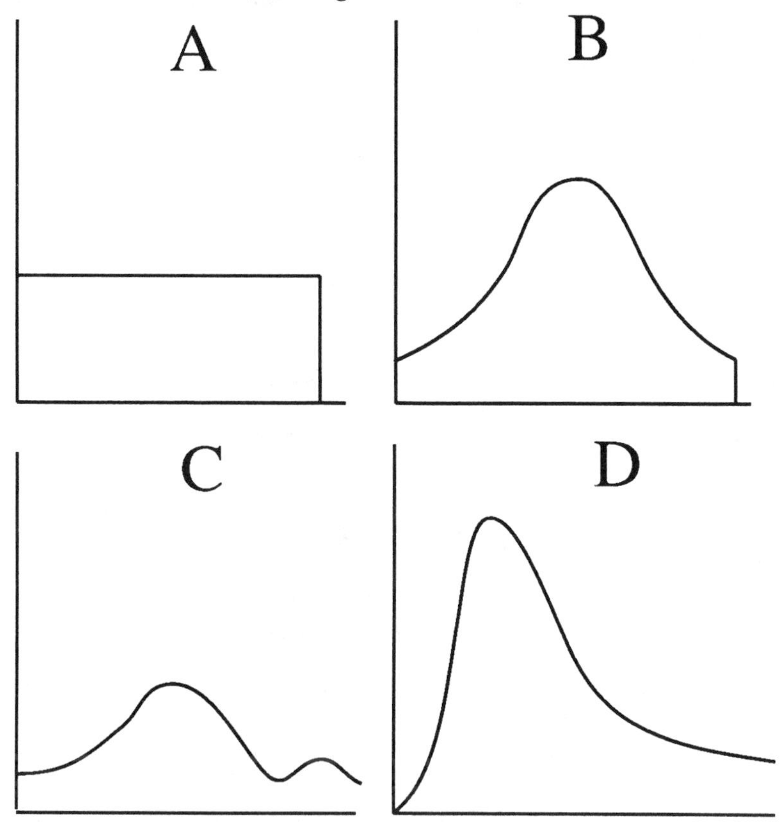

of common distributions in statistics), the question now is whether it is possible to choose a more limited class of distributions on the basis of theoretical arguments or otherwise. If so, is it possible to describe this class with a suitable mathematical function? In short, is it likely that alcohol consumption should obey some kind of distribution law?

Distribution "Laws" for Other Human Attributes

It is known from other branches of life sciences that some human attributes have distribution laws, at least in an approximate sense. Some, such as body length and intelligence, are approximately normal (Anastasi, 1958), whereas others such as income are very skew and resemble a theoretical curve known as the Pareto distribution (Blinder, 1974). Still other examples are known from the study of human consumption behavior. It is known that the consumption of many types of commodities roughly follows a mathematical curve known as the lognormal distribution (Aitchison & Brown, 1969), which is unimodal and skew with a long tail towards high consumption levels.

Individual differences are the result of differential mechanisms and processes. The shape of a distribution often mainly depends on certain structural properties of these processes. Hence, form, and not content, is often the main determinant of the distribution shape. This is the main reason for expecting that certain attributes should obey distribution laws, at least approximately. For instance, if a trait is determined by many independent factors of roughly equal importance, and if these factors combine additively, then it can be demonstrated mathematically (the so-called central limit theorem) that a Gaussian normal distribution would obtain. In this case the nature of the factors involved is immaterial.

Many human attributes, including body length and intelligence, are in fact determined by a multiplicity of genes and environmental factors that are believed to combine additively. (*Additivity* means that the removal of a factor has the same effect on the amount of the attribute that one gets, independently of whether one has much or little of this attribute to start with.) This fact explains the Gaussian character of these distributions. However, since additivity is only approximate and since a few rare factors may have a much larger impact on the outcome than others (say, the gene for dwarfism, or chromosomal defects in Down's syndrome), the fit between data and theoretical models is never perfect. Nevertheless, the theoretical curves still catch some essential features of the actual distributions; and as long as one does not forget that the curves are approximations, they can be useful for many purposes.

The reason why the Gaussian distribution plays so prominent role in statistics is that many traits are in fact determined by many different factors that combine additively, or almost additively. However, there are also important classes of phenomena for which additivity does not hold. Consumption behavior is a case in point.

As in the case of intelligence, individual differences in consumption behavior may be the result of a large number of independent physiological, psychological, and environmental factors. The reason that consumption distributions are nevertheless skew, rather than Gaussian, has to do with the fact that the factors do not combine additively. Removing or reducing one factor generally has a larger impact on a high-consuming individual than on a low-consuming individual, and typically, the reduction may be roughly proportional to the initial consumption level. This pattern would suggest that the factors combine multiplicatively, rather than additively.

The Law of Proportionate Effects

The law of proportionate effects says that the change in behavior elicited by a certain stimulus is approximately proportional to the initial behavior. When behavior follows the law of proportionate effects, the resulting distribution will be skew to the right. The reason why proportionate effects produce skew distributions is illustrated in Figure 2. Consider two equally large substrata that are identical except with respect to one factor. Those who have much of this factor drink more than those who have little. If the effect of this factor had been additive, the distributions of the two substrata would simply be shifted with respect to each other (left side, second row). With multiplicative effects, those with much of the factor will be stretched out along a larger segment of the consumption scale. Hence, their distribution curve will be lower (because the areas under the curves correspond to population size, which are equal for the two strata), and the overall distribution becomes skewed to the right (right side, third row).

When consumption behavior is determined by a large number of independent factors that combine multiplicatively, the distribution should resemble the so-called lognormal curve (Aitchison & Brown, 1969). The lognormal distribution is closely related to the Gaussian normal distribution, and the former can be transformed into a normal distribution by using a logarithmic consumption scale, rather than a linear one. A logarithmic consumption scale is stretched out at low levels and compressed at high levels; thus, the distance between 1 and 2 is the same as the distance between 10 and 20 and between 100 and 200, and so on.

There is some evidence that factors influencing people's consumption of alcoholic beverages tend to have proportional effects, that is, that they combine multiplicatively (Skog, 1979). Sex differences in drinking patterns give a simple illustration of this fact. It is well known that sex differences in alcohol consumption are quite large in most cultures and most substrata of each population. The difference between males and females is not an absolute one, however, as additivity would imply. Typically, the absolute difference between male and female consumption levels varies across cultures; and the greater the overall consumption level of the population, the larger the difference. However, the ratio of male to female consumption is much less variable. Typically males drink 3 to 4 times as much as females. Similar results have been reported for many other sociodemographic factors influencing drinking (Skog, 1979).

Naturally occurring changes in individual drinking give another opportunity for testing multiplicativity versus additivity. If changes are additive, then they should be of the same order of magnitude at all consumption

FIGURE 2
Additive (Left) and Multiplicative (Right) Change and Its Effect on the Distribution of the Population. Top: Initial Distributions. Middle: Half of the Drinkers Increase Their Intake by a Fixed Amount (Left) or by a Fixed Percentage (Right). Bottom: Resulting Distributions.

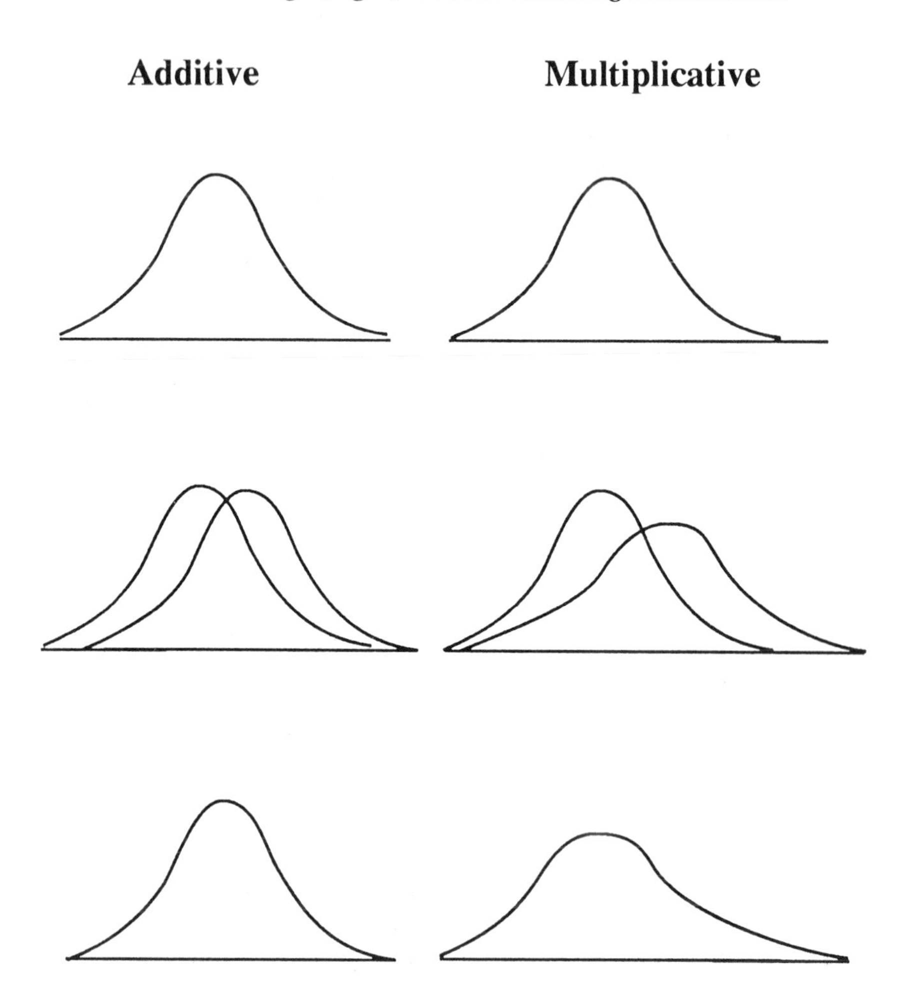

levels. If, however, changes are multiplicative, then the higher the initial consumption level, the larger the absolute changes should be. Hence, in the terminology of regression analysis, homoscedasticity should apply in an additive context, and heteroscedasticity would be expected in a multiplicative one. In the latter case, a logarithmic transformation of the consumption variables would produce homoscedasticity, since this would mean that the relative, rather than the absolute, changes are of the same order of magnitude at all consumption levels.

Data from a panel study of drinking in the general population in three Norwegian towns (Nordlund, 1978; Skog, 1979) are displayed in Figure 3. The subjects have been divided into 14 groups according to consumption level at Time 1, and for each of these groups the mean consumption level and an approximate 70% consumption band have been computed for Time 2. As can be seen from the diagram with linear scale (top), the absolute changes are not at all of the same order of magnitude at different consumption levels. In the diagram with logarithmic scales (bottom) it is seen that the relative changes are of approximately the same order of magnitude. Apparently, a multiplicative model is a much better approximation than an additive one.

Individual Change and Aggregate Stability

While attributes like intelligence and body height remain fairly constant throughout adult life, consumption behavior typically varies considerably. An abstainer may start to drink, a moderate drinker may start to drink somewhat less moderately, and an alcoholic may become abstinent. These individual changes may be fairly unpredictable and highly irregular. Intuitively, one might perhaps expect that such individual instabilities would easily destroy regularities in the distribution pattern.

However, they do not. Individual instability does not contradict aggregate stability. This fact is exemplified in Table 1, which shows empirical distributions of a sample according to self-reported annual intake of alcoholic beverages in two successive years (in the first two columns). (The third column shows the steady state distribution, which is explained below). The distributions are practically identical, in spite of the fact that large individual changes have occurred. The correlation between consumption levels for the 2 years was only 0.67, after logarithmic transformation of variables to correct curvilinearity and heteroscedasticity.

The structure of individual changes can be studied in Table 2, which for each consumption category at Time 1 shows the transition rates, that is, the fraction of the subjects who had decreased their intake, had remained in the same consumption category, or had increased their intake at Time 2.

This phenomenon—stability of the distribution of a population in spite of large individual changes—is not at all uncommon, and it is well known in many different branches of science. Such a distribution is called a steady state. Its characteristics obviously must reflect some crucial aspects of the process of change. This idea suggests that there are important links between the shape of the distribution of alcohol consumption and certain structural aspects of individual drinking careers.

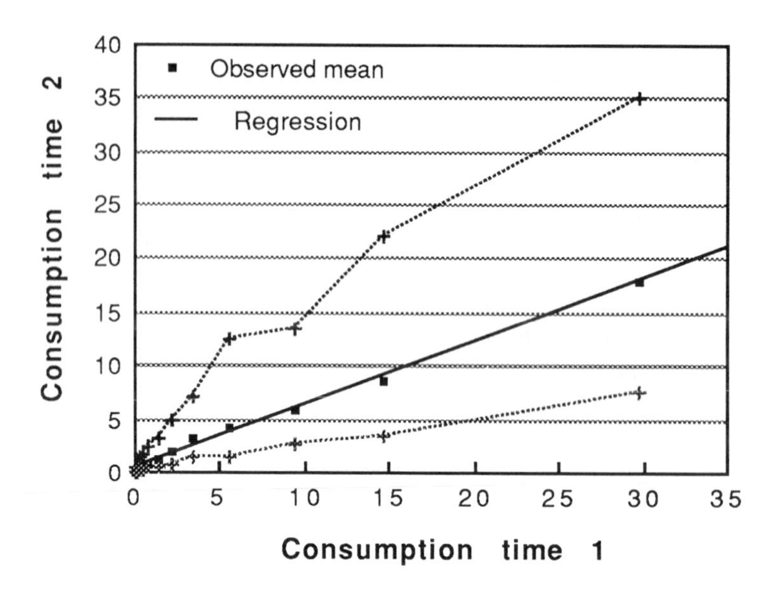

FIGURE 3
Changes in Consumption of Alcoholic Beverages, Displayed on Linear (above) and Logarithmic (below) Scales. Empirical Regression Curves (solid) and Conditional Standard Deviations (broken).

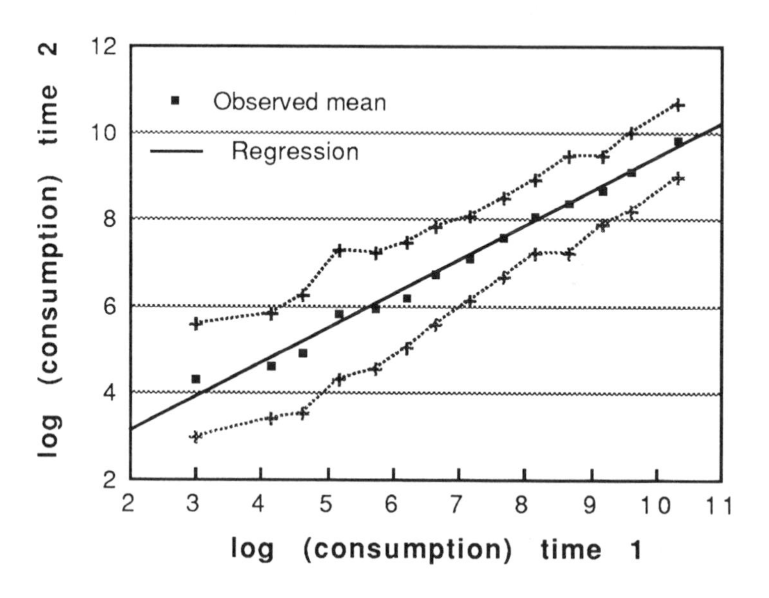

TABLE 1
Distribution of a Sample (%) According to Annual Intake of Alcoholic Beverages (Liters per Year) in Two Successive Years.

Consumption level	Time 1	Time 2	Steady state
Abstention	7.6	6.7	6.2
0 – 4.6	15.3	15.3	14.4
4.7– 9.1	14.7	13.1	13.3
9.2–18.3	18.0	19.6	20.0
18.4–36.5	16.8	19.3	20.2
36.6–73.0	16.5	15.0	15.0
73.1+	11.0	11.0	10.8
Total	99.9	100.0	99.9

Note: From O.-J. Skog and F. Duckert, *Journal of Studies on Alcohol* (forthcoming).

For any given matrix of transition rates there is one and only one steady state distribution. The steady state distribution corresponding to the transition rates in Table 2 is given in Table 1; and as one can see, it matches the observed distributions. The reader can easily verify by direct calculation that if the initial distribution had been different from the steady state distribution, a change in distribution would have occurred from Time 1 to Time 2.

The skewness of the steady state distribution in this case is due to the fact that the transition rates in Table 2 are in approximate accordance with the law of proportionate effects. The foregoing is seen from the fact that the relative changes are roughly the same at low and high levels. Note that the width of successive consumption categories increases so as to make the scale logarithmic.

The highest probabilities in each line of the transition matrix are typically the diagonal probabilities. Hence, those who have the largest chance of being in a certain state at a given time are typically those who occupied that state in the preceding period. However, the probability that a substantial change has occurred is quite significant at all levels. When changes occur, subjects typically move to neighbor consumption categories. Very large and dramatic jumps are in effect unlikely. Hence, the data suggest that processes of change are reasonably smooth. Light drinkers tend to remain light drinkers, and heavy drinkers tend to remain heavy drinkers. However, a certain drift can be observed, since light drinkers have a slightly higher probability for increasing than for decreasing, whereas the pattern is reversed for heavy drinkers, who have a larger probability for decreasing than for increasing. We shall discuss these floor and ceiling effects later.

TABLE 2
Transition Rates Between Different Consumption Levels (Liters per Year) from One Year to the Next.

Year 1 Consumption level	Year 2 Consumption Level							
	Abst.	0–4.6	4.7–9.1	8.2–18.3	18.4–36.5	36.6–73.0	>73.1	Total
Abst.	.40	.24	.04	.16	.08	—	.08	1.00
0 – 4.6	.12	.46	.22	.06	.06	.08	—	1.00
4.7– 9.1	.02	.27	.23	.31	.17	—	—	1.00
9.2–18.3	.03	.08	.22	.31	.27	.08	—	.99
18.4–36.5	.04	.04	.09	.24	.35	.15	.11	1.02
36.6–73.0	.02	.02	.02	.15	.22	.33	.24	1.00
73.1 +	—	—	.03	.08	.08	.39	.42	1.00

Note: From Skog and Duckert (forthcoming).

Some readers would perhaps at this point argue that although this pattern may be valid for normal drinkers it may not fit very well for alcoholics. For instance, according to the loss-of-control, or trigger, hypotheses of the classical disease model (Jellinek, 1960), moderate intake should rapidly develop into heavy drinking for an alcoholic. The fact is, however, that the pattern in Table 2 definitely is valid for alcoholics. The data are taken from a longitudinal study of a clinical population of chronic alcoholics (Skog & Duckert, forthcoming). This issue is addressed more closely in chapter 29 in this volume.

Summary

The theoretical and empirical facts outlined above lead one to expect certain regularities in the distribution of alcohol consumption. In particular, approximate proportionality of effects strongly suggests a considerable skewness towards high consumption levels.

It is important to realize that this skewness cannot be explained by the presumption that the population consists of two or more normally distributed substrata, say alcoholics and normal drinkers, as some writers have suggested (Tuck, 1980). Even if such a differentiation in substrata was meaningful, the law of proportionate effects would imply that the distribution within each substratum was very skew and strongly overlapping. Hence, one cannot remove the skewness by disaggregating the population into more homogeneous substrata.

The skewness of the distribution of alcohol consumption is mainly the product of mechanisms—namely, proportionality of effects—that have nothing to do with the beverage alcohol's addictive properties. Of course, this fact does not preclude the idea that addiction may contribute even further to the skewness of the distribution, but it is not the main cause.

It has been suggested that the lognormal distribution should give a good description of alcohol consumption (de Lint & Schmidt, 1968; Ledermann, 1956). As was mentioned above, this distribution is closely related to the law of proportionate effects. However, one could not expect a perfect fit between actual distributions and this theoretical curve. Some factors affecting individual drinking have a much larger impact than others, and this fact may sometimes produce deviations from the theoretical expectancy.

Regularities in Empirical Distributions

The Shape of Empirical Distributions

Distribution data from population surveys in many different countries have become available over the last two decades. Admittedly, these data are not ideal for studying distribution patterns, because the reliability and validity of these reports are not perfect. People have a tendency for underreporting their true consumption (Midanik, 1982; Pernanen, 1974), and the samples may be somewhat biased owing to higher-than-average rates of nonresponse from heavy drinkers and alcoholics. These limitations should be kept in mind when one is interpreting results. (It should be noted that, unless otherwise stated, abstainers are always excluded in the distribution data reported below and in subsequent sections. The reasons for this exclusion are mainly conventional.)

Available data confirm that alcohol distributions are very skew, in both high- and low-consumption cultures. Two typical examples from Norway and France, respectively, are shown in Figure 4. Norway's consumption level is one of the lowest in the Western world; France's is the highest.

As expected, the skewness prevails in homogeneous substrata of the population. This pattern is illustrated in Figure 5, which shows distributions for two substrata of the Norwegian population. One substratum, boys aged 15 years, is practically void of alcoholics, but the distribution is nevertheless very skew—even more so than the distribution of the whole population. The second stratum is a sample of alcoholics before treatment. Even in this group the distribution is quite skew, and a substantial fraction of the subjects had been drinking fairly moderate amounts for several months. This pattern is not at all uncommon, and during some periods chronic alcoholics may in fact drink more or less as "normal" drinkers. In fact, alcoholics and normal drinkers are not clearly separable in terms of, say, annual intake, because the distributions overlap to a considerable extent.

It is not easy to see from histograms whether the empirical distributions actually conform with the lognormal expectancy or not. A so-called log-probability plot, in which the cumulative distribution is plotted in a diagram with logarithmic abscissa and Gaussian ordinate (i.e., probits), is a useful aid when one wants to investigate the shape of the distribution in more detail. If the distribution is lognormal, the result will be a straight line. The slope of the line is determined by the variance of the distribution: the smaller the variance, the steeper the line becomes (cf. Aitchison & Brown, 1969).

Log-probability plots for some representative examples are shown in Figure 6. In some cases the plots are fairly close to straight lines, whereas in other cases a slightly curved pattern is observed. In these cases the plots are typically curved upwards, which implies that these distributions are somewhat less skew than the lognormal distribution. At best, therefore, the lognormal distribution gives only a rough approximation of actual distributions of alcohol consumption. Because the deviations from

FIGURE 4
Empirical Distributions of Annual Intake of Alcohol According to Surveys in France and Norway (Source: Skog, 1986).

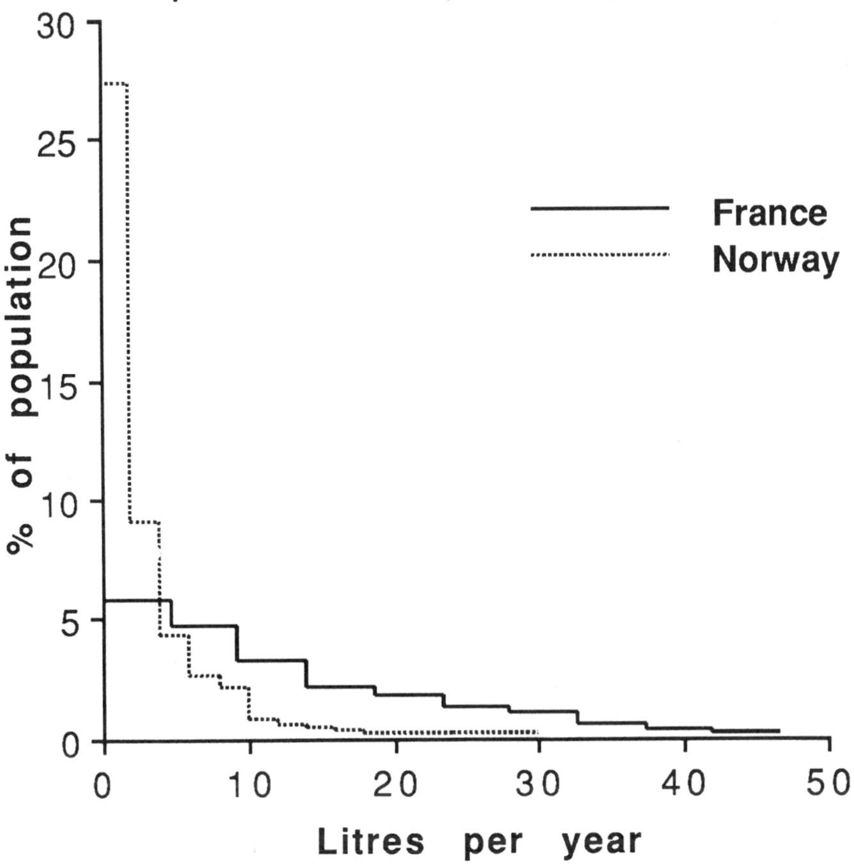

lognormality seem to vary from one population to another, it does not appear likely that all discrepancies can be explained by measurement error or the like.

It has been suggested that the so-called gamma distribution, which is slightly less skew than the lognormal, may often give a better fit to empirical distributions. However, one should not forget that it is unlikely that a single mathematical model will give an adequate description for all empirical distributions. After all, whereas there are theoretical reasons for expecting a certain degree of regularity in the distribution pattern, one cannot expect that actual distributions have exact mathematical properties.

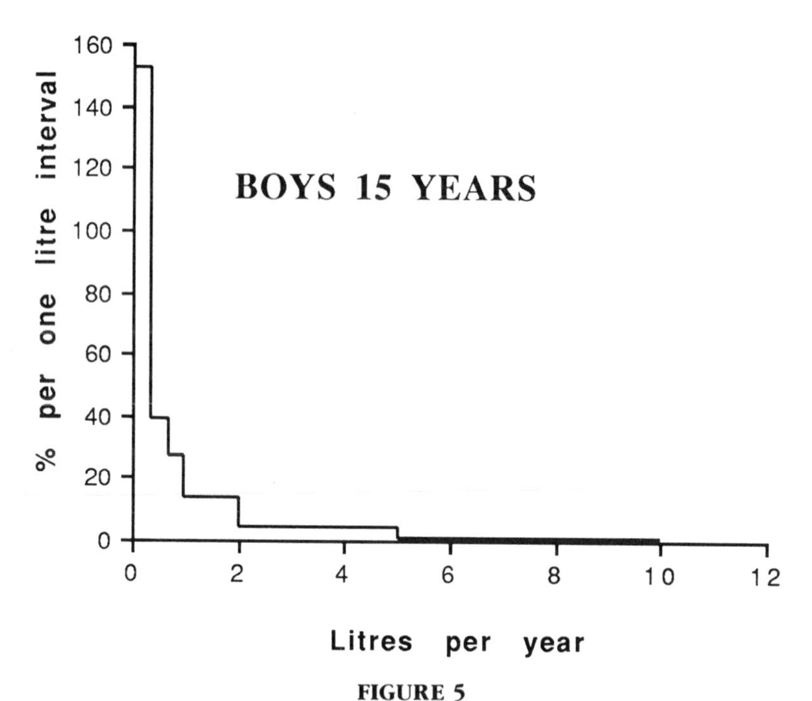

FIGURE 5
Empirical Distributions of Annual Intake of Alcohol According to Surveys in Two Substrata of the Norwegian Population. Boys Aged 15 Years (above) and Alcoholics Before Coming to Treatment (below).

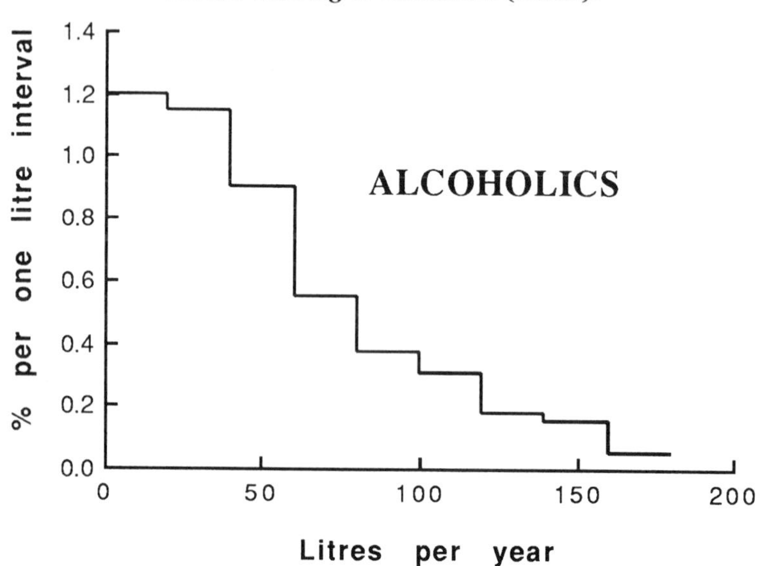

The discrepancies between actual distributions and the theoretical model are probably not larger in the case of alcohol consumption than for other human behaviors for which "distribution laws" have been found. Whether or not the discrepancies are so large that the lognormal curve (or the gamma curve) should not be used as a model is difficult to say. Obviously, the determination will depend on the purpose.

Description of the Skewness of the Distribution

A so-called Lorentz diagram is a useful tool for describing inter-individual differences in alcohol consumption. After ordering the sample according to the individuals' consumption level from lowest to highest, one calculates the fraction of the total consumption for which groups of different sizes are responsible. A Lorentz diagram is exemplified in Figure 7, with the use of data from France and Norway.

In Norway, the 50% of the drinkers with the lowest intake are responsible for less than 10% of all the alcohol being consumed. In France, that analogous group drinks about 20%. In Norway, 90% of the population drinks about 50% of the alcohol, leaving the other 50% to the remaining 10% of the population. In France, however, 90% of the population drinks about 70% of the alcohol and the remaining 10% thus drinks about 30%. Hence, in both countries a small percentage of the population drinks a disproportionately large percentage of the alcohol, and this pattern is even more pronounced in Norway than in France.

This difference exemplifies a general pattern. In low-consumption countries, consumption is typically more strongly concentrated in a small segment of the population, and this concentration is gradually relaxed as per capita consumption increases. This pattern exemplifies a kind of ceiling effect. However, the concentration is nevertheless quite pronounced even in countries with a very high consumption level.

Because of the skewness of the distribution, the mean consumption level is typically substantially higher than the median consumption level. Hence, the fraction of the population drinking more than the mean consumption is typically considerably smaller than 50%. This fact is illustrated in Table 3, showing the results from surveys of drinking habits in different countries. It is worth noting that the ratio of mean to median is typically larger in cultures with a low consumption level than in cultures with a high consumption level. This finding is closely related to the fact already mentioned, that is, that the concentration of consumption is largest in low-consumption countries.

The segment of the distribution above twice the mean can be considered the proper tail region of the distribution. The fraction of the population belonging to this region is surprisingly stable across cultures (cf.

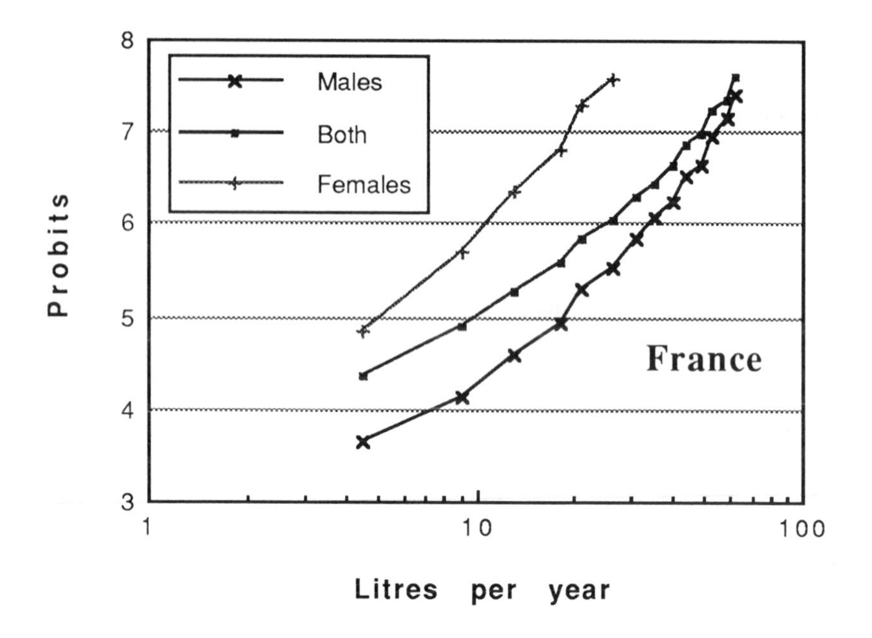

FIGURE 6
Log-Probability Plots of Alcohol Consumption Distributions in Selected Countries.

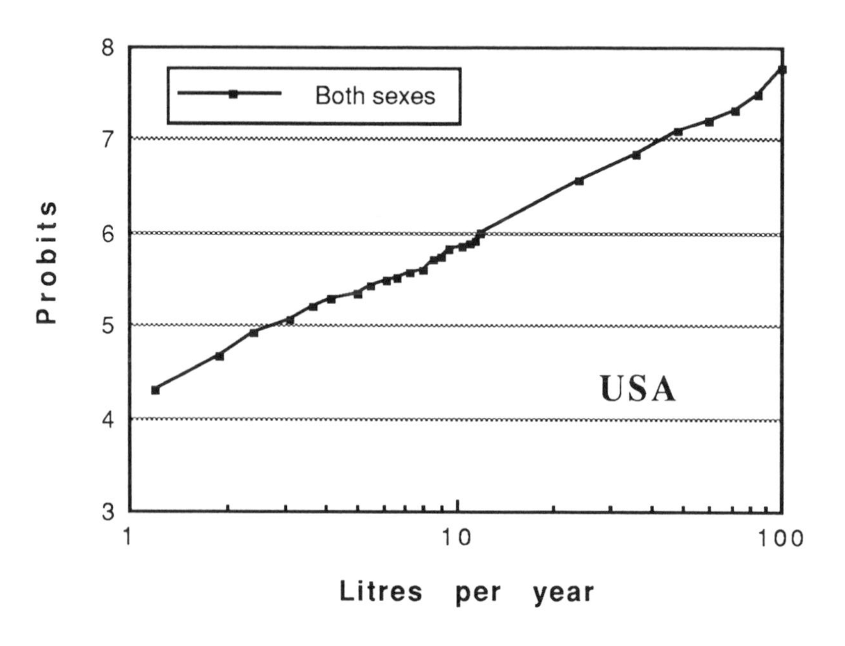

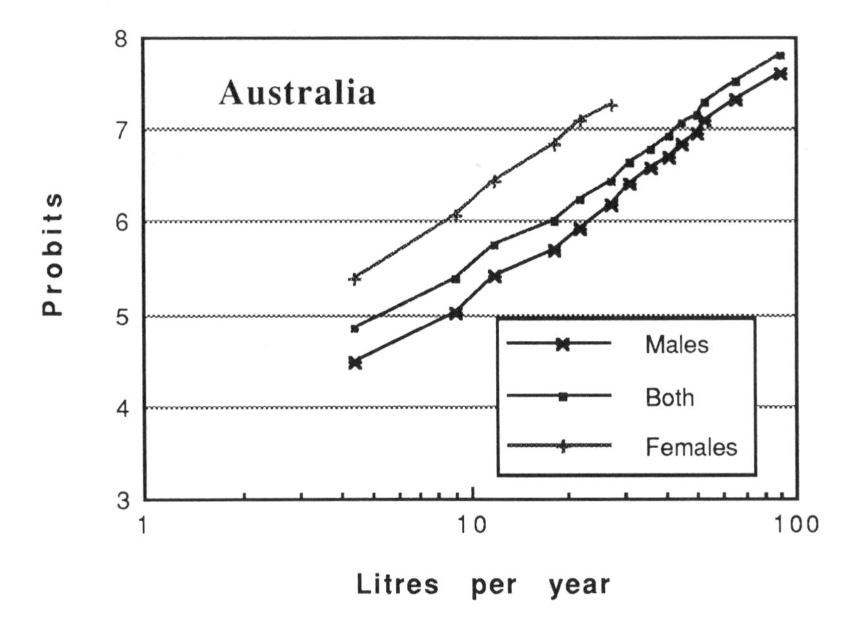

FIGURE 6 (continued).

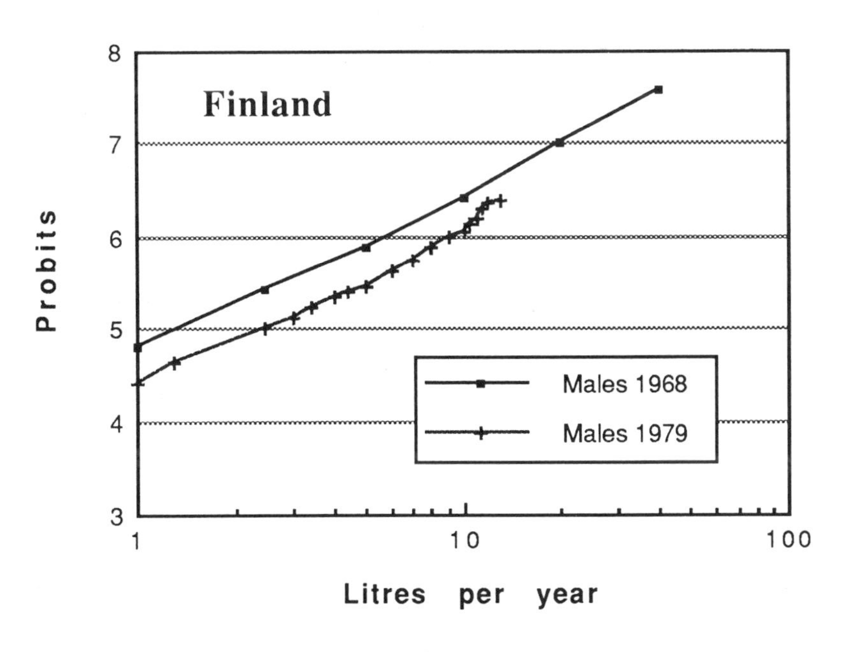

FIGURE 7
Lorentz Diagram of Concentration of Consumption in Norway and France.

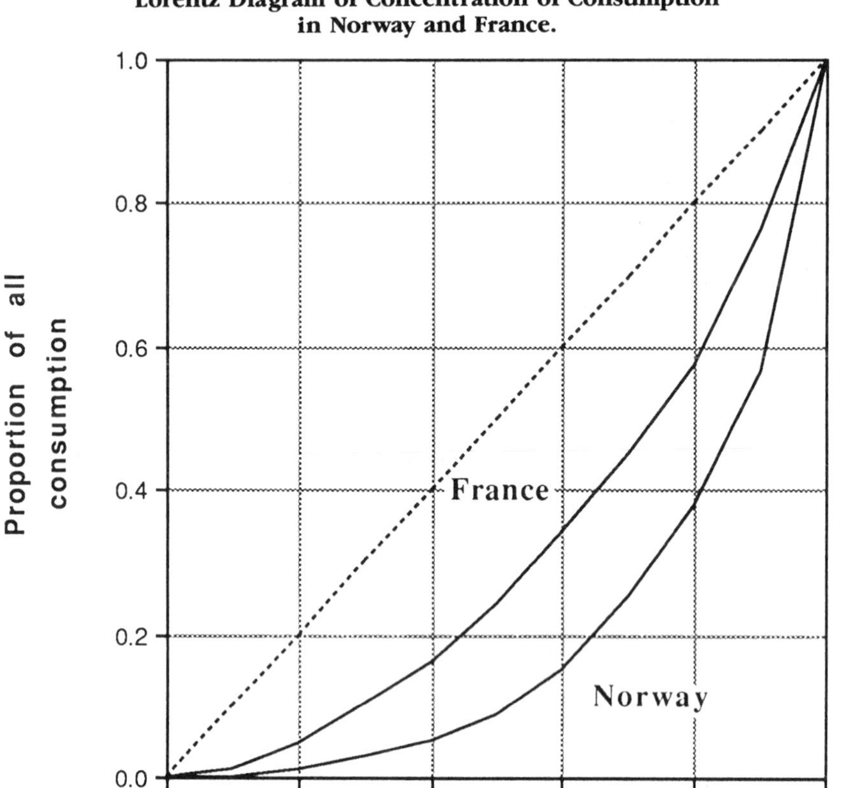

Table 3). This fraction is typically in the range of 10-15%, whatever the mean consumption level of the population. Because the prevalence of alcoholics is typically much smaller than this range, and because the prevalence of alcoholics also varies from one culture to the next (Popham, 1970), this is yet another piece of evidence that the skewness of the distribution cannot be explained solely in terms of the alcoholics.

Summary and Conclusions

Historically, the study of the distribution of alcohol consumption was initiated by the French demographer Sully Ledermann in the late 1940s.

TABLE 3
Estimated Mean and Median Consumption and Their Ratio, and the Proportion Drinking More than Twice the Mean in Surveys from Different Countries.

Region	Mean (l/yr)[a]	Median (l/yr)	Ratio	%over 2 × mean
Italy	11.7	8.4	1.4	9
U.S. (Italians)	5.1	2.6	2.0	11
France	18.6	13.5	1.4	13
Norway/Students	4.0	2.5	1.6	14
Norway/Youth 71	3.0	1.5	2.0	14
Norway/Youth 73	3.9	2.0	2.0	14
Sweden	2.4	0.9	2.7	16
Ille-et-Vilaine	14.0	9.9	1.4	14
Australia	9.6	5.6	1.7	15
USA	9.2	3.5	2.6	11
Switzerland	10.0	5.3	1.9	14
London 1965	5.0	2.8	1.8	12
London 1974	8.2	3.5	2.3	12
Scotland	8.2	3.5	2.3	13
St. Etienne	28.4	16.9	1.7	13
Marseille	20.7	12.8	1.6	12
Gironde	46.8	30	1.6	12
Gard	28.2	22.1	1.3	10
Savoie	47.5	32.5	1.5	14
Cotes-du-Nord	22.1	14.1	1.6	14
Vendee	29.2	17.5	1.7	16
Finland 79	3.8	1.3	2.9	13
Iceland 79	2.8	1.2	2.3	13
Norway 79	3.3	1.3	2.5	12
Sweden 79	3.5	1.5	2.3	13

Note: From Skog (1985).
[a] l/y = liters per year.

On the basis of a series of epidemiological studies of alcohol-related mortality in France, he had reached the conclusion that variations in mortality rates—both in time and space—closely matched variations in the per capita alcohol consumption level (for an overview, see Skog, 1982). In order to explain this conclusion, he proposed his theory of the distribution of alcohol consumption (Ledermann, 1956).

Although the Ledermann theory in its original form has been rightfully questioned (Duffy, 1986; Pittman, 1980; Skog, 1982), Ledermann was right in his basic presumption, namely, that there are fairly strong regularities in the distribution of alcohol consumption. That is, on the basis of the available data, it can be concluded that the distribution of alcohol consumption is very skew and resembles the lognormal curve. Furthermore, the concentration of consumption is typically largest in low-consumption countries. I shall briefly recapitulate some of the findings discussed above:

- The law of proportionate effects is approximately valid for alcohol consumption. Individual drinking is affected by many different factors, and these factors tend to combine multiplicatively.
- In spite of considerable individual fluctuations in consumption, the distribution may remain the same, provided that the overall consumption level in the population remains stable. Such fluctuations are best described by multiplicative models.
- These diachronic and synchronic mechanisms produce a substantial skewness in the distribution, which resembles (without exactly matching) the lognormal curve. Even homogenous substrata have very skew distributions due to multiplicativity, and the skewness may consequently not be explained as resulting from aggregation effects.
- The tail of the distribution (above twice the mean consumption level) typically contains 10-15% of the population, and the highest consuming 10% of the population typically drinks almost half of all the alcohol.

What are the implications of these findings for the prevalence of heavy drinking? Do they imply that high-consumption countries have a very much higher prevalence of heavy drinkers than low-consumption countries? Or does the weaker concentration of consumption in high-consumption countries counteract this effect? And what are the effect of changes in the general level of consumption in a culture—should one expect changes in the prevalence of heavy drinking as well?

By itself, lognormality (or approximate lognormality) has no strong implications for the issue of heavy drinking. The lognormal distribution has two parameters—the mean and the variance—and the prevalence of heavy use depends on both of them. Hence, unless something is known about the variance of the distribution or some related measure, not much can be said.

Ledermann (1956) not only proposed that alcohol consumption follows a lognormal distribution law, but he also proposed that the variance of the distribution was determined by the mean and hence that populations with the same mean consumption level also have the same prevalence of heavy drinking. Furthermore, he hypothesized that the logarithmic standard deviation decreases only slightly when the mean increases and that the decrease does not suffice to prevent the prevalence of heavy drinking from increasing. However, this hypothesis was not founded on any theory of drinking and was a purely technical statistical assumption.[1]

In order to address the questions raised above about implications of the distribution of consumption theory for heavy drinking, one has to discuss the formation of individual drinking habits in more detail. Our argument has so far been centered on purely structural aspects of the factors influencing drinking behavior. We have assumed that individual drinking habits are the result of an interplay of many different factors that tend to combine in a certain way and that produce changes of a certain type. From this starting point, certain predictions about the overall shape of the distribution have been deduced. However, these structural aspects have limited implications for the questions now raised. In order to address the issue of heavy drinking, we will have to take into consideration the kinds of factors that influence people's drinking behavior. These factors will be discussed in chapter 29.

Notes

1. Ledermann argued that nobody can drink more than 1 liter of pure alcohol a day. Therefore, he argued, the theoretical proportion above this limit according to the lognormal model must be very small. He presumed this theoretical proportion to be the same in all populations and estimated it from data of doubtful quality to be 0.03%. This presumption created a fixed relationship between the mean and the variance. The argument can hardly be taken seriously.

References

Aitchison, J., & Brown, J.A.C. (1969). *The lognormal distribution.* Cambridge: Cambridge University Press.

Anastasi, A. (1958) *Differential psychology.* New York: Macmillan.

Blinder, A.S. (1974) *Towards an economic theory of income distribution.* Cambridge, Massachusetts: MIT Press.

de Lint, J., & Schmidt, W. (1968). The distribution of alcohol consumption in Ontario. *Quarterly Journal of Studies on Alcohol, 29,* 968–973.

Duffy, J.C. (1986). The distribution of alcohol consumption—30 years on. *British Journal of Addiction, 81,* 735–774.

Jellinek, E.M. (1960). *The disease concept of alcoholism.* New Brunswick, NJ: Rutgers Center of Alcohol Studies.

Johnson, N.L., & Kotz, S. (1970). *Continuous univariate distributions* (Vols. 1 & 2). Boston: Houghton Mifflin.

Ledermann, S. (1956). *Alcool, alcoolism, alcoolisation* (Vol. 1). Paris: Presses Universitaries de France.

Midanik, L. (1982). The validity of self-reported alcohol consumption and alcohol problems: A literature review. *British Journal of Addiction,* 77, 357–382.

Nordlund, S. (1978). *Lokale Olmonopol—Virkninger pa Drikkevanene.* Oslo: Universitetsforlaget.

Pernanen, K. (1974). Validity of survey data on alcohol use. In R.J. Gibbens, Y. Israel, H. Kalant, R.E. Popham, W. Schmidt, & R.G. Smart (Eds.), *Research advances in alcohol and drug problems* (Vol, 1, pp. 355–374). New York: Wiley.

Pittman, D.J. (1980). *Primary prevention of alcohol abuse and alcoholism: An evaluation of the control of consumption policy.* St. Louis: Social Science Institute, Washington University.

Popham, R.E. (1970). Indirect methods of alcoholism prevalence estimation: A critical evaluation: In R.E. Popham (Ed.), *Alcohol and alcoholism* (pp. 294–306). Toronto: University of Toronto Press.

Skog, O.-J. (1979). *Modeller for drikkeatferd.* Oslo: National Institute for Alcohol Research.

Skog, O.-J. (1982). *The distribution of alcohol consumption: Part I. A critical discussion of the Ledermann model.* Oslo: National Institute for Alcohol Research.

Skog, O.-J. (1985). The collectivity of drinking cultures. A theory of the distribution of alcohol consumption. *British Journal of Addiction, 80,* 83–99.

Skog, O.-J., & Duckert, F. (forthcoming).The development of alcoholics' and heavy drinkers' consumption: A longitudinal study. *Journal of Studies on Alcohol.*

Tuck, M. (1980). *Alcoholism and social policy: Are we on the right lines?* London: Her Majesty's Stationery Office.

CHAPTER 8

Changes in American Drinking Patterns and Problems, 1967–1984

MICHAEL E. HILTON AND WALTER B. CLARK

The single distribution theory of alcohol consumption is one of the best-known theses in the alcohol studies literature. Based on the contributions of Ledermann (1956), Skog (1982, 1983) and others, the theory asserts that there is a relationship between the average per capita alcohol consumption of a population and the proportion of heavy drinkers in that same population. Expressed dynamically, the theory predicts that an increase in the average per capita volume of consumption will be accompanied by an increase in the proportion of heavy drinkers. Furthermore, it predicts that the increase in the proportion of heavy drinkers will be greater in magnitude than the increase in mean consumption volume. Presumably, there would also be an increase in the proportion of people reporting drinking problems. Collecting data for the consumption side of this argument is relatively easy because government statistics on alcohol consumption are readily available. For example, the rise in per capita alcohol consumption between 1967 and 1984 was 11.8% (National Institute on Alcohol Abuse and Alcoholism [NIAAA], 1985, 1986). According to the theory, this increase should have been accompanied by an even larger increase in the proportion of drinkers who can be classified as heavy drinkers. However, it is much more difficult to observe changes in the relative proportion of heavy drinkers or in the proportion of drinkers who report problems due to drinking. Making such an observation would require that two general population surveys using precisely the same items for measuring intake and problems be administered, separated by a span of several years. Given the cost and time involved, such observations are understandably rare. However, the 1984 national survey undertaken by the Alcohol Research Group did collect data, about both intake and prob-

Michael E. Hilton is with the Prevention Research Branch, National Institute on Alcohol Abuse and Alcoholism, Rockville, Md. Walter B. Clark is with the Alcohol Research Group, Medical Research Institute of San Francisco, Berkeley, Calif. This chapter is reprinted with permission from *Journal of Studies on Alcohol, 48,* pp. 515-522, 1987.

lems, which are precisely comparable to those collected in a 1967 national survey (Cahalan, 1970).

Here, we present some data from these two surveys on whether there have been any changes in either the prevalence of heavy drinking or the prevalence of self-reported drinking problems. Our purpose is not so much the analytic one of testing the single distribution hypothesis as the descriptive one of providing evidence about possible increases in the rates of heavy drinking and drinking problems at a time when per capita consumption is known to have increased. Because these consumption increases have created the expectation that prevalence rates for heavy drinking and drinking problems have been rising, it might be argued that greater resources should be directed at prevention and treatment efforts. Such policy discussions should be informed by any available evidence regarding changes in prevalence rates, and here we intend to provide just such information.

Similar studies of trends in drinking patterns have generally tended to report that there have been few changes over time. Fitzgerald and Mulford (1981) found few changes in the prevalence of heavy drinking in Iowa between 1961 and 1979, although they noted changes in light and moderate drinking. Most national studies of trends stem from Johnson, Armor, Polich, and Stambul (1977), a compilation of eight nationwide polls (mostly Harris polls) taken between 1971 and 1976. Clark and Midanik (1982) added data from the 1979 Alcohol Research Group survey to this sequence of studies and more recently Wilsnack, Wilsnack, and Klassen (1984, 1986) added their 1981 results, which focus on drinking patterns among women. The series of data accumulated in all three studies show little change in drinking patterns between 1971 and 1981. Additional literature containing trend comparisons, including those by Room (1983), Room and Beck (1974), and Hilton (1986), seems to confirm this general conclusion.

The present study differs from these previous studies in three important respects. First, it covers a much longer time span than most of the others, assessing changes that have taken place over a 17-year interval. Second, this study does not measure consumption on the basis of responses about typical or usual amounts of alcohol consumed, as did Wilsnack et al. (1984, 1986) and Johnson et al. (1977). As Room (1977) has noted, high-intake occasions can be masked by responses about typical drinking occasions, and this masked data will yield inaccurate volume estimates. Also, estimates based on volume alone fail to make use of the amount per occasion as a second dimension of the respondent's drinking pattern. Third, the present study examines trends in drinking problems as well as in drinking patterns. Drinking problem items tend to be much less standardized than drinking pattern items; therefore, studies in drinking problem trends that use strictly comparable items are especially rare. Finally, we give a caveat to the reader: our purpose is simply to describe changes that have taken place without regard to supplying explanations for why these changes have occurred.

Method

Both the 1967 and the 1984 surveys used strict area-probability sampling methods and asked a set of identical questions about drinking patterns and drinking problems. In both cases, face-to-face interviews of about an hour's length were conducted with adults living in households within the 48 contiguous states. By comparing the data from these surveys, we will be able to describe the changes that have occurred between 1967 and 1984.

The 1967 survey was the second wave of a longitudinal study, the first wave having been conducted in 1964 and the results published as *American Drinking Practices* (Cahalan, Cisin, & Crossley, 1969). The second wave was conducted in 1967, and the results were published as *Problem Drinkers: A National Survey* (Cahalan, 1970). The 1964 survey did not ask any questions about drinking problems and therefore is not discussed here. Only persons aged 21 years or older were interviewed in 1964; therefore, all respondents in the 1967 survey were at least 23 years old. In analyzing the 1984 data, we similarly limit the analysis to those who were at least 23 years old. This exclusion of younger respondents is unfortunate because it is well known that the youngest age categories include relatively large numbers of drinkers and problem drinkers.

The interview completion rate in the 1967 sample was 72% (allowing for losses in both 1964 and 1967). Those selected for reinterview in 1967 were disproportionally sampled from among heavy drinkers. However, the selection of these heavy drinkers was made according to probability methods and thus permitted reweighting to represent the adult population. In the present study we have weighted the 1967 data set for this differential selection and for the number of adults living within each selected household. These weights were constructed in such a way that the resultant *N* equals the actual number of people interviewed. Thus, all 1967 figures in the analysis to follow are weighted figures, but the *N*s are not inflated beyond the overall number of people interviewed.

The 1984 survey was designed to yield a representative sample of all U.S. adults and also to provide separate black and Hispanic samples of sufficient size for cross-ethnic analyses. Accordingly, approximately 1,500 blacks and 1,500 Hispanics were sampled in addition to about 2,000 nonblacks and non-Hispanics. These unequal selection probabilities require that we "downweight" the minority samples when looking at the U.S. population as a whole, and we have done so in this chapter. Other weights were applied to take into account the number of adults in a selected household and the varying completion rates for men and women, for those over versus those under age 35, and among residents of the four census regions—Northeast, North Central, South and West—of the U.S. mainland. The downweighting process produced an *N* of just over 1,900 cases, a number reflective of the actual number of non-black and non-Hispanic respondents who were interviewed plus an appropriately proportional number of black and Hispanic respondents. Hence, the number of cases is not inappropriately inflated when one is conducting tests of significance. All 1984 figures given below are based on these weighted data.

In both surveys respondents were asked to record the frequency with which they drank wine, beer, and distilled spirits along with the frequency of drinking any kind of alcoholic beverage. Respondents who never drink or who drank less often than once a year were defined as abstainers. All others were asked a series of questions about their drinking. For instance, respondents who reported drinking wine as often as once a year were asked: "Think of all the times you had wine recently. When you drink wine, how often do you have as many as five or six glasses?" Answer categories provided were: "nearly every time," "more than half the time," "less than half the time," "once in a while," and "never." Respondents were then asked about the proportion of occasions in which they consumed three of four glasses of wine and one or two glasses of wine. Similar questions were asked for beer and for distilled spirits. From the responses, we computed the number of drinks consumed per month of each individual beverage and of all beverages combined. To replicate the work done by Cahalan et al. (1969, Appendix I), we used a typology of drinking that takes into account not only the number of drinks consumed per month but also the amounts per drinking occasion. We call this measure Volmax (Figure 1).

Results

Drinking Patterns

Table 1 serves two purposes. The sample distributions of selected demographic characteristics are compared with census figures for 1970 and 1980, and demographic trends in the population that could affect our analysis insofar as they are related to drinking patterns and drinking problems are shown. The two samples were representative of census distributions on most characteristics including sex, region, race, and education. One concern, however, is that the 1967 sample underrepresented younger people by 3.2%. This underrepresentation might have the effect of lower-

FIGURE 1
Description of Volmax Measures

1. Abstainers are those who drink less often than once a year or who never drink.
2. Low volume–low maximum drinkers are those who drink fewer than 17.5 drinks per month and who never have more than 2 drinks per occasion.
3. Low volume–medium maximum drinkers are those who drink fewer than 17.5 drinks per month and who drink 3 or 4 drinks per occasion at least "once in a while" but who never drink 5 or more drinks per occasion.
4. Low volume–high maximum drinkers are those who drink fewer than 17.5 drinks per month and who drink 5 or more drinks per occasion at least "once in a while."
5. Medium volume–low maximum drinkers report drinking between 17.5 and 44.9 drinks per month and never have more than 2 drinks per occasion.
6. Medium volume–medium maximum drinkers are those who drink between 17.5 and 44.9 drinks per month and who drink 3 or 4 drinks per occasion at least "once in a while" but who never drink 5 or more drinks per occasion.
7. Medium volume–high maximum drinkers are those who drink between 17.5 and 44.9 drinks per month and who drink 5 or more drinks per occasion at least "once in a while."
8. High volume–low maximum drinkers report drinking 45 or more drinks per month and never have more than 2 drinks per occasion.
9. High volume–medium maximum drinkers are those who drink 45 or more drinks per month and who drink 3 or 4 drinks per occasion at least "once in a while" but who never drink 5 or more drinks per occasion.
10. High volume–high maximum drinkers are those who drink 45 or more drinks per month and who drink 5 or more drinks per occasion at least "once in a while."

ing the intake and problem estimates for that year inasmuch as younger people are known to be heavy drinkers and to have a high prevalence of problems. Also, the 1967 survey underrepresented men by 4.9% (and overrepresented women). Because the results presented here are controlled for sex, this underrepresentation should not alter the conclusions reached.

Several demographic trends in the underlying population will have to be kept in mind during the analysis. For instance, levels of educational attainment have increased, with substantially more people having completed high school in 1984 than had in 1967. The age distributions have also changed. The 1984 sample had a larger percentage of respondents in their 30s than the earlier sample, accurately reflecting population changes due to the maturation of those born during the baby boom. The 1984 sample also had a smaller percentage of respondents in their 40s, who would have been born during the late depression and war years. Since these characteristics are associated with drinking patterns, discovered changes in per capita consumption may reflect only a shift in the age or education distribution of the population. However, by looking at changes that have occurred within various categories of education and age, we should be able to distinguish between trends that have occurred within these categories and trends that have occurred because of changes in the

TABLE 1
Comparisons of the Survey Samples with Census Data on Selected Demographic Characteristics, in Percent[a]

	1967 Survey (N = 1359)	1970 Census	1984 Survey (N = 1904)	1980 Census
Sex				
Male	43.8	48.7	46.9	48.8
Female	56.2	51.3	53.1	51.4
Age				
23-29	14.0	17.2[b]	19.0	19.8[b]
30-39	21.0	19.5	24.0	23.5
40-49	24.0	20.8	17.0	16.1
50-59	19.0	18.2	14.0	15.1
60+	22.0	24.7	26.0	25.4
Region				
Northeast	24.0	24.1	22.0	21.1
North Central	31.0	27.8	26.0	25.2
South	32.0	30.1	33.0	34.0
West	13.0	17.1	19.0	19.7
Race[c]				
White	92.0	89.3	88.0	86.6
Black	7.0	9.8	10.0	10.9
Other	1.0	1.2	2.0	2.5
Education[d]				
Less than 4 years of high school	46.0	47.7	26.0	27.9
High school graduate	27.0	31.1	35.0	37.7
Some college or more	27.0	21.3	39.0	34.4

[a]The survey data are weighted figures as described in the text.
[b]Census figures for ages 20-29 adjusted to survey age categories of 23-29 by multiplying census figures by 0.6 (Department of Commerce, 1982).
[c]Census figures for race are ages 18 and older. Survey data include those 23 years old and older (Department of Commerce, 1982).
[d]Census figures for education are based on those 25 years old and older. Survey data include those 23 years old and older (Department of Commerce, 1982).

distribution of respondents among categories. Both kinds of changes are real, of course, but they call for different interpretations.

The data indicated that beverage preferences have changed. For wine, the mean ($\pm$ *SD*) drinks increased from 3.95 ± 14.34 per month to 6.76 ± 22.81 ($p < .05$); for beer, drinks increased from 19.36 ± 55.07 to 24.89 ± 64.05 ($p < .05$). A substantial decline in mean drinks of distilled spirits consumed per month was reported: 14.93 ± 44.38 in 1967 versus 9.33 ± 30.19 ($p < .05$) in 1984. (All figures were calculated on the basis of current drinkers.) Combining the three types of beverages into a measure of total drinks per month revealed an overall increase from 38.25 ± 77.36 in 1967 to 40.92 ± 81.52 in 1984, a difference that was nonsignificant.

Although these findings correctly indicate that beverage preferences have changed between 1967 and 1984, they do not reflect all of the increase in per capita consumption that took place. The per capita consumption of absolute alcohol from all beverages combined was approximately 2.37 gallons in 1967 and 2.65 gallons in 1985 (NIAAA, 1985, 1986). This was an 11.8% increase, whereas our survey figures show only a 7% increase. We suspect that the questions on quantity of drinks per occasion are to blame for this undercoverage. In the 1967 survey "five or six" drinks per occasion was the highest quantity asked about. No doubt many drinkers exceeded this amount per occasion, and for these drinkers the volume measurements obtained must have been inaccurate, in turn causing inaccurate aggregate estimates.

The distribution of the Volmax measure among various age categories and for each sex is shown in Table 2. There were no significant differences between the overall drinking patterns in 1967 and 1984. Among women the drinking patterns in 1967 and 1984 also did not differ significantly. However, the difference between the proportion of men who reported no current alcohol use in 1984 (25%) and the proportion in 1967 (20%) was significant ($p < .05$).

There were a few other differences between the findings of the two surveys, but these did not seem to form a pattern. For instance, in the 23-29 age group for each sex, the proportion of abstainers was greater in 1984 than it was in 1967. Also, among men aged 60 and over, the proportion who reported themselves as abstainers was greater in 1984 than in 1967. Among women aged 40-49 the proportion of abstainers was significantly greater in the later survey. Because these findings do not seem to form a meaningful pattern and because they are at variance with those of Hilton (1986), we are not inclined to place much emphasis on these differences. We think it more important to note the commonplace finding in both surveys that relatively more women than men in each age category were abstainers and that the proportion of abstainers tended to increase with age among both sexes.

We cross-tabulated drinking patterns as measured by Volmax against various other demographic categories (data not shown). As above, the few statistically significant differences found did not seem to be patterned. Among men in the Northeast, the proportion of abstainers was greater in 1984 (16%) than in 1967 (9%), but not so among women (25% in 1984 vs. 26% in 1967). In the North Central region, relatively more women reported abstinence in 1984 (35%) than in 1967 (25%),

TABLE 2
Quantity-Frequency of Drinking (Volume-Variability) by Sex and Age for 1967 and 1984, in Percent[a]

	Male age groups										Female age groups															
	23-29		30-39		40-49		50-59		60+		23-29		30-39		40-49		50-59		60+		Total men		Total women		Total	
	1967 (N = 104)	1984 (193)	1967 (156)	1984 (220)	1967 (180)	1984 (136)	1967 (143)	1984 (123)	1967 (168)	1984 (221)	1967 (92)	1984 (193)	1967 (129)	1984 (243)	1967 (157)	1984 (175)	1967 (106)	1984 (141)	1967 (124)	1984 (258)	1967 (751)	1984 (892)	1967 (608)	1984 (1010)	1967 (1359)	1984 (1902)
Abstainers	8	17*	13	13	18	23	25	29	32	42*	17	27*	34	31	26	38*	41	38	56	51	20	25*	36	38	29	32
Low Volume																										
Low Max	12	9	15	12	13	24*	14	11	23	21	25	18	22	24	30	26	30	29	29	32	16	15	27	26	22	21
Medium Max	8	8	15	9	13	10	8	10	5	5	21	15	20	10*	11	7	8	10	3	4	10	8	12	9	11	9
High Max	14	13	12	13	10	6	8	11	3	2	15	12	7	11	8	7	5	4	1	1	9	9	7	7	8	8
Medium Volume																										
Low Max	4	1	1	2	2	2	5	2	6	2	0	1	2	1	6	1*	1	3	4	3	3	2	3	2	3	2
Medium Max	3	3	5	6	3	2	2	4	3	4	10	8	2	5	2	4	2	5	0	2	3	4	3	4	3	4
High Max	19	17	8	14	15	6*	7	5	3	2	7	11	4	7	7	5	2	2	2	1	10	9	4	5	7	7
High Volume																										
Low Max	0	1	1	—[b]	1	—[b]	7	1*	11	8	1	1	—[b]	1	3	1	2	2	3	3	4	2	2	2	3	2
Medium Max	3	—[b]	1	1	4	—[b]	7	5	5	4	1	—[b]	2	—[b]	2	3	4	3	1	2	4	2	2	2	3	2
High Max	28	30	29	29	23	27	19	22	10	10	4	8	7	10	5	7	5	4	2	1	21	23	5	6	12	14
Total	99	99	100	99	102	100	102	100	101	100	101	101	100	100	100	99	100	100	101	100	100	99	101	101	101	101

[a]Percentages and Ns are weighted figures.
[b]Less than 0.5%
*$p < .05$. Difference of proportions test.

but men did not (17% in 1984 vs. 19% in 1967). In the South, more men reported high volume–high maximum drinking in 1984 (19%) than in 1967 (11%); no difference was found for women (3% in 1984 vs. 4% in 1967). There were regional differences in drinking patterns in both surveys, notably a greater proportion of abstainers in the South than in other regions. Within categories of education, relatively more of both sexes who had less than a high school education were abstainers in 1984 (men, 38%; women, 56%) than in 1967 (men, 27%; women, 43%). Men who had completed high school included relatively more abstainers in 1984 (24%) than in 1967 (15%), but the same was not true for women (1984, 35%; 1967, 33%). The relation of marital status to drinking patterns was much the same in 1984 as in 1967. One significant difference noted was that 26% of the married men reported themselves to be abstainers in 1984 and 19% in 1967. Among women, the only significant difference was that relatively fewer of those who never married were abstainers in 1984 (29%) than in 1967 (48%). There were few differences in religious affiliation between the surveys. Relatively more males grouped under the umbrella term *conservative Protestants* (see Table 4) reported being abstainers in 1984 (40%) than in 1967 (30%), yet a greater proportion of male conservative Protestants were high volume–high maximum drinkers in 1984 (17%) than in 1967 (11%). We dichotomized total family income at the median in both surveys to provide a comparable measure for both years. The only significant income difference between the two surveys was found among women—the proportion of abstainers was higher in 1984 (28%) than in 1967 (19%) among those with incomes higher than the median.

We have focused only on differences between the two surveys and not on the relationships between various sociodemographic variables and drinking patterns. What we found regarding the latter (relatively few systematic differences between 1967 and 1984) suggests that the relationships between these demographic characteristics and patterns of alcohol use have not changed very much over the years.

Drinking Problems

The 1984 survey precisely duplicated 14 questions on drinking problems that originally appeared in the 1967 survey. We grouped the questions into two categories: dependence items and problem consequences. The four dependence problem items were: "I have skipped a number of regular meals while drinking," "I have awakened the next day not being able to remember some of the things I had done while drinking," "I stayed intoxicated for several days at a time," and "Once I started drinking it was difficult for me to stop before I became completely intoxicated." The 10

TABLE 3
Drinkers Reporting Occurrence of Each Problem in Previous 12 Months, in Percent

	Men		Women	
	1967 (*N* = 476)	1984 (666)	1967 (492)	1984 (630)
Dependence problems				
Skipped meals	4.1	10.9*	2.1	6.1*
Loss of memory	4.9	11.8*	3.5	4.8
Couldn't stop until intoxicated	2.4	4.3	0.4	1.1
Binge drinking	1.6	1.1	0.3	0.9
Problem consequences				
Harmed friendships or social life	2.3	6.4*	2.3	2.7
Harmed marriage or home life	2.8	6.1*	0.6	2.6*
Harmed health	5.8	6.3	6.3	6.1
Harmed work or employment opportunities	0.9	2.9*	0.1	1.2*
Harmed finances	2.2	3.2	0.8	1.5
Had an accident	0.3	0.8	0.0	0.1
Legal trouble other than DUI	0.6	1.1	0.0	0.0
Lost or nearly lost a job	0.0	0.8*	0.3	0.4
People at work told me to cut down	1.0	1.0	0.0	0.0
Physician told me to cut down	3.4	2.0	2.9	0.9*
Any of four dependence problems	8.2	18.8*	5.2	8.2*
Any of nine problem consequences	10.8	13.3	8.0	7.1

*$p < .05$. Difference of proportions test. All figures based on weighted data.

problem consequences included items relating to trouble with spouse, employer, friends, police, and so on, that were, in the respondent's judgment, due to drinking. The problem areas covered by these items are shown in Table 3. In Cahalan's (1970) work these items were categorized as "tangible consequences" of drinking. Our analysis will concentrate on two problem measures constructed from the 14 items: any of the four dependence problems and any of nine problem consequences reported in the previous year (one item was dropped from the second measure for reasons discussed below). Problems had to have occurred within the previous 12 months to be counted.

The overall rates of both individual and aggregated problems in 1967 and in 1984 are shown in Table 3. Between 1967 and 1984 there were statistically significant increases (for either one sex or the other) in the rates of almost half of the individual problem items. The rate for any individual problem was small, however—usually under 5%. Increases in the rates of problems were therefore sometimes quite small in absolute terms. In relative terms, however, a doubling or tripling of problem rates can be found in Table 3. As in other surveys, "skipped meals" and "blackouts" were the most frequently reported experiences, and the greatest relative increases also occurred on these items. For "skipped meals" the differences were significant for both sexes; for "blackouts" the difference was significant only for men.

Other differences between the 1967 and 1984 problem rates were that "harmed marriage" and "harmed work" were significantly higher among both sexes in 1984, and that "harmed friendships" and "lost a job" (the latter very rare) were significantly greater in 1984 for men only. In both survey years, the problem rates were lower for women than for men. This finding is consistent with the data on drinking patterns, in which relatively more men than women were heavy drinkers in both surveys.

An exception to the general pattern of increasing rates occurred in response to the item "A physician suggested I cut down on drinking." Here, a lower proportion reported the problem in 1984 than in 1967 for both men and women, the difference being significant for women only. This was the only item for which a significant decrease was noted. Because of the unusual behavior of this item in relation to the other problem items, we excluded it from the problem consequences index, not wanting the interpretation of the index as a whole to be confounded by a strongly deviant item. Although conclusions based on a single item are necessarily weak, the findings here raise the question whether physicians were less likely to advise their patients to reduce alcohol intake in 1984 than in 1967 even though per capita consumption had increased.

A greater proportion of both men and women reported experiencing one or more of the four dependence problems during the previous year ($p < .05$). However, there were no statistically significant differences in the proportion of either men or women who reported experiencing one or more of the nine problem consequences. Thus, the overall picture appears to be one of higher rates of dependence problems in 1984 but of stable rates of problem consequences.

Since the proportion of respondents reporting any of four dependence items increased between 1967 and 1984, it would seem useful to examine the distribution of this change among various demographic categories (Table 4). A higher percentage of women reported one or more of the dependence items in 1984 than in 1967. However, when this difference was examined within various demographic categories, with consequent reduction in *N*, no significant differences were seen. The only exception to this was in the western region, where the rate of dependence problems changed from 1.2% in 1967 to 17% in 1984.

Among men, significant differences were found in the younger age categories and the lower educational categories. Of men with either a high school or less than a high school education, a greater proportion reported

TABLE 4
Drinkers in Selected Demographic Categories Reporting Any of Four Dependence Problems in the Past Year

	Men				Women			
	1967		1984		1967		1984	
	N	%	*N*	%	*N*	%	*N*	%
Total	474	8.2	665	18.8*	492	5.2	630	8.2*
Age								
23-29	73	13.5	158	31.0*	96	9.5	142	18.1
30-39	108	7.5	188	18.0*	108	6.0	168	9.3
40-49	120	7.6	105	21.9*	131	5.2	108	8.4
50-59	86	9.1	87	8.8	85	1.2	88	0.5
60 +	88	3.1	127	9.2	72	3.3	125	0.6
Education								
Less than high school	192	8.4	145	21.0*	204	6.6	115	12.1
High school graduate	122	7.8	222	22.2*	152	2.2	235	5.7
Some college	75	9.9	125	16.3	75	7.2	155	8.3
College graduate or more	84	6.9	174	14.4	61	5.5	125	9.0
Marital status								
Married	418	7.7	486	14.9*	389	4.6	426	5.7
Divorced, separated, never been married	49	13.7	161	29.0*	49	15.8	143	18.5

Region								
Northeast	135	9.5	163	14.1	133	7.1	169	5.3
North Central	145	6.5	196	20.5*	182	3.5	168	7.1
South	121	9.8	179	18.6*	119	7.6	165	5.4
West	73	6.5	128	22.4*	57	1.2	127	17.0*
Religion								
Catholic	144	12.2	197	17.5	188	7.0	209	7.9
Liberal Protestant	92	4.1	132	14.9*	90	4.9	128	6.6
Conservative Protestant[a]	197	6.7	240	21.1*	180	4.3	215	8.0
Income								
Above Median	208	8.6	247	20.4*	222	5.9	263	9.0
Below Median	259	8.1	381	17.7*	261	4.8	334	7.6
Urbanization								
Large metropolitan	170	12.7	229	22.3*	188	6.5	217	11.1
Small metropolitan	140	6.3	273	15.2*	145	2.6	261	5.7
Nonmetropolitan	164	5.1	163	19.9*	158	6.2	152	8.2

*$p < .05$. Difference of proportion test.

[a]Conservative Protestant includes the following denominations: Baptist, Methodist, United Brethren, Pentecostal, Assembly of God, Church of God, Nazarene, Holiness, Apostolic, Evangelical, Sanctified, Disciples of Christ, United Church of Christ, Christian Reformed, Jehovah's Witness, Congregational, Seventh Day Adventist, Latter-Day-Saint (Mormon), Brethren, Spiritual, Mennonite, Moravian, and Salvation Army.

a dependence problem in 1984 than had in 1967. A greater proportion of both married and single men reported a dependence problem in 1984 than had in 1967. Regionally, differences in the proportions were significant in the north central, southern and western states. In 1984, higher rates were found among both categories of Protestant men, but differences were especially large among men from conservative Protestant backgrounds. Differences were larger for those with higher incomes, a finding inconsistent with the finding that greater differences can be found at lower educational levels. Differences were significant for men in all three categories of urbanization. However, whereas rural areas formerly had substantially lower rates than large metropolitan ones, they later came to have similar rates of these problems. Thus, these data suggest a general pattern of increase in the proportion of men reporting one or more of the dependence indicators but not in the proportion of women reporting one or more of them.

A parallel analyses of problem consequences yielded few findings of interest (data not shown). Significant decreases in the proportions reporting these problems were found among women in the South and women who were divorced or separated or had never been married. Significant increases were found among men from small metropolitan areas and among men who were conservative Protestants.

Discussion

We have searched for evidence of changes in alcohol consumption and drinking problems between the years 1967 and 1984 by analyzing data from two general population surveys. Because only items that were identically worded in both surveys were included in the analysis, our focus was restricted to a small set of problem items, and these may not be indicative of the full range of drinking problems that are typically experienced. Another limitation is that the survey items did not ask the respondents how often they drank more than five or six drinks per occasion; we suspect that this omission is an important limitation on our ability to account for the total volume of alcohol consumed as estimated from published consumption statistics. The findings should be viewed with these limitations in mind.

Beverage preferences did change between 1967 and 1984. According to survey responses, Americans consumed more wine and beer but less distilled spirits in 1984 than they had in 1967. For total alcohol from all beverage types, however, the volume of drinks consumed did not change significantly. There were few significant differences in drinking patterns between 1967 and 1984, as measured by the Volmax index. An exception to this finding was a small increase in the percentage of men who were abstainers. Other scattered differences were observed when the data were analyzed according to various demographic breakdowns, but these did not fit into a recognizable pattern of change.

With regard to drinking problems, mixed findings were obtained. Little difference was found over time in the proportion of respondents experiencing any of the nine problem consequences, but there was an increase between 1967 and 1984 in the proportion who reported experiencing one or more of the four dependence problems. More specifically, increases were found for skipping meals because of drinking and experiencing memory losses after drinking. Among men, increases in rates of dependence problems were unevenly distributed across demographic categories, especially across age (with greater increases reported among younger respondents), educational attainment (with greater increases reported among the less-educated respondents), and religious affiliation (with greater increases reported among Protestants, especially conservative Protestants).

Finally, we must comment on the status of these findings as evidence for or against the single distribution theory. According to that theory, increases of greater than 11.8% would have been expected in the prevalence of heavy drinking. Our finding was that there was not a significant increase in the prevalence of heavy drinking between 1967 and 1984. At first glance this finding might seem evidence that the relationship between increases in per capita consumption and increases in heavy drinking is not necessarily what it has been specified to be under the single distribution hypothesis. But on closer inspection, we realize that the observed increase—from a prevalence of 12% to one of 14% among the total population—was of the expected magnitude. This increase may have failed to be significant only because the sample sizes employed were insufficiently large. Thus, the present findings do not necessarily count as counter-evidence against the single distribution theory. Nevertheless, this analysis has been of value since a known increase in consumption has created the expectation that heavy drinking and drinking problems are rising substantially. Though some increases do seem to be afoot, the present findings suggest that they have not been very large.

Acknowledgment

This research was supported by National Institute on Alcohol Abuse and Alcoholism grant AA05595.

References

Cahalan, D. (1970). *Problem drinkers: A national survey.* San Francisco: Jossey-Bass.

Cahalan, D., Cisin, I.H., & Crossley, H.M. (1969). *American Drinking practices: A national study of drinking behavior and attitudes* (Rutgers Center of Alcohol Studies Monograph No. 6). New Brunswick, New Jersey: Rutgers Center of Alcohol Studies.

Clark, W.B., & Midanik, L. (1982). Alcohol use and alcohol problems among U.S. adults: Results of the 1979 national survey. In National Institute on Alcohol Abuse and Alcoholism (Ed.), *Alcohol consumption and related problems* (Alcohol and Health Monograph No. 1, DHHS Publication No. ADM 82-1190, pp. 3-52). Washington DC: Government Printing office.

Department of Commerce. (1982). *Statistical abstract of the United States, 1982-1983.* Washington, DC: Government Printing Office.

Fitzgerald, J.L., & Mulford, H.A. (1981). The prevalence and extent of drinking in Iowa, 1979. *Journal of Studies on Alcohol, 42,* 38-47.

Hilton, M.E. (1986). Abstention in the general population of the U.S.A. *British Journal of Addiction, 81,* 95-112.

Johnson, P., Armor, D.J., Polich, M., & Stambul, H. (1977). *U.S. adult drinking practices: Time trends, social correlates and sex roles.* Santa Monica, CA: Rand.

Ledermann, S. (1956). *Alcool, alcoolisme, alcoolisation.* Paris: Presses Univarsitaires de France.

National Institute on Alcohol Abuse and Alcoholism. (1985). *U.S. alcohol epidemiological data reference manual: Vol. 1. U.S. apparent consumption of alcoholic beverages based on state sales tax, taxation, or receipt data.* Vol. 1. Washington, DC: Government Printing Office.

National Institute on Alcohol Abuse and Alcoholism. (1986) *Alcohol epidemiologic data system. Apparent per capita alcohol consumption, national state and regional trends, 1977-1984* (Surveillance Report No. 2). Washington DC: Government Printing Office.

Room R. (1977). Measurement and distribution of drinking patterns and problems in general populations. In G. Edwards, M. Gross, M. Keller, J. Moser, & R. Room (Eds.), *Alcohol related disabilities* (Offset Publication No. 32, pp. 6-87). Geneva: World Health Organization.

Room R. (1983). Region and urbanization as factors in drinking practices and problems. In B. Kissin & H. Begleiter (Eds.), *The pathogenesis of alcoholism: Psychosocial factors* (pp. 555-604). New York: Plenum Press.

Room, R., & Beck, K. (1974). Survey data on trends in U.S. consumption. *Drinking and Drug Practices Surveyor, 9,* 3-7.

Skog, O.-J. (1982). *The distribution of alcohol consumption: Part I. A critical discussion of the Ledermann model* (SIFA-Mimeograph No. 64). Oslo: Statens institutt for alkoholforskning (National Institute for Alcohol Research).

Skog, O.-J. (1983). *The distribution of alcohol consumption: Part II. A review of the first wave of empirical studies.* (SIFA-Mimeograph No. 67). Oslo: Statens institutt for alkoholforskning (National Institute for Alcohol Research).

Wilsnack, R.W., Wilsnack, S.C., & Klassen, A.D. (1984). Women's drinking and drinking problems: Patterns from a 1981 national survey. *American Journal of Public Health, 74,* 1232-1238.

Wilsnack, S.C., Wilsnack, R.W., & Klassen, A.D. (1986). Epidemiological research on women's drinking, 1978-1984. In National Institute on Alcohol Abuse and Alcoholism (Ed.), *Women and alcohol: Health related issues* (Research Monograph No. 16, DHHS Publication No. ADM 86-1139, pp. 1-68). Washington DC: Government Printing Office.

SECTION II

Observations on the Modern Setting

B. SOCIAL PROCESSES IN DRINKING

Introductory Note

Drinking behavior, like many forms of acquired human action, is learned behavior and most frequently occurs in a social context. In U.S. society the majority of adults classify themselves as drinkers, although their patterns of consumption may vary. Conversely, a significant minority classify themselves as abstainers, from 33% to 37% (National Institute on Alcohol Abuse and Alcoholism, 1990). Furthermore, the overwhelming number of drinkers cannot be classified as dependent on alcohol.

Helene Raskin White, Marsha E. Bates, and Valerie Johnson, in the first of the two chapters in this subsection (chapter 9), examine the relevance of a social learning model (Akers, 1977; Bandura, 1977) to the initiation and maintenance of alcohol use in adolescence (see also chapter 12, by Akers and La Greca, in this volume). They discuss the role of social reinforcers, observational learning, and alcohol expectancies in shaping patterns of drinking. In addition, they review and contrast family and peer influences and consider the effects of the media. The chapter suggests that environmental factors have a strong influence on the development of drinking behaviors.

White and her colleagues indicate that peers play an important role in determining adolescent drinking behaviors. Included as chapter 10 is an original research study by James Orcutt, which also examines the role of peer relationships in drinking contexts. Although the study focuses on the use of all beverage alcohol among subjects in this college sample, as the title implies ("... Beers and Peers"), the beverage of choice among college students is beer. Orcutt's research focus is on the important topic of how peer relationships and personal consumption patterns affect alcohol use and intoxication during the onset of drinking occasions on various days of the week. The respondents in this survey report on drinking events that occurred in the normal course of a week's activity; thus, consumption, intoxication, and reported alcohol-related problems can be viewed in terms of a situational context in which interpersonal influences of peers are present.

These two chapters demonstrate the importance of social influences on drinking and intoxication. Later in the book (section IV) we present chapters that discuss the importance of early socialization effects on later alcoholism.

References

Akers, R. (1977). *Deviant behavior: A social learning approach* 2nd ed. Belmont: Wadsworth.

Bandura, A. (1977). *Social learning theory.* Englewood Cliffs, NJ: Prentice-Hall.

National Institute on Alcohol Abuse and Alcoholism. (1990). *Seventh special report to the U.S. Congress on alcohol and health.* (DHHS Publication No. ADM 90-1656). Rockville, MD: Author.

CHAPTER 9

Learning to Drink: Familial, Peer, and Media Influences

HELENE RASKIN WHITE, MARSHA E. BATES, AND VALERIE JOHNSON

Drinking alcoholic beverages, like other acquired human behaviors, is learned and usually performed in a social context. There are many socially acceptable reasons and occasions for alcohol use, many social rewards for drinking, and many role models of drinking behaviors; these factors have resulted in the integration of a wide variety of drinking practices into our day-to-day life. The focus of this chapter is on social rewards and incentives for drinking that are dependent upon peoples' experiences within the context of specific social environments. We discuss the most proximal social determinants of drinking, including family and friends, who are powerful agents of social influence. In addition, we consider noninteractional models of drinking that are widely observed in the electronic and print media, both in an advertising context and during the recreational viewing of television and movies. Although this chapter emphasizes social learning processes during adolescence, it should be noted that these same processes operate throughout the life span (see White, Bates, & Johnson, 1990) and are applicable to the full range of drinking outcomes (see Abrams & Niaura, 1987). Before discussing social influences on the initiation and development of drinking behaviors, we provide a brief orientation to the major features of a social learning approach to the question of why people drink.

Helene Raskin White, Marsha E. Bates, and Valerie Johnson are with the Center of Alcohol Studies, Rutgers University, New Brunswick, N.J. This chapter is an abridged and updated version of "Social Reinforcement and Alcohol Consumption," from *Why People Drink: Parameters of Alcohol as a Social Reinforcer,* W.M. Cox, Ed. (New York: Gardner Press, 1990), pp. 233–261, and is published here with permission.

A Social Learning Perspective

Social reinforcers, that is, reinforcements provided by other people, are important determinants of many human behaviors. If others' reactions are perceived as approving (providing positive reinforcement) or consisting of the withdrawal of disapproval (providing negative reinforcement), they will serve to increase the likelihood that the behavior will be performed again. If the reactions of others are perceived as disapproving (providing, in effect, punishment), they will decrease the probability that the behavior will be engaged in again. The idea that consequences of behavior determine whether new behaviors will be acquired and existing ones will be modified is central to most learning theories. In addition, social learning theories emphasize that symbolic mental events (e.g., attitudes, expectancies) and observing others also influence learning. Hence, theories of social learning integrate learning principles derived from observable behavior with constructs based on cognitive processes that are not directly observable (see Bandura, 1969; Rotter, 1982; Skinner, 1971).

According to Bandura (1977), "Social learning theory approaches the explanation of human behavior in terms of a continuous reciprocal interaction between cognitive, behavioral and environmental determinants" (p. vii). Hence, a social learning approach seems well suited to examining why people drink, because drinking behaviors and effects appear to be determined by complex interactions among alcohol, the person who drinks it, and the environment (see, for example, Akers, 1985; Sher, 1985). [Also see chapter 12, by Akers and La Greca, in this volume.] From this perspective, variables in the person, environment, and behavioral domains are thought to affect and be affected by one another. Note that this model of *reciprocal determinism* is quite different from the view that drinking behavior is a unidirectional function of personal and environmental determinants (see Bandura, 1977).

The Person

The social learning model depicts the individual as bringing to the environment a unique endowment that (1) imposes certain limits on what can be learned and how it is learned and (2) influences his or her selection of various environments. Learning experiences transform one's genetic potentialities into behavioral repertoires (Doby, 1966). Because the foci of this chapter are important environmental influences on drinking behaviors, we will make little mention of relevant intrapersonal characteristics, including "self-efficacy" (Bandura, 1982; Love, Ollendick, Johnson, & Schlesinger, 1985), coping styles (e.g., Cooper, Russell, & George, 1988), and other intrapersonal characteristics that mediate the impact of social stresses and pressures on the individual's subsequent behavior and help determine whether alcohol will be used to regulate negative, socially induced, internal states. Understanding the scope and limits of environmental influences on drinking requires an appreciation that a person's characteristics will modify and be modified by drinking behaviors themselves as well as models and reinforcers in the relevant social environments. (See Cox, Lun, & Loper, 1983, for a review of personality characteristics that place adolescents at risk for the subsequent development of problem drinking.) Personality characteristics also help determine the impact of environmental influences. For instance, an adolescent who has strong needs for affiliation and approval will likely be more vulnerable to peer pressure to drink than one who does not.

The Environment

The environment imposes limitations on the nature and content of learning experiences. The distal environment consists of the larger cultural, religious, and other groups that help determine to what extent drinking behaviors will be reinforcing for the individual. Humans' capacity to represent symbolically and transmit information to one another, and thus to succeeding generations, provides avenues for the transference of integrated patterns of behavior whereby actions can be regulated through thought and language (Bandura, 1977). The learning of behavior patterns through experiencing common social reinforcers contributes to certain commonalities observed within large social groups (such as ethnic and religious groups, societies, and cultures) with respect to their attitudes, expectations, and patterns of alcohol use (White, 1982). These large groups typically have at their disposal legal, ethical, or educational institutions, or all of these, that function to promote those patterns of drinking that are consistent with the interests, beliefs, and goals of the group.

In addition to social motivations for drinking that are rather common in a society, family members and peers also prescribe why, when, and how much to drink. These smaller social groups reinforce or punish certain drinking behaviors by both their example and their expressed or implied attitudes toward drinking. Such groups also teach members the appropriate manner in which to comport themselves while intoxicated (Falk, Schuster, Bigelow, & Woods, 1982). In this chapter we will focus on these proximal environmental as well as media influences on drinking.

Observational Learning

Much social learning occurs in the absence of direct experience. That is, human thought and behavior may be markedly influenced by observing other people engage in behavior that is reinforced (vicarious or anticipated reinforcement). People model the behavior of others because of the expectation that their own behavior will produce the same positive consequences that they observed in the model experience (Bandura, 1977). Observational learning can occur without actual performance because of humans' ability to represent symbolically, store, and covertly rehearse behaviors. Thus, in the case of drinking, we suspect that observational learning of one's parents' drinking behaviors can occur long before those behaviors are actually performed. Note that children, by age 6, already hold beliefs regarding normative drinking behavior for men, women, and children (Spiegler, 1983). The child or adolescent may not actually engage in any learned drinking behaviors until some later age or until an appropriate social setting, very likely involving peers, occurs.

The terms *modeling, imitation,* and *observational learning* have been used to refer to the same phenomenon, but we prefer the distinction put forth by Bandura (1977), Rimm and Masters (1979), and others, who clearly differentiate between the temporary copying of a model's behavior (imitation) and the acquiring of more enduring behaviors (observational learning). Studies examining imitative phenomenon under controlled conditions have experimentally demonstrated the strength of modeling influences. During the past 10 years, these studies have confirmed that, under certain conditions, individuals imitate the specific characteristics of drinking behavior (e.g., the rate of consumption) that they observe among confederate models in laboratory settings (e.g., Collins, Parks, & Marlatt, 1985), "real world" bar settings (e.g., Reid, 1978), and semi-naturalistic bar settings (e.g., Caudill & Lipscomb, 1980). Modeling effects appear to be influenced by gender and drinking history (Lied & Marlatt, 1979), as well as by the degree of prior social interaction between subject and model (Hendricks, Sobell, & Cooper, 1978). It has been demonstrated that a negative social situation may lead to heavy drinking by subjects, regardless of the model's range of behavior drinking (Collins et al., 1985). Reid's (1978) study, however, in which subjects exposed to an unfriendly model were able to leave the drinking situation, indicates that a subject's perception of behavioral alternatives to drinking may be a significant consideration in his/her decision to use alcohol to cope with unpleasant social conditions.

Studies of concurrent modeling effects on drinking behaviors do not address the processes by which early observational learning experiences, which occur before the onset of drinking, come to bear on adolescents' later decisions to actually use alcohol. The study of alcohol expectancies, their development and subsequent impact on drinking decisions, bridges this gap by providing a theoretical concept to link early life experiences to later patterns of use.

Cognitions and Expectancies

Throughout the socialization process, symbolic representations of experience, including internalized values, beliefs, and expectancies, are formed and modified first through exposure to parents (or other primary caretakers) and later through the influence of other meaningful social agents (e.g., peers, teachers). Basic experimental research has demonstrated that differing beliefs about how often a behavior is likely to be reinforced (Kaufman, Baron, & Kopp, 1966) and why a particular behavioral consequence has occurred (Dulany, 1968) significantly modify how different individuals will behave in response to identical stimuli in the environment.

Note that alcohol is a psychoactive drug that produces powerful and relatively immediate effects and can thus serve as a potent primary reinforcer. Primary reinforcing effects are the result of "intrinsic" pharmacological actions of alcohol, and the strength of these effects may be genetically determined in part (see Goodwin, 1985). Increasing evidence suggests, however, that socially mediated expectancies and cognitions strongly influence effects of alcohol that, in the past, have been attributed to its pharmacological actions. For example, MacAndrew & Edgerton (1969) presented compelling anthropological evidence that alcohol effects on behavior vary greatly across different cultures and time periods. More recent laboratory studies have used the balanced placebo design to separate alcohol's pharmacological effects on social behaviors from effects due to subjects' belief that they have ingested alcohol (see Goldman, Brown, & Christiansen, 1987; Leigh, 1989; Marlatt & Rohsenow, 1980). Overall, these studies have confirmed that an individual's belief that she or he has received alcohol can be a determinant of alcohol effects, especially on somewhat deviant social behaviors (Hull & Bond, 1986).

The alcohol expectancy literature has shown that individuals' beliefs about how alcohol affects their thoughts, behaviors, and emotions are indeed related to drinking habits. Studies of the development and later impact of such beliefs are important because they provide information about the mechanisms by which early life experiences influence later drinking behaviors. Measures of outcome expectancies regarding the anticipated consequences of alcohol are thought to tap individuals' internal representation of their acculturation and socialization experiences before the onset of drinking (Smith, 1989). In general, the reasons for drinking most frequently endorsed by nonproblem drinkers involve the enhancement of already pleasurable internal states and social occasions. In contrast, alcoholics and problem drinkers expect more global positive changes, social assertiveness, social and physical pleasure, relaxation, and tension reduction from alcohol use than other drinkers do. Overall, the findings in this area suggest that outcome expectancies are good predictors of the onset and maintenance of adult and adolescent drinking styles, including problem use and alcoholism (Goldman et al., 1987). Recent studies of both concurrent (Christiansen & Goldman, 1983) and longitudinal (Christiansen, Smith, Roehling, & Goldman, 1989) relations between adolescent expectancies and drinking status show that those most involved with alcohol believe that alcohol use strongly facilitates cognitive and motor functioning as well as social interaction. The counterintuitive expectancies of enhanced cognitive and motor functioning have also been found both in high school students with a family history of alcoholism and among those with personality characteristics known to put people at risk of future problem use and thus may be of special, prognostic significance (Brown, Cramer, & Stetson, 1987; Mann, Chassin, & Sher, 1987).

The expectancy research also suggests that individuals at risk for or currently experiencing alcohol problems may have more social and nonsocial reinforcers connected with drinking, or more generalized stimulus conditions that trigger drinking behavior, or both, than those who are not experiencing problems. Generalization of stimulus conditions for drinking or generalized expectations of reinforcement from drinking may evoke (1) more intensive drinking behaviors and (2) nonadaptive drinking responses to stressful social stimuli. These suggestions parallel Smith's (1989) situational insensitivity hypothesis, which suggests that individuals for whom drinking is a problem hold situationally insensitive expectancies, value alcohol-related reinforcers more highly, and eventually drink in more contextually indiscriminate ways than others.

It thus seems likely that (1) the reinforcing properties of alcohol are perceived differently by various individuals in accordance with their previous exposure to specific others and social contexts for drinking and (2) resultant expectations about the effects of alcohol will actually modify how people experience and react to alcohol. Reasons for drinking, drinking behavior, and behavioral effects may thus be influenced by meaningful others even when others are not present.

In summary, a social learning approach that emphasizes the reciprocal interaction among the person, the environment, and behavior has been offered as a framework for examining why people drink. It is important to note that a social learning framework for studying the interindividual processes that influence drinking behaviors is not advanced here as an alternative to pharmacological, genetic, or other more molecular explanations of drinking phenomena and effects. Rather, it should be considered as providing a complementary descriptive and predictive level of analysis for what is clearly a complex, multidimensional causal process. Environmental influences will be examined in detail in the remainder of this chapter.

Proximal Environmental Influences

This section examines some of the empirical findings regarding proximal environmental influences on drinking behavior in adolescence. Although the majority of these studies do not test a social learning model per se, variables that represent various components of the model have been investigated.

There are three ways in which interpersonal influences have been postulated to operate: (1) directly (i.e., setting an example, providing reinforcement, or as a result of the quality of the relationship); (2) indirectly (i.e., influencing the development of values, attitudes, and behaviors and the formation of ties); and (3) conditionally (i.e., influencing the susceptibility to be influenced by another individual). Observational learning and reinforcement are the two primary social learning processes by which these influences are transmitted. The ways in which these processes operate depend upon many conditions, including the quality and nature of the relationship between the individuals as well as situations (Kandel, Kessler, & Margulies, 1978). The following sections focus primarily on the ways in which interpersonal influences affect the development of drinking behavior.

The Family

Parents. During infancy and early childhood, the family is the primary facilitator of the socialization process. Hence, parents (or other caretakers) have the primary responsibility for providing status, nurturance, training, and other elements of socialization. Basic research in modeling efficacy has found that models with higher status or power over the life events of an individual are more effective behavioral models than others are (e.g., Bandura, 1977). These findings suggest that, at least during childhood, parents are potentially powerful models of drinking behavior because of their initially exclusive control over the daily life of their children.

Kandel et al. (1978) have suggested that initiation to alcohol use results from direct modeling in which adolescents imitate the behavior of significant others. Similarly, Barnes (1977) has made a case for the "model building" theory of alcohol use, in which various patterns of alcohol use are learned by imitation of the example set by parents. (We have defined this more enduring learning as observational learning.) Barnes has claimed that the best predictors of youthful drinking habits are the attitudes and behaviors of parents in regard to alcohol use. She presented evidence that abstainers tend to come from homes in which parents abstain, moderate drinkers from homes in which parents drink moderately, and a disproportionate number of heavy drinkers from homes with parents who drink heavily. Much of the empirical research on alcohol use supports Barnes' conclusions. That is, parental drinking has been found to be strongly related to adolescent initiation (e.g., Barnes, Farrell, & Cairns, 1986; Fontane & Layne, 1979; Hawkins, Fine, & Sweeny, 1986). The majority of these studies, however, have been atheoretical and not designed to investigate the processes of social learning.

Research on parental influence has relied, for the most part, on adolescents' perceptions of parental use rather than parent self-reports, which could inflate the strength of observed relationships. One exception was a study by Newcomb, Huba, and Bentler (1983), who compared "direct modeling" and "cognitive mediational" explanations of maternal influence on adolescent alcohol and drug use. The former implies that direct observation and modeling are the essential processes for the acquisition of a behavior. The latter stresses that modeling effects are important but only to the extent that they influence the child's cognitive mediation, assimilation, and perceptions, which, in turn, motivate drinking. Newcomb et al. (1983) found that cognitive mediations best explained the acquisition by both sons and daughters of the drinking behavior of their mother. Hence, the study's data suggest that it is more the perceived than the actual parental behavior that children acquire. Another study relying on parent self-reports found that parental attitudes about use significantly contributed to initiation of use by their adolescent children (Hawkins et al., 1986).

Most research suggests that parents who drink teach their children to drink; but whether social learning occurs primarily through direct or vicarious social reinforcement has not been adequately addressed. There are several plausible ways in which parents may differentially reinforce, either intentionally or unintentionally, their children's use of alcohol. One way is to forbid or punish use. Studies have found that when parents forbid the use of alcohol, their adolescent offspring are less likely to use alcohol than offspring of parents who permit use (Hawkins et al., 1986; Mandell, Cooper, Silberstein, Novick, & Koloski, 1963). Alternatively, individuals may drink because they perceive that their parents expect them to drink. In fact, parents are generally the first ones to introduce alcohol to their children (Mandell et al.). Fontane and Layne (1979) found that the second most frequently identified situation in which college students felt that drinking was expected was the family context. (The first was at parties.) Rachal et al. (1975) also reported that a majority of adolescents consume alcohol with some regularity in the home on special occasions or with dinner. Thus, youth may learn to drink because drinking on special occasions with their family is associated with positive consequences. Note that factors that predict whether or not a student will initiate drinking are different from those that predict the frequency and degree of drinking after it has started. For example, in one study parental modeling appeared to be an important predictor of initiation, whereas contextual variables (e.g., drinking in cars) were the most important predictors of heavy drinking. Lack of parental knowledge of the child's drinking was the best predictor of drunkenness (Smart, Gray, & Bennett, 1978).

Although a full discussion of the quality of the parent-child relationship as it relates to drinking behavior is beyond the scope of this chapter, note that this variable influences social learning processes (see Zucker, 1976). Because family members are the primary agents of socialization, they set up initial reinforcement contingencies and, thus, influence all stages of development. The quality of the interaction between the adolescent and his or her parents and the quality of the home environment have been shown to influence adolescents' drinking regardless of what the parents' habitual use might be. Positive family relationships, attachment to family, communication, discipline, conflict, parental love, parental control, and family management techniques have been significant influences (Johnson & Pandina, 1991; Simons, Conger, & Whitbeck, 1988; Vicary & Lerner, 1986). (For a review of the way in which family variables affect alcohol and other drug use, see Kandel et al., 1978, and Glynn, 1981.)

Likewise, family background variables, including the personality and child-rearing practices of the parents, are strongly related to later problem drinking and alcoholism in the offspring (Barnes, 1977; McCord, McCord, & Gudeman, 1960; Zucker, 1976). Being reared in a home with an alcoholic parent can also contribute to offsprings' developing alcoholism in adulthood, although such an upbringing does not appear to affect early drinking patterns (see, for example, Alterman, Searles, & Hall, 1989; Johnson, Leonard, & Jacobs, 1989; Pandina & Johnson, 1989). Attitudes about alcohol and motivations to use it have been shown to be important determinants of alcohol problems (see White, 1982). Through observational learning within the home, children of alcoholics may learn inappropriate reasons for, and patterns of, alcohol use. The alcoholic parent may convey the attitude that heavy drinking is positively reinforcing. Parents are also models for the development of coping strategies (Knight, 1980). A child may learn to use alcohol to escape from problems rather than develop more positive coping mechanisms if he or she is exposed to alcohol use as a predominant coping mechanism and views the strategy as effective. (See Knight, 1980; Orford, 1979; and West & Prinz, 1987, for a discussion of additional characteristics of a home with an alcoholic parent that could contribute indirectly to the later development of alcoholism in the children.)

Parents are not the exclusive adult agents of influence over children and adolescents; other adults play influential roles as well. A young person's perceptions of adults in general and the transitions into adulthood can also affect how his or her drinking develops. According to Jessor and Jessor (1975), the initiation to drinking alcohol marks a transition in the life cycle that for many adolescents represents a symbolic means of dissolving their adolescent status and identifying with adults. In addition, some research suggests that adolescents develop drinking styles to conform to or reject traditional sex role standards (Wilsnack & Wilsnack, 1979).

Siblings. Another potential agent of interpersonal influence on patterns of alcohol use is the sibling. A few studies have identified an association between siblings' use of illicit drugs (e.g., Brook, Whiteman, Nomura, Gordon, & Cohen, 1988; Clayton & Lacy, 1982). When alcohol alone has been analyzed, some researchers have found a significant relationship (e.g., Needle et al., 1986), whereas others have not (e.g., Clayton & Lacy, 1982: Coombs & Paulson, 1988). In general, this research suggests that siblings exert greater influence than parents on overall substance use (Brook et al., 1988; Needle et al., 1986).

The research presented thus far provides a strong case for associations among family members' alcohol use.[1] However, the specific learning processes involved remain unclear. In the next section we explore peer influences on use and the relative strength of parent peer influence.

Peers

In addition to familial models of alcohol use, research indicates that both the actual or perceived number of peers using alcohol and peer attitudes regarding use of alcohol substantially influence adolescent use. (For a review, see Kandel, 1978; Radosevich, Lanza-Kaduce, Akers, & Krohn, 1980.) Although research has established that peers influence one another's use, it has not been determined to what extent the behavior is learned from the peer group and to what extent adolescents select friends whose drinking behaviors seem attractive. Recent research indicates that the relationship is reciprocal (Downs, 1987; Kandel, 1985).

In addition to friends' use and attitudes regarding alcohol, the degree to which adolescents are involved in peer activities (e.g., driving around in cars, attending parties) is an important factor predicting initiation (Kandel et at., 1978). Similarly, peer activities provide the context in which drinking most often occurs. A study that examined situations in which adolescents felt most tempted to drink alcohol found the situation almost always involved two or more individuals and usually occurred in the context of a party or other unstructured social occasions (Stetson, Brown, & Beatty, 1985). Either direct or indirect social pressures to drink were reported in almost all of these situations (see also Harford & Grant, 1987; Johnson, Marcos, & Bahr, 1987).

A few recent studies have operationalized and tested Akers' (1985) social learning model using parental and peer behavior and attitudes as predictors of adolescent drinking. Akers' social learning theory combines aspects of Sutherland's (1947) differential association perspective (i.e., the importance of significant others in learning "deviant" behavior) with Bandura's (1977) principles of social learning. According to Akers' formulation, differential associations (significant others) provide the environment in which exposure to definitions (beliefs and attitudes), imitation of models, and reinforcement take place. The definitions, in interaction with models and anticipated reinforcers, produce initial use. After initial use, imitation becomes less important, whereas definitions and reinforcements become more important.

Akers, Krohn, Lanza-Kaduce, and Radosevich (1979) investigated the relative importance of imitation, alcohol use by significant others, beliefs about alcohol, and reinforcers of alcohol use for explaining adolescent alcohol use and abuse. Their data demonstrated that friends' use of alcohol was the best predictor of alcohol use and abuse. Imitation, defined as the number of admired models the respondent had observed engaging in the behavior, was less related to drinking behavior than the other social learning variables. Whereas beliefs were the second-best predictor for use, reinforcements outranked beliefs in accounting for abuse. Thus, these results suggest that, over time, alcohol abusers begin to respond to direct reinforcement (from the effects of the drug itself) and definitions come to play a less significant role.

Lanza-Kaduce and Radosevich (1980) also tested this model and found strong evidence of one general learning process that operates across all stages and substances. Definitions appeared to be the most powerful set of predictor variables, followed by differential reinforcement and differential association variables. Again, the effect of imitation was weakest. In addition, stages of use were less differentiated by consequences of use than by normative definitions of use. Applying causal modeling techniques, Johnson (1988) tested a social learning model of continued alcohol use. Like Akers et al. (1979), she found that the proportion of friends who use, and friends' opinions about use, were the best predictors of continued use of alcohol. Contrary to the two previously discussed studies however, attitudes had a negligible effect.

Researchers with different theoretical orientations disagree about the relative influence of parents and peers during adolescence. In general, research indicates that peers exert a significantly stronger influence on adolescent alcohol and drug use than parents (or other adults) do, (e.g., Hansen et al., 1987; Huba & Bentler, 1980; White, Johnson, & Horwitz, 1986). Newcomb & Bentler (1986), however, found that rates of exposure and susceptibility to modeling effects differed among ethnic groups.

Similarly, Kandel and Andrews (1987), who collected data from best friends and parents, found that the socialization process varied depending upon the source of influence and the specific drug. Whereas parental influence was most important before initiation, peer imitation became most influential after initiation.

Most of the research points to an interaction of parent and peer influences. For example, Fontane and Layne (1979) suggested that the peer culture both reinforces drunkenness if parents display it and co-opts the parental influence of those parents who do not drink to drunkenness. Similarly, Maddox (1966) argued that although peer groups are important for learning how to drink and in providing occasions and support for drinking, teenagers first learn to drink from adult models. The peer group may thus provide support for drinking that has already been learned in the home (Barnes, 1977). Further, peer influences can be ameliorated by protective family factors (Brook, Whiteman, Gordon, Nomura, & Brook, 1986). In addition, parents can influence their children's values and choices of friends and thus can make their children susceptible to the influence of either friends who use drugs or those who do not (e.g., Hansen et al., 1987; Kline, Canter, & Robin, 1987; Simons et al., 1988). Zucker (1976) has suggested that, at the very least, the family indirectly influences drinking by defining the child's socioeconomic status, ethnic and cultural background, and milieu.

In his review of the literature on parental and peer influences on alcohol and other drug use, Glynn (1981) concluded that "there does not appear to be any point at which the drug [use] behavior of most adolescents is wholly influenced by either family or peers... [and that] parent and peer influences on adolescent alcohol use appear to be relatively equal... [and that] the most effective family influences appear to be those that are developed in advance of adolescence" (pp. 68–69).

In the previous sections we have examined proximal environmental influences of family and peers on drinking behaviors.[2] There are other potentially influential models of drinking behaviors in the environment with whom one does not interact directly. These models are often observed in the print and electronic media and are the focus of the next section.

Media Influences

The media provide an array of additional social reinforcements for alcohol use. According to Bandura (1977), the abundant and varied symbolic models provided by television and films strongly influence social learning. Research has demonstrated that media play an influential role in shaping social attitudes and behaviors for both children and adults (Liebert, Neals, & Davidson, 1973). Symbolic modeling, provided especially by television, can influence the development of moral judgments by portraying conduct as acceptable or reprehensible and by applying sanctions to such behavior. With the increasing use of symbolic modeling, proximal role models may come to occupy less prominent roles in social learning.

The extent of the role that the media play in socially reinforcing alcohol use has been a topic of increasing controversy over the past decade. Learning theorists suggest that media may (1) affect viewers' expectations concerning the use of alcohol, (2) affect viewers' attitudes about the acceptability or appropriateness of use, and (3)motivate people to model drinking behaviors (Greenberg, Fernandez-Collado, Graef, Korzenny, & Atkin, 1979). The media provide the means for the learning of drinking behavior by making models for observational learning, including vicarious reinforcement, accessible. Media provide models for observational learning through advertisements for alcoholic beverages via television, radio, newspapers, magazines, or billboards and by demonstrating the consumption of alcoholic beverages by characters in television shows and movies. Although it is generally accepted that the media provide information to people about the goals of drinking (i.e., to have fun, to socialize, and to feel better) and about brand choices, the role that media play in influencing attitudes and motivations regarding alcohol use is less clear. The causal connection between exposure to mass media and individual drinking behavior has yet to be systematically explored. The research approaches employed in the last decade to examine this issue have included: content analyses of TV and print media, experiments on program and advertising effects and advertising bans, econometric studies on alcohol sales, and correlational studies on media exposure and drinking behavior (see Atkin, 1987; Blane, 1988, and Smart, 1988 for reviews).

Overall, results of program content analyses suggest that television portrayals of alcohol use lead to favorable attitudes toward drinking, tend to normalize drinking, or emphasize only desirable concomitances such as wealth, success, and social approval. It has been found that whereas television characters at times exhibit signs of problem drinking, such problems are typically portrayed as short-lived and in response to a crisis, and characters often make speedy and unrealistic recoveries. Movies appear to have done a better job of realistically presenting the negative aspects of alcohol use than television has (see Herd, 1986). Viewers have been exposed to the punishing consequences of heavy drinking in such classics as *The Lost Weekend* and *Days of Wine and Roses* and, more recently, *Clean and Sober.* However, other movies have depicted drunkenness as a means of promoting euphoria and humor (e.g., *Arthur* and *Cocktail*) and accomplishing the seduction of women (e.g., *That Touch of Mink*).

In addition, people may be vicariously reinforced from watching characters in advertisements who consume alcohol and are also depicted as attractive, sexually desirable, healthy, and seemingly incapable of having anything but a good time. To combat criticism, in 1978 the Wine Institute adopted a new code of advertising standards, which includes eliminating (1) the use of athletes, (2) the implication of health benefits of use, (3) sexual exploitation, and (4) explicit appeals to youth. However, our causal viewing of beer and wine advertisements prompts us to question the extent to which the advertisers actually adhere to all of these standards (see Jacobson, Atkins, & Hacker, 1983). Alcohol advertisers, in general, have long claimed that their advertisements affect only brand choice and that advertising therefore influences only the market shares of the commodity (Mosher & Wallack, 1981). To date, some econometric studies have concluded that alcohol advertising only modestly increases sales of hard liquor and that restrictions on alcohol advertising do not decrease consumption (e.g., Ogborne & Smart, 1980). Recent experiments on the effects of alcohol advertising in television have concluded that there is generally a lack of evidence supporting a causal relationship between advertising and consumption rates (e.g., Kohn and Smart, 1984; Sobell et al., 1986).

There are also conflicting results about the influence of the media on the development of patterns of heavier drinking. Whereas data from content analyses have shown that many of the televised "drinking scenes" involve the consumption of at least five drinks and only mild disapproval of alcohol abuse among characters (Breed & De Foe, 1984), other authorities assert that it is a variety of risk factors (including a predisposition to heavy drinking) that are salient in an individual's decision to drink heavily (Pittman & Lambert, 1978; Rossiter & Robinson, 1980). Several factors have been found to mediate the influence of the media on alcohol consumption, including people's communication with significant others and their expectations of alcohol. Strickland and Wilson (1982) reported that high levels of communication within families moderated the influence of media on adolescents' consumption. In addition, we would expect that variables such as repetitiveness of exposure, attention to the message, acceptance of the message, and motivation to act on the message influence use (see Strickland, 1982).

Mass media campaigns to prevent alcohol problems have solicited air time to publicize the negative aspects of alcohol use. Typically, these media campaigns have concentrated on preventing problems associated with use (e.g., drunk driving, child abuse and neglect, fetal alcohol syndrome) or on promoting responsible use and increasing awareness of the potential problems and warning signs of alcoholism. Hewitt and Blane (1984) reviewed evaluations of 15 media campaigns conducted from 1971 to 1982. They found that the majority of campaigns promoted desired change in the degree of knowledge about alcohol problems but that attitudes and behaviors did not significantly change as a result of the exposure to these media prevention programs (see also Nathan & Niaura, 1987). Others have found that parents, friends, experts, or former users were considered much more credible in relaying information about drugs than the media were (e.g., Strickland, 1983).

In sum, although it seems clear that the media provide viewers with vicarious, positive social reinforcers for drinking, there is, as yet, no strong empirical evidence that exposure to alcohol use in the media actually increases the number of new drinkers, the overall consumption level, or the number of alcoholics. However, the role of media influences has not been thoroughly studied. Research needs to focus on children and adolescents because they are in their formative years when influences of models are most salient and social reinforcements and consequences of alcohol use appear to be more powerful than nonsocial reinforcers (Christiansen, Goldman, & Inn, 1982; Strickland & Pittman, 1984). The need for more research on this age group is underscored by the fact that from childhood through high school years young people spend more time in watching television than on any other leisure or work activity (Rubenstein, 1978).

Conclusion

In this chapter we explored the development of alcohol-using behaviors within the context of a social learning perspective. The direct social reinforcements provided by family and peers, as well as the symbolic processes involved in observational learning and noninteractional media influences on drinking behaviors, were examined. Recall that within this conceptual framework, patterns of human functioning are seen as arising from a continual interplay of personal, behavioral, and environmental influences. As succinctly explained by Bandura (1977), "Although the reciprocal sources of influence are separable for experimental purposes, in everyday life two-way control operates concurrently. In ongoing interchanges, one and the same event can be a stimulus, a response, or an environmental reinforcer depending upon the place in the sequence at which analysis arbitrarily begins" (p. 204). Thus, the complexity of any behavioral model embracing the concept of reciprocal determinism presents an enormous challenge for survey researchers. This challenge is especially evident for behaviors such as drinking, in which the potential for harmful behavioral repertoires makes issues such as the temporal ordering of influences a paramount concern.

Acknowledgment

The writing of this manuscript was supported, in part, by grants from the National Institute on Alcohol Abuse and Alcoholism (AA05823), the National Institute on Drug Abuse (DA03395), and the Alcoholic Beverage Medical Research Foundation.

Notes

1. An examination of family influences among adults would also include spouse influences. In general, the research in this area demonstrates an effect of husbands' on wives' drinking patterns (see White et al., 1990).
2. A complete discussion of peer influences in adulthood should include the influence of peers from the neighborhood and house of worship, as well as that of co-workers (see White et al., 1990). In addition to the effect of co-workers, other aspects of the work environment can also serve as reinforcements for alcohol use and heavy drinking (see Plant, 1979).

References

Abrams, D.B., & Niaura, R.S. (1987). Social learning theory. In H.T. Blane & K.E. Leonard, *Psychological theories of drinking and alcoholism* (pp. 131–178). New York: Guilford Press.

Akers, R.L. (1985). *Deviant behavior: A social learning approach.* Belmont, CA: Wadsworth.

Akers, R.L., Krohn, M.D., Lanza-Kaduce, L., & Radosevich, M. (1979). Social learning and deviant behavior: A specific test of a general theory. *American Sociological Review, 44,* 636–655.

Alterman, A.I., Searles, J.S., & Hall, J.G. (1989). Failure to find differences in drinking behavior as a function of familial risk for alcoholism: A Replication. *Journal of Abnormal Psychology, 98,* 50–53.

Atkin, C.K. (1987). Alcohol-beverage advertising. *Control issues in alcohol abuse prevention: Strategies for states and communities. Advances in substance abuse* (Supp. 1, 267–287).

Bandura, A. (1969). *Principles of behavior modification.* New York: Holt, Rinehart and Winston.

Bandura, A. (1977). *Social learning theory.* Englewood Cliffs, NJ: Prentice-Hall.

Bandura, A. (1982). Self-efficacy mechanism in human agency. *American Psychologist, 37,* 122–147.

Barnes, G.M. (1977). The development of adolescent drinking behavior: An evaluative review of the impact of the socialization process within the family. *Adolescence, 12,* 571–591.

Barnes, G.M., Farrell, M.P., & Cairns, A. (1986). Parental socialization factors and adolescent drinking behavior. *Journal of Marriage and the Family, 48,* 27–36.

Blane, H.T. (1988). Research on mass communications and alcohol. *Contemporary Drug Problems,* Spring, 7–20.

Breed, W., & De Foe, J.R. (1984). Drinking and smoking on television, 1950–1982. *Journal of Public Health Policy, 5,* 257–270.

Brook, J.S., Whiteman, M., Gordon, A.S., Nomura, C., & Brook, D.W. (1986). Onset of adolescent drinking: A longitudinal study of intrapersonal and interpersonal antecedents. *Advances in Alcohol and Substance Abuse, 5* (3), 91–110.

Brook, J.S., Whiteman, M., Nomura, C., Gordon, A.S., & Cohen, P. (1988). Personality, family and ecological influences on adolescent drug use: A developmental analysis. *Journal of Chemical Dependency Treatment, l*(2), 123–161.

Brown, S.A., Cramer, V.A., & Stetson, B.A. (1987). Adolescent alcohol expectancies in relation to personal and parental drinking patterns. *Journal of Abnormal Psychology, 96,* 117–121.

Caudill, B.D., & Lipscomb, T.R. (1980). Modeling influences on alcoholics' rates of alcohol consumption. *Journal of Applied Behavior Analysis, 13,* 355–365.

Christiansen, B.A., & Goldman, S. (1983). Alcohol-related expectancies versus demographic/background variables in the prediction of adolescent drinking. *Journal of Consulting and Clinical Psychology, 51,* 249–257.

Christiansen, A., Goldman, S., & Inn, A. (1982). Development of alcohol-related expectancies in adolescents: Separating pharmacolgical from social learning influences. *Journal of Consulting and Clinical Psychology, 50,* 336–344.

Christiansen, B.A., Smith, G.T., Roehling, P.V., & Goldman, M.S. (1989). Using alcohol expectancies to predict adolescent drinking behavior after one year. *Journal of Consulting and Clinical Psychology, 57,* 93–99.

Clayton, R.R., & Lacy, W.B. (1982). Interpersonal influences on male drug use and drug use intentions. *International Journal of the Addictions, 17,* 655–666.

Collins, R.L., Parks, G.A., & Marlatt, G.A. (1985). Social determinants of alcohol consumption: The effects of social interaction and model status on the self-administration of alcohol. *Journal of Consulting and Clinical Psychology, 53,* 189–200.

Coombs, R.H., & Paulson, M.J. (1988). Contrasting family patterns of adolescent drug users and nonusers. *Journal of Chemical Dependency Treatment, 1,*(2), 59–72.

Cooper, M.L., Russell, M., & George, W.H. (1988). Coping, expectancies and alcohol abuse: A test of social learning formulations. *Journal of Abnormal Psychology, 97,* 218–230.

Cox, W.M., Lun, K.S., & Loper, R. (1983). Identifying prealcoholic personality characteristics. In W.M. Cox (Ed.), *Identifying and measuring alcoholic personality traits* (pp. 5–19). San Francisco: Jossey-Bass.

Doby, J.T., (1966). *Introduction to social psychology.* New York: Appleton.

Downs, W.R. (1987). A panel study of normative structure adolescent alcohol use and peer alcohol use. *Journal of Studies on Alcohol, 48,* 167–175.

Dulany, D.E. (1968). Awareness, rules and propositional control: A confrontation with S-R behavior theory. In T.R. Dixon & D.L. Horton (Eds.), *Verbal behavior and general behavior theory.* Englewood Cliffs, NJ: Prentice-Hall.

Falk, J.L., Schuster, C.R., Bigelow, G.E., & Woods, J.H. (1982). Progress and needs in the experimental analysis of drug and alcohol dependence. *American Psychologist, 37,* 1124–1127.

Fontane, P.E., & Layne, N.R., Jr. (1979). The family as a context for developing youthful drinking patterns. *Journal of Alcohol and Drug Education, 24,*(3), 19–29.

Glynn, T.J. (1981). From family to peer: Transitions of influence among drug-using youth. In D.J. Lettieri & J.P. Lundford (Eds.), *Drug abuse and the American adolescent (National Institute on Drug Abuse Research Monograph, Serial No. 38). Washington, DC: U.S. Government Printing Office.*

Goldman, M.S., Brown, S.A., & Christiansen, B.A. (1987). Expectancy theory: Thinking about drinking. In H.T. Blane & K.E. Leonard (Eds). *Psychological theories of drinking and alcoholism* (pp. 181–226). New York: Guilford Press.

Goodwin, D.W. (1985). Genetic determinants of alcoholism. In J.H. Mendelson & N.K. Mello (Eds.), *The diagnosis and treatment of alcoholism* (2nd ed.). New York: McGraw-Hill.

Greenberg, B.S., Fernandez-Collado, C., Graef, D., Korzenny, F., & Atkin, C.K. (1979). Trends in use of alcohol and other substances on television. *Journal of Drug Education, 9,* 243–253.

Hansen, W.B., Graham, J.W., Sobel, J.L., Shelton, D.R., Flay, B.R., & Johnson, C.A. (1987). The consistency of peer and parental influences on tobacco, alcohol, and marijuana use among young adolescents. *Journal of Behavioral Medicine, 10,* 559–579.

Harford, T.C., & Grant, B.F. (1987). Psychosocial factors in adolescent drinking contexts. *Journal of Studies on Alcohol, 48,* 551–557.

Hawkins, J.D., Fine, D.N., & Sweeny, S.L. (1986). *The effects of parental attitudes on teenagers' use of gateway drugs.* Seattle, WA: University of Washington, Center for Social Welfare Research.

Hendricks, R.D., Sobell, M.B., & Cooper, A.M. (1978). Social influences on human ethanol consumption in an analogue situation. *Addictive Behavior, 3,* 253–259.

Herd, D. (1986). Ideology, melodrama, and the changing role of alcohol problems in American films. *Contemporary Drug Problems, 13,* 213–247.

Hewitt, L.E., & Blane, H.T. (1984). Prevention through mass media communication. In P.M. Miller & T.D. Nirenberg (Eds.), *Prevention of alcohol abuse.* New York: Plenum.

Huba, G.H., & Bentler, P.M. (1980). The role of peer and adult models for drug-taking at different stages in adolescence. *Journal of Youth and Adolescence, 9,* 449–465.

Hull, J.G., & Bond, C.R. (1986). Social and behavioral consequences of alcohol consumption and expectance: A meta-analysis. *Psychological Bulletin, 99,* 347–360.

Jacobson, M., Atkins, R., and Hacker, G. (1983). *The booze merchants.* Washington, DC: CSPI Books.

Jessor, R., & Jessor, S.L. (1975). Adolescent development vs. the onset of drinking: A longitudinal study. *Quarterly Journal of Studies on Alcohol, 36,* 27–51.

Jessor, R., & Jessor, S.L. (1977). *Problem behavior and psychological development: A longitudinal study of youth.* New York: Academic Press.

Johnson, R.E., Marcos, A.C., & Bahr, S.J. (1987). The role of peers in the complex etiology of adolescent drug use. *Criminology, 25,* 323–339.

Johnson, S., Leonard, K.E., & Jacobs, T. (1989). Drinking, drinking styles and drug use in children of alcoholics, depressives and controls. *Journal of Studies on Alcohol, 50,* 427–431.

Johnson, V. (1988). Adolescent alcohol and marijuana use: A longitudinal assessment of a social learning perspective. *American Journal of Drug and Alcohol Abuse, 14,* 419–439.

Johnson, V., & Pandina, R.J. (1991). Effects of the family environment on adolescent substance use, delinquency and coping styles. *American Journal of Drug and Alcohol Abuse, 17,* 71–88.

Kandel, D. (1978). Convergences in prospective longitudinal surveys of drug use in normal populations. In D. Kandel (Ed.), *Longitudinal research in drug use: Empirical findings and methodological issues* (pp. 3–38). Washington, DC: Hemisphere-Wiley.

Kandel, D.B. (1985). On the processes of peer influences in adolescent drug use: A developmental perspective. *Advances in Alcohol and Substance Abuse, 4*(3/4), 139–163.

Kandel, D.B., & Andrews, K. (1987). Process of adolescent socialization by parents and peers. *International journal of the Addictions, 22,* 319–342.

Kandel, D., Kessler, R.C., & Margulies, R.S. (1978). Antecedents of adolescent initiation into stages of drug use: A developmental analysis. In D. Kandel (Ed.), *Longitudinal research in drug use: Empirical findings and methodological issues*(pp. 73–99). Washington, DC: Hemisphere-Wiley.

Kaufman, A., Baron, A., & Kopp, R.E. (1966). Some effects of instructions on human operant behavior. *Psychonomic Monograph Supplements, 1* (pp. 243–250).

Kline, R.B., Canter, W.A., & Robin, A. (1987). Parameters of teenage alcohol use: A path analytic conceptual model. *Journal of Consulting and Clinical Psychology, 55,* 521–528.

Knight, J.A. (1980). The family in the crisis of alcoholism. In S.E. Gitlow & H.S. Peyser (Eds.), *Alcoholism: A practical treatment guide.* New York: Grune and Stratton.

Kohn, P.M., & Smart, R. (1984). The impact of television advertising on alcohol consumption: An experiment. *Journal of Studies on Alcohol, 45,* 295–301.

Lanza-Kaduce, L., & Radosevich, M. (1980). *Can a processual theory discriminate among stages of substance use? A test of social learning theory.* Unpublished manuscript. University of Florida, Gainesville.

Leigh, B.C. (1989). In search of the seven dwarves: Issues of measurement and meaning in alcohol expectancy research. *Psychological Bulletin, 150,* 361–373.

Liebert, R.M., Neale, J.M., & Davidson, E.S. (1973). *The early window effects of television on children and youth.* New York: Pergamon Press.

Lied, E.R., & Marlatt, G.A. (1979). Modeling as a determinant of alcohol consumption: Effect of subject sex and prior drinking history. *Addictive Behavior, 4,* 47–54.

Love, S.Q., Ollendick, T.H., Johnson, C., & Schlesinger, S.E. (1985). A preliminary report of the prediction of bulemic behaviors: A social learning analysis. *Bulletin of the Society of Psychologists in Addictive Behaviors, 4,* 93–101.

MacAndrew, C.R., & Edgerton, B. (1969). *Drunken comportment: A social exploration.* Chicago: Aldine.

Maddox, G.L. (1966). Teenagers and alcohol: Recent research. *Annals of the NY Academy of Science, 133,* 856–865.

Mandell, W., Cooper, A., Silberstein, R.M., Novick, J., & Koloski, E. (1963). *Youthful drinking : New York State, 1962.* Staten Island: Wakoff Research Center.

Mann, L.M., Chassin, L., & Sher, K.J. (1987). Alcohol expectancies and the risk for alcoholism. *Journal of Consulting and Clinical Psychology, 55,* 411–417.

Marlatt, G.A., & Rohsenow, D.J. (1980). Cognitive processes in alcohol use: Expectancy and the balanced placebo design. In N.K. Mello (Ed.), *Advances in substance abuse: Behavioral and biological research* (pp. 159–199). Greenwich, CT: JAI Press.

McCord, W., McCord, J., & Gudeman, J. (1960). *Origins of alcoholism.* Palo Alto: Stanford University Press.

Mosher, J.R., & Wallack, L.M. (1981). Government regulation of alcohol advertising: Protecting industry profits versus promoting public health. *Public Health Policy, 2,* 333–353.

Nathan, P.E., & Niaura, R.S. (1987). Primary and secondary prevention of alcohol problems. In W.M. Cox (Ed.), *Treatmant and prevention of alcohol problems: A resource manual* (pp. 333–354). New York: Academic Press.

Needle, R., McCubbin, H., Wilson, M., Reinbeck, R., Lazar, A., & Mederer, H. (1986). Interpersonal influences in adolescent drug use: The role of older siblings, parents and peers. *International Journal of the Addictions, 21,* 739–766.

Newcomb, M.D., & Bentler, P.M. (1986). Substance use and ethnicity: Differential impact of peer and adult models. *Journal of Psychology, 120,* 83–95.

Newcomb, M.D., Huba, G.J., & Bentler, P.M. (1983). Mothers' influence on the drug use of their children: Confirmatory tests of direct modeling and mediational theories. *Developmental Psychology, 19,* 714–726.

Ogborne, A.C., & Smart, R.G. (1980). Will restrictions on alcohol advertising reduce alcohol consumption? *British Journal of Addictions, 75,* 293–296.

Orford, J. (1979). Alcohol and the family. In M. Grant & P. Gwinner (Eds.), *Alcoholism perspectives,* Baltimore: University Park Press.

Pandina, R.J., & Johnson, V. (1989). Family drinking history as a predictor of alcohol and drug consumption among adolescent children. *Journal of Studies on Alcohol, 50,* 245–253.

Pittman, D.J., & Lambert, M.D. (1978). *Alcohol, alcoholism and advertising: A preliminary investigation of asserted associations.* St. Louis: Social Science Institute, Washington University.

Plant, M. (1979). *Drinking careers.* London: Tavistock.

Rachal, J.V., Williams, J.R., Breham, M.L., Cavanaugh, B., Moore, R.P., & Edkerman, W.C. (1975). *A national study of adolescent drinking behavior, attitudes and correlates.* Springfield, VA: National Institute on Alcohol Abuse and Alcoholism (NTIS No. PB-246-002).

Radosevich, M., Lanza-Kaduce, L., Akers, R.L., & Krohn, M.D. (1980). The society of adolescent drug and drinking behavior: A review of the state of the field, Pt. 2. *Deviant Behavior, 1,* 145–69.

Reid, J.B. (1978). Study of drinking in natural settings. In G.A. Marlatt & P.E. Nathan (Eds.), *Behavioral approaches to alcoholism.* New Brunswick, NJ: Rutgers Center of Alcohol Studies.

Rimm, D.C., & Masters, J.C. (1979). *Behavior Therapy.* New York: Academic Press.

Rossiter, J.R., & Robertson, T.S. (1980). Children's dispositions toward proprietary drugs and the role of television drug advertising. *Public Opinion Quarterly, 44,* 316–329.

Rotter, J.B. (1982). *The development and application of social learning theory.* New York: Praeger.

Rubenstein, E. (1978). Television and the young viewer. *American Scientist, 66,* 685–693.

Sher, K.J. (1985). Subjective effects of alcohol: The influence of setting and individual differences in alcohol expectancies. *Journal of Studies on Alcohol, 46,* 137–146.

Simons, R.L., Conger, R.D., & Whitbeck, L.B. (1988). Multistage social learning model of the influences of family and peers upon adolescent substance abuse. *Journal of Drug Issues, 18,* 293–315.

Skinner, B.F. (1971). Social behavior. In E. McGinnies & C.B. Ferster (Eds.), *The reinforcement of social behavior.* Boston: Houghton Mifflin.

Smart, R. (1988). Does alcohol advertising affect overall consumption? A review of empirical studies. *Journal of Studies on Alcohol, 49,* 314–323.

Smart, R.G., Gray, G., & Bennett, C. (1978). Predictors of drinking and signs of heavy drinking among high school students. *International Journal of the Addictions, 13,* 1079–1094.

Smith, G.T. (1989). Expectancy theory and alcohol: The situational insensitivity hypothesis. *Psychology of Addictive Behaviors, 2,* 108–115.

Sobell, L.C., Sobell, M.B., Riley, D.M., Klajner, F., Leo, G.L., Pavan, D., & Cancilla, A. (1986). Effect of television programming and advertising on alcohol consumption in normal drinkers. *Journal of Studies on Alcohol, 47,* 333–340.

Spiegler, D.S. (1983). Children's attitudes towards alcohol. *Journal of Studies on Alcohol, 44,* 545–548.

Stetson, B.A., Brown, S.A., & Beatty, P.A. (1985, August). *Coping with high-risk drinking situations: Abusing versus nonabusing adolescents.* Paper presented at the 93rd Annual Meeting of the American Psychological Association, Los Angeles.

Strickland, D.E. (1982). *Parents, peers, and problem drinking: An analysis of interpersonal influences on teenage alcohol abuse in the U.S.* Paper presented at the 28th International Institute on the Prevention and Treatment of Alcoholism, Munich, Germany.

Strickland, D.E. (1983). Advertising exposure, alcohol consumption and misuse of alcohol. In M. Grant, M. Plant, & A. Williams (Eds.), *Economics and alcohol: Consumption and controls.* New York: Gardner Press.

Strickland, D.E., & Pittman, D.J. (1984). Social learning and teenage alcohol use: Interpersonal and observational influences within the sociocultural environment. *Journal of Drug Issues, 14,* 137–150.

Strickland, D.E., & Wilson, J.B. (1982). *Social learning, family communication, and teenage drinking patterns.* St. Louis: Washington University Social Science Institute Working Paper Series.

Sutherland, E.H. (1947). *Principles of criminology.* Philadelphia: Lippincott.

Vicary, J.R., & Lerner, J. (1986). Parental attributes and adolescent drug use. *Journal of Adolescence, 9,* 115–122.

West, M.O., & Prinz, R.J. (1987). Parental alcoholism and childhood psychopathology. *Psychological Bulletin, 102,* 204–218.

White, H.R. (1982). Sociological theories of the etiology of alcoholism. In E.L. Gomberg, H.R. White, & J.A. Carpenter (Eds.), *Alcohol, science and society revisited.* Ann Arbor: University of Michigan Press and Rutgers Center of Alcohol Studies.

White, H.R., Bates, M.E., & Johnson, V. (1990). Social reinforcement and alcohol consumption. In W.M. Cox (Ed.), *Why people drink: Parameters of alcohol as a social reinforcer.* New York: Gardner Press.

White, H.R., Johnson, V., & Horwitz, A. (1986). An application of three deviance theories to adolescent substance use. *International Journal of the Addictions, 21,* 347–366.

Wilsnack, S.C., & Wilsnack, R.W. (1979). Sex roles and adolescent drinking. In H.T. Blane & M.E. Chafetz (Eds.), *Youth, alcohol, and social policy.* New York: Plenum Press.

Zucker, R.A. (1976). Parental influences upon drinking patterns of their children. In M. Greenblatt & M.A. Schuckit (Eds.), *Alcoholism problems in women and children.* New York: Grune & Stratton.

CHAPTER 10

The Social Integration of Beers and Peers: Situational Contingencies in Drinking and Intoxication

James D. Orcutt

For most people at most times and places, drinking is a thoroughly social activity. A number of surveys show that the bulk of alcohol consumption occurs in group contexts (e.g., Cahalan, Cisin, & Crossley, 1969; Harford, 1979; Room, 1972). Ethnographic and experimental studies demonstrate that drinking behavior and even its subjective effects are heavily influenced by the presence of fellow drinkers (Cavan, 1966; MacAndrew & Edgerton, 1969; Pliner & Cappell, 1974; Sher, 1985). Yet social scientific explanations of drinking patterns and alcohol problems mainly focus either on abstract cultural variables some distance removed from drinking situations (Room, 1976) or on individual characteristics of drinkers analytically isolated from their drinking partners (see Blane & Leonard, 1987). The middle ground between culture and personality—the interpersonal context of alcohol use—is poorly understood theoretically.

This study looks at an especially important element in the social fabric of drinking contexts: peer relationships. The drinking practices of close friends are consistently one of the most powerful predictors of the onset and patterning of alcohol use and abuse among adolescents and young adults (Kandel, 1980; Radosevich, Lanza-Kaduce, Akers, & Krohn, 1980; White, Johnson, & Horwitz, 1986). Unfortunately, the global measures of differential association or drinking networks that are typically employed in survey research fail to provide much insight into contextual effects of peer relationships on behavior and experience in specific drinking situations. This is not a trivial issue for alcohol studies or for deviance theory. Social learning theorists since Sutherland (1956; also see Akers, 1985; Akers, Krohn, Lanza-Kaduce, & Radosevich, 1979) have generally argued that the causal influence of primary group associations on deviant behav-

James D. Orcutt is with the Department of Sociology, Florida State University, Tallahassee, Fla. This chapter was written especially for this book.

ior is psychologically mediated by definitions, attitudes, and motives that the individual internalizes through a process of socialization. However, a number of multivariate analyses of adolescent drinking and drug use show that the mediating effects of subjective definitions or attitudes are quite modest in comparison to the strong, direct effects of friends' usage patterns (e.g., Jaquith, 1981; Johnson, Marcos, & Bahr, 1987; Strickland, 1982). To account for this theoretically problematic finding, Johnson et al. (pp. 336; cf. Jensen, 1972) speculated that survey measures of peer networks may reflect respondents' differential exposure to immediate situational "pressures" to drink or use drugs rather than "association with people, behavior, or definitions that alter one's views of the behavior in question." This line of argument is consistent with one of the few studies of peer relationships as a contextual factor in the development of adolescent drinking patterns. Harford and Spiegler (1983) found that the characteristic transition toward more frequent and heavier drinking from early to late adolescence appears only among teenagers who drink in peer (or both peer and home) contexts, as contrasted to those who drink exclusively in family settings.

Surveys of adult drinking practices indicate that the contextual influences of close friends extend well beyond the notorious "peer pressures" of adolescence. Cahalan et al. (1969, pp. 84–86) found that the only category of social circumstances under which both men and women consistently reported drinking "more than usual" was time spent with close friends. Furthermore, the tendency to drink more liberally in this particular social context increased steadily as a function of respondents' own drinking patterns and was most pronounced among "heavy" drinkers—roughly half of whom reported consuming even more than usual while in the company of close friends. In a later study, Cahalan and Room (1974) included an index of "heavy-drinking context" (e.g., proportion of close friends who "drink quite a bit"; proportion of get-togethers with close friends where alcohol is served) along with numerous individual and background variables in multiple regression analyses of heavy drinking and problematic consequences in a national sample of adult men. Their conclusion closely parallels the results of survey research on adolescent drinking behavior:

> Perhaps the most important finding from these analyses is the strength of prediction shown by heavy-drinking context; for both tangible consequences and problematic intake it shows even a stronger prediction than do the respondent's own attitudes toward drinking. (p. 139)

However, a possible qualification about the contextual influences of close friends emerged in a more recent factor analytic study of contextual and motivational factors in alcohol consumption and drinking problems among male participants in a large investigation of aging (Glynn, LoCastro, Hermos, & Bosse, 1983). Although a factor scale reflecting participation in "masculine activities" (e.g., drinking with friends of the same sex, drinking in bars) was positively related to the respondents' usual level of consumption and self-reported drinking problems, another factor labeled "social settings" (e.g., drinking with friends of both sexes, drinking with friends of the opposite sex) was negatively related to both dependent variables in multiple regression analyses. In short, the contextual effects of friends' drinking appear to vary according to the sex composition of situations in which these respondents tend to drink.

The most systematic attempt to overcome the crucial limitation of all of these studies—the absence of direct evidence on peer relationships in specific drinking situations—is Harford's (1979; 1983) research on daily drinking events reported by a sample of Boston adults over a 4-week period. Taking the most recent drinking event as his unit of analysis, Harford (1983) used analysis of variance to assess the main effects and interactions of physical location, number of drinking companions, and type of drinking companion (i.e., spouse or other relatives, or both; friends only; all other combinations) on the total amount of alcohol that respondents consumed during the event. Among male respondents, main effects for both location and type of companion pointed to higher consumption in bars and, across physical locations, in the presence of friends. Harford did not find main effects for the drinking companion variables among female respondents, but a significant interaction effect—location by type of companion—specified drinking in bars with friends as a relatively high-consumption event among these women. Harford cautioned that these data do "not permit causal interpretations of the effects of context on drinking since context variables are confounded with drinker status [and] heavier drinking adults may . . . prefer friends as drinking companions" (1983, p. 832). Nonetheless, his situational analysis offers a much closer view of the immediate link between peer relationships and increased consumption than previous studies of drinking networks and alcohol use offered.

The study of student drinking in this chapter attempts to extend Harford's approach by focusing on how peer relationships and individual consumption patterns not only separately but also jointly affect specific drinking events. To resolve the interpretive problem that confronted Harford, I employed an independent measure of "drinker status"—usual quantity of consumption—that was obtained before the collection of situational data on drinking events during a 3-day period. Using the additional leverage provided by this measurement strategy, I reexamined

Harford's basic findings on type of drinking companion and the "wetness" of drinking events: Do the positive effects of close friends on situational consumption persist when respondents' usual "drinker status" is taken into account? If so, such a finding would add credence to the argument that the "immediate pressure to [drink] at the risk of social discomfort or rejection" by peers generally increases consumption during drinking events (Johnson et al., 1987, p. 337; also see Rogers, 1970). However, other research I reviewed above suggests that more complex patterns will emerge from these situational data than a simple main effect of type of companion. Most notably, findings presented by Cahalan et al. (1969, pp. 84–86) on social contexts in which adults report drinking "more than usual" clearly take the form of a statistical interaction between context and drinker status; that is, the influence of one particular context—drinking with close friends—increases substantially as respondents' typical level of consumption increases. If this general pattern holds for specific drinking events among the young adults sampled in the present study, then it should yield a significant, nonadditive increment in situational consumption on those occasions where relatively heavy drinkers happen to be in the company of their close friends.

I also examined a possible "dysfunctional" consequence of drinking "in situations where problem behavior is expected and supported" by peers (Norem-Hebeisen & Hedin, 1981, p. 26). As a product of its direct impact on students' behavior, situational peer pressure should indirectly affect drinkers' subjective state of intoxication—occasionally forcing them to exceed the rate or quantity of consumption they are accustomed to handling in other contexts. Therefore, whereas the degree of intoxication should be strongly related to situational consumption in all drinking contexts, a peer pressure explanation implies that the presence of close friends will produce an "excessive" degree of intoxication at higher levels of consumption. This line of reasoning, then, suggests a statistical interaction between social context and situational consumption similar in form to the positive joint effect of peer context and usual quantity on drinking behavior. In this case, the dependent variable is a subjective measure based on students' estimates of how intoxicated they became during specific drinking events. Over and above the main effect of number of drinks consumed, do these estimates of intoxication increase even more sharply in those situations where close friends push drinkers beyond the limits characterizing other social contexts? Although such a question has not been raised in previous surveys of drinking contexts, it is highly relevant to theoretical issues and problematic consequences that flow from the pervasive integration of alcohol use into the peer group relations and activities of young people and adults.

Methods

Survey Design

My analysis of peer relationships and other situational contingencies in drinking and intoxication is based on a project originally designed to investigate temporal variations in affect and alcohol use (see Orcutt & Harvey, 1989). This survey employed a combination of questionnaire and diary methods. Student respondents were initially contacted during regular meetings of undergraduate classes at Florida State University and administered a seven-page questionnaire. This in-class questionnaire included summary measures of the respondents' typical drinking behavior (e.g., "usual" quantity and "average" frequency of use) as well as conventional measures of sociodemographic (sex, age), subjective (attitudes toward drinking), and interpersonal (friends' use of alcohol) variables that have been related to patterns of alcohol use in previous surveys of adolescents and adults.

After completing the questionnaire, the respondents were given a sealed "situational record" to be taken home and returned during a subsequent class meeting (where they would receive a payment of $2). Instructions on the outside cover of this diary-like instrument asked them to open and complete it at a specific time a little more than 3 days later. Half of the classes—and, thus, respondents—completed situational records covering 3 weekdays (Tuesday, Wednesday, Thursday); the other half filled out weekend records (Friday, Saturday, Sunday). Each page of this 15-page instrument referred to a different 4-hour block of time during the 3-day period, beginning with 8:00 a.m. to noon of the 1st day (Tuesday or Friday) and continuing through 4:00 p.m. to 8:00 p.m. of the 3rd day (Thursday or Sunday). For each 4-hour block, the respondents were asked to write a brief description of their "main activities and the situations [they] were in during this time" and to rate on 9-point scales the overall degree to which they perceived these activities and situations as stressful, boring, and so forth. Most important for present purposes, the respondents were instructed that if they drank "any alcoholic beverages" during a particular time block, they were to provide the following details about the drinking event: (1) the number of drinks of beer, liquor, or wine, or all of these, they had consumed during the 4-hour period; (2) how intoxicated they had become (from 1 [*not at all intoxicated*] to 9 [*extremely intoxicated*]); and (3) who, if anyone, had been drinking with them. Using a multiple-option checklist for the latter item ("check all that apply"), respondents could specify several types of close friends ("closest friend of same sex," "closest friend of opposite sex," "other close friends"), family members ("spouse," "other family"), or "acquaintances"—as well as solitary drinking ("by myself").

Sample

In all, 328 respondents completed both the in-class questionnaire and the situational record. The respondents who received the weekday version of the situational record came from six classes that met on a Monday-Wednesday-Friday schedule. Out of 257 students who completed an in-class questionnaire on a Monday meeting of these classes, 164 returned a usable weekday situational record the following Friday—a completion rate of 63.8%. Students from five Tuesday-Thursday classes received the weekend record on Thursday and were asked to return it the following Tuesday. Quite by coincidence, 164 of these 212 respondents returned a usable weekend record, for a completion rate of 77.4%.

These weekday and weekend subsamples were very similar on virtually all background characteristics and drinking variables included on the in-class questionnaires. Although over twice as many weekday respondents fell in the youngest age category (18 and under, 22.4% vs. 8.5%, $p < .01$), the age means for the weekday ($M = 20.0$, $SD = 2.54$) and weekend ($M = 20.2$, $SD = 2.54$) subsamples did not differ significantly. Furthermore, these subsamples did not differ significantly on sex (overall, 65.6% women), race (84% white, 14.5% black), or other status characteristics (marital status, religious affiliation, parents' education and annual income). The marginal distributions for "average" frequency of drinking and "usual" quantity consumed were quite similar across subsamples. Relatively few respondents had completely abstained from alcohol use during the previous year (7.3% weekday vs. 8.5% weekend), whereas less than a third of each subsample reported drinking more than once a week on the average (29.2% vs. 28.9%) or usually consuming four or more drinks *per* occasion (31.3% vs. 30.7%).

More interesting similarities between these subsamples emerged in analytical comparisons based on global measures of two theoretically relevant variables. The mean scores on a 5-point Likert measure of negative-to-positive attitude toward alcohol were identical in the weekday ($M = 3.5$, $SD = 1.17$) and weekend ($M = 3.5$, $SD = 1.14$) subsamples. Moreover, responses to a network measure of peer relationships—the number of the respondent's four closest friends who were said to "use alcohol at least once a month"—also showed identical means of 3.2 in each subsample ($SD = 1.15$ vs. 1.24). Within each subsample I performed multiple regression analyses to estimate the structural effects of peer relationships on respondents' drinking patterns while controlling for sex and attitude toward alcohol. The unstandardized partial coefficients for number of close friends (net of the significant main effects of sex and attitude) as a predictor of respondents' frequency of alcohol use were highly significant ($p < .001$) and almost equal in magnitude for the weekday ($b = .38$) and weekend ($b = .40$) subsamples. A parallel multiple regression analysis of usual quantity of consumption revealed that the unstandardized partial effect of number of close friends on quantity was exactly the same in each subsample ($b = .29$, $p < .001$). Thus, this preliminary analysis not only shows that the weekday and weekend subsamples are comparable in some very fundamental ways but also replicates the direct structural effect of differential association with close friends who drink on personal patterns of alcohol use that Johnson et al. (1987) interpreted as a reflection of situational peer pressure.[1]

Situational Measures

The data on situational consumption and intoxication come from two pairs of time blocks on the situational records that span the "prime time" hours for the daily onset of drinking among young adults (see Orcutt & Harvey, 1989; Sinnett & Morris, 1977). Specifically, the data are from weekday respondents who began drinking between the hours of 4:00 p.m. and midnight on Tuesday or Wednesday, or both, and from weekend respondents who began drinking during these same hours on Friday or Saturday, or both. For each of these 4 days, I constructed a set of independent cases (drinking events) suitable for regression analysis by combining data from two adjacent time blocks (4:00 p.m. to 8:00 p.m. and 8:00 p.m. to midnight) after excluding those respondents who had been drinking in the time block immediately preceding. For instance, my analysis of drinking events during Tuesday evening is based on the following 21 cases from the weekday subsample: (1) 10 respondents who drank during the 4-hour block between 4:00 p.m. and 8:00 p.m. on Tuesday but did not drink during the prior 4-hour block from noon to 4:00 p.m.; and (2) another 11 respondents who drank between 8:00 p.m. and midnight but did not drink between 4:00 p.m. and 8:00 p.m. Similarly, my analysis of Wednesday drinking events is based on 45 independent cases involving respondents who drank between 4:00 p.m. and 8:00 p.m. or, alternatively, between 8:00 p.m. and midnight on Wednesday. However, it is important to note that these 45 cases include 13 from respondents who also drank on Tuesday evening, for I excluded cases only when drinking occurred in an immediately adjacent time block. Likewise, I selected 83 independent cases of Friday drinking events and 71 independent cases of Saturday drinking events from the situational records completed by the weekend subsample, but 47 of the latter cases come from respondents who also

TABLE 1
Respondent Characteristics and Situational Measures for Sets of Drinking Events between 4:00 and 8:00 p.m. or 8:00 p.m. to Midnight—Means (and Standard Deviations) by Subsample and Day of Event

		WEEKDAY SUBSAMPLE		WEEKEND SUBSAMPLE	
		Tuesday (N = 21)	**Wednesday (N = 45)**	**Friday (N = 83)**	**Saturday (N = 71)**
Questionnaire Measures					
SEX	1 = Male 0 = Female	.62 (.50)	.47 (.50)	.29 (.46)	.37 (.49)
QUANT	1 = One drink 2 = 2–3 drinks 3 = 4–6 drinks 4 = More than 6	2.67 (.73)	2.50 (.73)	2.30 (.73)	2.39 (.67)
Situational Measures					
DRINKS	n = Total drinks during four-hour time block	3.48 (3.33)	3.76 (2.70)	3.59 (2.83)	3.90 (2.67)
INTOX	Range = 1 (not at all intoxicated) to 9 (extremely intoxicated)	2.95 (2.42)	3.20 (1.96)	3.63 (2.37)	3.91 (2.33)
PEER	1 = Closest friend of same sex and/or other close friends present; 0 = Both absent	.67 (.48)	.76 (.43)	.67 (.47)	.69 (.47)
OPPSEX	1 = Closest friend of opposite sex present; 0 = Absent	.24 (.44)	.40 (.49)	.47 (.50)	.49 (.50)
Means for Other Contexts (1 = Present; 0 = Absent)					
CLOSEST FRIEND OF SAME SEX		.43	.53	.38	.41
OTHER CLOSE FRIENDS		.52	.58	.54	.58
ACQUAINTANCES		.24	.27	.41	.44
SPOUSE		.05	.04	.05	.01
OTHER FAMILY		.05	.07	.07	.08
BY MYSELF		.24	.04	.02	.07

drank on Friday evening. Table 1 presents distributional data on the respondents involved in these four sets of cases and summarizes the situational measures from the 4-hour time blocks in which they began to drink.

As is shown in the first row of Table 1, male students—who constitute only a third of the weekday subsample—were overrepresented among those respondents who drank on Tuesday or Wednesday evenings. Also, in comparison to Friday or Saturday drinkers, the respondents involved in weekday drinking events tended to score higher on the questionnaire measure of "usual" quantity of consumption (QUANT). However, the measure of situational consumption (DRINKS) in the third row of Table 1 reveals that the actual number of drinks consumed was lowest for Tuesday drinking events ($M = 3.48$) and the highest for Saturday drinking events ($M = 3.90$). Furthermore, respondents who participated in weekend drinking reported higher levels of intoxication (INTOX) during the 4-hour period than respondents who drank on weekday evenings.

The bottom half of Table 1 describes the two contextual measures employed in my analysis—PEER and OPPSEX—and reports the means for other contextual options included on the situational record. The binary PEER context variable is a composite measure based on the specific options of drinking with "closest friend of same sex" or with "other close friends," or with both. If a respondent checked either or both of these options, the drinking event was scored as 1 on the PEER variable (or 0 if neither was checked). As the mean scores for this variable indicate, two thirds or more of the drinking events for each evening occurred in the company of peers. The other contextual measure I used in my analysis, OPPSEX, is simply the option for drinking with "closest friend of opposite sex." Although this type of drinking companion was present in only a fourth of the Tuesday drinking events, slightly less than half of the drinking events on other evenings included the respondent's closest friend of the opposite sex. Finally, the means for other contexts show that drinking with acquaintances was fairly common, especially on weekends, but that drinking with relatives was relatively rare among these students. Overall, very few respondents reported solitary drinking, although five of the Tuesday cases ($M = .24$) did involve that activity.

Results

I begin by addressing the basic questions of how peer relationships and personal consumption patterns affect alcohol use and intoxication during the onset of drinking events on different days of the week. First, for each daily subset of drinking events, I regressed the measure of situational consumption—number of DRINKS—on two predictors: QUANT, the respondent's usual quantity of consumption, and PEER, the presence (vs. absence) of close friends in the drinking situation. Pertinent to Harford's (1983) concern about the confounding of contextual factors with drinker status in his analysis, these main effects equations estimate the immediate influence of PEER context on DRINKS while controlling the status variable, QUANT.

TABLE 2
Unstandardized Coefficients (Standard Errors in Parentheses) for Regression Analyses of Situational Consumption (DRINKS) and Intoxication (INTOX) by Day of Drinking Event

A. Drinks Regressed on Quant, Peer, and Significant Quant × Peer Interaction

	TUESDAY		WEDNESDAY		FRIDAY		SATURDAY	
	Main	Interact	Main	Interact	Main	Interact	Main	Interact
QUANT	2.22**	−.35	1.40**		1.47**	.25	1.74**	
	(.83)	(1.22)	(.47)		(.37)	(.67)	(.44)	
PEER	2.54*	−7.82*	.40		1.96**	−1.95	.02	
	(1.26)	(4.11)	(.78)		(.58)	(1.89)	(.64)	
QUANT × PEER		3.95**		N.S.		1.74**		N.S.
		(1.51)				(.80)		
CONSTANT	−4.13	2.46	−.23		−1.13	1.60	−.16	
	(2.37)	(3.25)	(1.41)		(.95)	(1.57)	(1.15)	
R^2	.41	.58	.18		.27	.31	.19	
F	6.17**	7.72**	4.42**		14.72**	11.84**	7.70**	
[*df*]	[2/18]	[3/17]	[2/41]		[2/80]	[3/79]	[2/65]	

B. Intox Regressed on Drinks, Peer, and Significant Drinks × Peer Interaction

	TUESDAY		WEDNESDAY		FRIDAY		SATURDAY	
	Main	Interact	Main	Interact	Main	Interact	Main	Interact
DRINKS	.47**		.40**	.69**	.49**	.90**	.70**	.94**
	(.11)		(.11)	(.20)	(.07)	(.25)	(.06)	(.12)
PEER	1.38*		.48	1.88*	.81*	1.85**	.40	1.69**
	(.78)		(.59)	(1.01)	(.45)	(.76)	(.37)	(.64)
DRINKS × PEER		N.S.		−.40*		−.45*		−.33**
				(.24)		(.26)		(.14)
CONSTANT	.41		1.35	.35	1.31	.43	.85	−.06
	(.61)		(.63)	(.86)	(.38)	(.64)	(.39)	(.53)
R^2	.63		.26	.31	.44	.46	.65	.68
F	15.56**		7.29**	6.04**	31.74**	22.61**	61.30**	45.96**
[*df*]	[2/18]		[2/41]	[3/40]	[2/80]	[3/79]	[2/65]	[3/64]

*$p < .10$
**$p < .05$

Next, I estimated DRINKS with an "interaction" equation that included a third predictor: a product term for the joint, nonadditive effect of QUANT × PEER. In line with findings in Cahalan et al. (1969), I expected this interaction, if significant, to be positive—indicating that the situational impact of PEER on DRINKS is stronger among respondents with heavier drinking patterns. Panel A of Table 2 presents unstandardized coefficients from the four main effects equations for DRINKS and from two interaction equations that yielded a significant product term.

In the two simplest cases, the main effects equations for Wednesday and Saturday evening drinking events reveal that PEER context had very little effect on DRINKS net of the significant main effect of respondents' scores on QUANT. The unstandardized coefficients for the latter variable indicate that estimated consumption increased on these two evenings by roughly one and one-half DRINKS for each step on the 4-point questionnaire measure of QUANT (Wednesday $b = 1.40$; Saturday $b = 1.74$).

In contrast, the estimates for Tuesday evening show that most of the influence of QUANT is absorbed by a significant QUANT × PEER interaction. Specifically, QUANT was virtually unrelated to estimated DRINKS for seven respondents in nonpeer contexts ($b = -.35$ when PEER = 0), five of whom drank by themselves and none of whom consumed more than two DRINKS in the 4-hour drinking period. However, among 14 respondents who drank with their best friend of the same sex or other close friends, each step on QUANT (observed range = 2–4) yields an estimated increase of four DRINKS during these Tuesday evening time blocks ($b = 3.95$ when PEER = 1).

The interaction equation for Friday is based on a relatively large number of respondents and provides a more stable estimate of the positive nonadditive effect of QUANT × PEER on situational consumption. Among respondents in the lowest category of drinker status (QUANT = 1), the estimated number of DRINKS consumed with close friends ($\hat{Y} = 1.64$ when PEER = 1) is slightly smaller than the comparable estimate in nonpeer contexts ($\hat{Y} = 1.85$ when PEER = 0). From that point on, situational consumption with close friends increasingly exceeds consumption in nonpeer contexts by a relative increment of $b = 1.74$ for each step of increase in QUANT. For Friday respondents with the heaviest drinking pattern (QUANT = 4), the estimate of situational DRINKS is three times higher in peer contexts ($\hat{Y} = 7.61$ when PEER = 1) than in nonpeer contexts ($\hat{Y} = 2.60$ when PEER = 0).

I used a similar hierarchical regression strategy to estimate the situational ratings of INTOX for each day as a function of the main effects of PEER and DRINKS and, subsequently, the nonadditive effect of DRINKS × PEER. Of course, I expected the respondents' levels of intoxication to be strongly related to the number of drinks they consumed during the 4-hour time blocks. The key question here is whether the subjective experience measured by INTOX was in some way contingent on the presence—or pressures—of close friends. The results in Panel B of Table 2 provide some interesting answers to this question. Owing to the theoretical importance of these findings, I employed a liberal criterion ($p < .10$) for the inclusion of statistically significant interactions.

Tuesday evening drinking events constitute the only case in which the DRINKS × PEER product term was nonsignificant as a predictor of INTOX. A strong main effect of DRINKS ($b = .47$) predicts a half-point increase on the 9-point scale of INTOX for each drink the respondents consumed during Tuesday time blocks. In addition, a modest main effect of PEER elevates the estimated level of INTOX by slightly more than a point ($b = 1.38$) net of situational consumption.

For the other three evenings, the nonadditive effect of DRINKS × PEER on INTOX was significant and *negative* in direction. The interaction equations containing this effect are very similar in structure for all 3 days. Therefore, to aid interpretation of these findings, Figure 1 graphs the observed mean ratings of INTOX within categories of DRINKS and PEER (vs. "NON-PEER" when PEER = 0) during Saturday drinking events against linear estimates calculated from the Saturday interaction equation, in which the DRINKS × PEER product term is significant at $p < .05$.

As Figure 1 makes clear, the positive main effect of number of DRINKS on ratings of INTOX—the upward trend in estimated and observed values for both PEER and NON-PEER conditions—remains strong and interpretable in the interaction equations for Saturday and, by the same token, for Wednesday and Friday. However, the negative DRINKS × PEER interaction captures a difference in the slopes of this relationship for the PEER and NON-PEER contexts. In relation to the conditional slope for NON-PEER drinking events, the initially higher but flatter slope of INTOX on DRINKS in the PEER condition models an "excess" of intoxication that is *specific to lower levels of situational consumption.* These conditional slopes converge as the number of DRINKS increases, and they eventually cross—specifically, at five drinks for the Wednesday and Saturday estimates, and at four drinks for the Friday estimates. Thus, the presence of close friends appears to have intensified the experience of intoxication when, and only when, individuals consumed a relatively small amount of alcohol during drinking events on these three evenings.[2]

Whereas PEER context is an influential situational contingency in either the amount of drinking or the level of intoxication for every evening in this study, parallel regression analyses of the effects of OPPSEX—drinking with the closest friend of the opposite sex—yielded few significant results. The most noteworthy findings for this contextual variable emerged in the Wednesday evening time blocks. Controlling for the main effect of QUANT ($b = 1.37, p < .01$), I found that OPPSEX significantly increased situational consumption by one and two-thirds DRINKS ($b = 1.67, p < .05$). Furthermore, a significant QUANT × OPPSEX interaction ($b = 1.51, p = .08$) for Wednesday indicated that the presence of a friend of the opposite sex had a more pronounced positive effect on DRINKS among respondents with heavier drinking patterns. In short, the situational effects of OPPSEX on

FIGURE 1
Estimated (line) and observed (column) relationship between INTOX and DRINKS by PEER (vs. NON-PEER) context for Saturday drinking events

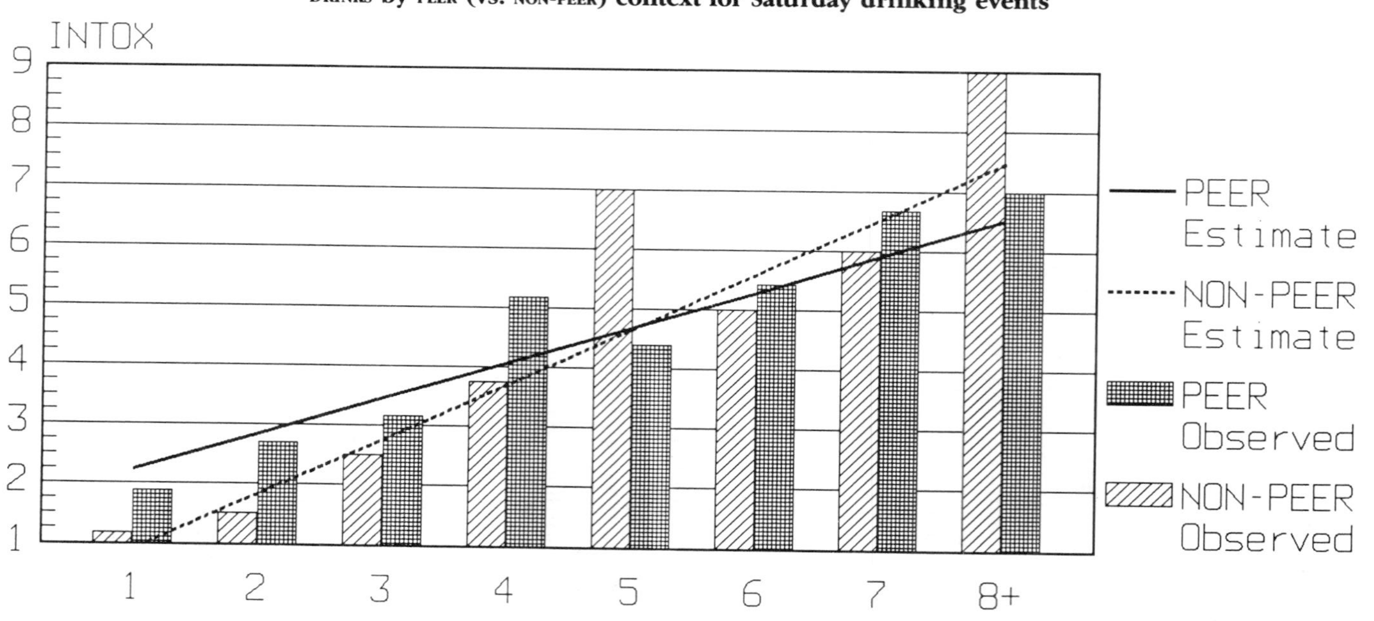

Wednesday consumption closely resemble the main and interaction effects of PEER context on Tuesday and, especially, Friday consumption. OPPSEX also had a modest negative main effect ($b = -.73$, $p = .07$) on INTOX net of the strong main effect of DRINKS ($b = .55$, $p < .01$) during Friday drinking events; but, aside from that finding, there was little evidence of the moderating effects of opposite sex contexts that Glynn et al. (1983) noted in their study of adult male drinkers.

This last point raises a final question: Is the sex composition of respondents reporting drinking events somehow implicated in the contextual effects of PEER or OPPSEX on situational consumption and intoxication? For the most part, the answer is no. The inclusion of a dummy variable for SEX of respondent in both the main effects and interaction equations for PEER context did not substantially alter the significant results in Table 2 for Tuesday, Friday, or Saturday drinking events. In fact, there was no instance in which SEX had a significant effect on DRINKS net of the main effect of QUANT, or on INTOX net of the main effect of DRINKS. Even so, a control for SEX reduced the interaction of DRINKS × PEER on INTOX to nonsignificance ($p = .11$) for Wednesday drinking events.

Likewise, the addition of SEX to the main effects and interaction equations for OPPSEX had little impact on the original results for this variable. That is, OPPSEX still had a significant positive main effect on DRINKS for Wednesday ($p < .05$) net of QUANT and SEX, and the positive nonadditive effect of QUANT × OPPSEX remained significant ($p = .07$) with SEX included in the interaction equation. The negative main effect of OPPSEX on Friday INTOX was also unaffected ($p = .06$) by simultaneous controls for DRINKS and SEX.

Conclusion

The results show that peer relationships play an important part in the situational patterning of consumption and intoxication among student drinkers. However, the significant interaction effects involving the measures of social context are difficult to reconcile with a simple model of situational peer pressure. These effects indicate that situational consumption by lighter drinkers—respondents who scored lower on QUANT—was generally unresponsive to any pressures from their close friends. Yet the presence of close friends of the same or opposite sex had a relatively strong effect on the number of drinks consumed by heavy drinkers, at least during the evening hours of Tuesday, Wednesday, and Friday. If anything, then, these results suggest that heavier drinkers, both here and in the Cahalan et al. (1969) study, feel freer to exercise their preferences in the company of close friends than in other, more restrained contexts (see Burns, 1980).

The significant DRINKS × PEER interactions in the analyses of INTOX are even more problematic for a peer pressure explanation of drinking behavior. Rather than contributing to inordinate intoxication at higher levels of consumption, the presence of close friends appears to enhance the intoxicating effects of a relatively few drinks. This pattern, which emerges in the data for 3 different evenings, resembles and extends the findings of experimental research on the ability of social settings to alter the subjective effects of alcohol (Pliner & Cappell, 1974; Sher, 1985). Whereas laboratory studies typically compare solitary settings to group settings composed of previously unacquainted subjects, the present results indicate that a particular type of natural social setting—the peer context—facilitates the experience of intoxication up to a certain point, where higher doses of alcohol override this contextual effect.

The mechanistic concept of situational pressure might be of some relevance in explaining the initiation of alcohol or drug use among young people from proscriptive backgrounds or with strongly negative attitudes toward the activity (Rogers, 1970; Orcutt, 1987), but it is clearly inadequate as an account of more routine forms of interpersonal influence and shared experience among drinking companions. Other investigations of the structure of drinking groups (Bruun, 1962; Storm & Cutler, 1981), modeling influences on drinking behavior (Caudill & Marlatt, 1975; DeRicco & Niemann, 1980), and situational norms or expectancies toward the effects of alcohol (Roizen, 1983; Sher, 1985) point to more promising explanations for the contextual patterning of consumption, intoxication, and alcohol problems. In concert with observational and experimental work, survey research can play a useful role in this theoretical enterprise by moving beyond static measures of drinking networks and focusing more directly on the integration of alcohol use into the daily activities that reaffirm friendship and other social relationships.

Notes

1. The standardized coefficients (*betas*) for these preliminary analyses are of some interest in light of previous research that has focused on the relative strength of social network and attitudinal predictors of drinking behavior (e.g., Akers et al., 1979; Jaquith, 1981; Johnson et al., 1987). The following standardized partial effects of friends' use, attitude toward alcohol, and sex on frequency and quantity of alcohol use in each subsample can be meaningfully compared within equations (all coefficients significant at $p < .05$ except Sex for Frequency/Weekday):

Dependent Variable and Subsample	Friends' Use	Attitude toward Alcohol	Sex
Frequency/Weekday	*beta* = .38	.39	.11
Frequency/Weekend	*beta* = .44	.33	.12
Quantity/Weekday	*beta* = .36	.43	.16
Quantity/Weekend	*beta* = .37	.37	.18

2. By basing these analyses on situational measures within fixed, 4-hour time blocks during which respondents first began drinking on a given evening, I have truncated (right-censored) some drinking episodes that extended over two or more adjacent time blocks in the situational records. Although this within-block analytical design avoids the complexities of changing values for contextual measures and varying durations of drinking events across time blocks, it may not be sensitive to longer-run positive effects of peer pressure at higher levels of consumption on the eventual degree of intoxication attained in drinking episodes that extend across time blocks. To obtain at least a rough assessment of these long-run effects, I reestimated the Saturday INTOX equations using data from 12 respondents who began drinking on that day between 4:00 p.m. and 8:00 p.m. and who continued to drink during the next 4-hour block, 8:00 p.m. to midnight. I summed the total number of DRINKS they consumed in these two time blocks, based the PEER variable on a contrast between eight respondents who drank with peers in both time blocks (PEER = 1) versus the other four respondents (PEER = 0), and used these combined measures to predict the degree of INTOX attained in the 8:00 p.m.-to-midnight time block. The results of this 8-hour analysis were very similar to the Saturday INTOX estimates reported in Table 2 and shown in Figure 1. Most important, the negative nonadditive effect of DRINKS × PEER on INTOX ($b = -37$) was similar in form to the 4-hour interaction term and, even in this analysis of 12 cases, was nearly significant at $p = .11$.

References

Akers, R.L. (1985). *Deviant behavior: A social learning approach* (3rd ed.). Belmont, CA: Wadsworth.

Akers, R.L., Krohn, M.D., Lanza-Kaduce, L., & Radosevich, M. (1979). Social learning and deviant behavior: A specific test of a general theory. *American Sociological Review, 44,* 636–655.

Bland, H.T., & Leonard, K.E. (Eds.). (1987). *Psychological theories of drinking and alcoholism.* New York: Guilford.

Bruun, K. (1962). The significance of roles and norms in the small group for individual behavioral changes while drinking. In D.J. Pittman & C.R. Snyder (Eds.), *Society, culture, and drinking patterns.* New York: Wiley.

Burns, T.F. (1980). Getting rowdy with the boys. *Journal of Drug Issues, 10,* 273–286.

Cahalan, D., Cisin, I.H., & Crossley, H.M. (1969). *American drinking practices.* New Brunswick, NJ: Rutgers Center of Alcohol Studies.

Cahalan, D., & Room, R. (1974). *Problem drinking among American men.* New Brunswick, NJ: Rutgers Center of Alcohol Studies.

Caudill, B.D., & Marlatt, G.A. (1975). Modeling influences in social drinking: An experimental analogue. *Journal of Consulting and Clinical Psychology, 43,* 405–415.

Cavan, S. (1966). *Liquor license: An ethnography of bar behavior.* Chicago: Aldine.

DeRicco, D.A., & Niemann, J.E. (1980). In vivo effects of peer modeling on drinking rate. *Journal of Applied Behavior Analysis, 13,* 149–152.

Glynn, R.J., LoCastro, J.S., Hermos, J.A., & Bosse, R. (1983). Social contexts and motives for drinking in men. *Journal of Studies on Alcohol, 44,* 1011–1025.

Harford, T.C. (1979). Ecological factors in drinking. In H.T. Blane & M.E. Chafetz (Eds.), *Youth, alcohol and social policy.* New York: Plenum.

Harford, T.C. (1983). A contextual analysis of drinking events. *International Journal of the Addictions, 18,* 825–834.

Harford, T.C., & Spiegler, D.L. (1983). Developmental trends of adolescent drinking. *Journal of Studies on Alcohol, 44,* 181–188.

Jaquith, S.M. (1981). Adolescent marijuana and alcohol use: An empirical test of differential association theory. *Criminology, 19,* 271–280.

Jensen, G.F. (1972). Parents, peers and delinquent action: A test of the differential association perspective. *American Journal of Sociology, 78,* 562–575.

Johnson, R.E., Marcos, A.C., & Bahr, S.J. (1987). The role of peers in the complex etiology of adolescent drug use. *Criminology, 25,* 323–340.

Kandel, D.B. (1980). Drug and drinking behavior among youth. In A. Inkeles, N.J. Smelser, & R.H. Turner (Eds.), *Annual review of sociology* (Vol. 6). Palo Alto, CA: Annual Reviews.

MacAndrew, D., & Edgerton, R.B. (1969). *Drunken comportment: A social explanation.* Chicago: Aldine.

Norem-Hebeisen, A., & Hedin, D.P. (1981). Influences on adolescent problem behavior: Causes, connections, and contexts. In *Adolescent peer pressure: Theory, correlates, and program implications for drug abuse prevention* (National Institute on Drug Abuse, DHHS Publication No. ADM 83-1152). Washington, DC: U.S. Government Printing Office.

Orcutt, J.D. (1987). Differential association and marijuana use: A closer look at Sutherland (with a little help from Becker). *Criminology, 25,* 341–358.

Orcutt, J.D., & Harvey, L.K. (1989). *Stress and alcohol use: A situational analysis of weekday and weekend drinking.* Paper presented at the annual meeting of the Midwest Sociological Society.

Pliner, P., & Cappell, H. (1974). Modification of affective consequences of alcohol: A comparison of social and solitary drinking. *Journal of Abnormal Psychology, 83,* 418–425.

Radosevich, M., Lanza-Kaduce, L., Akers, R.L., & Krohn, M.D. (1980). The sociology of adolescent drug and drinking behavior: A review of the state of the field: Part II. *Deviant Behavior, 1,* 145–169.

Rogers, E.M. (1970). Group influences on student drinking behavior. In G.L. Maddox (Ed.), *The domesticated drug: Drinking among collegians.* New Haven: College & University Press.

Roizen, R. (1983). Loosening up: General-population views of the effects of alcohol. In R. Room & G. Collins (Eds.), *Alcohol and disinhibition: Nature and meaning of the link* (National Institute on Alcohol Abuse and Alcoholism Research Monograph No. 12, DHSS Publication No. ADM 83-1246). Washington, DC: U.S. Government Printing Office.

Room, R. (1972). Drinking patterns in large U.S. cities: A comparison of San Francisco and national samples. *Quarterly Journal of Studies on Alcohol* (Suppl. No. 6, 28–57).

Room, R. (1976). Ambivalence as a sociological explanation: The case of cultural explanations of alcohol problems. *American Sociological Review, 41,* 1047–1065.

Sher, K.J. (1985). Subjective effects of alcohol: The influence of setting and individual differences in alcohol expectancies. *Journal of Studies on Alcohol, 46,* 137–146.

Sinnett, E.R., & Morris, J.B. (1977). Temporal patterns of the use of non-prescribed drugs. *Perceptual and Motor Skills, 45,* 1239–1245.

Storm, T., & Cutler, R.E. (1981). Observations of drinking in natural settings: Vancouver beer parlors and cocktail lounges. *Journal of Studies on Alcohol, 42,* 972–997.

Strickland, D.E. (1982). "Social learning and deviant behavior: A specific test of a general theory": A comment and critique. *American Sociological Review, 47,* 162–167.

Sutherland, E.H. (1956). A statement of the theory. In A. Cohen, A. Lindesmith, & K. Schuessler (Eds.), *The Sutherland papers.* Bloomington: Indiana University Press.

White, H.R., Johnson, V., & Horwitz, A. (1986). An application of three deviance theories to adolescent substance use. *International Journal of the Addictions, 21,* 347–366.

SECTION III

Social Structure, Subcultures, and Drinking Patterns

A. AGE, GENDER, AND SEXUAL ORIENTATION

Introductory Note

Drinking patterns in contemporary society have been studied in relation to basic structural variables of age and gender and to a limited extent in reference to sexual orientation. Alcoholism is known to vary strikingly in incidence between men and women and to find fullest expression in the middle and later years of life. Moreover, there is substantial variation in the sex ratios of alcoholism from one sociocultural milieu to another. Given the pronounced variation in drinking patterns along the spectrum of age, gender, and sexual orientation, sociological inquiry into these differences has proceeded at an accelerating pace over the last three decades. Mention should be made here of the pioneering efforts of the original California social research group of Ira Cisin, Wendell Lipscomb, and Genevieve Knupfer to collect systematic epidemiological information on the drinking patterns of Americans (Wiener, 1981). Their work has been continued by the Alcohol Research Group, in Berkeley, California, whose director is the sociologist Robin Room.

American society in the 20th century has evidenced a strong protective tendency toward its youth as evidenced by such legislation as child labor laws, marriage laws, and compulsory school regulations, and by various states enacting a minimum purchase age for alcoholic beverages of 21 years of age. The drinking patterns of adolescents and young adults have been studied more extensively that those of any other parts of the age pyramid, and no one chapter can do justice to the scope and complexity of these investigations. However, in chapter 11, Robert J. Pandina, Helene Raskin White, and Gail Gleason Milgram present an overview and assessment of the drinking practices of youth in the United States. These authors focus on the changing drinking patterns of young people over the last two decades and on the emergence of extensive recreational drug use by young people since the late 1960s. Furthermore, they address the crucial questions of societal intervention into the youthful drinking behavior.

At the other end of the age spectrum, attention to drinking patterns among the elderly began to receive extensive scientific attention only in the last decade. Ronald Akers and Anthony La Greca in their original research, based on in-home interviews with elderly individuals in four different types of communities and reported in chapter 12, continue the process of mapping the topography of drinking behavior by older adults. Their data are analyzed within the theoretical framework of social learning (see chapter 9 in this volume) and they test the relevance of various psychosocial theories of alcohol use among older individuals. Their finding that life event stresses do not cause nondrinkers to begin drinking or drinkers to increase their intake in this population stands in bold contrast to that of some reports in the popular press.

Mention of female drinking patterns in the 1962 edition of *Society, Culture, and Drinking Patterns* was basically relegated to a footnote referring to the work of Edith Lisansky on the female alcoholic. The intervening three decades have witnessed a major growth of empirical data on the drinking behavior of women in U.S. society, but this information in comparison with that for men is still rather modest. Edith Lisansky Gomberg in chapter 13 reviews the current state of knowledge on female drinking, problem drinking, and alcoholism, as well as some of the research findings from her study of women who are being treated for alcoholism and their matched controls.

Some areas of alcohol research remain relatively unexplored. Sexual orientation in reference to drinking patterns, both nonproblematic and problematic, falls in this category. Peter Nardi's original article on the posited relationships between alcoholism and homosexuality is reprinted as chapter 14 with an update prepared by him in 1990. Research data on the drinking patterns of gay men and lesbians still remain sparse, although some information has been obtained in the course of research on the AIDS pandemic.

The four chapters in the subsection highlight the need for studies of drinking behavior in various segments of the population subdivided by age, gender, and sexual orientation. In the next subsection we explore other relevant subdivisions.

References

Wiener, C. (1981). *The politics of alcoholism: Building an arena around a social problem.* New Brunswick, NJ: Transaction Books.

CHAPTER 11

Assessing Youthful Drinking Patterns

ROBERT J. PANDINA, HELENE RASKIN WHITE, AND
GAIL GLEASON MILGRAM

American society has long maintained a strong investment in the manner in which its youth are raised. Part of that investment is reflected in formal and informal social regulation of many behaviors that emerge in that period loosely characterized as adolescence. Of special concern are those behaviors that are perceived as potential threats to the development of competencies believed to be essential for the assumption and performance of adult roles. Often behaviors closely regulated in this developmental epoch are considered, paradoxically, normative behaviors and, in some cases, "rights" in adulthood. Drinking alcoholic beverages is clearly one of those adult norms our post–World War II society has attempted to understand and regulate among its youth. Whether we have been successful in either regard remains as open and important a question for us, today, as it was for Maddox (1962) who wrote on this topic in the original version of this volume nearly 30 years ago.

In today's social climate, the perceived threat that youthful drinking poses to successful transition to adulthood has taken expression in several important action trends including enactment of nationwide legal restrictions prohibiting drinking behavior among those under the age of 21, continued refinement and expansion of education programs aimed at prevention, emergence of specialized treatment programs for youthful problem drinkers, and, of course, continuation of a broad research agenda aimed both at understanding the nature and extent of youthful drinking practices and, to a lesser extent, at determining whether present interventions are successfully coping with any threats that arise from such practices.

Robert J. Pandina, Helene Raskin White, and Gail Gleason Milgram are with the Center of Alcohol Studies, Rutgers University, New Brunswick, N.J. This chapter is a revised and updated version of "Methods, Problems, and Trends in Studies of Adolescent Drinking Practices," by Robert J. Pandina, which appeared in *Annals of Behavioral Medicine, 8* (2-3), pp. 20-26, 1986.

The purpose of this chapter is to provide an overview and assessment of a limited segment of information provided by the current research agenda. Specifically, we wish to identify trends in use patterns during the decades of the 1970s and 1980s in order to examine the scope and status of problematic drinking among youth at present. Further, we wish to identify and comment about the types of interventions currently in vogue, which are intended to reduce the size of the threat and to deal with the casualties. The scope of our review is, of necessity, limited. Such important themes as etiological factors and mechanisms are not addressed in detail here. Further, these issues are of such scope, complexity, and import (literally, dozens of subtopics encompassing thousands of relevant citations) that separate consideration is warranted. Overviews of these topics can be found elsewhere (e.g., Forney, Forney, & Ripling, 1989; Halebsky, 1987; Jessor & Jessor, 1977; Kandel, 1978, 1980; Lettieri, 1975; Lettieri, Sayers, & Wallenstein, 1980; Mayer, 1988; Pandina & Schuele, 1983).

Challenges in Characterizing Drinking Practices

Selecting a Database

Constructing a reliable and valid topological map of drinking practices among youth across time requires a careful distillation of information from a range of complimentary sources each having advantages and weaknesses. No single study or, for that matter, study type can provide a complete picture in itself. For example, relying upon school-based surveys or clinical samples introduces selective biases unintentionally distorting the picture. School-based surveys cannot reflect the patterns of dropouts and absentees (estimated to average about 15–20% nationally but to be possibly as high as 30–35% in some locales) who have been shown to be more extreme in their use (e.g., Clayton & Ritter, 1985; Fagan & Pablon, 1989; Kandel, 1975a: Mensch & Kandel, 1988). Clinical samples may differ as a function of programmatic biases (e.g., eligibility criteria, referral source, program sponsor, and philosophy) and limited sample size, and they typically represent young people at more extreme ends of the use continuum, some of whom may display other confounding pathological conditions. Complexity may be compounded by the need to consider key subsamples that may not appear in either school or clinical samples (e.g., special ethnic populations such as American Indian reservation youth, ghetto or barrio youth gangs, or newly arrived Asian immigrants). Further, it is important to recognize that for a given individual or for a class of individuals drinking may not remain constant and, thus, it may be valuable to observe individuals at varying time intervals so as to capture the dynamic element missing in the snapshot view provided by single-assessment cross-sectional studies.

There are, of course, the typical methodological problems of comparing across studies that use different assay techniques (e.g., survey versus personal interview), varying format of salient probes, varying definitions of key constructs (e.g., of *problem drinking*), varying determination of the reliability and validity of self-report data, and so on. Although not all these can be addressed at length in this review, two key problems will be discussed, namely, the reliability and validity of self-report data and techniques for assessing problem drinking status.

In painting our composite portrait of youthful drinking we have drawn from several key data sources. To provide a broad framework, we rely on relatively large-scale, nationally based, cross-sectional studies employing probability samples of youths enrolled in schools or of those residing in households. These studies include an ongoing effort to assess high school seniors (between 16,000 and 18,000 each year, drawn from approximately 140 schools, with a subsample of 2,400 from each class being reassessed periodically) beginning with the class of 1975 and continuing to the present (e.g., Johnston, O'Malley, & Bachman, 1989) and a series of cross-sectional national household surveys (e.g., Fishburne, Abelson, & Cisin, 1980; Gallup, 1985, 1987; National Institute on Drug Abuse, 1986, 1989) that include subsamples of youth. This class of study provides both continuity across time and an element of representativeness to our portrait. This view is supplemented further by other studies employing national sampling frames but more limited in time. Included in the time-limited studies are those of Rachal and colleagues (e.g., 1980) conducted in 1974 and 1978, involving a school-based national probability sample. We have also taken advantage of heroic summaries of earlier surveys in documenting baseline patterns (e.g., Maddox & McCall, 1964; Blane & Hewitt, 1977).

The picture provided by these types of cross-sectional studies are enriched by consideration of results of a number of specially designed ongoing longitudinal studies of drinking and drug-taking behavior. Overviews of a number of these studies are available (e.g., Jessor & Jessor, 1977; Kandel, 1978; Newcomb & Bentler, 1988; Pandina, Labouvie, Johnson, & White, 1988; Pandina, Labouvie, & White, 1984; Rachal et al., 1980; Robins, 1984). Given the ongoing nature of these studies, results continue to emerge that provide important clues to changing trends.

Validity of Self-Report Measures

The vast majority of cross-sectional and longitudinal studies of youthful drinking rely upon the self-reports of individuals about their current and past behavior. The question arises, of course, what the utility of such estimates in providing a valid and reliable record of true drinking practices is. A number of investigations have been aimed specifically at this problem and, for the most part, have found encouraging results (Rouse, Kozel, & Richards, 1985). Even among difficult samples (e.g., arrestees, juvenile delinquents, clients in treatment) reports appear to be quite reliable (Ball, 1967; Single, Kandel & Johnson, 1975). Studies specifically aimed at judging the veracity of drug use reports have confirmed self-reports through use of urine assays of drug metabolites. Problems, of course, are not completely absent. Memory distortions for events, especially remote happenings, occur. Descriptions of consequences of use may be subject to individual interpretations. Nonetheless, it appears that, within reasonable and specifiable limits, individuals are both willing and able to provide useful reports of their drinking behavior (Elliott & Huizinga, 1988).

Defining "Problem Drinking" among Adolescents

Perhaps the most difficult methodological problem encountered in providing an accurate assessment in this area is the lack of consensus regarding operational definitions or conceptual guidelines about what constitutes an adolescent "problem drinker" or "alcoholic." Recent estimates suggest that there are over 300,000 teenage alcoholics (ages 13 to 17) in the United States (Kinney & Leaton, 1987). At the same time, however, many clinicians believe that alcoholism among teens characterized by physical dependence and withdrawal phenomena is quite rare.

The fact that a young person is referred to or treated at an alcohol treatment facility does not necessarily indicate that he or she displays an adult form of alcoholism. It should be kept in mind that adolescents may encounter problems with alcohol because of a single acute episode rather than as the result of a chronic condition. For example, problems that occur because of drunkenness, such as drunk driving or having a fight with a friend, even though not indicative of a repetitive pattern of problems, are often precursors to referral. Many of these problems, while potentially serious, result from inexperience and carelessness rather than a chronic disease state. Young people drink less often than adults, but when they drink, they tend to drink in larger amounts (Harford & Mills, 1978). Thus, they are at risk for suffering from acute effects (e.g., blackouts and hangovers) as well as for displaying behavioral concomitants of intoxication (e.g., belligerence). Also, it should be kept in mind that the risk of social and interpersonal consequences is heightened because adolescents are legally under age and their drinking often violates the law as well as social norms.

For these reasons, it would seem more fruitful to focus on teenage problem drinking than teenage alcoholism. Traditional measures, however, have not been adequate (see Harrell, Kapsak, & Cook, 1986; Mayer & Filstead, 1980; O'Gorman, Stringfield, & Smith, 1977; White, 1989). Identifying problem drinking in adolescents requires departure from several aspects of the widely used model of adult alcoholism. The progressive nature of the disease, medical complications, physical dependence, and other chronic symptoms are less clearly associated with adolescent alcohol problems. However, although physiological addiction during adolescence may be rare (Filstead, 1982), with many adult signs and symptoms absent, under no circumstances should it be concluded that adolescent drinking problems do not require treatment.

Measures of frequency, quantity, and variability of alcohol use may be necessary but are not sufficient indicators to diagnose the problem status of adolescent drinking accurately. Information about negative consequences attributable to drinking, in addition to information on drinking patterns, would seem most appropriate for diagnosing problem drinking (see also Sadava, 1985; White, 1987). Especially important to identify are characteristics of drinking and its consequences that are dysfunctional for satisfactory growth and resolution of key developmental tasks (Harrell & Wirtz, 1989).

Until recently, screening tools for diagnosing problem drinking among adolescents were virtually absent. The Adolescent Alcohol Involvement Scale (AAIS) (Mayer & Filstead, 1980) represents the only instrument with clinical applicability that had been developed specifically for adolescents (Jacobson, 1983). Yet it has been criticized for its lack of conceptual clarity, which is desirable to distinguish between problem and nonproblem drinking. Also, although the test has been found to be multidimensional, it uses scores as if it were unidimensional (Moberg, 1983). Other adolescent scales (e.g., Donovan & Jessor, 1978) had been developed primarily for research purposes and had limited clinical utility, especially because of their arbitrary cutoffs and incomplete lists of consequences (White, 1989).

Owing to this scarcity of appropriate and easily administered assessment tools, health professionals have had to rely on unstandardized interview techniques or tests designed for adults (Harrell et al., 1986; Owen & Nyberg, 1983). For example, the DSM-III (American Psychiatric Association, 1980) criteria for alcohol dependence among adults are typically used by clinicians without giving any particular consideration to special problems of adolescents (MacFarland, 1983). Other adult assessment instruments, such as the AUI (Wanberg & Horn, 1983) and the ADS (Edwards & Gross, 1976; Skinner, 1981), were constructed on clinical samples of adult alcoholics and do not have documented applicability for teenagers in their early years of problem drinking. Some researchers have studied the validity of using adult assessment instruments such as the MacAndrew scale with young problem drinkers, but they have found high rates of both false positives and negatives (Moore, 1984).

Recently a few screening instruments have been developed specifically for adolescents, but they require further evaluation. A few examples are the RAPI (Rutgers Alcohol Problem Index) (White & Labouvie, 1989), a 23-item paper-and-pencil instrument dealing with symptoms or consequences (or both) of alcohol use; the PIC (personal involvement with chemicals scale) (Winters & Henly, 1987), a 29-item paper-and-pencil instrument that assesses psychological involvement in alcohol and drug use; and Drinking and You (Harrell & Wirtz, 1989), a 24-item screening instrument covering loss of control and psychological, social, and physical symptoms of problem drinking.

Emergence of Illicit Drug Use: A Special Confound

The spread in use of marijuana, cocaine, and other illicit drugs among Americans, especially young Americans, observed over the past 20 years represents one of the more striking and well-documented examples of social change since the World War II era (Kandel, 1980). Although the exact date of onset of the "epidemic" is difficult to pinpoint, many believe that 1965 (± 2 years) is a reasonable bench mark (O'Donnell, Voss, Clayton, Slatin, & Room, 1976). Since 1965, drug use patterns have undergone rapid and often erratic transition. Both epoch and cohort effects are apparent and have been a source of continuing interest (Johnston et al., 1989). Initially, the epidemic begun in the mid-1960s was confined substantially to older adolescents and young adults (17–24 years of age) but by the mid-to-late 1970s the phenomenon had spread to younger adolescents. The high-water mark in terms of greatest exposure and prevalence rates for the more popular illicit drugs (e.g., marijuana) was reached in the late 1970s and early 1980s (1979 ± 2 years) although some substances (e.g., cocaine) reached their peak rates more recently.

Brief consideration of prevalence rates among school-attending high school seniors provides an instructive example of the manner in which the so-called drug epidemic may affect our understanding of drinking patterns of contemporary youth (Johnston et al., 1989). The use of marijuana rose from a base line annual prevalence rate of less than an estimated 10% of high school seniors in 1965 to a high of slightly over 50% in 1979, and it currently stands at about 33%. Intensive use, defined as daily use, was less than 2% in 1969, rose to a high of 10.7% in 1978, and currently stands at about 3%. Similar nonlinear trends are observed for many, but not all, drugs between 1975 and 1988; exceptions include the steady increase in cocaine use and the steady decrease in hallucinogen use.

Several further observations are necessary. Virtually all marijuana users are also alcohol users, although the degree of overlap among user groups displaying different intensity patterns for each substance (e.g., once-a-month vs. daily use) may vary (e.g. Pandina & Schuele, 1983; Pandina et al., 1988; Pandina, White, & Yorke, 1981) Of course, not all alcohol users are necessarily marijuana users; likewise heavy users of marijuana may be moderate in their alcohol ingestion. Generally, alcohol use (and often tobacco use) precedes marijuana use, and use of other substances, if it occurs, typically follows alcohol and marijuana use, a fact that has led to a well-replicated developmental stage model of substance use (Kandel, 1975b).

These observations lead to several conclusions. First, many adolescent alcohol users may also be concurrently classified as drug users, although the intensity patterns for the substances may differ markedly. Thus, attempts at categorizing alcohol use patterns and delineating use-related sequelae must carefully weigh interactive influences of other drug use. Consequences attributed to a given alcohol use pattern may be, in fact, related to concurrent marijuana use or to interactive effects. This problem extends beyond epidemiological projections of use to studies of etiological factors and creates difficulties in achieving appropriate diagnoses and selecting appropriate treatment strategies.

Historic fluctuations in use patterns of various drugs occur with regularity, although the practical significance of these changes is difficult to evaluate. Increases or decreases in use of any given substance *X* may exacerbate or ameliorate estimates of problems attributed to substance *Y.* For example, the dramatic increase in marijuana use observed in the early 1970s was believed by some public health officials to have been paralleled by a net decrease in alcohol use, and attention was deflected away from alcohol. In the mid-1970s and again in the mid-1980s public perception again focused on the alcohol epidemic among youth. Careful inspection of the alcohol use patterns of those epochs demonstrated rather stable pictures of use practices. In addition, mini-epidemics, short in duration but intensive in impact (e.g., the recent PCP scare and the current "crack" cocaine phenomenon) may be superimposed on more stable use patterns. These are, of course, tip-of-the-iceberg issues and should serve as illustrations of the complexity engendered in studying use practices and problems. These factors also play havoc with those charged with making and executing policy as well as those counting and treating the casualties.

Contemporary Youth: Where Do We Stand?

The task of presenting a capsule view of alcohol use practices of American adolescents is facilitated by the many thorough reviews, cited above, which summarize and contrast research results from cross-sectional and longitudinal studies. The results summarized by these extensive reviews and national survey reports, supplemented by more in-depth analyses permitted by longitudinal data, provide a broad basis for summarizing current use practices. Historical trends can also be usefully abstracted.

The onset of drinking for the majority of young people occurs before their completing the first year of high school, the average age of first use being about 13. By the end of high school, peak exposure levels are observed, with lifetime prevalence figures exceeding 90%. After age 13, the greatest increases in exposure rate are observed between ages 14 and 15, with rates reaching asymptotic levels at about age 17 (cf.: Grade 11). Exposure levels are similar for males and females. The age of onset and rates of exposure have been relatively stable since the mid-to-late 1960s.

By the time peak exposure rates are achieved, about 85% of a given sample have used alcohol at least once in the previous year and 65–70% drink at least monthly. Males exceed females in monthly use (by about 7–9%). These rates have fluctuated somewhat during the past 10 years, the greatest changes being seen in the steady increase in proportion of females in this category. Of the older monthly users, the majority drink on two to four occasions a month, with females lagging behind males in use frequency. Younger drinkers tend to drink with somewhat less frequency, the figures for younger males approximating those for older females.

Differences are more apparent as one examines more intensive alcohol use patterns, with larger proportions of older male adolescents exhibiting heavier drinking. A number of indicators have been used to characterize intensive or heavy drinking among youth. Each measure adds a useful dimension to attempts to delineate the nature and estimate the magnitude of extreme drinking. The most recent estimates indicate that about 4–5% of high school seniors drink on a daily basis, males exceeding females by a ratio of nearly 2.5:1.

Intensity measures that employ quantity-frequency per unit times estimates are preferable in that these methods permit an indication of degree and length of intoxication episodes. While drinking by youths itself, irrespective of quantity, is considered a problem by many (a point of view that has merit), it can be argued that impairment is more accurately estimated by drinking that leads to higher levels of intoxication, such as those associated with blood alcohol concentrations (BACs) in excess of .10%. It is useful to note that a 150-lb male who ingests five 12-oz servings of standard beer (or the equivalent) over a 2-hour period would achieve a peak blood alcohol concentration in the range of .10%. Equivalent ingestion by a 130-lb female would result in a concentration of .14%; in both cases acute intoxication would result. A similar point can be made for other substances with relatively large effective dose ranges (e.g., marijuana), which, to some extent, differentiates them from other drugs (e.g., hallucinogens, PCP), the ingestion of even small doses of which leads to marked and lengthy acute episodes and dramatic impairment.

Although such indices of heavy use are useful, difficulties are unavoidable. The major problem is capturing a truly representative picture of use intensity across the use continuum in a single characterization (Pandina et al., 1981). For example, Rachal et al. (1980) classified about 15% of 15–18-year-olds as heavier users, defined as individuals who drank at least once a week and had five or more drinks on each occasion. These researchers characterized as moderate drinkers those who drank (1) at least once a week and had one drink or less on each occasion; (2) three to four times a month and had two to four drinks each time; or (3) no more than once a month and had five or more drinks each time. About 17% fell into the moderate category. This category is more heterogeneous than the heavier users classification and contains individuals who on occasion drink heavily.

In addition to daily drinking, Johnston et al. (1989) probed episodic heavy drinking, using a criterion measure similar to that selected by Rachal's group. High school seniors were asked whether they had consumed five or more drinks in a row (indicating probable intoxication) in the two weeks before the survey. Thirty-five percent answered positively; again, males exceeded females (43% vs. 27%) in such episodes. Note that the more recent data indicate a modest decline (about 3-6%) from the high-water marks of 1978 to 1982. Thirty percent of all subjects also indicated that most or all of their friends experienced frequent intoxication (at least weekly); 56% reported that they were frequently around people who often became intoxicated or high. The latter trends of perceived peer use have remained relatively consistent across time.

Students in Johnston's surveys were also asked to rate the perceived level of risk associated with various use patterns. About 40–45% reported risks associated with weekend binge drinking (having five or more drinks in a row on both days); about 25–30% judged drinking a few drinks daily risky, and over two thirds found that daily heavy drinking was risky. Interestingly, these results were in contrast to student approval of various use behaviors. Almost three quarters disapproved of moderate daily drinking (one to two drinks a day), and about 60% disapproved of weekend binge drinking. About 90% disapproved of daily heavy drinking.

Extrapolations from these "softer" indicators to hard estimates of high-risk drinking behaviors and beliefs are themselves risky. However, these data, viewed within the context of firmer data about drinking patterns, provide an arena for speculation about the relative size and nature of the at-risk youthful population.

It is tempting to characterize the at-risk population in terms of a three-dimensional cone with three plateaus that become increasingly smaller as the cone reaches its peak, the broadest and lowest plateau being composed of occasional and relatively regular drinkers but with intensity levels that vary considerably. The second, and smaller, level consists of about 30–40% of the population who are regular drinkers whose drinking episodes are more frequent, who are more likely to experience higher levels of intoxication, and whose closest friends share similar drinking styles. The smallest, but highest, plateau (the crown of the cone encompassing the peak) includes relatively heavy drinkers (about 5–10%) of the population who have frequent episodes of intoxication and who are most likely to be weekend drinkers. The probability of experiencing negative consequences across several life domains increases as one moves closest to the crown of the cone. To continue the image, it should be recognized that the microstructure (surface area, topography of the plateau, etc.) of the cone's regions may vary across time but that the macrostructure (a three-tiered cone) appears to remain relatively constant.

This description, though highly speculative, does fit reasonably well with information provided from several studies across the years, many of which tapped different populations and used different mapping devices and strategies. For example, Rachal et al. (1980) classified about 25–30% of 10th to 12th graders as alcohol misusers. This classification included those who experienced frequent intoxication or negative consequences, or both, in at least one life area. When misusers were restricted to only those experiencing both intoxication and negative consequences, the pool of misusers was reduced to about 18–20%. When a more restrictive definition was used, the pool dropped to about 5–10% of the sample.

The broad picture presented above for drinking patterns has remained relatively stable for the last 10–15 years and, in many ways, parallels many observations made for adult drinking patterns (Cahalan, 1970). Historical trends, especially in the area of daily drinking and attitudes toward heavy drinking, are apparent. At present, daily drinking is at its lowest point in the last 10 years, and more youths see extreme forms of drinking as risky and disapprove of such behavior. A noticeable downward trend is apparent in the more recent survey data; however, the stability of the deceleration remains to be seen. The year 1979 appears to have been the high-water mark, for example, 7% of high school seniors reported daily drinking; this figure is in contrast to the just under 6% in 1975 and 1982 and the about 4–5% characteristic between 1984 and 1989.

Although the trends reported above are consistent for the majority of the adolescent population, sociodemographic differences are evident. For example, decreases in several prevalence rates indicators (e.g., yearly, monthly, daily) are more apparent for large and small metropolitan compared to nonmetropolitan areas. Regional differences are also leveling. The southern and western regions of the country are beginning to catch up to the northeast and north central regions, where moderate and heavy use have traditionally been higher.

The gender gap may also be narrowing; however, differences in intensive forms of use are still apparent. Ethnic differences appear to be maintaining the trends observed over the past 10–15 years, namely black and Hispanic youths (especially females) lag behind white youths in many use indicators including intensive drinking.

Interestingly, differences on a number of drinking parameters are apparent when one contrasts individuals on the basis of academic aspirations. College-bound students lag behind non-college-bound students on many markers of use and problem intensity. As a grouping, they may be imagined as a constricted dip on each plateau of the cone. Equally interesting, monitoring of these individuals across time indicates that once they enter college they begin to look like their non-college-attending peers (Johnston et al., 1989).

No appreciable changes in the well-established beverage preferences of youth are noteworthy—beer continues to be the overwhelming favorite of youth. Wine and distilled spirits are distant seconds, although the margin of preference between these two has narrowed.

The stability of trends in alcohol use stands in marked contrast to the more erratic picture observed for illicit drugs over the course of the past 10–15 years. Although it may be too early to present a thorough evaluation of the impact of illicit drug use upon alcohol patterns, it does appear that most illicit drug use is superimposed upon a rather stable base of alcohol use. This important observation is further supported by studies of developmental sequencing and multiple use patterns (Clayton & Ritter, 1985; Pandina et al., 1981; Pandina et al., 1988).

Prevention and Education

Currently interest in preventing alcohol problems in the United States has been renewed; Gusfield (1982) has attributed this new activism to changes in alcohol policies. If one defines *prevention* in terms of forestalling the occurrence of alcohol problems, then environmental conditions, situational opportunities and restraints, normative standards, and legal controls are all significant factors for consideration (Gusfield, 1982). Five models for prevention have been delineated: (1) the prohibition model, which uses legal coercive controls to prohibit manufacture, distribution, and consumption; (2) the regulation model, which focuses on reducing alcohol consumption by regulating economic and physical availability; (3) the alternative model, which emphasizes alternative sources of satisfaction; (4) the deterrence treatment model, which develops systems for early identification of individuals who are exhibiting problem behavior; and (5) the information and education model, which disseminates information about the consequences of using alcohol, with the goal being moderate use or abstinence (Mandell, 1982).

Dembo (1984) discussed several of the limitations of current alcohol and drug prevention models, including the following: (1) programs often overlook the critical factors of the individual's social and cultural experiences; (2) many programs take place in institutional settings rather than in neighborhood locations; (3) programs often focus on white, middle-class youths and do not target non-white nonyouths; (4) programs emphasize nonuse, which is counter to the fact that most young people have been introduced to alcohol and drugs and that many continue to use them; and (5) the majority of programs are designed for large audiences and use one-way communication techniques such as lectures and films.

Since the many programs designed toward control of, education about, or elimination of alcohol problems have not prevented their occurrence (Moskowitz, 1989; Nathan 1983), many observers have called for more comprehensive approaches. For example, Perry and Jessor (1985) offer a three-dimensional conceptual model for preventing drug abuse and promoting health. One dimension includes the domains of health (i.e., physical, psychological, social, and personal); the second dimension consists of two health promotion strategies (i.e., weakening health-compromising behavior and strengthening health-enhancing behavior); and the third dimension involves three foci of intervention (i.e., behavior, personality, and the environment).

There are several models for prevention, but the information and education model is the most utilized model of prevention in the United States; however, alcohol and drug education has not lived up to society's expectation as an effective method for reducing alcohol-related problems. A major reason for the disappointing results may lie in the fact that prevention objectives have been poorly defined and most programs have failed to clarify their educational goals. Weisheit, Hopkins, Kearney, and Mauss (1984) investigated the appropriateness of the school setting for prevention programming. Their results indicated that although students might not be receptive to much of what occurs in school, they do not "tune out" alcohol education. But although the school may be an appropriate setting, unless several important features are present, a program cannot be successful. First, a statement of philosophy and program policies, along with the goals to be achieved by the program, needs to be developed and made known to all school personnel. In addition, educators need to be trained in content and strategies to conduct the program effectively. Similarly, parents need to be informed of, and involved in, the alcohol and drug education program and policies. An innovative program in the Seattle schools focuses on training teachers and parents in identifying risk factors in alcohol and drug abuse and in developing skills to deal with these issues. In this program, the relationship between children and the significant adults in their lives is considered more important than the given curriculum (Hawkins, 1986).

The needs of students are also critical to the success of any program, and yet they are often overlooked. Regardless of whether the student is a drinker or a nondrinker, alcohol will be involved in some, if not most, of the activities in which adolescents participate. Therefore adolescents should have the opportunity to discuss these events, clarify their position, and follow through with appropriate decisions. The educational atmosphere needs to facilitate this process and motivate the discussion of alternatives in an open and nonthreatening atmosphere. Since the peer group is also part of the classroom, it is essential that the students be given the opportunity to hear and react to each other. Perry's (1987) review of adolescent prevention programs suggests that peer-led strategies may be an efficient and effective method for drug abuse and health promotion programs.

Alcohol and drug education programs must be comprehensive and integrated throughout a child's years in school. Too often these programs have been one-shot events that occur on the day of the prom or the day before graduation. Alcohol and drug education should occur every year and address the needs of the students at the various levels of development. It is also important that these programs be conducted by educators in the school. When outsiders are brought in to do 1- or 2-day programs, they do not have a rapport with the students and they leave when they are finished, creating a dilemma for the young person who has a personal or family problem and would like to speak with the lecturer. Thus, having the program rooted in the school allows for easier access to sources of help.

The Student Assistance Program (SAP) is experiencing dramatic growth. The SAP is grounded in the school system's policy and is designed to assist students who are experiencing problems. Information, advice, assistance, and referral are all part of the SAP. In order for SAPS to be effective, parents and community members need to be aware of the program components. In addition, school systems need to tackle the issue of teachers and administrators with alcohol or drug problems because they serve as potential role models for students.

Given the complexity of programmatic issues, evaluation of alcohol and drug education programs is often not conducted; and when an evaluation is carried out, it is frequently deficient. Hewitt's (1981) review of alcohol education programs concluded that few of the programs evaluated effects on behavior. Because making an impact on behavior is a major goal of the education, little can be stated regarding the effectiveness of alcohol education programs if this variable is not measured. Further, in order to measure behavioral effects, the follow-up needs to last at least 2 to 3 years to allow these effects to occur (Bry, 1978; Staulcup, Kenward, & Frigo, 1979). In her review of alcohol and drug education programs, Milgram (1987) noted that too often programs are expected to produce dramatic effects in relatively short periods of time; when they fail to do so, support for the programs is often withdrawn.

Adolescent Treatment

Before the 1970s, adolescent treatment facilities were almost nonexistent. The prevailing view was that adolescents who had problems with alcohol or drugs, or both, were experiencing difficulties adjusting and would "outgrow" the problem. Adolescents who received treatment did so in adult facilities with a small group of peers. However, at many facilities individuals under 18 were not admitted for several reasons: facilities were concerned about working with what was perceived as a difficult population, legal issues complicated matters, and the special needs of this population (i.e., educational, social, recreational) drained the facilities' resources. The growth of adolescent facilities during the 1970s and 1980s has been noteworthy. Most private treatment programs have started adolescent units, and many adolescent facilities are freestanding. Though some programs have taken adult treatment models and superimposed them on adolescents, the need to focus on the adolescent is motivating many programs to develop specific strategies for this population.

Adolescent treatment must take into account the differences between adolescent and adult problem drinkers. For example, most treatment personnel recognize that the adolescent problem drinker is probably a multiple substance user. Also, adolescents have higher energy levels than adults and often require a different form and time period of treatment than adults (Coar, 1984). Family involvement in the treatment process is often encouraged because it increases the likelihood that the adolescent will remain in treatment and also motivates the family to work on issues and relationships that have been affected by the alcohol or drug (or both) problem (Weidman, 1987).

Adolescent treatment also needs to be concerned with the important developmental issues of adolescents. For young adults, therapy is involved with socialization: learning to deal with identity issues, family issues, and the development of interpersonal skills and alternative methods of coping (Gomberg, 1986). Education is a primary factor in adolescent treatment. Since most adolescents are still attending school, it is essential that the facility provide a remedial education program to assist individuals in learning the content and skills that were missed during the period of drinking or drugging and to accommodate the ongoing educational needs of the students. Recreational and social activities also need to be built into the program so that the young patient may develop healthy strategies for fun and enjoyment. Methods of coping with problems also need to be incorporated into the program. Since alcohol and drugs often were used as coping mechanisms, the adolescent alcohol or drug patient needs to learn communication skills, techniques for dealing with anger, strategies for saying no, and so forth.

Most adolescent facilities have individual and group counseling, as well as family sessions. They also conduct an intake physical and provide other medical services to assist the young person in achieving good health. Most also incorporate self-help fellowship programs into their programs and refer patients to them after treatment, because the majority of programs view treatment as ongoing and do not consider a person cured upon release. Abstinence from the use of substances during treatment is required by virtually all treatment programs, and abstinence from all psychoactive drugs for life is seen as the ideal goal (Hoffmann, Sonis, & Halikas, 1987).

Selecting and integrating treatment components (i.e., outpatient, inpatient, day-care, etc.) for an adolescent requires an adequate knowledge about the person and advance treatment planning. Adolescents who require detoxification or pose other acute medical problems require medically supervised inpatient treatment before being placed into less-supervised programs. When a person's desire for treatment is strong, when his or her acute medical or psychiatric problems are nonexistent or minimal, when he or she is willing to abstain from all mood-altering drugs, and when the appropriate family members can be engaged in treatment, then it is likely that outpatient treatment will be an appropriate alternative. Day treatment programs and alternative school day programs are also important components in the complete services profile. Because many facilities are located in remote areas or otherwise difficult to reach, transportation is often a key factor for the adolescent client.

Though most would agree that adolescent treatment is beneficial, there are debates regarding key issues; for example, whether lifetime abstinence is a realistic goal, what the criteria for successful treatment should consist of, what the adequate duration of treatment is (relatively short to quite lengthy time frames have been suggested), whether it is appropriate to use locked units with this population, whether siblings who use alcohol or drugs, or both, should be required to enter treatment, and so on. The role of the family and that of the school and the responsibilities of each institution are also being addressed. The need for outcome studies of the young people who have completed treatment is apparent. Blum (1987) has noted that though short-term therapy may be beneficial, the long-term impact is uncertain. Though most professionals accept the fact that adolescents constitute a more difficult treatment population than adults, a clear understanding of impediments to treatment is lacking. Program evaluation (including process assessments) and comparisons across program types need to be implemented.

Conclusions

Crafting an adequate description of the topography of drinking among adolescents is as difficult today as it was for Maddox (1962) 3 decades ago. Presently, we have the benefit of a broader range of studies, cross-sectional and longitudinal in methodology, encompassing varying subpopulations, and utilizing the latest in computer-assisted technology. We still struggle, however, with the same types of issues, not the least of which is how to achieve a consensus view of what constitutes an adolescent "problem drinker" or "alcoholic." We have made progress in this very important area. Current concepts need to be tested with both normal and clinical populations and adapted for use in diagnosis and treatment.

The current apparent downward shift in size of what appears to be the at-risk pool of problematic users should be carefully monitored. Current lower levels may represent a temporary decline only. Further, a number of school, household, and community-based studies may miss tapping important subpopulations (e.g., dropouts or absentees, minorities), the inclusion of which could add a significant topographical feature to our three-dimensional cone. Of course, the inclusion of drug use patterns, largely unknown to and unanticipated by Maddox at the time, makes any characterization just that much more difficult.

Two very important and lasting trends merit special mention. First, efforts at prevention of alcohol-related phenomena—running the gamut from drinking, itself, to situational drinking (e.g., drinking and driving), to problematic drinking—are currently receiving more attention than at any other time in the past 30 years. Intervention modalities have been expanded beyond the traditional passive education mode to include peer-oriented methods and have extended beyond concern about agent alcohol, alone, to concern about the general behavioral health of the teenager. This trend, in itself, is a significant leap forward and should be encouraged.

The second trend is the movement toward conceptualization of treatment of the adolescent problem drinker as more than a simple linear extension of that provided for the adult. We fully expect this healthy situation to mature in the near term. Especially important are carefully designed programs tailored specifically to meet the unique needs of the adolescent.

Acknowledgments

The preparation of this manuscript was supported in part by National Institute on Alcohol Abuse and Alcoholism Grant No. 5 ROI AA05823 and National Institute on Drug Abuse Grant No. 5 ROI DA03395.

References

American Psychiatric Association Task Force on Nomenclature and Statistics. (1980). *Diagnostic and statistical manual of mental disorders (DSM-III)*, Washington, DC: American Psychiatric Association.

Ball, J.C. (1967). The reliability and validity of interview data obtained from 59 narcotic drug addicts. *American Journal of Sociology, 72,* 650–654.

Blane, H.T., & Hewitt, L.E. (1977). *Alcohol and youth: An analysis of the literature, 1960–1975.* Washington, DC: National Institute on Alcohol Abuse and Alcoholism.

Blum, R.W. (1987). Adolescent substance abuse: diagnostic and treatment issues. In P.D. Rogers (Ed.), *Pediatric clinics of North America* (pp. 523–537). Philadelphia: W.B. Sanders.

Bry, B.H. (1978). Research design in drug abuse prevention: Review and recommendations. *International Journal of the Addictions, 13,* 1157–1168.

Cahalan, D. (1970). *Problem drinkers.* New Brunswick, NJ: Rutgers Center of Alcohol Studies.

Clayton, R.R., & Ritter, C. (1985). The epidemiology of alcohol and drug abuse among adolescents. *Advances in Alcohol and Substance Abuse, 4*(3/4), 69–97.

Coar, J.M. (1984). Treatment of adolescent alcoholics. In M.J. Goby (Ed.), *Alcoholism treatment and recovery* (pp. 59–81). St. Louis, MO: Catholic Health Association of the United States.

Dembo, R. (1984). Substance abuse prevention programming and research: A partnership in need of improvement. In S. Eiseman, J.A. Wingard, & G.J. Huba (Eds.), *Drug abuse: Foundation for a psychosocial approach* (pp. 88–107). Farmingdale, NY: Baywood.

Donovan, J.E., & Jessor, R. (1978). Adolescent problem drinking: Psychosocial correlates in a national sample study. *Journal of Studies on Alcohol, 39,* 1506–1524.

Edwards, G., & Gross, M.M. (1976). Alcohol dependence: Provisional description of a clinical syndrome. *British Journal of Medicine, 1,* 1058–1061.

Ehrlich, P. (1987). 12-step principles and adolescent chemical dependence treatment. *Journal of Psychoactive Drugs, 19(3),* 311–317.

Elliott, D.S., & Huizinga, D. (1988, November). *Establishing temporal order.* Paper presented at the American Society of Criminology 40th Annual Meeting, Chicago, IL.

Fagan, J., & Pablon, E. (1989). *Contributions of delinquency and substance use to school drop out among inner city youths.* New York: John Jay College of Criminal Justice.

Filstead, W.J. (1982). Adolescence and alcohol. In E.M. Pattison & E. Kaufman (Eds.), *Encyclopedia handbook on alcoholism* (pp. 769–778). New York: Gardner Press.

Filstead, W.J. & Anderson, C.L. (1983). Conceptual and clinical issues in the treatment of adolescent alcohol and substance misusers. In R. Isralowitz & M. Singer (Eds.), *Adolescent substance abuse: A guide to prevention and treatment* (pp. 103–116). New York: Haworth Press.

Fishburne, P.M., Abelson, H.T., & Cisin, I. (1980). *National survey on drug abuse, main findings, 1979.* Rockville, MD: National Institute on Drug Abuse.

Forney, M.A., Forney, P.D., & Ripling, W.K. (1989). Predictor variables of adolescent drinking. *Advances in Alcohol and Substance Abuse, 8* (2), 97–117.

Friedman, A.S., & Beschner, G.M. (1985). *Treatment services for adolescent substance abusers* (Treatment Research Monograph Series, DHHS Publication No. ADM 85-1342). Washington, DC: U.S. Government Printing Office.

Gallup, G. Jr., (1985). Alcohol use and abuse in America. *The Gallup Poll* (The Gallup Report No. 242).

Gallup, G. Jr., (1987). Alcohol use and abuse in America. *The Gallup Poll* (The Gallup Report No. 265).

Gomberg, E.S.L. (1986). Some issues in the treatment of the young male alcoholic. *Alcoholism Treatment Quarterly, 3*(1), 109–118.

Gusfield, J.R. (1982). Prevention: Rise, decline and renaissance. In E.L. Gomberg, H.R. White, & J.A. Carpenter (Eds.), *Alcohol, Science and Society Revisited* (pp. 402–425). Ann Arbor, MI: University of Michigan Press and Rutgers Center of Alcohol Studies.

Halebsky, M.A. (1987). Adolescent alcohol and substance abuse: Parent and peer effects. *Adolescence, 22,* 961–967.

Harford, T.C., & Mills, G.S. (1978). Age-related trends in alcohol consumption. *Journal of Studies on Alcohol, 39,* 207–210.

Harrell, A., Kapsak, K., & Cook, R. (1986). *Screening for adolescent drinking problems: Phase I. Development* (Final Report, Publication No. PB 86-220654), Springfield, Va.: National Technical Information Service.

Harrell, A., & Wirtz P. (1989). Screening for adolescent problem drinking: Validation of a multidimensional instrument for case identification. *Psychological Assessment: A Journal of Consulting and Clinical Psychology, 1,* 61–63.

Hawkins, J.D., & Lam, T. (1986). Teacher practices, social development, and delinquency. In J.D. Burchard (Ed.), *The prevention of delinquent behavior* (pp. 000–000. Newberry Park, CA: Sage.

Hennecke, L., & Gitlow, S.E. (1983). Alcohol use and alcoholism in adolescence. *New York State Journal of Medicine, 83*(7), 936–940.

Hewitt, L.E. (1981). Current status of alcohol education programs for youth. In National Institute on Alcohol Abuse and Alcoholism (Ed.), *Special population issues (Alcohol and Health Monograph No. 4)* (pp. 000–000). Rockville, MD: National Institute on Alcohol Abuse and Alcoholism.

Hoffmann, N.G., Sonis, W.A., & Halikas, J.A. (1987). Issues in the evaluation of chemical dependency treatment programs for adolescents. In P.D. Rogers (Ed.), *Pediatric clinics of North America* (pp. 449–459). Philadelphia: W.B. Saunders.

Jacobson, G.R. (1983). Detection, assessment, and diagnosis of alcoholism: Current techniques. In M. Galanter (Ed.), *Recent developments in alcoholism: Vol. 1* (pp. 377–413). New York: Plenum Press.

Jessor, R., & Jessor, S.L. (1977). *Problem behavior and psychosocial development: A longitudinal study of youth.* New York: Academic Press.

Johnston, L.D., O'Malley, P.M., & Bachman, J.G. (1989). *Drug use, drinking, and smoking: National survey results from high school, college, and young adult populations 1975–1988.* Rockville, MD: National Institute on Drug Abuse.

Kandel, D. (1975a). Researching the hard-to-reach: Illicit drug use among high school absentees. *Addictive Diseases, 1,* 465–480.

Kandel, D. (1975b). Stages in adolescent involvement in drug use. *Science, 190,* 912–914.

Kandel D. (1978). *Longitudinal research on drug use: Empirical findings and methodological issues.* New York: Academic Press.

Kandel, D. (1980) Drug and drinking behavior among youth. *Annual Review of Sociology, 6,* 235–285.

Kinney, J., & Leaton, G. (1987). *Loosening the grip: A handbook of alcohol information* (3rd ed.). St. Louis, MO.: C.V. Mosby.

Lettieri, D.J. (Ed.). (1975). *Predicting adolescent drug abuse: A review of issues, methods, and correlates.* Rockville, MD: National Institute on Drug Abuse.

Lettieri, D.J., Sayers, M., & Wallenstein, P.H. (Eds.). (1980). *Theories on drug abuse: Selected contemporary perspectives* (Research Monograph 30). Rockville, MD: National Institute on Drug Abuse.

MacFarland, B.J. (1983). Adolescent chemical dependency assessment. In D.H. Niles (Ed.), *Proceedings of the winter institute: Working with adolescents* (pp. 45–55). Wisconsin: University of Wisconsin-Extension, Center for Alcohol and Other Drug Studies.

Maddox, G.L. (1962). Teenage drinking in the United States. In D.J. Pittman & C.R. Snyder (Eds.), *Society, culture, and drinking patterns* (pp. 230–245). New York: Wiley.

Maddox, G.L., & McCall, B.C. (1964). *Drinking among teen-agers.* New Brunswick, NJ: Rutgers Center of Alcohol Studies.

Mandell, W. (1982). Preventing alcohol-related problems and dependencies through information and education programs. In E.M. Pattison & E. Kaufman (Eds.), *Encyclopedic handbook of alcoholism* (pp. 468–492). New York: Gardner Press.

Mauss, A.L., Hopkins, R.H., Weisheit, R.A., & Kearney, K.A. (1988). The problematic prospects for prevention in the classroom: Should alcohol education programs be expected to reduce drinking by youth? *Journal of Studies on Alcohol, 49*(1), 51–61.

Mayer, J.E. (1988). The personality characteristics of adolescents who use and misuse alcohol. *Adolescence, 23,* 383–404.

Mayer, J.E., & Filstead, W.J. (1980). Empirical procedures for defining adolescent alcohol misuse. In W.J. Filstead, & J.E. Mayer (Eds), *Adolescence and alcohol* (pp. 51–68). Cambridge, MA: Ballinger.

Mensch, B.S., & Kandel, D.B. (1988). Dropping out of high school and drug involvement. *Sociology of Education, 61,* 93–113.

Milgram, G.G. (1982). Youthful drinking: Past and present. *Journal of Drug Education, 12*(4), 289–308.

Milgram, G.G. (1987). Alcohol and drug education programs. *Journal of Drug Education, 17*(1), 43–57.

Moberg, D.P. (1983). Identifying adolescents with alcohol problems: A field test of the adolescent alcohol involvement scale. *Journal of Studies on Alcohol, 44:* 701–721.

Moore, R.H. (1984). The concurrent and construct validity of the MacAndrew Alcoholism Scale among at-risk adolescent males. *Journal of Consulting and Clinical Psychology, 40,* 1264–1269.

Moskowitz, J.M. (1989). The primary prevention of alcohol problems: A critical review of the research literature. *Journal of Studies on Alcohol, 50,* 000–000.

Nakken, J.M. (1989). Issues in adolescent chemical dependency assessment. In P.B. Henry (Ed.), *Journal of Chemical Dependency Treatment, 2*(1), 71–93. New York: Haworth Press.

Nathan, P.E. (1983). Failures in prevention: Why we can't prevent the devastating effect of alcoholism and drug abuse. *American Psychologist, 38,* 459–467.

National Institute on Drug Abuse (1986). *National household survey on drug abuse: Main findings, 1985.* Rockville, MD: National Institute on Drug Abuse.

National Institute on Drug Abuse (1989). *National household survey on drug abuse: 1988 population estimates.* Rockville, MD: National Institute on Drug Abuse.

Newcomb, M.D., & Bentler, P.M. (1988). *Consequences of adolescent drug use: Impact on young adults.* Newberry Park, CA: Sage.

O'Donnell, J.A., Voss, H.L., Clayton, R.R., Slatin, G.T., & Room, R.G. (1976). *Young men and drugs: A nationwide survey* (Research Monograph 5). Rockville, MD: National Institute on Drug Abuse.

O'Gorman, P., Stringfield, S., & Smith, I. (Eds.). (1977). Defining adolescent alcohol use: Implications toward a definition of adolescent alcoholism. New York: National Council on Alcoholism.

Owen, P.L., & Nyberg, L.R. (1983). Assessing alcohol and drug problems among adolescents: Current practices. *Journal of Drug Education, 13,* 249–254.

Pandina, R.J., Labouvie, E.W., Johnson, V., & White, H.R. (1988). The impact of prolonged marijuana use on personal and social competence in adolescence. In G. Chesher, P. Consroe, & R. Musty (Eds.), *Proceedings of the Melbourne Symposium on Cannabis* (pp. 183–200). Canberra, Australia: ANCADA, Australian Department of Health.

Pandina, R.J., Labouvie, E.W., & White, H.R. (1984). Potential contributions of the life span developmental approach to the study of adolescent alcohol and drug use. The Rutgers Health and Human Development Project: A working model. *Journal of Drug Issues, 14,* 253–268.

Pandina, R.J., & Schuele, J.A. (1983). Psychosocial correlates of alcohol and drug use of adolescent students and adolescents in treatment. *Journal of Studies on Alcohol, 44,* 950–973.

Pandina, R.J., White, H.R., & Yorke, J. (1981). Estimation of substance use involvement: Theoretical considerations and empirical findings. *International Journal of the Addictions, 16,* 1–24.

Perry, C.L. (1987). Results of prevention programs with adolescents. *Drug and Alcohol Dependence, 20,* 13–19.

Perry, C.L., & Jessor, R. (1985). The concept of health promotion and the prevention of adolescent drug abuse. *Health Education Quarterly, 12*(2), 169–184.

Rachal, J.V., Guess, L., Hubbard, R., Maisto, S.A., Cavanaugh, E.R., Waddell, R., & Benrud, C. (1980). *The nature and extent of adolescent alcohol and drug use: The 1974 and 1978 National Sample Studies.* Research Triangle Park, NC: Research Triangle Institute.

Robins, L.N. (1984). The natural history of adolescent drug use. *American Journal of Public Health, 74,* 656–657.

Rouse, B.A., Kozel, N.J., & Richards, L.G. (Eds.). (1985). *Self-report methods of estimating drug use: Meeting current challenges to validity.* (National Institute on Drug Abuse Research Monograph No. 57, DHHS Publication No. ADM 85-1402). Washington, DC: Government Printing Office.

Sadava, S.W. (1985). Problem behavior theory and consumption and consequences of alcohol use. *Journal of Studies on Alcohol, 46,* 392–397.

Single, E., Kandel, D.B., & Johnson, B. (1975). The reliability and validity of drug use responses in a large scale longitudinal survey. *Journal of Drug Issues, 5,* 426–443.

Skinner, H.A. (1981). Primary syndromes of alcohol abuse: Their measurement and correlates. *British Journal of Addiction, 76,* 63–76.

Staulcup, H., Kenward, K., & Frigo, D. (1979). A review of federal primary alcoholism preventon projects. *Journal of Studies on Alcohol, 40*(11), 943–968.

Wanberg, K.W., & Horn, J.L. (1983). Assessment of alcohol use with multidimensional concepts and measures. *American Psychologist, 38,* 1055–1069.

Weidman, A. (1987). Family therapy and reductions in treatment dropout in a residential therapeutic community for chemically dependent adolescents. *Journal of Substance Abuse Treatment, 4,* 21–28.

Weisheit, R.A., Hopkins, R.H., Kearney, K.A., & Mauss, A.L. (1984). The school as a setting for primary prevention. *Journal of Alcohol and Drug Education, 30*(1), 27–35.

White, H.R. (1982, April). *Adolescent "problem" drinking: Defining use, misuse, and abuse in youthful populations.* Paper presented at the 13th Annual Medical-Scientific Conference of the National Alcoholism Forum, Washington, DC.

White, H.R. (1987). Longitudinal stability and dimensional structure of problem drinking in adolescence. *Journal of Studies on Alcohol, 48,* 541–550.

White, H.R. (1989). Relationship between heavy drug and alcohol use and problem use among adolescents. In S. Einstein (Ed.), *Drug and alcohol use: Issues and factors* (pp. 61–71). New York: Plenum Press.

White, H.R., & Labouvie, E.W. (1989). Towards the assessment of adolescent problem drinking. *Journal of Studies on Alcohol, 50,* 30–37.

Wilcox, J.A. (1985). Adolescent alcoholism. *Journal of Psychoactive Drugs, 17*(2), 77–85.

Winters, K.C., & Henly, G.A. (1987). Advances in the assessment of adolescent clinical dependency. Development of a chemical use problem severity scale. *Psychology of Addictive Behavior, 1,* 146–153.

CHAPTER 12

Alcohol Use among the Elderly: Social Learning, Community Context, and Life Events

RONALD L. AKERS AND ANTHONY J. LA GRECA

Over the past decade, alcohol use and abuse among the elderly has come to occupy the attention of a growing number of social scientists, health and social service providers, and policymakers. A body of knowledge is developing, but we are still at the beginning of providing answers to key questions about the way alcohol does or does not fit into the lives of the elderly. Much remains to be learned about basic prevalence, distribution, and correlates of drinking and alcohol abuse in this age group. To the sociologist, however, the most noticeable lacuna is the absence of a coherent, systematic, and empirically supported theory of drinking behavior among the elderly. It is this issue to which this chapter is devoted. We report findings here from the first major study explicitly designed to assess the importance of key psychosocial theories explaining alcohol use among older adults. Specifically, we report findings on a social learning model and a life events–stress model of alcohol use and abuse by older adults. The research also tested social bonding, anomie, and locus of control models, but space limitations preclude our reporting on them here. Fuller presentation of these models and results have been reported elsewhere. What we offer here are summaries of some of those findings along with new findings. (For additional reports from the research, see Akers & La Greca, 1988; Akers, La Greca, Cochran, & Sellers, 1989; Akers, La Greca, & Sellers, 1988; La Greca, Akers, & Dwyer, 1988; Sellers & Akers, 1988.)

Alcohol Use among Older Adults

Although there are methodological and other difficulties that qualify some of the conclusions, the major finding from past survey research is

Ronald L. Akers and Anthony J. La Greca are with the Department of Sociology, University of Florida, Gainesville, Fla. This chapter was written especially for this book.

that the percentage of respondents who report drinking and heavy drinking decreases with age. Surveys of local, regional, and national samples have consistently found drinking behavior to be negatively associated with age. The elderly are less likely than younger persons and middle-aged persons to be drinkers, heavy drinkers, and problem drinkers (Barnes, 1979; Borgatta, Montgomery, & Borgatta, 1982; Cahalan & Cisin, 1968; Department of Health and Human Services [DHHS], 1981; Fitzgerald & Mulford, 1981; Holzer et al., 1984; Knupfer & Room, 1964; Meyers, Goldman, Hingson, Scotch, & Mangione, 1981–1982; Smart & Liban, 1982; Weschler, Demone, & Gottlieb, 1978; Wilsnak & Cheloha, 1987). However, studies of populations under treatment and in institutions have suggested that there is unrecognized and underreported abuse of alcohol by the elderly (Gorwitz, Bahn, Warthen, & Copper, 1970; Maddox, Robins, & Rosenberg, 1984; McCusker, Cherubin, & Zimberg, 1971; National Institute in Alcohol Abuse and Alcoholism [NIAAA], 1982; Petersen & Whittington, 1977; Zimberg, 1974a; 1974b).

Neither societal nor sociological interest in alcohol use among the elderly is predicated on how serious or widespread the problem is currently. There is some evidence that in the elderly alcohol has greater toxicity and is metabolized more slowly than in the young and middle-aged and thus that less of it is needed to produce the same effects in the elderly as in younger populations. Lower lean body mass and body water of older adults combine to compound the effects of ethanol, even at low levels of consumption. Also, alcohol exacerbates health problems associated with aging and interacts harmfully with several types of medication taken by the elderly (Hartford & Samorajski, 1982; NIAAA, 1982; Vogel-Sprott & Burrett, 1984).

If the lower levels of drinking among the elderly were due entirely to aging effects, then those who are young or middle-aged now will drop down from their higher levels of drinking to take on the patterns typical for the present elderly. However, we know that some of the difference by age is the result of a cohort effect. The lower prevalence of alcohol use among the elderly today is partly a function of the fact that they are of a generation that came of age during Prohibition and the Great Depression. This era reflected a less substance-oriented and more alcohol-proscriptive culture. The "new" elderly from the baby boom period after World War II do not share this culture. As the more substance-use accepting generation of the 1960s ages we can expect increased prevalence of drinking and alcohol problems for the elderly (Fitzgerald & Mulford, 1981; Glantz, 1982; Holzer et al., 1984; Meyers, Goldman, Hingson, Scotch, & Mangione, 1981–82). Sociologically we are concerned with explaining whatever pattern of drinking there is in the older population and relating it to social psychological and social structural variables.

Research on older adults and alcohol consumption has yet to establish confirmed and replicated relationships between key socioeconomic and demographic factors and drinking by older adults. The strongest correlates are age and gender, with women drinking less than males (Borgatta et al., 1982; Holzer et al., 1984; Smart & Liban, 1981). There is little difference in the use or abuse of alcohol between black and white elderly (Borgatta et al., 1982; Holzer et al., 1984). Only education has been consistently found to have a clear relationship to drinking done by older adults. Borgatta et al. (1982), Holzer et al. (1984), and Shuckit and Miller (1975) all found that drinking by older adults increases as education increases. There are either mixed or inconsistent findings regarding the correlation of drinking with occupational status, employment status, and income (Barnes, 1979; Borgatta et al., 1982; Holzer et al., 1984; Shuckit & Miller, 1975; Smart & Liban, 1981). Similarly, there is a lack of consensus on the relationship between marital status and drinking by older adults (Holzer et al., 1984; Shuckit & Miller, 1975; Smart & Liban, 1981). With a few exceptions (Borgatta et al., 1982; Smart & Liban, 1981), past research has not utilized a multivariate analysis to assess the relative importance of these correlates. But probably the major shortcoming of this research is the one just mentioned, the lack of theoretical focus. Our recently concluded research was designed to address the theoretical issues.

Methods

Data Collection

We conducted in-home interviews of 1,410 older adults (age 60 and above) in four communities: two retirement communities (age homogeneous) and two age-integrated communities. The retirement community in New Jersey was located near residents' families and preretirement friends; the retirement community in Florida was not. One of the age-integrated communities was in Pinellas County, Florida (St. Petersburg, Clearwater), where the concentration of older adults is over two and one-half times the national average. The other age-integrated community was in Alachua County (Gainesville, Florida), where the proportion of elderly is about two thirds that of the national average. These four sites represent community settings typical of noninstitutionalized older adults (Golant, 1975). A random sample was generated for Pinellas and Alachua counties, with interview completion rates of 67% and 82.1%. (A lack of funds prevented a follow-up of first refusals in Pinellas County). Entry into the two retirement communities involved the usual restrictions associated with such entry (see Streib, Folts, & La Greca, 1984), and so no recontacts of any type were allowed. Therefore, 40% of those initially contacted in the Florida retirement community, and 25% of those initially contacted in the New Jersey community, were interviewed. The age, gender, and race distribution of our two random samples closely approximates that of the census data on Pinellas and Alachua counties. The retirement communities had no communitywide tabulations of these demographics.

Measures of Alcohol Use

The dependent measures were designed to cover the continuum of alcohol behavior ranging from abstinence to heavy drinking, as well as measures assessing problem drinking. We also retrospectively assess continuity or change in drinking behavior after age 60.

YFREQALC, frequency of alcohol consumption in the past year, reflects alcohol consumption (beer, wine, or liquor) during the previous 12 months with responses on a 9-point scale (*no drinking, once or twice, less than once a month, once a month, two to three times a month, once or twice a week, three to four times a week, nearly every day,* and *daily*). For tabular presentation, the nine categories were collapsed into four: *no drinking/abstainers,* including those who did no drinking, or drank only once or twice, in the previous year; *monthly drinking*—those who drank less than once a month up to two to three times a month; *weekly drinking*—those who drank once or twice a week up to three to four times a week; and *daily drinking*—those who drank nearly every day or daily. A similar measure was constructed for frequency of alcohol consumption in the past month (MFREQALC).

QUANTALC, typical quantity of alcohol consumption, is a measure of the quantity of beer, wine, or liquor consumed by the respondent on a typical drinking day. The quantity is measured by the number of cans or bottles of beer, or glasses of wine, or drinks consisting of approximately one ounce of liquor. Responses were categorized as abstainers, or as one, two, three, four, or five or more bottles, cans, or glasses of alcohol.

YQFINDX, the quantity-frequency index of alcohol consumption, is a cross-tabulation of the frequency of alcohol consumption with the typical quantity of alcohol consumption in the past year (QUANTALC), ordinally categorized into four levels: *no drinking/abstainers; light drinking* (from one to six drinks consumed less than once a month up to one or two drinks taken once or twice a week); *moderate drinking* (from 7 to 12 drinks less than once a month or one to two drinks daily); and *heavy drinking* (from six or more drinks once or twice a week to seven or more drinks daily). These values were adapted from Cahalan, Cisin, and Crossley (1967). This index was also constructed for the month immediately past (MQFINDX).

ELDONSET (elderly onset, increase, decrease, or maintenance of drinking) is measured by respondents' retrospective report of their drinking patterns in the decade before age 60 compared with responses for the years after their turning 60. It is an ordinal scale ranging from *no drinking/abstiners* to *no change in drinking* to *increased drinking* to *first began drinking* to *first began heavy drinking* (after age 60).

YPROBALC, alcohol-related problems in the past year, is an assessment of whether respondents experienced any alcohol-related problems in the past year. The list of problems covers physiological, psychological, and social consequences related to problem drinking (e.g., developing liver disease or yellow jaundice as a consequence of drinking; wanting but being unable to stop drinking; neglecting usual or typical daily responsibilities as a result of drinking). The list represents the range of problems identified by the National Institute of Mental Health (see Robins, Helzer, Croughan, & Ratcliff, 1981).

Table 1 shows the distribution of alcohol-drinking behavior for our sample. Nearly a third (62%) of our sample reported some drinking in the previous year, with over a fifth reporting daily drinking or nearly daily drinking. A little over 6% of our sample (10% of the drinkers) had been heavy drinkers in the previous year or month. About 3.1% of our sample (about 6% of our drinkers) had experienced one or more alcohol-related problems in the previous year. These rates of drinking and problem drinking are comparable to those found in other survey research on this age group (Borgatta et al., 1982; Cahalan & Cisin, 1968; DHHS, 1981; Holzer et al., 1984; Meyers et al., 1981–82). The typical pattern of drinking after one turns 60 is, not surprisingly, to continue essentially as one was doing before that, but 5% of our respondents started drinking or increased their drinking after reaching the age of 60 (ELDONSET).

Sociodemographic Correlates of Drinking

We included seven measures of sociodemographic variables in this analysis: *age, race, sex, marital status, education, income,* and *occupational prestige* (current if the respondent was still working or previous if he or she was retired). Education is measured in years of schooling completed, income is measured by monthly household income from all sources, and occupational prestige is measured by the Nam scores from the 1970 U.S. Census status scores (Nam & Powers, 1983). The drinking variables are significantly related to all of the sociodemographic variables, with the magnitude of the relationships ranging from weak (r = .08) to moderate

TABLE 1
Percentage Distribution of Drinking Behavior among the Elderly

(*N* = 1410)		
Frequency of Drinking during Past Year (YFREQALC)		
No drinking/abstainers		38.2%
Monthly drinking (up to 2–3 times a month)		19.6
Weekly Drinking (up to 3–4 times a week)		21.1
Daily drinking		21.1
Frequency of Drinking in Past Month (MFREQALC)		
No drinking/abstainers		41.3%
1–3 times		16.9
1–4 times a week		20.6
Daily drinking		21.0
Quantity/Frequency of Drinking during Past Year (YQFINDX)		
	Total Sample	Drinkers Only
No drinking/abstainers	38.2%	—
Light drinking	30.5	49.3%
Moderate drinking	25.0	40.4
Heavy drinking	5.7	9.2
Excessive drinking	.7	1.1
Alcohol-Related Problems during Past Year (YPROBALC)		
Total sample		3.1%
Drinkers only		5.5
Drinking after Age 60 (ELDONSET)		
No drinking/abstainers		36.7%
No change in drinking		54.6
Decreased drinking		3.9
Increased drinking		2.9
First began drinking		1.1
First began heavy drinking		.8
Percent Elderly Drinking by Community		
New Jersey Retirement (N = 216)		67.6%
Florida Retirement (N = 516)		76.4
Alachua County (N = 352)		41.5
Pinellas County (N = 326)		57.1

($r = .32$). A regression model with these variables accounts for 30% of the variance in YFREQALC (tables not shown). We turn now to social learning and life event explanations of elderly drinking, measures of concepts, and findings on models derived from the theories.

Social Learning Theory and Alcohol Use among the Elderly

Social learning has been used to refer to a number of social-behavioral formulations in psychology (see especially Bandura, 1977; Stumphauzer, 1986) as well as to some explanations of alcohol use (White, Bates, & Johnson, 1990). [See also chapter 9, by White, Bates, & Johnson, in this volume.] But the theoretical model most often given this label in sociology (particularly in the sociology of deviance) is the one first proposed by Burgess and Akers (1966) and further developed and tested by Akers over several years. This theory ties general behavioral learning principles to a sociological theory of association and reference group influence (Sutherland, 1947) and has been used to account for various forms of deviant behavior, drug addiction, and alcohol use (Akers, 1985).

As stated by Akers (1985, pp. 39–61; Akers, Krohn, Lanza-Kaduce, & Radosevich, 1979) this social learning theory proposes that behavior is acquired both through direct behavioral conditioning ("differential reinforcement") and through imitation or modeling of others' behavior. Behavior is strengthened and sustained through reward (positive reinforcement) and avoidance of punishment (negative reinforcement) or weakened by aversive stimuli or loss of reward. Which behavior is acquired and persists depends on the balance of past, present, and anticipated rewards or punishments for the behavior and those attached to alternative behavior. The reinforcement can be both nonsocial (as in the direct physiological effects of alcohol) and social. The social rewards may be monetary, symbolic, or intangible and may involve self-reinforcement. Social groups with which one is in interaction, or with which one identifies, influence or control sources of social reinforcement.

It is in interaction (or identification) with these significant groups in their lives that individuals learn normative attitudes toward certain behavior as good or bad, right or wrong. These are termed *definitions;* and the more individuals positively define the behavior or define it as justified (with *neutralizing definitions*), the more likely they are to engage in it. The process whereby one interacts with others and is exposed to positive or negative norms is labeled *differential association* (Sutherland, 1947). The most important of these groups are the primary groups of family and friendship groups, but they also include work groups, schools, churches, community organizations, and other secondary groups.

The theory proposes, then, that abstinence or drinking and the pattern of drinking that develops (light, moderate, or heavy) result from greater rewarding consequences, on balance, over aversive contingencies for that behavior (both social consequences and the effects of the alcohol), that is, from *differential reinforcement.* Further, one's definitions of, or attitudes toward, drinking are conducive to consumption of alcohol when, on balance, the positive and neutralizing definitions of drinking offset the negative. The greater the extent to which one associates with drinkers and persons who hold favorable attitudes toward drinking (*differential reinforcement*) and who provide drinking models to imitate than with abstainers and persons who hold negative attitudes toward drinking, the more likely one is to drink and to drink more frequently. Therefore, drinking of alcohol can be expected to the extent that one has been differentially associated with other drinkers, that drinking has been differentially reinforced over abstinence, and that one defines drinking as more desirable than abstinence or, at least, as justified. Linear causal models are tested here, but it should be remembered that they are approximations of an underlying process that involves nonrecursive relationships. Once drinking patterns develop, they have an effect on choice of friends, interaction patterns, others' reactions, and other variables (Akers, 1985, pp. 39–61, 152–163).

The principal concepts in the theory, then, are: *definitions* of drinking, *differential association, imitation* of drinking and nondrinking models, and *differential reinforcement* for a given pattern of drinking or abstinence. We exclude imitation from the model here because concepts and measures of modeling are less applicable, and even more difficult to measure in this age group, than in adolescent groups, on which the model has been previously tested. We test the theory here through regression analysis of the overall model, examining the combined total effects of the measures of these main concepts from the theory on alcohol use among the elderly, as well as the net effects of each variable and each subset of variables. Earlier research has found that social learning models account for substantial amounts of variance in teenagers' use of marijuana and other drugs, alcohol, and tobacco. (Akers, 1985; Akers & Cochran, 1985; Dembo, Grandum, LaVoie, Krohn, Skinner, Massey, & Akers, 1985; Lanza-Kaduce, Krohn, & Akers, 1984; White, Johnson, & Horwitz, 1986). Our aim is to see how well the model also fits elderly substance use.

Social Learning Variables

All five of the *definitions* variables are measured by responses to Likert-type questions about positive, negative, or neutralizing attitudes toward alcohol: (1) BALDEF (balance of positive-negative definitions of alcohol use), (2) NEUTDEF (neutralizing definitions justifying drinking), (3) NEGDEFA (negative definitions of alcohol), (4) PRESDEF (prescriptive definitions of alcohol use under certain circumstances) and (5) general adherence to conventional beliefs about religion and the law). The *normative* dimension of *differential association* is measured by respondents' perceptions of the normative attitudes toward drinking held by their primary groups:

(6) SPNORMS (spouse's drinking norms), (7) FAMNORMS (other family's drinking norms), (8) FRNORMS (drinking norms of best friends, friends with whom the respondent most frequently associates, and longest-time friends). The *interactive* dimension, on the other hand, is measured by respondents' report of the drinking behavior of members of their primary groups; (9) SPASSN (spouse's drinking), (10) CHIASSN (respondent's adult children's drinking), (11) OFASSN (drinking by other family members), (12) BFASSM (best friend's drinking), and (13) PFRASSN (proportion of best, longest-time, and most-frequently-associated-with friends who are drinkers). *Differential reinforcement* is measured by respondent's reports of positive and negative reactions from others and consequences of drinking with nondrinkers being asked about anticipated reactions and consequences: (14) SREINBA (balance of reinforcement for drinking or abstinence in social relationships), (15) SPREAC (spouse's reaction), (16) FAMREAC (other family's reactions), (17) FREAC (friends' reactions), (18) PREINBAL (balance of perceived physical effects of alcohol), and (19) BALRC (overall balance of perceived rewards minus costs of drinking). For more information on measures of these variables see Akers et al. (1989).

Frequency, Quantity, and Late Onset of Alcohol Use

Table 2 presents the bivariate correlation (r) and the standardized partial regression coefficients (B) for YFREQALC, YQFINDX, and ELDONSET by each social learning variable. As this table shows, all of the drinking measures are significantly related to each of the social learning variables, with r equaling from .25 to .65. When the social learning variables are entered into regression equations, substantial amounts of variances are explained. Although there is some multicollinearity in the model, diagnostic tests revealed it to be within acceptable limits. These findings, then, support social learning theory; the variables measuring concepts from the theory sufficiently distinguish elderly drinkers from nondrinkers and different quantity-frequency patterns of alcohol use among drinkers, as well as patterns of drinking after entrance into the elderly years.

The social learning model applies very well to the drinking and abstinence behavior of elderly men and women, just as it has been shown to do with adolescents (Akers et al., 1979). The social learning model accounted for 55% of the variance in alcohol use in the Akers et al. (1979) study of adolescents. The model here accounts for 59% of the variance in YFREQALC and also accounts for high proportions of variance in YQFINDX (R^2 = .52) and ELDONSET (R^2 = .51).

Separate regression equations for the three subsets of social learning variables (definitions, association, reinforcement) were run (tables not shown); the equations show clearly that any one of the subsets of major

TABLE 2
Correlation and Partial Regression Coefficients for Alcohol Variables by Social Learning Variables

Social Learning Variables	Alcohol Variables YFREQALC		YQFINDX		ELDONSET	
I. *Definitions*	$r =$	$B =$	$r =$	$B =$	$r =$	$B =$
1. BALDEF	.59**	.25**	.55**	.22**	.59**	.12**
2. NEUTDEF	.34**	.05*	.34**	.07**	.30**	.03
3. NEGDEFA	.47**	.02	.44**	.02	.39**	-.04
4. PRESDEF	.44**	.05	.41**	.03	.43**	-.04
5. GENDER	.19**	.05	.19**	.06*	.15*	.03
II. *Differential Association*						
6. SPNORMS	.40**	.08**	.35**	.09**	.30**	-.04
7. FAMNORMS	.35**	.04	.33**	.04	.25**	-.04
8. FRNORMS	.48**	.05	.44**	.06*	.39**	-.06*
9. SPASSN	.45**	.17**	.41**	.15**	.33**	.07*
10. CHIASSN	.33**	.01	.31**	.01	.28**	.01
11. OFASSN	.32**	.01	.30**	.01	.27**	-.02
12. BFASSN	.57**	.18**	.54**	.17**	.46**	.04
13. PFRASSN	.53**	.03	.50**	.03	.50**	.08*
II. *Differential Reinforcement*						
14. SREINBAL	.51**	.07*	.50**	.02	.55**	.04
15. SPREAC	.43**	.06*	.39**	.04	.36**	-.02
16. FAMREAC	.46**	.03	.43**	.04	.46**	.02
17. FREAC	.54**	.07*	.51**	.08**	.62**	.15**
18. PREINBAL	.61**	.24**	.58**	.21**	.62**	.25**
19. BALRC	.65**	.18**	.62**	.17**	.65**	.23**
adj. R^2 =	.59		.52		.51	

** $p < .01$

* $p < .05$

Variables coded so that positive relationships are in expected direction.

social learning variables, by itself, has a powerful relationship to elderly drinking. The subsets of variables account for 27% to 50% of the variance in the drinking variables. Including the sociodemographic variables in the regression equations makes essentially no difference in the relative effects of the social learning variables and adds little to explained variance (tables not shown).

Elderly Problem Drinking

As Table 1 shows, only 44 persons in this sample (3.1% of the total sample and 5.5% of the drinkers) reported one or more problems of personal or interpersonal functioning and control during the previous year. Nevertheless, the few elderly in our sample who had been engaged in deviant behavior or had been arrested, convicted, or jailed come disproportionately from among this small number of problem drinkers. Only 3.4% of the nondrinkers and 6.3% of the drinkers in our study had any kind of contact with the criminal justice system in the past year, but 15% of the problem drinkers had such contact. Similarly, only 4.6% of the abstaining elderly in the study self-reported illegal behavior in the past year, whereas 5.2% of the drinkers and 20% of the problem drinkers reported involvement in illegal behavior. Thus, the elderly problem drinker runs nearly five times the nondrinker's risk of criminal sanction and four times the chance of illegal behavior.

Any survey among the general, noninstitutionalized elderly population will find relatively few problem drinkers, and our findings must be qualified by the small number of problem drinking cases. Nevertheless, we believe that those identified in our research as problem drinkers are indeed those with alcohol-related difficulties, including the same alcohol-crime link that has been found for younger age groups. (For a further report on alcohol, contact with the legal system, and illegal behavior among the elderly, see Akers and La Greca, 1988.)

Given the skewed distribution and low number along with the dichotomous dependent variable, logistic regression is more appropriate to analyze the data on problem drinking than ordinary least squares (OLS) regression (Hanushek & Jackson, 1977). A logistic regression social learning model for problem drinking was tested; it produced standardized and unstandardized regression coefficients and instantaneous rate-of-change coefficients (table not shown).

The differences in the predicted probabilities in the logistic regression model indicate that the additive effects of the social learning variables make a very substantial difference in the probability of problem drinking. Those drinking respondents in our sample who are interacting in the most extreme anti-heavy drinking environment (wherein the respondents have a spouse, family members, and friends, all of whom are light or moderate drinkers, as well as differential reinforcement and definitions favoring moderate drinking) have essentially a zero chance of being one of the 44 problem drinkers, p (Y = 1) being .001. Those respondents at the other extreme, holding attitudes and interacting in an environment most supportive of heavy drinking, have a very high likelihood of being among the problem drinkers, p (Y = 1) being .683.

The analysis indicates that the problem drinker is more likely than the average drinker to report negative effects from alcohol and to hold negative attitudes toward heavy drinking. Other things being equal, these negative outcomes and attitudes should be associated with less, rather than more, heavy drinking. This is, of course, the age-old question of why it is that with all of the difficulties connected with alcohol abuse, problem drinkers do not naturally curtail drinking. At some point many do—when they "hit bottom." But short of this they do not. The key to why such drinkers continue to consume alcohol in the face of these unpleasant outcomes and attitudes lies in the balancing concept in social learning theory. Problem drinkers in our study score high in the positive direction on our measure of the overall balance of rewards versus costs of drinking. Some of this is negative reinforcement in which the heavy drinking helps the person to "escape" from other unpleasant or aversive conditions. Some of it is the social reinforcement that comes from associating with other heavy drinkers. We found that problem drinkers differentially associate with heavy drinkers and that the balance of social reinforcement for heavy drinking was positive. Problem drinking is sustained by the support from friends with similar drinking habits and the fact that problem drinkers experience the inducements and rewards for drinking as considerably outbalancing the costs and aversive consequences of drinking.

Community Context and Social Learning in Elderly Drinking

As Table 1 shows, we found noticeable differences in levels of drinking by community of residence. In addition to the differences in percentage of drinkers shown in the table, we found that 30% of the respondents in the Florida retirement community reported drinking every day (or nearly every day), which is more than twice the level of daily drinking in Alachua County and in the retirement community in New Jersey. Clearly, although all respondents were 60 years of age and older, the community in which they resided made a difference in the chances of drinking and in how frequently they were likely to drink. Theoretically, we view each community as providing a general social structural context within which the social learning process operates to affect the drinking behavior of the elderly residents. There has been a resurgence of interest in the "micro-macro link" in sociology (see Alexander, Giesen, Munch, & Smelser, 1987). There is little agreement about what constitutes a micro- or macro-level variable. But if the community context can be defined as *macro* and social learning as *micro,* the design and findings of our study allow for an exploration of this issue with regard to alcohol use among the elderly. The differential association–social learning tradition provides a good model for the micro-macro connection because, from the beginning, it has conceptualized learning as the process by which social structure has an impact on individual behavior (Akers, 1985; Cressey, 1960; Sutherland, 1947).

We tested a theoretical model that viewed the community context as the structural, or macro-level, explanation for variations in drinking patterns among the elderly respondents in our sample and viewed social learning as the mediating process, or set of micro-level variables, by which the community context had an effect on drinking behavior. We knew that there were some differences in drinking customs in the communities. For instance, the Florida retirement community had a golf course and community club with a daily "happy hour," whereas the New Jersey community had a community club, but one with no bar or provision for serving liquor. Alachua County is a smaller, less urbanized area and is less affected by in-migration than the St. Petersburg area of Pinellas. It still has more of the religious and other southern community traditions.

However, we had no way of systematically measuring differences in drinking traditions that might have built up in each community. We did know that even though the respondents were all in the older age group the communities differed systematically in how densely concentrated the residents were among other elderly. The most age-dense community was the New Jersey retirement community (a separate retirement community but one set in the midst of and surrounded by other retirement communities). The Florida retirement community is separate from but not far from an age-integrated community. Although the most age-dense community did not have the highest level of drinking (primarily because its population was somewhat less affluent than the Florida retirement community's), it had the second highest proportion of drinkers, well above that of both Alachua and Pinellas counties. Of the two age-integrated communities, Pinellas County, with 35% of the population 60 and over, had a higher concentration of elderly people than Alachua County, with only 10% of the population 60 and over. And it was the latter that had the fewest drinkers among the elderly respondents. Thus, although not perfectly related, the age density of the community appears to be a factor in drinking among the elderly living in it, and we used it as measure of overall community characteristics.

But the effect of community context also reflects differences in the social background, drinking experiences, traditions, and social characteristics through the selective process by which the people come to live in the different communities. That is, we view community context as involving both a "global" dimension reflecting a unitary characteristic of the community and a dimension reflecting the aggregated socioeconomic characteristics of the individual residents. We measured the first dimension by ranking and assigning an ordinal score to the communities on age density and the second dimension by the education and income of the residents of the communities (the two sociodemographic variables most strongly related to drinking, r = .28 and .32, respectively).

We tested a LISREL model with these measures of community context as structural variables and social learning as the micro-level variables. Theoretically, the social learning constructs (definitions, association, reinforcement) are themselves different indicators or dimensions of a single underlying larger construct of the social learning process. Therefore, we ran a LISREL model with a second-order factor as a higher-level social learning construct. That model had a good fit to the data ($X^2 = 3.78$, $df = 3$, $p = .44$, and adjusted goodness-of-fit index = .99).[1]

Next we ran a structural model with the indicators and constructs from this social learning model as exogenous, the second-order social learning construct as the first endogenous variable, and frequency of drinking as the second endogenous construct (the dependent variable). Again, there was a good fit to the data (a significant X^2 at 45.16 with $df = 12$, but with a Wheaton index of 3.9). The squared multiple correlation for the drinking construct (interpretable as explained variance) = .75.

Then we ran a good-fitting structural model, including the social learning constructs and a community context construct with age density, education, and income as the indicators. Without social learning effects taken into account in the model, community context has a direct effect of .409. When the social learning constructs are in the model, that effect reduces to zero and goes slightly beyond that to a negative coefficient. Community context retains an indirect relationship to drinking through social learning that is, in fact, somewhat stronger than the original effect.

Although they are not directly interpretable as least squares standardized betas, LISREL does produce coefficients, betas, which show net direct and indirect effects. There is a strong direct effect of community on the second-order social learning construct ($\beta = .41$). The social learning construct has an even stronger direct effect on frequency of drinking among the elderly ($\beta = 1.30$), but community context has very little direct effect on drinking ($\beta = -.14$). The impact of community context on drinking patterns among the elderly is primarily indirect through the social learning construct. The second-order social learning construct explains .74 of the variance in drinking on its own. The community context construct accounts for .24 explained variance but adds only .02 explained variance when placed in the structural model with social learning constructs (see Figure 1).

The conclusion to be drawn, then, is that, as theoretically expected, the structural variable of community context is mediated through the micro-level process of social learning. Living in a community with certain age and socioeconomic characteristics does affect how much one is likely to drink alcohol. But this effect comes not so much from those characteristics directly producing the behavior. Rather, the community context has a direct effect on the social learning process, on one's choice of people

FIGURE 1
LISREL (Unstandardized Solution) Model of Social Learning and Community Context in Elderly Drinking

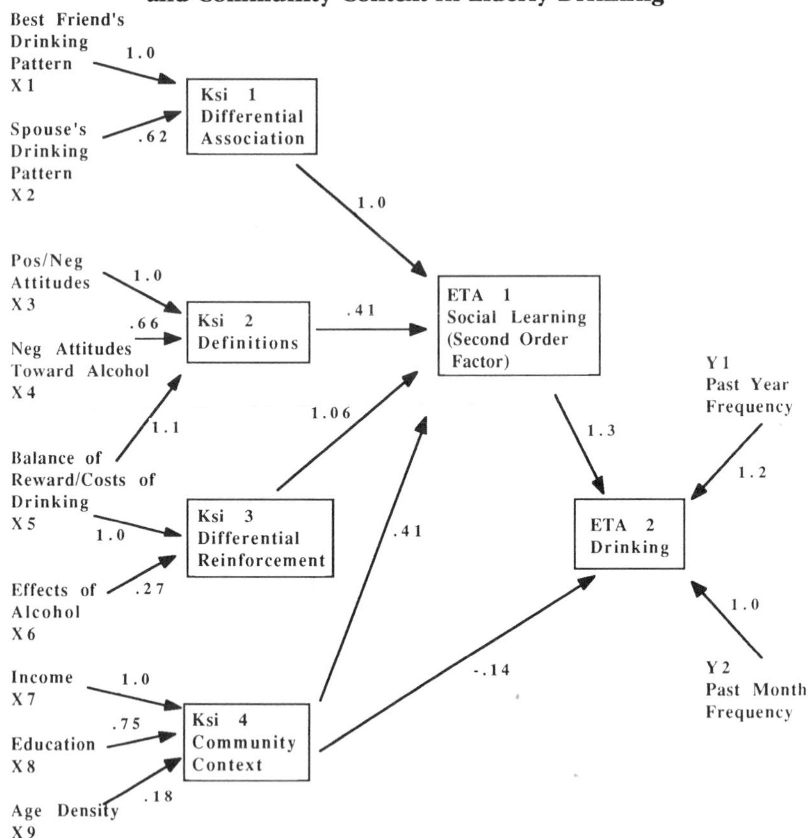

with whom to associate, on the normative orientations to which one is exposed and which one internalizes, and on the degree to which one finds alcohol use (and perhaps other variables not included in the model) to be reinforcing—a degree that, in turn, affects one's drinking behavior.

Social learning theory is a general behavioral approach to alcohol behavior that is compatible with theories of social structure and is applicable to adults as well as adolescents (Akers, 1985; Akers & Cochran, 1985; Elliott, Huizinga, & Ageton, 1985; Pearson & Weiner, 1985). It accounts for variations in the full range of drinking from abstinence to moderate and problem drinking among the elderly. Both the general social learning model and the structure-process model of elderly alcohol use are supported.

Life Event Stress and Alcohol Use

A persistent hypothesis in research on older adults and alcohol use is that the elderly either initiate or increase drinking as a means of coping with stress. This phenomenon of "late onset" has been supported by research on clinical populations (Glatt, Rosin, & Jauhar, 1978; Hubbard, Santos, & Santos, 1979; & Schuckit, 1977) but not by survey research on general populations of older adults (Borgatta et al., 1982; Meyers et al., 1981–82). As we have seen, there is very little late onset or increase in alcohol use among the elderly in our study, and the persistence or change in drinking patterns as one becomes elderly is related to social learning variables. The stress associated with the late onset hypothesis is typically seen in other research as stemming from life events common to older adults. These include loss of spouse, changes in health, retirement, and other potentially negative happenings that often adversely affect the well-being of older adults (Cohen, Teresi, & Holmes, 1985a, 1985b; Krause, 1986).

However, the late onset hypothesis is not tested in isolation from other themes in gerontological research. These themes include the personal hardiness of older adults whose individual coping resources negate the use of escapist means to cope (McCrae, 1982) as well as the theme that sources of social support mitigate the effects of stressful life events, thus diminishing the need for escapist behavior (Hultsch & Plenons, 1979; Eckenrode & Gore, 1981).

In spite of this history of research on the late onset hypothesis, no study has yet collected survey data specifically intended to test this model. From the outset, one of the specific aims of our research was to test the two central hypotheses associated with the late onset model: (1) that the experience of life events is positively related to alcohol consumption and problem drinking; and (2) that this relationship is more likely to occur for those with less individual coping resources and less social support.

For our research, the number of life events is based on a checklist of events having occurred in the past year (death of spouse, death of family member or close friend, own health problem, family health problem, family separation, change of residence, retirement, financial difficulty, being fired from job, marriage, and divorce) (see Holmes & Rahe, 1967; Paykel, Prusoff, & Uhlenhuth, 1971). Nearly three out of four of our respondents (73.6%) had experienced at least one life event in the previous year, the most frequent event being the death of a family member or close friend.

Individual *coping ability* is a summed average of four items measuring reported ability to deal effectively with health problems, loss of spouse, loss of friends, or financial difficulties. *Importance of religion* is a one-item measure of the extent to which religious beliefs play an important role in the daily life of the respondent.

Social support variables are *satisfaction with contact,* which is a summed average of three items measuring respondents' satisfaction with the closeness of contact with their children, close friends, and other family members. *Social support in time of need* and *social support in time of depression* are each a summed average of four items measuring how strongly respondents feel they can count on the support of friends and family members in time of need or during those times when they feel depressed. We also included two objective measures of social support: *marital status* and *living arrangement.*

Our findings show no support for either of the hypotheses. Bivariate correlations show almost no association between the occurrence of life events and the frequency, quantity, or problems of drinking behavior among older adults in our sample. Indeed, the direction of the relationships was opposite theoretical expectation—the more life events the elderly experience, the lower the chance of their drinking. Regression analyses show that although the already low negative effects of life events on YFREQALC and YQFINDX are reduced even further when the social support and coping variables are entered they do not become positive. The greater the social support and coping skills, the less likely the elderly are to drink; but stress from life events has essentially no effect (tables not shown).

Transition from lifelong abstention to problem drinking as a means of coping with the stress of old age is a very rare phenomenon. If late onset drinking does ever occur, it is most likely to occur in those older adults who enter later life with a predisposition for heavy drinking. It is not likely to be provoked by stressful life events. The elderly are, by and large, quite capable of dealing with them without recourse to alcohol.

Conclusion

This chapter has reported recent findings on alcohol use and nonuse among older adults. As in past research, alcohol use and abuse was found to be lower for this age group than for younger populations. The main focus of our research, however, was the testing of key psychosocial theories of alcohol use among the elderly. Social learning theory, long proven to be of value in furthering our understanding of adolescent drinking, also proved to be a powerful tool to aid our understanding of the continuum of nondrinking to drinking by older adults. Social learning theory is also useful in advancing our understanding of the process by which community context has an effect on drinking behavior. Finally, our cross-sectional test of the life events–stress model showed that such events do not initiate or increase drinking by older adults. However, further explication of our theoretical models is continuing with panel data.

Acknowledgment

The research reported here was supported by National Institute on Alcohol Abuse and Alcoholism grant P50AA05793.

Note

1. LISREL is a computer software package that performs structural equation modeling. The statistical procedure combines path analytic and factor analytic concepts and performs a simultaneous estimation of the unknown parameters in a set of linear structural equations including measurement structure, correlated measurement errors, and correlated equation disturbances. The program provides a maximum likelihood goodness-of-fit (chi-square) test of the overall fit of the model to the data. For greater detail, see Joreskog and Sorbom (1981).

References

Akers, R.L. (1985). *Deviant behavior: A social learning approach* (3rd ed.). Belmont, CA: Wadsworth.

Akers, R.L., & Cochran, J.K. (1985). Adolescent marijuana use: A test of three theories of deviant behavior. *Deviant Behavior, 6,* 323–346.

Akers, R.L., Krohn, M.D., Lanza-Kaduce, L., & Radosevich, M.J. (1979). Social learning and deviant behavior: A specific test of a general theory. *American Sociological Review, 44,* 635–655.

Akers, R.L. & La Greca, A.J. (1988). Alcohol, contact with the legal system, and illegal behavior among the elderly. In B. McCarthy & R. Langworthy (Eds.), *Older offenders: Perspectives in criminology and criminal justice* (pp. 35–50). New York: Praeger Press.

Akers, R.L., La Greca, A.J., Cochran, J., & Sellers, C. (1989). Social learning theory and alcohol behavior among the elderly. *Sociological Quarterly, 30,* 625–638.

Akers, R.L., La Greca, A.J., & Sellers, C. (1988). Theoretical perspectives on deviant behavior among the elderly. In B. McCarthy & R. Langworthy (Eds.), *Older offenders: Perspectives in criminology and criminal justice.* New York: Praeger Press.

Alexander, J.C., Giesen, B., Munch, R., & Smelser, N.J. (1987). *The micro-macro link.* Berkeley: University of California Press.

Bandura, A. (1977). *Social learning theory.* Englewood Cliffs, NJ: Prentice-Hall.

Barnes, G.M. (1979). Alcohol use among older persons: Findings from a western New York State general population survey. *Journal of the American Geriatric Society, 17,* 244–250.

Borgatta, E.R., Montgomery, R., & Borgatta, M.L. (1982). Alcohol use and abuse, life crisis events, and the elderly. *Research on Aging, 4,* 378–408.

Burgess, R.L., & Akers, R.L. (1966). A differential association-reinforcement theory of criminal behavior. *Social Problems, 14,* 128–147.

Cahalan, D., & Cisin, I.H. (1968). American drinking practices: Summary of findings from a national probability sample. I. Extent of drinking by population subgroups. *Quarterly Journal of Studies on Alcohol, 29,* 130–151.

Cahalan, D., Cisin, I.H., & Crossley, H.S. (1967). *American drinking practices.* Washington, D.C.: George Washington University.

Cohen, C.I., Teresi, J., & Holmes, D. (1985a). Social networks, stress, and physical health. *Research on Aging, 7,* 409–431.

Cohen, C.I., Teresi, J., & Holmes, D. (1985b). Social networks, stress, adaptation, and health: A longitudinal study of an inner-city elderly population. *Journal of Gerontology, 40,* 478–486.

Cressey, D.R. (1960). Epidemiology and individual conduct: A case from criminology. *Pacific Sociological Review, 3,* 47–58.

Dembo, R., Grandum, G., LaVoie, L., Schmeidler, J., & Burgos, W. (1986). Parents and drugs revisited: Some further evidence in support of the social learning theory. *Criminology, 24,* 85–104.

Department of Health and Human Services. (1981). *Alcohol and health.* Washington, DC: United States Government Printing Office.

Dupree, L.W., Broskowski, H., & Schonfeld, L. (1984). The gerontology alcohol project: A behavioral treatment program for elderly alcohol abusers. *The Gerontologist, 24,* 510–516.

Eckenrode, J., & Gore, S. (1981). Stressful events and social supports: The significance of context. In B. Gottlieb (Ed.), *Social networks and social support.* Beverly Hills, CA: Sage.

Elliott, D.S., Huizinga, D., & Ageton, S.S. (1985). *Explaining delinquency and drug use.* Beverly Hills, CA: Sage.

Fitzgerald, J.L., & Mulford, H.A. (1981). The prevalence and extent of drinking in Iowa, 1979. *Journal of Studies on Alcohol, 42,* 38–47.

Glantz, M. (1982). Predictions of elderly drug abuse. In D.M. Petersen & F.J. Whittington (Eds.), *Drugs, alcohol and aging* (pp. 7–16). Dubuque, IA: Kendall/Hunt.

Glatt, M.M., Rosin, A.J., & Jauhar, P. (1978). Alcoholic problems in the elderly. *Age and Aging, 7,* 64–71.

Golant, S.M. (1975). Residential concentrations of the future elderly. *The Gerontologist, 15,* 16–23.

Gorwitz, K., Bahn, A., Warthen, F.J., & Copper, M. (1970). Some epidemiological data on alcoholism in Maryland. *Quarterly Journal of Studies on Alcohol, 31,* 423–443.

Hanusheck, E.A., & Jackson, J.E. (1977). *Statistical methods for social scientists.* New York: Academic Press.

Hartford, J.T., & Samorajski, T. (1982). Alcoholism in the geriatric population. *Journal of the American Geriatric Society, 30*(1), 18–24.

Holmes, T.H., & Rahe, R.H. (1967). The social readjustment rating scale. *Journal of Psychosomatic Research, 11,* 213–218.

Holzer, C., Robins, L., Meyers, J., Weissman, M., Tischler, G., Leaf, P., Anthony, J., & Bednarksi, P. (1984). Antecedents and correlates of alcohol abuse and dependence in the elderly. In G. Maddox, L. Robins & N. Rosenberg (Eds.), *Nature and extent of alcohol abuse among the elderly* (DHHS No. ADM 84–1321, pp. 217–244). Washington, DC: U.S. Government Printing Office.

Hubbard, R.W., Santos, J.F., & Santos, M.A. (1979). Alcohol and older adults: Overt and covert influences. *Social Casework: The Journal of Contemporary Social Work, 60,* 166–170.

Hultsch, D.F., & Plemons, J.K. (1979). Life events and life-span development. In P.B. Baltes & O.G. Brim (Eds.), *Life-span development and behavior.* New York: Academic Press.

Joreskog, K.G., & Sorbom, D. (1981). LISREL V: *Analysis of structural relationships by maximum likelihood and least squares methods.* Chicago, IL: National Educational Resources.

Knupfer, G., & Room, R. (1964). Age, sex and social class as factors in amount of drinking in a metropolitan community. *Social Problems, 12,* 224–240.

Krause, N. (1986). Social support, stress, and well-being among older adults. *Journal of Gerontology, 41,* 512–519.

Krohn, M.D., Lanza-Kaduce, L., & Akers, R.L. (1984). Community context and theories of deviant behavior. *Sociological Quarterly, 25,* 353–371.

Krohn, M.D., Skinner, W.F., Massey, J., & Akers, R.L. (1985). Social learning theory and adolescent cigarette smoking: A longitudinal study. *Social Problems, 32,* 455–471.

La Greca, A.J., Akers, R.L., & Dwyer, J.W. (1988). Life events and alcohol behavior among older adults. *The Gerontologist, 28*(4), 552–558.

Lanza-Kaduce, L., Krohn, M.D., & Akers, R.L. (1984). Cessation of alcohol and drug use among adolescents. *Deviant Behavior, 5,* 79–96.

Lauer, R.M., Akers, R.L., Massey, J., & Clarke, W. (1982). The evaluation of cigarette smoking among adolescents: The Muscatine study. *Preventive Medicine, 11,* 417–428.

Maddox, G.L., Robins, L.N., & Rosenberg, N. (Eds.). (1984). *Nature and extent of alcohol problems among the elderly* (NIAAA Publication 84–1321). Rockville, MD: U.S. Department of Health and Human Services, Public Health Service, Alcohol, Drug Abuse, and Mental Health Administration.

McCrae, R.R. (1982). Age differences in the use of coping mechanisms. *Journal of Gerontology, 37,* 454–460.

McCusker, J., Cherubin, C.E., & Zimberg, S. (1971). Prevalence of alcoholism in general municipal hospital population. *New York State Journal of Medicine, 71,* 751–754.

Meyers, A.R., Goldman, E., Hingson, R., Scotch, N., & Mangione, T. (1981–82). Evidence of cohort and generational differences in drinking behavior of older adults. *International Journal of Aging and Human Development, 14,* 31–44.

Mulford, A. (1964). Drinking and deviant behavior, USA, 1963. *Quarterly Journal of Studies on Alcohol, 25,* 634–650.

Nam, C.B., & Powers, M.G. (1983). *The socioeconomic approach to status management.* Houston: Cap and Gown.

National Institutes on Alcohol Abuse and Alcoholism (1982). Alcohol problems in the elderly compounded by many factors. Information and Feature Service 103 (December), 1.

Paykel, E.C., Prusoff, B.A., & Uhlenhuth, E.H. (1971). Scaling of life events. *Archives of General Psychiatry, 25,* 340–347.

Pearson, F.S., & Weiner, N.A. (1985). Toward an integration of criminological theories. *Journal of Criminal Law and Criminology, 76,* 116–150.

Petersen, D.M., & Whittington, F.J. (1977). Drug use among the elderly: A review. *Journal of Psychedelic Drugs, 9,* 25–37.

Robins, L.N., Helzer, J.E., Croughan, J., & Ratcliff, K.S. (1981). National Institute of Mental Health diagnostic interview schedule: Its history, characteristics, and validity. *Archives of General Psychiatry, 38,* 381–389.

Rosin, A.J., & Glatt, M.M. (1971). Alcohol excess in the elderly. *Quarterly Journal of Studies on Alcohol, 32,* 530–539.

Sellers, C., & Akers, R.L. (1988, November). *Alcohol and the elderly: A new test of social bonding theory.* Paper presented at the Annual Meetings of the American Society of Criminology, Chicago.
Schuckit, M.A. (1977). Geriatric alcoholism and drug abuse. *The Gerontologist, 17,* 168–174.
Schuckit, M.A., & Miller, P.L. (1975). Alcoholism in elderly men: A survey of a general medical ward. *Annals of the New York Academy of Sciences, 273,* 558–571.
Smart, G., & Liban, C.B. (1982). Predictors of problem drinking among elderly, middle-aged and youthful drinkers. In D.M. Petersen & F.J. Whittington (Eds.), *Drugs, alcohol and aging* (pp. 43–53). Dubuque, IA: Kendall/Hunt.
Streib, G.F., Folts, W.E., & La Greca, A.J. (1984). Entry into retirement communities: Process and related problems. *Research on Aging, 6,* 257–270.
Stumphauzer, J.S. (1986). *Helping delinquents change.* New York: Haworth Press.
Sutherland, E.H. (1947). *Principles of criminology.* Philadelphia: J.B. Lippincott.
Vogel-Sprott, M., & Barrett, P. (1984). Age, drinking habits and the effects of alcohol. *Journal of Studies on Alcohol, 45,* 517–521.
Warheit, G.J., Wobie, H.K., & Black B. (1986). *Alcohol behaviors, abuse, and dependence among age and sex groups.* Unpublished manuscript, University of Florida.
Wechsler, H., Demone, H.H., & Gottlieb, H. (1978). Drinking patterns of greater Boston adults: Subgroup differences on the QFV index. *Journal of Studies on Alcohol, 39,* 1158–1165.
White, H.R., Bates, M.E., & Johnson, V. (1990). Social reinforcement and alcohol consumption. In W.M. Cox (Ed.), *Why people drink.* New York: Gardner.
White, H.R., Johnson, V., & Horwitz, A. (1986). An application of three deviance theories for adolescent substance use. *International Journal of the Addictions, 21,* 347–366.
Wilsnack, R.W., & Cheloha, R. (1987). Women's roles and problem drinking across the life span. *Social Problems, 34*(3), 231–248.
Zimberg, S. (1974a). The elderly alcoholic. *The Gerontologist, 14,* 221–224.
Zimberg, S. (1974b). Two types of problem drinkers: Both can be managed. *Geriatrics, 29*(8), 135–138.

CHAPTER 13

Women and Alcohol: Psychosocial Aspects

EDITH S. LISANSKY GOMBERG

There has been a continuous production of reviews of literature about women and alcoholism for more than 30 years (Beckman, 1975; Blume, 1986; Burtle, 1979; Corrigan, 1980; Curlee, 1967; Gomberg, 1976a; Gomberg, 1976b; Gomberg, 1989a; Kinsey, 1966; Leland, 1982; Lindbeck, 1972; Lisansky, 1957; McCrady, 1984; National Institute on Alcohol Abuse and Alcoholism [NIAAA], 1980, 1986; Roman, 1988; Schuckit, 1972; Wilsnack & Beckman, 1984). This list does not even include the many popular writings, annotated bibliographies, and articles in journals and in books on alcohol-related problems. Reviews have been published in Canada (e.g., Kalant, 1980), in the United Kingdom (e.g., Camberwell Council on Alcoholism, 1980), and in other countries. A complete list of all the works published in the last 30 years about women's drinking behaviors and women's alcohol-related problems would look impressive until compared with a similar list about male drinking behavior and problems. In examining treatment outcome research, for example, it has been pointed out that over a 30-year period, women constituted approximately 7% of the research subjects (Vannicelli, 1984). Although the research literature has improved somewhat in this regard, it is still unfortunately true that many studies fail to distinguish male and female data.

The point that women are underrepresented in the scientific literature about alcoholism has been made many times. The original *Society, Culture, and Drinking Patterns* (Pittman & Snyder, 1962) did contain a section called "Age and Sex," which dealt with adolescents and college students, and each of the three chapters did indeed note the existence of female adolescent and college students. The chapter by Robins, Bates, and O'Neal (1962) on adult drinking patterns of former problem children also noted "sex differences." It was Jackson (1962) in her chapter on alcoholism and the family who was closest to recognition of women's alcohol

Edith S. Lisansky Gomberg is with the Department of Psychiatry, University of Michigan, Ann Arbor, Mich. This chapter was written especially for this book.

problems; she discussed research findings about wives of alcoholics and about women alcoholics as well. The editors, in a footnote, referred the reader elsewhere: "A review of the state of knowledge regarding the woman alcoholic, a subject treated only incidentally in this volume... may be found in Lisansky" (Pittman & Snyder, 1962, p. 226).

In dealing with women's drinking behaviors and alcoholism, one is confronted by interface of two challenging research issues: female biology, psychology, and social role on the one hand, and the effects of alcohol, intoxication, and heavy drinking on the other. The psychology of women was described by Freud as "a dark continent," and there is currently Sturm und Drang about Gilligan's (1982) view of female development and psychological characteristics as different from men's. If alcoholism or any other psychopathological behavior is the same for men and women, why study women at all? Does one need to argue that there are major differences in biology, psychology, and social role of the sexes? As for the effects of alcohol, intoxication, and heavy drinking, these are research questions fraught with conflicting views (Gomberg, 1989c) and we are still arguing about the definition and diagnosis of alcoholism and whether alcoholism is or is not a disease. [See chapters 21, 22, and 39, by Fingarette, Hore, and Roman & Blum, respectively, in this volume.] In this chapter I will review the current state of knowledge on female drinking, problem drinking, and alcoholism. Then I will present recent findings from a Michigan study of women in treatment for alcoholism and matched controls.

Female Drinking

Although there is an occasional challenge, it is generally agreed that more men drink than women and that men drink in larger quantities. Such a generalization, however, becomes complicated because there are differences across the life span (Fillmore, 1987; Wilsnack & Cheloha, 1987; Wilsnack, Klassen, & Wilsnack, 1986) and there are differences in the drinking behaviors of women related to variations in education, marital status, income, employment, race, and the drinking status of the woman's spouse or lover. More drinking, that is, higher frequency or quantity (or both), is associated with the young adult years, ages 21–34 (Wilsnack, Wilsnack & Klassen, 1984), with nonmarried status—including being single, divorced, separated, or cohabiting—and with higher educational achievement (Celentano & McQueen, 1984). Ethnicity and race are also related to women's drinking: when Mexican-American women are compared with Anglo women, more abstainers are reported among the former (Holck, Warren, Smith, & Rochat, 1984); and a recent survey comparing black and white women shows black women in all age groups (except 30–39) with a higher proportion of abstainers than white women (Herd, 1988). (See chapters 16 and 17, by Harper & Saifnoorian and Neff, respectively, in this volume.)

The role of employment outside the home and its relationship to women's drinking is not clear. Table 1 presents some of the mixed results obtained in comparison of workplace with at-home women. Obviously these findings are related not only to workplace status but to age, marital status, the nature of employment, and a host of other factors. Earlier work indicated that there was more heavy drinking among workplace women than among married housewives, but Wilsnack, Wilsnack, and Klassen (1984) found more heavy drinking among housewives than among employed women.

The last study listed in Table 1 is not a survey of working women and housewives but rather one of alcoholic women and age-matched nonalcoholic women: fewer of the alcoholic women were workers, and more of them were unemployed (seeking work) or housewives (Gomberg, 1986). This is a complex issue indeed because whereas being unemployed and staying at home might relate to the woman's alcoholic drinking career, it is also true that the alcoholic and nonalcoholic women in this study, who came from similar socioeconomic backgrounds, were not comparable in educational achievement (Gomberg, forthcoming). Perhaps the lack of clarity about workplace status and drinking behavior is linked to the heterogeneity of women's motivations to enter the workplace. But if the workplace is indeed conducive to heavier drinking to a greater extent than staying at home, the finding suggests that there will be no decrease of heavy or problem drinkers among women in the future since the percent of women in the workplace keeps increasing. It also suggests the importance of trying to reach women in the workplace through education and case finding.

There is a convergence hypothesis: rates of drinking and problem drinking have increased faster among women than among men, and drinking patterns of men and women are becoming alike. This hypothesis was examined and rejected by Ferrence (1980) who stated: "There is no clear evidence to support the contention that rates of problem drinking for women and men have converged" (p. 114). A more recent study also rejects the convergence hypothesis (Bell, Havlicek, & Roncek, 1984). Several studies, however, report no sex difference whatsoever in alcohol consumption (Richman & Flaherty, 1986) or a "dramatic increase" in substance use among women aged 18 to 34 (Kaestner, Frank, Marel, & Schmeidler, 1986). [See chapter 11, by Pandina, White, & Milgram, in this volume.] Johnston, O'Malley, and Bachman (1989) report that in their annual survey of high school students' drug use:

TABLE 1
Social and Heavy Drinking by Women In the Workplace and Other Women

Author(s)	Sample	Proportion of Drinkers among Working Women and Others	Proportion of Heavy Drinkers among Working Women and Others
Siassi et al., 1973	937 UAW members	23% of working women 60% of "national sample"	38% (of those who drink) were heavy drinkers 8% of "national sample"
Johnson, 1982	NIAAA national survey of 1141 women, 18 and over	59% of working women 50% of women at home	27% of married workers 26% "separated" workers 17% unemployed, separated women 17% of those "not in workforce"
Celentano & McQueen, 1984	1100 Baltimore women	72% of employed women 56% of "not employed" women	9.4% of employed women 6.8% of nonemployed women.
Wilsnack et al., 1984	917 women, national survey	62% of married employed women 57% of married housewives	4% of married employed women 6% of married housewives
Shore, 1985	147 women in business/ professional jobs	97% of business women 62% of women in "national surveys"	11% heavy drinking rate
Gomberg, 1986	301 alcoholic women in treatment and 137 matched control women		Work status Working: 53% of alcoholics 77% of controls Unemployed: 19% of alcoholics 6% of controls Housewives: 27% of alcoholics 17% of controls

> There remains a quite substantial sex difference among high school seniors in the prevalence of occasions of heavy drinking (27% for females vs. 43% for males in 1988) but this difference has been diminishing very gradually since the study began over a decade ago. There also remain very substantial sex differences in alcohol use among college students and young adults generally, with males drinking more. However, there has been little change in the differences between 1980 and 1988. (p. 13)

What needs to be distinguished is a male-female comparison of alcohol and drug use, on the one hand, and, on the other, changes in alcohol and drug use among women, particularly young women, over time. Among college students there are a number of illicit substances whose patterns of use show that differences between the sexes are diminishing. Generally, males more frequently use alcoholic beverages and consume more of them than females in all age groups—although the gender differences are less than they were a generation ago. For example, annual studies of male and female college students from 1980 to 1988 (Johnston, O'Malley, & Bachman, 1989) show the prevalence (use within the last year) to be almost identical for the sexes.

In examining the changes in alcohol use among women over time, one concludes that whereas the percentage of U.S. women who drink has remained quite steady for the last 20 years or more—most national surveys finding that 60–61% of the female population report they drink—questions may be raised about the make-up of the 60%. It is more than possible that two trends combine to keep the percentage steady: the proportion of older women in the female population, who are less likely to report themselves as drinkers, grows larger, but this potential drop is offset by an increasing proportion of younger drinkers. When Straus and Bacon (1953) investigated college students' drinking, 79% of the men and 61% of the women students reported that they drank. When college students were surveyed 20–25 years later, the male percentage had risen a little; the female percentage had risen to 75% (Fillmore, 1987).

The changes are also reflected in alcohol-related problems. There is evidence that the proportion of young women drivers' involvement in alcohol-related fatal crashes has increased, particularly in the age group 21–24 (Popkin, 1989). Comparing various age cohorts of women, Fillmore (1984) concluded that "the youngest cohort, who in 1979 were ages 21–29, seem to show a much higher rate of heavy-drinking than past cohorts measured at the same age, particularly among the employed" (p. 31).

When men and women's drinking and alcohol-related problems are compared with each other over time, one may conclude that rates are not converging so that gender differences disappear but that these same gender differences have grown smaller. And when we look at women's use of alcohol and drugs over time, it is clear that problems associated with heavier drinking have increased among women, primarily as a function of younger drinkers, those under 40.

Female Drinking: Motives and Effects

There are several interesting issues raised in the research literature about motives and effects. First, recent studies of adolescents (Windle & Barnes, 1988) and college students (Ratliff & Burkhart, 1984) suggest that pleasure seeking as a motive for drinking is very similar for the two sexes. A search for excitement and for sensation is a characteristic reason for drinking among the young, and although some young women may drink for escapist reasons, hedonistic reasons are more characteristic. At the same time, there are several reports suggesting that some young women may drink alcoholic beverages to minimize feelings of depression. However, alcohol may serve to minimize depression in the short run but increase feelings of depression in the long run (Aneshensel & Huba, 1983; Birnbaum, Taylor, & Parker, 1983). Second, an interesting sex difference has been observed in the effects of alcohol on self-disclosure (Caudill, Wilson, & Abrams, 1987). When young women subjects believe they are intoxicated, they display significantly lower levels of self-disclosure than other young women who have been told their beverage is nonalcoholic. This finding relates to other findings about women's expectations and beliefs about the effects of alcohol; for example, women report increased sexual arousal as their blood alcohol level rises, although physiological measures indicate the opposite is true (Wilson & Lawson, 1978). Other experiments show that whereas males show less anxiety when drinking with women, women show more anxiety when drinking with men (Marlatt, 1986). Thus, in spite of the pleasure-seeking motive for drinking, young women believe that drinking makes them "sexier" and probably more vulnerable: hence the increase in anxiety when drinking with men and the drop in self-disclosure. The anxiety is not unrealistic, because the work of George, Gournic, and McAfee (1988) has indicated that among college students drinking women are viewed differently (e.g., as more sexually available) than their nondrinking counterparts.

Epidemiology of Female Problem Drinking and Alcoholism

One needs to separate alcohol-related problems, the consequences of drinking—usually of heavy drinking—from clinical alcoholism. The mean age of women in treatment for alcoholism was reported in 1978 as the mid-40s (Armor, Polich, & Stambul, 1978). More recently, clinical reports suggest that the average age of women in treatment is more likely to be in the 30s. Fillmore (1987), comparing men's and women's problem drinking across the life span, found men's rates to remain fairly even over the adult life course but women to be at the highest risk for "serious" alcohol-related problems while they are in their 30s.

The numbers of men and women patients in any alcoholism treatment facility varies widely, depending on which population of alcohol abusers is treated. Women are apparently more likely to appear at middle-class and upper-class directed facilities, such as the offices of physicians, private hospitals, and some outpatient facilities. They are far less likely to appear in medical centers for veterans, correctional facilities, and Skid Row shelters.

Who are the women at high risk for problem drinking and alcoholism? The following list, assembled from a number of sources, indicates that prevention efforts would profitably be directed towards the following groups.

1. Women raised in alcoholic families, in which a parent or older sibling is drinking heavily during the woman's growing-up years.
2. High school girls who are not college-bound and not interested in school, and who are sexually precocious, and marked by early use of marijuana.
3. Young, nonmarried women in the workplace or in search of employment.
4. Women whose spouse or lover or close friends drink heavily.
5. Recently divorced or separated women, particularly those heading a one-parent family with young children.
6. Women in crisis situations, for example, women who have recently experienced a traumatic loss like a death, hysterectomy, divorce, and so on.
7. Women who use alcohol to cope with depressed affect.
8. Lesbians.
9. Women in the military.
10. Women in the criminal justice system.

The question of a history of childhood abuse as antecedent to female alcoholism has been raised; alcoholic women in treatment frequently report childhood abuse. There are several intervening conditions involved in the linkage of childhood abuse and alcoholism, one of those being alcoholism or drug use in a parent. Interestingly enough, sexual abuse was rarely reported as perpetrated by a parent, but the absence of ordinary parental protection and the isolation of a child growing up in an alcoholic family appear to be relevant (Miller, Downs, Gondoli, & Keil, 1987). Spousal abuse is another question but is probably a concomitant rather than a cause of male or female drinking problems.

Wilsnack and Cheloha (1987) have summarized findings about women at risk at different points in the life span: among younger women (aged 21–34), those who have not been married and are not employed full time are most at risk. Women aged 35–49 are most at risk when their marriages are disrupted, they are not employed outside the home, or they have no children living at home (or all three). Older women (50–64) who are married but have no job or children living at home are most at risk. At all ages, risk is multiplied if the woman's "significant other" is a heavy drinker.

Female Problem Drinking and Alcoholism: Comparisons of Female and Male Alcoholics

Several research studies have investigated differences between male and female alcoholics in clinical populations.

First, in clinical populations, women alcoholics report a family history positive for alcoholism more frequently than male alcoholics do (Armor, Polich, & Stambul, 1978; Lisansky, 1957; Winokur & Clayton, 1968). A survey of professional people who were members of Alcoholics Anonymous produced a report of positive family history by 29% of the men but 41% of the women (Bissell & Haberman, 1984). The high frequency with which women alcoholics report having grown up with one or more siblings who drank heavily or became problem drinkers later has been consistent (Gomberg, 1989b; Lisansky, 1957).

Second, although there is a trend toward convergence, males continue to begin drinking earlier and to become problem drinkers at an earlier age than women. Women experience their first drink and first intoxication later than men. They also develop problem drinking later and come to treatment facilities with a shorter duration of alcoholism than men do (Lisansky, 1957; Ross, 1989). This pattern also apparently holds true when the drug of choice is other than alcohol; it appears that women take less time to become addicted to narcotics (Anglin, Hser, & McGlothlin, 1987).

Third, women are more likely than men to cite a stressful event as a precipitant to alcoholic drinking. This finding has produced a number of questions; for example, are these events cited because women are more defensive? more likely to use denial? more vulnerable to environmental stress? more likely to report consequent stresses as antecedent to alcoholism? Various reports have interpreted this gender difference in various ways (e.g., Allan & Cooke, 1985; Lisansky, 1957). One difficulty in studying this question lies in the various definitions of *stressful event,* which have included specific traumatic loss, prolonged depressed mood, and loneliness. The recent emphasis by practitioners on the alcoholics' reports of physical and sexual abuse in their family of origin may be a variant of trauma-as-precipitant theories, because the time interval between the painful event and the development of problem drinking has never been specified.

Fourth, men in treatment facilities less frequently report a spouse or lover who is a heavy or problem drinker than women do. The importance or the social environment in facilitating heavy drinking among women has been emphasized by the epidemiological findings of Wilsnack, Wilsnack, and Klassen who in 1984 reported "the strong associations that women's use of alcohol had with drinking by close companions (spouses, siblings, and friends)" (p. 1237). It appears, then, to be true of both clinical populations and the general population that women are likely to have drinking patterns that resemble those of significant others in their social world. A history of female alcoholism that includes antecedent heavy drinking by the husband is quite common, but the reverse—alcoholism à deux begun with the woman's drinking—is not reported often. Women alcoholics admitted to a treatment facility were reported as three to five times more likely to have an alcoholic spouse than male alcoholics (Hesselbrock, et al., 1984).

Fifth, at-home versus public drinking also differentiates male and female alcoholism. Although women vary by age, (that is, younger women are more likely to be drinking in public places than older women), taken in the aggregate, women do more drinking at home—alone or with someone close to them—than men do (Wanberg & Knapp, 1970). Women have fewer and shorter drinking bouts, and do less morning drinking, than men (Rimmer, Pitts, Reich, & Winokur, 1971).

Sixth, Cloninger (1987) has a typology of alcoholism that includes Types I and II. Type I, "milieu-limited," is associated with a positive family history, relatively later onset, milder symptoms, and rare involvement in criminal activity. Type I, although characterized by genetic predisposition, requires an environmental releaser. Type II is seen only among males and is characterized by early onset, antisocial behavior, and greater severity. Genetic predisposition for Type II is independent of environmental influence. Cloninger writes:

> Women develop loss of control (Type I) alcoholism predominantly, with a later onset and more rapid progression of complications associated with guilt, depression, and medical complications from sustained high blood levels of alcohol, such as cirrhosis and liver abnormalities. Both types of alcoholism are common in men but Type II alcoholism is characteristic of most men in hospital treatment samples. (p. 411)

Seventh, when men and women alcoholics in a treatment facility are compared on marital status, there appears to be a consistent difference:

women alcoholics report more marital disruption than the men do. Whether this is true because there is less tolerance for the alcoholic women than for the men, we do not really know. Certainly there are different social attitudes toward the wife of an alcoholic man than toward the husband of an alcoholic woman, although the nature of these attitudes has rarely been investigated.

Eighth, the extent to which other drugs are used and abused by alcoholics varies by sex, by drug, and by age. Generally, males more frequently use illegal drugs (e.g., heroin), and females more frequently use prescription drugs (e.g., minor tranquilizers, sedatives, hypnotics). What is true of the general population is also true of alcoholics. Some authors will describe the use of illegal substances as "drug abuse" and the use of prescribed substances as "drug use," but the fact remains that whether the drugs are "street drugs" or prescribed medications there are gender differences in the patterns of drug use or abuse. Age differences also complicate the picture: at this time, younger men as well as younger male alcoholics are more likely to be using illegal substances than older men or older alcoholics (Gomberg, 1982). One finds the same thing when comparing younger and older women and younger and older female alcoholics.

Ninth, research results on comparative prognoses for male and female alcoholics in treatment are mixed. Earlier in the 20th century, women were seen as having poorer prognoses (Karpman, 1948), but it has never been clear whether that perception meant that (1) they behaved in ways that caretakers described as more difficult to manage and "worse" than male patients (Ellinwood, Smith, & Vaillant, 1966), (2) they were more likely to manifest other psychiatric disorders in addition to alcoholism, or (3) they dropped out of treatment earlier. Although women alcoholics (like women narcotic addicts or women schizophrenics) have been described as more difficult patients, recent work yields a mixed picture about prognosis. Comparatively low recovery rates for women in the past may have been the consequence of a self-fulfilling prophecy, but analysis of prognosis studies comparing results with male and female patients yields ambiguous results (Annis & Liban, 1980). We have no doubt that there are gender differences in therapeutic concerns. For example: "The dominant area in the life events world of alcoholics seems to vary according to sex, with men attributing greater meaning to work events, women to private life events" (Remy, Soukup-Stepan, & Tatossian, 1987, p. 49).

Tenth, on a number of other comparisons, male-female differences among problem drinkers are not clear. Jellinek (1952) described a phasic model for symptom progression, based on a sample of alcoholic males.

[See chapter 20, by Jellinek, in this volume.] This phasic model has, in recent years, been disregarded. Two research reports, however, have dealt with this symptom progression among women alcoholics (James, 1975; Piazza, Peterson, Yates, & Sundgren, 1986). It is interesting to note that women alcoholics report morning drinking earlier in the symptom sequence than male alcoholics do, whereas they report guilt feelings and persistent remorse later.

A typology of female alcoholics has differentiated primary from secondary alcoholism (Schuckit, Pitts, Reich, King, & Winokur, 1969), secondary alcoholism being the type that follows from depression. Many people in alcohol studies would agree that women alcoholics show more diagnosable depressive disorders than male alcoholics, but the sequence of these disorders is still in question.

This line of thought raises the question of comorbidity, which is really three questions: (1) Do men and women alcoholics differ in the extent to which they present "dual diagnoses"? (2) Are those diagnoses different, that is, do men and women present as alcoholics with similar accompanying disorders? (3) Do the other diagnostic disorders antedate the alcoholism—do both disorders develop simultaneously—or does the disorder result from the drinking?

Results are ambiguous about whether men and women differ in the extent to which they present dual diagnoses. In a large epidemiological sample, Helzer and Pryzbek (1988) found a male-to-female ratio of alcoholism of 5:1 but added that

> since it is so much more deviant for women to be alcoholic than men, it is not surprising that the association of alcoholism with other diagnoses is stronger in women. In fact, while 44% of the male alcoholics have a second diagnosis, 65% of the female alcoholics do. (p. 222)

This is consistent with an earlier work reporting more psychiatric treatment history in a group of women alcoholics than in a group of male alcoholics in the same facility. However, a recent study of 260 male and 241 female alcoholic patients concluded that women do not show higher rates of psychiatric disorder than men (Ross, Glaser, & Stiasny, 1988).

There is more agreement on whether the diagnoses are different. Helzer and Pryzbek (1988) reported more antisocial personality among male than female alcoholics. Bedi and Halikas (1985) found a lifetime rate of affective disorder of 43% in females and 29% in males. Hesselbrock, Meyer, and Keener (1985) reported that 18% of hospitalized male alcoholics and 38% of female hospitalized alcoholics were depressed and that the women also showed larger percentages of phobic, obsessive-compulsive, and panic disorders. The findings of Ross, Glaser, and Stiasny (1988) agree in noting that women alcoholics indeed showed more neurotic disorders and men were more likely to be diagnosed as having antisocial personality, but they disagree with other findings: no significant gender differences were found in the prevalence of affective disorders.

Do the other diagnostic disorders antedate the alcoholism? Do both disorders develop simultaneously? Or does the disorder result from the drinking? The distinction between primary and secondary alcoholism among women was one attempt to answer this question. A fair amount of research effort has been expended recently in experimental studies of women's drinking and depression to demonstrate that female depression, noted clinically at admission, is indeed a result of the biochemical effects of alcohol. There are also some reports suggesting that depression is antecedent to problem drinking and that the heavy intake of alcohol is part of an attempt to cope with depressed feelings. The Michigan study of alcoholic women (described below) found no significant age differences in depression among alcoholic women: in the age range compared, they showed significantly more depression than an age-matched sample of nonalcoholic women, but alcoholic women who began drinking early in life and those who began later showed little difference in depressed affect. Where age differences were significant, there was an association of younger age at onset of alcoholism and younger age at admission with impulsivity and acting-out behaviors (Gomberg, 1982, 1986; Hesselbrock, et al., 1984). The association of earlier onset and antisocial personality held for both sexes.

Eleventh, and finally, sex differences in the consequences of alcoholism (medical, familial, legal, and occupational) are inevitable when one considers the differences in biology, drinking behaviors, and social roles. Health problems differ, with males reporting more episodes of delirium tremens, blackouts, and alcohol-related accidents (Rimmer, et al., 1971) and women more frequently reporting the development of hepatic disorder (Hill, 1986). Gender differences appear in the frequency with which job-related and legal consequences are reported. This finding is probably related to social role, that is, men being more likely to be in the workplace, to drink in public places, to act out belligerence. In familial consequences, we have noted the greater frequency of marital disruption reported by alcoholic women, and it is likely that guilt, related to parenting, is greater among alcoholic women. Although much has been written about the greater stigma attached to female alcoholism, some questions on this point have been raised. A Finnish survey (International Council on Alcohol and Addictions, 1985) suggests that "for a given amount of drinking, men are more likely to have been criticized by others than women" (p. 26)!

Michigan Study: Alcoholic Women in Treatment

The subjects of the study, 301 women in 21 different treatment facilities in southeastern Michigan, were interviewed after 1–2 weeks of detoxification. The interview sites included inpatient, outpatient, small-town, and urban facilities, and the women interviewed had been variously referred by themselves, physicians, family members, the courts, or friends. All consecutive admissions between the ages of 20 and 50 were asked to participate; there was a 7% refusal rate. Women were divided into three age groups: 20 to 29 ($N = 99$), 30 to 39 ($N = 108$), and 40 to 49 ($N = 94$). A woman's age at onset was, inevitably, related to her current chronological age: the mean age of alcoholic women in their 20s at onset was 19.58 years (SD = 3.18), that of those in their 30s (mean age) at onset was 28.34 years (SD = 5.35), and that of those in their 40s (mean age) at onset was 36.52 years (SD = 5.93). A control group, matched for age and socioeconomic background, was obtained by nomination, either by study participants or by female members of local chapters of Alcoholics Anonymous. Nonalcoholic women included 34 women in their 20s, 53 in their 30s, and 50 in their 40s. The white-collar or blue-collar status of respondents' parents was compared with a random sample of women between the ages of 20 and 50 interviewed in a national study (Veroff, Douvan, & Kulka, 1981). The socioeconomic status (SES) of the national study women was lower than that of the alcoholic and control groups in this study, but SES differences between alcoholic and control groups were small and nonsignificant.

The interview schedule, which was modified only in the section on drinking for the control group, included items related to childhood and adolescence, adult roles, social supports, psychiatric health symptomatology, and drinking history.

Results

Antecedent or prealcoholic events. In a finding consistent with past findings, a positive family history of heavy or problem drinking appeared significantly more often with the alcoholic women; the differences were greatest when drinking by the father and drinking by one or more siblings (while the respondent was growing up) were compared for alcoholic and nonalcoholic women. Interestingly enough, the presence of a heavy or problem drinker in the family did not distinguish between different age groups of alcoholic women. Although results with male alcoholics have indicated that more family history of alcoholism is associated with earlier onset, the same was not found for this sample of female alcoholics (Gomberg, 1986).

When alcoholic and nonalcoholic groups were queried about negative events in early life (e.g., not living with both real parents), few differences distinguished one from the other. When the group members were asked about their affective response to early life events (e.g., feeling unloved and unwanted), however, differences were significant (Gomberg, 1989b). Alcoholic women reported a significantly higher degree of childhood depression. The two groups differed in early life emotional problems and in problem behaviors linked with impulse control, such as temper tantrums, enuresis, running away from home, and so on (Gomberg, 1989b). Clearly there is a link between the more frequent presence of heavy or problem drinking in the family of origin, the report of more parental conflict, and conflict with parents on the part of the prealcoholic women; it is interesting that there were significant differences between alcoholics and nonalcoholics in reported closeness to their mother, but not in reported closeness to their father, who was usually the heavy or problem drinker.

The best predictors of early onset of alcohol problems in females included age at first intoxication, the use of drugs other than alcohol in early adolescence, childhood temper tantrums, and reported unhappy childhood (Gomberg, 1988a).

Role behaviors. Although they came from similar social backgrounds, the alcoholic women reported leaving school, leaving home, and getting married at an earlier age than the control women (Gomberg, 1989a). Inevitably, this pattern has led them to lower occupational status, indicated in the lower part of Table 2.

As Table 2 shows, the alcoholic women were more likely to be unemployed or to be homemakers, and the control women were more frequently working outside the home. The differences are most striking for the women in their 40s.

In findings consistent with those of previous studies, alcoholic women reported more marital disruption and were more often married to a heavy or problem drinker. Significant differences also appeared in frequency of miscarriage and hysterectomy. Not only did the alcoholic women agree significantly more frequently than the controls that an intoxicated woman is more "obnoxious and disgusting than an intoxicated man," but more of them believed that the mother's alcoholism has a worse effect on the children than the father's alcoholism (Gomberg, 1988a).

Depression. Two indices of depression, one reflecting low self-esteem and the other current mood, were both associated with earlier age at on-

TABLE 2
Michigan Study: Alcoholic Women in Treatment
Employment status and occupations of alcoholic and nonalcoholic women, in percentages

	Working	Unemployed	Homemaker
Alcoholic women:			
19–29 years	68.9	22.2	8.9
30–39 years	58.5	15.7	27.8
40–49 years	39.6	17.6	42.9
Nonalcoholic women:			
21–29 years	78.1	9.4	12.5
30–39 years	81.1	5.7	13.2
40–49 years	71.4	4.1	24.5

OCCUPATIONS OF EMPLOYED ALCOHOLIC AND NONALCOHOLIC WOMEN*

	Professional	Managers	Clerical	Sales	Craft/ operative	Service labor
Alcoholic women	24.5	8.8	22.0	5.7	19.5	19.5
Nonalcoholic women	33.0	16.5	27.2	5.8	2.9	14.6

* $\chi^2 = 19.6$, $df = 5$, $p = <.001$

set (Turnbull & Gomberg, 1988). The frequency with which a suicide attempt was reported was significantly higher with alcoholic women than with controls and, as Table 3 shows, there were striking age differences among the alcoholic women, the youngest group being most at risk (Gomberg, 1989d).

Social supports. The alcoholic women reported less prealcoholic support during childhood and adolescence than the controls reported. Although the alcoholic women saw themselves as giving as much support to others as the nonalcoholic women, they saw themselves as receiving less current social support (Schilit & Gomberg, 1987). In view of the state of their lives when they enter treatment, it is not an unrealistic view.

Why do alcoholic women enter treatment? The most frequently cited circumstances driving the alcoholic women into treatment were: deepening depression (87.8%), increasing alcohol-related medical problems (67.4%), and difficulty with spouse or lover (65.4%). There were age differences in alcohol-related problems. The youngest alcoholic women entered treatment under pressure from job-related stresses, from alcohol-related symptoms like blackouts and episodes of delirium tremens, and from trouble with the police. The oldest group reported significantly more problems with a child or difficulties when a child moves out of the home (Gomberg & Turnbull, 1990).

Conclusion

Looking back over more than 30 years of interest in the drinking behaviors and alcohol-related problems of women, we can see that we have made progress. The glass that is half-full contains information about drinking in the general population and the beginning of knowledge of the factors that influence a choice of abstinence, moderate-drinking, or heavy drinking. That glass also contains information about the onset, development, manifestations, consequences, and treatments of female alcoholism. The glass that is half-empty contains questions about the contexts of social drinking and what encourages more or less drinking, interactions in alcoholic families that facilitate or inhibit the development of alcohol problems in the next generation, treatment characteristics that are more or less effective with women patients, and effective ways to do primary prevention.

In spite of progress, one continues to ask: Why does the study of women, their development and their problems, lag—like women's paychecks—so far behind in spite of feminism, women in the workplace, and changing sex role definitions?

TABLE 3
Michigan Study: Alcoholic Women in Treatment
Percentages of women reporting suicide attempts

	All cases	20–29	30–39	40–49
Alcoholic	40.0	50.5	43.0	25.5
Control	8.8	8.8	9.4	8.0
χ^2	43.5	18.3	18.4	6.4
Significance	.0000	.0000	.0000	.0000
Cramer's phi	.32	.37	.34	.21

Age differences (20s vs. 30s vs. 40s)

1. Alcoholic women:	
χ^2	13.15
Significance	.0014
Cramer's phi	.21
2. Control women:	
χ^2	.66
Significance	.9673
Cramer's phi	.02

Acknowledgments

Thanks are due to those who worked in different stages of the Michigan research project, which was supported by grants from the National Institute on Alcohol Abuse and Alcoholism (AA04143), The University of Michigan Rackham School of Graduate Studies, and the National Institutes of Health Biomedical Research Grant to the Vice President for Research, University of Michigan. Contributing to the research project were: Elizabeth M. Douvan, Mary Ellen Colten, Susan Goff Timmer, David M. Klingel, Joanne E. Turnbull, and Rebecca Schilit.

References

Allan, C.A., & Cooke, D.J. (1985). Stressful life events and alcohol misuse in women: A critical review. *Journal of Studies on Alcohol, 46,* 147–52.

Aneshensel, C.S., & Huba, G.J. (1983). Depression, alcohol use and smoking over one year: A four-wave longitudinal causal model. *Journal of Abnormal Psychology, 92,* 134–50.

Anglin, M.D., Hser, Y.I., & McGlothlin, W.H. (1987). Sex differences in addict careers: 2. Becoming addicted. *American Journal of Drug and Alcohol Abuse, 13,* 59–71.

Annis, H.M., & Liban, C.B. (1980). Alcoholism in women: Treatment modalities and outcomes. In O. Kalant (Ed.), Research advances in alcohol and drug problems: Vol. 5. *Alcohol and drug problems in women* (pp. 385–422). New York: Plenum Press.

Armor, D.J., Polich, J.M., & Stambul, H.B. (1978). *Alcoholism and treatment.* New York: Wiley.

Beckman, L.J. (1975). Women alcoholics: A review of social and psychological studies. *Journal of Studies on Alcohol, 36,* 797–824.

Bedi, A.R., & Halikas, J.A. (1985). Alcoholism and affective disorder. *Alcoholism: Clinical and Experimental Research, 9,* 133–134.

Bell, R., Havlicek, P.L., & Roncek, D.W. (1984). Sex differences in the use of alcohol and tranquilizers: Testing a role convergence hypothesis. *American Journal of Drug and Alcohol Abuse, 10,* 551–561.

Birnbaum, I.M., Taylor, T.H., & Parker, E.S. (1983). Alcohol and sober mood state in female social drinkers. *Alcoholism: Clinical and Experimental Research, 7,* 362–368.

Bissell, L., & Haberman, P.W. (1984). *Alcoholism in the professions.* New York: Oxford University Press.

Blume, S.B. (1986). Women and alcohol: A review. *Journal of the American Medical Association, 256,* 1467–1470.

Burtle, V. (1979). *Women who drink: Alcoholic experience and psychotherapy.* Springfield, IL: Charles C. Thomas.

Camberwell Council on Alcoholism (1980). *Women and alcohol.* London: Tavistock.

Caudill, B.D., Wilson, G.T., & Abrams, D.B. (1987). Alcohol and self-disclosure: Analysis of interpersonal behavior in male and female social drinkers. *Journal of Studies on Alcohol, 48,* 401–409.

Celentano, D.D., & McQueen, D.V. (1984). Alcohol consumption patterns among women in Baltimore. *Journal of Studies on Alcohol, 45,* 355–358.

Cloninger, C.R. (1987). Neurogenetic adaptive mechanisms in alcoholism. *Science, 236,* 410–416.

Corrigan, W.M. (1980). *Alcoholic women in treatment.* New York: Oxford University Press.

Curlee, J. (1967). Alcoholic women: Some considerations for further research. *Bulletin of Menninger Clinic, 31,* 154–163.

Dorris, M. (1989). *The broken cord.* New York: Harper & Row.

Ellinwood, E.H., Jr., Smith, W.G., & Vaillant, G.E. (1966). Narcotic addiction in males and females: A comparison. *International Journal of the Addictions, 1,* 33–45.

Ferrence, R.G. (1980). Sex differences in the prevalence of problem drinking. In O. Kalant (Ed.), Research advances in alcohol and drug problems: Vol. 5. *Alcohol and drug problems in women* (pp. 69–124). New York: Plenum Press.

Fillmore, K.M. (1984). "When angels fall": Women's drinking as cultural preoccupation and as reality. In S.C. Wilsnack & L.J. Beckman (Eds.), *Alcohol problems in women: Antecedents, consequences and intervention* (pp. 7–36). New York: Guilford.

Fillmore, K.M. (1987). Women's drinking across the adult life course as compared to men's. *British Journal of Addiction, 82,* 801–811.

George, W.H., Gournic, S.J., & McAfee, M.P. (1988). Perceptions of postdrinking female sexuality: Effects of gender, beverage choice, and drink payment. *Journal of Applied Social Psychology, 18,* 1295–1317.

Gilligan, C. (1982). *In a different voice: Psychological theory and women's development.* Cambridge, MA: Harvard University Press.

Gomberg, E.S.L. (1976a). Alcoholism in women. In B. Kissin & H. Begleiter (Eds.), *The biology of alcoholism: Vol. 4. Social Aspects of Alcoholism* (pp. 117–166). New York: Plenum Press.

Gomberg, E.S.L. (1976b). The female alcoholic. In R.E. Tarter & A.A. Sugarman (Eds.), *Alcoholism: Interdisciplinary approaches to an enduring problem* (pp. 603–636). Reading, MA: Addison-Wesley.
Gomberg, E.S.L. (1982). The young male alcoholic: A pilot study. *Journal of Studies on Alcohol, 43,* 683–701.
Gomberg, E.S.L. (1986). Women and alcoholism: Psychosocial issues. In National Institute on Alcohol Abuse and Alcoholism (Ed.), *Women and alcohol: Health-related issues* (Research Monograph 16) (DHHS Publication No. ADM 86-1139, pp. 78–120). Washington, DC: Government Printing Office.
Gomberg, E.S.L. (1988a). Alcoholic women in treatment: The question of stigma and age. *Alcohol and Alcoholism, 23,* 507–514.
Gomberg, E.S.L. (1988b). Predicting age at onset for alcoholic women. *Alcoholism: Clinical and Experimental Research, 12,* 337. (Abstract).
Gomberg, E.S.L. (1989a). *Alcohol and women* (Center of Alcohol Studies Pamphlet Series). New Brunswick, NJ: Center of Alcohol Studies.
Gomberg, E.S.L. (1989b). Alcoholic women in treatment: Early histories and early problem behaviors. *Advances in Alcohol and Substance Abuse, 8,* 133–147.
Gomberg, E.S.L. (1989c). On terms used and abused: The concept of "codependency." Current Issues in alcohol/drug studies. *Drugs and Society, 3* (3/4), 113–132.
Gomberg, E.S.L. (1989d). Suicide risk among women with alcohol problems. *American Journal of Public Health, 79,* 1363–1365.
Gomberg, E.S.L. (forthcoming). Alcoholic women in treatment: Education and occupation.
Gomberg, E.S.L. Women and alcohol: Primary prevention of problems. Unpublished manuscript.
Gomberg, E.S.L., & Turnbull, J.E. (1990). Alcoholism in women: Pathways to treatment. *Alcoholism: Clinical and Experimental Research, 14,* 312 (abstract).
Helzer, J.E., & Pryzbeck, T.R. (1988). The co-occurrence of alcoholism with other psychiatric disorders in the general population and its impact on treatment. *Journal of Studies on Alcohol, 49,* 219–224.
Herd, D. (1988). Drinking by black and white women: Results from a national survey. *Social Problems, 35,* 493–505.
Hesselbrock, M.N., Hesselbrock, V.M., Babor, T.F., Stabenau, J.R., Meyer, R.E., & Weidenman, M.A. (1984). Antisocial behavior, psychopathology and problem drinking in the natural history of alcoholism. In D.W. Goodwin, K.T. Van Dusen, & S.A. Mednick (Eds.), *Longitudinal research in alcoholism* (pp. 197–214). Boston: Kluwer-Nijhoff.
Hesselbrock, M.N., Meyer, R.E., & Keener, J.K. (1985). Psychopathology in hospitalized alcoholics. *Archives of General Psychiatry, 42,* 1050–1055.
Hill, S.Y. (1986). Physiological effects on alcohol in women. In National Institute of Alcohol Abuse and Alcoholism (Ed.), *Women and alcohol: Health-related issues* (Research Monograph No. 16) (DHHS Publ. No. ADM 86–1139, pp. 199–214). Washington, DC: Government Printing Office.
Holck, S.E., Warren, C.W., Smith, J.C., & Rochat, R.W. (1984). Alcohol consumption among Mexican American and Anglo women: Results of a survey along the U.S.-Mexico border. *Journal of Studies on Alcohol, 45,* 149–154.
International Council on Alcohol and Addictions, Alcohol Epidemiology Section, 10th Annual Meeting (1985). Discussion session on "Women, gender, and alcohol." *Drinking and Drug Practices Surveyor, 20,* 1, 25–32.

Jackson, J.K. (1962). Alcoholism and the family. In D.J. Pittman & C.R. Snyder (Eds.), *Society, culture, and drinking patterns* (pp. 472–492). New York: Wiley.
James, J.E. (1975). Symptoms of alcoholism in women: A preliminary survey of A.A. members. *Journal of Studies on Alcohol, 19,* 1564–1569.
Jellinek, E.M. (1952). Phases of alcohol addiction. *Quarterly Journal of Studies on Alcohol, 13,* 673–684.
Johnston, L.D., O'Malley, P.M., & Bachman, J.G. (1989). *Drug use, drinking and smoking: National survey results from high school, college and young adults populations, 1975–1988* (DHHS Publ. No. ADM 89-1638). Rockville, MD: National Institute on Drug Abuse.
Kaestner, E., Frank, B., Marel, R., & Schmeidler, J. (1986). Substance use among females in New York State: Catching up with the males. *Advances in Alcohol and Substance Abuse, 5,* 29–49.
Kalant, O. (Ed.). (1980). *Research advances in alcohol and drug problems:* Vol. 5. *Alcohol and drug problems in women.* New York: Plenum Press.
Karpman, B. (1948). *The alcoholic woman.* Washington, DC: Linacre Press.
Kinsey, B.A. (1966). *The female alcoholic: A social psychological study.* Springfield, IL: Charles C. Thomas.
Leland, J. (1982). Gender, drinking, and alcohol abuse. In A. Ihsan (Ed.), *Gender and psychopathology* (pp. 201–235). New York: Academic Press.
Lindbeck, J. (1972). The woman alcoholic: A review of the literature. *International Journal of Addictions, 7,* 567–580.
Lisansky, E.S. (1957). Alcoholism in women: Social and psychological concomitants: I. Social history data. *Quarterly Journal of Studies on Alcohol, 18,* 588–623.
Marlatt, G.A. (1986). Sex differences in psychosocial alcohol research: Implications for prevention. In National Institute on Alcohol Abuse and Alcoholism (Ed.), *Women and alcohol: Health-related issues* (Research Monograph 16) (DHHS Publication No. ADM 86-1139, pp. 260–271). Washington, DC: U.S. Government Printing Office.
McCrady, B.S. (1984). Women and alcoholism. In E.A. Blechman (Ed.), *Behavior modification with women* (pp. 428–448). New York: Guilford.
Miller, B.A., Downs, W.R., Gondoli, D.M., & Keil, A. (1987). The role of childhood sexual abuse in the development of alcoholism in women. *Violence and Victims, 2,* 157–172.
National Institute on Alcohol Abuse and Alcoholism (Ed.). (1980). *Alcohol and women* (Research Monograph 1) (DHEW Publication No. ADM 80–835). Washington, D.C.: U.S. Government Printing Office.
National Institute on Alcohol Abuse and Alcoholism (Ed.). (1986). *Women and alcohol: Health-related issues* (Research Monograph 16) (DHHS Publication No. ADM 86–1139). Washington, DC: U.S. Government Printing Office.
Pittman, D.J., & Snyder, C.R. (Eds.). (1962). *Society, culture, and drinking patterns.* New York: Wiley.
Piazza, N.J., Peterson, J.S., Yates, J.W., & Sundgren, A.S. (1986). Progression of symptoms in women alcoholics: Comparison of Jellinek's model with two groups. *Psychological Reports, 59,* 367–370.
Popkin, C.L. (1989, October). Drinking and driving by young females. *33rd Annual Proceedings, Association for the Advancement of Automotive Medicine* (pp. 29–40) Baltimore, MD.
Ratliff, K.G., & Burkhart, B.R. (1984). Sex differences in motivations for and effects of drinking among college students. *Journal of Studies on Alcohol, 45,* 26–32.

Remy, M., Soukup-Stepan, S., & Tatossian, A. (1987). For a new use of life event questionnaires: Study of the life events world of a population of male and female alcoholics. *Social Psychiatry, 22,* 49–57.

Richman, J.A., & Flaherty, J.A. (1986). Sex differences in drinking among medical students: Patterns and psychosocial correlates. *Journal of Studies on Alcohol, 47,* 283–289.

Rimmer, J., Pitts, F.N., Reich, T., & Winokur, G. (1971). Alcoholism: II. Sex, socioeconomic status and race in two hospitalized samples. *Quarterly Journal of Studies on Alcohol, 32,* 942–952.

Robins, L.N., Bates, W.M., & O'Neal, P. (1962). Adult drinking patterns of former problem children. In D.J. Pittman & C.R. Snyder (Eds.), *Society, culture, and drinking patterns,* (pp. 395–412). New York: Wiley.

Roman, P.M. (1988). *Women and alcohol use: A review of the research literature* (DHHS Publication No. ADM 88–1574). Rockville, MD: National Institute on Alcohol Abuse and Alcoholism.

Ross, H.E. (1989). Alcohol and drug abuse in treated alcoholics: A comparison of men and women. *Alcoholism: Clinical and Experimental Research, 13,* 810–816.

Ross, H.E., Glaser, F.B., & Stiasny, S. (1988). Sex differences in the prevalence of psychiatric disorders in patients with alcohol and drug problems. *British Journal of Addiction, 83,* 1179–1192.

Schilit, R., & Gomberg, E.S.L. (1987). Social support structures of women in treatment for alcoholism. *Health and Social Work, 12,* 187–196.

Schuckit, M.A. (1972). The woman alcoholic: A literature review. *Psychiatry in Medicine, 3,* 37–42.

Schuckit, M.A., Pitts, F.N., Jr., Reich, T., King, L.J., & Winokur, G. (1969). Alcoholism: I. Two types of alcoholism in women. *Archives of General Psychiatry, 20,* 301–306.

Straus, R., & Bacon, S.D. (1953). *Drinking in college.* New Haven, CT: Yale University Press.

Turnbull, J.E., & Gomberg, E.S.L. (1988). Impact of depressive symptomatology on alcohol problems in women. *Alcoholism: Clinical and Experimental Research, 12,* 374–381.

Vannicelli, M. (1984). Treatment outcome of alcoholic women: The state of the art in relation to sex bias and expectancy effects. In S.C. Wilsnack & L.J. Beckman (Eds.), *Alcohol problems in women* (pp. 369–412). New York: Guilford.

Veroff, J., Douvan, E., & Kulka, R. (1981). *The inner American and mental health in America.* New York: Basic Books.

Wanberg, K.W., & Knapp, J. (1970). Differences in drinking symptoms and behavior of men and women alcoholics. *British Journal of Addiction, 64,* 347–355.

Wilsnack, R.W., Wilsnack, S.C., & Klassen, A.D. (1984). Women's drinking and drinking problems: Patterns from a 1981 national survey. *American Journal of Public Health, 74,* 1231–1238.

Wilsnack, R.W., & Cheloha, R. (1987). Women's roles and problem drinking across the life span. *Social Problems, 34,* 231–248.

Wilsnack, R.W., Klassen, A.D., & Wilsnack, S.C. (1986). Retrospective analysis of lifetime changes in women's drinking behavior. *Advances in Alcohol and Substance Abuse, 5,* 9–28.

Wilsnack, S.C., & Beckman, L.J. (1984). *Alcohol problems in women.* New York: Guilford.

Wilson, G., & Lawson, D.M. (1978). Expectancies, alcohol, and sexual arousal in women. *Journal of Abnormal Psychology, 87,* 358–367.

Windle, M., & Barnes, G.M. (1988). Similarities and differences in correlates of alcohol consumption and problem behavior among male and female adolescents. *International Journal of the Addictions, 23,* 707–728.

Winokur, G., & Clayton, P.J. (1968). Family histories: IV. Comparison of male and female alcoholics. *Quarterly Journal of Studies on Alcohol, 29,* 885–891.

CHAPTER 14

Alcoholism and Homosexuality: A Theoretical Perspective

Peter M. Nardi

Despite some indication of a high estimated rate of alcoholism among homosexuals (Fifield, 1975; Lohrenz, Connelly, Coyne, & Spare, 1978; Saghir & Robins, 1973), relatively little research focuses directly on the subject. This is due in part to the difficulty in delineating a representative sample of homosexual alcoholics. Alcoholism is difficult enough to define, and finding a cross-section of homosexuals (both open gays as well as those still repressed) is unrealistic. It is also due in part to the traditional resistance in social science to studying homosexuality. In either event, the available material is severely limited and of questionable reliability, as the following analysis reveals.

Journal of Studies on Alcohol is one of the most important publications in the alcohol field, publishing original articles and reviews of current alcohol literature. The *Journal's* index serves as an important resource for the field of alcohol studies, with thousands of citations each year. From 1951 through 1980, however, there were only 42 references under the heading of homosexuality.[1]

As shown in Table 1, only 10 of those references (24% of the total 42) were actual studies of alcohol use among homosexuals. The remainder were either (1) references to latent homosexuality and its undocumented causal relationship to alcoholism (29%), (2) studies using homosexuality as one of several pieces of demographic data, but not focusing on it (40%), or (3) research using homosexuals as a control group to compare with alcoholics, under a psychoanalytic research model (7%). In short, less than one tenth of 1% of all available references in 30 years are on

Peter M. Nardi is with the Department of Sociology, Pitzer College, Claremont, Calif. This chapter is an updated version of the author's article that appeared under the same title in *Journal of Homosexuality, 1* (4), pp. 9–25, 1982, and *Gay and Sober*, T. Ziebold and J. Mongeon, Eds. (New York: Harrington Press, 1982), and is published here with permission.

TABLE 1
Trends in Research on Alcoholism and Homosexuality Reported in *Journal of Studies on Alcohol*

Year	Psycho-analytic studies	Homosexuality as a demographic	Actual studies of gay men and lesbians	Total
1951–60	12[a]	5	2[b]	19
1961–70	0	5	1[c]	6
1971–80	0	7	10	17
Total	12	17	13	42

Notes: [a] Three of these were not supportive of the psychoanalytic model.
[b] Both used gays as a control group only.
[c] A case study of an alcoholic with homosexual tendencies.

alcohol use among homosexuals, despite high estimates of alcoholism among them and much anecdotal concern over their drinking patterns.

The relationship between alcoholism and homosexuality has had an unusual history. For decades, followers of Freudian thought have sought to explain alcoholism simply in terms of latent homosexuality. Today, more complex sociocultural theories are often offered to describe the relationship between problem drinking and homosexuality. Somewhere between the early psychoanalytic ideas and the current sociocultural views is a vast chasm of misguided research, theories, and unanswered questions. Although causal relationships have not been established, the myths and assumptions that surround this issue are numerous. This chapter is a beginning step in organizing what is known about the relationship between alcoholism and homosexuality. It is an analysis and exploration of underlying assumptions, focusing on several theoretical perspectives and concluding with suggestions for much-needed further investigation using a sociological approach.

No attempt is made in this chapter to define alcoholism specifically. Rather, the focus is on drinking patterns in general, with particular emphasis on problem drinking or alcoholism. The words *gay* and *lesbian* refer to those men and women who are open about their homosexual identity either to themselves or to others, although *gay* will be used for brevity to include both men and women. *Repressed homosexuals* are those who have not accepted their sexual identity.

The current trend, however, is clearly in the direction of research on homosexual populations and away from a psychoanalytic model. This trend is illustrated by the recent appearance of articles on alcoholism in gay magazines and newspapers (such as Abramson, 1979; Anderson, 1979; Bowring, 1979; Shilts, 1976; Ziebold, 1979), by special sessions of professional meetings (such as the 1980 National Council on Alcoholism in

Seattle), by the formation of the National Association of Gay Alcoholism Professionals, and by the publication of gay-oriented materials by several alcohol agencies (such as Alcoholics Anonymous; Alcoholism Center for Women in Los Angeles; Comp-Care [Michael, 1976, 1977]; *Do It Now Foundation* [Schwartz, 1980]; Hazelden Educational Services; and Wisconsin Clearinghouse for Alcohol and Other Drug Information). This trend coincides with major social and intellectual advances with respect to homosexuality issues between 1950 and 1980. However, many of the recent scholarly publications on alcoholism and homosexuality are anecdotal or contain serious methodological problems. In fact, the figures usually introduced to support excessive alcoholism rates among homosexuals come from studies with limited generalizability.

The most quoted figure is from Fifield's (1975) study conducted for Los Angeles County by the Gay Community Services Center. Gathering self-report data from 200 gay bar users, 98 bartenders and owners, 53 recovered gay alcoholics in treatment programs, and 132 users of the Services Center, Fifield estimates problem drinking rates in the gay population as three times higher than rates in the general population. Using bartenders' estimates, Fifield concludes that 10.4% of the total adult gay population in Los Angeles County is a "primary target group of gay men and women in crisis or danger stages of alcohol consumption and in need of alcoholism services" (p.8). A secondary target group at high risk of needing future treatment is estimated as 21% of the gay population. Thus, 31.4% of the Los Angeles County gay and lesbian population shows signs of alcoholism or heavy drinking. Given the dependency on bar goers and bartenders, the use of self-reported problem drinking, the lack of a control group of heterosexual bar goers, and the derivation of figures from previously estimated percentages, this study has limited generalizability and reliability.

Another highly quoted finding comes from a study of 145 homosexual men selected from two Kansas university towns and two Kansas cities (Lohrenz et al., 1978). Most of the subjects were college educated, were current students or professionals, and were from urban areas. As a result of their scores on the Michigan Alcoholism Screening Test (MAST), 29% of the subjects were categorized as alcoholic. The failure to obtain a control group, to verify the MAST scores with more detailed questions, and to survey a more representative cross-section of gay people also limits the reliability and validity of these findings.

Saghir and Robins (1973) included a few questions on drinking behavior on their three to four hour long interviews with 89 male and 57 female homosexuals. Their sample was selected mostly from gay political or social (or both) organizations (about 65%), by word of mouth (about 25%), and from gay bars (between 5% and 10%) in Chicago and San Francisco. Among lesbians, 35% described their drinking behavior as excessive or alcohol dependent at some point in their lives. Among a control group of 43 heterosexual women, 5% described their drinking as excessive or dependent at some point in their lives.

Among gay men, 30% said they were excessive or dependent drinkers at some point in their lives. Among a control group of 35 heterosexual men, 20% described their drinking as excessive or alcohol dependent at some point in their lives. Problems with the study are its small sample size, its limited representativenesss of the gay population, and its dependence on only a few questions about drinking behavior.

Besides these studies, no other research has been published with statistical data on drinking rates among gay and lesbian populations. Weinberg and Williams (1974) in their survey of over 2,000 gay men included exactly *one* question on drinking in their 145-question survey ("Do you ever drink more than you should?" the choices being "never," "hardly ever," "sometimes," and "many times"). The poor and biased phrasing of that question; the dependence on Mattachine members and gay bar users in New York, San Francisco, Amsterdam, and Copenhagen; and the low response rate (averaging about 30%) severely limit the reliability of the finding (27.9% responded *sometimes* and 11.4% said *many times*).

The highly publicized Bell and Weinberg (1978) "Kinsey study" of almost 1,000 San Francisco homosexuals fails to include a single question on drinking behavior. Even Read's (1980) ethnography of a male gay bar does not include any detailed descriptions or discussions of drinking styles, attitudes, or patterns.

There have recently appeared several descriptive articles focused on gay alcoholics and oriented toward the counseling profession. Small and Leach (1977) discuss case histories of 10 gay male alcoholics and offer suggestions for counseling; Diamond and Wilsnack (1978) report descriptive findings from 10 lesbians alcohol abusers, revealing strong dependency needs, low self-esteem, and depression among their clients; and Beaton and Guild (1976) describe their group counseling sessions with three gay men and two lesbians and offer strategies for nongay therapists.

Other published articles on the topic include a survey of nongay alcoholism treatment agencies and the services offered for gay men and women (Judd, 1978), a discussion of lesbianism and alcoholism (Hawkins, 1976), and an interview with a lesbian alcoholic (Sandmaier, 1979).

In sum, the direction of research in recent years is toward understanding alcohol use from within the gay community. However, the current literature on homosexuality and alcoholism does not yet include a systematic, rigorous epidemiological survey of alcoholism or of drinking patterns among the gay subculture. The attention that is growing and the interest in the topic that is developing are positive signs for future quality research. Understanding the assumptions underlying prior research on the subject and sharpening the conceptual frameworks for new research are important first steps in achieving the quality that is much needed in this area.

Theoretical Perspectives

The shift from the historical dominance of the psychoanalytic model to the current trend in studying gay alcoholics from a sociocultural perspective represents not only a major change in society's knowledge about and attitude toward homosexuality but also a shift in theoretical perspectives and assumptions about alcoholism. An analysis of each of the major theoretical viewpoints and how it may relate to homosexuality will clarify the myths and assumptions many people hold about the connection between alcoholism and homosexuality and will enhance the ability to evaluate past research and to set the foundation for quality research. Several theoretical viewpoints exist concerning the etiology of alcoholism: biological-genetic, psychoanalytic, learning, and sociocultural (see Buss, 1966).

Biological-Genetic Arguments

Various researchers have viewed alcoholism as a function of anatomy, physiology, metabolic abnormalities, tissue chemistry, allergies, enzyme defects, or genetic transmission (see Kessel & Walton, 1965). One of the major studies supporting a genetic model is Goodwin's (1976) work with Danish children. He demonstrated that male offspring of alcoholics, separated from their biological parents in early infancy and raised by nonrelatives, had "nearly twice the number of alcohol problems and four times the rate of alcoholism as the children whose parents had no record of hospitalization for alcoholism" (p.73). The finding suggested to Goodwin a genetic predisposition to severe alcohol abuse. Goodwin (1979) has also argued that what is inherited is not a predisposition to alcohol but a lack of intolerance for alcohol; those with the allergy to alcohol are the ones who avoid it and remain nonalcoholic.

These studies tend to locate the etiology of alcoholism in biological or genetic abnormalities. Deficiencies or pathologies in the physiological system affect reactions and responses to alcohol. Although biological or genetic explanations cannot be ruled out entirely, no conclusive evidence has yet been presented to establish a dominant biological or genetic trait for alcoholism.

Similarly, some have presented biological or genetic abnormalities to explain homosexuality, usually emphasizing hormonal imbalances (Glass, Deuel, & Wright, 1940; Lang, 1940) or heredity (Kallman, 1952). Wilson (1978), from a sociobiological position, even discusses a genetic predisposition to be homosexual. However, no substantial evidence has been forthcoming to support these early studies in establishing a dominant biological or genetic basis for homosexuality (see West, 1967).

Although no one has made the connection explicit or published research on the subject, some dedicated biologically oriented theorists could easily attempt to search for biological or genetic variables to explain both homosexuality and alcoholism simultaneously. Perhaps, they might argue, the reason for high incidences of drinking problems among the homosexuals is some shared gene or hormonal imbalance. More important, the assumptions underlying a nature-oriented form of inquiry involve a "what-went-wrong" philosophy: that is, deviations from "normal" or "natural" genetic/biological structures are used to describe why the individual ended up both homosexual and alcoholic. By locating the source of behavior in biological or genetic factors, such theorists absolve or overlook oppressive social conditions.

Although genetic or biological theories might combine with other theories in explaining the issue and, thus, contribute to our understanding of alcoholism among homosexuals, the potential misuse of them as sole explanations could prove more detrimental to our understanding of behavior.

Psychoanalytic Tradition

Nowhere is the relationship between alcoholism and homosexuality more distorted than in psychoanalytic theories. As was demonstrated in Table 1, most of the published material cited in 30 years of the *Journal of Studies on Alcohol* abstracts, especially during the 1950s, derives from a psychoanalytic perspective and emphasizes a causal relationship between latent homosexuality and alcoholism. In fact, the link between the two has been explicitly made since the early 1900s. As Buss (1966) states: "Classical psychoanalytic theory emphasizes orality and homosexuality in the genesis of alcoholism" (p. 445).

Alcoholics are seen to be fixated in either the oral or anal stage, to overidentify with the father, to be anxious about masculine inadequacy and incompleteness, to have emotionally absent fathers and overindulgent mothers, to have experienced traumatic weaning, to exhibit penis envy, or to have an irrational fear of being heterosexual (Buss, 1966; Roebuck & Kessler, 1972; Small & Leach, 1977). These same phrases are often used by psychiatrists to describe the etiology of homosexuality (see Bieber et al., 1962; Socarides, 1968).

Modern psychoanalytic theory less often insists that homosexuality is basic to alcoholism, emphasizing more that homosexual urges are controlled by drinking behavior. Yet, erroneous assumptions and myths about homosexuality persist. For example, Hatterer (1970) responds as follows to a client troubled by his homosexual experiences: "You're right about your homosexual activities being directly triggered by your drinking. I'd go so far to say now it's possible you'd stop all of it if you were sober" (p. 249). Levy (1958) also draws on faulty reasoning when he writes:

> Transient overt homosexual contacts while drinking... are not uncommon among patients. In most cases this represents the lack of concern with, and clear perception of, the nature and needs of the sexual object. In most cases a woman or a sheep might have done as well. (p. 656)

Similar sexist assumptions have guided research conducted with homosexuals as control groups. Rather than study alcoholism among homosexual populations, researchers have tended to look for homosexuality (as defined by masculinity-femininity [M-F] scales) among alcoholics. A 1959 study by Machover, Puzzo, Machover, and Plumeau, using M-F scales and projective tests, confirmed a hypothesis that remitted alcoholics would exhibit more homosexual tendencies than unremitted alcoholics. They based this hypothesis on "the clinical impression that tendencies toward feminine, including maternal, identification were more frequent among the remitted alcoholics" (p. 529).

Gibbins and Walters (1960) also assumed a psychoanalytic model to test the relationship between latent homosexuality and alcoholism, using a control group of "self-confessed manifest homosexuals" who had been "arrested for homosexual offenses." They reason that

> if there is any substance to the belief that alcoholic males are latent homosexuals, one might expect that they and nonalcoholic manifest homosexuals would perceive certain stimuli in their environment in a somewhat similar manner, i.e., that in some respects alcoholics would differ from normal subjects and resemble homosexuals. (p. 618)

However, their findings are mixed, failing to show any differences between subjects in one experiment and, in two other experiments, showing alcoholics as scoring between homosexuals and "normals." Yet Gibbins and Walters conclude: "While the experiments as a whole do not provide strong evidence for the psychoanalytic theory, the results suggest that this theory should not be lightly discarded" (p. 618). Interpreting results to conform to firmly held assumptions is also evident in one of Tahka's (1966) conclusions from his study of 50 male alcoholics in Stockholm: "All the subjects preferred men to women as drinking companions. This might be interpreted as an indication of latent homosexual needs" (p. 179).

The relationship between latent homosexuality and alcoholism is even expressed in artistic form in Tenneessee Williams' 1955 play *Cat on a Hot Tin Roof.* Brick's alcoholism is linked to his frustrating relationship with his wife Maggie and his repressed homosexual feelings about his dead friend Skipper. Lolli (1956) sees in this play an illustration of the psychoanalytic refrain: "Oral frustrations contribute both to alcoholism and homosexuality. Therefore, the presence of latent homosexual traits in alcoholics is neither unusual nor mild" (p. 550).

Critics of this perspective attack the overemphasis on oral aspects of homosexuality. The typical psychoanalytic approach usually ignores the range of sexual practices and the emotional-love dimensions of same-sex relationships. Furthermore, it does not account for lesbians, for the repressed homosexuals who are not alcoholic, for the open gays and lesbians who are not alcoholic, and for the open gays who are alcoholic (Small & Leach, 1977). Although repression of fundamental characteristics of self can often lead to destructive behavior, the focus of psychoanalysis is of particular relevance here. The relationship between latent homosexuality and alcoholism assumes that learning to overcome one's repressed homosexual feelings and to live heterosexually is the best "cure" for alcoholism. Thus, the focus of therapy is on one's sexuality, not on the drinking or the repression. In other words, when a pathology is linked to repressed homosexuality in psychiatry, the link is made to the homosexuality instead of to the repression (Adam, 1978). Szasz (1970) similarly has criticized psychiatry for manufacturing mental illness and labeling homosexuality as a disease, thereby obscuring the fact that homosexuals are medically stigmatized and socially persecuted.

Connecting a disease concept of homosexuality with a disease concept of alcoholism, however, makes sense from a psychoanalytic therapeutic viewpoint (see Conrad & Schneider, 1980). Each disease can then be cured with similar techniques, psychoanalysis or aversion therapy, and only by medical experts. Adam (1978) has described the historical role of therapeutic ideologies in perpetuating inferiorization and domination of gay people. The psychoanalytic model has served to maintain internalization of negative self-images. As Adam states:

> The psychiatric literature provides a compendium of the responses to domination; it is a vast document of the contempt internalized by an inferiorized people from a society which stands opposed to its self-realization. The "cure" is the disease. Acting as an agent of inferiorization, the therapeutic establishment cannot but compound the malaise of the oppressed. (pp. 41–2)

In short, psychoanalytic theory has been the major source of false assumptions concerning the relationship between alcoholism and homosexuality. Psychoanalytic research on alcoholism and psychoanalytic treatment of alcoholics exhibiting homosexual behavior, as a result, have suffered from the use of oppressive and sexist concepts and methodologies.

Learning Theory

If psychoanalytic theory focuses on inner thought and feelings, a learning approach emphasizes more overt behavior. From this perspective, alcoholism is seen as a learned behavior resulting from reinforcement of pleasurable experiences or from avoidance of negative ones. Tension reduction, relaxation, peer approval, and feelings of power have all been listed as associated with drinking alcohol (Schuckit & Haglund, 1977).

Akers (1973) believes that drinking patterns are a "function of the social rewards attached to drinking and the reinforcing effects of the alcohol" (p. 127). Negative reinforcement occurs when anxiety is reduced and stressful situations are avoided. Positive reinforcement occurs when social groups reward drinking behavior and when pleasurable effects are experienced. When negative sanctions are applied to what begins to seem like excessive drinking, these punishing sanctions become weak, irregular, and infrequent compared with the numerous rewards of drinking behavior (Trice, 1966). Imitation of peers' and elders' drinking patterns is also viewed as a dominant characteristic of initial learning experiences with alcohol (Maddox, 1962). [See chapter 9, by White, Bates, & Johnson, in this volume.]

Thus, learning theorists view social sanctions and norms concerning drinking behavior as important components in explaining drinking patterns and alcoholism. Since learning theorists posit reinforcement schedules, some also believe that unlearning behavior can occur through similar means. Aversion techniques, ranging from nausea-inducing drugs, such as Antabuse, to electric shock treatment, are often prescribed for alcohol dependent people (Steffen, Steffen, & Nathan, 1977).

Similar reinforcement arguments have been offered to explain homosexual behavior as learned behavior. Some believe it also can be unlearned. The misguided use of electric shock therapy to unlearn homosexuality and reinforce heterosexuality is well documented (see Adam, 1978). Others see homosexual behavior as a gradual process of reinforcement and rationalization (Akers, 1973). Inhibitions toward homosexuality decrease as positive rewards accrue from direct homosexual experiences, from masturbatory imagery, and from relatively less positive heterosexual experiences (Akers, 1973). As participation in a gay subculture increases, often at first through bars, emerging gay people develop rationalizations and justifications for their behavior. Social rewards and minimization of anxiety over being different increase, thus leading to the continuation of homosexual behavior (Akers, 1973).

A learning model explanation of excessive drinking among gays and lesbians stresses tension reduction and the more positive hedonistic aspects of the openly and visibly gay community. For some gay people who are just coming out, getting involved sexually with another of the same biological sex is possible only while intoxicated (Chafetz, Blane, & Hill, 1970). The tension, anxiety, and guilt feelings generated in the context of a society that does not condone homosexual behavior are reduced by increased alcohol use. The resultant strength and feelings of power allow emerging gay people to make sexual contacts and overcome social resistances.

Another possible explanation is the positive reinforcements of an openly gay lifestyle, which stresses bar life and drinking. The emergence of gay bars as a common institution for introduction into a gay community derives from their history of permissiveness and protectiveness (Achilles, 1967). Gay bars emphasize the leisure time dimensions of one's gay identity: they provide some anonymity and segregation from the dominant culture and permit sexual contacts to be made with relative safety and respectability (Achilles, 1967). Hooker (1965) has also described gay bars as centers for communication among the gay community and as free markets retailing both leisure time activities, such as entertainment and drinking, and sexual services. That gay bars serve a wide range of needs, interests, and goals in often highly ritualized ways is documented in Read's (1980) ethnography of a working-class gay male bar. Whatever the motives of gay bar patrons, and whatever needs are being filled, the importance and availability of alcohol in achieving their goals are strongly evident in the gay bar subculture.

Other social activities within the gay community, as well as the gay media, constantly encourage a fashionable social scene replete with alcohol (Warren, 1974). This aspect of gay culture mirrors the dominant culture's emphasis on drinking as an acceptable component of a successful social event. As Warren states: "There is pressure to drink alcohol.... Getting drunk in gay bars, like getting drunk at home gatherings, is normal trouble in the gay community, rather than deviance" (p. 58).

This idea might explain why some open gays and lesbians drink heavily: not because of anxiety or low self-esteem but because of the acquired habit of drinking in gay settings. The positive aspects of being a part of a gay community reinforce drinking patterns. Drinking is not used to escape from something; rather, it is used to join something. An emerging gay person's initial socialization into a gay community often occurs through his or her going to gay bars and enacting the drinking roles perceived as essential for a gay identity. Whether heavy drinking may be for some the outcome of continually reinforced positive feelings engendered by the gay lifestyle and not due to negative feelings has yet to be systematically explored.

As with all subcultures, there exists a diversity of people within the group. There are many different types of homosexuals; there are many different types of alcoholics and alcoholisms. Searching for a single etiology to explain all drinking by homosexuals or to explain all forms of alcoholism is a misguided task. For some open gays, a pleasure-seeking explanation is probably a more accurate learning model. For others just coming out, a tension-reduction model may serve as a sharper explanation, especially in the context of emerging positive rewards as contact is made with a gay community. For those still "in the closet" and repressing their identity, a tension-reduction model may also be an appropriate framework for an understanding of their drinking behavior. Thus, learning theory can offer several avenues of research for analyzing drinking behavior patterns among various types of gay people at various stages of identity formation.

A Sociocultural Approach

Variations in rates of alcoholism from one society to another, and from one subculture to another, have led many to see a sociocultural explanation for drinking patterns and behaviors. Kessel and Walton (1965) emphasize such factors as incitement (money, leisure time, advertising), opportunity (social class, occupation, number of area bars, laws), and example (peers and parents) as contributors to alcohol use and abuse. Trice (1966) has focused on the social values, rules, and meanings a particular group gives to alcohol, and Bales (1946) has offered a model of three interacting sets of contributors: dynamic factors, alternative factors, and orienting factors. Dynamic factors include acute psychic tensions at the group level; alternative factors are culturally defined patterns of behavior other than heavy drinking but functionally equivalent in relieving acute psychic tensions; and orienting factors involve the group's traditional ways of defining the norms and attitudes about drinking.

Some scholars have developed anomie theories for explaining the source of tension and anxiety experienced by the incipient alcoholic. Snyder (1964) emphasizes the low rate of alcoholism among culturally integrated and cohesive societies and the higher rate among societies with more social disorganization. Ullman (1958) similarly stresses the importance of inconsistent drinking norms and unintegrated drinking customs in the production of ambivalent feelings about drinking. Room (1976), on the other hand, rejects the concept of ambivalence as an explanation for problematic drinking behavior.

Others have argued that drinking behavior is related to definitions emerging out of social interaction, emphasizing the power of labels and socially constructed meanings within a culture. MacAndrew and Edgerton (1969), for example, refute the universality of the disinhibiting effects of alcohol. They observed, in numerous cross-cultural settings, that a person's social behavior while drunk is highly variable and situationally defined. How a society defines drinking and drunkenness, what meanings are constructed for behavior "under the influence," and what situational factors and social norms are relevant—all affect drinking patterns and definitions of alcoholism.

Furthermore, some argue that definitions of drinking problems and behaviors are imposed by those in power to make and enforce rules (Conrad & Schneider, 1980). The emergence of the temperance movement as a crusade symbolic of the power of the ruling class illustrates the process of socially defining, transmitting, labeling, and controlling drinking behavior (Gusfield, 1963). Duster (1970) also describes the relationship among social class, power, and definitions of substance abuse as immoral or illegal.

Changing social structural conditions also have been demonstrated to affect drinking behavior. Brenner (1973) shows that "psychiatric hospitalization of persons diagnosed as having psychosis with alcoholism increases sharply during economic downturns and decreases during upturns" (p. 225).

In short, a sociocultural view of alcohol use emphasizes the norms and values of the society toward drinking, the meanings people attach to drinking, and the definitions and laws imposed by those in power to enact and control the norms.

Similarly, one can develop a sociocultural analysis of homosexuality, focusing on a society's definitions, norms, and attitudes toward it. By viewing the social context in which an individual is socialized (for example, social class, ethnic background, religious influences, family dynamics), one can gain a clearer understanding of how a person acquires a sexual identity and of what is done with it by society and by the individual. Instead of defining homosexuality as an illness or pathology, therefore, a sociocultural perspective emphasizes the meanings given by those in power to homosexual behavior. Stigma, oppression, and individual rage and anxiety are seen to be created by the social context and cannot be fully understood apart form the dominant culture's values and beliefs (Hills, 1980). The fact that homosexuality has been variously regarded through time as a sin, a sickness, a moral issue, a legal issue, and a lifestyle illustrates the importance of social definitions in how gay people are treated and how they perceive themselves (Conrad & Schneider, 1980).

An analysis of drinking behavior within gay and lesbian subcultures, then, employing a sociocultural model, leads to a more complex and less reductionistic understanding of the issue. Unless the problem is viewed from the perspective and social context of gay people, inappropriate research methodologies and misguided assumptions will persist. As Robinson (1976) states in his call for a sociological study of alcoholism,

> We must not take for granted at the outset what "alcoholism," or "drinking problem," or "being an alcoholic" is. What is needed instead is an understanding of what these things mean to particular people in particular situations. (p. 8)

If one focuses on the social context in which gay people find themselves, how they define reality and perceive their situation, and what symbols and values they hold with respect to alcohol use, one begins to develop a more complete picture of the relationship between homosexuality and alcoholism.

Emphasizing the point that gay people congregate in bars is too simple an explanation for understanding their drinking patterns. It is only one factor ("opportunity" in Trice's perspective, or "orienting factors" in Bales') among many that interact. A necessary starting point in understanding homosexual drinking patterns is to focus on

> the inner world of alcoholism—with the view from inside out rather than outside in.... We must, in effect, put aside our own frame of reference and be willing to enter into that of another to see how he makes sense of his experiences and reacts to them. (Wallace, 1977, p. 6)

Understanding how certain gay individuals manage and control their feelings in an oppressive social context illustrates this phenomenological perspective. A homophobic society instills in those coming to terms with their sexuality a variety of feelings about the immorality and deviant nature of homosexuality. A typical response is to deny to oneself (and often loudly to others) that one is homosexual. Self-hatred, fear of being different, and lowered self-esteem often lead to strong ego defenses and rigid denial (Ziebold, 1978). Hiding one's feelings, sexual and otherwise, becomes normative. Thus, homosexuals trying to come out find they must struggle not only against society's expectations but also against their own perceptions. Some may give up, becoming alienated from their own feelings. They hide behind the closet door, locked securely by the illusion of safety while slowly destroying their own identity and mental health.

These dynamic factors may also lead some to increase their consumption of alcohol to aid in their coming-out process or to maintain their concealed identity. As Ziebold (1979) writes, "Homosexual individuals who have been forced to develop rigid defenses against social reaction to their sexual and affectional orientation may unknowingly let these same reflexes reinforce a budding dependency on alcohol" (p. 39). Given the orienting factors of socially approved drinking settings, such as parties and gay bars, alcohol can easily become, for those coming out, one means of coping with the perceived oppressive social situation and personal psychic confusion. The absence of significant, subculturally valued alternatives to drinking settings, especially in the smaller, less urban centers, contributes to the dependency on alcohol as an acceptable solution to feelings of anxiety, alienation, and low self-esteem. This role, of course, is not much different from the role of alcohol in heterosexual society, except that for many heterosexuals there are numerous socially sanctioned, positive alternative sources of dependency. Social interaction does not always depend on singles bars; family bonds may often be stronger, and work-related friendships may be closer. For those heterosexuals who also find these factors absent, vulnerability to the alcohol alternative increases. Although the absence of these factors can often be replaced by newer social ties to gay groups and "families" (Nardi, 1982), the gay person just coming out or the repressed homosexual would probably not as yet have made these new connections. For some open gays for whom the absence of these factors may be acute, for those homosexuals not yet involved in a gay subculture (still "in the closet"), and for those still repressing their identity, alcohol can easily become a source of dependency and strength. Ziebold (1979) clearly illustrates this sociocultural model when he writes:

> A high level of psychic stress, polarization towards bars and cocktail parties as the primary basis for social interaction, and a closing off of alternative modes of relief in everyday living: these are the forces of oppression, and alcoholism is one of the resulting symptoms. (p. 40)

The dilemma faced by gay and lesbian alcoholics is further heightened by the additional stigma of alcoholism. Concealing from oneself and from others that one is an alcoholic is a common practice. Whether this concealment is a function of an unconscious denial process, is due to a society's imposing negative labels on certain forms of drinking behavior, or is a result of epistemological confusion (i.e., difficulty in coming to know oneself because of conflicting labels, social comparisons, and mixed outcomes with alcohol use [Wallace, 1977]), there is clearly an attempt to hide oneself behind another closed door. Alcoholism has historically been treated as a crime, a sin, a moral issue, a legal matter, or, more recently, a disease (Conrad & Schneider, 1980). But even today it is kept quiet in families, concealed at work, and hidden from others. Being labeled alcoholic has been as stigmatizing as being labeled homosexual (see Goffman, 1963).

Confined by double closet doors, gay and lesbian alcoholics must work on opening both. As alcoholics, homosexuals must hide their drinking from other gays for fear of rejection; as homosexuals, alcoholics must hide their sexuality from heterosexual alcoholics or therapists for fear of rejection (Ziebold, 1979). The problems are further intensified if the gay alcoholic is a member of another minority also oppressed and stigmatized by society. Women, blacks, Hispanics, and Native Americans are some of those who face additional blocks to finding positive alternative sources of identity and support. The social system reinforces the alienation and low self-esteem of gay and lesbian alcoholics and labels them morally weak, thereby increasing their vulnerability to addictive behavior. How society defines and regulates interactions and roles for homosexuals, alcoholics, and other minorities must be analyzed first. Studying how these people define their situation and attempt to express their feelings in this social context will lead to a fuller understanding of the complex, dynamic relationship between homosexuality and alcoholism.

Summary

As with any social phenomenon, adopting a particular theoretical framework structures the kinds of questions asked and limits the range of answers offered. A biological or genetic perspective tends to focus on deviations from what is assumed to be natural. Anomalies, illnesses, and pathologies are described, and variations in "normal" behavior become so labeled. Overlooked or deemphasized are the social structural factors and environmental strains that may be contributing to the "deviant" behavior.

Psychoanalytic paradigms focus attention on latent homosexuality, thereby failing to account for problem drinking among open gays and lesbians. Furthermore, assumptions underlying this perspective emphasize the "deviant" sexuality and not the societal conditions leading to the repression.

Learning theory models best contribute to our understanding of why some of those openly gay and involved in a gay subculture may become alcoholic. Socialization into a hedonistic, positively reinforcing lifestyle revolving around bars and other alcohol-oriented social functions is offered as an explanation by this perspective. Future research needs to focus more on the positive dimensions many perceive while learning to become a member of a gay subculture of bars and parties.

A more encompassing viewpoint is the sociocultural one, emphasizing labeling theories, conflict models, and interactionist perspectives. Understanding drinking behavior and patterns among gay populations necessitates analysis of the meanings and definitions of alcohol use that people within a subculture evolve. How those in power structure the roles of gay people and alcoholics, how they define problem drinking and alcoholism, and how gay people in turn respond to these structures and definitions are the issues this perspective emphasizes.

Through clarification of both the models for research and the assumptions underlying various theoretical perspectives, our understanding of the relationship between homosexuality and alcoholism will be enhanced and our strategies for prevention and treatment can only improve.

Alcoholism and Homosexuality: An Update in 1990

Although there has been some increase in attention devoted to the topic of gay and lesbian drinking in the last 10 years (see Israelstam & Lambert, 1983 & 1986; Mosbacker, 1988; Swallow, 1983), it is still a topic that is severely under-studied. To this date, no major epidemiological data are available about drinking patterns among gay men and lesbians. And several recent texts on alcohol (such as Engs, 1987; Kinney & Leaton, 1987; Royce, 1989) continue to ignore the topic totally.

In a recent article on the topic, Lewis and Jordan (1989) continue to report the questionable and dated findings of the 1975 Fifield study criticized earlier in this chapter for its methodological problems of sampling. They claim: "Some research seems to indicate a significantly higher percentage of lesbian and gay alcoholism than in the general population.... When one compares [the Fifield findings of 1 in 3] with the one in 10 in the general population, the large numbers become more obvious" (p. 175).

That current articles on homosexuality and alcoholism can do no better than quote 15-year-old data collected in bars is disturbing enough. That no one has published good research on the prevalence and patterns of drinking among gay men and lesbians is even more disturbing. Unfortunately it has taken the acquired immune deficiency syndrome (AIDS) pandemic to focus some renewed attention on the issue.

Despite the assumption that human immunodeficiency virus (HIV) status and substance abuse may be related, the data on this relationship remain sparse. Lewis and Jordan (1989, p. 187) claim that "today the number of substance abusers who are also HIV+ (the virus that causes the immunosuppression) causes counselors to be concerned." Yet, no data are provided about that number.

What is known is best summarized by Jerrells, Marietta, Bone, Weight, and Eckardt (1988, p. 173):

> A number of investigators have suggested that prolonged and excessive consumption of ethanol results in alterations of host defense mechanisms manifested ultimately in increased susceptibility to infections from pathogenic and opportunistic organisms.... Neither the extent of this effect nor the mechanisms have been clearly elucidated.

MacGregor (1986, p. 37) argues that alcoholism impairs cell-mediated immunity and predisposes to infection: "Enumeration of T-lymphocytes in alcoholics shows lower levels of these cells than in the normal population." Alcohol, in other words, is a moderate immunosuppressant and more markedly so in patients who drink chronically (MacGregor, 1986). This is not to say that alcohol abuse or other substance abuse causes AIDS, but it clearly has some cofactor effect on how a person's immune system may be functioning. How this effect may relate to making some individuals more susceptible to getting HIV and how it may contribute to the progression from testing HIV-positive to exhibiting full-blown AIDS have yet to be detailed.

Apart from the physiological aspects of alcohol abuse and AIDS, a contemporary discussion on alcohol use in the gay community must deal with a variety of sociocultural dimensions. Several areas need to be addressed by research, beginning with the response to a diagnosis of seropositive HIV. This stressful information can result in some people's responding by using excessive amounts of alcohol. What if the person is a recovering alcoholic? What needs to be investigated is how people react to information that they are HIV-positive or have developed full-blown AIDS.

Another issue that needs to be studied is the relationship between alcohol consumption and risk-taking sexual behaviors. One of the statements included in many "safer sex" documents is the necessity of engaging in "sober sex." The disinhibiting effects of alcohol and the impaired judgment that often results can lead to participation in unsafe sexual practices.

The role of treatment facilities for alcoholics who may be HIV-positive or have AIDS needs to be understood. Petrakis (1986) reports of a case in which a treatment facility refused someone admission because of his HIV-positive status. Can these facilities cope with such patients? What kinds of training are their personnel receiving about AIDS? How do counselors, 12-step programs, physicians, and other sources of help deal with clients who are HIV-positive, who have AIDS, or are simply part of the "worried well"?

Finally, Petrakis (1986) reports a few cases of alcoholics who were suicidal and sought to get AIDS as a form of self-destruction. Conversely, some people with AIDS are suicidal and may choose "drinking themselves to death" as a solution. What this relationship is between AIDS, suicide, and alcohol has yet to be fully understood.

Whether alcohol abuse is seen as suppressing the immune system, disinhibiting risky sexual behaviors, helping someone cope with the psychic stress of learning of his or her HIV-positive status, or being used as a form of suicide, research on gay men and lesbians can no longer afford to ignore the role of substance abuse in their lives. For it is with AIDS that the biological, psychological, and sociocultural theories come together to more fully explain the complex relationships between alcoholism and homosexuality.

Acknowledgments

The author wishes to thank Barry Adam, Andrew Berner, Stephen Murray, Farrell Webb, and Thomas Ziebold for their comments and suggestions.

Note

1. Abstracting has not always been as systematic or comprehensive as in recent years. There is, however, no indication that references to homosexuality have been systematically omitted. If errors exist, they probably are randomly distributed throughout the years and across various subtopics. Thus, the numbers in Table 1 should be regarded as not absolute but rather as suggestive of the relative proportion of articles on the subject.

References

Abramson, M. (1979, August). *Loving an alcoholic. Christopher Street*, pp. 15–17.
Achilles, N. (1967). The development of the homosexual bar as an institution. In J. Gagnon & W. Simon (Eds.), *Sexual deviance*. New York: Harper & Row.
Adam, B. (1978). *The survival of domination*. New York: Elsevier.
Akers, R. (1973). *Deviant behavior: A social learning approach*. Belmont, CA: Wadsworth.
Anderson S. (1979, June 28). Beating the bottle the gay way. *Advocate*, pp. 25–27.
Bales, R. (1946). Cultural differences in rates of alcoholism. *Quarterly Journal of Studies on Alcohol*, *6*, 480–499.
Beaton, S., & Guild, N. (1976). Treatment for gay problem drinkers. *Social Casework*, *57*, 302–308.
Bell, A., & Weinberg, M. (1978). *Homosexualities: A study of diversity among men and women*. New York: Simon & Schuster.
Bieber, I., Dain, H., Dince, P., Drelich, M., Grand, H., Gundlach, R., Kremer, M., Rifkin, A., Wilbur, C., & Bieber, T. (1962). *Homosexuality: A psychoanalytic study*. New York: Basic Books.
Bowring, D. (1979, December 8). HATS helps gay alcoholics. *Gay Community News*, pp. 10–12.
Brenner, M. H. (1973). *Mental illness and the economy*. Cambridge: Harvard University Press.
Buss, A. (1966). *Psychopathology*. New York: Wiley.
Chafetz, M., Blane, H., & Hill, M. (1970). *Frontiers of alcoholism*. New York: Science House.
Conrad, P., & Schneider, J. (1980). *Deviance and medicalization: From badness to sickness*. St. Louis: Mosby.
Diamond, D., & Wilsnack, S. (1978). Alcohol abuse among lesbians: A descriptive study. *Journal of Homosexuality*, *4* (2), 123–142.
Duster, T. (1970). *The legislation of morality: Law, drugs, and moral judgment*. New York: Free Press.

Engs, R. (1987). *Alcohol and other drugs: Self-responsibility*. Bloomington, Indiana: Tichenor.

Fifield, L. (1975). *On my way to nowhere: Alienated, isolated, drunk*. Los Angeles: Gay Community Services Center and Department of Health Services.

Gibbins, R., & Walters, R. (1960). Three preliminary studies of a psychoanalytic theory of alcohol addiction. *Quarterly Journal of Studies on Alcohol, 21*, 618–641.

Glass, S. J., Deuel, H. J., & Wright, C. A. (1940). Sex hormone studies in male homosexuality. *Endocrinology, 26*, 590–594.

Goffman, E. (1963). *Stigma: Notes on the management of spoiled identity*. Englewood Cliffs, NJ: Prentice-Hall.

Goodwin, D. (1976). *Is alcoholism hereditary?* New York: Oxford University Press.

Goodwin, D. (1979). Alcoholism and heredity. *Archives of General Psychiatry, 36*, 57–61.

Gusfield, J. (1963). *Symbolic crusade: Status politics and the American temperance movement*. Urbana: University of Illinois Press.

Hatterer, L. (1970). *Changing homosexuality in the male*. New York: McGraw-Hill.

Hawkins J. (1976). Lesbianism and alcoholism. In M. Greenblatt & M. Schuckit (Eds.), *Alcoholism problems in women and children*. New York: Grune & Stratton.

Hills, S. (1980). *Demystifying social deviance*. New York: McGraw-Hill.

Hooker, E. (1965). Male homosexuals and their "worlds." In J. Marmor (Ed.), *Sexual inversion*. New York: Basic Books.

Israelstam, S. & Lambert, S. (1983). Homosexuality as a cause of alcoholism: A historical perspective. *International Journal of the Addictions, 18*, 1085–1107.

Israelstam, S., & Lambert, S. (1986). Homosexuality and alcohol: Observation and research after the psychoanalytic era. *International Journal of the Addictions, 4/5*, 509–537.

Jerrells, T., Marietta, C., Bone, G., Weight, F., & Eckardt, M. (1988). Ethanol-associated immunosuppression. In T. Bridge, A. Mirsky, & F. Goodwin (Eds.), *Psychological, neuropsychiatric, and substance abuse aspects of AIDS*. New York: Raven Press.

Judd, T. D. (1978). A survey of non-gay alcoholism treatment agencies and services offered for gay women and men. In D. Smith, S. Anderson, M. Buxton, N. Gottlieb, W. Harvey, & T. Chung (Eds.), *A multicultural view of drug abuse*. Cambridge, MA: G. K. Hall/Shenkman.

Kallmann, F. J. (1952). Comparative twin study of the genetic aspects of male homosexuality. *Journal of Nervous and Mental Disease, 115*, 283–298.

Kessel, N., & Walton, H. (1965). *Alcoholism*. Baltimore: Penguin.

Kinney, J., & Leaton, G. (1987). *Loosening the grip: A handbook of alcohol information* (3rd ed.). St. Louis: Times Mirror/Mosby.

Lang, T. (1940). Studies in the genetic determination of homosexuality. *Journal of Nervous and Mental Disease, 92*, 55–64.

Levy, R. (1958). The psychodynamic functions of alcohol. *Quarterly Journal of Studies on Alcohol, 19*, 649–659.

Lewis, G., & Jordan, S. (1989). Treatment of the gay or lesbian alcoholic. In G. Lawson & A. Lawson (Eds.), *Alcoholism and substance abuse in special populations*. Rockville, MD: Aspen.

Lohrenz, L., Connelly, J., Coyne, L., & Spare, K. (1978). Alcohol problems in several midwestern homosexual communities. *Journal of Studies on Alcohol, 39*, 1959–1963.

Lolli, G. (1956). Alcoholism and homosexuality in Tennessee Williams' "Cat on a Hot Tin Roof." *Quarterly Journal of Studies on Alcohol, 17*, 543–553.

MacAndrew, C., & Edgerton, R. (1969). *Drunken comportment*. Chicago: Aldine.

MacGregor, R. (1986). Alcohol and the immune system. In P. Petrakis (Ed.), *Acquired immune deficiency syndrome and chemical dependency*. Washington, DC: National Institute on Alcohol Abuse and Alcoholism, U.S. Government Printing Office.

Machover, S., Puzzo, F., Machover, K., & Plumeau, F. (1959). Clinical and objective studies of personality variables in alcoholism: III. An objective study of homosexuality in alcoholism. *Quarterly Journal of Studies on Alcohol, 20*, 528–542.

Maddox, G. (1962). Teenage drinking in the United States. In D. Pittman & C. Synder (Eds.), *Society, Culture, and Drinking Patterns*. New York: Wiley.

Michael, J. (1976). *The gay drinking problem: There is a solution*. Minneapolis: CompCare.

Michael, J. (1977). *Sober, clean and gay!* Minneapolis: CompCare.

Mosbacher, D. (1988). Lesbian alcohol and substance abuse. *Psychiatric Annals, 18*, (1), 47–50.

Nardi, P. M. (1982). Alcohol treatment and the non-traditional "family" structures of gays and lesbians. *Journal of Alcohol and Drug Education, 27*,(2), 83–89.

Petrakis, P. (1986). *Acquired immune deficiency syndrome and chemical dependency*. Report of symposium sponsored by the American Medical Society on Alcoholism and Other Drug Dependencies, Inc., and the National Council on Alcoholism. Washington, DC: National Institute on Alcohol Abuse and Alcoholism, Government Printing Office.

Read, K. (1980). *Other voices: The style of a male homosexual tavern*. Novato, CA: Chandler & Sharp.

Robinson, D. (1976). *From drinking to alcoholism: A sociological commentary*. New York: Wiley.

Roebuck, J., & Kessler, R. (1972). *The etiology of alcoholism*. Springfield, IL: Charles C. Thomas.

Room, R. (1976). Ambivalence as a sociological explanation: The case of cultural explanations of alcohol problems. *American Sociological Review, 41*, 1047–1065.

Royce, J. (1989). *Alcohol problems and alcoholism: A comprehensive review* (rev. ed.). New York: Free Press.

Saghir, M., & Robins, E. (1973). *Male and female homosexuality*. Baltimore: Williams & Wilkins.

Sandmaier, M. (1979). *The invisible alcoholics: Women and alcohol abuse in America*. New York: McGraw-Hill.

Schuckit, M., & Haglund, R. (1977). An overview of the etiological theories on alcoholism. In N. Estes & E. Heinemann (Eds.), *Alcoholism: Development, consequences and interventions*. St. Louis: Mosby.

Schwartz, L. (1980). *Alcoholism among lesbians/gay men: A critical problem in critical proportions*. Phoenix: Do It Now Foundation.

Shilts, R. (1976, February 25). Alcoholism: A look in depth at how a national menace is affecting the gay community. *Advocate*, pp. 16–19, 22–25.

Small, E., & Leach, B. (1977). Counseling homosexual alcoholics. *Journal of Studies on Alcohol, 38*, 2077–2086.

Snyder, C. (1964). Inebriety, alcoholism and anomie. In M. Clinard (Ed.), *Anomie and deviant behavior*. New York: Free Press.
Socarides, C. (1968). *The overt homosexual*. New York: Grune & Stratton.
Steffen, J., Steffen, V., & Nathan, P. (1977). Behavioral approaches to alcohol abuse. In N. Estes & E. Heinemann (Eds.), *Alcoholism: Development, consequences and interventions*. St. Louis: Mosby.
Swallow, J. (1983). *Out from under: Sober dykes and our friends*. San Francisco: Spinsters, Inc.
Szasz, T. (1970). *The manufacture of madness*. New York: Harper & Row.
Tahka, V. (1966). *The alcoholic personality: A clinical study*. Helsinki: Finnish Foundation for Alcohol Studies.
Trice, H. (1966). *Alcoholism in America*. New York: McGraw-Hill.
Ullman, A. (1958). Sociocultural backgrounds of alcoholism. *Annals of the American Academy of Political and Social Science, 315*, 48–54.
Wallace, J. (1977). Alcoholism from the inside out: A phenomenological analysis. In N. Estes & E. Heinemann (Eds.), *Alcoholism: Development, consequences, and interventions*. St. Louis: Mosby.
Warren, C. (1974). *Identity and community in the gay world*. New York: Wiley.
Weinberg, M., & Williams, C. (1974). *Male homosexuals*. New York: Oxford University Press.
West, D. J. (1967). *Homosexuality*. Chicago: Aldine.
Williams, T. (1955). *Cat on a hot tin roof*. New York: New Directions.
Wilson, E. (1978). *On human nature*. New York: Bantam Books.
Ziebold, T. (1978). *Alcoholism and the gay community*. Washington D.C.: Whitman-Walker Clinic and Blade Communications.
Ziebold, T. (1979, January). Alcoholism and recovery: Gays helping gays. *Christopher Street*, pp. 36–44.

SECTION III

Social Structure, Subcultures, and Drinking Patterns

B. RELIGION, RACE, AND ETHNICITY

Introductory Note

It is firmly established that different ethnic groups within the United States exhibit strikingly different rates of drinking pathologies, ranging from simple inebriety to the extremes of alcoholism and the alcoholic psychoses. There is evidence also that, despite a common core of opposition to drunkenness, the varieties of religious groups and traditions with which ethnicity is so often intertwined are differentially effective in thwarting drinking pathologies. One cannot say for certain what the absolute rates of alcoholism or other drinking pathologies are for particular ethnic, religio-ethnic, or racial groups. It is well documented, nevertheless, that the Irish and Scandinavians, to take two clear-cut examples, have contributed disproportionately to the total fund of drinking pathologies in this country, whereas the yield from certain other groups such as the southern Italians and the Jews has been far below chance expectations. Moreover, relative differences among such groups persist with surprising tenacity over the generations (Greeley & McCready, 1978). Collectively, these facts lend a special significance to the study of ethnic- and religious-group drinking patterns in the context of a concern to understand alcoholism. More broadly, they offer a challenge to sociological inquiry analogous to the challenge offered by Durkheim in connection with varying group rates of suicide.

Chapter 15, by Barry Glassner, continues the seminal work of Charles Snyder (1958) in understanding the conception of the content sources and vitality of Jewish normative attitudes towards drinking beyond the nexus of the ritual drinking. Glassner bases his conclusions on the research he conducted with Berg in a northeastern American community in the late 1970s. Their analysis states that the insulation of Jewish people from drinking problems is grounded in four factors: (1) Jewish perception of alcohol-related problems as non-Jewish, (2) moderation practices from childhood, (3) insulation by peers, and (4) avoidance repertoire. Glassner and Berg are among the few scholars over the last three decades who have explored the salience of religious affiliation on drinking patterns of individuals. Particular religious patterns, whether that of Islam (Badri, 1976), the Latter-day Saints, or Christian Scientists, have received scant attention in reference to the relation between their religious belief systems and their attitudes toward drinking. Measures of religiosity constructed on frequency of church attendance or similar indices are not the best measures for determining the impact of religious belief systems on drinking practices. Ideally the researcher should be cognizant of the specific religious group's attitude toward the use of alcoholic beverages in different contexts.

In the original *Society, Culture, and Drinking Patterns,* negligible reference was made to drinking practices of various racial groups. The absence of these materials was primarily based upon the paucity of research studies that had been conducted on different racial groups in U.S. society. In this volume the deficiency is partially corrected. Frederick Harper and Elaheh Saifnoorian, in chapter 16, "Drinking Patterns among African-Americans," present information on black drinking patterns derived both from historical and more recent research investigations. However, it should be pointed out that systematic investigations of drinking patterns and pathologies are still relatively limited for this significant group in the population. The same can be said in reference to other racial groups in the society, such as Asians (Bennett & Ames, 1985).

Lastly, chapter 17 is James Neff's original study of the relationships among race, ethnicity, and drinking patterns. These data are drawn from a random sample of approximately 1,300 Anglo, black, and Mexican-American male and female drinkers in a community in the southwestern United States. His findings indicate that there are significant differences in beverage preferences by race and ethnicity; moreover, the study points out the crucial importance of sociocultural differences in the perceived role of alcohol and drinking norms.

In conclusion, one is struck by the lack of empirical investigations by U.S. researchers on the crucial importance of race, ethnicity, and religion in the examination of drinking patterns. Moreover, studies of social and economic class differences, as well as those of various occupational categories (Plant, 1979), in reference to both drinking patterns and pathologies, are extremely limited in the contemporary alcohol studies literature.

References

Badri, M.B. (1976). *Islam and alcoholism.* Tacoma Park, MD: International Graphic Printing Service.

Bennett, L.A., & Ames, G.M. (Eds.). (1985). The American experience with alcohol: Contrasting cultural perspectives. New York: Plenum Press.

Greeley, A., & McCready, W. (1978). A preliminary reconnaissance into the persistence and explanation of ethnic subcultural drinking patterns. *Medical Anthropology, 2,* 31–51.

Plant, M. (1979). *Drinking careers: Occupations, drinking habits, and drinking problems.* London: Tavistock.

Snyder, C. (1958). *Alcohol and the Jews.* Glencoe, IL: Free Press.

CHAPTER 15

Jewish Sobriety

BARRY GLASSNER

In his reviews of historical accounts of the drinking patterns of Jews cross-culturally, Charles Snyder suggested that sobriety and temperance have been the rule for centuries (Snyder, 1958). Snyder's own research on Jewish sobriety offered what remains the seminal statement on this topic. Following upon the work of Bales (1944), Snyder (1958) explored the proscription of drunkenness and hedonistic consumption of alcohol in Jewish communities, as well as the prescriptive use of alcohol in religious ceremonies and rites of passage (see Mizruchi & Perruchi, 1970).

My work, in collaboration with Bruce Berg (Glassner & Berg, 1980, 1984), was undertaken a quarter-century after Snyder's and in the light of considerable evidence that Jews had drifted away from the Orthodox religious affiliations Snyder had found to be so important. "The noteworthy sobriety of the Jews appears to be primarily associated with the culture of Orthodox Judaism—a religious culture with a ritualistic emphasis, prescribing frequent drinking which is integrated with familial religious practices," Snyder (1978, p. 96) contended. He joined Bales (1944) in predicting that the decline of Orthodoxy would bring the end of low Jewish alcoholism rates.

Since Jewish alcohol problem rates have remained low—1% of the adult Jewish population versus more than 7% of the adult U.S. population (Glassner & Berg, 1980; Unkovic, Adler, & Miller, 1975)—we looked beyond Orthodox religious practice for explanations of Jewish sobriety.

Some theoretical constructions have pointed to special positions of the ethnic group in society. Glazer (1952) built upon Kant's (1789) argument that Jews are in vulnerable positions and that they respond with an emphasis upon self-control and propriety. "It is the consequences of the siege, passed down from generation to generation, and including such elements as the desire to hold on to one's senses and a distaste for the irrational, that sets a limit to Jewish drinking," Glazer (1952, p. 185) con-

Barry Glassner is with the Department of Sociology, University of Southern California, Los Angeles, Calif. This chapter was written especially for this book.

cluded. Similar arguments hold that Jews try to avoid the public censure that might arise from affective drinking or that Jews have come to emphasize education and rational self-control (Snyder, 1978). Although this sort of discussion is descriptive of some Jewish value styles, it is by itself insufficient to explain low rates of alcohol-related problems. Even if one is willing to grant causal power to values, the problem arises that these values somehow become substance specific. Jews do exhibit high rates of obesity and overeating, high rates for neuroses and some psychoses and, within subpopulations, high rates of drug usage.

What seem to be operative, then, are alcohol-specific protective characteristics in Jewish communities. Among those frequently cited are sacramental and family drinking (King, 1961; Snyder, 1978), which have depended, of course, upon traditional religious practices. (One must wonder about their continuation and vitality amid declines in traditional religiosity.)

A contrasting approach to the question of relatively low Jewish rates of alcohol-related problems seeks characteristics usually found among substance-abusing groups but absent from nonabusing groups. Perhaps the most often confirmed sociological generalization on this issue is that involvement with marijuana and alcohol is highly contingent upon usage of these substances by peers. If an explanation of Jewish drinking based upon this generalization is not to be seen as begging the question, however, it must address important issues. Do Jews, in fact, view members of their ethnic group as their principal drinking peers, rather than turn to other peers, drawn from, for instance, neighborhood, workplace or school? How could Jews maintain such peer relationships when other ethnic groups do not? Ample evidence suggests, for instance, that after the first generation in America, Italians and Irish do accommodate themselves to the drinking patterns of additional peers outside their own groups, even where ethnic group members cannot be properly said to be assimilated at the structural or primary group levels.

Protection from Alcoholism

The methodology employed in our study is described in detail elsewhere (Glassner & Berg, 1980). In short, we developed a stratified random sample of 88 Jews—variously self-defined as Orthodox, Conservative, Reform, and nonpracticing—in a central New York State city that is consistently listed among the ideal northeastern cities for market and social scientific research because it offers "mean demographics" on such variables as population size, ethnic composition, age, and income distributions. Each person in our sample was interviewed in depth by means of a structured, open-ended instrument containing 144 questions.

From the transcript material, we have identified four processes that seem to protect Jews from alcohol-related problems:

1. Definition of Alcohol-Related Problems as Non-Jewish

There appears to be a widespread belief among members of the sample that alcoholism is something that does not happen to Jews. A substantial majority of the respondents indicated this belief, even though we did not ask any questions about it. In most cases the respondents directly stated this association. An Orthodox counselor in his 60s reported:

> Nobody ever drank to get drunk. I mean that wasn't, isn't a Jewish concept. Liquor and wine is part of Jewish, you know, holiday and tradition. More sociability at parties.

A Reform college professor in his 40s stated:

> [A Christian friend] drinks a lot. I don't think he's an alcoholic, but his idea of socializing is to drink, so I buy a bottle of Scotch to have in the house, just in case, because he likes J & B. So when he comes over, I'll always offer him a drink. He and his wife are very drink-oriented people. They drink, and I associate that with non-Jewishness myself. I always associate Jewish tradition with food, as most people do, I think.

A Reform housewife in her 30s has long held this association:

> We were exposed to liquor plenty. Actually, my father worked for a beer company. It sounds like a stupid generalization, but non-Jewish people drink more heavily than Jewish people. That's a generalization I've been brought up with . . . and I still think it's true.

In some cases, the association is made, but in less direct statements. A Reform housewife in her 50s noted:

> I've been a guest at [a local Jewish Country Club]. . . . You know, they never spend money on their bar there like they do in most Gentile country clubs.

Our finding that Jews define alcoholism as an out-group characteristic is a reconfirmation of the finding within earlier studies (Kramer & Leventman, 1961; Snyder, 1958). Indeed, Zimberg (1976) went so far as to claim that "the sociocultural attitudes equating Jewish identity with sobriety are perhaps the major factors accounting for low alcoholism in Jews."

That Jews define alcohol-related problems as not Jewish is customarily interpreted within the literature as one way in which the pious Jew maintains his sense of moral superiority (Snyder, 1978). Our data support this interpretation among the Orthodox, about whom the earlier literature concentrated, but verbalization of the belief is not less among other Jews in our sample, for whom it seems to be taken as a learned fact.

To the extent that defining one's group as nonalcoholic helps to prevent members of the group from becoming problem drinkers, recent campaigns promoting the idea that an increasing number of Jews are alcoholics, may, ironically, increase alcohol abuse (see Steinhardt, 1988). That the campaigns are having an effect is evident in several interviews. One finds questioning and confusion about the association between alcoholism and non-Jews among several of the younger respondents.

This first protective process thereby turns out to be two-sided. A newsletter quotes a psychologist's (Spencer, 1979) claim that "the Jewish alcohol abuser is . . . the most difficult to work with. They 'know' they don't exist. They believe they have the same reality as unicorns. This makes it very difficult for them to identify themselves as alcohol abusers." Other clinicians report that Jewish alcoholics use the defense mechanism of defining *alcoholic* as non-Jewish ("Since I'm Jewish, I can't be alcoholic").

Thus, on the one hand, the defining of alcohol-related problems as non-Jewish serves as an eligibility control that can contribute to avoidance of alcohol-related problems within the large Jewish group by making more difficult the moves we have noted as necessary for substance abuse. If extensive drinking is considered non-Jewish, it becomes more difficult for a Jew to believe that problems are solved through the use of alcohol, to learn how to use alcohol to solve problems, or to rationalize to significant others such deviant use of alcohol. On the other hand, for those Jews who despite the deterrents do find their way to alcohol-related problems, the Jewish community's definition can hinder the work of personnel from the helping services by promoting denial on the part of the alcoholic.

2. Moderation Practices from Childhood

Snyder (1958) argued that Jewish sobriety "depends upon the continuity and vitality of the Orthodox religious traditions." He emphasized the importance of Orthodox religious affiliation per se for his respondents, but within his work one also sees that it would be possible to extend the implications of the protective characteristics of Orthodoxy. He argued, for instance, that "the thread of Orthodox life may be woven into many regional fabrics without losing its sobering influence." Our data suggest that such protection has been extended even further into the contemporary secularized community. Through religious and ceremonial usage of alcohol, Jews continue to learn "prescriptive" (Mizruchi & Perruchi, 1970) drinking norms. In contrast to the expectations of those authors (Bales,

1944; Malzberg, 1940; Zimberg, 1976) who expected the loss of traditional protections with the loss of Orthodoxy, we have detected a good deal of tenacity and adaptability of traditions and their latent symbolism.

Children continue to have many opportunities to learn to associate drinking of alcohol primarily with special (dare we say *sacred*?) occasions (see Knupfer & Room, 1967, p. 690). On the last page of his book, Snyder (1978) listed several conditions we find are currently met in new ways:

> Where drinking is an integral part of the socialization process, where it is interrelated with the central moral symbolism and is repeatedly practiced in the rites of a group, the phenomenon of alcoholism is conspicuous by its absence. (p. 202)

Sobriety amid drinking continues to be the norm in Jewish homes. Our interviews suggest that Jews perpetuate this association and its connection with ritual not only by affiliation with religious life but also by restricting drinking to special secular occasions and by cataloguing drinking as a symbolic part of festive eating.

Almost without exception, our interviewees reported that drinking in their homes during childhood was predictable and that drunkenness was never condoned or was condoned only on very rare occasions. A Conservative secretary in her 60s said:

> Parents don't usually put a drink in front of themselves or their children. Mine didn't, and I wouldn't with my kids. Some kids will say, "You do it, why can't I do it?" We never did drink except for a special occasion.

A nonpracticing medical researcher in her 30s noted:

> It was a ritual in our daily life. It was a family routine. [My father] would sit and read the paper for a little while and have the drink and then we would have dinner and my mother would have a drink with him.

Nearly half of those interviewed could not remember their first drink but did recall that all their early drinking was during childhood in the home as part of religious ceremonies. Or those in the remainder of the sample, 89% recalled taking their first drinks in a recreational, family-supervised context, in the home or at the synagogue, and doing so before the age of 13. Thus, only about 5% of those in the sample recalled taking their first drinks outside the family and at a time later than childhood.

Our respondents thus grew up experiencing the conditions that several studies have found to result in adults' having low rates of alcohol-related problems: parents who agree about drinking norms for themselves and their children, formative drinking experiences within the family, and controlled drinking at an early age (McGonegal, 1972; Wilkinson, 1970). Jews are sheltered from what Pittman (1967) calls the ambivalent drinking culture of the United States, which consists of conflicting attitudes toward alcohol consumption. The Jewish family maintains instead what Pittman calls a permissive drinking culture, in which drinking is permitted, but excessive drinking is not.

Not only is Jewish drinking limited primarily to predictable family, social, and religious occasions, but the drinking is closely associated with eating (Hill, 1977). Most of the interviewees alluded to the relative importance of food over drink and the absence of drinking without eating. A nonpracticing businessman in his 40s described the balance between eating and drinking that was modeled for him:

> I can remember more my mother arranging platters of cold cuts and potato salad or baking because she was having people [to the house] than I can ever remember her serving drinks. The emphasis was always on the food, and the liquor was deemphasized. If it was there it was because there were people who expected it to be there.

A Conservative schoolteacher in her 30s observed:

> Drinking was just never entered as an activity. It's part of a thing like eating. It's OK with foods. It goes, to me, it goes along with eating.

The moderating effect of associating drinking with eating was first noted by Feldman (1923) in a comparison of the contrasting practices of English Gentiles and English Jews. In recent decades, several studies (reviewed in Plaut, 1967; and Zucker, 1976) have suggested that groups characterized by widespread drinking but few alcohol-related problems have integrated their drinking into a variety of basic activities. In the case of Jews, this integration is taken a significant step further, toward making drinking a symbolic act. Drinking serves as a symbolic punctuation mark that helps to separate certain positive events (religious services, weddings, dinners, etc.) from all other events (Mandelbaum, 1965). By being a symbolizer, the drinking takes on a symbolic character itself.

3. Insulation by Peers

If drinking patterns learned in childhood are to be maintained, and if the self-definition of Jews as moderate drinkers is to remain believable, Jews must collectively reiterate moderate drinking patterns in adulthood. This reiteration is accomplished in part through family practices, as we have already noted. However, the serious threat to moderate drinking occurs outside the home, at parties and other public occasions, when considerable drinking is often encouraged or expected.

In large part, the ability of our respondents to avoid excessive drinking contexts seems to be due to their socializing mainly with other Jews. Thus, 76% of those in the sample reported that all or nearly all their friends are Jewish, although most said that their work associates are primarily non-Jews. The respondents are not, as a group, ideologically opposed to integrating. When asked if it mattered one way or the other whether they spent recreational time with Jews or with non-Jews, 64% answered negatively, 25% affirmatively, and 14% (including two individuals who answered affirmatively) described types of situations in which it would or would not matter.

A gravitation toward Jewish things and people (90% see "Jewishness" as important in their lives) protects Jews from excessive drinking in obvious and not so obvious ways. First, a Jew is unlikely to find others who would encourage, cooperate with, or even condone excessive drinking. Thus, any rationalization for extra drinking becomes difficult. This is particularly evident in those cases when Jews drink primarily among other Jews. The process goes further, however, to the selection of non-Jews who drink like Jews (see Alexander & Campbell, 1970). A nonpracticing college professor in his 50s said:

> Most people at the party were not Jewish, actually. This one guy was making a real ass of himself. He'd had too much to drink, and it made everybody uncomfortable. I guess our friends just are not heavy drinkers. I mean, we carry drinks around at parties, but when this guy got drunk he really stood out.... I think he eventually got the message, because he was one of the first to leave.

Socializing with Jews and other moderate drinkers makes inconvenient, or removes as an option, the use of alcohol to various practical ends; in contrast to members of other cultural groups, such as the Oaxacans of Mexico, in which intoxication provides persons with license to shout insults and attend social gatherings without invitation (Dennis, 1975), Jews who become intoxicated are likely to be scorned or pitied.

A net result of the moderate drinking environment is that Jews maintain images of normal and preferred social interactions over their life spans that do not include excessive drinking. This point is evident, for example, in the case of an Orthodox businessman in his 40s who answered the question "When did people drink when you were growing up?" with "My family just didn't drink. People we are with now just don't seem to drink." A further illustration is the Conservative male college student who answered a question about parents' drinking patterns with, "My parents rarely drink at all. Myself, I went through the drinking phase with my friends, but we've turned the page now."

Group avoidance is thus a dynamic process. By modeling consistent group drinking practices and norms in adulthood, Jews perpetuate the drinking associations into the next generation. Also, because unique alcohol consumption practices serve to emphasize ethnic differences in this multiethnic nation (Cisin, 1978), adult drinking patterns contribute not only to perpetuation of the first process (defining drunkenness as non-Jewish), but also to group boundary maintenance.

4. Avoidance Repertoire

Despite the social interaction patterns previously described, drinking histories collected in the interviews reveal that at least 46% of the respondents found themselves in social situations during their adulthood in which they were pressured to drink more than they wished to drink. Here they utilized what we have termed *avoidance repertoire,* or interaction techniques appropriate to the social context, which permitted them to control their alcohol consumption. We recorded several exchanges similar to the following discussion with a nonpracticing businessman in his 40s:

Q: Does it matter if other people drink with you?
A: No, it seems to matter more whether I drink with them.
Q: What do you mean?
A: My not drinking tends to make more people uncomfortable than their drinking makes me.

The drinking histories, symbolic associations, and uniformity of group opinion we have noted thus far seem to result in these respondents' feeling they can offer an assertive no when encouraged to drink more than they wish. When we asked respondents how they avoid excessive drinking during various pressure situations, many were curious about why we would suspect that such a situation might prove to be a problem. They perceive their avoidance repertoire as nonaction. A Conservative housewife in her 40s said:

I don't care one way or another whether I have a drink. If everybody is drinking and I feel like having a drink I'll have a drink. If everybody is drinking and I don't want a drink, I don't drink.

Nevertheless, upon further probing, many respondents did reveal a variety of techniques used in such avoidance. Several reported that they begin a party with a mixed drink, but refill the glass throughout the night with only the mixer. Other respondents reported that their family as a unit will act to prevent unwanted drinking. Some told of plans whereby the wife will reprimand the husband for drinking too much at a party, when actually he is still nursing his first or second drink. Other couples develop reputations as nondrinkers by telling jokes about drinkers or by avoiding drinking situations.

Other respondents reported specific avoidance repertoires they have worked out for particular situations. For example, a Conservative photographer in his 40s told of the problems he used to experience at weddings, where sometimes he was "on the verge of being hit because they felt it wasn't showing proper respect for the bride and groom by toasting them [i.e., his abstention from toasting was seen as an insult]." This photographer reported that since those early years he has consistently utilized a series of "convincing lines" to indicate that he cannot drink while working. He couples this with taking more pictures at times when others are involved in heavy drinking. "Being Jewish, your reflexes and everything else sharpen you, and that expedience staves off any future attempts [by] people to push drinks at you," he explained.

In summary, this research suggests four processes that protect Jews from alcohol abuse: (1) defining alcoholism as something that happens to non-Jews; (2) subscribing to practices and symbolism learned early, which integrate sober drinking into daily life; (3) restricting most primary relations to others with similar drinking norms and practices; and (4) developing a repertoire of avoidance techniques.

Images of Alcoholism

In his later work on Jewish drinking patterns, Charles Snyder examined the nature of alcohol-related problems among Israeli Jews (Snyder, Palgi, Elgar, & Elian, 1982). In that study, Snyder and his colleagues compared three ethnic communities—Ashkenazi, Sephardi, and Oriental—and found higher rates of alcoholism and related problems among the second and third of these communities. By way of those comparisons, Snyder was able to suggest how the avoidance of alcohol-related problems among Jews is rooted in the cultural patterns of European (Ashkenazi) Jewry.

In our American research, we found relatively low rates of alcohol-related problems among all four self-defined communities of Jews we studied: Orthodox, Conservative, Reform, and nonpracticing. Nearly all of these Jews were, of course, Ashkenazi. Despite their similarities, however, we found marked differences between the groups in perceptions of alcoholism and alcoholics.

The four social processes just discussed operate to lessen the possibility of Jews' developing alcohol-related problems. These processes appear to operate in all subgroups of Jews. In contrast, we found marked differences among the affiliation categories of Jews regarding perceptions of alcoholism and alcoholics. Let us look at how each affiliation group defines alcoholism.

Orthodox

Of the available definitions of alcoholism, we suspected Orthodox Jews would select a highly moralistic version that blamed secular society. Alcoholism has long and widely been considered a moral weakness, and only in the past few decades has a major effort been made to counter this view by educating the public that alcoholism is a disease (Conrad & Schneider, 1980; Ries, 1971). Because the Orthodox are the most traditional Jewish affiliate group and the least accepting of their fellow Jews' accommodation to American ways, one would anticipate that they would be the least likely to accept such "newfangled" notions as disease definitions of alcoholism. Instead, our findings suggest that 63% of the Orthodox (compared to 25% of Reform and nonpracticing) Jews offer disease definitions of alcoholism in response to our open-ended question "What do you think alcoholism is?" A teacher in her 30s replied, "It's a kind of sickness. It can be cured, at least in the beginning." An engineer in his 40s responded, "I think it is a disease. People that are alcoholic, a lot of them, can't; they need some sort of professional help, plus family help together."

As these quotations attest, Orthodox respondents frequently talked about the alcoholic's being cured, whether by conventional medicine or by a combination of medicine and determination.

Half of the Orthodox respondents said they have never known anyone who was even accused of drinking too much, and the persons cited by the other half of the sample as possible alcoholics were in every case very distant from the respondent (e.g., an aunt's second husband and "someone I once met in college"). None of the Orthodox respondents reported knowing a Jew who was considered to have a drinking problem.

In short, the Orthodox respondents appear to have stayed a great distance away from alcoholism and to have picked up a readily available definition of the phenomenon. The respondents' disease definition is externalistic and simplifying. It treats alcoholism as something that happens to someone, much like a bacterial infection. The conviction is probably not held very deeply. Indeed, because the Orthodox find little practical need for defining alcoholism, many respondents had not consciously thought of the issue before we asked about it.

Conservative

Alone among the affiliate groups, Conservative respondents rambled in their discussions about the key issues of concern in our study. When Conservatives were asked to define alcoholism, their median response was 80 words long, well over the median of 45 words for the remainder of the sample. About 33% of the Conservatives' definitions, but only 5% of those from the remaining sample, could not be coded into a single category. Transcripts of the Conservatives' responses frequently display hesitations and ambiguities, in contrast to comments made by members of other affiliation groups. A writer in his 30s (who did not falter in response to other questions) answered:

> I don't think it is a disease. I think it is too much drinking. You know, I don't think, you know, every day, with somebody, somebody very well. It's nobody's, it's not anybody's fault that it's a disease. I think it's a character weakness of a person. I think that when people say it's a disease, I think it's crazy.

A repairman in his 40s said:

> I think it's a, um, it's a need to drink and to consume alcohol, on a regular basis. Not just a need to drink, but it's, it's habit forming to the point where you can't do without it. And, it doesn't necessarily mean that a person has to get drunk. But it's being hooked on something that, uh, you need to use as a crutch, or uh, filler for other things.

The latter excerpt includes the sole consistent characteristic within the Conservative respondents' interpretations of alcoholism: the idea that drinking is habit forming. Forty-six percent of the Conservatives (but only 10% of the other respondents) include habit as a defining characteristic: "It's habit forming. It's a drug. Your body demands a certain amount." "You get the habit and up and use more and more." "There is a compulsive need, even when nobody else is around." This emphasis on habit is the only common feature of the Conservatives' responses.

The idea of alcoholism as a response to life's troubles appears in Conservatives' discussions, but seldom in the pseudopsychological form we will note below for the Reform and nonpracticing. Among Conservatives, this image is enmeshed with other impressions. For a student in his 20s, it is a characteristic of an illness: "Alcoholism is a sickness, I think, it is a, you have to seek refuge through alcohol. You really can't deal with the world as it is or you are having trouble facing your own reality. An escape."

Thus, the Conservatives are midway between Orthodox and Reform and nonpracticing in several respects. As a subjective affiliation, Conservatism is viewed as less religious than Orthodoxy and more religious than Reform or nonpracticing. Conservatives generally see themselves and are seen by other Jews as balancing traditional Jewish lifestyles and values with involvement in American society. Sklar (1972, p. 207) has called Conservatism a "halfway house" between Reform and Orthodox, "a type of Judaism which, while not Orthodox, derives from traditional sources; while not completely Reform, it is sufficiently advanced so as to be 'Modern.' " His research suggests that a difficulty for the Conservative movement has been its lack of a distinct body of religious and secular beliefs and practices (pp. 199–245).

Their confusion and emphasis regarding definitions of alcoholism coincide with these positions. Conservatives are divided into those (29%) who accept the disease definitions prevalent among the Orthodox and others (33%) who—like the Reform and the nonpracticing—adopt psychological versions. But the largest group of Conservatives (38%) is unclear about how to define the phenomenon. The major commonality in Conservatives' definitions—habit—is itself a notion midway between the disease definitions of the Orthodox and the psychologistic definitions of the Reform and nonpracticing. In common usage, ***habit*** is a vague term implying something that happens to a person's mind and body and hence is unintentional or outside the person's control but also involves a person's own acts.

Conservatives as a group are in the middle of a continuum also with regard to their experiences with alleged alcoholics. A third of the Conservatives (50% of the Orthodox, and 7% of the Reform and nonpracticing) claimed to have never known an alcoholic. Of those Conservatives who have known someone suspected of alcoholism, a third have known a Jewish alcoholic. Most of the alleged alcoholics with whom the Conservatives are familiar are distant acquaintances, but unlike those known by the Orthodox, some of the suspected alcoholics are personal friends or in-laws.

Reform and Nonpracticing

In contrast, the Reform and nonpracticing interviewees tended to define alcoholism in psychological terms as a psychological dependence and weakness. A housewife in her 50s said that an alcoholic is

> somebody who would need a drink to face the day, or face anything in the day. Just to get himself going, or herself going. In other words, would just really need it all through the day. To get up, to feel relaxed, to feel self-confident.

A college student in his 20s called alcoholism

> a dependency on alcohol. In other words, it's not just going out and having a drink, but needing that drink to relax. If you can't relax without that drink, you're an alcoholic, whether it is one drink or a bottle.

Seventy-one percent of the Reform and nonpracticing respondents defined alcoholism as a dependency with psychological overtones.

Reform Jews have accommodated themselves to and been active in American society for many years, a state of affairs that has provided them with social involvements different from those of more traditional Jews. One result is that Reform and nonpracticing respondents differ from the Orthodox not only in how they view alcoholism but, concurrently, in their contact with alcoholism and drinking. The dramatic difference is demonstrated by the fact that 93% report they have known an alleged alcoholic and 37% say they have known a Jewish alcoholic. Unlike those known to the Orthodox, many of the presumed alcoholics are close acquaintances, including aunts and uncles, parents' friends, in-laws, parents, and personal friends. The difference is also apparent in the role drinking occupies in the respondents' daily lives. Reform and nonpracticing Jews regularly engage in social drinking. Most respondents said they drink every couple of days, either at business or at social gatherings, whereas Orthodox Jews reported nonceremonial drinking as rare.

The Reform and nonpracticing respondents consider drinking to be a part of social life in two primary ways. First they see alcohol as a natural ingredient in social gatherings. For instance, a retired manufacturer in his 70s said he drinks

> to make the people that we are with more comfortable when they are at our house. The other way around, when I'm out, I don't want them to feel that I am not social.

A business owner in his 30s said:

> It is very social to drink. It's a common ground either for discussion or as a starting point to being together.

These sorts of comments are totally unlike those of the Orthodox respondents, as is the case with the second social version of drinking, in which Reform and nonpracticing Jews talked of drinking as useful in its own right. Many respondents described the importance of drinking with business associates, but for several others the secular symbolism of drinking went still further. A college professor in his 40s explained:

> I think the reason why I keep a store of liquor in the house and why I put wine on the table [is that] it's sort of a symbol of, well, we're doing the elegant

thing . . . it was always the thing to do if you were rich. Since none of us were rich, this is why I think we do it now.

Summary

The study of Jewish drinking patterns and beliefs continues to provide important information about a variety of alcohol-related issues. We have suggested four protective processes that appear to contribute to the avoidance of alcohol abuse by American Jews. In addition, we have discussed some significant variations among subgroups of Jews in beliefs about alcoholism and alcoholics.

It will be particularly instructive if further research can focus upon changes in drinking patterns and beliefs among Jews, because these occur in contrasting types of communities or over different historical periods. In addition, comparative research with other ethnic groups might reveal patterns of protective processes and beliefs similar to those we have described for American Jews.

References

Alexander, D.N., & Campbell, E.Q. (1970). Normative milieux and social behaviors. In G.L. Maddox (Ed.), *The domesticated drug* (pp. 268–69). New Haven, CT: College and University Press.

Bales, R.F. (1944). The "fixation factor" in alcohol addiction. Unpublished doctoral dissertation. Harvard University.

Cahalan, D., & Cisin I.H. (1968). American drinking practices. *Quarterly Journal of Studies on Alcohol, 29,* 130–151.

Chafetz, M.E., & Demone, H.W. (1962). *Alcoholism and society.* New York: Oxford University Press.

Cisin, I.H. (1978). Formal and informal social controls over drinking. In L. Ewing & B. Rouse (Eds.), *Drinking alcohol in American society* (pp. 145–158). Chicago: Nelson-Hall.

Conrad, P., & Schneider, J. (1980). *Deviance and medicalization.* St. Louis, MO: Mosby.

Dennis, P. (1975). The role of the drunk in an Oaxacan village. *American Anthropologist, 77,* 856–863.

Feldman, W.M. (1923). Racial aspects of alcoholism. *British Journal of Inebriety, 21,* 1–15.

Glassner, B., & Berg, B. (1980). How Jews avoid alcohol problems. *American Sociological Review, 45,* 647–664.

Glassner, B., & Berg, B. (1984). How define alcoholism. *Journal of Studies on Alcohol, 45,* 16–25.

Glazer, N. (1952). Why Jews stay sober. *Commentary, 13,* 181–186.

Greeley, A.M., & McCready, W.C. (1978). *Societal influences on drinking behavior.* NORC paper presented to the International Medical Advisory Conference of the Brewing Associations, Toronto.

HEW. Department of Health, Education and Welfare. (1972). *Alcohol and health.* First Special Report to the U.S. Congress from the secretary of Health, Education and Welfare. Washington, D.C.: U.S. Government Printing Office.
Hill, T.M. (1977). Survey of Jewish drinking patterns. *Military Chaplain's Review,* 65–77.
Kant, I. (1940–1941). *Anthropologie.* Cited in E.M. Jellinek, Immanuel Kant on drinking. *Quarterly Journal of Studies on Alcohol, 1,* 777–778. (Original work published 1789)
Keller, M. (1970). The great Jewish drink mystery. *British Journal of Addiction, 64,* 287–295.
King, A.R. (1961). The alcohol problem in Israel. *Quarterly Journal of Studies on Alcohol, 22,* 321–324.
Knupfer, G., & Room, R. (1967). Drinking patterns and attitudes of Irish, Jewish and White Protestant American men. *Quarterly Journal of Studies on Alcohol, 28,* 676–699.
Kramer, J.R., & Leventman, S. (1961). *Children of the gilded ghetto.* New Haven, CT: Yale University Press.
Levy, L. (1973). Drug use on campus. *Drug Forum, 2,* 141–171.
Lowenthal, U., Wald, D., & Klein, H. (1975). Hospitalization of alcoholics and the therapeutic community. *Harefuah, 89,* 316–320.
McGonegal, J. (1972). The role of sanction in drinking behavior. *Quarterly Journal of Studies on Alcohol, 33,* 692–697.
Malzberg, B. (1940). *Social and biological aspects of mental disease.* Utica, NY: State Hospital Press.
Mandelbaum, D.G. (1965). Alcohol and culture. *Current Anthropology, 6,* 281–288.
Mizruchi, E.H., & R. Perruchi (1970). Prescription, proscription and permissiveness: Aspects of norms and deviant drinking behavior. In G. Maddox (Ed.), *The domesticated drug* (pp. 234–53). New Haven, CT: College and University Press.
Mulford, H.A. (1964). Drinking and deviant behavior. *Quarterly Journal of Studies on Alcohol, 25,* 634–650.
Pittman, D.J. (1967). International overview: Social and cultural factors in drinking patterns, pathological and non-pathological. In D.J. Pittman (Ed.), *Alcoholism* (pp. 3–20). New York: Harper & Row.
Plaut, T.F. (1967). *Alcohol problems: A report to the nation by the cooperative commission on the study of alcoholism.* New York: Oxford University Press.
Ries, J. (1971). Public acceptance of the disease concept of alcoholism. *Journal of Health and Social Behavior, 18,* 338–344.
Riley, J.W., & Marder, C. (1947). The social pattern of alcoholic drinking. *Quarterly Journal of Studies on Alcohol, 8,* 265–273.
Roberts, B.H., & Myers, J.K. (1967). Religion, national origin, immigration and mental illness. In S.I. Weinberg (Ed.), *Sociology of mental disorders* (pp. 68–72). Chicago: Aldine Press.
Room, R. (1968). Cultural contingencies of alcoholism. *Journal of Health and Social Behavior, 8,* 99–113.
Schmidt, W., & Popham, R.E. (1976). Impressions of Jewish alcoholics. *Journal of Studies on Alcohol, 37,* 931–939.
Sklar, M. (1972). *Conservative Judaism.* New York: Schocken.
Snyder, C. (1978). *Alcohol and the Jews.* Carbondale, IL: Southern Illinois Press (Reprint. Original work published 1958 in Glencoe, IL, by Free Press).

Snyder, C., Palgi, P., Elgar, P., & Elian, B. (1982). Alcoholism among the Jews in Israel: A pilot study. *Journal of Studies on Alcohol, 43,* 623–654.
Spencer, J.M. (1979, Spring). West Coast seminar lifts veil of denial. *Alcoholism and the Jewish Community Newsletter,* pp. 1–3.
Steinhardt, D. (1988). Alcoholism: The myth of Jewish immunity. *Psychology Today, 22,* p. 10.
Unkovic, C.M., Adler, R.J., & Miller, S.E. (1975). A contemporary study of Jewish alcoholism. *Alcohol Digest, 9,* vi–xiii.
Wilkinson, R. (1970). *The prevention of drinking problems.* New York: Oxford University Press.
Zimberg, S. (1976). Socio-psychiatric perspective on Jewish alcohol abuse. Paper presented to the Task Force on Alcoholism of the Commission on Synagogue Relations, New York.
Zucker, R.A. (1976). Parental influences on the drinking patterns of their children. In M. Greenblatt & M. Schuckit (Eds.), *Alcoholism problems in women and children* (pp. 211–38). New York: Grune & Stratton.

CHAPTER 16

Drinking Patterns among Black Americans

Frederick D. Harper and Elaheh Saifnoorian

This chapter focuses on the following topics related to black drinking patterns: (1) historical and etiological factors; (2) drinking patterns and consequences; (3) special populations and subgroups (youth, women, and families); (4) treatment, prevention, and alcohol education; and (5) issues and recommendations.

Historical and Etiological Factors

As early as the Atlantic slave trade, alcohol was employed as a commodity of barter in the trade of African slaves (Mannix, 1962). During American slavery, alcohol was very much a part of the weekend and holiday social life of many plantations—often thought by slaveowners to be a means of release for slaves from long hard hours of work as well as a method of diverting blacks from thoughts of escape to freedom (Franklin, 1967). On the contrary, some slaveowners refused to allow their slaves access to alcohol for fear that such activity would lead to rebellion and slave revolt on the plantation.

Subsequent to the era of American slavery, blacks found themselves isolated from the diverse social and recreational activities that whites enjoyed; thus, many found relaxation and recreation in the form of heavy drinking at public taverns and private homes, especially on weekends and during holidays. Herd (1987) states that migration of blacks to urban northern cities in search of jobs during the early 1900s appears to have resulted in increased use of alcoholic beverages.

For generations, black civil rights and social protest leaders have warned that heavy and problem drinking is a destructive activity and a

Frederick D. Harper is with the School of Education, Howard University, Washington, D.C. Elaheh Saifnoorian is a doctoral student, School of Education, Howard University. This chapter was written especially for this book.

barrier to black freedom and self-improvement. As early as the 1800s, Frederick Douglass scolded slaves about their heavy weekend and holiday drinking (Douglass, 1892). More recently, black leaders from various circles (e.g., civil rights, religion, and athletics) have spoken out against alcohol abuse and misuse—leaders such as Jackie Robinson, Jesse Jackson, W.E.B. DuBois, Marcus Garvey, and Paul Robeson.

Etiological Factors

Blacks have long maintained drinking patterns characterized by a preference for group drinking, weekend and holiday drinking, and a tendency to either drink heavily or abstain altogether. Moreover, blacks have suffered disproportionately from negative consequences and social problems associated with heavy and inappropriate drinking (e.g., homicides, industrial accidents, and alcohol-related health problems).

Within the context of these drinking patterns and consequences, theorists and students of alcoholism have come forth with etiological explanations. Earlier explanations for heavy drinking or nondrinking among blacks tended to rely on racially stereotypical reasons such as deficits or weakness, matriarchal families, compensation for social inadequacies, low self-esteem, and as a means of sexual seduction (e.g., see Maddox & Williams, 1968; Strayer, 1961).

While the above explanations tended to blame blacks for their own alcohol plight (i.e., by focusing on personality attributes and lifestyle orientation), there are other etiological explanations of social factors external to the drinker—factors such as racial discrimination (Larkins, 1965) and social hardship of living (Sterne & Pittman, 1972).

Harper (1976) attempts to identify a number of varied factors (e.g., historical, social, psychological, and economic) that serve to explain drinking patterns of blacks. Such explanations of heavy drinking patterns include (1) historical drinking patterns established as a result of the hardship of slavery and the isolation of racial segregation that followed; (2) the accessibility of liquor stores and dealers in black residential neighborhoods; (3) acceptance among peers for heavy drinking, group drinking, and weekend drinking; (4) frustrations resulting from racism, job problems, and financial responsibilities; and (5) an attempt to fulfill psychological needs or escape unpleasant emotional moods.

Nondrinking or abstention among a large proportion of black women has been explained in terms of family responsibility and lack of money (no time and extra money for drinking), fundamental religious upbringing, a dislike for the bitter and dry taste of many alcoholic beverages, the need to stay sober in order to manage their children and often their problem drinking men (especially husbands/lovers who may be prone to trouble while drinking), and the lack of access to alcohol due to a tendency to stay at home or be constantly at work (Benjamin, 1976; Cahalan & Cisin, 1968; Harper, 1976).

Recent etiological propositions tend to focus on racial consciousness and pride as being associated with drinking attitude. For example, blacks with little racial consciousness and pride are more likely to be alcohol abusers compared to those with a sense of racial consciousness and pride (Caution, 1986; Gary & Berry, 1986). Gary and Berry (1986) concluded that racial consciousness is a reliable predictor of attitudes about alcohol use, and that those conscious of racial issues are less likely to be tolerant of alcohol abuse. Moreover, Caution (1986) posits that Afro-centric values can assuage the alcohol-misuse problem of blacks and be a basis for prevention (i.e., values of oneness with nature, survival of the group, kinship, spirituality, and black history and culture).

Drinking Patterns and Consequences

National and regional studies of probability samples have very often found no significant differences in drinking rates and/or alcohol consumption between blacks and whites. However, when racial subgroups are compared, black women have consistently shown a significantly greater proportion of abstainers than white women (Cahalan & Cisin, 1968; Fernandez-Pol & Bluestone, 1986; Herd, 1988; Neff, 1986). Moreover, a number of studies indicates that black youth have a lower rate of drinkers and/or heavy drinkers compared to white youth (Dawkins, 1976; Globetti, 1970; NIAAA, 1983).

There appears to be a lack of consensus on the drinking patterns of blacks as a group and when compared to whites. Some within-race and between-race variations may be due to a number of factors including differences in populations studied, sampling error, changes in drinking practices over the years, method of data collection, and validity of data collected. On the other hand, black drinking practices, as with any social group, can differ across socioeconomic lines—i.e., across age group, geographic residence (e.g., urban vs. rural), social class, educational level, or gender. These differences and inconsistencies suggest the need for a national survey that compares black drinking patterns by socioeconomic groups within race as well as matched samples between races (e.g., black compared to white Americans).

The social concern of black drinking seems not to focus on rates and amount of consumption but rather the grave consequences and problems associated with alcohol abuse and misuse. Such alcohol-related social problems, especially among urban blacks have been documented repeatedly over the years in the forms of alcohol-related arrests, crime, homicide, family disruption, health problems, and industrial accidents.

For example, Gary (1986) documents the association of alcohol drinking and homicidal events among black men. Along the same line, Goodman et al.'s (1989) study of homicidal victims (aged 15–64) revealed that 49% of black victims were diagnosed with alcohol in the bloodstream compared to only 39% of white victims. In an earlier study, Harper (1976) found comparable or higher percentages of alcohol incidence in black homicidal victims in urban areas of Atlanta, Cleveland, Miami, and Washington, D.C.

In terms of additional alcohol-related problems, Sterne and Pittman's (1972) urban case study of black families documents negative consequences associated with alcohol misuse such as marital disruption, loss of money, sexual infidelity, gambling, lack of child support, child neglect by heavy drinking mothers, family fights and arrests, and a practice of allowing babies and small children to sip alcohol.

Drinking in public places and arguments during drinking often combine to yield a higher rate of alcohol arrests for urban blacks compared to whites (Harper, 1989). In general, for a large proportion of blacks, heavy drinking in groups on weekends, when combined with poverty, urban living, stress, poor health, and violence, often results in negative consequences.

Special Subgroups and Populations

Youths

National and local surveys suggest no significant differences in drinking rates for black and white youths (see Dawkins, 1976; Globetti, 1970; Harper, 1988; NIAAA, 1983). Some studies even indicate slightly higher rates (but not significant) of drinking (Dawkins, 1976) and heavy drinking (Globetti, 1970) by white rural youths.

In regards to urban youths, Dawkins (1981) surveyed 1,095 subjects (93% black, 7% white; 51% female) in Washington, D.C. on their drinking practices and drug use. Dawkins found that this predominantly black urban sample was (1) most likely to drink during weekends, on holidays, and at night, (2) most likely to have taken the first drink between ages 11 and 15, (3) very likely to receive their first drink from parents or friends at a party, (4) very likely to drink with another person or group, and (5) drank mainly to celebrate an occasion or have a "good time." Eighty percent reported taking a drink within the previous year and 56% reported drinking during the previous month.

Dawkins and Dawkins (1983) carried out a study on alcohol-related offenses involving 342 subjects (black, Hispanic, and white youth offenders). Their results yielded a significant correlation between drinking rate/level and juvenile offenses for all groups. However, drinking rate/level served to be a predictor of criminal behavior for black youths but not so for the other two racial groups.

A large proportion of urban youths who experiment with alcohol tend to have parents who are heavy drinkers. Sterne and Pittman's (1976) study of drinking families in urban St. Louis revealed that black youths who grew up in heavy-drinking families and neighborhoods tend to develop similar drinking patterns themselves, and a predisposition for alcohol-related problems. On the same topic, NIAAA (1983) delineates the following consequences for youths with heavy drinking parents: youths of heavy drinkers tend to have lower self-esteem and higher levels of anxiety, aggression, and psychosomatic symptoms compared to the general population of youths; youths of heavy drinkers are more likely to have school problems and display antisocial behavior; and high rates of separation and divorce among alcoholic parents often result in stress for their children.

Women

There appear to be greater differences in drinking patterns between black and white women than their male counterparts. Earlier (Cahalan & Cisin, 1968) and recent (Herd, 1988) surveys indicate a significantly larger proportion of black women are abstainers compared to white women. Cahalan and Cisin's (1968) national survey found a slight majority of black women (51%) to be abstainers; however, at the other extreme, the rate of black female heavy drinkers was higher than that for white women. Herd's (1988) more recent survey confirms Cahalan and Cisin's findings of black women as being more likely than white women to be abstainers; nevertheless, Herd found white women to have a ***higher*** rate of heavy drinkers than black women (just the opposite of the Cahalan and Cisin survey published 20 years earlier).

Even more revealing, Herd (1988) examined subgroups from a national probability sample of women (1,224 blacks and 1,034 whites) and found the proportion of black female abstainers and the degree of differences between black and white abstainers greater in elderly or aged subgroups. For example, the following is the percentage of nondrinkers in various subgroups of older black women: those aged 60 and greater (69%), widowed (70%), and retired (75%). Among those categorized as low-income black women (regardless of age), the nondrinker rate was 53%.

In regard to drinking setting, preferences, etiological cause, and practices, Dawkins and Harper (1983) examined self-report data from a representative sample of 201 alcoholic women. Although no significant differences were found as to "cause" of heavy drinking, differences did occur on other criteria. For example, black alcoholic women were significantly more likely to prefer gin and sweet wines, drink in a group or with friends, drink in public (including "on the street"), drink in the morning, and drink on holidays. White women were significantly more likely to prefer bourbon or mixed drinks, drink at home or alone, and start with heavy drinking at a later age. The reader must be reminded that these findings were based on data of drinking patterns of diagnosed alcoholic women; and, therefore, are not generalizable to the drinking behaviors of the general population of black and white women.

It appears from the limited empirical research that black and white women are less alike in drinking patterns than black and white men; the latter tend to be relatively similar, especially in rates of drinking behavior.

Families

A review of the recent literature and an analysis of black families and substance abuse yielded the following observations and conclusions (Harper, 1986):

- Heavy and inappropriate use of alcohol seems to be associated with instability of and negative consequences for the black family.
- A large proportion of urban black families have contributed to their own heavy drinking and problem drinking via (1) drinking lifestyles; (2) lack of knowledge, false knowledge, and inappropriate attitudes about alcohol use and alcoholism; and (3) outright irresponsibility, such as child neglect, financial mismanagement, family conflict, and sharing alcohol with babies and youths.
- Heavy drinking tends to contribute to black family disruption and a number of alcohol-related problems: gambling, sexual infidelity, loss of money, unstable work habits, family violence, neglect of parental responsibilities, damage to household property, and police arrest.
- Heavy drinking and problem drinking of black family members are frequently influenced by or associated with (1) intergenerational family drinking patterns dating back to racial segregation and, seemingly, American slavery; (2) the high prevalence of liquor stores accessible to family households (i.e., in residential neighborhoods); (3) peer pressure to drink, especially in heavy drinking neighborhoods and/or housing projects; and (4) social stresses (e.g., racism and racial discrimination) and boredom that drive family members to drink for convivial reasons and as a means of escape.

Moreover, there is evidence that heavy drinking among black families can result in family grief and loss due to alcohol-related homicide (Gary, 1986) and suicide (Smith & Carter, 1986); children's embarrassment by the public drunkenness of their parents; family fights, neighborhood disturbances, frequent accidents, and alcohol-related illnesses; and victimization of a drinking family member by criminal elements (Gary, 1983).

Treatment, Prevention, and Alcohol Education

In the late 1970s, NIAAA commissioned a report to survey and analyze the literature and examine programs throughout the country concerning the latter's practices and problems. The resulting report, *Alcoholism Treatment and Black Americans,* generated the following observations and conclusions (Harper, 1979):

1. Based on hospital and clinical data, it appears that black alcoholics in treatment tend to be younger than their white counterparts.
2. Alcoholics Anonymous (AA) groups that are predominantly black by membership and/or those located in black neighborhoods often prove to be more successful for the black alcoholic compared to those AA groups that are located outside the black community with predominantly white memberships. (Nevertheless, any AA group, regardless of racial composition, is recommended for black alcoholics.)
3. Outreach counseling is very important in the identification and subsequent treatment of black alcoholics since they are not likely to reach treatment via self-referral or referral by a family member.
4. Much of the research literature and most treatment programs relate to the alcoholism treatment of urban and rural lower-income blacks.
5. Racial discrimination and racism have had and apparently continue to have an effect on the etiology and effective treatment of black alcoholics.
6. Treatment efforts with black alcoholics should be comprehensive, culture-specific, community-based, and inclusive of family support and involvement.

The same report (Harper, 1979) identified and ranked a number of problems associated with the alcoholism treatment of blacks, as follows:

Rank (importance of)	Nature of Problem/Concern
1	Continued funding of programs serving black patients.
2	Finding employment for black patients.
3	Support for the training of staff.

4	Inappropriate attitudes by white policy-makers, program administrators, and researchers toward black alcoholism and blacks in general.
5	Outreach efforts to get black alcoholics into treatment.
6	The ability to reach middle-class black alcoholics.
7	Gaining community acceptance of a new alcoholism treatment program.
8	Poor race relations among staff.
9	Poor race relations between staff and clients.
10	Getting black alcoholics interested in AA.
11	Lack of adequate treatment for black women.
12	Gaining family cooperation during treatment.
13	Breaking through the resistance and defenses of the black alcoholic during the counseling process.
14	Motivating the black client to continue and cooperate with treatment.
15	Minimizing heavy drinking during weekends and evenings (in general, when treatment programs are usually closed and, thus, therapeutic support is unavailable).
16	Difficulties in writing proposals and evaluation reports.
17	Getting black churches to help black alcoholics and their families.

Very recent works continue to focus on the dynamics of black culture, community, and family in the effective treatment of the black alcoholic (Brisbane, 1987; Prugh, 1987; Thompson & Simmons-Cooper, 1988). For example, Prugh (1987) calls upon the clergy of black churches to play a significant role in helping both the black alcoholic and his/her family members through the power of spiritual motivation. Thompson and Simmons-Cooper (1988) posit that, as differentiated from white alcoholism, those who treat black alcoholics need to address the ethnically oriented cultural, psychological, and historical issues and factors that influence drinking patterns and alcoholism treatment. Lastly, Brisbane (1987) investigated the family dynamics of 40 black alcoholic women in treatment who all had alcoholic parents (subjects were adult children of alcoholics). The results indicated these alcoholic women revealed significantly more punitive, negative, and blaming attitudes toward the alcoholic mother compared to the alcoholic father.

Prevention and Education

Alcohol education and alcohol problem prevention are often discussed within the same context or same breath. However, when it comes to blacks, the literature is extremely limited.

Any meaningful prevention programs for blacks would have to consider their culture, lifestyle, drinking patterns, and attitudes. Dawkins (1988) takes the view that in planning education and prevention programs, one has to understand two forces that facilitate heavy drinking and problem drinking among blacks. Dawkins dichotomizes these into *external forces* (factors outside the black community such as racial discrimination and racism) and *internal forces* (factors inside the black community that include negative role models, peer pressure to drink, and an attitude for heavy drinking).

One of the few reported prevention and education projects involving predominantly black youths was carried out in Washington, D.C. with school children in Kindergarten and first grade (Washington Area Council on Alcoholism and Drug Abuse, 1982). The project incorporated youth-parent training workshops and involved teachers and community leaders as models and adjunct participants. The first year's outcome (1980–81) suggested positive gains in knowledge and attitudes about alcohol and smoking. Moreover, the changed attitudes of the children had a positive impact on their parents' attitudes about heavy drinking, smoking, and health in general.

Alcohol education and prevention should certainly be a required part of the school curriculum, church/religious activities for youths, and social programs related to children and adolescents. Urban and heavy drinking black communities are definitely at risk for alcoholism and problem drinking; therefore, insights from research and etiological theory should be employed to assist in planning prevention. Such prevention programs must address unrealistic attitudes about drinking and alcohol-related social problems such as family crises/disruption, homicide, crime, illness, arrests, and assaults.

Issues and Recommendations

Many issues have evolved in the literature and at professional meetings on the topic of drinking among blacks. Since a thorough discussion of "issues" would require a full-blown paper, we will limit the following to a few pro-and-con examples of unresolved issues or dilemmas:

- Are blacks more likely than whites to be heavy drinkers vis-à-vis true alcoholics?

- Black alcoholics are no different from white alcoholics—"an alcoholic is an alcoholic."
- Alcohol has been a necessary means of black survival versus it has been a means of self-destruction.
- Blacks are primarily the cause of their own alcohol problems versus the influence of external forces (e.g., racial discrimination).
- Heavy drinking among blacks is motivated mainly by a conviviality function versus that of a utilitarian function.
- How do we explain the large variation between the drinking patterns of black males and females as well as the large proportion of black women who are nondrinkers?

Recommendations

There is surely a need for a greater number of studies on black drinking behaviors. The empirical literature raises numerous questions concerning internal and external validity when one considers issues such as (1) small or unrepresentative sample size, (2) a lack of studies on middle-class and upper-class blacks, (3) a lack of studies on the drinking patterns of blacks in the general population (i.e., versus captive research subjects such as treatment outpatients, hospital inpatients, criminal offenders, homicide victims, tavern drinkers, and other accessible subjects), and (4) a dearth of studies on prevention, alcohol education, and alcohol and safety.

When we talk about the drinking patterns of blacks, we must realize that blacks as a population have a number of socioeconomic groups whose drinking patterns can vary significantly according to age, social class, educational level, occupational group, gender, geographic residence (e.g., urban vs. rural or Southeast versus Northeast), etc. When we talk about black drinking patterns, we must be cognizant of these critical differences. Finally, we must increase our knowledge of black drinking behaviors via improved and expanded empirical research and literature analyses, and focus more attention on public education.

References

Benjamin R. (1976). Rural black folk and alcohol. In F.D. Harper (Ed.), *Alcohol abuse and black America.* Alexandria, VA: Douglass.

Brisbane, F.L. (1987). Divided feelings of black alcoholic daughters: An exploratory study. *Alcohol, Health and Research World, 11,* 48–50.

Cahalan, D., & Cisin, I.H. (1968). American drinking practices: Summary of findings from a national probability sample. *Journal of Studies on Alcohol, 29,* 103–151.

Caution, G.L. (1986). Alcoholism and the black family. In R.J. Ackerman (Ed.), *Growing in the shadow: Children of alcoholics.* Pompano Beach, FL: Health Communications, Inc.

Dawkins, M.P. (1976). Alcohol use among black and white alcoholics. In F.D. Harper (Ed.), *Alcohol abuse and black America.* Alexandria, VA: Douglass.
Dawkins, M.P. (1981). *Alcohol use among black and white adolescents in Washington, D.C.* Washington, DC: Washington Area Council on Alcoholism and Drug Abuse.
Dawkins, M.P., & Dawkins, R.L. (1983). Alcohol use and delinquency among black, white, and Hispanic adolescent offenders. *Adolescence, 18,* 799–809.
Dawkins, M.P., & Harper, F.D. (1983). Alcoholism among women: A comparison of black and white problem drinkers. *International Journal of the Addictions, 18,* 333–349.
Dawkins, M.P. (1988). Alcoholism prevention and black youth. *Journal of Drug Issues, 18,* 15–20.
Douglass, F. (1892). *Life and times of Frederick Douglass.* Washington, DC: Author. (Reprinted in 1962 by Macmillan.)
Fernandez-Pol, & Bluestone, H. (1986). Drinking patterns of inner-city black Americans and Puerto Ricans. *Journal of Studies on Alcohol, 47,* 156–160.
Franklin, J.H. (1967). *From slavery to freedom: A history of Negro Americans* (3rd ed.). New York: Vintage.
Gary, L.E. (1983). The impact of alcohol and drug abuse on homicidal victims. In T.D. Watts & R. Wright, Jr. (Eds.), *Black alcoholism: Toward a comprehensive understanding.* Springfield, IL: Charles C. Thomas.
Gary, L.E. (1986). Drinking, homicide, and the black male. *Journal of Black Studies, 17,* 15–31.
Gary, L.E., & Berry, G.L. (1986). Predicting attitudes toward substance use in a black community: Implications for prevention. *Community Mental Health Journal, 21,* 42–51.
Globetti, G. (1970). The drinking patterns of Negro and white high school students in two Mississippi communities. *Journal of Negro Education, 39,* 60–90.
Goodman, R.A., Mercy, J.A., & Rosenberg, M.L. (1989). Alcohol use and homicide victimization: An examination of racial/ethnic differences. In D. Spiegler, D. Tate, S. Aitken, & C. Christian (Eds.), *Alcohol use among U.S. ethnic minorities.* Rockville, MD: NIAAA.
Harper, F.D. (Ed.), (1976). *Alcohol abuse and black America.* Alexandria, VA: Douglass.
Harper, F.D. (1979). *Alcoholism treatment and black Americans* (DHHS Publication No. ADM 79–853). Washington, DC: U.S. Government Printing Office.
Harper, F.D. (1986). *The black family and substance abuse.* Detroit: Detroit Urban League.
Harper, F.D. (1988). Alcohol and black youth: An overview. *Journal of Drug Abuse, 18,* 7–14.
Harper, F.D. (1989). Alcoholism and blacks: An overview. In T.D. Watts & R. Wright, Jr. (Eds.), *Alcoholism in minority populations.* Springfield, IL: Charles C. Thomas.
Herd, D. (1987). Rethinking black drinking. *British Journal of Addiction, 82,* 219–223.
Herd D. (1988). Drinking by black and white women: Results from a national survey. *Social Problems, 35,* 493–504.
Larkins, J. (1965). *Alcohol and the Negro: Explosive issues.* Zebulon, NC: Record Publishing.
Maddox, G., and Williams, J. (1968). Drinking behavior of Negro collegians. *Journal of Studies on Alcohol, 29,* 117–129.

Mannix, D.P. (1962). *Black cargoes: A history of the Atlantic slave trade.* New York: Viking.

National Institute on Alcohol Abuse and Alcoholism (NIAAA), (1983). *Fifth special report to the U.S. Congress on alcohol and health* from the secretary of Health and Human Services. Rockville, MD: U.S. Department of Health and Human Services.

Neff, J.A. (1986). Alcohol consumption and psychological distress among U.S. Anglos, Hispanics, and blacks. *Alcohol and Alcoholism, 21,* 111–119.

Prugh, T. (1987). Black church: A foundation for recovery. *Alcohol, Health and Research World, 11,* 52–54.

Smith, J.A., & Carter, J.H. (1986). Suicide and black adolescents: A medical dilemma. *Journal of the National Medical Association, 78,* 1061–1064.

Sterne M., & Pittman, D.J. (1972). *Drinking patterns in the ghetto.* Unpublished report. St. Louis: Washington University, Social Science Institute.

Sterne, M., & Pittman, D.J. (1976). Alcohol abuse and the black family. In F.D. Harper (Ed.), *Alcohol abuse and black America.* Alexandria, VA: Douglass.

Strayer, R. (1961). A study of the Negro alcoholic. *Journal of Studies on Alcohol, 22,* 111–123.

Thompson, T., & Simmons-Cooper, C. (1988). Chemical dependency and black adolescents. *Journal of Drug Issues, 18,* 21–31.

Washington Area Council on Alcoholism and Drug Abuse (1982). Final report of an evaluation of the "Alcohol/Smoking Risk–Reduction Project." Unpublished paper. Washington, D.C.: Author.

CHAPTER 17

Race, Ethnicity, and Drinking Patterns: The Role of Demographic Factors, Drinking Motives, and Expectancies

JAMES ALAN NEFF

This chapter presents preliminary data from the first wave of a prospective study of drinking patterns among Anglo, black, and Mexican-American regular drinkers in San Antonio, Texas. The ultimate objective of the larger study is to assess the nature of racial and ethnic and sex differences in drinking patterns and to identify psychosocial factors associated with patterns of heavy or problem drinking, both cross-sectionally and longitudinally. In general, the project seeks to assess the extent to which racial and ethnic differences in drinking patterns may be attributed to socioeconomic or sociocultural factors, or both.

Although 12-month follow-up interviews are currently being conducted to assess changes in drinking patterns, data are examined here for the sample of 1,286 adult regular drinkers (412 Anglos, 239 blacks, and 635 Mexican-Americans) interviewed in wave 1. Extensive information is available on these respondents regarding drinking patterns, drinking motives and expectancies, and drinking contexts. In addition, because our overarching conceptual model posits that alcohol consumption cannot be explained without consideration of the context in which drinking takes place, information is collected regarding sociodemographic characteristics, stresses and strains in various role domains, typical coping patterns, and the nature and extent of social networks. Sociocultural dimensions are assessed by examining racial and ethnic differences in "value orientations" such as fatalism, familism, and religiosity (Chandler, 1979; Kluckhohn, 1951).

James Alan Neff is with the Department of Psychiatry, University of Texas Health Science Center, San Antonio. The research for this study was supported by grant AA06723 from the National Institute on Alcohol Abuse and Alcoholism. This chapter was written especially for this book.

In this chapter, racial and ethnic differences in basic quantity, frequency, and composite Quantity × Frequency measures of alcohol consumption are examined. To account for observed differences in drinking patterns, this chapter examines the effects of statistical controls for sociodemographic, drinking motive, and expectancy variables. The first step of the analysis is to determine whether observed differences in drinking patterns result simply from the socioeconomically disadvantaged statuses of blacks and Mexican-Americans in our society. Beyond sociodemographic controls, the second step of the analysis is to examine the influence of drinking motives and alcohol expectancies. Although drinking motives and expectancies are manifest at the individual level, differences in motives across racial and ethnic groups are viewed as reflecting underlying sociocultural variation in orientations toward alcohol.

Background

Epidemiologic interest in the relationship of race and ethnicity to drinking patterns dates back at least to Jessor, Graves, Hanson, and Jessor (1968). In this tri-ethnic study, Hispanics had a somewhat higher mean quantity-frequency of alcohol use and were more likely (at 10%) than Anglos (at 3%) to be classified as heavy drinkers (i.e., those drinking more than five drinks at a sitting). Hispanics in Cahalan's national study (Cahalan, Cisin, & Crossley, 1969) were slightly more likely to abstain than U.S. whites, though among drinkers, the proportion of Hispanic heavy drinkers was greater than among whites (30% vs. 11%, respectively). Cahalan found similar abstinence rates among black and white males (23% vs. 21%, respectively); among drinkers, black and white males were about equally likely to be classified as heavy drinkers according to quantity-frequency-variability criteria (24% vs. 29%, respectively). Among females, abstinence was much more common among Hispanics. Heavy drinking was less common among Hispanic females, and more common among black females, than among white females.

More recently, Caetano has examined drinking patterns, problems, and attitudes toward alcohol among Anglos, blacks, and Hispanics in two data sets. In the San Francisco Bay area (Caetano, 1984), Hispanic males had slightly higher rates of "frequent heavy drinking," though racial and ethnic differences in drinking patterns were not marked in the total sample. However, among regular drinkers (i.e., those drinking at least once a month), both black and Hispanic males had higher rates of heavy drinking than Anglos did. More recently, Caetano (1987) has presented national data on 1,453 Hispanics, 1,821 whites, and 1,947 blacks. Examining drinking patterns and settings, the study found that whites were more likely to drink in bars and clubs, and blacks were most likely to drink in public settings like parks, streets, or parking lots. Although Hispanic males did not appear to drink any more often in public than other groups, they did tend to report heavier public consumption than either whites or blacks. Heavier drinking was generally associated with younger age, though heavier patterns continued until middle age among blacks and Hispanics (Caetano & Herd, 1988). Among Hispanic males and females, acculturation level was positively associated with liberal attitudes toward alcohol use (Caetano, 1987).

In our earlier study of 164 Anglo, 168 black, and 149 Mexican-American males from San Antonio who drank on a regular basis (at least two times a month), striking racial and ethnic differences in drinking patterns had been observed (Neff, Hoppe, & Perea, 1987). Mexican-Americans drank slightly less frequently than either Anglos or blacks in this study but drank significantly higher quantities of alcohol in each drinking occasion. There were no differences, however, in average quantity and frequency of alcohol consumed. Differences in quantity of alcohol consumption were not explained by demographic controls (Neff, Hoppe, & Perea, 1987).

Unfortunately, most recent research on racial and ethnic differences has been descriptive and has not actually attempted to explain observed differences. Although Caetano has extensive national data, most of his analyses have been noncomparative, focusing only upon trends among Hispanics. Where racial and ethnic comparisons have been made (cf. Caetano & Herd, 1988), adjustments have not been made for critical sociodemographic differences between groups.

The premise of the present research is that it is crucial to differentiate empirically between socioeconomic and sociocultural influences upon drinking patterns. Blacks and Mexican-Americans occupy similarly disadvantaged socioeconomic positions in society, though they may differ considerably on cultural dimensions relevant to drinking. Both sociodemographic and sociocultural influences on racial and ethnic differences in drinking patterns are examined in this chapter. Although a variety of sociocultural influences could be examined (cf. Chandler, 1979), the present analyses focus upon the perceived role of alcohol use, reflected by self-reported drinking motives and alcohol expectancies.

The possibility that racial and ethnic differences in drinking practices may reflect underlying differences in drinking motives or expectancies of alcohol is suggested by research suggesting high rates of "escape" or "personal effects" drinking among blacks and Hispanics compared with the general population (Cahalan et al., 1969; Jessor et al., 1968). Caetano's (1984) data also indicate greater endorsement of motives such as drinking to "blow off steam" and to "have fun" among blacks and Hispanics than among Anglos. Johnson and Matre (1978) found Hispanics more likely than Anglos (59% vs. 32%, respectively) to report "drinking to unwind."

In Neff, Hoppe, and Perea's (1987) study of 481 Anglo, black, and Mexican-American male regular drinkers, Mexican-Americans were more likely than Anglos to endorse "negative personal" drinking motives (similar to "escape" motives) such as drinking "to forget problems," "to alleviate boredom," or "to feel more satisfied." Among Mexican-Americans, endorsement of these motives was greatest among the least acculturated. Although data on blacks were not presented in that publication, both Anglos and blacks were less likely than Mexican-Americans to endorse negative personal motives (Neff, Hoppe, & Perea, 1986).

Such findings are suggestive, though extant research has not explicitly examined differences in motives as a possible explanation of differences in drinking patterns between groups. The present chapter presents new data from an ongoing study in San Antonio, Texas, addressing the influence of drinking motives at two levels. Motives will be examined in terms of, on the one hand, individual drinking motives (i.e., reasons for drinking) and, on the other, global alcohol expectancies (i.e., the individual's evaluation of the acceptability of drinking and drunkenness). Global positive expectancies are considered to provide a reflection of subgroup norms regarding alcohol use (Rohsenow, 1983).

Methods

Sampling Procedures

Standardized household interviews were conducted with 1,286 regular drinkers (i.e., those drinking at least two to three times a month) aged 20-60, all of whom were residents of San Antonio. Multistage area probability sampling techniques were used to stratify census tracts in urban San Antonio by median household income and by percentage of black or Hispanic origin. These two stratification factors were imposed to reflect socioeconomic status variation within and between ethnic groups and to reflect ethnic heterogeneity of census tracts. Within strata, tracts and blocks were randomly drawn for study, and interviews were allocated to tracts and blocks in proportion to the population of the respective racial and ethnic groups in those areas. Randomly drawn blocks within tracts were assigned to enumerators who contacted each residence, listing eligible drinkers within each. A random sample of male and female respondents was drawn from these household enumerations. Informed consent was obtained from all respondents, the refusal rate for the study being approximately 32%. Field work was conducted during 1988.

Demographic characteristics of the resulting sample are presented in Table 1. Blacks and Mexican-Americans in the sample were younger, on average, than Anglos. Minority respondents were significantly more disad-

TABLE 1
Demographic Characteristics of Wave I Respondents

	Male			Female			Test statistics		
	Anglo (n = 200)	Black (n = 123)	Mexican-American (n = 324)	Anglo (n = 212)	Black (n = 116)	Mexican-American (n = 311)	Ethnic(E)	Sex(S)	ExS
Mean age (SD)	37.52 (11.44)	36.55 (11.76)	35.91 (10.81)	38.86 (10.80)	36.38 (10.72)	35.83 (9.93)	$F = 5.93^{**}$	.45	.58
Mean education (SD)	14.52 (2.8)	12.76 (2.3)	11.75 (3.27)	14.00 (2.75)	12.75 (1.94)	11.51 (2.73)	$F = 108.78^{**}$	2.10	1.25
Mean income (SD)	$29,815 (13,012)	$19,398 (11,570)	$23,052 (12,087)	$28,152 (13,022)	$17,513 (11,978)	$19,990 (11,799)	$F = 67.08^{**}$	10.38**	1.25
% Married	58%	40%	70%	58%	28%	53%	$\chi^2 = 52.77^{**}$	10.53**	6.94
% Unemployed	7%	21%	12%	6%	20%	10%	$\chi^2 = 27.81^{*}$	.54	0.0
% Managerial/professional/administrative	52%	16%	24%	50%	16%	20%	$\chi^2 = 109.47$	.02	.01

$^{*}p < .05$. $^{**}p < .01$.

vantaged in income, education, occupational status, and employment than Anglos. Blacks, in particular, had the lowest average income and the highest prevalence of unemployment. Mexican-Americans were most likely to be married at present; blacks were least likely to be married.

The Measurement of Drinking Patterns

Information regarding quantity, frequency, and variability of beer, wine, and liquor consumption was obtained in the interview. Because of evidence that the nature of racial and ethnic differences in alcohol consumption may vary by the alcohol consumption dimension examined (Neff, Hoppe, & Perea, 1987), quantity (i.e., the reported number of drinks consumed in a typical drinking occasion) and frequency (i.e., the typical number of drinking occasions in a week) are examined separately. Quantity and frequency were assessed for beer, wine, and liquor independently, and total quantity and total frequency measures were computed by summing across beverage types. Multiplicative quantity-frequency measures of total drinks of beer, wine, and liquor for a typical week were constructed, in addition to measures of total quantity-frequency summed across beverages. For the present analysis, results are presented for the total quantity, frequency, and quantity-frequency measures. Beverage-specific findings are presented where they clarify overall trends.

The Measurement of Dimensions of Drinking Motive and Expectancy

The individual's motives or reasons for drinking were assessed with 20 items drawn from Mulford and Miller (1960), Cahalan et al. (1969), and Cutter and O'Farrell (1984). Items were worded in the first person (e.g., "I drink because it helps me to relax") and followed traditional distinctions between "social" and "personal" or "escape" drinking motives (cf. Mulford and Miller, 1960).

Global expectancies regarding alcohol were assessed with 23 items tapping general positive expectancies of the sociability and tension-reducing functions of alcohol (Caetano, 1984; Paine, 1977). This item set included several items constructed especially for the present study and dealing with the acceptability of drinking and drunkenness for men and women in differing contexts (e.g., at home, in public). These items were generally worded in the third person (e.g., "Getting drunk is just an innocent way of having fun") to tap broader cultural norms or beliefs about alcohol and its use (Rohsenow, 1983) as opposed to personal motives.

Exploratory factor and reliability analyses were conducted to clarify the dimensionality of individual motive and global expectancy item sets. Unrotated principal factor analyses of the 20 personal motive and 23 general motive item sets were conducted for the total sample as well as within Anglo, black, and Mexican-American subgroups to identify similar factors across groups. Statistical Analysis System software (SAS, 1988) was used.

Unrotated principal factor analyses (using squared multiple correlations as communality estimates) for the individual motive item set yielded two consistent factors across subgroups—the first (14 items) dealing with personal or escape motives such as "drinking to feel less self-conscious" and the second (5 items) dealing with sociability motives such as "drinking because people I know drink." All items had factor loadings of at least .30 on their respective factors. Information regarding item content and factor loadings is available from the author. The internal consistency reliability (Cronbach's alpha) for the personal motive measure was .89 for Anglos, .87 for blacks, and .90 for Mexican-Americans. The reliability for the social motive measure was .63 for Anglos, .48 for blacks, and .59 for Mexican-Americans.

Application of similar factor analytic procedures to the motive and expectancy item set yielded a global first factor (17 items) dealing with general positive expectancies or acceptance of alcohol use and drunkenness. Although further rotated analyses showed subfactors of this global factor (e.g., an "acceptance of drunkenness" factor), the magnitude of loadings on the first factor and the consistency of loadings across racial and ethnic subgroups suggests a global construct of acceptance of alcohol use. The reliability for this measure was .91 for Anglos, .85 for blacks, and .88 for Mexican-Americans.

Whereas each of these scales measure attitudes regarding alcohol at some level, it is important to note that the global expectancy measure correlates .51 ($p < .001$) with personal and .28 ($p < .001$) with social drinking motives. There is some overlap among dimensions, though the measures tap different aspects or domains of drinking motives.

The Need for Social Desirability Controls

Because Ross and Mirowsky (1984) have shown that Mexican-Americans are more prone than Anglos to report socially desirable responses in surveys similar to this one, it is important to control for possible confounding effects of response bias factors. A 29-item social desirability measure was used here, with items drawn from the original Crowne-Marlowe (1964) scale. This measure had an internal consistency reliability (Cronbach's alpha) of .77 for Anglos, .74 for blacks, and .76 for Mexican-Americans in this data set.

Analysis and Results

Race and Ethnicity, Sex, and Drinking Motives

The mean scores on the composite drinking motive measures are presented for racial/ethnic and sex subgroups in Table 2. Significant race and ethnicity main effects were obtained for all three motive measures. Black males and females had higher personal motive scores than their Anglo or Mexican-American counterparts. Higher social drinking motive scores were found among Anglos—particularly among males—with the lowest social motive mean scores being found among Mexican-Americans. On both personal and social motive measures, males had higher scores than females, though differences on social motives were only marginally significant. Positive expectancies were highest among Mexican-American and black males, and among black females.

Analysis of Drinking Pattern Indices

Given significant racial and ethnic differences on both demographic factors and drinking motives, the analysis shifts to examine racial and ethnic and sex differences in drinking patterns. Also examined is the extent to which such variation can be accounted for by sociodemographic or drinking motive and expectancy differences between groups. Analyses are presented for total quantity, frequency, and quantity-frequency of alcohol consumption summed across beer, wine, and liquor. Where trends vary for specific beverages, these trends are noted.

A series of analysis of covariance models were conducted, beginning with a straightforward examination of main and interaction effects of race and ethnicity and of sex, without adjustments for covariates. Three additional covariance analyses were then conducted, the first looking at the effect of adding demographic and social desirability controls, and the second looking at the addition of drinking motives, including demographic controls, to the model. The third analyses incorporated tests for interactions between, on the one hand, each of the three drinking motive measures and, on the other, race and ethnicity. Interaction effects were tested simultaneously in the same model. The results of these analyses are presented in Table 3.

Total quantity consumption. Significant main effects of race and ethnicity and of sex emerged with regard to quantity. Sex differences were attenuated somewhat by demographic and particularly by drinking motive controls; males drank higher quantities in all racial and ethnic groups. For race and ethnicity, the general pattern showed the highest quantity of consumption to be among Mexican-Americans (both male and female),

TABLE 2
Mean Scores on Drinking Motive and Expectancy Measures by Race/Ethnicity and by Sex

	Males			Females			*F* Statistics		
	Anglo	Black	Mexican-American	Anglo	Black	Mexican-American	Ethnic (E)	Sex (S)	ExS
Individual motives									
Personal	30.07	31.05	30.10	29.30	30.29	29.41	3.28*	4.45*	0.18
Social	12.78	12.57	12.42	12.48	12.42	12.31	3.16*	3.35	0.42
Global expectancies									
+ Expectancies	33.94	34.41	34.55	32.18	34.68	33.38	8.69**	8.02**	1.55

*$p < .05$. **$p < .01$.

TABLE 3
Observed and Adjusted Mean Scores on Drinking Indices by Race/Ethnicity and by Sex

	Males			Females			*F* Statistics		
	Anglo	Black	Mexican-American	Anglo	Black	Mexican-American	Ethnic (E)	Sex (S)	ExS
Total quantity									
Observed	6.32	5.46	7.38	4.54	4.48	5.84	13.9*	25.5*	0.6
Adjusted^1	6.45	5.00	7.22	4.57	4.17	5.84	13.8*	22.1*	1.0
Adjusted^2	6.33	4.77	7.16	4.83	4.13	6.14	16.1*	12.1*	0.6
Adjusted^3	6.33	4.94	7.13	4.90	4.25	6.16	2.4	0.4	0.5
Total frequency									
Observed	4.25	3.87	3.03	2.83	3.04	1.95	29.1*	59.0*	1.2
Adjusted^1	4.11	3.75	3.07	2.69	3.01	2.03	15.4*	48.6*	1.5
Adjusted^2	4.05	3.64	3.02	2.81	3.07	2.07	14.5*	33.1*	1.3
Adjusted^3	4.05	3.60	3.03	2.82	3.06	2.05	0.5	0.1	1.3
Total quantity-frequency									
Observed	17.20	15.60	16.90	7.26	9.12	8.38	.07	58.87**	.70
Adjusted^1	18.21	12.77	15.86	7.54	8.09	7.87	1.32	57.17**	2.45
Adjusted^2	17.68	11.66	15.39	8.38	8.10	8.61	2.10	38.02**	2.12
Adjusted^3	17.94	11.77	15.31	8.45	8.32	8.68	2.65	1.80	2.33

Notes:
Observed = Model including ethnicity and sex.
Adjusted^1 = Controls added for demographic factors and social desirability.
Adjusted^2 = Controls for demographics, social desirability, and drinking motives.
Adjusted^3 = Controls for above *and* Motive × Factor interactions.
*$p < .05$. **$p < .01$.

with the lowest quantity of consumption being among blacks. Both the racial and ethnic and the sex differences became insignificant when interactions of motives and expectancies with race and ethnicity and with sex were included.

Total frequency consumption. The main effects of not only race and ethnicity but also sex were significant, though the magnitude of these effects was attenuated by demographic and motive controls. Males were generally more frequent drinkers than females and, among both males and females, Anglos and blacks drank more frequently than Mexican-Americans. Among females, blacks tended to be slightly more frequent drinkers than Anglos. As with quantity, differences in race and ethnicity and in sex were eliminated by adjustments for interactions between motives, race and ethnicity, and sex.

Total quantity-frequency. Sex differences on total quantity-frequency were highly significant, with males reporting up to two times the amount of alcohol consumption of females. However, because Anglos and blacks tended to be more frequent, lower-quantity drinkers and Mexican-Americans tended to be less frequent, higher-quantity drinkers, the combination of these dimensions averaged out drinking pattern differences. In short, the total volume of alcohol consumed in a typical week by Anglos, blacks, and Mexican-Americans is similar, though their specific drinking patterns differ.

Variation in Consumption Patterns by Beverage Type

Overall trends are reflected by the total consumption measures, though some variation in findings by beverage type are worth noting. For beer, Anglos and blacks were more frequent, lower-quantity drinkers, and Mexican-Americans were less frequent, higher-quantity drinkers. For wine and liquor, however, Anglos and blacks tended toward more frequent, higher-quantity consumption and Mexican-Americans were less frequent, lower-quantity drinkers. Males generally drank beer more frequently and in higher quantity than females, though sex differences were reversed for wine. However, after statistical controls for demographics, drinking motives, and Motive × Ethnicity interaction effects, race and ethnicity effects became insignificant for all drinking dimensions except liquor frequency and both beer and liquor quantity-frequency.

Drinking Motives and Expectancies and Drinking Patterns

The significance of Drinking Motive × Race and Ethnicity interactions in the previous analyses suggests that measures of drinking motive and expectancy relate differentially to drinking pattern measures across racial and ethnic subgroups. Before an examination of relationships within subgroups, however, it is useful to consider the nature of the relationships between drinking motive and expectancy measures and drinking pattern measures for the total sample. These coefficients were computed in models that included the three motive measures simultaneously, in addition to race and ethnicity, sex, demographic factors, and social desirability. Only the drinking motive coefficients are discussed here.

The only significant motive and expectancy predictor of the total quantity of consumption was the global positive expectancy measure, which was associated with higher-quantity consumption ($\beta = .16$, $p < .01$). For total frequency of consumption, both personal and social drinking motives were significantly related to frequency. Specifically, personal motives were associated with more frequent drinking ($\beta = .08$, $p < .01$); social motives were associated with less frequent drinking ($\beta = -.18$, $p < .01$). For total quantity-frequency, personal motives ($\beta = .52$, $p < .01$) and positive expectancies ($\beta = .31$, $p < .01$) were associated with higher total consumption; social motives were associated with lower total consumption ($\beta = -1.24$, $p < .01$).

A look at trends for specific beverages shows that beer quantity, frequency, and quantity-frequency followed the general trends, being positively associated with personal motives and global positive expectancies and negatively associated with social drinking motives. In contrast, personal motives were unrelated to both wine and liquor dimensions, and general positive expectancies were positively related only to liquor quantity ($\beta = .05$, $p < .05$). Social drinking motives were significantly related only to wine quantity, the endorsement of social drinking motives being associated with higher wine quantity ($\beta = .08$, $p < .05$).

Regression coefficients were estimated for each motive and expectancy variable within racial and ethnic subgroups and are presented in Table 4. Tests for the significance of slope differences across groups are indicated in Table 3. Table 4 indicates the significance of individual subgroup regression coefficients. Those which differ significantly from zero are marked with an asterisk.

In these analyses, personal motives were most strongly and consistently related to total quantity and total quantity-frequency among Anglos, with personal motives being associated with heavier consumption. Global positive expectancies were positively related to total frequency among Mexican-Americans, less strongly among Anglos. Social drinking motives were negatively related to total frequency and total quantity-frequency among blacks.

With regard to beverage-specific trends, personal motives were positively related to beer and liquor quantity and to beer quantity-frequency

TABLE 4
Standardized Partial Regression Coefficients Predicting Drinking Indices by Drinking Motive and Expectancy Measures for Racial and Ethnic Subgroups

	Anglo	Black	Mexican-American
Total quantity			
Individual motives			
Personal	.21*	.09	.01
Social	−.12	−.18	.04
+ Expectancies	.14*	−.01	.22*
Total frequency			
Individual motives			
Personal	.04	.05	.05
Social	−.17	−.34*	−.11
+ Expectancies	.04	.05	.003
Total quantity-frequency			
Individual motives			
Personal	.63*	.45	.18
Social	−.88	−2.10*	.13
+ Expectancies	.32	.01	.30

* $p < .05$. ** $p < .01$.

among Anglos. Positive expectancies were positively related to beer and liquor quantity and to liquor quantity-frequency among Mexican-Americans and Anglos. Social drinking motives were negatively related to beer and wine frequency and beer quantity-frequency for blacks. Beyond these general trends, expectancies were positively related to beer frequency among Anglos, personal motives were positively associated with beer frequency among Mexican-Americans, and social motives were positively related to liquor quantity among Mexican-Americans.

It is substantively important to note that significant variation in the nature of motive and expectancy effects upon alcohol consumption does exist across racial and ethnic groups. When such interactions are added to the statistical model predicting alcohol consumption, main effects of race and ethnicity become largely nonsignificant. The interpretation of these trends will be considered below.

Discussion

This chapter has presented preliminary analyses of data from the first wave of a longitudinal tri-ethnic study of drinking patterns. The principal objectives of the chapter have been to document the existence of differences across groups in drinking patterns and to assess the role of demographic and of drinking motive and expectancy factors as they contribute to the explanation of such differences. Before a discussion of the findings, however, it is important to note that our data pertain only to regular drinkers—not to the general population. Thus, the reader should keep in mind that references to Anglos, blacks, or Mexican-Americans actually mean Anglo, black, or Mexican-American drinkers.

A major finding of the study is that there are no racial or ethnic differences in the total amount of alcohol consumed in a typical week in our sample of drinkers. However, rather striking racial and ethnic differences in the total quantity and frequency of consumption exist, and these patterns differ somewhat by type of beverage. In general, Anglo and black drinkers tend to be more frequent, lower-quantity drinkers. Mexican-American male and female drinkers, in contrast, tend to be less frequent, higher-quantity drinkers. These general trends are most characteristic of beer consumption. For wine and liquor, however, Anglos are more frequent, higher-quantity drinkers, and Mexican-Americans manifest the lightest drinking patterns. Sex differences are dramatic as well, with males being uniformly heavier beer drinkers than females. However, females are higher-quantity, more frequent wine drinkers than males. No sex differences were found for liquor consumption.

In the context of the overall study, these trends are generally attenuated, though not completely eliminated, by additive demographic controls and motive and expectancy controls. However, it is interesting to note that the effects of these controls vary by beverage type. Specifically, additive demographic and motive controls eliminate effects of race and ethnicity only for wine and liquor consumption, not for beer. Although controls generally reduce the magnitude of race and ethnicity differences, such differences remain significant for quantity, frequency, and quantity-frequency of beer consumption. Further, for wine and liquor, additive controls eliminate racial and ethnic differences only for quantity, not frequency. Differences in the quantity of wine and liquor consumption are largely explained by demographic factors.

Interestingly, the effects of demographic controls appear most salient for beverages consumed in higher quantity by Anglos than by Mexican-Americans. Such specificity may, in part, reflect distinctive types of drinkers or drinking styles, rooted in social class differences. Such types or styles are suggested by demographic profiles associated with higher-quantity consumption of different beverages. High-quantity beer consumption is more common among younger, less-educated male respondents. In contrast, higher-quantity wine consumption is more typical of more highly educated individuals and females. Higher-quantity liquor consumption, as well, is associated with higher educational levels, though no sex differences are observed here.

Because drinking styles are related to social class dimensions, it is possible that sociodemographic controls may eliminate observed racial and ethnic differences in quantity of consumption. Why a similar explanation does not apply to beer consumption is unclear, though it is notable that wine and liquor are clearly more costly than beer. Perhaps because of cost differentials, beer may be more universally accessible across social class lines, in contrast to wine and liquor, which would be much more expensive to consume in high quantities.

As the most popular beverage in our sample, beer and its consumption trends dominate the findings for the total consumption measures. As a result, in contrast to the wine and liquor findings, racial and ethnic differences in beer quantity and frequency are not explained by either demographic or additive drinking motive and expectancy controls. The fact that high-quantity beer and total alcohol consumption among Mexican-Americans is not explained by differing drinking motives is intriguing. It is especially interesting that blacks in our data most commonly report personal or "escape" drinking motives and Mexican-Americans report the lowest levels of social drinking motives. Thus, in the present data, Mexican-Americans are neither escape nor social drinkers as previous research has suggested (cf. Cahalan et al., 1969; Gilbert and Cervantes, 1986). However, the global motive findings do indicate higher levels of positive expectancies and greater acceptance of drinking and drunkenness among both Mexican-American and black males than among Anglos. Males generally have more positive expectancies than females, though this tendency is reversed among blacks, black females having somewhat higher levels of positive expectancies than black males.

The fact that Mexican-American males view drinking more positively than Anglos does not in itself account for their high-quantity consumption (particularly of beer). Rather, the analyses indicate that relationships between motive dimensions and drinking patterns vary across groups and that these differences in slopes account for apparent racial and ethnic differences. Specifically, global positive expectancies relate positively to quantity dimensions most strongly and consistently among Mexican-Americans, thus suggesting that acceptance of drinking and drunkenness may serve to facilitate higher-quantity consumption in this group. In contrast, the interaction analyses show that the endorsement of personal or escape motives may facilitate high-quantity consumption among Anglos and that endorsement of social drinking motives is associated with less frequent consumption and lower total quantity-frequency consumption among blacks. This latter finding is of interest given Neff, Prihoda, and Hoppe's (in press) observation of a relatively high prevalence of solitary drinking, associated with heavy consumption patterns, among black drinkers. Although we have not specifically examined drinking contexts in the present analyses, the endorsement of social drinking motives among blacks may well counter tendencies toward heavier drinking associated with solitary drinking in this group. This possibility will require further examination.

In a larger context, these analyses support a distinction between individual and global drinking motives and expectancies. Individual motives and global expectancies regarding alcohol appear to have effects on alcohol consumption independent of each other and independent of demographic factors. It does not appear that motives and expectancies are so highly interdependent that, for example, the effect of global expectancies is explained away by the individual's personal drinking motives (or vice versa). The findings further suggest that not all motives are equally important as predictors of alcohol use across racial and ethnic subgroups. Thus, it is important to note, in terms of a sociological perspective on race and ethnicity and on drinking patterns, that Mexican-American patterns appear much more strongly and consistently influenced by global drinking norms and that black and Anglo patterns appear more influenced by individual drinking motives.

In conclusion, the present study raises an intriguing theoretical and substantive issue. That is, Mexican-Americans and blacks are both socioeconomically disadvantaged in comparison to Anglos, and might be expected to drink more heavily, according to existing "stress" formulations of minority drinking (Alcocer, 1982; Larkins, 1965). However, in the present study, black patterns are more similar to those of Anglos than to those of Mexican-Americans. These analyses suggest that such distinctive patterns result from differing relationships between alcohol consumption dimensions and individual or global drinking motives across groups. These relationships appear to facilitate or inhibit heavier consumption patterns. The present analyses highlight the importance of drinking motives and expectancies and further challenge notions regarding the importance of escape drinking motives among minority drinkers. However, the proportion of variance in drinking patterns explained here is admittedly low (ranging from 5% to 24%), and more detailed analyses are clearly needed. This chapter represents but a preliminary step in our plan to identify psychosocial influences upon drinking patterns of Anglos, blacks, and Mexican-Americans.

Acknowledgment

The research was supported by National Institute on Alcohol Abuse and Alcoholism grant AA06723.

References

Alcocer, A.M. (1982). *Alcohol use and abuse among the Hispanic American population* (Alcohol and Health Monograph 4, Special Population Issues, pp. 361–382). Washington, DC: U.S. Government Printing Office.

Caetano, R. (1984). Ethnicity and drinking in northern California: A comparison among whites, blacks, and Hispanics. *Alcohol and Alcoholism, 19*, 31–44.

Caetano, R. (1987). Alcohol use and depression among U.S. Hispanics. *British Journal of Addiction, 82*, 1245–1251.

Caetano, R., & Herd, D. (1988). Drinking in different social contexts among white, black, and Hispanic men. *Yale Journal of Biology and Medicine, 61*, 243–258.

Cahalan, D., Cisin, I., & Crossley, H. (1969). *American drinking practices*. New Brunswick, NJ: Rutgers Center of Alcohol Studies.

Chandler, C. (1979). Traditionalism in a modern setting: A comparison of Anglo and Mexican-American value orientations. *Human Organization, 38*, 153–159.

Crowne, D., & Marlowe, D. (1964). *The approval motive*. New York: Wiley.

Cutter, H., & O'Farrell, T. (1984). Relationship between reasons for drinking and customary drinking behavior. *Journal of Studies on Alcohol, 45*, 321–325.

Gilbert, M.J., & Cervantes, R. (1986). Patterns and practices of alcohol use among Mexican Americans: A comprehensive review. *Hispanic Journal of Behavioral Sciences, 8*, 1–60.

Jessor, R., Graves, T., Hanson, R., & Jessor, S. (1968). *Society, personality, and deviant behavior: A study of a tri-ethnic community*. New York: Holt, Rinehart and Winston.

Johnson, L., & Matre, M. (1978). Anomie and alcohol use. *Journal of Studies on Alcohol, 39*, 894–902.

Kluckhohn, C. (1951). Values and value orientations in the theory of action: An exploration in definition and classification. In T. Parsons and E. Shils (Eds.), *Toward a general theory of action*. New York: Harper & Row.

Larkins, V. (1965). *Alcohol and the Negro*. Zebulon: Record Publishing.

Mulford, H.A., & Miller, D.E. (1960). Drinking in Iowa: III. A scale of definitions of alcohol related to drinking behavior. *Quarterly Journal of Studies on Alcohol, 21*, 267–278.

Neff, J.A., Hoppe, S.K., & Perea, P. (1986). *Alcohol consumption and drinking motives among Anglo, black, and Mexican-American males.* Paper presented at the Annual Meeting of the American Public Health Association.

Neff, J.A., Hoppe, S.K., & Perea, P. (1987). Acculturation and alcohol use: Drinking patterns and problems among Anglo and Mexican American male drinkers. *Hispanic Journal of Behavioral Sciences, 9*, 151–181.

Neff, J.A., Prihoda, T.J., & Hoppe, S.K. (1991). "Machismo," self-esteem, education and high maximum drinking among Anglo, black and Mexican-American male drinkers. *Journal of Studies on Alcohol,* 52, 458–463.

Paine, H.J. (1977). Attitudes and patterns of alcohol use among Mexican Americans: Implications for service delivery. *Journal of Studies on Alcohol, 38*, 544–553.

Rohsenow, D.J. (1983). Drinking habits and expectancies about alcohol's effects for self versus others. *Journal of Consulting and Clinical Psychology, 51*, 752–756.

Ross, C.E., & Mirowsky, J. (1984). Socially-desirable response and acquiescence in a cross-cultural survey of mental health. *Journal of Health and Social Behavior, 25*, 189–197.

Statistical Analysis System Institute Inc. (1988). *SAS/STAT™ User's guide, release 6.03 Edition*. Cary, NC: SAS Institute.

SECTION III

Social Structure, Subcultures, and Drinking Patterns

C. DRINKING-CENTERED INSTITUTIONS

Introductory Note

The actual contexts within which drinking occurs have received extensive analysis by social scientists over the last three decades. That there are institutions in Western society, such as public drinking houses, bars, taverns, or cocktail lounges, that are organized around the act of drinking and that many individuals spend considerable time in these settings are undeniable facts.

In chapter 18, Mary Ann Campbell presents a brief summary of various research studies over the last four decades of primarily American public drinking places and provides the reader with a detailed tavern taxonomy based on cocktail lounges, drink-and-dine establishments, neighborhood taverns, Skid Row places, and nightclubs. Furthermore, from her extensive participant observation of tavern social life in a working-class neighborhood undergoing gentrification in a large midwestern city, she is able to map the various social networks that develop in various tavern types. This analysis allows Campbell to focus on the potential of these networks to handle alcohol-related problems among regular patrons.

However, public drinking places have assumed decreased significance in U.S. society. Trends of suburbanization, the paucity of mass transit along with the reliance upon the automobile as the primary mode of individual transportation, and citizens' concerns about driving while their judgment is impaired by alcohol, as well as potential legal liability problems faced by tavern owners, have accelerated the drift away from drinking in commercial contexts. In a related study (1989) Oldenburg's trenchant analysis links the decay of U.S. public life to the decline of gathering places (cafes, pubs, and so forth) that are within walking distance of home or work. Such a trend has led to an increased privatization of drinking; most drinking occasions are now in noncommercial contexts (see chapter 6, by Klein, in this volume).

Chapter 19, James Rooney's provocative research on patterns of alcohol use in Spanish society, stands in sharp contrast to studies of patterns found in culturally pluralistic U.S. society. Basic to this Mediterranean society is the fact that alcohol use is integrated into general beverage consumption and alcohol occupies no special moral category, as it does in the United States, Canada, and Scandinavian countries. To a remarkable extent, alcohol consumption occurs in the context of dining, meals, and snacks. One is struck by Rooney's observations on how public drinking places in the form of beverage and food shops, as well as all-male taverns, serve as the public places that Oldenburg finds withering away in the United States. Spain is also characterized by numerous youth bars as well as canteens in educational institutions that have students 14 years of age and over; social policy makers seem to have little concern about the access of youth to beer, because alcohol use by people in this age category is viewed as nonproblematic. In sum, Rooney's study sharply demonstrates the crucial role that cultural meaning and context place on alcohol use.

References

Oldenburg, R. (1989). *The great good place.* New York: Paragon House.

CHAPTER 18

Public Drinking Places and Society

Mary Ann Campbell

When Marshall B. Clinard wrote "The Public Drinking House and Society" (1962) over a quarter of a century ago, he noted correctly that there was a dearth of studies on this institution in the sociological literature. At the conclusion of his article, Clinard suggested the need for studies of social interaction in various types of public drinking places. This need has been met by a proliferation of studies on a variety of tavern types. Though some recent research, notably that of Nusbaumer, Mauss, and Pearson (1982), has attempted to demonstrate the influence of patronage on problem drinking, most studies of public drinking places focus on sociability, rather than the serving of beverage alcohol, as the main social fact to be examined (Anderson, 1976; Cavan, 1966; Katovich & Reese, 1987; Kotarba, 1984; LeMasters, 1975). Generally, these studies have challenged the stereotype of the public drinking place as a non-normative behavior setting that fosters alcohol abuse and other social problems.

The main question to be addressed in this chapter asks what the specific elements of the culture of public drinking places in the United States are. Certainly this culture shares characteristics of the dominant culture of U.S. society. The salience of beverage alcohol use to interaction differentiates the public drinking place culture from other subcultural groups. Rubington has explicated the "alcohol language" (1971, p. 721) utilized in this subculture.

Beyond this distinction, the tavern as public drinking establishment has been defined and differentiated from other social contexts where beverage alcohol use is involved, such as street drinking (see for example Rubington, 1971; Spradley, 1970) and drinking in private clubs; (see Adler & Adler, 1983; Roebuck & Frese, 1976). Research has further explicated the definition of the tavern as public drinking establishment by differenti-

This chapter was written especially for this book. It is based in part on the author's doctoral dissertation, "Tavern Culture and Neighborhood Networks: Implications for Alcohol Problem Prevention," Washington University, 1988.

ating among types of taverns (Cavan, 1966; Clinard, 1962; Dinitz, 1951; Macrory, 1952; Richards, 1963–1964).

I focus first on questions regarding the general perspective on drinking found in the tavern subculture, the marginal status of the tavern in society, and the debate on the role of the tavern with regard to deviant behavior, particularly deviant drinking behavior. This line of investigation is followed by an explanation of the techniques used to classify taverns in the present study. Finally, I discuss tavern-centered social networks and suggest propositions concerning the likelihood of personal networks developing in each type of tavern.

Characteristics of Public Drinking Establishments

Taverns are "establishments whose business consists mainly of selling and serving beer, wine or other intoxicating liquors for consumption on the premises" (Clinard, 1962, p. 271). Following Clinard's analysis, this chapter defines the tavern as a public establishment, thereby differentiating it from the bar or cocktail lounge of a private club.[1] Although the public tavern and the private club are both drinking-centered institutions, the democratic nature of the former sets it apart from member- and guest-only establishments and justifies a separate analysis of the social context of tavern drinking, which, theoretically at least, is open to all members of the public of legal drinking age with the financial means to participate (Clinard, 1962; LeMasters, 1975; Roebuck & Frese, 1976).

Taverns are found in rural as well as urban areas (Clinard, 1962; Macrory, 1952) and in a variety of forms, from Skid Row dives to posh nightclubs (Cavan, 1966; Clinard, 1962; Richards, 1963–1964). Though there is significant variation among them (to be discussed later in this chapter) all taverns share certain features in common. Clinard outlines six characteristics of the tavern as a public institution. In addition to its commercial and public nature, and the necessity of serving beverage alcohol (as distinguished from serving food, which is optional), the tavern has a physical structure, a set of norms, and a functionary role—that of tavern-keeper or bartender. Finally, "the tavern is a setting for *group drinking* [italics in original] in the sense that it is done in the company of others in a public place" (p. 271).

Social Control and Deviance: The Marginal Status of the Tavern in Society

As a type of "unserious behavior setting" (Cavan, 1966, p. 8), the tavern provides an accessible space for taking time out from the pressures of everyday work and home life (Gottlieb, 1957; LeMasters, 1975; Macrory, 1950). Thus, in addition to the consumption of beverage alcohol, the activities of play and sociability are important features the tavern has in common with other types of recreational settings, such as athletic events, picnics, or parties (Byrne, 1978; Bott, 1955; LeMasters, 1975). Although the tavern may also be the site of serious activities, its social milieu can most generally be described as nonserious, informal, and recreational (Bissonette, 1977; Byrne, 1978; Cavan, 1966).

Specifically because it is characterized by this type of social milieu, which by its very nature indicates an acceptance of all members of the public and the permission of unconventional behavior, the tavern is a potential setting for deviance. The deviant stereotype has plagued the tavern throughout history (Clinard, 1962; Koren, 1899; Moore, 1897) and has led to its "marginal status" (Gottlieb, 1957, p. 559) in society.

Liquor control laws vary temporally and geographically, as the involvement of moral entrepreneurs does. Observers indicate that the development of informal normative constraints within the group drinking context of a tavern may also be expected to vary temporally and geographically, and with the particular characteristics of the clientele (Cavan, 1966; Gottlieb, 1957; LeMasters, 1975). Thus, several levels of social control can be expected to contribute to the normative structure of any public drinking establishment: the formal rules and attendant sanctions of laws governing the sale of beverage alcohol for consumption on the premises, the moral stance of the community in which the tavern is situated, and the informal norms and sanctions that emerge in the group drinking context of the tavern. The interrelationships among the normative constraints of the small-group, legal, and moral orders that operate to regulate behavior in this unserious, drinking-centered setting bear exploration.

The significance of the tavern in society is nowhere better illustrated than in the continuing debate over its social value. Historically, the contradiction lies between the temperate view of the tavern as "contributing to drunkenness, unhappy home life, loss of jobs, mental trouble, neglect of children by parents, crime and personal demoralization and misery" (Macrory, 1950, p. 613) and the alternative position that "primarily the saloon is a social center" (Moore, 1897, p. 4). Sociological research on the public drinking establishment has not resolved this debate but has produced some relevant findings.

In an attempt to answer the question whether the tavern contributes to drunkenness and alcoholism, Dinitz (1951) surveyed 197 white male recovering alcoholics (members of Alcoholics Anonymous) about their perceptions of the impact of the tavern setting on their drinking patterns. Although he had no control group with which to compare these data,

Dinitz concluded that "initiation into drinking has very little to do with tavern going" (p. 362). The findings of Clark (1981) support this conclusion.

However, there is evidence indicating that the perception that the tavern contributes to various forms of behavior considered deviant still remains but exists in conflict with the perception that the tavern serves positive functions for the community. A survey of 1,441 regular tavern patrons, occasional patrons, and nonpatrons on attitudes toward the role of the tavern in the community concluded that "there is a marked lack of consensus relative to the role and functions of the tavern in the community" (Macrory, 1950, pp. 465–466).

Such a lack of consensus in attitudes toward the tavern is in line with the sociocultural analysis of the United States as having an ambivalent drinking culture, characterized by conflict between or among coexisting value structures (Pittman, 1967a). Although some groups, such as certain religious ones, advocate abstinence from beverage alcohol, positive sentiments toward drinking are a feature of other religious groups, certain ethnic groups, and the beverage alcohol industry (Pittman, 1967b, p. 8). The value structure of the tavern may generally be typed as an overly permissive subculture, one in which "the cultural attitude is permissive toward drinking, to behaviors which occur when intoxicated, and to drinking pathologies" (Pittman, 1967b, p. 5).[2]

The view of tavern subculture as overly permissive is implicitly supported by Kotarba's (1984) research on tavern drinking and driving. As home territory bars (Cavan, 1966), the four taverns in Kotarba's study provided the kind of deviant subculture that supported patrons' justification accounts of their drinking and driving as acceptable behavior (Scott & Lyman, 1968). Further, among the taverns in Kotarba's study, some were known to "cater to many off-duty policemen" (1984, p. 158). Patrons befriended these policemen in the hope or with the plan that dropping a policeman's name would enable them to avoid arrest for driving while intoxicated (DWI) if they were stopped. Kotarba notes that this practice was fairly successful in enabling tavern patrons to avoid two deviant labels: the official one, *DWI*, as well as the informal one, *alcoholic*.

Tavern Types

Many taverns offer much more than beverage alcohol by the drink, including packaged liquor, entertainment, games, music, food, and even sports activities and contests. In efforts to differentiate and classify taverns with regard to their functions, researchers have employed these tavern characteristics as well as other variables, such as size, decor, and price and selection of beverage alcohol, as well as patron characteristics, such as occupation, age, sex, race, and interaction patterns. The main contributions of these studies have been the development of ideal types (Cavan, 1966; Clinard, 1962; Dinitz, 1951; Macrory, 1952, 1950; Richards, 1963–1964).

The data analyzed in this chapter are drawn from previous studies, as well as from extensive field work I did in a 2-year study of taverns in a large midwestern city. The purpose of the ethnography was to explore and classify public drinking places in order to find out which types of taverns might be conducive to the development of personal networks of support. A taxonomy (or set of inclusionary categories) of taverns was developed, and prototypical taverns in each category were studied in depth. My study replicates previous findings on types of taverns and goes on to analyze the social networks found among their patrons, proprietors, and staff. These findings are presented in a discussion of each tavern type and summarized at the conclusion of this chapter in a set of propositions concerning tavern-centered networks.

To develop a taxonomy of taverns, I selected some variables from previous classification schemes and formulated others during the course of field work, using observations and respondents' descriptions of differences among the taverns in their neighborhood (see Spradley & McCurdy, 1972, pp. 63–77). Variables fall into three categories: physical, commercial, and social. I expected to find that cases observed in the field would approximate ideal types for certain categories of taverns as constructed from the literature: the Skid Row tavern, the drink-and-dine tavern, the nightclub, the cocktail lounge, and the neighborhood tavern (see Table 1).

Using my own observations, information provided by proprietors and management and employees, informal interviews with patrons, and media advertisements placed by the establishments, I typed each of 24 observed taverns on each variable. I then categorized each tavern according to the type into which it fit on the largest number of variables. Most cases fit into one tavern category on a preponderance of the 12 variables. This pattern was particularly true of the 12 neighborhood taverns, 8 of which exhibited the characteristics of that category on all 12 variables. The following discussion classifies tavern-centered social networks.

The Cocktail Lounge

Cocktail lounges are frequently located in commercial areas, such as business districts or shopping centers (Clinard, 1962; Gottlieb, 1957). "They usually have long bars, booths, and attractive decorations" (Clinard, 1962, p. 276). Entertainment sometimes is provided by a professional performer but more typically involves a television, juke box, or electronic game (Clinard, 1962; Gottlieb, 1957; Richards, 1963–1964).

TABLE 1
Tavern Taxonomy: Tavern Types

Index Variables	Cocktail lounge	Drink-and-dine	Neighborhood tavern	Skid Row	Nightclub
A. *Physical*					
1. Location	Main street, usually on outer fringes of neighborhood.	Main street, usually on outer fringes of neighborhood.	Side street, within neighborhood.	Isolated on side street.	Downtown or near an airport or major highway, often in a hotel.
2. Floor Plan	Bar and stools, tables and chairs, stage, game room.	Tables and chairs, bar and stools.	Bar and stools, table and chairs, electronic games or pool table.	Bar and stools.	Tables and chairs, stage.
B. *Commercial*					
1. Beverage Alcohol	Wide selection featuring mixed drinks, house specialties, brandies. No packaged liquor.	Wide selection featuring mixed drinks, house specialties and wine. No packaged liquor.	Limited selection, bar brands, perhaps pints and six-packs to go.	Limited selection, no wine or brandy. Half-pints, pints, and six-packs to go.	Limited selection featuring mixed drinks. No packaged liquor.
2. Food	Limited menu: lunch or dinner, or both. Food sales smaller than liquor sales.	Full menu: lunch and dinner, maybe breakfast. Food sales as much as, or more than, liquor sales.	Occasional luncheon or single-item food special. Packaged snacks.	No food usually, perhaps pre-packaged snacks.	No food usually.
3. Prices	High liquor and food prices.	High liquor prices. Food prices range from moderate to high.	Low liquor prices, cheap bottled or draft beer, or both. Moderate food prices.	Moderate-to-low prices on liquor by the drink and packaged liquor.	High prices, cover charge, and often a two-or-more-drink minimum.

4. Entertainment	Live entertainment 3 or more nights a week, juke box, electronic games.	Sometimes live entertainment on weekends.	Juke box, electronic games, perhaps a pool table.	Juke box, perhaps a pinball machine or pool table.	Live entertainment.
5. Hours of Peak Business	Evenings on weekends and weeknights.	Lunch and dinner hours, weekends.	Early morning, late afternoon (after work), weekends.	Early morning to early evening.	Weekend evenings, sometimes weeknight evenings.
C. *Social*					
1. Staff	Proprietor/manager, bartenders, cocktail servers, bar helpers.	Proprietor, managers, cocktail waitresses, bus persons, host/hostess, cooks, food waitresses.	Proprietor, bartender, bar helper.	Proprietor, bartender.	Manager, service bartender(s), cocktail waitresses.
2. Clientele	Large and varied: dates and singles come from various areas of the city for entertainment and to meet people.	Large and varied: dates and singles come from various areas of the city for food and entertainment, local businessmen for lunches.	Neighborhood residents, large group of regulars who come to be with friends.	One-time customers, there for a drink only, alcoholics, homeless men, prostitutes.	Large and varied: dates come from various areas of the city for entertainment.
3. Social Functions	Marketplace, musical entertainment.	Fine dining, perhaps marketplace and musical entertainment.	Gathering place, information exchange, support networks.	Refuge for social isolates.	Musical entertainment.
4. Reputation	Metropolitan, with substantial advertising.	Metropolitan, with substantial advertising.	Local.	Local notoriety.	Metropolitan, with substantial advertising.

The literature on the cocktail lounge specifies that "primarily mixed drinks" (Clinard, 1962, p. 276; Gottlieb, 1957, p. 559) are served; and food may be served, with "house specialties frequent" (Richards, 1963–1964, p. 267). Prices of beverage alcohol, food, and entertainment are generally higher than in other types of public drinking establishments, with the exception of the nightclub (Richards). The staff usually includes bartenders, waiters or waitresses, and sometimes a cook (Richards). The reputation of a cocktail lounge may be "city or district wide" (Richards, p. 267).

A typical cocktail lounge "is open for business in the early afternoon and will cater to afternoon and early-evening clientele" (Gottlieb, 1957, p. 559) or to "late evening patrons" (Clinard, 1962, p. 276). From a study of 22 cocktail lounges in the Chicago metropolitan area, Gottlieb concludes that most lounge patrons are members of the upper–middle class (see also Clinard, 1962). These patrons are characterized as transients (Gottlieb, p. 560), in that they typically do not reside in the area of the cocktail lounge but frequent it mainly for convenience and marketplace uses, as described by Cavan (1966). Thus the ecology, patronage, and main functions of the cocktail lounge do not dispose this type of tavern to the development of social networks among its clientele.

The Drink-and-Dine Tavern

Clinard locates the drink-and-dine tavern "in business districts or . . . along main highways" (1962, p. 277). His description of a public drinking establishment where "patrons are most frequently businessmen" and "many business deals are transacted over cocktails and steaks" (p. 277) carries with it a social class connotation that does not apply to all drink-and-dine taverns. These establishments range on a continuum from the restaurant-bar, which features fine foods, to the bar-and-grill, which serves up an inexpensive and limited menu.

In operationalizing the concept of the drink-and-dine tavern for my research, I utilized the applicable legal licensing standards for restaurants and beverage alcohol: that is, an establishment in the area studied must do a certain volume of total business and a minimum proportion of its business in food sales in order to maintain its license to sell beverage alcohol. All such licensed establishments that sell beverage alcohol for consumption on the premises are therefore classified as drink-and-dine taverns. It remains the case that some taverns in this category function differently at varying times of the day or week. For example, some restaurant-bars close early in the evening; others operate as cocktail lounges after their kitchens close, particularly on weekends. Using the legal licensing standard applicable locally provides a way to categorize each establishment by its major function.

The drink-and-dine tavern is patronized for the convenience either of its location (Cavan, 1966) or of its ability to provide beverage alcohol with food (Cavan), and as a marketplace (Clinard, 1962). Clinard posits that "there is little interaction between patrons [because] they tend to come in small individual groups" (p. 277). Thus, it would not be expected that social networks among patrons would have their loci in such taverns.

However, an establishment in the drink-and-dine category might have a core group of regular patrons among whom the development of a tavern-centered social network would be possible. This situation could occur in a modest establishment with only a neighborhoodwide reputation and clientele, as well as in a drink-and-dine tavern located near centralized, work-related activities and regularly patronized by a group of professionals (for instance, a restaurant-bar situated in the vicinity of a municipal courts building and frequented by lawyers). Nonetheless, the major function of the drink-and-dine tavern in combining restaurant and bar service does not generally make it conducive to the establishment of personal networks.

The Nightclub

The main feature that distinguishes the nightclub from other types of taverns is that it exists mainly to provide live entertainment. Cavan (1966, pp. 155–156) refers to this form of entertainment as "the programmed production," describing it as being "in some general fashion scripted, rehearsed, and presented to the patrons." This is usually musical entertainment provided by a band or singer or both, a duo, a trio, or an orchestra; however, it may also be a comedy act, or two of the above may share one billing.

Related to the programmed production is the temporal aspect of patronage. Patrons do not come and go on a fluid basis as in other types of public drinking places; rather, they line up outside the establishment, are allowed to enter at a specified time before the show when doors open, and are expected to leave at the conclusion of a show. This policy is especially firmly in effect when there are two shows a night—the usual situation, at least on weekends. The twice nightly format allows a maximum profit to be made by the club, as well as by the act, in the shortest possible time frame.

Related to the nightclub's primary functions are its spatial differences from other types of public drinking places. Unlike other taverns, which may have small rooms offering pool tables, pinball machines, and videogames with which their patrons can actively entertain themselves, the nightclub consists of only one large room, offering entertainment that its patrons experience passively—as spectators rather than game players (Katovich & Reese, 1987). Cavan asserts that the nightclub is more respectable than other types of public drinking places—"respectable enough to warrant the patronage of respectable people" (1966, p. 166).

Sales of beverage alcohol, but not of food (except on certain special occasions such as New Year's Eve), are ancillary. However, they usually provide a higher profit margin than sales of tickets for the show, and so sales of beverage alcohol play an important role in this type of public drinking place. In conjunction with this money-making function, it is common for nightclubs to have a two-drink-minimum rule, so as to insure the desired overall profit margin when ticket sales are slow.

The only bar is the service bar, generally hidden from the view of the patrons and reached through swinging restaurant doors at the rear of the room. Unlike bartenders in other types of public drinking places, the service bartenders at the nightclub do not interact with patrons, but they do need to cooperate with the cocktail servers, on whom they depend for tips. Both bartenders and servers have the goal of making and selling as many drinks as fast as possible. The amount of tips is related to volume of sales, though the ratio is usually lower than at cocktail lounges or drink-and-dine establishments, both because the typical patrons at such establishments are not known as big tippers and because the bartenders and servers in nightclubs spend a shorter amount of time serving and interacting with the patrons than in other types of public drinking places. Observations indicate that although the structure and function of this type of public drinking place militate against the development of social networks among patrons and staff members, the nightclub does present fertile ground for social networks to develop among the staff members.

The Skid Row Tavern

If the nightclub represents the elite public drinking setting, the Skid Row tavern epitomizes the opposite extreme.[3] Dumont paints a vivid description of a case in this category:

> The Star Tavern is in the shadow of an "el" in an area soon to be demolished by an urban renewal program.... It is a typical lower class tavern with large bottles of third-rate whiskey and brandy, illuminated beer displays, unstable stools, and a bar that is a mosaic of stains and cigarette burns. (1967, pp. 938–39)

Also in contrast to a nightclub, the Skid Row tavern offers no entertainment to its customers, with the possible exception of a pinball machine

(Richards, 1963–1964). Beverage alcohol is the major drawing card of such an establishment, whose "patronage is largely single and homeless men, migrant laborers, and alcoholics" (Clinard, 1962, p. 276).

The cost of beverage alcohol is low, as is the cost of whatever food may be available. If food is served, it is typically prepackaged or meets only minimal legal health requirements. The bartender is the only staff required; frequently the position is filled by the individual or family who owns the tavern (Richards, 1963–1964).

The characteristic motivation for patronizing a Skid Row tavern is what Cavan terms convenience use:

> For some patrons the convenience use may be characterized in terms of a setting that provides a place where drinks can be purchased or consumed when the need or desire arises, and the activity of patrons within this setting may be limited to this alone. (1966, p. 145)

Although close proximity to other settings could result in the convenience use of a restaurant-bar in a mall by middle-class shoppers, or of a hotel lounge by tourists, the meaning of convenience for typical patrons of a Skid Row tavern centers on the easy availability of beverage alcohol to the virtual exclusion of other considerations, such as food, decor, and entertainment.

Dinitz (1951) found that for drinkers in general not only patronizing taverns but patronizing the tavern closest to home increased significantly between the social drinking phase and the excessive phase and between the excessive and alcoholic drinking phases (1950, pp. 286–87).[4] The social stigma attached to habitues of the Skid Row tavern imbues this type of public drinking establishment with a reputation of "local notoriety" (Richards, 1963–1964, p. 267), which Clinard asserts has "contributed much toward the development of the stereotype of all taverns" (1962, p. 276).

Though any sort of tavern might be used as a marketplace, this function has been characterized as virtually endemic to the Skid Row tavern. Of taverns in this category, Clinard writes:

> Although their primary function is to provide a place for cheap drinking, they are often the site of gambling and soliciting for prostitution. There are frequent violations of state and municipal laws relating to taverns, as well as drunk and disorderly conduct and gambling. In this type of tavern, violations of regulations which are strictly enforced in other places are often permitted; for example, closing hours are widely disobeyed, and many establishments virtually operate on a 24-hour basis. (1962, p. 276)

The above composite picture of the Skid Row tavern might suggest that in such taverns an anomic atmosphere would prevail, precluding the development of relationships among patrons. However, this image is challenged by Dumont:

> The regulars at the Star Tavern form a cohesive and durable social system. They are their own and their only reference group. They are almost universally alienated from their families. Except for occasional spurts of unskilled labor they have no identity as part of a work force. Most importantly, they are a group without a future. Most of them have no idea what they will do or where they will go when their shabby rooms and the tavern fall to the bulldozer. For these men the tavern provides the only opportunity for socialization. It is a distortion to say that they spend their days in a barroom only to drink. The barroom hangout of homeless men does not exist only to exploit and aggravate social pathology. It performs a life-sustaining function for men who have literally nothing else. It may provide their only opportunity for a tolerant and supportive environment, for socialization, for rest and warmth. (1967, pp. 942–943)

This observation is echoed in Spradley's study of homeless men (1970, p. 118).

Despite great advancement made in the decriminalization of public drunkenness in recent decades, Skid Row areas, taverns, and habitues retain the stigma associated with the lifestyle of urban nomads. Drawing on the reports of his informants, Spradley concludes, as Dumont does, that "Skid-Row bars are not simply places to drink, they are institutions where strangers with spoiled identities can meet and find security in their common humanity as tramps" (1970, p. 256). It is precisely because their social and economic resources are so scarce that a network of relationships among patrons of the Skid Row tavern is crucial; it is the last thread by which many hang on to the fabric of social integration.

The Neighborhood Tavern

A study comparing 24 neighborhood taverns and 22 cocktail lounges in the Chicago area found that the neighborhood tavern is a phenomenon both of the lower social classes and of the immediate neighborhood (Gottlieb, 1957). The image of the neighborhood tavern as the public drinking domain of the working class is reflected in other studies. Macrory (1950) estimated that about two thirds of the patrons of the neighborhood taverns he studied in Wisconsin were laborers, and LeMasters (1975) stated that most of the habitues of the Oasis Tavern, the establishment he studied, were construction workers.

More characteristic of the neighborhood tavern than a local clientele is the presence of a group of patrons who frequent the tavern on a regular basis (Clinard, 1962; Gottlieb, 1957; LeMasters, 1975). These regulars may live in the neighborhood, as observed by Gottlieb (1957), or they may live some distance away (Kotarba, 1984).[5] Whether they are former residents of the neighborhood or are connected to the neighborhood tavern through their workplace, church, or sports team, these regulars can typically be found in the neighborhood tavern frequently and at roughly the same time of day and day of the week (Gottlieb, 1957; Chandler, 1948).

It is apparent that the neighborhood tavern serves an integrative function for its regulars, fulfilling social needs that perhaps would otherwise be unsatisfied. These are summarized by Clinard (1962):

> Other than [simply providing a place for] drinking there appear to be three chief functions of the neighborhood tavern: (a) as a meeting place where social relationships with other persons can be established, (b) as a place for recreation such as games, and (c) as a place to talk over personal problems with the tavernkeeper or others. (p. 279)

Cavan calls the neighborhood tavern, as well as the English pub, which it resembles,[6] "prototypes of the home territory use of the public drinking place," while specifying that "the lines along which the patrons of the home territory are drawn are not necessarily limited to residential areas" (1966, p. 206). That is to say, any type of tavern might be used as a home territory by patrons who "share one or more features of their social identity, and this common bond forms the basis of defining those who are welcome in the establishment and those who are not" (p. 206). Neighborhood taverns studied by Gottlieb included four that catered to Polish-Americans, three to German-Americans, five to southern whites, and three to blacks (1957, p. 560). Cavan found that racial and ethnic collectivities including blacks, Native Americans, Mexican-Americans, Irish, Italians, Russians, Chinese, Filipinos, and British subjects used public drinking places in San Francisco as home territories (pp. 206–207).

Though any type of tavern may be used by any social group as a home territory (Cavan, 1966), this function is the hallmark of the neighborhood tavern. Data from participant-observation studies of neighborhood taverns (Cavan, 1966; Gottlieb, 1975; LeMasters, 1975) best fit the definition of the home territory use of the tavern:

> Some public drinking places derive their special character from the fact that they are used as though they were not public places at all, but rather as though they were the private retreat for some special group. Those who use the public drinking places in this way frequently designate them as a "home," a "second home," or a "home away from home." (Cavan, 1966, p. 205)

The privatization of the public drinking place by the regular patrons establishes boundaries for social interaction with those outside the core group.

The homelike entertainment provided by the neighborhood tavern does not detract from the primary function it serves for its clientele: "providing a meeting place for friends and neighbors where they may share like interests and problems, or relax and enjoy visiting together" (Clinard, 1962, p. 278). Strangers whose barstools are juxtaposed can become "bosom buddies" in a few beers' time. Regardless of whether a regular is close friends with other patrons who are in the bar at any particular time, he or she recognizes them as familiar faces and knows that a number of these regulars will certainly be in the tavern at any time of the day or night. And sometimes, for some regular patrons, these familiar faces are the most substantial personal community they have. So they keep coming in to the place which, by virtue of its everyday familiarity, is the center of their social life.

We can understand this familiarity among strangers by distinguishing between two types of regulars.[7] I refer first to regulars who spend a substantial and predictable amount of time in the neighborhood tavern as the *hard-core regulars*. Their social networks are characterized by intensity (closeness of interaction) and frequency of interaction. Spatial as well as temporal markers are used to identify them. These regulars always sit at the bar, unless they are in a family group, in which case it is typical for couples that the husbands gather at the bar while the wives and children sit at a nearby table. Each group carries on its own conversation, but these conversations are punctuated by remarks exchanged between the groups. An elderly couple, however, usually sit together, either at the bar or at a table.

There is also a spatial differentiation among hard-core regulars according to the function the tavern serves for them. Those who are primarily interested in sociability usually gather at the end of the bar closest to the front door. Doing so allows them to meet and greet friends as they arrive, as well as to communicate through glances or words (or both) the lack of welcome for walk-in traffic off the street. Functions of sociability and normative boundary maintenance are thus associated with positioning at the bar, whether or not the regulars themselves are aware of it. The hard-core regulars who engage in heavy drinking can be found at the opposite end of the bar. They have few social or other resources and spend most of their day each day in the neighborhood tavern. They have no jobs or families, so the sociability function of the tavern is important to them. But drinking is their primary interest.

Though the hard-core regulars have interpersonal networks that include the proprietors and staff of the neighborhood tavern, a second group of regulars is not so closely connected to the focal role played by the proprietor or to the hard-core regulars. I refer to this group simply as *regular patrons*, because they patronize the neighborhood tavern on a regular basis. Their networks are characterized by frequency of interaction but not by intensity. It is possible for a regular patron to move to the hard-core group after a time, having demonstrated acceptability to its members.

The booths and most of the tables are the least desirable spaces in the status system of the neighborhood tavern subculture. The elite group of patrons, the hard-core regulars, rarely use the booths, except perhaps for courting or confidential business discussions. Otherwise, tables and booths are the spaces to which nonelite regular patrons are relegated when the bar is full. Even when food is served at the neighborhood tavern, which is a special occasion, the most desirable place for eating among hard-core regulars is at the bar.

As in any subculture where people engage in everyday face-to-face interaction in a particular environment, the participants share a set of values and have an understanding of which types of behavior are permissible, as well as of which acts constitute nonnormative behavior. Although there are slight differences in the composition of the "regular" daytime and evening crowds, they have more in common than not. Three main factors distinguish the hard-core regulars from the rest of the neighborhood tavern clientele. First, whether or not they drink beverage alcohol, their primary purpose in patronizing the neighborhood tavern is to belong to a home territory. The second distinguishing factor is that the hard-core regulars of the neighborhood tavern tend to name as their three best friends other hard-core regulars, forming a nearly closed subculture. The final factor distinguishing the hard-core regulars from other neighborhood tavern patrons is their close relationship to the proprietor, the person playing the focal role in the neighborhood tavern subculture. A tightly knit, intensive network of interpersonal relationships revolve around this proprietor-bartender role. I observed that help seeking through this natural network took place on many levels and that the proprietor responded with a range of helping responses, from providing small loans to helping in getting social services to trying to help with drinking-related problems.

The neighborhood tavern has two main functions: the first may be considered generic to the class of institutions structured around the recreational use of alcohol; the second is specific to the particular subclass of neighborhood tavern and further informed by the character of the particular neighborhood social structure. The second function, that of sociability, is the focus of my research. A close analysis of the everyday interaction of hard-core regulars reveals that there exists a complex web of relationships. Certain pivotal figures—the proprietors and their close associates—are the focal points of these networks of relationships.

Among the neighborhood tavern regulars, there are two types of friendship networks. Networks of relationships among patrons with substantial social and material resources focus on the tavern as a social gathering place or club. Sociability is the primary concern and alcohol is of secondary concern. In contrast are the networks of patrons for whom alcohol is of paramount concern. Their networks are important sources of survival, as each network member has few resources and many social strains. But sociability is important here as well. The networks are surrogate families for people who have lost their families or jobs, or both. Because of the reputation of the neighborhood tavern proprietors in the community, both types of patrons center their social lives in the tavern. High-risk members of the neighborhood tavern society are suffering from poverty, unemployment and underemployment, poor housing and illness. For them, seeking help is a process contained in the small, tightly knit, intensive networks that articulate around neighborhood leaders.

Conclusion

Mitchell's tyopology of social networks (1969) suggests three variations, which range from incidental to primary involvement among network members. Categorical networks operate at a superficial and routine level of everyday exchange. Structural networks develop among incumbents of roles in a formally organized structure, such as a workplace. Personal networks link together significant others in primary relationships.

Field observations indicate that tavern patron networks vary by tavern type. Although categorical networks can be found in and among all types of public drinking establishments, they do not form the basis of primary relationships. In cocktail lounges, nightclubs, and drink-and-dine taverns, structural networks are present among employees. In cocktail lounges and drink-and-dine establishments, these networks may also include regular luncheon or cocktail hour patrons. But these networks are framed by the context of the situation and the time limitation involved; they do not extend beyond co-worker or employee-patron relationships.

Though taverns typically serve multiple functions and are patronized for a wide variety of reasons, certain primary functions and typical clientele characterize different types of taverns. Data from this project as well as previous participant-observation research suggest four propositions regarding the functions of the five types of taverns:

1. The nightclub and the cocktail lounge are less likely than other types of taverns to possess a core group of regular patrons and to serve an integrative function.

2. The secondary use of a drink-and-dine tavern as a social center, to the extent that it may occur, varies ecologically and temporally.
3. The Skid Row tavern, though characterized in usage by convenience in providing beverage alcohol, serves an integrative function for its habitues.
4. The neighborhood tavern is more likely than any other type of tavern to cater to a regular clientele and to serve the integrative function of a social center.

Any tavern may be a locus of structural and categorical networks among patrons, employees, and proprietors. Many taverns do not fit readily into an ideal type; rather, there exist "marginal establishments of each type" (Gottlieb, 1957, p. 559). To the extent that a public drinking establishment approximates either the neighborhood tavern or Skid Row tavern, it is likely to have a relatively homogeneous clientele and to be a locus of personal networks among its patrons, employees, and proprietors. Patrons may utilize such networks in seeking help for a variety of problems, including alcohol problems.

The development of tavern-centered personal networks takes place primarily in the setting of the neighborhood tavern, or perhaps occasionally in the Skid Row tavern and develops from the lack of social distance as well as the frequency and regularity of interaction between proprietor or bartender and the clientele. Observations lead to the generalization that the greater the social similarities between the proprietor or bartender and the patrons, the more likely the development of a hard-core regular clientele and tavern-centered personal networks.

Implications for Further Research

The findings and analysis presented here suggest several implications for further research. First, it is important that tavern-centered natural helping network patterns such as those found in this exploratory study be further documented through replication. Sociometric analysis could be used to test whether the existence of a hard core of regulars tends to be common to public drinking establishments that fit the neighborhood tavern type.

Second, this study could be improved upon. Research on help seeking and help giving in tavern-centered networks could have a more explicit focus on alcohol-related problems. This is a very sensitive area, one I found subjects reluctant to discuss. Researchers should anticipate difficulty in documenting help seeking and help giving with regard to alcohol-related problems, but the goal is worth the heroic effort required for immersing themselves totally in field work.

Notes

1. See Adler and Adler (1983) for an incisive study reporting that private clubs function virtually as public taverns in dry counties.
2. Pittman (1967) notes further that "this type, the Over-Permissive Culture, does not occur completely in societies, but only approximations in certain nonliterate societies, in those cultures undergoing considerable change, and those in which there are strong economic vested interests in the production and distribution of alcoholic beverages" (p. 5).
3. For a discussion of the origin of the term *Skid Row* and an explication of the stigmatization of habitues of such areas, as well as an analysis of prospects for alcoholism intervention in such areas, see "Homeless men," by D.J. Pittman, in *Poor Americans: How the White Poor Live*, M. Pilisuk and P. Pilisuk, Eds. (Chicago: Aldine, 1971), pp. 112–116.
4. However, it should also be noted that some subjects reported that they never patronized taverns, even during their alcoholic drinking phase (Dinitz, 1950, p. 283).
5. The term *regular* is used popularly by and about patrons who patronize a particular tavern on a regular basis. Chandler distinguished a "hard core of regulars" from other regular patrons (1948, p. 73).
6. Clinard makes a similar comparison between these working-class public drinking establishments (1962, pp. 278–279).
7. Katovich and Reese (1987) distinguish among urban tavern patrons, using four categories: regulars, regular irregulars, irregular regulars, and nonregulars. I follow Chandler (1948, p. 73) instead, and divide neighborhood tavern patrons into three groups: hard-core regulars, regular patrons, and occasional patrons.

References

Adler, P.A., & Adler, P. (1983). Dry with a wink: Normative clash and social order. *Urban Life, 12*, 123–139.

Anderson, E. (1976). *A place on the corner*. Chicago: University of Chicago Press.

Barker, J.M. (1970). *The saloon problem and social reform*. New York: Arno Press.

Bissonette, R. (1977). The bartender as a mental health service gatekeeper: A role analysis. *Community Mental Health Journal, 13*, 92–99.

Bott, E. (1957). *Family and social network: Roles, norms, and external relationships in ordinary urban families* (2nd ed.). New York: Free Press.

Byrne, N. (1978). Sociotemporal considerations of everyday life suggested by an empirical study of the bar milieu. *Urban Life, 6*, 417–438.

Cahalan, D., Cisin, I.H., & Crossley, H.M. (1969). *American drinking practices: A mental study of drinking behavior and attitudes* (Rutgers Center of Alcohol Studies Monograph No. 6). New Brunswick, NJ: Rutgers Center of Alcohol Studies.

Campbell, M.A. (1989, August). Alcohol and identity: The case of the neighborhood tavern. Paper presented to the Society for the Study of Social Problems, San Francisco.

Campbell, M.A. (1988, October). Applying the natural helping network model in tavern-centered alcoholism prevention programs. Paper presented to the Society for Applied Sociology, Oak Lawn, IL.

Campbell, M.A. (1988). *Tavern culture and neighborhood networks: Implications for alcohol problem prevention* (Unpublished doctoral dissertation, Washington University).
Cavan, S. (1966). *Liquor license: An ethnography of bar behavior.* Chicago: Aldine.
Chandler, M.K. (1948). *The social organizations of workers in a rooming house area* (Unpublished doctoral dissertation, University of Chicago).
Clark, W. (1981). The contemporary tavern. In Y. Israel, F.B. Glaser, H. Kalant, R. Popham, W. Schmidt, & R. Smart (Eds.), *Research advances in alcohol and drug problems* (Vol. 6, pp. 425–470). New York: Plenum.
Clinard, M.B. (1962). The public drinking house and society. In D.J. Pittman & C.R. Snyder (Eds.), *Society, culture, and drinking patterns* (pp. 270–292). New York: Wiley.
Cloyd, J. (1976). The marketplace bar. *Urban Life, 5*, 293–312.
Cromwell, W.O. (1935). *The return of the saloon.* Chicago: Juvenile Protective Association.
Cromwell, W.O. (1940a). *The tavern: A social problem.* Chicago: Juvenile Protective Association.
Cromwell, W.O. (1940b). *The tavern in community life.* Chicago: Juvenile Protective Association.
Cromwell, W.O. (1948). *The tavern in relation to children and youth.* Chicago: Juvenile Protective Association.
Dinitz, S. (1951). *The relation of the tavern to the drinking phases of alcoholics* (Unpublished doctoral dissertation, University of Wisconsin).
Duis, P.R. (1943). *The saloon: Public drinking in Chicago and Boston, 1880–1920.* Chicago: University of Illinois Press.
Dumont, M.P. (1967). Tavern culture: The sustenance of homeless men. *American Journal of Orthopsychiatry, 3*, 938–945.
Gans, H.J. (1962). *The urban villagers: Group and class in the life of Italian-Americans.* New York: Free Press.
Gomberg, E.L., White, H.R., & Carpenter, J.A. (Eds.), (1985). *Alcohol, science and society revisited.* New Brunswick, NJ: Rutgers Center of Alcohol Studies.
Gottlieb, D. (1957). The neighborhood tavern and the cocktail lounge: A study of class differences. *American Journal of Sociology, 62*, 550–562.
Howe, L.K. (1977). *Pink collar workers: Inside the world of women's work.* New York: Avon Books.
Hutt, C. (1973). *The death of the English pub.* London: Arrow Books.
Katovich, M.A., & Reese, W.A., II. (1987). The regular: Full-time identities and memberships in an urban bar. *Journal of Contemporary Ethnography, 16*, 308–343.
Koren, J. (1899). *Economic aspects of the liquor problem.* Boston and New York: Houghton Mifflin.
Kotarba, J.A. (1984). One more for the road: The subversion of labeling within the tavern subculture. In J.D. Douglas (Ed.), *The sociology of deviance* (pp. 152–162). Boston: Allyn and Bacon.
LeMasters, E.E. (1975). *Blue-collar aristocrats: Life styles at a working-class tavern.* Madison: University of Wisconsin Press.
Liebow, E. (1967). *Tally's corner: A study of Negro street corner men.* Boston: Little, Brown.
Listiak, A. (1974). Legitimate deviance and social class: Bar behavior during Grey Cup week. *Sociological Focus, 7*, 13–43.

Macrory, B.E. (1952). The tavern and the community. *Quarterly Journal of Studies on Alcohol, 13*, 609–637.

Macrory, B.E. (1950). *A sociological analysis of the role and function of the tavern in the community* (Unpublished doctoral dissertation, University of Wisconsin).

Mass Observation. (1943). *The pub and the people*. London: Victor Gollantz, Ltd.

Mitchell, J.C. (1969). *Social networks in urban situations*. Manchester: Manchester University Press.

Moore, E.C. (1897). The social value of the saloon. *American Journal of Sociology, 3*, 1–12.

Mosher, J.F. (1979). Dram shop liability and the prevention of alcohol-related problems. *Journal of Studies on Alcohol, 40*, 773–798.

Nusbaumer, M.R., Mauss, A.L., & Pearson, D.C. (1982). Draughts and drunks: The contributions of taverns and bars to excessive drinking in America. *Deviant Behavior, 3*, 329–357.

Pittman, D.J. (1971). Homeless men. In M. Pilisuk & P. Pilisuk (Eds.), *Poor Americans: How the white poor live* (pp. 112–116). Chicago: Aldine.

Pittman, D.J. (Ed.). (1967a). *Alcoholism*. New York: Harper & Row.

Pittman, D.J. (1967b). International overview: Social and cultural factors in drinking patterns, pathological and nonpathological. In D.J. Pittman (Ed.), *Alcoholism* (pp. 3–20). New York: Harper & Row.

Pittman, D.J., & Snyder, C.R., (Eds.). (1962). *Society, culture, and drinking patterns*. New York: Wiley.

Prus, R., & Irini, S. (1980). *Hookers, rounders and desk clerks*. Toronto: Gage.

Read, K.E. (1980). *Other voices*. Novato, CA: Chandler and Sharp.

Richards, C.E. (1963–1964). City taverns. *Human Organization, 22*, 260–268.

Roebuck, J.B., & Frese, W. (1976). *The Rendezvous: A case study of an after-hours club*. New York: Free Press.

Rubington, E. (1971). The language of 'drunks.' *Quarterly Journal of Studies on Alcohol, 32*, 721–740.

Scott, M.B., & Lyman, S.M. (1968). Accounts. *American Sociological Review, 33*, 46–62.

Shaffer, D.R., & Sadowski, C. (1978). This table is mine: Respect for marked barroom tables as a function of gender of spatial marker and desirability of locale. *Sociometry, 38*, 408–419.

Spradley, J.P. (1970). *You owe yourself a drunk: An ethnography of urban nomads*. Boston: Little, Brown.

Spradley, J.P., & Mann, B.J. (1975). *The cocktail waitress: Woman's work in a man's world*. New York: Wiley.

Spradley, J.P., & McCurdy, D.W. (1972). The cultural experience: Ethography in complex society. Chicago: Science Research Associates.

Stevens, W.B. (1973). The Missouri tavern. *Missouri Historical Review*. Columbia, MO: The State Historical Society of Missouri.

Whyte, W.F. (1955). *Street corner society* (2nd ed.). Chicago: University of Chicago Press.

Zien, H. (1987). Meeting the liability crisis. *Wisconsin Beverage Journal, 45*, 12.

CHAPTER 19

Patterns of Alcohol Use in Spanish Society

James F. Rooney

This analysis takes as its focus the major patterns of alcohol consumption in Spanish society, with only small attention given to differences among the various regions of the country. These patterns of behavior are analyzed in terms of social institutions, behavior that is standardized and expected within a society. The author rejects the chemicalistic fallacy: that the effect of a drug is due entirely to the chemical effect on the organism. Although there are some fairly uniform effects of given quantities of alcohol on the human organism, this fact in no way determines the quantity or the pattern of consumption that is customary in any particular culture. Rather, this analysis utilizes the paradigm of symbolic interaction: that humans interact principally in terms of meanings and that meanings are culturally defined. Alcohol use accordingly is the object of social definition and control, which can vary greatly among societies.

Research Procedures

This report is based chiefly on observational data the author gathered while traveling and residing in Spain and is supplemented with published data from population surveys as well as census data and alcohol consumption records. The author has visited Spain six times. Five times he was a tourist, staying 3 to 4 weeks on each visit. Over the course of these trips, he visited 14 of the 15 regions of the Spanish mainland, most of them at least twice. From the beginning, he recorded notes regarding patterns of alcohol consumption in various types of public establishments. The sixth visit to Spain was of a year's duration, during which the author resided in the Province of Huelva in the region of Andalucia and had the opportunity both to become a regular client in one tavern in an urban working-class

James F. Rooney is with the Division of Behavioral Science, Penn State University at Harrisburg. This chapter was written especially for this book.

barrio and to be integrated into its social network. The author also made systematic observations in other types of eating and drinking establishments in the City of Huelva. He conducted focused interviews with bar owners to clarify issues of fact and of interpretation.

Findings

At the institutional level, alcohol use is thoroughly integrated into beverage use in Spanish culture. To make this point clear to an Anglo-Saxon readership, one must realize that there are no coffee shops as such in Spain. Rather there are beverage and food shops. According to the governmental regulatory code, no special license or permit is required to sell alcoholic beverages. Rather, establishments are required to obtain a general license to sell food and beverages. Thus any licensed establishment can sell any combination of coffee, tea, hot chocolate, colas, ginger ale, fruit-flavored sodas, beer, wine, and brandy. Larger bars will also carry an inventory of whiskey, cordials, and liqueurs in addition to all the previously mentioned beverages. All establishments will also carry some foods, although the variation is considerable, ranging from three or four light snacks up to full meals. This principle of the integration of beverages is also seen in coin-operated vending machines for dispensing beverages, often located in public places such as train and bus stations as well as many art museums. Machines typically contain seven varieties of canned beverages, one of which is beer, the others being colas, ginger ale, and fruit-flavored sodas.

The undifferentiated beverage and food shops flourish not only in the community, but also in high schools and technical schools, which have students generally between the ages of 14 and 18. Such educational centers usually have a *cantina* (a bar or saloon) which closely duplicates the products sold in bars of the outside community; snacks, lunches, coffee, tea, sodas, beer, wine, and brandies are available. Policies regarding the dispensing and control of alcoholic beverages vary widely between regions and between individual schools. Beer is generally available to students in all educational centers. However, a policy may be mandated that beer be the only alcoholic beverage available to students under 18 years of age, or that no alcohol be sold before noon, or that there be a two-drink limit for each person. These regulations may or may not be enforced, however. Observations in high school cafeterias reveal that the majority of students consume coffee or soft drinks and fewer than 20% take beer either separately or with lunch.

Clearly, alcohol is not placed in a separate moral category in the Spanish cognitive map but rather constitutes one class of beverages among others, all of which are sold in the same establishment and generally have some degree of association with food consumption. Martinez and Martin (1987, p. 46) well summarize the integral position of alcohol in Spanish culture: "The consumption of alcohol is [as] integrated into common behaviors as sleeping and eating." Furthermore, alcohol consumption has a positive social sanction as a practice most closely linked to sociability and for the obligatory celebration of all types of festive and ritual activities.

The key variable that influences the institutionalized social contexts and patterns of alcohol use in Spanish society is the degree to which alcohol is associated with food consumption. Basically, settings in which food consumption is more salient involve lesser focus upon alcohol and generally include a greater proportion of women as customers. The sex composition in turn exerts a tremendous influence on the general ambience and norms of behavior operating in any particular beverage shop. In contrast, settings in which food consumption is less frequent involve a greater focus upon alcohol consumption and simultaneously include fewer women. This generalization holds true throughout all regions of Spain for adults. Youth bars patronized principally by persons under the age of 25 operate on different norms, which will be explained separately.

Basically there are three patterns of relationship between alcohol and food consumption. Alcohol (1) can form part of a major meal, (2) can be taken with a snack or light lunch, and (3) can be consumed apart from food. Each of these patterns has been in existence for over a century, and each deserves a separate examination.

Alcohol and Major Meals

The principal meal is consumed either around 2:00 p.m. or between 9:00 p.m. and 12:00 p.m. and typically consists of four to six courses plus dessert and coffee. When the principal meal is eaten at home, wine frequently accompanies it. To serve this domestic market, nearly all larger food markets sell low-priced wines, and there are wine merchants in most neighborhoods who dispense wine directly from large barrels into bottles brought by customers. For the more affluent, liquor stores sell vintage wines and a full array of liquors.

The consumption of wine in a restaurant differs among social classes. In the more expensive restaurants, wine is not included in the price of a meal but must be ordered from the wine menu separately. This policy gives diners the opportunity to select a vintage wine to their liking. Wine can also be ordered by the glass.

Whether taken at home or in a restaurant, the major meal of the day usually is not consumed hurriedly. Lunch periods in Spain typically are at least 2 hours long, providing ample time for diners to eat and drink slowly while focusing on conversations with companions. There is a danger, however, for those who not only regularly consume several drinks with the major meal but also drink regularly outside of mealtime. A study of employed males in the Galicia region revealed that moderate drinkers consumed approximately equal quantities of alcohol with meals and outside of meals, whereas excessive drinkers consumed a greater quantity with meals (Lorenzo Lago, Fariñas Vasco, Carrera Machado, Belo Gonzales, & Rodriguez Lopez, 1987).

In working-class restaurants two major choices are presented to the diner: one can select items à la carte, or order the "special of the day," a set package that typically consists of soup, meat and fish, two vegetables, six slices of bread, a bottle of wine, and dessert. The wine is served in unlabeled bottles filled from a barrel in the back room, having been purchased either from a wine merchant or directly from a local farmer. The special of the day constitutes a complete package; there is no opportunity for substitution of items or reduction of the price. When one to three persons together order the meal of the day, one bottle of wine is brought to the table. If four to six persons sit together, two bottles are brought to the table. Observation in restaurants in all regions reveals that most persons consume a quarter to a third of a bottle of wine with a meal. No one considers it necessary to drink all of the wine, just as scarcely anyone eats all six slices of bread. Clearly, wine consumption at meals is regarded in a very utilitarian manner.

That wine is perceived as a beverage to accompany a meal is shown all the more clearly by the frequent practice of diluting. It should be understood that despite the fact that excellent vintage wines are produced in Spain, *vino comun,* or common wine, sold at low prices in cardboard packages in grocery stores, from the barrel in wine shops and in taverns, frequently is of low quality and may have a bitter taste. In fact, many people refer to common white wine as *vino peleón,* literally "wine that fights," the connotation being that wine of this quality combats the person who consumes it. Public drunkenness is actually rare. But when it occurs, it often is attributed to this fighting wine. If a man is seen intoxicated while walking home, people are likely to say that he is "going by, drunk with vino peleón."

To make this wine more palatable, many drinkers buy a bottle of seltzer water and add it to the wine with a meal. In some localities, customers may ask that a type of sweet ginger ale be added to the wine "to make it sweeter" or "to make the wine taste better." In some working-class restaurants and taverns, the owner adds spiced soda water to the bulk wine to improve the taste. As a result, all persons who order the house wine are served a diluted and flavored wine. This practice anticipated and long preceded the introduction of wine coolers in the United States. Some establishments also add ginger ale to draft beer, thus reducing the alcohol content to approximately 2%. This combination drink is called a *clara.* Clearly the purpose of these practices is to produce a palatable, pleasant beverage to accompany a meal; seeking the euphoric effect is not a consideration.

Alcohol and Snacks

Light lunches and snacks of some kind are served in all beverage shops and in most restaurants. Snacks, called *tapas,* are small individual servings of a wide variety of foods, such as various types of meat and poultry, shellfish, fresh or salted fish, snails, olives, fruit and vegetable salads, egg dishes, and various types of pasta salads. Typically, large serving plates of the various tapas of the day are set on the counter of establishments in the late morning. Persons desiring only a snack at any time of the day will choose a single serving of one item. Those desiring a light lunch will choose approximately three different tapas. A glass of wine or of beer very frequently is consumed along with the food. But beverage shops differ considerably in the number and quality of the tapas they serve. Some may have as many as 15 or 20 different foods of excellent quality, whereas others may have only three or four selections, and these may not be very appealing. Inasmuch as there is a greater abundance and quality of food than of alcohol, customers are drawn because of the food, and alcohol consumption is subsidiary. Those beverage shops with a wide and attractive assortment of food draw a clientele composed of both men and women. On the other hand, inasmuch as there is a less varied and appealing food offering in other shops, customers are drawn to such establishments because of the availability of alcoholic beverages and the opportunity for interaction with other customers. These establishments are peopled almost exclusively by men.

This association of food and the sex ratio of the clientele is also related to location, by whether the establishment is situated near the client's place of work or near his or her home. Those shops situated near his or her place of work and offering an attractive selection of foods become popular for lunchtime patronage as well as for midmorning snacks and for early morning coffee and rolls. Because they serve a utilitarian function, they are patronized by both men and women employees for coffee breaks, snacks, and lunch.

Over the past 40 years, the sex ratio of customers in the beverage shops that serve attractive food has shifted greatly owing to the entrance of more women into the labor force. From 1950 to 1989, the percentage of the labor force that is female has increased steadily from approximately 16% to 34% (Instituto Nacional de Estadistica, 1987, 1989). Consequently, beverage and food shops that focus upon the breakfast and lunch business have experienced a large increase in female clientele.

Alcohol Consumed apart from Food: All-Male Taverns

Beverage shops located near one's home, however, need not be patronized for utilitarian purposes; one can, of course, eat at home. It must be emphasized that in contrast to Italy, where alcohol in the form of wine is consumed almost exclusively with food (Lolli, Serianni, Golder, & Luzzato-Fegiz, 1958), Spain, whereas it does share that pattern, has also had a secondary consumption pattern separate from food intake for many years. Although many establishments in residential areas serve a wide variety of attractive snacks and some serve full meals, others have only a skimpy supply of food. The latter type of establishment in residential areas is generally older, less well maintained, and less attractive, offers mostly beer and wine at lower prices, and is patronized principally by males whose focus is alcohol consumption and sociability. All-male taverns serve as open clubs and social centers.

Such businesses are rarely patronized by women in that most women in Spain prefer to drink alcohol less frequently than men, and because the behavior of the men in such settings often is not consistent with the cultural definition of propriety in the presence of women. Sex role behavior in Spain is more sharply demarked than in the United States. To a greater extent, women are defined as possessing beauty and delicacy plus a sense of shame in being witnesses to crude behavior.

The working-class beverage shop in a residential area serves a large number of social functions, but above all it is, for men of the neighborhood, a social center and haven of respite from responsibility and hierarchical control. Participation in the culture of the tavern is very particularistic in the sense that a person may choose any of a large number of modes of interaction according to his or her particular desires. A patron can sit alone at a table in the back and relax privately, and even doze off without pressure to buy drinks; a group of friends may sit together at a table and converse or play cards; a person may stand at the counter and be socially available to all who enter. The majority of taverngoers choose the latter mode of participation.

Regulars in a bar have told me that a man should drink principally in only one establishment so that he can become known and have friends there. In other words, the tavern serves as a social network in which a man can find a niche. It is a place to meet friends, to relax in a noncompetitive atmosphere with peers, discuss the events of the day, and especially to leave responsibility and many inhibitions behind. In fact, the sex segregation of the tavern assures that the establishment serves as a haven of respite from domestic concerns. Since women are not allowed access, men can be virtually guaranteed of not being "pestered" by wives. In an unusual circumstance, a wife may go to the tavern to fetch her husband. She does not, however, enter the establishment, but rather remains outside while a young child is sent inside to inform the father of the urgent domestic situation. The tavern is also a place to sit and read the newspaper, especially in the morning. In fact, most taverns have a "house newspaper" available for those who have not brought their own. Here a man can catch up on the news at his leisure without the intrusion of children or spouse.

Wylie (1974) observed in a French village in Provence that taverns or cafes served similar social functions: they were social centers for men and havens from wives and children. Likewise, case studies of beer gardens in Rhodesia (Wolcott, 1974) and of workingmen's taverns in the United States (LeMasters, 1975) report that their principal function is to serve as gathering places for male friendship groups and that drinking enhances social solidarity and gaiety.

The majority of customers of men's bars are both married and employed. Unmarried younger men do not patronize all-male bars regularly. They prefer to go to the youth bars, where they can meet young women. Although the social institutions of machismo and male dominance in the household are widespread in Spain, nevertheless in the noncompetitive fellowship of males in the beverage shop they can relax from their usual responsibilities of upholding standards of decorum in employment and domestic situations. Savater (1983) notes that in the hospitable orbit of the tavern one can set aside one's usual personality and construct another one to share with associates. At times men engage in quite uninhibited behavior, such as using strong language in jokingly denigrating or razzing associates about their ineptitude, stupidity, ugliness, laziness, or sexual performance, as well as occasionally spitting on the floor, playing tricks, and engaging in boisterous horseplay, which can include poking a companion in a tender spot or giving him the hotseat with a cigarette lighter. [The functions of these drinking groups are comparable to those described for American ones by Clinard, 1962, and Trice, 1966—Eds.]

Clearly the basic trait of interaction in the all-male tavern is that of spontaneity and quite uninhibited emotional expression. This interaction occurs in an atmosphere of openness and social access among regulars, both formally and interpersonally. In the formal sense, Max Weber (1947) defines an open relationship as one available to anyone who wants to participate. Any adult male is free to participate in barroom activity. Everyone in the tavern is free to speak to anyone else. The lack of social distance between participants can be explained by four factors that contribute to social homogeneity. All participants are male, with similar educational and occupational status, and all live in the same barrio under similar living conditions. In addition, the regulars see each other in the bar very frequently and have a large number of associates in common. Although clearly there are sociability preferences, all regulars in the establishment feel free to talk with anyone else.

The one form of competitive behavior that takes place in men's bars is verbal dueling. The dueling consists of an exchange of an interrelated series of insults in which each contestant uses his opponent as a foil to demonstrate adroitness and wit to an appreciative audience, some of whose members may contribute comments to the duel. Octavio Romano (1960) has noted that adeptness at verbal dueling is a means of gaining prestige in a Mexican-American community. In Spain, the duel inevitably ends with both parties asserting the strength of their friendship and each other's integrity. Goffman (1955) has noted the necessity for a culture to have reconstructive behaviors for repair or ruptures in social relations following a faux pas. Although a duel is not a faux pas, reconstructive devices of asserting friendship and solidarity are employed to repair any possible rupture or ill feeling that might have been engendered during the verbal assault.

In summary, the major trait of social interaction among the patrons in an all-male tavern with an atmosphere of equality is that of spontaneity and uninhibited emotional expression, often with grandiose hyperbole—all exhibited without fear of recrimination or of jeopardizing the standards of propriety required of them as employees and heads of households. Such beverage shops are very similar to workingmen's bars in the United States except that the degree of emotional expressiveness is notably more ample and free in Spanish bars.

It is convenient to divide the customers of workingmen's taverns into three general types according to their frequency of patronage.

Visitors are persons who come to a particular tavern only occasionally and are not personally known there.

Regulars are those who usually patronize a particular tavern at least twice a week and are known and recognized there. One distinguishing mark of being integrated into the social network is that one receives messages from other patrons relayed by the bartender. Most frequently, regulars come in during the late afternoon or early evening hours after work but before dinner and sometimes during the morning or at midday on Saturdays and Sundays. They usually consume a few drinks at each visit, talk to companions, perhaps watch part of a soccer game on television, and leave within 30 to 60 minutes. Their participation in the life of the tavern is only one of the many types of social activities they regularly engage in. Of all persons who patronize a men's tavern during the course of a week, well over two thirds are regulars. In general, these persons are high-frequency, low-quantity drinkers. Systematic surveys of the Spanish population indicate that this is the modal consumption pattern in Spain, occurring among 46% of the adult male population in Madrid (Martinez & Martin, 1987), among 47% of males in Seville (Gili, Giner, Lacalle et al., 1989), among 49% in Santiago (Lorenzo Lago et al., 1987), and among 48% to 54%, according to various surveys, in the Basque country (Elzo, 1987).

Residents are persons who generally spend most of their free time at the neighborhood tavern. Some may come in for one or two early morning drinks before going to work. After work, all return to spend several hours at the tavern, which constitutes their principal sphere of social activity and focus of community. Friends who wish to locate them go to the tavern; if they do not find the residents there, they leave word with the bartender. During their long visits in the drinking establishment, most residents drink wine and drink it slowly, thereby avoiding a high blood alcohol concentration. Similarly, Wylie (1974) observed in a French village that although many men spent long hours in the tavern very few ever became intoxicated. Wolcott (1974) reports that in the African beer gardens of Rhodesia, the focal point of customers is sociability with friends and acquaintances. Beer is usually consumed slowly, amidst much conversation. Some might enter and have one drink and sit for hours, as Blum (1969) also observed in cafes in Greece. At times, a regular might be passing by and notice a particular friend in the bar, enter, and pass an hour talking without buying anything at all. Others might sit and not drink when using the bar as a meeting place for an appointment with an associate.

Bar owners make no efforts to speed up or control the pace of drinking on the part of customers and thereby to increase sales and profits. From a study of a pueblo in Andalucia, Pitt-Rivers (1961) concludes that economic values are subjected to control by moral and social values. The acquisition of money is not regarded as inherently good but rather is subject to moral judgement according to how it is acquired and used. Bar owners have told me that success in the business is dependent upon an owner's developing a personal relationship with customers by being open, communicative, and tolerant. Thus, economic transactions in taverns take place within a particularistic context of individual preferences and community values.

Alcohol Consumption and Sociability

Regardless of the quantity and pace of alcohol consumption, the focus of these men is principally upon interaction with their fellows. However, the interaction at times may lapse, especially on hot Sunday afternoons in July and August, when some men seated at tables fall asleep during the usual time for siesta. Clearly the bar is their social, recreational, and relaxation center and, as has been stated, can be utilized for a variety of modes of participation according to their particular needs.

Alcohol consumption, however, constitutes the medium for social participation and for relating to friends; there is considerable buying of rounds of drinks and also considerable sharing of cigarettes among both the residents and the regulars. Pitt-Rivers (1961) notes that in Spain people like to make a gesture of generosity toward a friend but also like to make a show of their generosity. Money generates prestige only if it is employed in a morally approved manner, and sharing among friends is regarded as a requirement in the system of friendship. Although most residents do not consume drinks at a rapid pace, over the course of 3 to 6 hours spent in the beverage shop, they usually consume from 5 to 10 drinks nearly every day. Clearly, this cadre of residents constitutes high-frequency, high-quantity drinkers. Systematic surveys of the Spanish population indicate that such drinkers constitute 16% of the male population in Madrid (Martinez & Martin, 1987), 7% in Seville (Lacalle, Giner, Gili, & Franco, 1989), between 12% and 14% (according to various surveys) in the Basque country (Elzo, 1987), and 26% in Cantabria, along the northern coast of Spain (Diez Manrique & Peña, 1989).

Although the data are presented somewhat differently, surveys conducted in the United States yield results about the frequency of visiting a tavern that are within the range of the Spanish surveys. Data from both a national sample and a sample of an urbanized county in the Pacific Northwest reveal that nearly 12% of adults go to taverns at least weekly. Furthermore, 10% of the Pacific Northwest adults sampled consume eight or more drinks on a particular day at least once a month (Nusbaumer, Mauss, & Pearson, 1982). In Spain, however, this quantity of alcohol on a particular day is consumed at least once a week by a similar proportion of adults.

In Spain, however, even for heavy drinkers, sociability appears to be more closely integrated with alcohol consumption than in the United States. Bacon (1958) notes that deliberate seeking of euphoric effects is the dominant style of alcohol use among U.S. alcoholics. Furthermore, achieving euphoric effects is institutionalized or standardized behavior, particularly among certain groups of young men in the United States who engage in deliberate heavy drinking, especially on weekends, with the objective of intoxication (Globetti, 1970; Globetti & Windham, 1967).

Drunkenness

Despite the high quantity they consume, drunkenness is infrequent among tavern regulars in Spain. They do, however, run the risk of becoming maintenance-level, or delta, alcoholics, as described by Jellinek (1960). Reinterpreting Jellinek's formulation, Alonso Fernandez (1981) states that persons who exhibit this type of daily drinking pattern but generally avoid drunkenness are better described as "regular excessive drinkers." Alonso Fernandez further points out that although many drinkers who follow this pattern need not be alcoholic, they nevertheless run considerable health risks and have a high probability of developing physical dependency after many years. A number of daily high-volume drinkers have told me that they usually wake up each morning with slight tremors. This phenomenon is not interpreted as alarming or pathological but rather as a condition that can easily be alleviated with one morning drink. For them, alcohol is the pleasurable cure to possible problems.

The infrequent occurrence of obvious drunkenness even among habitual excessive drinkers may be explained in terms of the type of beverage usually consumed, the focus of interaction in drinking settings, and the attitudes toward drunkenness in Spanish culture. First of all, the majority of the residents usually consume vino comun, which has an alcohol content of approximately 11%, in contrast to the 40% to 50% alcohol in distilled spirits; and this wine often is diluted with soda water. Second, the focus of interest in the bar is, again, upon interaction with companions. The bar is a social center, and drinking serves as the common activity by which residents relate to one another. Hence, most men talk much and drink slowly. Representative surveys of the population also reveal the strong association of drinking and sociability. In Seville, 93% of both males and females endorsed the statement that drinking is a means of expressing friendship (Gili, Giner, & Lacalle, 1989), which statement was endorsed by 58% of males in Madrid (Martinez & Martin, 1987). The latter authors note that these survey data reflect the social nature of what drinking signifies in Spanish society, where a great part of social relations takes place in bars and beverage shops. Primarily, alcohol accompanies sociability rather than vice versa.

In the United States and Canada, by far the most frequent use of alcohol occurs in the context of social relationships. In a national survey in the United States, Cahalan, Cisin, and Crossley (1969) found that of persons drinking at least monthly the majority most frequently drank with friends, and next most often with family members; only a small minority usually drank alone.

Drunkenness in Spain is further limited by widely held censorial norms against it. The few persons who do become intoxicated in drinking establishments commit a breech of a major norm: they become unable to participate in the social interaction. Although they are usually not outrightly chided or rejected when intoxicated, they are usually ignored by their associates. This pattern constrasts to that found in a study of a French village in which the few persons who became drunk were looked upon as amusingly childish (Wylie, 1974).

Survey results closely parallel these participant observations. In the Province of Seville, there is a high degree of rejection of drunkenness as shameful among men and as completely unacceptable among women (Lacalle et al., 1989) and widespread agreement that drunkenness would marginalize a person from social groups (Gili et al., 1989). A survey in Madrid reports that over 90% of both men and women support the statement that it is disagreeable to see a person drunk, and 87% disagree with the statement that some people feel better to get drunk once in a while (Martinez & Martin, 1987).

The Chiquiteo

Although this analysis of drinking patterns has been concerned chiefly with males, it should be noted that there is an institutionalized pattern of drinking outside of mealtime followed by some women as well as men in urban areas, particularly in the Basque country in northeastern Spain. The name, as well as the origin, of the practice is Basque, *txikiteo,* which has been transliterated into Castilian as *chiquiteo* and is a ritualized pattern by which sociability and interpersonal relations are linked to the consumption of alcohol. The chiquiteo is a form of "doing the rounds" in which each participant in a drinking group buys a round of drinks for the group. The larger the drinking group, the greater the number of drinks required to be consumed by each participant. However, norms dictate that each serving be small. Although doing the rounds is common among men in barrooms in the United States, the rules of the chiquiteo require that the participants move on to a different bar for each round of drinks. A group that regularly engages in this practice usually establishes a fixed itinerary. Ramirez Goicoechea (1984) points out that making regular visits to a number of bars gives group members the opportunity to extend their range of social contacts.

The rules further require that the group be composed of members of only one sex. Although this pattern originated among men, it often occurs among employed women, and after working hours, but before supper. After visiting as many bars as there are group members, the chiquiteo often ends with the group going to dinner. In contrast to English pub-crawling or American bar-hopping, the Spanish chiquiteo usually includes emphasis on eating much as well as drinking much. The less frequent appearance of the chiquiteo among men probably is due to the fact that men most often become either regular patrons or residents in one bar and practice celebratory drinking with their friends there.

Youth Bars

Drinking on the part of youth and young adults has departed from the tradition of their elders in a number of important ways: (1) substituting beer for wine as the initial beverage used, (2) initiating drinking at a younger age, (3) initiating alcohol consumption in the social context of peers rather than of the family, and (4) of separating alcohol use from food consumption to a greater degree. However, young persons are similar to their elders in that when alcohol is consumed apart from meals the principal focus is upon social interaction rather than upon alcohol consumption.

Bars catering to teenagers and persons in their 20s have been in existence for many years. Disco clubs have become popular in major cities since the introduction of rock music in the mid-1960s. Although the minimum legal age for purchasing alcohol in Spain is 16 years, no one is concerned with formalities of the law. This nonconcern for the age law has been mentioned in relation to the availability of beer, even in a government-sponsored public institution, the secondary schools. Pitt-Rivers (1961) points out that Spaniards sharply distinguish legality from morality. The penal code originates from the central government, whereas the code of moral behavior comes from the norms of the people. Consequently, there is a large part of the penal code to which the citizenry is morally indifferent. With this in mind, it is easy to understand the results of a survey of school students in grades 7 through 12 in the Basque country, in which 91% report that it is very easy to obtain alcohol (Orrantia & Fraile, 1984). My own observations reveal that youngsters of 10 and 12 years are able to buy liter bottles of beer in grocery and convenience stores if they choose.

Although the minimum purchase age law is not enforced, taverns for adults are rarely patronized by teenagers of 14 or 15 years nor by those of the legal age of 16 to 20 years. Persons of this age group have no interest in socializing with older men. Rather, young people of 14 years and over and of both sexes gather at bars for young persons primarily to socialize with members of their own age group and to meet members of the opposite sex. Youth bars in most cities tend to be concentrated in close proximity. The area where such bars are situated serves simply as a gathering point for persons of this age group. The focus is so much upon socializing that the majority of youngsters who congregate in the bars and in the street outside are not consuming drinks but rather are occupied in conversing and in meeting people, as well as in walking from one bar to another to see if perhaps more action is taking place and to participate generally in this iridescent scene. The bars themselves, in addition to the adjacent sidewalks and streets, are transformed into an area of *open social participation,* Goffman's (1971) term for a condition of sanctioned freedom from structured expectations and the accompanying freedom to be personally expressive.

The primary attraction of the youth bar is the availability of large numbers of persons of one's age group in a setting free from adult control. Of the minority who are consuming a beverage, nearly all drink beer, with a small proportion drinking sodas or mixed drinks with distilled liquor. Very rarely does anyone take wine. Nor is food of any consideration; only a very small minority eat an occasional snack. Thus, youth are very similar to patrons of men's taverns in that their focus is primarily upon social interaction. They differ, however, in that many will consume no alcohol; and for those who do, beer is far and away their beverage of choice.

Age Differences in Drinking Patterns

The shift in primary beverage choice between Spanish adolescents and young adults on the one hand and their parents on the other contrasts with survey results in the United States. Barnes (1977), through a study of both adolescents and their parents in one metropolitan area in the United States, found that the drinking patterns of adolescents very closely paralleled that of their elders, both in frequency and in type of beverage used. Spanish young people, by contrast, are not following the beverage choice of their elders but have shifted from wine to beer as their beverage of choice. In the United States beer is the beverage of choice of both adults and young people.

It is important to note that beer has become increasingly more popular over the past 35 years. This assertion is borne out both by production data and by population surveys regarding consumption. Production data reveal that beer consumption has experienced more than a 25-fold increase, moving from 2.15 liters per capita in 1950 up to 58.67 liters per capita in 1985 (Instituto Nacional de Estadistica 1953, 1987). In contrast, wine has enjoyed only a 64% increase, moving from 51.5 liters to 84.5 liters per capita over the same years.

The reasons for this great increase in beer consumption are found in the results of population surveys. A survey of the population 18 years and over in the metropolitan area of Madrid reveals that wine as the beverage of initial use in childhood or in youth has declined progressively over the years, having been the first beverage used for approximately two thirds of those aged 50 and over, for slightly less than half of those in their 40s, for a third of those in their 30s, and for only a fifth of those 18 to 29 years old (Martinez & Martin, 1987). On the other hand, the percentage of those taking beer as their initial beverage has increased from 14% to 59% over these same age groups.

Furthermore, the age at which alcohol use is initiated has declined considerably over the past generation. Of those aged 40 or more, approximately 20% first use alcohol by the age of 16. This proportion increases to 28% among those in their 30s and stands at 50% among those 18 to 29 years of age (Martinez & Martin, 1987).

The progressive substitution of beer for wine as the beverage of choice and the phenomenon of initiating drinking at progressively younger ages are shown also in a national survey of school children in the 6th and 8th grades. The aggregate data for both grades and both sexes show that the proportion of these school children who have ever used beer is 64%, in contrast to the 43% who have ever taken wine. Furthermore, analysis of frequency of use reveals that 10% use beer at least weekly and an additional 18% use it occasionally, in contrast to 6% who drink wine at least weekly and 9% who drink it occasionally (Mendoza, 1987). Thus, school children in the 6th and 8th grades who use alcohol are nearly twice as likely to drink beer rather than wine, thus following the established pattern among older youth.

Further analysis of the data regarding school children reveals that their beer consumption is linked with social relations in the peer group. Among those students who go out with their friends every day after school, 79% drink beer daily; the proportion of regular beer drinkers declines proportionally with the frequency of seeing friends outside of school (Mendoza, 1989). These results are highly similar to those of Jessor and Jessor (1977), who found, through a longitudinal study of young persons aged 12 to 17 in the United States, that becoming a drinker—and especially a heavier drinker—was associated with greater involvement in the peer group, an attenuation of attachment to norms of the parents, and a greater desire for freedom from parental control.

Future Drinking Patterns

Martinez and Martin (1987) pose the question whether this new pattern among the young of drinking beer unrelated to food consumption constitutes a major change in Spanish drinking patterns. Will those now in their 20s carry this drinking pattern with them throughout their lives and establish it as the dominant pattern of alcohol use, or will the contemporary young adults revert to the traditional use of wine with food as the dominant pattern as they mature and take on adult roles? But the pattern of behavior of the young is not without precedent in Spanish culture. These researchers do not take into account the fact of the existence of a well-established secondary pattern of alcohol use in Spain that developed in workingmen's taverns. Here alcohol consumption is integrated within a pattern of social relations and is separated from food consumption. This appears to be the pattern the majority of young people are following. Whether this pattern will become dominant in Spanish culture in future years is a question well worth asking about a process well worth monitoring.

References

Alonso Fernandez, F. (1981). *Alcohol-dependencia.* Madrid: Ediciones Piromides.

Bacon, S. (1958). Alcoholics do not drink. *Annals of the American Academy of Political and Social Science, 315,* 55–64.

Barnes, G. (1977). The development of adolescent drinking behavior: An evaluative review of the impact of the socialization process within the family. *Adolescence, 12,* 571–591.

Blum, R. (1969). *Society and drugs.* San Francisco: Jossey-Bass. Publication of the Institute for the Study of Human Problems.

Cahalan, D., Cisin, I.H., & Crossley, H.M. (1969). *American drinking practices: A national study of drinking behavior and attitudes.* New Brunswick, NJ: Rutgers Center of Alcohol Studies.

Clinard, M. (1962). The public drinking house and society. In D.J. Pittman & C.R. Snyder (Eds.), *Society, culture, and drinking patterns* (pp. 270–92). New York: Wiley.

Diez Manrique, J.F., & Peña, C. (1989). Respuesta de la comunidad a los problemas relacionados con el alcohol en Cantabria. In *Problemas relacionados con el consumo de alcohol* (pp. 99–108). Junta de Andalucia, Consejeria de Salud y Servicios Sociales, Comisionado para la Droga.

Elzo, J. (1987). La investigacion epidemiologica y sociologica de la drogadiccion en Euskadi (1978–1986). In Gobierno Vasco, *Libro blanco de las drogodependencias en Euskadi, 1987* (pp. 71–89). Vitoria: Servicio Central de Publicaciones.

Gili, M., Giner, J., & Lacalle, J.R. (1989). Actitudes y normas culturales de la poblacion frente a la bebida. In *Problemas relacionados con el consumo de alcohol* (pp. 51–61). Junta de Andalucia. Consejeria de Salud y Servicios Sociales, Comisionado para la Droga.

Gili, M., Giner, J., Lacalle, J.R., Franco, D., Perea, E., & Dieguez, J. (1989). Patterns of consumption of alcohol in Seville, Spain: Results of a general population survey. *British Journal of Addiction, 84,* 277–285.

Globetti, G. (1970). The drinking patterns of Negro and white high school students in two Mississippi communities. *Journal of Negro Education, 30,* 60–69.

Globetti, G., & Windham, G.O. (1967). The social adjustments of high school students and the use of beverage alcohol. *Sociology and Social Research, 51,* 148–157.

Goffman, E. (1955). On facework: An analysis of ritual elements in social interaction. *Psychiatry, 18,* 213–231.

Goffman, E. (1971). *Relations in public: Microstudies of the public order.* New York: Basic Books.

Instituto Nacional de Estadistica. (1953). *Anuario estadistico de España, año 1953.* Madrid: Edicion Manual.
Instituto Nacional de Estadistica. (1987). *Anuario estadistico de España, año 1987.* Madrid: Edicion Manual.
Instituto Nacional de Estadistica. (1989). *Encuesta de poblacion activa: Resultados detallados. Primer trimestre de 1989.* Madrid: Edicion Manual.
Jellinek, E.M. (1960). *The disease concept of alcoholism.* New Brunswick, NJ: Rutgers Center of Alcohol Studies.
Jessor, R., & Jessor, S.L. (1977). *Problem behavior and psychosocial development: A longitudinal study of youth.* New York: Academic Press.
Lacalle, J.R., Giner, J., Gili, M., & Franco, D. (1989). Patrones de consumo y problemas relacionados con el alcohol. In *Problemas relacionados con el consumo del alcohol* (pp. 33–49). Junta de Andalucia, Consejeria de Salud y Servicios Sociales, Comisionado para la Droga.
LeMasters, E. (1975). *Blue collar aristocrats: Life-styles at a working-class tavern.* Madison, WI: University of Wisconsin Press.
Lolli, G., Serianni, E., Golder, G.M., & Luzzatto-Fegiz, P. (1958). *Alcohol in Italian culture.* Glencoe, IL: Free Press.
Lorenzo Lago, A., Fariñas Vasco, E., Carrera Machado, I., Belo Gonzalez, C., & Rodriguez Lopez, A. (1987). Habitos de consumo de alcohol en una empresa gallega. *Revista Española de Drogodependencia, 12,* 167–176.
Martinez, R.N., & Martin, L. (1987). Patrones de consumo de alcohol en la comunidad de Madrid. *Comunidad y Drogas, 5–6,* 39–62.
Mendoza, R. (1987). Consumo de alcohol y tabaco en los escolares espanoles. *Comunidad y Drogas, 5–6,* 83–102.
Mendoza, R. (1989). El consumo de alcohol en los escolares españoles: Datos del estudio Europeo sobre los habitos de los escolares en relacion con la salud. In *Problemas relacionados con el consumo del alcohol* (pp. 173–90). Junta de Andalucia, Consejeria de Salud y Servicios Sociales, Comisionado para la Droga.
Nusbaumer, M.R., Mauss, A.L., & Pearson, D.C. (1982). Draughts and drunks: The contributions of bars and taverns to excessive drinking in America. *Deviant Behavior, 3,* 329–358.
Orrantia, I., & Fraile, A. (1984). *El consumo de drogas de los alumnos del 7° y 8° de EGB, BUP, COU y FP de Vizcaya.* Reporte al Gobierno Vasco. Bilbao: Multicopia.
Pitt-Rivers, J. (1961). *The people of the Sierra.* Chicago: University of Chicago Press.
Ramirez Goicoechea, E. (1984). Cuadrillas en el Pais Vasco: Identidad local y revitalizacion etnica. *Revista Española de Investigaciones Sociologicas, 25,* 213–220.
Romano, O. (1960). Donship in a Mexican-American community in Texas. *American Anthropologist, 62,* 966–976.
Savater, F. (1983). Elogio de la taberna. In *Sobras Completas* (pp. 131–37). Madrid: Ediciones Literarias.
Trice, H. (1966). *Alcoholism in America.* New York: McGraw-Hill.
Weber, M. (1947). *Theory of social and economic organization* (translated by A.M. Henderson and Talcott Parsons). Glencoe, IL: Free Press.
Wolcott, H.F. (1974). *The African beer gardens of Bulawayo: Integrated drinking in a segregated society.* New Brunswick, NJ: Rutgers Center of Alcohol Studies.
Wylie, L.W. (1974). *Village in the Vaucluse* (3rd ed.) Cambridge, MA: Harvard University Press.

SECTION IV

The Genesis and Patterning of Alcoholism and Alcohol-Related Problems

A. DEFINITION, PATTERNING, AND EXTENT

Introductory Note

A major line of investigation in clinical medicine has always been concerned with the "natural history" of disease. This approach assumes that typically each disease has a sequence of symptoms that extends from the prodromal to the terminal phases of the illness. In short, the disease, whether alcoholism or diabetes, is expected to have a patterned symptomatology that can best be discerned if the patient is followed over a number of years. For certain diseases, it is possible for therapeutic intervention to alter the "natural history" of the disease, and the impact of this intervention must be carefully evaluated in each case.

E.M. Jellinek, in one of the most significant contributions to the alcoholism field, applies the natural-history-of-disease assumptions to a questionnaire study of the drinking habits of male alcohol addicts. This study, "Phases of Alcohol Addiction" (which is reprinted here as chapter 20 from, coincidentally, chapter 20 in the original volume), is based on the assumption that there are two categories of alcoholics—alcohol addicts and habitual symptomatic excessive drinkers (nonaddicts). The crucial differentiating factor is the addicts' "loss of control after drinking begins."

Members of Jellinek's questionnaire group were presented with a set of symptoms associated with alcoholism and were instructed to indicate the sequence in which these features appeared in their own drinking histories. Jellinek's findings are that the symptoms occur in the typical case in sequence but that "not all symptoms . . . occur necessarily in all alcohol addicts, nor do they occur in every addict in the same sequence." These symptoms are patterned into four sequential phases, namely, the prealcoholic symptomatic, the prodromal, the crucial, and the chronic. Thus, Jellinek has documented concrete stages through which alcoholism progresses. This ingenious research study by Jellinek has perhaps had more impact on the alcoholism field than any other and forms the basis for the disease concept of alcoholism. It should be noted, however, that Jellinek's methodology has been extensively criticized (see chapter 21, by Fingarette).

In spite of Jellinek's claim, the question whether alcoholism is a disease remains one of the most controversial issues in the alcohol field. Although it is not the intention of this book to support one viewpoint over another, we present two chapters, 21 and 22, which discuss the advantages and disadvantages of a disease model. In chapter 21, Herbert Fingarette argues that "the idea that alcoholism is a disease is a myth" and that the current scientific evidence clearly challenges all the beliefs behind the disease concept. He claims that although treatment staffers and alcoholics may profit from the economic benefits that accrue to a disease concept, the average taxpayer suffers. He also claims that a disease model hinders early intervention and that medical treatments have not been successful in helping alcoholics.

Brian Hore, a clinical psychiatrist from the United Kingdom, makes a similar argument while presenting primarily the British view on the disease concept of alcoholism. In chapter 22, written especially for this book, Hore first gives a history of the disease concept. He claims that although it is better to have alcoholism considered within a health sphere as opposed to a morality sphere, the disease model removes individual responsibility and prevents early recognition and treatment of the problem. After reviewing the empirical research relating to the disease concept, Hore concludes that genetic, psychological, and biological factors should be linked into a theory of alcoholism. Later in this section we present chapters that discuss the role of social and environmental factors in such a multidisciplinary theory.

CHAPTER 20

Phases of Alcohol Addiction[1]

E.M. Jellinek

Only certain forms of excessive drinking—those that in the present report are designated as alcoholism—are accessible to medical-psychiatric treatment. The other forms of excessive drinking, too, present more or less serious problems, but they can be managed only on the level of applied sociology, including law enforcement. Nevertheless, the medical profession may have an advisory role in the handling of these latter problems and must take an interest in them from the viewpoint of preventive medicine.

The conditions that have been briefly defined by the Alcoholism Subcommittee of the World Health Organization as alcoholism are described in the following pages in greater detail, in order to delimit more definitely those excessive drinkers whose rehabilitation primarily requires medical-psychiatric treatment. Furthermore, such detailed description may serve to forestall a certain potential danger that attaches to the disease conception of alcoholism, or, more precisely, of addictive drinking.

With the exception of specialists in alcoholism, the broader medical profession, representatives of the biological and social sciences, and the lay public use the term *alcoholism* as a designation for any form of excessive drinking instead of as a label for a limited and well-defined area of excessive drinking behaviors. Automatically, the disease conception of alcoholism becomes extended to all excessive drinking irrespective of whether or not there is any physical or psychological pathology involved in the drinking behavior. Such an unwarranted extension of the disease conception can only be harmful, because sooner or later the misapplication will reflect on the legitimate use too and, more important, will tend to weaken the ethical basis of social sanctions against drunkenness.

E.M. Jellinek (1890–1963), co-founder, along with Howard W. Haggard, of the then Yale Center of Alcohol Studies, is the author of *The Disease Concept of Alcoholism* (New Brunswick, N.J.: Rutgers Center of Alcohol Studies, 1960). This chapter is reprinted from *Society, Culture, and Drinking Patterns,* D.J. Pittman and C.R. Snyder, Eds. (New York: Wiley, 1962), pp. 356–368.

The Disease Conception of Alcohol Addiction

The subcommittee has distinguished two categories of alcoholics, namely, "alcohol addicts" and "habitual symptomatic excessive drinkers." For brevity's sake the latter will be referred to as nonaddictive alcoholics. Strictly speaking, the disease conception attaches to the alcohol addicts only, and not to the habitual symptomatic excessive drinkers.

In both groups the excessive drinking is symptomatic of underlying psychological or social pathology, but in one group after several years of excessive drinking "loss of control" over the alcohol intake occurs, whereas in the other group this phenomenon never develops. The group with the loss of control is designated as "alcohol addicts." (There are other differences between these two groups, and these will be seen in the course of the description of the "phases.")

The disease conception of alcohol addiction applies not to the excessive drinking but solely to the loss of control, which, again, occurs in only one group of alcoholics and then only after many years of excessive drinking. There is no intention to deny that the nonaddictive alcoholic is a sick person; but his ailment is not the excessive drinking, but rather the psychological or social difficulties from which alcohol intoxication gives temporary surcease.

The loss of control is a disease condition per se that results from a process that superimposes itself upon those abnormal psychological conditions of which excessive drinking is a symptom. The fact that many excessive drinkers drink as much as or more than the addict for 30 or 40 years without developing loss of control indicates that in the group of alcohol addicts a superimposed process must occur.

Whether this superimposed process is of a psychopathological nature or whether some physical pathology is involved cannot be stated as yet with any degree of assurance, the claims of various investigators notwithstanding. Nor is it possible to go beyond conjecture concerning the question whether the loss of control originates in a predisposing factor (psychological or physical), or whether it is a factor acquired in the course of prolonged excessive drinking.

The fact that this loss of control does not occur in a large group of excessive drinkers would point towards a predisposing *X* factor in the addictive alcoholics. On the other hand this explanation is not indispensable, in that the difference between addictive and nonaddictive alcoholics could be a matter of acquired modes of living—for instance, a difference in acquired nutritional habits.

The Meaning of Symptomatic Drinking

The use of alcoholic beverages by society has primarily a symbolic meaning, and secondarily it achieves "function." Cultures that accept this custom differ in the nature and degree of the functions, which they regard as legitimate. The differences in these functions are determined by the general pattern of the culture—for example, the need for the release and for the special control of aggression, the need and the ways and means of achieving identification, and the nature and intensity of anxieties and the modus for their relief. The more the original symbolic character of the custom is preserved, the less room will be granted by the culture to the functions of drinking.

Any drinking within the accepted ways is symptomatic of the culture of which the drinker is a member. Within that frame of cultural symptomatology there may be in addition individual symptoms expressed in the act of drinking. The fact that a given individual drinks a glass of beer with his meal may be the symptom of the culture that accepts such a use as a refreshment, or as a "nutritional supplement." That this individual drinks at this given moment may be a symptom of his fatigue, his elation, or some other mood, and thus an individual symptom; but if his culture accepts the use for these purposes, it is at the same time a cultural symptom. In this sense even the small or moderate use of alcoholic beverages is symptomatic, and it may be said that all drinkers are culturally symptomatic drinkers or, at least, started as such.

The vast majority of the users of alcoholic beverages stay within the limits of the culturally accepted drinking behaviors and drink predominantly as an expression of their culture. Although an individual expression may be present in these behaviors, its role remains insignificant.

For the purpose of the present discussion the expression *symptomatic drinking* will be limited to the predominant use of alcoholic beverages for the relief of major individual stresses.

A certain unknown proportion of these users of alcoholic beverages, perhaps 20%, are occasionally inclined to take advantage of the functions of alcohol, which they have experienced in the course of its "cultural use." At least at times, the individual motivation becomes predominant, and on those occasions alcohol loses its character as an ingredient of a beverage and is used as a drug.

The "occasional symptomatic excessive drinker" tends to take care of the stresses and strains of living in socially accepted—that is, "normal"—ways, and his drinking is most of the time within the cultural pattern. After a long accumulation of stresses, however, or because of some particularly heavy stress, his tolerance for tension is lowered and he takes recourse to heroic relief of his symptoms through alcoholic intoxication.[2] Under these circumstances the "relief" may take on an explosive character, and thus the occasional symptomatic excessive drinker may create serious problems. No psychological abnormality can be claimed for this type of drinker, although he does not represent a well-integrated personality.

Nevertheless, within the group of apparent occasional symptomatic excessive drinkers there is a certain proportion of definitely deviating personalities who after a shorter or longer period of occasional symptomatic relief take recourse to a constant alcoholic relief, and drinking becomes with them a "mode of living." These are the "alcoholics" of whom, again, a certain proportion suffer loss of control—that is, become addictive alcoholics.

The proportion of alcoholics (addictive and nonaddictive) varies from country to country, but does not seem to exceed in any country 5% or 6% of all users of alcoholic beverages. The ratio of addictive to nonaddictive alcoholics is unknown.

The Chart of Alcohol Addiction

The course of alcohol addiction is represented graphically in Figure 1. The diagram is based on an analysis of more than 2,000 drinking histories of male alcohol addicts. Not all symptoms shown in the diagram occur necessarily in all alcohol addicts, nor do they occur in every addict in the same sequence. The "phases" and the sequences of symptoms within the phases are characteristic, however, of the great majority of alcohol addicts and represent what may be called the average trend.

For alcoholic women the phases are not as clear-cut as in men, and the development is frequently more rapid.

The phases vary in their duration according to individual characteristics and environmental factors. The lengths of the different phases on the diagram do not indicate differences in duration but are determined by the number of symptoms that have to be shown in any given phase.

The chart of the phases of alcohol addiction serves as the basis of description, and the differences between addictive and nonaddictive alcoholics are indicated in the text.

1. The Prealcoholic Symptomatic Phase

The very beginning of the use of alcoholic beverages is always socially motivated in the prospective addictive and nonaddictive alcoholic. In contrast to the average social drinker, however, the prospective alcoholic (together with the occasional symptomatic excessive drinker) soon experiences a rewarding relief in the drinking situation. The relief is strongly

FIGURE 1

The phases of alcohol addiction. The large bars denote the onset of major symptoms that initiate phases. The short bars denote the onset of symptoms within phases. Reference to the numbering of the symptoms is made in the text, where the numbers appear in parentheses.

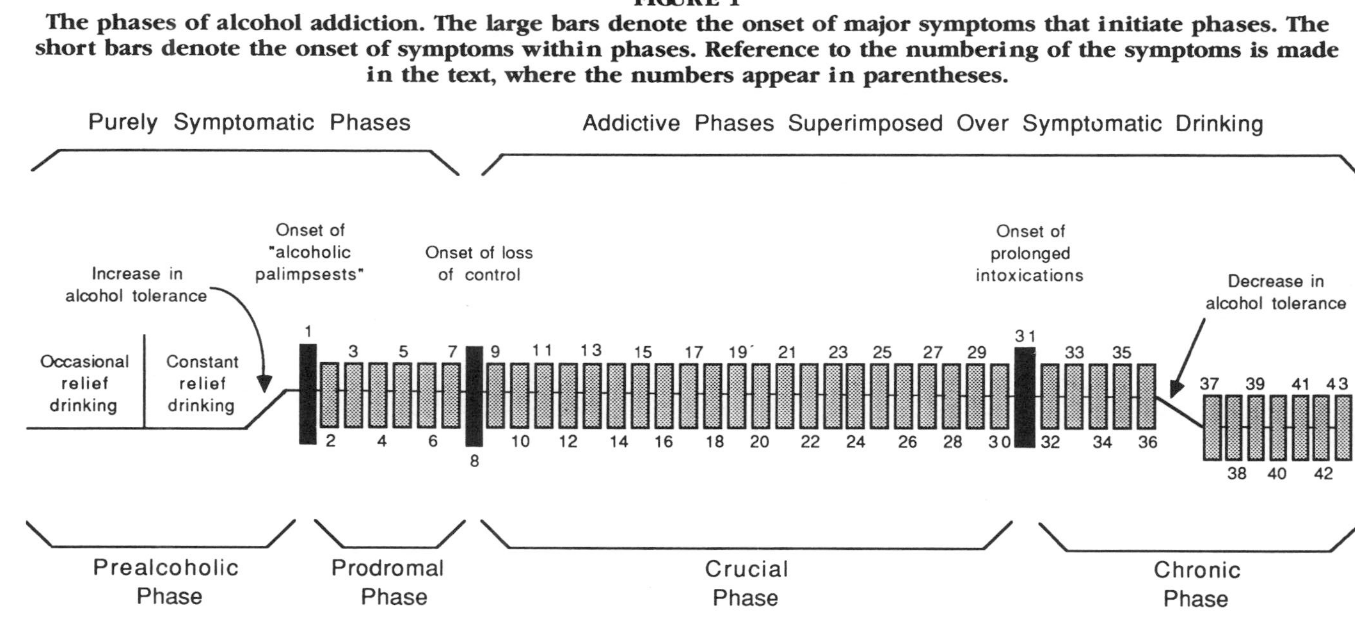

marked in his case because either his tensions are much greater than in other members of his social circle, or he has not learned to handle those tensions as others do.

Initially this drinker ascribes his relief to the situation rather than to the drinking, and he seeks therefore those situations in which incidental drinking will occur. Sooner or later, of course, he becomes aware of the contingency between relief and drinking.

In the beginning he seeks this relief occasionally only, but in the course of 6 months to 2 years his tolerance for tension decreases to such a degree that he takes recourse to alcoholic relief practically daily. Nevertheless, his drinking does not result in overt intoxication, but he reaches toward the evening a stage of surcease from emotional stress. Even in the absence of intoxication this process involves fairly heavy drinking, particularly in comparison to the use of alcoholic beverages by other members of his circle. The drinking is, nevertheless, not conspicuous either to his associates or to himself. After a certain time an increase in alcohol tolerance may be noticed; that is, the drinker requires a somewhat larger amount of alcohol than formerly in order to reach the desired stage of sedation.

This type of drinking behavior may last from several months to 2 years, according to circumstances, and may be designated as the prealcoholic phase, which is divided into stages of occasional relief drinking and constant relief drinking.

2. The Prodromal Phase

The sudden onset of a behavior resembling the blackouts in anoxemia marks the beginning of the prodromal phase of alcohol addiction. The drinker who may have had not more than 50 or 60 grams of absolute alcohol and who is not showing any signs of intoxication may carry on a reasonable conversation or may go through quite elaborate activities without a trace of memory the next day, although sometimes one or two minor details may be hazily remembered. This amnesia, which is not connected with loss of consciousness, has been called by Bonhoeffer the "alcoholic palimpsests," with reference to old Roman manuscripts superimposed over an incompletely erased manuscript.

Alcoholic palimpsests (1) may occur on rare occasions in an average drinker when he drinks intoxicating amounts in a state of physical or emotional exhaustion.[3] Nonaddictive alcoholics, of course, also may experience palimpsests, but infrequently and only following rather marked intoxication. Thus, the frequency of palimpsests and their occurrence after medium alcohol intake are characteristic of the prospective alcohol addict.

This would suggest heightened susceptibility to alcohol in the prospective addict. Such a susceptibility may be psychologically or physiologically determined. The analogy with the blackouts of anoxemia is tempting. Of course, an insufficient oxygen supply cannot be assumed, but a malutilization of oxygen may be involved. The present status of the knowledge of alcoholism does not permit of more than vague conjectures which, nevertheless, may constitute bases for experimental hypotheses.

The onset of alcoholic palimpsests is followed (in some instances, preceded) by the onset of drinking behaviors indicating that for this drinker beer, wine, and spirits have practically ceased to be beverages and have become sources of a drug he "needs." Some of these behaviors imply that this drinker has some vague realization that he drinks differently from others.

Surreptitious drinking (2) is one of these behaviors. At social gatherings the drinker seeks occasions for having a few drinks unknown to others, because he fears that if it were known that he drinks more than the others he would be misjudged. Those to whom drinking is only a custom or a small pleasure would not understand that because he is different from them alcohol is for him a necessity, although he is not a drunkard.

Preoccupation with alcohol (3) is further evidence of this need. When he prepares to go to a social gathering his first thought is whether there will be sufficient alcohol for his requirements, and he has several drinks in anticipation of a possible shortage. Because of this increasing dependence upon alcohol, the onset of *avid drinking* (4) (gulping of the first drink or first two drinks) occurs at this time.

As the drinker realizes, at least vaguely, that his drinking is outside the ordinary, he develops *guilt feelings about his drinking behavior* (5), and thus in turn he begins to *avoid reference to alcohol* (6) in conversation.

These behaviors, together with an *increasing frequency of alcoholic palimpsests* (7), foreshadow the development of alcohol addiction; they are premonitory signs. This period may be called the prodromal phase of alcohol addiction.

The consumption of alcoholic beverages in the prodromal phase is heavy, but not conspicuous, in that it does not lead to marked, overt intoxications. The effect is that the prospective addict reaches towards evening a state that may be designated as emotional anesthesia. Nevertheless, this condition requires drinking well beyond the ordinary usage. The drinking is on a level that may begin to interfere with metabolic and nervous processes, as evidenced by the frequent alcoholic palimpsests.

The "covering-up" which is shown by the drinker in this stage is the first sign that his drinking might separate him from society, although initially the drinking may have served as a technique to overcome some lack of social integration.

Because in the prodromal phase rationalizations of the drinking behavior are not strong and there is some insight into, as well as fear of, possible consequences, it is still feasible to intercept incipient alcohol addiction at this stage. In the United States of America, the publicity given to the prodromal symptoms begins to bring prospective alcoholics to clinics as well as to groups of Alcoholics Anonymous. It goes without saying that even at this stage the only possible modus for this type of drinker is total abstinence.

The prodromal period may last anywhere from 6 months to 4 or 5 years according to the physical and psychological makeup of the drinker, his family ties, vocational relations, general interests, and so forth. Eventually the prodromal phase ends, and the crucial or acute phase begins with the onset of loss of control, which is the critical symptom of alcohol addiction.

3. The Crucial Phase

Loss of control (8) means that any drinking of alcohol starts a chain reaction felt by the drinker as a physical demand for alcohol. This state, possibly a conversion phenomenon, may take hours or weeks for its full development; it lasts until the drinker is too intoxicated or too sick to ingest more alcohol. The physical discomfort following this drinking behavior is contrary to the object of the drinker, which is merely to feel "different." As a matter of fact, the bout may not even be started by any individual need of the moment, but by a "social drink."

After recovery from the intoxication, it is not the loss of control—that is, the physical demand, apparent or real—which leads to a new bout after several days or several weeks. The renewal of drinking is set off by the original psychological conflicts or by a simple social situation that involves drinking.

The loss of control is effective after the individual has started drinking, but it does not give rise to the beginning of a new drinking bout. The drinker has lost the ability to control the quantity once he has started, but he still can control whether he will drink on any given occasion or not. This capability is evidenced in the fact that after the onset of loss of control the drinker can go through a period of voluntary abstinence ("going on the water wagon").

The question why the drinker returns to drinking after repeated disastrous experiences is often raised. Although he will not admit it, the alcohol addict believes that he has lost his will power and that he can and must regain it. He is not aware that he has undergone a process that makes it impossible for him to control his alcohol intake. To "master his will" becomes a matter of the greatest importance to him. When tensions rise, a drink is the natural remedy for him and he is convinced that this time it will be one or two drinks only.

Almost simultaneously with the onset of loss of control the alcohol addict begins to *rationalize his drinking behavior* (9): he produces the well-known alcoholic *alibis*. He finds explanations that convince him that he did not lose control, that on the contrary he had a good reason to get intoxicated, and that in the absence of such reasons he is able to handle alcohol as well as anybody else. He needs these rationalizations primarily for himself and only secondarily for his family and associates. The rationalizations make it possible for him to continue with his drinking, and this possibility is of the greatest importance to him in that he knows no alternative for handling his problems.

This is the beginning of an entire "system of rationalizations" that progressively spreads to every aspect of his life. Although this system largely originates in inner needs, it also serves to counter *social pressures* (10) which arise at the time of the loss of control. At this time, of course, the drinking behavior becomes conspicuous, and the parents, wife, friends, and employer may begin to reprove and warn the drinker.

In spite of all the rationalizations there is a marked loss of self-esteem, and this loss of course demands compensations that in a certain sense are also rationalizations. One way of compensation is the *grandiose behavior* (11) the addict begins to display at this time. Extravagant expenditures and grandiloquence convince him that he is not as bad as he had thought at times.

The rationalization system gives rise to another system, namely, the "system of isolation." The rationalizations quite naturally lead to the idea that the fault lies not within himself but in others, and this delusion results in a progressive withdrawal from the social environment. The first sign of this attitude is a *marked aggressive behavior* (12).

Inevitably, this latter behavior generates guilt. Although even in the prodromal period remorse about the drinking arose from time to time, now *persistent remorse* (13) arises, and this added tension is a further cause of drinking.

In compliance with social pressures the addict now goes on *periods of total abstinence* (14). There is, however, another modus of control of drinking that arises out of the rationalizations of the addict. He believes that his trouble comes from his not drinking the right kind of beverages or not drinking in the right way. He now attempts to control his troubles by *changing the pattern of his drinking* (15) by setting up rules about not drinking before a certain hour of the day, drinking in certain places only, and so forth.

The strain of the struggle increases his hostility towards his environment and he begins to *drop friends* (16) and *quit jobs* (17). It goes without saying that some associates drop him and that he loses some jobs, but more frequently he takes the initiative as an anticipatory defense.

The isolation becomes more pronounced as his entire *behavior becomes alcohol centered* (18); that is, he begins to be concerned about how activities might interfere with his drinking instead of how his drinking is affecting his activities. This development, of course, involves a more marked egocentric outlook, which in turn leads to more rationalizations and more isolation. There ensues a *loss of outside interests* (19) and a *reinterpretation of interpersonal relations* (20), coupled with *marked self-pity* (21). The isolation and rationalizations have increased by this time in intensity and find their expression either in contemplated or actual *geographic escape* (22).

Under the impact of these events, a *change in family habits* (23) occurs. The wife and children, who may have had good social activities, may withdraw from them for fear of embarrassment; or, quite contrarily, they may suddenly begin intensive outside activities in order to escape from the home environment. This and other events lead to the onset of *unreasonable resentments* (24) in the alcohol addict.

The predominance of concern with alcohol induces the addict to *protect his supply* (25)—that is, to lay in a large stock of alcoholic beverages, hidden in the most unthought-of places. A fear of being deprived of the most necessary substance for his living is expressed in this behavior.

Neglect of proper nutrition (26) aggravates the beginnings of the effects of heavy drinking on the organism, and frequently the *first hospitalization* (27) for some alcoholic complaint occurs at this time.

One of the frequent organic effects is a *decrease of the sexual drive* (28), which increases hostility towards the wife and is rationalized into what he imagines to be her extramarital sexual activities, a process that in turn gives rise to the well-known *alcoholic jealousy* (29).

By this time remorse, resentment, struggle between alcoholic needs and duties, loss of self-esteem, and doubts and false reassurance have so disorganized the addict that he cannot start the day without steadying himself with alcohol immediately after arising or even before getting out of bed. This is the beginning of *regular matutinal drinking* (30), which previously had occurred on rare occasions only. This behavior terminates the crucial phase and foreshadows the beginning of the chronic phase.

During the crucial phase intoxication is the rule, but it is limited to the evening hours. For most of this phase drinking begins sometime in the afternoon and leads by the evening to intoxication. It should be noted that the physical demand involved in the loss of control results in continual rather than continuous drinking. Particularly the resorting to the "matutinal drink," which occurs toward the end of the crucial phase, shows the continual pattern. The first drink at rising, let us say at 7 a.m., is followed by another drink at 10 a.m. or 11 a.m., and another drink around 1 p.m., whereas the more intensive drinking hardly starts before 5 p.m.

Throughout, the crucial phase presents a great struggle of the addict against the complete loss of social footing. Occasionally the aftereffects of the evening's intoxication cause some loss of time, but generally the addict succeeds in looking after his job, although he neglects his family. He makes a particularly strong effort to avoid intoxication during the day. Progressively, however, his social motivations weaken, and the morning drink jeopardizes his effort to comply with his vocational duties, in that this effort involves a conscious resistance against the apparent or real physical demand for alcohol.

The onset of the loss of control is the beginning of the "disease process" of alcohol addiction, which is superimposed over the excessive symptomatic drinking. This disease process progressively undermines the morale and the physical resistance of the addict.

4. The Chronic Phase

The increasingly dominating role of alcohol, and the struggle against the demand set up by matutinal drinking, at last break down the resistance of the addict, and he finds himself for the first time intoxicated in the daytime and on a weekday and continues in that state for several days until he is entirely incapacitated. This is the onset of *prolonged intoxications* (31), referred to in the vernacular as "benders."

This latter drinking behavior meets with such unanimous social rejection that it involves a grave social risk. Only an originally psychopathic personality or a person who has later in life undergone a psychopathological process would expose himself to that risk.

These long-drawn-out bouts commonly bring about *marked ethical deterioration* (32) and *impairment of thinking* (33), which, however, are not irreversible. True *alcoholic psychoses* (34) may occur at this time, but in not more than 10% of all alcoholics.

The loss of morale is so heightened that the addict *drinks with persons far below his social level* (35) in preference to his usual associates—perhaps as an opportunity to appear superior—and, if nothing else is available, he will *take recourse to "technical products"* (36) such as bay rum or rubbing alcohol.

A *loss of alcohol tolerance* (37) is commonly noted at this time. Half of the previously required amount of alcohol may be sufficient to bring about a stuporous state.

Indefinable fears (38) and *tremors* (39) become persistent. These symptoms also occur sporadically during the crucial phase, but in the chronic phase they are present as soon as alcohol disappears from the organism. In consequence, the addict "controls" the symptoms through alcohol. The same is true of *psychomotor inhibition* (40), the inability to initiate a simple mechanical act—such as winding a watch—in the absence of alcohol.

The need to control these symptoms of drinking exceeds the need of relieving the original underlying symptoms of the personality conflict, and the *drinking takes on an obsessive character* (41).

In many addicts, approximately 60%, some *vague religious desires develop* (42) as the rationalizations become weaker. Finally, in the course of the frequently prolonged intoxications, the rationalizations become so frequently and so mercilessly tested against reality that the entire *rationalization system fails* (43) and the addict admits defeat. He now becomes spontaneously accessible to treatment. Nevertheless, his obsessive drinking continues, because he does not see a way out.

Formerly it was thought that the addict must reach this stage of utter defeat in order to be treated successfully. Clinical experience has shown, however, that this "defeat" can be induced long before it would occur of itself and that even incipient alcoholism can be intercepted. Since the incipient alcoholic can be easily recognized, it is possible to tackle the problem from the preventive angle.

5. The "Alcoholic Personality"

The aggressions, feelings of guilt, remorse, resentments, withdrawal, and so on, that develop in the phases of alcohol addiction are largely consequences of the excessive drinking. At the same time, however, they constitute sources of more excessive drinking.

In addition to relieving, through alcohol, symptoms of an underlying personality conflict, the addict now tends to relieve, through further drinking, the stresses created by his drinking behavior.

By and large, these reactions to excessive drinking—which have quite a neurotic appearance—give the impression of an "alcoholic personality," although they are secondary behaviors superimposed over a large variety of personality types that have a few traits in common, in particular a low capacity for coping with tensions. There does not emerge, however, any specific personality trait or physical characteristic which inevitably would lead to excessive symptomatic drinking. Apart from psychological and possibly physical liabilities, there must be a constellation of social and economic factors that facilitate the development of addictive and nonaddictive alcoholism in a susceptible terrain.

The Nonaddictive Alcoholic

Some differences between the nonaddictive alcoholic and the alcohol addict have previously been stated in this chapter. These differences will be recapitulated and elaborated, and additional differential features will be considered.

One may readily visualize the main difference by erasing the large bars of the diagram in Figure 1. This action results in a diagram that suggests a progressive exacerbation of the use of alcohol for symptom relief and the social and health consequences resulting from such use, but a progression without any clear-cut phases.

The prealcoholic phase is the same for the nonaddictive alcoholic as for the alcohol addict. That is to say, he progresses from occasional to constant relief of individual symptoms through alcohol.

The behaviors denoting that alcohol has become a drug rather than an ingredient of a beverage—symptoms (2) to (6)—occur also in the nonaddictive drinker, but, as mentioned before, the alcoholic palimpsests occur rarely, and only after overt intoxication.

Loss of control is not experienced by the nonaddictive alcoholic, and this is the main differentiating criterion between the two categories of alcoholics. Initially, of course, it could not be said whether the drinker had yet reached the crucial phase. However, after 10 or 12 years of heavy drinking without loss of control, though symptoms (2) to (6) were persistent and palimpsests were rare and did not occur after a medium intake of alcohol, the differential diagnosis is rather safe.

The absence of loss of control has many involvements. First of all, because there is no inability to stop drinking within a given situation, there is no need to rationalize the inability. Nevertheless, rationalizations are developed for justifying the excessive use of alcohol and some neglect of the family attendant upon such use. Likewise, there is no need to change the pattern of drinking, which in the addict is an attempt to overcome the loss of control. Periods of total abstinence, however, occur as responses to social pressure.

On the other hand, there is the same tendency toward isolation as in the addict, but the social repercussions are much less marked, because the nonaddictive alcoholic can avoid drunken behavior whenever the social situation requires it.

The effects of prolonged heavy drinking on the organism may occur in the nonaddictive alcoholic, too; even delirium tremens may develop. The libido may be diminished, and "alcoholic jealousy" may result.

Generally, there is a tendency toward a progressive dominance of alcohol, resulting in greater psychological and bodily effects. In the absence of any grave underlying psychopathies, a deteriorative process is speeded up by habitual alcoholic excess, and such a nonaddictive drinker may slide to the bottom of society.

Notes

1. This chapter presents a summary and refined view of the conception of phases in the drinking history of alcoholics originally set forth in Jellinek (1946). It incorporates those modifications suggested by extensive application and testing of the questionnaire upon which the earlier study was based. The present version was first published under the auspices of the Alcoholism Subcommittee of the World Health Organization (Second Report, Annex 2, World Health Organization Technical Report Series, No. 48, August 1952) and was subsequently reproduced in 1952 in *Quarterly Journal of Studies on Alcohol, 13,* 673–684. It is reprinted here with only slight modification.
2. This group does not include the regular "periodic alcoholics."
3. The numerals in parentheses following the designations of the individual symptoms represent their order as given in Figure 1.

References

Jellinek, E.M. (1946). Phases in the drinking history of alcoholics: Analysis of a survey conducted by the official organ at Alcoholics Anonymous. *Quarterly Journal of Studies on Alcohol, 7,* 1–88.

CHAPTER 21

Alcoholism: The Mythical Disease

HERBERT FINGARETTE

The idea that alcoholism is a disease is a myth, and a harmful myth at that. The phrase itself—"alcoholism is a disease"—is a slogan. It lacks definite medical meaning and therefore precludes one from taking any scientific attitude toward it, pro or con. But the slogan has political potency. And it is associated in the public consciousness with a number of beliefs about heavy drinking that do have meaning and do have important consequences for the treatment of individuals and for social policy. These beliefs lack scientific foundation; most have been decisively refuted by the scientific evidence (Fingarette, 1988). [See also chapter 22, by Hore, in this volume.]

This assertion obviously conflicts with the barrage of pronouncements in support of alcoholism's classification as a disease by health professionals and organizations such as the American Medical Association, by the explosively proliferating treatment programs, and by innumerable public service organizations. So it may seem that a sweeping challenge to the disease concept can be only hyperbole, the sensationalist exaggeration of a few partial truths and a few minor doubts.

To the contrary: The public has been profoundly misled and is still being actively misled. Credulous media articles have featured so many dramatic human-interest anecdotes by "recovering alcoholics," so many "scientific" pronouncements about medical opinion and new discoveries, that it is no wonder the lay public responds with trusting belief.

Yet this much is unambiguous and incontrovertible: The public has been kept unaware of a mass of scientific evidence accumulated over the

Herbert Fingarette is a professor emeritus, Department of Philosophy, University of California, Santa Barbara. This chapter is reprinted (with changes by the author) with permission from *The Public Interest* (Spring 1988), pp. 3–22. Supporting evidence and additional references for this chapter are found in his book *Heavy Drinking: The Myth of Alcoholism as a Disease* (Berkeley: University of California Press, 1988). Individuals wishing further information on the viewpoint presented here are strongly advised to consult the book.

past couple of decades, evidence familiar to researchers in the field, which radically challenges each major belief generally associated in the public mind with the phrase "alcoholism is a disease." I refer not to isolated experiments or offbeat theories but to massive, accumulated, mainstream scientific work by leading authorities, published in recognized journals. If the barrage of "public service" announcements leaves the public wholly unaware of this contrary evidence, shouldn't this state of affairs in itself raise grave questions about the credibility of those who assure the public that alcoholism has now been scientifically demonstrated to be a disease?

One may wonder why it is important whether or not alcoholism is a disease. To begin with, *disease* is the word that triggers provision of health insurance payments, employment benefits such as paid leave and workmen's compensation, and other government benefits. The direct cost of treatment for the "disease" of alcoholism is rapidly rising, already exceeding a billion dollars annually (Holden, 1987). Add in all related health costs and other kinds of benefits, and the dollar figure is well into the tens of billions annually (Miller & Hester, 1986). Alcoholism is, of course, profoundly harmful, both to the drinkers themselves and to others. But if it ceased to be characterized as a disease, all the disease-oriented methods of treatment and resulting expenditures would be threatened; this upheaval would in turn threaten the material interests of hundreds of thousands of alcoholics and treatment staffers who receive these billions in funds. The other side of the coin would be many billions in savings for taxpayers and those who pay insurance premiums.

It is not surprising that the disease concept of alcoholism is now vigorously promoted by a vast network of lobbies, national and local, professional and volunteer, ranging from the most prestigious medical associations to the most crassly commercial private, profit-making providers of treatment. This promotion is big politics and big business.

Use of the word *disease* also shapes the values and attitudes of society. The selling of the disease concept of alcoholism has led courts, legislatures, and the populace generally to view damage caused by heavy drinkers as a product of "the disease and not the drinker." The public remains ambivalent about this, and criminal law continues to resist excusing alcoholics for criminal acts. But the pressure is there, and of more practical importance, civil law has largely given in. Civil law now often mandates leniency or complete absolution for the alcoholic from the rules, regulations, and moral norms to which nondiseased persons are held legally or morally accountable. Such was the thrust of a recent appeal to the U.S. Supreme Court by two veterans, who claimed certain benefits in spite of their having failed to apply for them at any time during the legally specified 10-year period after discharge from the army. Their excuse: alcoholism, and the claim that their persistent heavy drinking was a disease entitling them to exemption from the regulations. The Court's decision was rendered in *Traynor v. Turnage* (1988).

Traynor v. Turnage (1988) has been the only major U.S. Supreme Court case dealing with the use of "the disease of alcoholism" as an excuse since the landmark case of *Powell v. Texas* (1968). In the latter case, the Court refused to excuse Powell from criminal responsibility. Fundamental to the Court's reasoning was that it found inadequate scientific evidence to support either the claim of disease or the claim that the alcoholic's drinking is involuntary. Again, in the more recent Traynor case, the Court found that "the authorities remain sharply divided" (p. 634) about whether alcoholism is a disease; and since this case did not require a decision on that point, the Court declined to take any position. However on the issue of whether the alcoholic's drinking is "willful" or is wholly involuntary, the Court accepted as reasonable the Veterans Administration's flat rule that the drinking is "willful." The Court said that such a rule is supported by "a substantial body of medical literature" (p. 633). The Court cited as its principal scholarly authority for this conclusion my detailed study of the legal and medical issues in the *Harvard Law Review* (Fingarette, 1970), a study whose general thrust is faithfully echoed in the more recent analysis presented in the text above.

What seems compassion when done in the name of "disease" turns out, when the facts are confronted, not only to subvert the drinker's autonomy and will to change but also to exacerbate a serious social problem. The reason is that the excuses and benefits offered heavy drinkers work psychologically as incentives for them to continue drinking. The doctrine that the alcoholic is "helpless" delivers the message that he might as well drink, because he lacks the ability to refrain. As for the expensive treatments, they do no real good. Certainly our current disease-oriented policies have not reduced the scale of the problem; in fact, the number of chronic heavy drinkers reported keeps rising. (It is currently somewhere in the range of 10 to 20 million, depending on the definitions one uses [Clark & Cahalan, 1976; Room, 1974, 1977; Schuckit, 1985].)

In the remainder of this discussion I will set out the major beliefs associated with the disease concept of alcoholism and then summarize the actual evidence on each issue. I will also sketch an alternative perspective on chronic heavy drinking that is warranted by the evidence we have today.

Conventional Wisdom

Science, according to the conventional view, has established that there is a specific disease that is triggered by the consumption of alcoholic beverages. Not everyone is susceptible; most people are not. But (the argument continues) a significant minority of the population has a distinctive biological vulnerability, an "allergy" to alcohol. For these people, to start drinking is to start down a fatal road. The stages are well defined and develop in regular order, as with any disease, with the symptoms accumulating and becoming increasingly disabling and demoralizing. First comes what looks like normal social drinking, but then, insidiously and inevitably, come heavier and more frequent drinking, drunken bouts, secret drinking, morning drinking, and, after a while, "blackouts" of memory from the night before. It begins to take more and more liquor to get the same effect—physical "tolerance" develops—and any attempt to stop drinking brings on the unbearable and potentially life-threatening "withdrawal" symptoms. Eventually, the crucial symptom develops: "loss of control." At that point, whenever the person takes a drink, the alcohol automatically triggers an inability to control the drinking, and drunken bouts become the rule. There follows an inevitable, deepening slavery to alcohol, which wrecks social life, brings ruin, and culminates in death. The only escape—according to this elaborate myth—is appropriate medical treatment for the disease.

The myth offers the false hope that as a result of recent "breakthroughs" in science we now basically understand what causes the disease—a genetic and neurophysiological defect. But fortunately, it is claimed, medical treatment is available, and generally produces excellent results. However, the argument continues, even after successful treatment the alcoholic can never drink again. The so-called allergy is never cured; the disease is in remission, but the danger remains. The lifelong truth for the alcoholic is, as the saying goes, "one drink—one drunk." The possibility of a normal life depends on complete abstinence from alcohol. Thus, there are no "cured" alcoholics, only "recovering" ones (Mann, 1950).

That is the classical disease concept of alcoholism. As I have said, just about every statement in it is either known to be false or (at a minimum) lacks scientific foundation.

Origins of the Myth

Before turning to the substance of the specific claims, one ought to be aware of the historical context. For it is important to recognize that the disease concept of alcoholism not only has no basis in current science but has never had any scientific justification.

The understanding of alcoholism as a disease first surfaced in the early 19th century. The growing popularity of materialistic and mechanistic views bolstered the doctrine that drinking problems stemmed from a simple malfunctioning of the bodily machinery. The new idea was popularized by Benjamin Rush (Levine, 1978; Rorabaugh, 1979), one of the leading medical theorists of the day.

Rush's claim was ideological, not scientific; neither Rush nor anyone else at that time had the experimental facilities or the biological knowledge to justify it. It seemed plausible because of its compatibility with the crude biological theories of the time, assumptions that we now know to be erroneous. Nevertheless, the idea seized the public imagination, in part because it appealed to the growing mercantile and manufacturing classes whose demand for a disciplining "work ethic" (especially among the working class) was supported by this new "scientific" indictment of drinking. One should realize that the 19th-century version of the doctrine, as advanced by the politically powerful temperance movement, indicted not just some but all drinking. Alcohol (like heroin today) was viewed as inherently addictive. The drinker's personal characteristics and situation were considered irrelevant.

The 19th-century temperance movement crested in 1919 with the enactment of the Prohibition Amendment; but by 1933 the idea of total prohibition had lost credibility, and the amendment was repealed. For one thing, the public no longer accepted the idea that no one at all could drink alcohol safely. In addition, the costs of prohibition—such as gangsterism and public cynicism about the law—had become too high. Most people wanted to do openly and legally, in a civilized way, what large numbers of people had been doing surreptitiously (Gusfield, 1963).

For the temperance impulse to survive, it had to be updated in a way that did not stigmatize all drinking on moral or medical grounds. Any new antialcohol movement had to be more selective in its target, by taking into account the desires of drinkers generally, as well as the interests of the now legal (and growing) alcoholic beverage industry.

A new sect arose with just the right formula. Alcoholics Anonymous (AA), founded in 1935, taught that alcohol was not the villain in and of itself and that most people could drink safely. (In this way the great majority of drinkers and the beverage industry were mollified.) A minority of potential drinkers, however, were said to have a peculiar biological vulnerability; these unfortunates, it was held, were "allergic" to alcohol, so that their drinking activated the disease, which then proceeded insidiously along the lines outlined earlier.

This version of the disease theory of alcoholism, along with subsequent minor variants of the theory, is often referred to now as the "classic" disease concept of alcoholism. Like the temperance doctrine, the new doctrine was not based on any scientific research or discovery. It was created by the two ex-alcoholics who founded AA: William Wilson, a New York stockbroker, and Robert Holbrook Smith, a physician from Akron, Ohio. Their ideas in turn were inspired by the Oxford religious movement, and by the ideas of another physician, William Silkworth. They attracted a small following, and a few sympathetic magazine articles helped the movement grow.

Alcoholism and Science

What AA still needed was something that would serve as a scientific authority for its tenets. After all, the point of speaking of a "disease" was to suggest science, medicine, and an objective malfunction of the body. The classic disease theory of alcoholism was given just such an apparent scientific confirmation in 1946. A respected scientist, E.M. Jellinek, published a lengthy scientific article consisting of 80+ pages impressively filled with charts and figures. He carefully defined what he called the "phases of alcoholism," which went in a regular pattern, from apparently innocent social drinking ever downward to doom. [See chapter 20, by Jellinek, in this volume.] The portrait, overall and in its detail, largely mirrored the AA portrait of the alcoholic. Jellinek's work and AA proselytizing generated an unfaltering momentum; the disease concept that they promulgated has never been publicly supplanted by the prosaic truth.

Jellinek's portrait of the "phases of alcoholism" was not an independent scientific confirmation of AA doctrine. For as Jellinek explicitly stated, his data derived entirely from a sampling of AA members, a small fraction of whom had answered and mailed back a questionnaire that had appeared in the AA newsletter. The questionnaire was prepared by AA members, not by Jellinek; Jellinek himself criticized it, finding it scientifically inadequate. In addition, many AA members did not even subscribe to the newsletter and so had no opportunity to respond. Jellinek obtained only 158 questionnaires but for various reasons could actually use just 98 of them. This was a grossly inadequate set of data, of course, but it was all Jellinek had to work with.

Predictably, the data from these 98 questionnaires generated a portrait of alcoholism that coincided with the AA portrait. Since Jellinek was a reputable scientist, it is not surprising that he pointed to the various limitations of the data base and the highly tentative nature of his conclusions. It is equally unsurprising that AA propagandists publicized the impressively charted and statistically annotated portrait drawn by Jellinek but glossed over his scholarly reservations about the hypotheses and data.

The "alcoholism movement," as it has come to be called among those familiar with the facts, has grown at an accelerating rate. Its growth results from the cumulative effect of the great number of drinkers indoctrinated by AA, people who passionately identify themselves with the AA portrait of "the alcoholic." AA has vigorously supported the idea of "treatment" for alcoholics; in turn, the rapidly proliferating "treatment" centers for the "disease of alcoholism" have generally supported AA. All this has generated a kind of snowballing effect.

By the 1970s there were powerful lobbying organizations in place at all levels of government. The National Council on Alcoholism (NCA), for example, which has propagated the disease concept of alcoholism, has been a major national umbrella group from the early days of the movement. Until 1982 the NCA was partially subsidized by the liquor industry, which had several representatives on its board. The alliance was a natural one: At the cost of conceding that a small segment of the population is allergic to alcohol and ought not to drink, the liquor industry gained a freer hand with which to appeal to the majority of people, who are ostensibly not allergic.

Health professionals further widened the net, and economic incentives came powerfully into play. Federal and local governments began to open their health budgets to providers of alcoholism treatment and also to alcoholism researchers. Insurance companies are increasingly required to do the same. Today, treatment aimed at getting alcoholics to stop drinking brings in over a billion dollars a year. Alcoholism researchers now rely on what is probably the second-largest funding source after defense—government health funds. And by now there are hundreds of thousands of former heavy drinkers who feel an intense emotional commitment; they supply a large proportion of the staffs of treatment centers.

Large and powerful health-professional organizations (such as the American Medical Association) now have internal constituencies whose professional power and wealth derive from their role as the authorities responsible for dealing with the "disease" of alcoholism. As usual, these interest constituencies lobby internally, and the larger organization is persuaded to take an official stand in favor of the meaningless slogan "Alcoholism is a disease." Thus, there are many health organizations that now endorse this slogan.

Judges, legislators, and bureaucrats all have a stake in the doctrine. They can now with clear consciences get the intractable social problems posed by heavy drinkers off their agenda by compelling or persuading these unmanageable people to go elsewhere—that is, to get "treatment." Why should these public officials mistrust—or want to mistrust—this safe-as-motherhood way of getting troublesome problems off their backs while winning popular approval? The ample evidence that these "treatment" programs are ineffective, and waste considerable amounts of money and resources, is ignored.

The "Phases of Alcoholism"

The "phases-of-alcoholism" portrait of the alcoholic has been examined in detail in a number of major studies dating back to the 1960s (Cahalan & Room, 1974). A recent summary of the scientific literature on this topic indicates that the typical drinking pattern is characterized by much fluctuating between levels of consumption (Rudy, 1986). Thus, many drinkers with numerous and severe problems are found later to have markedly improved, or to have developed different problems. Some also deteriorate. Individual drinkers do not develop in any consistent pattern, nor do they remain stable in a single pattern. Some claim "loss of control"; others do not. Many report no social problems associated with their drinking (and so, not surprisingly, many heavy drinkers are not recognized as such by friends, colleagues, or even family).

One of the leading scientists in the field, Marc Schuckit (1984, p. 59), summarizes the evidence on whether alcoholics drink persistently by pointing out that "in any given month, one half of alcoholics will be abstinent, with a mean of four months of being dry in any one-year to two-year period." In general, as George Vaillant (1983), has reported, the cumulative evidence is that during any reasonably long period (10 to 20 years), roughly a third of alcoholics "mature out" into various forms of moderate drinking or abstinence. The rate of "maturing out" for heavy problem drinkers—including those not diagnosed as alcoholics—is substantially higher.

A number of factors are associated with rates of "natural" improvement (i.e., improvement independent of any formal treatment): Higher socioeconomic class, more education, regular employment, and being married are positively associated with higher improvement rates. Those who "mature out" at rates lower than the average tend to be socially deprived and alienated. "None of this," says one specialist on the topic, "fits with the disease model of alcoholism insofar as that model implies keeping early symptoms and early problems and adding others as time passes" (Clark & Cahalan, 1976, p. 258). Certainly none of this fits with the concept of a disease whose pattern of development is uniform and essentially independent of individual social and cultural characteristics.

Biological Causes?

What does it mean to say that alcoholism is a disease? In public discussions in the news media, it is usually taken to mean that alcoholism has a single biological cause. "I believe [alcoholics] have a genetic predisposition and a certain kind of biochemistry that dooms [them] to be [alcoholics] if [they] use alcohol" (Bellows, 1988, p. 9). This is a characteristic remark, with what in this domain is a familiar kind of specious authority. The statement was printed in an alcoholism bulletin issued under the aegis of a University of California Extension Division alcoholism program. It appears in an interview with Kevin Bellows, a lay activist heading an international organization fighting alcoholism.

Lay activists are not alone in pressing this theme. When I was on a network talk show recently, the physician on the panel—a man high in government alcoholism advisory councils—devoted most of his time to running through a list of recent research discoveries about the biological peculiarities of alcoholics. His thesis was that alcoholism is unquestionably a disease, and he plainly implied that it has a biological cause. What the lay audience does not realize is that the newly discovered biological phenomena can rarely be regarded as *causes* of chronic heavy drinking; instead, they are merely *associated* with chronic heavy drinking or with intoxication. Nevertheless, the audience is led to infer that these phenomena play a causal role; in fact, we know that there simply are no decisive physical causes of alcoholism.

Long-term heavy drinking is undoubtedly an important contributing cause of bodily ailments—including major organ, nerve, circulatory, and tissue disorders. The illness and mortality rates of heavy drinkers are far higher than those of the population generally. Chronic heavy drinking is rivaled only by habitual smoking as a major contributor to the nation's hospital and morgue populations. But all this affliction is, it must be stressed, the effect of drinking; the drinking behavior itself is the cause. Stop the behavior and you stop its terrible physical effects.

Another abnormal physical condition associated with heavy drinking is the appearance of biological "markers." These metabolic and other physiological conditions—statistically abnormal but not necessarily ailments in and of themselves—may often be present among alcoholics. More significantly, some of them are present in persons who are not and have not been alcoholics but who have been identified on independent grounds as being at higher-than-average risk of eventually becoming alcoholics. Such "markers" can serve as warning signs for those at higher risk. It has been hypothesized that some of these biological "markers" may play a causal role in bringing about alcoholic patterns of drinking. The question is, What kind and what degree of causality are at issue?

One much discussed metabolic "marker" is the difference in the way those who are independently identified as being at higher risk oxidize alcohol into acetaldehyde and in turn metabolize the acetaldehyde. The toxic effects of acetaldehyde in the brain have led to speculation that it might play a key causal role in inducing alcoholism (Schuckit & Viamontes, 1979). Analogous claims have been made about the higher level of morphinelike substances that alcoholics secrete when they metabolize alcohol (Schuckit, 1984). As it happens (so often in these matters), there are serious difficulties in measuring acetaldehyde accurately, and the reported results remain inadequately confirmed. But these confirmation problems are problems of technique and not of fundamental importance.

The substantive point, generally obscured by the excitement of the new discovery, is that even if the existence of any such metabolic processes were confirmed they still would not cause alcoholic behavior, because the metabolism of alcohol takes place only when there is alcohol in the body. Therefore, these metabolic products cannot be present in alcoholics who have not been drinking for a period of time and in whom the total metabolic process in question is not presently taking place. Yet by definition these individuals return to drinking and do so recurrently, in spite of the intermittent periods of sobriety. The metabolic phenomena bear only on drinking that is done while the individuals are in a state of intoxication; the key question about alcoholism, however, is why a sober person, with no significant toxic product remaining in the body, should resume drinking when it is known to have such harmful effects.

The story of biological discoveries concerning alcoholism is always the same: Many unconfirmed results are unearthed, but no causal link to repetitive drinking is ever established. There is one exception, however: the recent discoveries in genetics. A study of these, and of how they have been reported to the public, is revealing.

Alcoholism and Genes

Several excellently designed genetic studies of alcoholism have recently come up with credible positive results (Cloninger, Bohman, & Sigvardsson, 1981; Goodwin, Schulsinger, Hermansen, Guze, & Winokur, 1973); thus, we have been hearing from activists, staff members at treatment centers, and physicians that "alcoholism is a genetic disease." The reality—as revealed by the data—is very different from what this slogan suggests. [See chapter 30, by Goodwin, in this volume.]

The course followed in these recent "decisive" studies have been simple: find children who were born of an alcoholic mother or father, who were put up for adoption very shortly after birth, and who thus spent little time with their biological parents. Then see whether this group of children shows a higher rate of alcoholism in later life than a comparable group of adoptees whose biological parents were not alcoholics. Controlling all other relevant conditions so that they are the same for both groups, one can infer that any eventual differences in the group rates of alcoholism are attributable to their heredity, the one respect in which they differ. In all these studies, the prevalence of alcoholism was significantly greater among the biological sons of alcoholics, especially the sons of alcoholic fathers. Doesn't this finding suggest that alcoholism is hereditary?

To answer the question, let us consider the first of these reports, the 1973 article by Donald Goodwin and his colleagues. They conclude that about 18% of the biological sons of an alcoholic parent themselves become alcoholics, whereas only 5% of the biological sons of nonalcoholic parents become alcoholics—a statistically significant ratio of almost four to one, which in all probability is ascribable to heredity. This apparent hereditary predisposition to alcoholism is what we typically hear about in the media, with or without the precise numbers.

Now let us look at the same data from a different angle, and in a more meaningful context. As simple arithmetic tells us, if 18% of the sons of alcoholics do become alcoholics, then 82%—more than four out of five—do not. Thus, to generalize from the Goodwin data, we can say that the odds are very high—better than four to one—that the son born of an alcoholic parent will not become an alcoholic. In other words, it is utterly false, and perniciously misleading, to tell people with a parental background of alcoholism that their heredity "dooms" them to become alcoholics, or even that their heredity makes it probable that they will become alcoholics. Quite the contrary. Their alcoholic heredity does make it more probable that they'll become alcoholics than if they had nonalcoholic parents, but the probability is still low. The point is that life circumstances are far more important than genes in determining how many people in any group will become heavy drinkers.

There is yet another important implication: Since 5% of the sons of nonalcoholic parents become alcoholics, and since there are far more nonalcoholic parents than alcoholic ones, that 5% ends up representing a far larger total number of alcoholic sons. This implication is consistent with what we know anyway: the great majority of alcoholics do not have alcoholic parents.

The most recent (and influential) adoptee genetic study, reported by Cloninger and his colleagues (1981, p. 867), concludes with these words: "The demonstration of the critical importance of sociocultural influences in most alcoholics suggests that major changes in social attitudes about drinking styles can change dramatically the prevalence of alcohol abuse regardless of genetic predisposition."

Given the possibly dramatic effect of social attitudes and beliefs, the media emphasis on genes as the cause of alcoholism has a pernicious, though unremarked upon, effect. As we have noted, only a minority of alcoholics have an alcoholic parent. Emphasis on heredity as the "cause" of alcoholism may give a false sense of assurance to the far greater number of people who are in fact in danger of becoming alcoholics but do not have an alcoholic parent. These potential alcoholics may feel free to drink heavily, believing themselves genetically immune to the "disease."

The Special Committee of the Royal College of Psychiatry (Edwards et al., 1979, p. 108) put the matter in perspective by saying the following in its book-length statement on alcoholism: "It is common to find that some genetic contribution can be established for many aspects of human attributes on disorders (ranging from musical ability to duodenal ulcers), and drinking is unlikely to be the exception."

Causes of Alcoholism

There is a consensus among scientists that no single cause of alcoholism, biological or otherwise, has ever been scientifically established. There are many causal factors, and they vary from drinking pattern to drinking pattern, from drinker to drinker. We already know many of the predominant influences that evoke or shape patterns of drinking. We know that family environment plays a role, as age does. Ethnic and cultural values are also important: the Irish, Scandinavians, and Russians tend to be heavy drinkers; Jews do not. [See chapter 15, by Glassner, in this volume.] The French traditionally drank modest amounts at one sitting but drank more regularly over the course of the day. Cultural norms have changed in France in recent decades, and so have drinking styles.

We have interesting anthropological reports about the introduction of European styles of drinking into non-European tribal societies. Among the Chichicastenango Indians of Guatemala, for example, there are two different ways of drinking heavily (Marshall, 1981). When drinking ceremonially, in the traditional way, men retain their dignity and fulfill their ceremonial duties even if they have drunk so much that they cannot walk unassisted. When they drink in the bars and taverns where secular and European values and culture hold sway, the men dance, weep, quarrel, and act promiscuously.

The immediate social setting and its cultural meaning are obviously important in our own society. The amount and style of drinking typically vary according to whether the drinker is in a bar, at a formal dinner party, a postgame party, or an employee get-together. It is known that situations of frustration or tension, as well as the desire for excitement, pleasure, or release from feelings of fatigue or social inhibitions, often lead people to drink. Much depends on what the individual has "learned" from the culture about the supposed effects of alcohol, and whether the person desires those particular effects at a particular moment. [See chapter 9, by White, Bates, & Johnson, in this volume.]

But do any of these many factors apply to alcoholics? The belief in a unique disease of alcoholism leads many to wonder whether the sorts of influences mentioned above can make much of a difference when it comes to the supposedly overwhelming craving of alcoholics. Once one realizes that there is no distinct group of "diseased" drinkers, however, one is less surprised to learn that no group of drinkers is immune to such influences or is vulnerable only to other influences.

Do Alcoholics Lack Control?

In fact, alcoholics do have substantial control over their drinking, and they do respond to circumstances. Contrary to what the public has been led to believe, this idea is not disputed by experts. Many studies have described conditions under which diagnosed alcoholics will drink moderately or excessively, or will choose not to drink at all (Fingarette, 1988). Far from being driven by an overwhelming "craving," they turn out to be responsive to common incentives and disincentives, to appeals and arguments, to rules and regulations. Alcohol does not automatically trigger uncontrolled drinking. Resisting our usual appeals and ignoring reasons we consider forceful are results not of alcohol's chemical effect but of the fact that the heavy drinker has different values, fears, and strategies. Thus, in their usual settings alcoholics behave without concern for what others regard as rational considerations.

But when alcoholics in treatment in a hospital setting, for example, are told that they are not to drink, they typically follow the rule. In some studies they have been informed that alcoholic beverages are available, but that they should abstain. Having decided to cooperate, they voluntarily refrain from drinking. More significantly, it has been reported that the occasional few who cheated nevertheless did not drink to excess but voluntarily limited themselves to a drink or two in order to keep their rule violation from being detected (Parades, Hood, Seymour, & Gollob, 1973). In short, when what they value is at stake, alcoholics control their drinking accordingly.

Alcoholics have been tested in situations in which they can perform light but boring work to "earn" liquor; their preference is to avoid the boring activity and forgo the additional drinking. When promised money if they drink only moderately, they drink moderately enough to earn the money. When threatened with denial of social privileges if they drink more than a certain amount, they drink moderately, as directed. The list of such experiments is extensive (Pattison, Sobell, & Sobell, 1977). One can easily confirm the conclusions by carefully observing one's own heavy-drinking acquaintances, provided one ignores the stereotypes of "the alcoholic."

Some people object that these experiments take place in "protected" settings and are therefore invalid. This objection gets things backwards. The point is that it is precisely settings, circumstances, and motivations that are the crucial influences on how alcoholics choose to drink. The alcohol per se—either its availability or its actual presence in the person's system—is not decisive.

Indeed, the alcohol per se or its ready availability seems to be irrelevant to how the alcoholic drinks. Among the most persuasive experiments demonstrating the irrelevance of alcohol to the alcoholic's drinking are several studies in which alcoholic subjects were deceived about whether they were drinking an alcoholic or nonalcoholic beverage. Alan Marlatt and his colleagues (Marlatt, Deming, & Reid, 1973), for example, asked a group of alcoholics to help them "taste-rate" three different brands of the same beverage. Each individual subject was installed in a private room with three large pitchers of beverage, each pitcher supposedly containing a different brand of the same beverage. Their task, of course, was phony. Without their knowledge, the subjects had been assigned to one of four groups. One group was told that the beverage in the three pitchers was tonic water—which was true. But a second group was told that the beverage was a tonic-and-vodka mix—though in fact it, too, was pure tonic water. Those in the third group were told that the beverage was tonic and vodka—which in fact it was. Those in the fourth group were told that it was simply tonic water—whereas in fact it, too, was tonic and vodka. The subjects were left alone (actually observed through a one-way window) and allowed to "taste" the drinks at will, which they did. The total amount drunk and the rapidity of sips were secretly recorded.

The results of this study (and several similar ones) are illuminating. First, none of the alcoholic subjects drank all the beverage, even though, according to the disease theory, those who were actually drinking vodka ought to have proceeded to drink uncontrollably. Second, all of those who believed they were drinking vodka—whether they really were or had been deceived—drank more and faster. Conversely, all of those who believed they were drinking pure tonic—though some were actually drinking vodka—drank less, and more slowly. The inference is unambiguous: The actual presence or absence of alcohol in the system made no difference in the drinking pattern; but what the alcoholics believed was in the beverage did make a difference—in fact, all the difference.

These results fit into a more general pattern revealed by similar experiments on other aspects of alcohol-related behavior in both alcoholics and nonalcoholics: Change the beliefs about the presence of alcohol (or the effect it is supposed to have), and the behavior changes. But the alcohol itself plays no measurable role.

Mark Keller (1972), one of the early leaders of the alcoholism movement, has responded to such evidence by redefining (or, as he would say, "reexplaining") the key concept of "loss of control." We are now told that this concept never connoted an automatically induced inability to stop drinking. Like other sophisticated advocates of the disease concept, Keller now means that one "can't be sure." The alcoholic who has resolved to stop drinking may or may not stand by his resolution. We are told that "loss of control" is compatible, though unpredictably, with temporary, long-term, or indefinite remission. Here medical terms such as *remission* provide a facade of scientific expertise, but the substance of what we are told is that "loss of control" is consistent with just about anything. This reasoning precludes prediction, and of course explains nothing. If the concept retains any empirical content at all, it amounts to a platitude: Someone who for years has relied on a certain way of handling life's stresses may resolve to change, but he or she "can't be sure" whether that promise will be fully kept. This is reasonable thinking. But it is not a scientific explanation of an inner process that causes drinking.

Similarly, the idea that "craving" causes the alcoholic to drink uncontrollably has been tacitly modified. It was plausible in its original sense, which is still the popular understanding: an inordinately powerful, "overwhelming," and "irresistible" desire. But the current experimental work regards "mild craving" as a form of "craving." Of course the whole point of "craving" as an explanation of a supposed irresistible compulsion to drink is abandoned here. But the word is retained—and the public is misled.

There have been other adjustments in response to new evidence, designed to retain the "disease" terminology at whatever cost. We now read that "of course alcoholism is an illness that consists of not just one but many diseases, having different forms and causes." We also hear—in pronouncements addressed to more knowledgeable audiences—that alcoholism is a disease with biological, psychological, social, cultural, economic, and even spiritual dimensions, all of them important. This is a startling amplification of the meaning of *disease,* to the point where it can refer to any human problem at all. It is an important step toward expanding the medicalization of human problems—a trend that has been deservedly criticized in recent years. [See chapter 39, by Roman & Blum, in this volume.]

A Useful Lie?

Even if the disease concept lacks a scientific foundation, might not it nevertheless be a useful social white lie, in that it causes alcoholics to enter treatment? This common—and plausible—argument suffers from two fatal flaws.

First, it disregards the effects of this doctrine on the large number of heavy drinkers who do not plan to enter treatment. Many of these heavy drinkers see their own conditions (often correctly) as not fitting the criteria of "alcoholism" under some current diagnostic formula. The inference they draw is that they are therefore not ill and thus have no cause for concern. Their inclination to deny their problems is thus encouraged. This encouragement can be disastrous, because persistent heavy drinking is physically, mentally, and often socially destructive.

Furthermore, since most people diagnosable as alcoholics today do not enter treatment, the disease concept insidiously provides an incentive for them to keep drinking heavily. For those many alcoholics who do not enter treatment and who (by definition) want very much to have a drink, the disease doctrine assures them that they might as well do so, in that an effort to refrain is doomed anyway.

Moreover, a major implication of the disease concept, and a motive for promoting it, is that what is labeled "disease" is held to be excusable because involuntary. Special benefits are provided alcoholics in employment, health, and civil rights law. The motivation behind this approach may be humane and compassionate, but what it does functionally is to reward people who continue to drink heavily. The policy is insidious: The only known way to have the drinker stop drinking is to establish circumstances that provide a motivation to stop drinking, not an excuse to continue. The U.S. Supreme Court has recently faced this issue in two cases (*McKelvey v Turnage* was merged with *Traynor v Turnage*; see the discussion of the outcome of these cases earlier in this chapter). And the criminal courts have thus far resisted excusing alcoholics from criminal responsibility for their misconduct. But it is difficult to hold this line when the American Medical Association insists the misconduct is involuntary.

The second flaw in the social white-lie argument is the mistaken assumption that use of the word *disease* leads alcoholics to seek a medical treatment that works. In fact, medical treatment for alcoholism is ineffective. Medical authority has been abused for the purpose of enlisting public faith in a useless treatment for which Americans have paid more than a billion dollars. To understand why the treatment does no good, we should recall that many different kinds of studies of alcoholics have shown substantial rates of so-called natural improvement. As a 1986 (Brownell, Marlatt, Lichtenstein, & Wilson, 1986, p. 766) report concludes, "The vast majority of [addicted] persons who change do so on their own." This "natural" rate of improvement, which varies according to class, age, socioeconomic status, and certain other psychological and social variables, lends credibility to the claims of success made by programs that "treat" the "disease" of alcoholism.

Many of the clients—and, in the expensive programs, almost all of the clients—are middle-class, middle-aged people, who are intensely motivated to change and whose families and social relations are still intact. Many, often most, are much improved by the time they complete the program. They are, of course, delighted with the change; they paid money and went through an emotional ordeal, and now receive renewed affection and respect from their family, friends, and coworkers. They had been continually told during treatment that they were helpless, and that only treatment could save them. Many of them fervently believe that they could never have been cured without the treatment.

The sound and the fury signify nothing, however; the rates of improvement in these disease-oriented treatment programs do not significantly differ from the natural rates of improvement for comparable but untreated demographic groups. That is to say, these expensive programs (which cost between $5,000 and $20,000) contribute little or nothing to the improvement. Even so, the claims that patients leave their programs improved are true; to the layman such claims are impressive. The reality, however, is less impressive; over half a dozen major studies in the past two decades have concluded that the money, time, and trust expended on these treatments are badly spent.

There is some disagreement about the effectiveness of more modest forms of treatment. Some reports—for example, a major study done for the Congressional Office of Technology Assessment (Saxe, Dougherty, Esty, & Fine, 1983)—conclude that no single method of treatment is superior to any other (a judgment made by all the major studies) (Saxe et al., 1983; Fingarette, 1988). But according to the Saxe study (p. 53), the data appear to show that "treatment seems better than no treatment." That is, some help-oriented intervention of any kind—it doesn't matter which—may contribute modestly to improvement. The now-classic British experiment led by Griffith Edwards showed that an hour or so of firm and sensible advice produced overall results as good as those produced by a full year of the most complete and sophisticated treatment procedures in a first-class alcoholism hospital and clinic. Such conclusions have led a number of authorities (including a World Health Organization committee in 1980) to argue for brief, informal counseling on an outpatient basis as the preferred method in most cases.

Note, however, that what is now recommended is not really medical treatment. Physicians may still control it, and the institutional setting may be "outpatient," but the assistance provided is merely brief, informal, common-sense advice. The medical setting merely adds unnecessary expense.

So much for the optimistic view about "treatment." A British report concludes (Orford & Edwards, 1977, p. 118) that "it seems likely that treatment may often be quite puny in its powers in comparison to the sum of [nontreatment] forces."

The more pessimistic reading of the treatment-outcome data is that these elaborate treatments for alcoholism as a disease have no measurable impact at all. In a review of a number of different long-term studies of treatment programs, George Vaillant (1983, p. 123) states that "there is compelling evidence that the results of our treatment were no better than the natural history of the disease." [See chapter 25, by Vaillant & Milofsky, in this volume.] Reviewing other major treatment programs with long-term follow-ups, he remarks that the best that can be said is that these programs do no harm.

New Approaches

In recent years, early evaluation studies have been reexamined from a nondisease perspective, which has produced interesting results. For example, it appears that the heaviest and longest-term drinkers improve more than would be expected "naturally" when they are removed from their daily routine and relocated, with complete abstinence as their goal. This group is only a small subset of those diagnosable as alcoholics, of course. The important point, though, is that it is helpful to abandon the one-disease, one-treatment approach and to differentiate among the many different patterns of drinking, reasons for drinking, and modes of helping drinkers.

Indeed, when we abandon the single-entity disease approach and view alcoholism pluralistically, many new insights and strategies emerge. For example, much depends on the criteria of success that are used. The disease concept focuses attention on only one criterion—total, permanent abstinence. Only a small percentage of alcoholics ever achieve this abolitionist goal. But a pluralistic view encourages us to value other achievements, and to measure success by other standards. Thus, marked improvement is quite common when one takes as measures of success additional days on the job, fewer days in the hospital, smaller quantities of alcohol drunk, more moderate drinking on any one occasion, and fewer alcohol-related domestic problems or police incidents. The Rand Report authors (Polich, Armor, & Braiker, 1980, pp. 60–63) found that about

42% of heavy drinkers with withdrawal symptoms had reverted to somewhat more moderate drinking with no associated problems at the end of four years. Yet, as nonabstainers, they would count as failures from the disease-concept standpoint.

The newer perspective also suggests a different conception of the road to improvement. Instead of the hopeful search for a medical magic bullet that will cure the disease, the goal here is to change the way drinkers live. One should learn from one's mistakes, rather than view any one mistake as a proof of failure or a sign of doom. Also consistent with the newer pluralistic, nondisease approach is the selection of specific strategies and tactics for helping different sorts of drinkers; methods and goals are tailored to the individual in ways that leave the one-disease, one-treatment approach far behind.

Much controversy remains about pluralistic goals. One of the most fiercely debated issues is whether so-called controlled drinking is a legitimate therapeutic goal. Some contend that controlled drinking by an alcoholic inevitably leads to uncontrolled drinking. Disease-concept lobbies, such as the National Council on Alcoholism, have tried to suppress scientific publications reporting success with controlled drinking and have excoriated them upon publication. Some have argued that publishing such data can "literally kill alcoholics." Authors of scientific studies, such as Mark and Linda Sobell, have been accused of fraud by their opponents (though expert committees have affirmed the scientific integrity of the Sobells' work) (Dickens, Doob, Warwick, & Winegard, 1982; Trachtenberg, 1984). Attacks like these have been common since 1962, when D.L. Davies (1962) merely reviewed the literature and summarized the favorable results already reported in a number of published studies—and was severely criticized for doing so. But since that time hundreds of similar reports have appeared. One recent study (Nordström & Berglund, 1987) concludes that most formerly heavy drinkers who are now socially adjusted become social drinkers rather than abstainers.

In any case, the goal of total abstinence insisted upon by advocates of the disease concept is not a proven successful alternative, since only a small minority achieves it. If doubt remains about whether the controversy over controlled drinking is fueled by nonscientific factors, that doubt can be dispelled when one realizes that opposition to controlled drinking (like support for the disease concept of alcoholism) is largely confined to the United States and to countries dominated by U.S. intellectual influence. Most physicians in the United Kingdom, for example, do not adhere to the disease concept of alcoholism. And the goal of controlled drinking—used selectively but extensively—is widely favored in Canada and the United Kingdom. British physicians have little professional or financial incentive to bring problem drinkers into their consulting rooms or hospitals. U.S. physicians, in contrast, defend an enormous growth in institutional power and fee-for-service income. The selling of the term *disease* has been the key to this vast expansion of medical power and wealth in the United States.

What should our attitude be, then, to the long-term heavy drinker? Alcoholics do not knowingly make the wicked choice to be drunkards. Righteous condemnation and punitive moralism are therefore inappropriate. Compassion, not abuse, should be shown toward any human being launched upon a destructive way of life. But compassion must be realistic: It is not compassionate to encourage drinkers to deny their power to change, to assure them that they are helpless and dependent on others, to excuse them legally and give them special government benefits that foster a refusal to confront the need to change. Alcoholics are not helpless; they can take control of their lives. In the last analysis, alcoholics must truly want to change and actively choose to change. To do so they must make many difficult daily choices. We can help them by offering moral support and good advice and by assisting them in dealing with their genuine physical ailments and social needs. But we must also make it clear that heavy drinkers must take responsibility for their own lives. Alcoholism is not a disease; the assumption of personal responsibility, however, is a sign of health, and needless submission to spurious medical authority is a pathology.

References

Bellows, K. (1988). Speaking woman to woman about alcohol. *Prevention File, 3,* 8–9.

Brownell, K.D., Marlatt, G.A., Lichtenstein, E., & Wilson, G.T. (1986). Understanding and preventing relapse. *American Psychologist, 41,* 765–782.

Cahalan, D., & Room, R. (1974). *Problem drinking among American men.* New Brunswick, NJ: Rutgers Center of Alcohol Studies.

Clark, W.B., & Cahalan, D. (1976). Changes in problem drinking over a four-year span. *Addictive Behaviors, 1,* 251–259.

Cloninger, C.R., Bohman, M., & Sigvardsson, S. (1981). Inheritance of alcohol abuse: Cross-fostering analysis of adopted men. *Archives of General Psychiatry, 38,* 861–868.

Davies, D.L. (1962). Normal drinking in recovered alcohol addicts. *Quarterly Journal of Studies on Alcohol, 23,* 94–104.

Dickens, B.M., Doob, A.N., Warwick, O.H., & Winegard, W.C. (1982). *Report of the Committee of Enquiry into allegations concerning Drs. Linda and Mark Sobell.* Toronto: Addiction Research Foundation, 1982.

Edwards, G., Bewley, T.H., Connell, P.H., Glatt, M.M., Milne, H.B., Murray, R.M., Oppenheim, A.N., & Walton, H.J. (1979). *Alcohol and alcoholism: Report of a special committee of the Royal College of Psychiatrists.* London: Tavistock.

Fingarette, H. (1970). The perils of Powell: In search of factual foundations for the "disease concept of alcoholism." *Harvard Law Review, 83,* 793–812.

Fingarette, H. (1988). *Heavy drinking: The myth of alcoholism as a disease.* Berkeley: University of California Press.
Goodwin, D.W., Schulsinger, F., Hermansen, L., Guze, S.B., & Winokur, G. (1973). Alcohol problems in adoptees raised apart from alcoholic biological parents. *Archives of General Psychiatry, 28,* 238–243.
Gusfield, J.R. (1963). *Symbolic crusade.* Urbana, IL: University of Illinois Press.
Holden, C. (1987). Alcoholism and the medical cost crunch. *Science, 235,* 1132–1133.
Jellinek, E.M. (1946). Phases in the drinking history of alcoholics. *Quarterly Journal of Studies on Alcohol, 7,* 1–88.
Keller, M. On the loss-of-control phenomenon in alcoholism. (1972). *British Journal of Addiction, 67,* 153–166.
Levine, H.G. (1978). The discovery of addiction: Changing conceptions of habitual drunkenness in America. *Journal of Studies on Alcohol, 39,* 143–174.
Mann, M. (1950). *Primer on alcohol: How people drink, how to recognize alcoholics, and what to do about them.* New York: Rinehart.
Marlatt, G.A., Deming, B., & Reid, J.B. (1973). Loss of control drinking in alcoholics: An experimental analogue. *Journal of Abnormal Psychology, 81,* 233–241.
Marshall, M. (1981). Four hundred rabbits: An anthropological view of ethanol as a disinhibitor. In R. Room & G. Collins (Eds.), *Alcohol and disinhibition* (pp. 186–211). Rockville, MD: National Institute on Alcohol Abuse and Alcoholism.
Miller, W.R., & Hester, R.K. (1986). Inpatient alcoholism treatment: Who pays? *American Psychologist, 41,* 794–805.
Nordström, G., & Berglund, M. (1987). A prospective study of successful long-term adjustment in alcohol dependence. *Journal of Studies on Alcohol, 48,* 95–103.
Orford, J., & Edwards, G. (1977). *Alcoholism: A comparison of treatment and advice, with a study of the influence of marriage.* Oxford: Oxford University Press.
Paredes, A., Hood, W.R., Seymour, H., & Gollob, M. (1973). Loss of control in alcoholism: An investigation of the hypothesis with experimental findings. *Quarterly Journal of Studies on Alcohol, 34,* 1146–1161.
Pattison, E.M., Sobell, M.B., & Sobell, L.C. *Emerging concepts of alcohol dependence.* New York: Springer.
Polich, J., Armor, D.J., & Braiker, H.B. (1980). *The course of alcoholism: Four years after treatment.* Santa Monica, CA: The Rand Corporation.
Powell v. Texas. 392 U.S. 514 (1968).
Room, R. (1977). Measurement and distribution of drinking patterns and problems. In G. Edwards, M.M. Gross, M. Keller, J. Moser, & R. Room (Eds.), *Alcohol-related disabilities* (pp. 61–87). Geneva: World Health Organization.
Rorabaugh, W.J. (1979). *The alcoholic republic.* Oxford: Oxford University Press.
Rudy, D.R. (1986). *Becoming alcoholic.* Carbondale, IL: Southern Illinois University Press.
Saxe, L., Dougherty, D., Esty, K., & Fine, M. (1983). *The effectiveness and costs of alcoholism treatment* (Health Technology Case Study 22). Washington, D.C.: Office of Technology Assessment, U.S. Congress.
Schuckit, M.A. (1984). *Drug and alcohol abuse.* New York: Plenum Press.
Schuckit, M.A. (1985). *Alcohol patterns and problems.* New Brunswick, NJ: Rutgers University Press.
Schuckit, M.A., & Viamontes, R. (1979). Ethanol ingestion: Differences in blood acetaldehyde concentrations in relatives of alcoholics and controls. *Science, 203,* 54–55.

Trachtenberg, R.L. (1984). *Report of the Steering Group to the Administrator: Alcohol, Drug Abuse, and Mental Health Administration regarding its attempts to investigate allegations of scientific misconduct concerning Drs. Mark and Linda Sobel.* Rockville, MD: Alcohol, Drug Abuse, and Mental Health Administration.

Traynor v. Turnage. 99 S. Ct. L. Ed. 2d 618 (1988).

Vaillant, G.E. (1983). *The natural history of alcoholism.* Cambridge, MA: Harvard University Press.

CHAPTER 22

The Disease Concept of Alcoholism

Brian D. Hore

Concern over those individuals who drink alcohol to excess goes back many centuries, this concern being directed towards the effect alcohol has both on the individual and on society. Over the centuries there have been various interpretations of the nature of such behavior, that is, why individuals drink in this manner.

Levine (1978) noted that the idea of alcoholism as a progressive disease with the key symptom of "loss of control" is nothing new, nor indeed did such a concept start with the foundation of Alcoholics Anonymous or the publication of Jellinek's monograph (1960); it is in fact at least 200 years old. During the last century there was much debate about whether individuals who drank to intoxication did so because they chose to or whether they could not really help losing control of their drinking. Levine (1978) emphasized that long before opium was regarded as an addictive substance, alcohol was. During the 19th century, the temperance movement, which was perhaps the most prominent of the forces concerned with alcohol-related problems, accepted the concept of addiction but emphasized alcohol as being the addictive agent. After Prohibition, the disease model of alcoholism with the essential causation being related not to the agent alcohol but to the host—that is, the individual—came more to the fore.

In the early years of American colonization drinking to intoxication was seen as a matter of choice; however, Rush (1810, 1814) designated it as a disease of the will and one of increasingly frequent bouts of intoxication. The views of individuals concerned with the temperance movement resembled those of Alcoholics Anonymous 100 years before AA was founded. Bemen (1829), for instance, stated that drunkenness in itself is a disease: "When the case is formed and the habit established no man is his own maker" (pp. 6,7).

In the ideology of the 19th century and before, the most important single feature of the prevailing concept of alcoholism was the apparent

Brian D. Hore is with the Alcoholism Treatment Unit, Withington Hospital, Manchester, United Kingdom. This chapter was written especially for this book.

involuntary nature of drinking, and in temperance writings this was the essence of the disease concept. The behavior of members of the temperance societies in some ways also resembled that of self-help groups such as AA. For example, members of the temperance societies were encouraged to produce reform in drunken individuals; and once reformed, such individuals were to help fellow sufferers in the same way as members of AA offer help to fellow alcoholics. It is interesting to note that when such individuals were regarded as suffering from a disease called alcoholism, efforts to develop asylums for the inebriated were in fact started and supported by the temperance organizations.

Rush (1810, 1814) had been the first to recommend a sober house where drunkards could get special treatment. Over 200 years later it is those who largely accept the disease concept of alcoholism who recommend that individuals be placed in special treatment centers. A disease model, however, was not the only view in the 19th century; Todd (1882), cited by Jellinek (1960), stated, "I consider it certain that the great multitudes of drunkards could stop drinking today and forever if they would, but they do not want to.... The man drinks simply because he likes to drink or because he likes to be drunk" (p. 207). Todd also commented, perhaps cynically, that many physicians, especially specialists who make the treatment of drunkenness a business and source of profit, are positive that it is a disease. Similar criticisms today (e.g., Fingarette, 1988) [see chapters 21 and 39, by Fingarette and Roman & Blum, respectively, in this volume] are currently made by antagonists of the disease model to those who recommend specialized treatment, particularly of an inpatient type.

Therefore, during the last century and up to the present day there has been a major debate about whether alcoholism is a disease. Much of this debate, though centering to a degree on what constitutes a disease and whether alcohol fits into such a definition, is also concerned about the issue of control of drinking, that is, whether such control is within the voluntary capacity of an individual or not. Debate in this field has been extremely heated, as was seen after the recent publications by Fingarette (1988) and Brody (1989).

It is interesting to speculate on why there has been so much antagonism and hostility relating to differing concepts of alcohol misuse, whereas there have been far fewer arguments in the field of addictions—for example, opiate addiction—and in the concepts of mental illness, such as schizophrenia. The vast majority of British psychiatrists believe (without any unequivocal evidence) that schizophrenia is the result of some physiological or biochemical disturbance of cerebral function. Apart from the views of Laing (1965) and others in the 1960s (which really had little influence on the mainstream of psychiatry in the United Kingdom) psychiatrists have continued to accept this theory. Laing (1965) believed that schizophrenic behavior was understandable as behavior adapted to an extremely adverse environment, particularly during development, and that such individuals therefore were not ill, much of their behavior being more properly seen as in fact protective of themselves. He felt also, as others have in the field of alcoholism, that because there was no evidence for a biochemical disorder in schizophrenia, it was not appropriate to treat such individuals as ill.

However, despite the parallels with alcoholism, this debate (at least in the United Kingdom) produced very little in the way of effect on the mainstream view of schizophrenia. Edwards (1985) commented that the effect of criticism of the disease concept has led to its being largely replaced by a view based on learning theory—one that sees alcoholism as a behavioral disorder. Without arguing whether this extreme degree of change has actually occurred, one can state that criticisms of the disease concept have without doubt had a major effect. If alcoholism is not viewed as a disease, then it is not a matter primarily to be dealt with by medical intervention. No such claims have been made by social scientists or psychologists in fields of mental illness such as schizophrenia. The reason may be that the principal treatment for schizophrenia has remained pharmacological; that is, there has been little place for behavioral techniques and for the role of other personnel, for example, psychologists. This explanation would fit with Edwards' view (1985) that perhaps psychologists' attitudes toward the disease concept have been influenced by their wishes to claim the treatment of alcoholism as their own province.

The Contribution of Jellinek

There is no doubt that although it must be seen in the context of the debate of the 19th century, Jellinek's (1960) publication of the disease concept of alcoholism, which attempted to extract a disorder called *alcoholism* from a variety of drinking patterns, provided a major landmark. Together with the views of Alcoholics Anonymous, the academic support of the Yale Summer School of Alcohol Studies, and the development of the National Council on Alcoholism, this study led in the United States to an almost universal acceptance of the disease concept. Jellinek's monograph is also of interest in that in examining attitudes in a variety of nations he found that not all of them accepted the disease concept. Indeed, concern about alcohol-related problems was primarily confined to the United States and to Scandinavian countries, although in the latter case it was more connected with puritan tradition and the strength of the temperance and Prohibition movements than with the acceptance of the disease concept. It is also interesting to look at the situation in the United Kingdom as it was then and is now.

Although the *British Journal of Addiction* (formerly the *British Journal of Inebriety*) was being published in 1892, Jellinek stated that the great majority of physicians in the United Kingdom did not believe in the disease concept. The British traditional view was that alcoholism was nothing but drunkenness, bad behavior that just did not occur among decent people. This ambivalence and in effect lack of interest in the disease concept of alcoholism may be the reason that (certainly in comparison with the situation in the United States) very little in the way of specialized facilities have been provided to help people with alcohol problems in the United Kingdom, and indeed currently there is both a strong move against such specialized facilities and an increasing emphasis that many people with alcohol problems are far better dealt with in a community setting by those without specialist training (Hore, 1988).

Jellinek particularly stressed two types of alcoholics, whose characteristics he considered to be indicative of a disease concept. These types he called *gamma* and *delta* alcoholics. The following features were included in both groups: acquired tissue tolerance to alcohol, adaptive cell metabolism, withdrawal symptoms, craving, and either loss of control or inability to abstain. Gamma alcoholics lost control immediately when they started drinking, the ingestion of one drink leading to a chain reaction, whereas delta alcoholics displayed the inability to abstain, even for a day or two, from alcohol.

Jellinek was fully aware that cultural and economic factors in different countries were important in influencing levels of alcohol consumption and also cautioned that an exclusive emphasis on the disease concept of alcoholism (as seen in gamma and delta patterns of behavior) would exclude many other problems due to alcohol [see chapter 20, by Jellinek, in this volume]. He was therefore aware that alcohol could produce a wider range of problems than those due to the behavior of individuals with the alleged disease of alcoholism. He also considered certain varieties of drinking behavior to be symptomatic—that is, engaged in to relieve underlying symptoms—and did not regard these varieties, lacking the features of gamma and delta alcoholism, as illnesses. Alcohol, he felt, was an agent capable of inducing addiction, and in those with gamma or delta alcoholism it was likely that there would be a progression from psychological to physical addiction. Jellinek also felt that in gamma and delta alcoholism there was an adaptation of cell metabolism that represented physiopathological changes analogous to those in drug addiction.

He agreed that the nature of these physiopathological changes was unknown despite various theories then current and drew close parallels between what he believed to be the addictive patterns of drinking in the gamma and delta subjects and the patterns of behavior in those who were regarded as addicted to opiates. He appeared to regard opiate addiction as a disease; and in that sense, he felt, alcoholism (of the gamma and delta type) could also be regarded as a disease. He also did not think that because an individual had a disease that was a form of addiction it was completely outside the individual's power to do something about the addiction; such a person was therefore responsible for taking the necessary steps to deal with the problem.

It is worth stressing the point discussed by others, including Heather and Robertson (1981) and Keller (1972), that labeling individuals as having the disease of alcoholism excludes them from responsibility and also stigmatizes them. [This point is also central to the arguments of American labeling theorists of deviant behavior, such as Becker and Lemert, among others.—Eds.] Jellinek clearly separated those learning processes that preceded addiction from the addiction itself. He agreed that learned behavior may be an important factor leading to drinking before the onset of the disease process; and though he understood psychological formulations that related to seeing drinking as a symptom of underlying disorder, he did not believe that these formulations explained alcoholism.

Essentially Jellinek's theory about the disease of alcoholism involves seeing an individual drinking for a variety of reasons (including social, psychological, and cultural) and believing that in certain individuals this drinking leads to a process (as yet not understood) of physiological and pathological changes, which explains the increased tissue tolerance (with subsequent increases in the amounts of alcohol required to produce the same effect), the adaptation of cell metabolism to alcohol (evidence of which is seen in the severe withdrawal symptoms that occur when alcohol is withheld from those individuals), and the patterns of loss of control and inability to abstain. Although Jellinek clearly understood that genetic factors might be important, he seems to have supported the view of Fleming (1937), who stressed that any individual, however healthy or well organized, who drinks heavily and long enough can become addicted. The presence of withdrawal symptoms was the only way that physical dependence could be definitely established. Jellinek drew on the experimental work of Isbell, Fraser, Wikler, Belleville, and Eisenman (1955) (particularly their experiments in which former morphine addicts were given large amounts of alcohol and their withdrawal symptoms were assessed), who also seemed to believe that the presence of a withdrawal syndrome indicated physical dependence.

In the early 1950s it was clear that considerable thought was going into concepts of addiction in general, and the World Health Organization (WHO) (1952) described drugs as falling into three types in relation to addiction. There were, for example, opiates, in which the pharmacological properties were of primary importance; habit-forming drugs, in which the psychological reaction of the user was of primary importance; and intermediate drugs, in which both factors were important. Alcohol was considered a drug whose pharmacological action was intermediate, between that of addictive and habit-forming drugs, and whose effects were seen to cause dependence in those individuals predisposed by their makeup to seek and find an escape in alcohol. This theory is different from that of Fleming (1937) and suggests that both personal and the pharmacological factors play an important part in the process. One has the impression that Jellinek was anxious to fit alcohol into the WHO classification of addiction even though it was clearly understood that only a minority of users of alcohol became addicted, certainly in comparison to users of opiates. Increased tissue tolerance was regarded as being clearly demonstrable in relation to alcohol through tests on individuals to measure the level of tolerance (*tolerance* being defined as the threshold of the blood concentration at which performance showed deterioration). In heavy drinkers a higher blood alcohol concentration was required to bring about that effect. Increased tolerance was regarded as a changed physiological response to a substance—for example, alcohol—and this increased tolerance would combine with an ability to compensate. Physiological experiments on humans had already been carried out demonstrating that tolerance could be induced, and also it was known that alcohol was cross-tolerant with other drugs. It was known in addition that changes in tolerance could not be explained by the change in the rate of absorption of alcohol from the gut or the rate of increase in oxidation or excretion. Another key element of the disease concept as emphasized by Jellinek was the withdrawal syndrome, which Isbell, Fraser, Wikler, Belleville, and Eisenman (1955) demonstrated could be produced.

Craving, in Jellinek's view, was of a psychological type in between bouts of drinking—thus he accounted for people's relapses after periods of abstinence—and of a physiological type or potentially physiological type during the time the people were drinking. Loss of control, to Jellinek, meant the loss of freedom following the first ingestion of alcohol in a new bout of drinking in those individuals who suffered from gamma alcoholism and was characterized by minor withdrawal symptoms in the presence of alcohol in the blood stream and the failure to achieve the desired euphoria for more than a few minutes. Inability to abstain, as mentioned above, meant that the individual was unable to abstain even for a day or two from alcohol.

Implications of the Illness Concept

In discussing the attitude of WHO, Jellinek (1960) stated that whether alcoholism was regarded as an illness or not was not crucial and that what was most essential was where alcoholism was to be placed—within the sphere of health or that of morality. He also stated that there are many conditions that physicians legitimately claim as in their province although these conditions do not constitute diseases. Heather and Robertson (1981) and nonmedical persons would presumably object strongly to the acceptance of this idea.

The main beneficial effect of the disease concept of alcoholism was placing the condition within the sphere of health rather than within the sphere of morality. Jellinek also stressed that ultimately the attitude of the public at large would be crucial, and that if that attitude were positive it would provide hope and incentive for the rehabilitation of the alcoholic.

In essence Jellinek selected from patterns of excessive drinking two types, the gamma and delta, which he considered to be two diseases because of the presence of tissue tolerance, adapted cell metabolism responsible for this, the presence of withdrawal symptoms, craving, and the abnormal patterns of drinking, and the loss of control or the inability to abstain.

Later Developments of the Disease Concept

In subsequent years (before major critiques of the disease concept) there were two important developments. First, Keller (1972) noted that the immediate loss of control implicit in the idea of one drink's leading inevitably to continuous drinking was in fact not an accurate description of what happened to alcohol-dependent individuals in their day-to-day existence. Although this immediate loss of control can happen in many individuals, they do not lose control on every occasion and in fact may control their drinking for certain periods. Keller (1972) emphasized that the key was the individual's lack of choice and a lack of consistency of control once the drinking begins, as well as a lack of choice about whether to drink in the first place. A second important development was that of Edwards and Gross (1976), who advanced the idea of the alcohol dependence syndrome. The syndrome was graded; that is, people could be at different levels of dependence and could move in either direction. Edwards and Gross also believed that biochemical factors as well as psychological ones were important in describing the phenomenon of the syndrome. However, as Heather and Robertson (1981) noted, with emphasis particularly on the immediate loss of control over drinking after a period of abstinence, in many ways the alcohol dependence syndrome was a more socially acceptable form of the disease concept.

The Term *Disease*

It is impossible to get agreement on what constitutes the term *disease;* it is not enough to say a disease is what doctors say a disease is, because this approach is not logically sound and can also change with different generations of physicians. Thus, there are multiple definitions of *disease.* In the United States., the National Council on Alcoholism (1972) defined *disease,* in relation to alcoholism, as a definite morbid process having (1) a characteristic train of symptoms, (2) the potential to affect the whole body or any of its parts, and (3) an etiology, a pathology, and a prognosis that may be known or unknown. Gitlow (1988) stated that the ultimate decision regarding classification of any illness or disease rests upon whether or not the signs and symptoms, prognosis, and so on, commonly associated with it are similar enough to describe a specific entity.

Hershon (1974) felt two essential features had to be present before a drinking disorder could be characterized as having a disease status. First, the disorder had to relate to an etiologically relevant underlying pathology and, second, that the individual could not by will avoid having this disorder.

Implications of the Disease Concept

The major implication of labeling a condition as a disease is that doing so places it within the health sphere. Two alternatives to the disease label are placing such a disorder among deviant behavior (similar to the sin model) or, as Heather and Robertson (1981) pointed out, classifying it as a learned behavioral disorder. There is of course a difference between a scientific and a public acceptance of such a labeling. The disease and deviance concepts of alcoholism are considerably easier for the public to accept because of their simplicity in comparison with the concept of a learned behavioral disorder. In summary, the main criticisms of labeling alcoholism as a disease are that doing so (1) removes responsibility from the individual for his or her own condition, (2) fosters unwillingness on the part of individuals to pay attention to their symptoms in the early stages of an alcohol problem, and (3) tends to encourage perpetuation of the notion of an irreversible drinking pattern. Keller (1972), it should be noted, has challenged this third criticism, stating that it is very difficult to imagine individuals who are showing alcoholic behavior as requiring encouragement to drink. A further criticism has been leveled at the domination of the medical profession over this field and at what is seen as a neglect of the wider problem of alcohol abuse.

The most thorough critical analysis of the disease concept of alcoholism, particularly as exemplified by Jellinek's classic description (1960), is that by Heather and Robertson (1981). Their monograph used the terminology of behavioral psychology and a knowledge of statistics. After clarifying the various components of the disease concept of alcoholism as described by Jellinek and others, Heather and Robertson drew attention to the important experimental evidence that had developed in recent years. They began by examining laboratory studies of intoxication in alcoholics. This work was predominately that of Mendelson and his associates (Mello, McNamee, & Mendelson, 1968; Mello & Mendelson, 1972; Mendelson & Mello, 1966). Using behavioral psychology terminology, Heather and Robertson stated that these experiments attempted to determine the environmental correlates of designated behavioral responses in an individual. By changing environmental consequences one could, they said, record changes in the responses. In relation to the subject under consideration, that is, abusive drinking, the underlying goal in operant methodology is to identify those variables that contribute to the maintenance of that drinking.

The first group of experiments involved alcoholics carrying out spontaneous tasks, the reward for which would be money or alcohol; in general, the harder the individual worked, the more reward he or she obtained. Within this type of situation (which of course is different from that outside the laboratory) individuals did not drink rapidly to levels of intoxication. Indeed, they clearly exerted control over their drinking. They often drank to reasonably high blood alcohol concentrations, becoming mildly but not totally intoxicated. Further, although they could obtain more alcohol by working harder, they frequently chose not to do so. They could also wait for a drink and did not have to drink as soon as it became available. In essence, Heather and Robertson stated, these factors suggested the presence of some type of control over their drinking behavior.

The second group of experiments described what happened to individuals when they were given priming doses of alcohol or when they believed that they had received alcohol and in fact had not. During these experiments craving was usually rated. Interestingly enough, as with experiments on aggression, the researchers found (excluding levels of dependence of an individual) that whether individuals craved alcohol was much more a function of whether they believed they were receiving alcohol than whether they actually had received alcohol. This finding supports a psychological rather than a physiological explanation of craving. Heather and Robertson (1981) felt that the priming experiments, therefore, provided no evidence for the theory of the physiological basis of loss of control. Studies by Hodgson, Stockwell, and Rankin (e.g., 1979) in recent years have repeated this type of experiment with the difference that they have measured severity of dependence. In the severely dependent group behaviorally measured craving was primarily determined by the alcohol content of the drink; in the moderately dependent group craving was primarily determined by the instructional set. However, Heather and Robertson (1981) said that a physiologically based craving is also predicted by a learning theory model of alcoholism as well as by the disease model. In learning theory terms, such craving relates to the differential reinforcement histories attaching to the two levels of dependence. In particular, the definition of severe dependence by which the groups used in the experiments were divided makes "clear that the severely dependent alcoholic had engaged in repeated and frequent consumption of alcohol in order to escape or avoid withdrawal symptoms" (p. 108).

Heather and Robertson (1981) discussed the continued utility of the concepts of loss of control and craving. They felt that the experiments by Mendelson and his associates (Mello, McNamee, & Mendelson, 1968; Mello & Mendelson, 1972; Mendelson & Mello, 1966) demonstrated quite clearly that the inevitable loss of control implicit in the frequently heard phrase "one drink, one drunk" was (certainly within that setting) not correct. Nor, they felt, did the experiments support the idea that alcoholics cannot abstain; both of these concepts are important in the treatment implication that alcoholics should always be given a totally abstinent treatment goal. Heather and Robertson commented as follows on the inconsistency of control described by Keller (1972) and the indeterminacy of control as put forth in the Edwards and Gross (1976) description of alcohol dependence syndrome:

> A conception of impaired control does not qualify as a scientific proposition because ... by the epistemological canons currently accepted by the majority of scientists (Popper, 1959) it is not falsifiable. The reason it is not falsifiable is that it cannot specify the conditions under which uncontrolled drinking will occur and the conditions under which it will not. For the purposes of scientific discourse therefore it is strictly meaningless. (p. 123)

Heather and Robertson (1981) maintained that all this concept of lack of consistency of control did was to describe what actually happened; it did not take researchers any further into explaining the pattern of drinking. They asserted that this concept of impaired control in no particular way achieved more than a restatement of the original problem to be solved.

Heather and Robertson (1981) considered the most important aspect of experimental drinking studies to be the illustration that alcoholic drinking is often behavioral; that is, the amount and patterning of alcoholic drinking is shaped by the environmental consequences of that drinking. The experiments showed that the drinking behavior of alcoholics was essentially modifiable in the same way as normal drinking. They stated:

"There is no sense in describing a specific disease of alcoholism and no sense in searching for the roots of the general and irreversible loss of control in the alcoholic" (p. 127).

Biological Aspects of Alcoholism

As stated above, Jellinek (1960) in his discussion of gamma and delta alcoholism described (in addition to the concepts of loss of control and craving) pharmacological tolerance and the presence of withdrawal symptoms. Heather and Robertson (1981) did not comment to any great degree on how these two types of phenomena could be explained in psychological terms. Further, it would appear that even if a neurophysiological basis (see below) were found for alcohol dependence that equated it to an addiction, this finding would not automatically establish alcohol dependence as a disease in their view. They used the case of smoking cigarettes, which in many smokers is embedded in pharmacological addiction and definitely results in physiological damage. But they did not view nicotine dependency as a disease, "partly because huge numbers of people would be thereby classified as sick and enormous legal and financial problems would arise" (p. 245).

Gross (1977) and his colleagues developed a method of measuring withdrawal syndrome in a quantifiable manner and also subjected withdrawal symptoms to factor analysis. Their research found that in alcoholics in an uninterrupted episode of drinking critical cumulative blood alcohol concentrations, rather than quantities or days of alcohol intake, were related to the development of most features of the withdrawal syndrome during a period of four to six days of alcohol intake; significant levels of the withdrawal syndrome tended to occur after drinking episodes in which the average daily peak concentrations were greater than approximately 200 mg per 100 ml and the average daily minimal concentrations were greater than approximately 50 mg per 100 ml. Significant correlations were obtained between blood alcohol concentration during drinking and the severity during withdrawal of the clinical components relating to hallucinogenesis and affective disturbance, including tremor. These findings are clearly related to similar ones between cumulative blood alcohol concentration and the severity of withdrawal in animals such as mice. Gross stated:

> The withdrawal syndrome may reasonably be viewed as an indirect toxic effect of the sustained critical concentrations of alcohol, presumably a consequence of the underlying changes which produce physical dependence during the intake period. With a sufficient number of such episodes or prolongation of drinking over a sufficient period of time and an insufficient number and dura-

> tion of non-drinking periods which could presumably permit recovery there is evidence that some residual manifestations of withdrawal may persist for months and might even become irreversible. (p. 111)

The concept of physical dependence, as Gross (1977) explained, attempts to account for the alcohol withdrawal syndrome. It implies an adaptation to alcohol at the tissue level (probably in the central nervous system), which requires the presence of alcohol to maintain homeostasis. One definition he stated (not dissimilar to that of Kalant, 1973) refers to a latent state of hyperexcitability that develops after prolonged alcohol-induced depression of the central nervous system as a result of compensatory adaptation to the presence of the depressant. Gross wrote that once physical dependence has been induced it can more easily be reintroduced. This phenomenon, designated the "carry-over" of physical dependence, suggests the persistence of long-lasting physical changes. Tolerance to alcohol also was part of Jellinek's original concept. Functional tolerance is the most important type of tolerance and was defined by Gross as "the diminished effect of the same blood alcohol concentration on neurophysiological and behavioural responses" (p. 114). Studies of Skid Row alcoholics have shown that such tolerance can be extreme (Kessel et al., 1974). Once functional tolerance has occurred it is more readily reacquired (Gross, 1977). This phenomenon, designated the "carry-over" of functional tolerance, suggests also the persistence of long-lasting changes involving the underlying mechanism of functional tolerance.

Kalant (1973), in reviewing biological models of alcohol tolerance and physical dependence, considered tolerance and physical dependence closely related phenomena that develop essentially in parallel in man, rat, monkey, and other species. In addition, he stated that there was also general acceptance of the view that the withdrawal reaction by which physical dependence is identified is essentially the mirror image of the pattern of acute actions of alcohol. Moreover, he added, this pattern is not unique to alcohol but appears to be common to the withdrawal state of all or most of the hypnosedative and minor tranquilizers. Kalant concluded that although there are various theoretical models of drug dependence they are all variants of the same theme.

Summary and Conclusions

There are always some individuals who continue to drink despite damage to themselves and others. Such individuals in normal life may stop drinking on occasion or control their drinking but clearly on other occasions will repeatedly drink to levels of intoxication likely to cause themselves considerable harm. They may show patterns of increased functional tolerance and also, at least in some cases, a characteristic disturbance of physiological function in any period of abstinence—that is, a withdrawal syndrome. But a description of the type of behavior alone is not sufficient; it is necessary to understand the best way to explain it and the implications of that explanation.

The disease concept of Alcoholics Anonymous—and indeed the concepts of irreversible loss of control, inability to abstain, and craving, as exemplified by Jellinek (1960)—would not seem to have survived the evidence of the experimental studies mentioned by Heather and Robertson (1981) or, in fact, the careful clinical observations of those who work with alcoholics. These individuals do on occasion control their drinking over periods and do not always exemplify the phrase "one drink, one drunk"; they can abstain on occasion and do not crave in a manner that validates the claim that this phenomenon has a purely biological cause. However, these individuals do show changes in tolerance and withdrawal symptoms, and these cannot be explained easily in terms of learning theory and are more explicable in biochemical terms (see below). In simplistic terms, therefore, it would seem that psychological, or sociopsychological, explanations can account for the patterns of drinking and craving in abusive drinkers but cannot explain such phenomena as functional tolerance or the withdrawal syndrome and the "carry-over" of these effects, phenomena that would suggest the persistence of long-lasting changes.

Studies focusing on biological changes related to tolerance and withdrawal symptoms have been reviewed by Wallace (1988). Some of these studies, Wallace reports, have examined genetic factors, and, in animals, factors relating to brain levels of serotonin and opioid peptides. Others have centered on neurotransmitters such as noradrenalin and gamma amino butyric acid. In both humans and animals, it is claimed, an increase in the availability of serotonin reduces consumption of alcohol. Further attempts in biological research include those examining the hypothesis of alcohol condensation products.

It is also suggested (Trachtenberg & Blum, 1987) that the physiological craving for alcohol may be the result of naturally occurring deficits of an opiatelike substance as well as other neurochemicals. These deficits may occur genetically or as a result of long-term drinking. Further psychological factors, such as stress, can produce a chronic deficiency of naturally occurring enkephalins and endorphins. These theories would link genetic, psychological, and biological factors into a theory of alcoholism. At present it has to be said that these theories are complex and cannot be regarded as proved. However, there does seem to be growing evidence of biochemical changes in fundamental areas of cerebral function, such as the level of neurotransmitters and opioidlike substances in animals and possibly also in humans after regular alcohol use. This evidence—together with the fact, already described by Gross (1977), that functional tolerance and physical dependence carry over—suggests that in those that consume alcohol over a period there is at least the existence of long-lasting biological changes. The illness and deviancy models of alcoholism have in the last 20 years been joined by a third model, that is, alcoholism perceived as a learned reversible behavioral disorder. Major implications regarding treatment goals and methods of treatment, particularly the goal of controlled drinking and treatment of a behavioral nature, have followed from this paradigm and have had considerable influence. It is questionable, however, in the United Kingdom, although these theories have had considerable impact, whether they have replaced the model of the alcohol dependence syndrome among those professionals who work in specialized treatment units, halfway houses, and other care facilities for alcoholics.

Thus, from the evidence and models presented in this chapter, it is not valid to assume that alcoholics never control their drinking, that they must be understood in the sense of a simplistic phrase such as "one drink, one drunk." Further, craving for alcohol would seem at least in part to be psychological in nature rather than of pure physiological function. However, the behavioral theory does not seem to explain the phenomena of functional tolerance and physical dependence (together with the associated withdrawal syndrome and the carry-over effect of tolerance and physical dependence). There is also suggestive evidence that genetic (although this would not be denied by the behaviorists) and biochemical abnormalities involving key cerebral systems may be altered in animals, and possibly in humans, after alcohol use.

Alcoholism is probably best regarded as more than a unidimensional illness involving more than a single factor. If this idea were accepted, it might lead to cooperation between professionals. Such cooperation might replace the apparent present disunity among the relatively few professionals (Hore, 1984), from whatever discipline, who wish to work with those who abuse alcohol.

Acknowledgment

The author would like to thank Nicola Thornton for her excellent preparation of the manuscript.

References

Beman, N.S. (1829). *Beman on intemperance* (rev. stereotype ed.). New York: John P. Haven.

Brody, A. (1989). Martian, animal and human chemical dependency. A critique of Fingarette's rejection of alcoholism as a disease. Paper presented at 17th International Institute on the Treatment and Prevention of Alcohol Problems, Pontault-Combault, France (available from ICAA, Case Postale 189, 1, Avenue du Tribunal-Federal, CH-1001 Lausanne, Switzerland).

Edwards, G. (1985). Paradigm shift or change in ownership? The conceptional significance of D.L. Davies' classic paper. *Drug and Alcohol Dependence, 15,* 19–34.

Edwards, G., & Gross, M.M. (1976). Alcohol dependence: A provisional description of a clinical syndrome. *British Medical Journal, 1,* 1058–1061.

Fingarette, H. (1988). *Heavy drinking: The myth of alcoholism as a disease.* Berkeley: University of California Press.

Fleming, R. (1937). The treatment of chronic alcoholism. *New England Journal of Medicine, 217,* 779–783.

Gitlow, S. (1988). An overview. In S.E. Gitlow & H.S. Peysor (Eds.), *Alcoholism: A practical treatment guide.* 2nd ed., pp. 1–18. Philadelphia: Grune and Stratton.

Gross, M.M. (1977). Psychobiological contributions to the alcohol dependence syndrome: A selective review of recent research. In G. Edwards, M.M. Gross, M. Keller, J. Moser, & R. Room (Eds.), *Alcohol-related disabilities* (WHO Offset Publications No. 32, pp. 107–131). Geneva: World Health Organization.

Heather, N., & Robertson, I. (1981). *Controlled drinking.* London: Methuen.

Hershon, H.I. (1974). Alcoholism and the concept of disease. *British Journal of Addiction, 69,* 123–131.

Hodgson, R.J., Stockwell, T.R., & Rankin, H.J. (1979). Can alcohol reduce tension? *Behavior Research and Therapy, 17,* 459–466.

Hore, B.D. (1984). Disunity amongst the few. *Alcohol and Alcoholism, 19,* 197–198.

Hore, B.D. (1988). Alcoholism treatment in the United Kingdom. In T. Kamada, K. Kuriyama, & H. Suwaki (Eds.), *Biomedical aspects of alcohol and alcoholism.* Tokyo, Japan: Aino Foundation, Gendaikik Kushitshu Publishing.

Isbell, H., Fraser, H.F., Wikler, A., Belleville, R.E., & Eisenman, A.J. (1955). An experimental study of the aetiology of "rumfits" and delerium tremens. *Quarterly Journal of Studies on Alcohol, 16,* 1–33.

Jellinek, E.M. (1960). *The disease concept of alcoholism.* New Brunswick, NJ: Rutgers Center of Alcohol Studies.

Kalant, H. (1973). Biological models of alcohol tolerance and physical dependence. In M.M. Gross (Ed.), *Alcohol intoxication and withdrawal, experimental studies: Advances in experimental medicine and biology* (Vol. 35, pp. 3–14). New York: Plenum Press.

Keller, M. (1972). On the loss of control phenomenon in alcoholism. *British Journal of Addiction, 67,* 153–166.

Kessell, W.I.N., Makenjuola, J.D.A., Rossall, C.J., Chand, T.G., Hore, B.D., Redmond, A.D., Rees, D.W., Gordon, M., & Wallace, P.D. (1984). The Manchester Detoxification Service. *The Lancet, 1,* 839–842.

Laing, R. (1965). *The divided self.* Bergenfield, NJ: Penguin U.S.A.

Levine, H.G. (1978). The discovery of addiction: Changing conceptions of habitual drunkenness in America. *Journal of Studies on Alcohol, 39,* 143–174.

Mello, N.K., McNamee, H.B., & Mendelson, J.H. (1968). *Drinking patterns of chronic alcoholics: Gambling and motivation for alcohol* (Psychiatric Research Report No. 24). Washington D.C.: Psychiatric Association.

Mello, N.K., & Mendelson, J.H. (1972). Drinking patterns during work-contingent and non-contingent alcohol acquisition. *Psychosomatic Medicine, 34,* 139–164.

Mendelson, J.H., & Mello, N.K. (1966). Experimental analysis of drinking behavior of chronic alcoholics. *Annals of New York Academy of Sciences, 133,* 828–845.

National Council on Alcoholism, Criteria Committee. (1972). Criteria for the diagnosis of alcoholism. *American Journal of Psychiatry, 129,* 127–135.

Popper, K.R. (1959). *The logic of scientific discovery.* London: Hutchinson.

Rush, B. (1810). *Medical inquiries and observations upon the disease of the mind.* New York: Hafner.

Rush, B. (1934). An inquiry into the effect of ardent spirits. In Y.A. Henderson. *A new deal in liquor: A plea for dilution* (8th ed., pp. 185–227). New York: Doubleday. (Original work published 1814)

Todd, J.E. (1960). Drunkenness a vice, not a disease. In E.M. Jellinek. (1960). *The disease concept of alcoholism* (Appendix A, pp. 207–210). New Haven, CT: College and University Press. (Original work published 1882)

Trachtenberg, M.D., & Blum, K. (1987). Alcohol and opioid: Peptides, neuropharmacological rationale for physical craving of alcohol. *American Journal of Drug Alcohol Abuse, 13,* 365–372.

Wallace, P. (1988). The relevance to clinical care of recent research in neurobiology. *Journal of Substance Abuse Treatment, 5,* 207–217.

World Health Organization. (1952). Expert Committee on drugs liable to produce addiction: Third report (WHO Technical Report Series No. 57). Geneva, Switzerland: Author.

SECTION IV

The Genesis and Patterning of Alcoholism and Alcohol-Related Problems

B. DEVELOPMENTAL CONTEXT

Introductory Note

Why, in grossly similar sociocultural settings, do certain persons become alcoholics while others do not? There is no definitive answer to this question, and perhaps there never will be, but the question itself has given rise to extensive speculation, as well as to a variety of studies concerned with the origins of alcoholism in childhood.

A cardinal difficulty with most research on the causes of alcoholism is that we have been forced to consider potential causes in the terminal stages of alcoholism rather than during or prior to its inception. In consequence, the findings seem to be almost hopelessly contaminated from the standpoint of unravelling etiology, unless we are to be content with clinical reconstruction resting upon a principle of coherence. Of course, this situation is by no means unique to the study of alcoholism. Nor is its recognition to be construed as implying that it is unimportant to learn what we can about the personality and social relationships of alcoholics in the later stages. Indeed, such studies not only may have important therapeutic implications but also may reveal durable features of the individual that contributed to alcoholism's onset. However, the possibility remains that this type of investigation will tell us more about the effects of 15 or 20 years of excessive drinking than it does about the causes of alcoholism. From an etiological standpoint, then, there is a need for studies that circumvent the dilemma that inheres in studying alcoholics in the later stages of alcoholism, that is, after they have reached Alcoholics Anonymous, the hospital, the clinic, or the jail. For this reason longitudinal studies have been hailed as the only research design in which the true causes of alcoholism can be uncovered. It has been argued that longitudinal research is invaluable because it can establish temporal connections between theoretically important variables (Kandel, 1978). The chapters in this subsection provide data from longitudinal research.

In the first of these studies (chapter 23), reprinted from the original volume, the authors—Lee Robins, William Bates, and Patricia O'Neal—investigate whether there is a higher incidence of childhood behavior disorders among persons who in later life become alcoholics than among other, supposedly normal people who do not. Further, their work attempts to specify the kinds of behavioral disorders characteristic of prealcoholics in childhood and to delineate gross features of the socialization setting that may be considered as plausible antecedents to the development of alcoholism. Their data indicate that childhood ethnic background, socioeconomic status, parental inadequacy, and behavior problems are the best predictors of later adulthood alcoholism. They also demonstrate that some of their predictors (i.e., sex and ethnicity) are related to heavy drinking and lose their ability to distinguish alcoholics from nonalcoholics, among heavy drinkers. Other predictors (i.e., parental inadequacy and socioeconomic status) remain powerful predictors even after one controls for the experience of heavy drinking.

Another classic longitudinal study in the alcoholism field is the study by William and Joan McCord (1960), which also appeared, in an abridged form, in the original volume (McCord & McCord, 1962). Joan McCord has conducted a follow-up to the original study, which is reprinted here as chapter 24. In this chapter she describes developmental pathways that lead to alcoholism in later adulthood (ages 45–53 years) and compares rates and predictors for sons of alcoholics and sons of nonalcoholics. She finds that sons of alcoholics, if the alcoholic father was held in high esteem by the mother, were more likely to model their fathers and also become alcoholics or, at the very least, to have some problems with alcohol and drink daily. For men whose fathers were not alcoholic, there was a greater likelihood of becoming an alcoholic if the child was not controlled in early adolescence. Hence, the data suggest that different paradigms describe the pathways to alcoholism for men who are sons of alcoholics and for those who are not. Although modeling may partially account for the higher incidence of alcoholism in sons of alcoholics (see also chapter 9, by White, Bates, & Johnson), Goodwin (see chapter 30) offers an alternative explanation.

The original Robins et al. (chapter 23) and McCord and McCord (1960) studies are classics in terms of their longitudinal design. The findings from these pioneer studies have been replicated in recent research, such as the prospective longitudinal study by George E. Vaillant and Eva S. Milofsky, which is reprinted here as chapter 25. Like Robins and her colleagues, Vaillant and Milofsky find that ethnicity, parental inadequacy as measured by the number of alcoholic relatives, and premorbid antisocial behavior are the risk factors best predicting later alcoholism. Their original article has been criticized by Zucker and Gomberg (1986) for inappropriate data analytic techniques and, especially, for misinterpretation of the results. Vaillant and Milofsky respond to this critique in a postscript to their chapter.

In the final chapter in this subsection, Robert Zucker (chapter 26) critiques longitudinal research on alcoholism. He points to the failure to assess all the salient risk domains, the neglect of interactional influences, and the failure to adequately conceptualize alcoholic processes as major flaws in current research designs. After reviewing the major longitudinal studies of early childhood predictors, Zucker summarizes the commonalities and noncommonalities in these studies. Zucker argues that there is strong evidence for a steady progression of problem behaviors from early childhood to adulthood but that there are factors in adolescence that can either vitiate or ameliorate earlier childhood factors. Although Zucker claims that the pathway into problematic alcohol involvement is by way of influences that are nonspecific to alcoholism, he also stresses that environmental exposure is important. In the following subsection we present chapters that examine environmental influences on alcoholism.

References

Kandel, D.B. (Ed.). (1978). *Longitudinal research on drug use*. New York: Wiley.

McCord, W., & McCord, J. (1962). A longitudinal study of the personality of alcoholics. In D.J. Pittman & C.R. Snyder (Eds.), *Society, culture, and drinking patterns* (pp. 413–30). New York: Wiley.

McCord, W., & McCord, J. (1960). *Origins of alcoholism*. Stanford: Stanford University Press.

Zucker, R.A., & Gomberg, E.S.L. (1986). Etiology of alcoholism reconsidered: The case for a biopsychosocial process. *American Psychologist, 41,* 783–793.

CHAPTER 23

Adult Drinking Patterns of Former Problem Children

LEE N. ROBINS, WILLIAM M. BATES, AND PATRICIA O'NEAL

There are at least three established ways of studying the etiology of a disease. The first of these is to attempt to induce the disease experimentally in disease-free individuals. In the case of diseases like alcoholism for which no reliable cure is available, this approach raises serious ethical problems and consequently has been avoided. The second approach is to try to reconstruct events that may have led to the occurrence of the disease by studying persons who currently suffer from it. Such an approach has often been used in research on the etiology of alcoholism, but it contains major and unavoidable drawbacks. Over and above the uncertainties that inhere in attempts to reconstruct the past are the difficulties of disentangling causes or antecedent factors from the effects of the disease. The third approach avoids these pitfalls. It involves studying disease-free cohorts of children or younger persons and studying them again in later life to see who got the disease without experimental intervention. It is this third approach that we have taken to the study of alcoholism.

Advantages and Limitations of the Present Approach

There are, however, problems involved in taking this approach to alcoholism research. The most serious of these is briefly as follows: When a disease occurs in only a relatively small proportion of a population, as alcoholism does, it becomes necessary—barring extremely time-consuming and expensive research—to select a cohort of younger persons whose probability of manifesting the disease in later life is high. This procedure is essential if enough cases are to be obtained to illuminate etiology in a systematic way. Yet a procedure of this sort sacrifices ran-

Lee N. Robins and Patricia O'Neal are with the Department of Psychiatry, Washington University School of Medicine, St. Louis, Mo. William M. Bates is with Loyola University of Chicago, Ill. This chapter is reprinted from the original D.J. Pittman and C.R. Snyder, Eds., *Society, Culture, and Drinking Patterns* (New York: Wiley, 1962), pp. 395–412.

domness of selection. The alcoholics, for instance, who emerge in the course of this follow-up study cannot be assumed to be typical of alcoholics in general. If there are alternative sets of factors that antedate the onset of alcoholism but occur only in prealcoholic populations selected by other criteria, they will not be uncovered by such a procedure. The findings from a study of this kind apply, therefore, to a subgroup of the alcoholic population. Although they may suggest hypotheses that may ultimately prove to have broader application, the findings would not be generalized to alcoholics as a whole.

The cohort actually selected for this study was the patient population of 30 years before at a child guidance clinic in St. Louis, a group chosen with the expectation of its yielding a fair number of alcoholics. The high rate of adult deviance—signified by arrests, divorce, mobility patterns, and psychiatric illnesses other than alcoholism—previously found (O'Neal & Robins, 1958; Robins & O'Neal, 1958a, 1958b) in this "patient group" suggested that it might exhibit an unusually high rate of alcoholism, because alcoholism may also be viewed both as socially disapproved behavior and as a psychiatric illness. As will be noted in detail later on, the findings fully justified this expectation.

The information about these patients as children that was utilized in this research comes from records of the child guidance clinic collected quite independently of considerations of which children might eventually become alcoholics. Of course the content and scope of this information is necessarily restricted to what the clinic saw fit to record, and it suffers from certain other limitations, which will be discussed. It has, however, the decided advantage of obviating the need for reconstruction from the memory of adults and permits the disentanglement of prealcoholic phenomena from the consequences of years of excessive drinking. Moreover, the follow-up investigation of a cohort so studied initially made it possible to identify and compare alcoholics, on the one hand, and a "control group" of persons who exhibited behavior disorders in childhood but did not develop drinking problems as adults, on the other. The follow-up also yielded useful information on spontaneous recovery from alcoholism of a sort that cannot be obtained in many other studies of alcoholism.[1]

Although the particular longitudinal study that we have undertaken has its advantages, it also has definite limitations apart from matters of the scope of the data and of the generality of the findings derived from a "disease-prone" population. In the first place, the information on childhood, which was collected quite independently of the follow-up study, is not as systematic as that obtained at the time of the follow-up. The clinic that made these records did use a standard outline of topics; thus, the same general kinds of data were collected in the great majority of cases. However, identical questions were not necessarily used to elicit information, and little attempt was made to record negative evidence. The early records, therefore, fail to meet the most rigorous standards of comparability.

In the second place, the profile of drinking constructed from our follow-up interviews is not as detailed as it might have been had our study been exclusively devoted to drinking patterns. The study of drinking patterns reported here represents only one phase of a broader investigation of the social and psychiatric adjustment of the patient group. Although it would have been desirable to ask many more questions about drinking than were actually asked in the interviews, time was limited by other considerations. In the third place, it would have been very desirable, given the tendency of alcoholics to rationalize their drinking, to have interviewed a member of the family in every case as a means of checking statements on drinking experience. This procedure again was beyond the scope of the present study but could well be a part of future follow-up studies of alcoholism.

Details of Method

The patient cohort that is the core of this study consists of every child seen at the St. Louis Municipal Psychiatric Clinic in the years 1924 through 1929 who was white, who had an intelligence quotient of at least 80, and who had been referred for behavior problems of some kind. There are 524 such former patients, 503 of whom have survived to age 25 and can therefore be thought of as having made some sort of life adjustment as adults. In addition, this study made use of a matched "comparison group" of normal children who provide a baseline against which the extent of alcoholism in the cohort with childhood behavior problems can be compared. The comparison group consists of students selected from public school records to match the patient group with respect to age, sex, race, IQ, and census tract lived in. Of 100 control subjects selected, 99 had survived to age 25.

The members of these two groups were personally interviewed with the use of a standardized questionnaire 30 years after their clinic experience or, in the case of the comparison group, school attendance. Respondents were asked when they began their drinking, how much and how often they drank in their youth, how much and how often they drink currently, and whether there was ever a period in their lives when they drank more than now. Respondents were asked, too, if they thought there was any period when they drank too much and whether any family member ever complained about their drinking. They were also asked whether they had ever been on a bender; lost a job or been arrested because of drinking; or had liver disease, delirium tremens, or any other medical complications of alcohol. In cases in which potential respondents had died, a close relative was interviewed in their stead. Information was also collected from police, hospital, social agency, credit bureau, Veterans Administration, and other records and from death certificates where relevant. At the time of writing, 75% of all subjects who survived to age 25 had been interviewed, the analysis of childhood clinic records was complete for 221 cases, and the collection of data from other records was nearly complete for all cases.

In studying the extent of alcoholism in a total population, it is necessary to set up criteria by which alcoholism can be said to be present or absent. This problem does not, of course, arise if one studies the etiology of alcoholism in a group of known alcoholics selected either by their own definition, as in the case of members of Alcoholics Anonymous, or by the definition of medical personnel, as in the case of hospitalized alcoholics. A review of some of the definitions offered in the literature indicates there is a general acceptance of the idea that an alcoholic is a person who gets into trouble through drinking, either socially or medically, but there are no clear-cut criteria for the degree or kinds of trouble he must have before he qualifies as an alcoholic.

For the purposes of this chapter, two sets of standards for classification of alcoholics were developed: first, criteria for "chronic alcoholics," which are probably rigid enough to include only cases most people working in the field would accept as such, and second, criteria for "probable alcoholics." The criteria upon which these distinctions are based are set forth in Figure 1, along with the other criteria and distinctions among drinkers used in this study.

The criteria established for heavy drinkers shown in Figure 1 require both the consumption of relatively large amounts of alcohol and chronicity. Consequently, some persons were classified in the "no excessive drinking" category who had occasionally gotten drunk, provided that this behavior had not been a chronic problem. The heavy drinkers, by definition, include no one for whom there are public records of excessive drinking, since such records would constitute evidence of social difficulties associated with drinking. As a result, the evidence for heavy drinking comes entirely from interviews. Standards for heavy drinking refer to levels reported as *usual* intake. It seems possible, considering the social disapproval of heavy drinking, that many respondents minimized their actual drinking in the interview statements. It may seem to some that the standards chosen for classifying persons as heavy drinkers are rather low. However, as will be seen in the results to follow, only one fifth of the comparison group reported drinking as much as the minimum standards for heavy drinking require. Therefore, whether or not it is unusual to con-

FIGURE 1
Criteria for the Classification of Types of Drinker

ALCOHOLIC

Chronic Alcoholic

1. Death from acute alcoholism or
2. Medical complications of alcohol (cirrhosis, neuropathy, DTs) or
3. Diagnosis of alcoholism in hospital, army, or prison or
4. Arrests, loss of jobs, absenteeism, fighting, serious family complaints[2] due to alcohol along with chronic unemployment or
5. Five or more arrests for drinking within 5 years, the last arrest being within the last 3 years.

Probable Alcoholic

1. Statement from patient that he cannot control his drinking, in the absence of above criteria or
2. Any of the complaints in 4, above, in the absence of chronic unemployment.

HEAVY DRINKER

None of the complications above, but over an extended period

1. Drinking at least three drinks three times a week or
2. Drinking at least seven drinks a sitting.

NO EXCESSIVE DRINKING

1. Having none of the complications of alcohol and
2. Never having drunk as much as the minimum above over any extended period.

sume this much, it is surely unusual for respondents to report what we have called heavy drinking.

Results

Alcoholism and Heavy Drinking in the Patient and Comparison Groups

The data in Table 1 indicate a significantly higher rate of alcoholism for the patient group than for the comparison group. In making these estimates, we assumed that all cases of alcoholism were identified through agency records, whether or not interviews took place.[3]

Of the entire sample, 89% were located, and there was sufficient information on the whereabouts of these cases to permit checks of the appropriate public records. Another 5%, who could not be located currently, had a known address within the last few years that made possible the examination of appropriate records in all locales where they were known previously to have lived. For the remaining 6%, there was little or no information covering the interval between appearance at the clinic as a child and the present. The estimates of alcoholism in Table 1 are therefore probably somewhat below the reality, but the increment to be expected if all cases had been located would be small. Moreover, the location of the remainder of the cases would probably only serve to enhance the differ-

TABLE 1
Types of Drinker in the Patient and Comparison Groups (in percent)

		Patients (n = 503)		Comparion (n = 99)
Alcoholic	Chronic 6 / Probable 9	15	None / 2	2
Heavy drinker (estimated)[a]		22		18
No excessive drinking		63		80
Total		100		100

χ^2 (2 df) = 33.18; $p < .001$

[a]Since evidence for heavy drinking comes only from the interview, an estimate was made on the basis of the proportion of heavy drinkers found in the 301 patients and 85 control subjects interviewed who were not alcoholics. Of these, 25% of the patients and 19% of the control subjects were heavy drinkers. The figures in the table assume that the same proportion of the uninterviewed are also heavy drinkers. If none of the uninterviewed are assigned to the heavy drinker category, there are 15% of the patients and 16% of the control subjects now definitely known to be heavy drinkers.

ence between patient and comparison groups, since 98% of the latter had been located but only 87% of the patients.

Sex differences. The differences reported in the literature on alcoholism between American men and women tend toward a ratio in the neighborhood of six male alcoholics for every female alcoholic. It is hardly surprising, therefore, that the male patients in our study included a significantly higher proportion of alcoholics than the females did. However, the difference between men and women was less striking than the differences reported by others. The proportion of male chronic alcoholics exceeded the proportion of female chronic alcoholics by less than 2 to 1, and the proportion of all male alcoholics exceeded the proportion of all female alcoholics by less than 3 to 1 (see Table 2).

The relatively high rate of alcoholism among the female patients probably reflects the fact that the clinic population was largely male—as is the case in most child guidance clinics (Ackerson, 1931, p. 91f.)—and that the girls referred to such a clinic, therefore, tend to be even more atypical of the total population of girls than the males are of boys in general.

Although more of the boy than the girl clinic patients become alcoholics as adults, the differences are largely accounted for by their greater experience of heavy drinking. Of all the males who were ever heavy drinkers, 55% are classified as alcoholics; among the women who were ever heavy drinkers, 44% are classified as alcoholics. This difference in

TABLE 2
Sex Differences among Types of Drinkers in the Patient and Comparison Groups (in percent)

		PATIENTS[a]				COMPARISON	
		Men (n = 367)		Women (n = 141)		Men (n = 70)	Women (n = 29)
Alcoholic	Chronic	7	21	4	8	–	–
	Probable	14		4		3	–
Heavy drinker			16		10	19	11
No excessive drinking			41		54	66	79
Uncertain (uninterviewed)			22		28	12	10
Total			100		100	100	100

[a]For the sex difference among patients (excluding the 83 men and 40 women whose drinking behavior was uncertain): χ^2 (2 df) = 16.08; $p < .001$.

the rate of alcoholics between men and women who had experienced heavy drinking is not statistically significant.

Rates of Recovery

For purposes of this study, an alcoholic was considered "recovered" from alcoholism if he claimed to have had no social or medical difficulties resulting from drinking within the three years before being interviewed and if there was no objective evidence to the contrary. Because alcoholism is characterized by periods of temporary remission, shorter periods of recovery were not counted. Heavy drinkers were considered recovered if they claimed not to have had a period within the last three years when they drank as much as is specified by the criteria for heavy drinking in Figure 1.

It is evident from Table 3 that the comparison group contains so few alcoholics and heavy drinkers that the differences in recovery between patient and comparison groups cannot be assessed statistically.

The alcoholic category contains all those who have ever had social or medical difficulties with drinking, that is, both the chronic and probable alcoholics. Most of the alcoholic patients had social or medical difficulties within the three years preceding the interview (or within three years of their death, if they are deceased). More than half of those who have been heavy drinkers decreased their drinking only as they grew older. The higher rate of recovery found among heavy drinkers may be partly a function of the kind of evidence available. Because evidence for heavy drink-

TABLE 3
Extent of Recovery from Excessive Drinking in the Patient and Comparison Groups (in percent)

	PATIENTS		COMPARISON	
Current Drinking	Alcoholic (*n* = 84)	Heavy Drinking Only (*n* = 74)	Alcoholic (*n* = 2)	Heavy Drinking Only (*n* = 16)
Alcoholic	59	–	(0)[a]	–
Heavy drinking	7	41	–	69(11)
Drinking mildly	19	53	(2)	31(5)
Abstaining	7	5	–	–
No information	8	1	–	–
Total	100	100		100

[a]Numbers of persons shown in parentheses.

ing comes from interviews, there was no opportunity to validate the statements of patients who claimed to have reduced their consumption.[4]

Few of the recovered alcoholics claim that they are now abstainers. This is an interesting observation in view of the common belief that alcoholics can never become social drinkers because the first drink leads inevitably to uncontrolled drinking. Yet 26% of those with previous medical or social troubles associated with drinking now claim to be drinking heavily or moderately without complications, and objective evidence does not contradict their claim. It is still possible, however, that they are in fact unable to limit their alcohol consumption but have learned to drink in more discreet settings so that there are no public records of their drinking.

Factors were sought in the childhood records of the patient cohort that might distinguish those alcoholics and heavy drinkers who recovered from those who did not. Approximately one hundred fifty relationships were examined. Of these, only seven were found to be statistically significant, and none of these was common to both recovered alcoholics and recovered heavy drinkers. Since this incidence of positive findings is no greater than chance, we conclude that available information about the childhood of these patients does not permit us to predict who will recover from excessive drinking and who will not.

Antecedents of Alcoholism in the Patient Cohort

Although the patient cohort has a high rate of alcoholism in contrast with the comparison group, we cannot account for these differences by comparing the childhood records of these two groups since, by definition, the comparison group did not attend the clinic and therefore has no comparable childhood records. We can compare the childhood clinic records of patients who became alcoholics both with the records of patients who became heavy drinkers but never alcoholics, and with the records of patients who never drank heavily. These comparisons will give some insight into the etiological factors in the histories of prealcoholic patients and may permit inferences about why children who come to a child guidance clinic have a high rate of alcoholism as adults.

At the time of writing, analysis of childhood records was complete for all the chronic alcoholics, for 85% of the probable alcoholics and heavy drinkers, and for 37% of those interviewed who were never heavy drinkers.[5]

Ethnic Background

Because of reports in the literature (Bales, 1946; Snyder, 1958) that the Irish are overrepresented among alcoholics and the Jews underrepresented, it was of interest to see whether ethnicity was an important variable in the patient group.

The ethnic background of patients, as shown in Table 4, was determined by the birthplaces of parents and grandparents and by the ethnicity, if recorded, of American-born grandparents. The German group was kept separate not because of any special hypothesis about its drinking behavior but because in this St. Louis population it was by far the largest ethnic group. Since ethnicity beyond the parental generation had not been previously analyzed, it was necessary to go back to the original records for this information. As a result, it was possible to include all known alcoholics and heavy drinkers, whether or not their records had been totally analyzed. As anticipated, the Irish show the highest rate of alcoholism and were the only group with a significantly different rate, whereas the Jews have the highest proportion who have never been excessive drinkers.

When only those who have ever been heavy drinkers are compared, however, the rate of alcoholism among the Irish is not significantly higher, nor is the rate for the Jews significantly lower than the rates for other groups. Whatever the ethnic group, between 48% and 56% of the heavy drinkers became alcoholics. Ethnic differences in alcoholism, therefore, appear to be functions of differences in experience of heavy drinking rather than of differences in predisposition to become alcoholic if exposed to such a pattern. These findings are entirely consistent with the conclusions of Bales and of Snyder in the works cited above that different cultural patterns for normal drinking account for the difference between Irish and Jewish rates of alcoholism.

TABLE 4
Type of Drinker, Patient Group, by Ethnicity (in percent)[a]

	Irish[b] (n = 28)	German (n = 64)	Jewish (n = 15)	Old American (n = 78)	Other[c] (n = 33)
Alcoholic	46	31	20	31	33
Heavy drinker	36	25	20	33	36
No excessive drinking	18	44	60	36	31
Total	100	100	100	100	100

[a]Ethnicity was determined for all known cases of alcoholism (n = 84) and heavy drinking (n = 74), even where other analysis was not complete, and for all analyzed cases with no heavy drinking (n = 84). Excluded are 13 alcoholics, 7 heavy drinkers, and 4 with no excessive drinking for whom no information about ethnicity was given in the childhood records.
[b]Irish versus all others: χ^2 (2 df) = 7.20; $p < .05$.
[c]Italian, Romanian, Austrian, English, French, Hungarian.

Childhood Social Status

The patient cohort was largely but not exclusively made up of children of lower socioeconomic status. Since referrals were often made by social agencies, many of the children came from families that were not self-supporting.[6]

Allocation of patients to social status categories was largely on the basis of their family breadwinners' occupations, because this information was uniformly present in the clinic records. Housing, parents' levels of education, grandparents' occupations, and the neighborhood lived in were used, in addition, to allocate patients when this information was available. The distribution of types of drinkers in the resulting status categories is shown in Table 5.

The lowest socioeconomic group, composed of children from families dependent on social agencies or supported by the criminal activities of the parents, yielded a very high rate of alcoholism. Professional and executive families produced few alcoholics. Unlike the findings of differences between men and women and between ethnic groups, these differences are not simply associated with differences in experience of heavy drinking, because a strikingly high proportion of the heavy drinkers from the lowest status group became alcoholics and very few of the heavy drinkers from white-collar families became alcoholics. Not only do more of the lowest-status group drink heavily; but when they drink heavily, they are apparently much more likely to have social or medical problems resulting from their drinking [χ^2 (with Yates' correction) = 11.64; $p < .05$].

TABLE 5
Type of Drinker, Patient Group, by Childhood Family Status (in percent)[a]

	Professional or Executive (n = 15)	Other White Collar (n = 36)	Self-Sustaining Labor (n = 92)	Marginal (n = 49)	Illegal or Chronically Dependent (n = 23)
Alcoholic	7	19	32	37	65
Heavy drinker	40	31	30	33	9
No excessive drinking	53	50	38	30	26
Total	100	100	100	100	100

χ^2 (with Yates' correction) = 16.40; $p < .05$

[a]Omitted are six cases for whom there was no information concerning family status; four of these were heavy drinkers, two were never heavy drinkers.

TABLE 6
Type of Drinker, Patient Group, by Parental Adequacy (in percent)[a]

	Both or Only Parent Adequate (n = 51)	Mixed or Somewhat Inadequate (n = 111)	Both or Only Parent Inadequate (n = 49)
Alcoholic	14	34	49
Heavy drinker	31	33	20
No excessive drinking	55	33	31
Total	100	100	100

χ^2 (2 at 4 df) = 18.80; $p < .001$

[a]Ten cases omitted for whom parental adequacy was unknown: one alcoholic, five heavy drinkers, four no excessive drinking.

Parental Adequacy

Evaluation of the "adequacy" of parents with whom the child lived was based on parents' performance of the basic obligations of physical care, financial support, supervision, and provision of a socially acceptable model. No consideration was given to the subtler aspects of interpersonal relations between parent and child. Yet, using these gross criteria of parental adequacy, inadequate parents were found to produce a higher rate of alcoholic offspring than adequate parents, as is evident in Table 6.

TABLE 7
Type of Drinker, Patient Group, by Parental[a] Antisocial Behavior (in percent)

	Both Parents Anti-social (n = 41)	Only Father Anti-social (n = 78)	Anti-social Father[b] (Total) (n = 119)	Only Mother Anti-social (n = 14)	Other Parental Problems Only (n = 42)	No Known Parental Problems (n = 46)
Alcoholic	37	41	40	21	24	22
Heavy drinker	31	29	29	29	29	32
No excessive drinking	32	30	31	50	47	46
Total	100	100	100	100	100	100

[a]Natural parents only, whether or not the child lived with the natural parent.

[b]For antisocial fathers versus others: χ^2 (2 df) = 9.94; $p < .01$.

Again, while children of adequate parents were less often heavy drinkers, their lower rate of alcoholism could not be explained simply by their lesser experience of heavy drinking. A higher percentage of the heavy drinkers who had inadequate parents became alcoholics than of the heavy drinkers who had adequate parents [χ^2 (2 df) = 9.43; $p < .01$].

Parents were judged inadequate if they did not perform the traditional parental role, regardless of whether their failure arose from irresponsibility or from some incapacitating physical or mental disease. Some parents were therefore judged inadequate because they were confined to hospitals or to their beds at home for long periods during the childhood of the patient. Others were so judged because they wantonly neglected or deserted the child, drank excessively, or committed flagrantly illegal acts with the child's knowledge. Some inadequate parents, therefore, showed antisocial behavior; others were inadequate in other ways. In some instances the child lived with the inadequate parent; in other instances the child was known to have had an inadequate parent but was brought up exclusively by the other parent, who was adequate, or by adequate permanent foster parents.

When the father had showed antisocial behavior, 40% of the children became alcoholic. When only the mother had exhibited antisocial behavior, or when the parents were inadequate in other ways, they were not more likely to produce an alcoholic child than if they had no known problems (Table 7).

There were only eight cases in which the father was known to have shown antisocial behavior but in which the child was brought up exclusively by an adequate mother or by adequate permanent foster parents. In none of these cases did the child develop alcoholism. Although there are too few cases to draw reliable inference, this finding suggests that the predisposition to alcoholism is not solely genetic but may require the experience of living with an antisocial father.[7]

It is interesting that the effect of the father's antisocial behavior does not appear to depend on excessive drinking by the father (that is, if our data on this point are in fair correspondence with the reality [Table 8]). Fathers reported to drink excessively were no more likely to produce alcoholic children than fathers who were arrested, were erratic workers, had deserted, were guilty of sexual misbehavior, or beat their wives or children. Apparently the presence of an antisocial father in the home predisposes a child to excessive drinking whether or not the father sets an example of such drinking.

There are many aspects of the child's experience in the home that are dependent on the adequacy of his parents—for instance, whether discipline is lax, how frequently the family moves, whether the home is broken, or whether the child suffers deprivations of food, shelter, clothes, and supervision. All these variables show statistically significant positive correlations with adult alcoholism, illustrating in many ways the relations between parental adequacy and adult drinking behavior.

Childhood Behavior Problems

In previous papers (O'Neal & Robins, 1958; Robins & O'Neal, 1958a, 1958b) based upon our larger study, it was reported that antisocial behavior in childhood, particularly the severe antisocial behavior that results in juvenile court appearance, is related to different measures of deviance in adult adjustment such as arrests, incarceration, and divorce. Antisocial behavior in childhood is also related to deviant adult drinking behavior (Table 9). Patients who appeared in juvenile court have a higher rate of alcoholism than patients without antisocial behavior do. Those with antisocial behavior but no juvenile court appearances fall in between, indicating that it is the severity of the antisocial behavior rather than the experience in court that is important in the juvenile-court cases' high rate. Also, the high rate of alcoholism in patients who appeared in juvenile court is not simply a reflection of their greater experience of heavy drinking, since significantly more of the heavy drinkers from the juvenile court group became alcoholic [χ^2 $(2\,df) = 8.01$; $p < .02$].

Childhood records were analyzed for all types of behavior problems reported, and such problems were categorized into 71 symptoms. Alcoholics had more kinds of childhood symptoms per child than those who were not heavy drinkers. The median number of symptoms of alcoholics was 12 out of the possible 71, and for those who were never excessive drinkers, 8

TABLE 8
Type of Drinker, Patient Group, by Nature of Father's Antisocial Behavior (in percent)

	Drinking (n = 76)	Nonsupport (n = 61)	Arrests (n = 20)	Sex (n = 37)	Cruelty (n = 50)	Desertion (n = 46)	None (n=81)
Alcoholic	38	46	70	46	42	48	23
Heavy drinker	34	26	15	32	30	26	26
No excessive drinking	28	28	15	22	28	26	51
Total	100	100	100	100	100	100	100

TABLE 9
Type of Drinker, Patient Group, by Juvenile Antisocial Behavior

	Juvenile Court Record (n = 105)	Antisocial Behavior (No Juvenile Court) (n = 57)	No Antisocial Behavior (n = 59)
Alcoholic	45	25	15
Heavy drinker	27	38	27
No excessive drinking	28	37	58
Total	100	100	100
	χ^2 (4 df) = 22.36; $p < .001$		

out of 71; but the excess among alcoholics is accounted for only by symptoms of antisocial behavior (such as running away, theft, and assault). Alcoholics did not have more symptoms of a neurotic type (mood disturbances, restlessness, and seclusiveness, for example) as children. For 30 out of 35 symptoms of antisocial behavior, the proportion of prealcoholics showing the symptom was higher than the proportion of those who would never drink excessively. The data on this point are summarized in Table 10.

Moreover, for half of the 30 childhood symptoms of antisocial behavior in which the proportion of alcoholics exceeded the proportion of those without heavy drinking, the differences were statistically significant. For none of the 5 antisocial symptoms in which the rate of alcoholics was lower was the difference significant. Out of 36 neurotic symptoms encountered, half were reported for a higher proportion of the prealcoholics and half for a higher proportion of those who would never drink heavily—exactly what one would expect by chance. For none of the specific neurotic symptoms was the proportion of alcoholics significantly greater than the proportion of those who never drank excessively. The childhood symptomatology of the alcoholics, then, is characterized by many symptoms of antisocial behavior. However, there is no evidence that alcoholics had more neurotic problems in childhood than those patients who never drank excessively as adults.

There are apparently no specific antisocial symptoms that enable us to predict alcoholism. In an examination of 15 symptoms of childhood antisocial behavior, significant differences were found in the proportions of alcoholics compared with those with no experience of excessive drinking. However, the higher rate of juvenile court appearances among subsequent alcoholics suggested that these symptoms might be typical of severely antisocial children rather than of prealcoholics in particular. When alcoholics who had appeared before juvenile courts were compared with other

TABLE 10
Relative Proportions in the Patient Group of Alcoholics and Persons with No Experience of Excessive Drinking, by Type of Childhood Symptoms (in percent)

	Antisocial Behavior Symptoms[a] (n = 35)	Neurotic Symptoms[b] (n = 36)
Higher proportion of alcoholics[c]	86	50
Higher proportion of no excessive drinking	11	47
Equal proportions	3	3
Total	100	100

[a]Incorrigibility, arrests, physical aggression, correctional institutionalization, delinquent associates, impulsive behavior, truancy, poor school achievement, school expulsion, diagnosed "psychopathy," vagrancy, pathological lying, irresponsibility, thievery, poor work history, staying out late at night, excessive drinking, vandalism, use of aliases, lack of guilt feelings, verbal aggression, sexual perversion, excess masturbation, initiating incest, rape, premarital sex experience, illegitimate pregnancy, exhibitionism, excess sex talk or play, hostility toward family, fighting with contemporaries, teasing, discipline problems, attending multiple schools, slovenliness. The first 15 are those in which alcoholics are significantly higher.

[b]Passivity, learning disabilities, poor coordination, faints, fits, eating problems, bizarre food preferences, eneuresis, sleepwalking, fears, seclusiveness, withdrawn, paranoid ideas, irritability, nausea, apparent stupidity, brooding, being unaffectionate, overdependence, being teased, odd school behavior, somatic symptoms, insomnia, nightmares, nailbiting, thumbsucking, ties, depression, "nervousness," diagnosed psychosis, oversensitivity, restlessness, daydreaming, suicidal tendencies, low energy, tantrums.

[c]For "higher proportion of alcoholics" versus others: χ^2 (1 df) = 12.5; $p < .001$.

patients who had appeared before juvenile courts, none of these symptoms, not even adolescent drinking, was significantly more frequent in the alcoholic group. Children with severe symptoms of antisocial behavior show a high rate of alcoholism as adults, but the knowledge of the choice of antisocial symptoms apparently does not assist us in predicting which of the antisocial children will become alcoholics.

Interrelationships between Childhood Factors

We have found three aspects of the childhood of these patients to be related to later alcoholism: the social status of the family, the adequacy of the parents, and the number of (antisocial) symptoms.[8] These variables, however, are also highly intercorrelated. Inadequate parents tend to be found frequently at the lower end of the socioeconomic scale and tend to produce children who exhibit much antisocial behavior. The question therefore is whether these variables are independent determinants of adult drinking. Measures of partial association indicate that childhood social status and the number of symptoms in childhood are independent of each other and of parental adequacy. Social status is related to alcoholism when one is controlling for number of symptoms ($\chi^2 = 11.31$; $p < .05$) and controlling for parental adequacy ($\chi^2 = 8.31$; $p < .05$). Number of symptoms is related to alcoholism when one is controlling for social status ($\chi^2 = 15.0$; $p < .01$) and controlling for parental adequacy ($\chi^2 = 9.48$; $p < .05$). Parental adequacy is related in the expected direction, but differences do not reach the 5% level of confidence when one is controlling for social status ($\chi^2 = 5.61$; $p < .20$) and number of symptoms ($\chi^2 = 6.72$; $p < .10$). Limiting our definition of parental inadequacy to the presence of an antisocial father in the home would probably delimit this variable sufficiently to obtain significant relationships on partial association.

Summary and Comment

The principal findings of this study may be summarized briefly as follows:

First, alcoholism was found by follow-up study to be significantly more prevalent among a cohort of former patients of a child guidance clinic than among an otherwise matched comparison group with no childhood clinic experience.

Second, some former alcoholics were found to be drinking heavily or mildly but without evidence of social or medical complications, the implication being that in certain instances alcoholics may be able eventually to drink socially without loss of control.

Third, although the extent of alcoholism was seen to differ by sex and ethnic categories, these differences lost significance when the experience of heavy drinking was controlled.

Fourth, antecedent factors evident in the childhood histories of the clinic patients and found to be significantly related to alcoholism in later life were: low family social status; parental inadequacy, in particular antisocial behavior on the part of fathers; and serious antisocial behavior, as evidenced by records of juvenile court appearances and a clinic record of a variety of symptoms of antisocial behavior.

It is important to note that the antecedent factors of low family status, parental inadequacy, antisocial fathers, and antisocial behavior in childhood significantly differentiated the alcoholics from nonalcoholics in later life even when only those who had experienced heavy drinking were considered. This finding is in contrast to the differentiating effect of sex and ethnic factors, which disappeared when only those persons with experience of heavy drinking were considered. This finding suggests that sex and ethnic groups are different kinds of variables from the others. Perhaps the difference is that sex and ethnic status are totally ascribed statuses, assigned at birth, whereas the other variables depend either on the behavior of the patient himself (antisocial behavior in childhood) or on the behavior of his parents (achieved class position, inadequacy, and antisocial behavior of the parents). If we make the assumption that heavy drinking can be normal or accepted social behavior in certain sociocultural groups but that alcoholism, at least in America, always violates the social norms, we can postulate that alcoholism requires both exposure to the mores of heavy drinking and a pathological individual.

The high rate of alcoholism observed in this and other studies in the Irish as contrasted with other ethnic groups and in men as contrasted with women does not indicate a higher rate of individuals pathologically susceptible to alcoholism in these groups. Rather, the fact that these groups accept heavy drinking as normal behavior exposes a larger proportion of their members to heavy drinking and thereby makes alcohol available to a larger proportion of the fixed number of alcohol-susceptible members of the group. The high rate of alcoholism among those who have low status, inadequate parents, and antisocial childhood behavior, on the other hand, reflects not only a greater experience of heavy drinking among these groups but also a higher proportion of alcoholism-susceptible members, as indicated by the higher proportion of alcoholics even when one is controlling for the amount of experience of heavy drinking.

A study of this kind that takes as its population a group of patients referred to a child guidance clinic does not permit generalizations about the etiology of alcoholism. Most alcoholics did not attend such clinics as children, and their alcoholism may have developed from different roots and been reached by different paths. This study indicates, however, that there is among alcoholics a group of still-undetermined size whose alcoholism develops in the context of both a long history of antisocial behavior dating from childhood and a childhood experience of grossly inadequate parental care and extremely low social status.[9] A striking finding of this study is that not only is the occurrence of alcoholism highly related to evidences in childhood of pathology in the subjects, and their parents, but the kind of pathology related to alcoholism can best be described as antisocial rather than neurotic behavior.

Acknowledgments

The research upon which this chapter is based was supported by grants from the Foundation Fund for Research in Psychiatry and the United States Public Health Service.

Notes

1. At the time of follow-up, some of those persons who formerly had difficulties with their drinking no longer evidenced such problems, and the rate of recovery could be studied. By contrast, if one studies a group of known alcoholics who either are by definition not recovered (e.g., those currently hospitalized or imprisoned for alcoholism) or are by definition in remission (members of Alcoholics Anonymous, for example), it is not possible to estimate the normal rate of recovery.
2. Excluded are those cases in which the only known social difficulties were complaints from families opposed on principle to drinking.
3. It is not anticipated that a significant number of those who are not now alcoholics will become alcoholics later on in life. The average age of the sample is now 43 years, and other research indicates that the first symptoms of alcoholism seldom appear after this age. For example, Amark (1951) found that 94% of Swedish alcoholics showed symptoms of alcoholism by age 43, 100% by age 54.
4. With the alcoholics, of course, objective evidence from public records sometimes discounted claims of recovery. In the case of heavy drinkers professing recovery, interviews with relatives would have been especially valuable as a check on their claims.
5. A far smaller proportion of the "no excessive drinking" group was analyzed because the 84 cases previously analyzed in the course of this study were enough to permit statistical analysis. Although we cannot guarantee that this is an unselected sample of all the cases interviewed which belong in this category, analysis had been previously completed without consideration of the drinking behavior of this group, and we have no reason to suspect that this is a biased sample.
6. For this reason, even though the comparison group was matched with the patient group on the basis of census tracts, the clinic patients were of lower social status than the comparison group.
7. Our findings support those of Roe and Burks (1945), who found that children of alcoholic parents reared in foster homes do not develop alcoholism.
8. Since differences in the number of symptoms between alcoholics and other patients depend entirely on the excess of symptoms of antisocial behavior in the alcoholics, differences in the total number of symptoms are equivalent to differences in the number of antisocial symptoms.
9. Several other studies suggest that this group may form a significant proportion of existing alcoholics. Amark's (1951) findings—a high rate of criminal behavior and alcoholism in the brothers of alcoholics and some evidence that the fathers also had a high rate of criminal activities—suggest the antisocial family as the origin of alcoholism. The McCords (1962) support this conclusion, although like this study, theirs does not have a representative sample of alcoholics. In Bleuler's (1955) study of the family background of 50 well-to-do hospitalized alcoholics the finding that even in this upper-class group 19 had "grossly unfavorable home environment" as children suggests parental inadequacy, and the finding that 29 had long-standing and intimate contact before the age of 20 with persons who had alcoholic problems suggests antisocial behavior in the parents.

   Sherfey (1955), in studying admissions to the Payne Whitney Clinic, found 11% of those she was able to diagnose to be "asocial psychopathic . . . with a life-long behavioral abnormality." Like Bleuler (1955), she was working with an upper-class group in which asocial psychopathy should be minimal. Of the many reports of the psychological test scores of alcoholics, one of the few reliable findings, by Chotlos and Deiter (1959), is that alcoholics have a high Pd (psychopathic deviate) score on the Minnesota Multiphasic Personality Inventory, which suggests the intimate relationship between alcoholism and other forms of antisocial behavior. Although personality test scores are influenced to an unknown extent by the long history of excessive alcohol intake, these studies suggest that antisocial personality characteristics are common in alcoholics and that many alcoholics were brought up in homes where male members of the family also exhibited severe antisocial behavior.

References

Ackerson, L. (1931). *Children's behavior problems.* Chicago: University of Chicago Press.

Amark, C.A. (1951). *Study in alcoholism.* Copenhagen: Ejnar Munksgaard.

Bales, R.F. (1946). Cultural differences in rates of alcoholism. *Quarterly Journal of Studies on Alcohol,* 6, 480–499.

Bleuler, M. (1955). Familial and personal background of chronic alcoholics. In O. Diethelm (Ed.), *Etiology of chronic alcoholism.* Springfield, IL: Charles C. Thomas.

Chotlos, J.W., & Deiter, J.B. (1959). Psychological considerations in the etiology of alcoholism. In D.J. Pittman (Ed.), *Alcoholism: An interdisciplinary approach.* Springfield, IL: Charles C. Thomas.

McCord, W., & McCord, J. (1962). A longitudinal study of the personality of alcoholics. In D.J. Pittman & C.R. Snyder (Eds.), *Society, culture, and drinking patterns,* (pp. 413–430). New York: Wiley.

O'Neal, P., & Robins L.N. (1958). The relation of childhood behavior problems to adult psychiatric status. *American Journal of Psychiatry, 114,* 961–969.

Robins, L.N., & O'Neal, P. (1958a). Mortality, mobility and crime. *American Sociological Review, 23,* 162–171.

Robins, L.N., & O'Neal, P. (1958b). The marital history of former problem children. *Social Problems, 5,* 347–358.

Roe, A., & Burks, B. (1945). *Adult adjustment of foster children of alcoholic and psychotic parentage and the influence of the foster home.* New Haven, CT: Journal of Studies on Alcohol.

Sherfey, M.J. (1955). Psychopathology and character structure in chronic alcoholism. In O. Diethelm (Ed.), *Etiology of chronic alcoholism.* Springfield, IL: Charles C. Thomas.

Snyder, C.R. (1958). *Alcohol and the Jews.* Glencoe, IL: Free Press.

CHAPTER 24

Identifying Developmental Paradigms Leading to Alcoholism

Joan McCord

Alcoholism, like crime and mental illness, seems to run in families. Few who have known an alcoholic are likely to argue that an alcoholic's behavior will have no impact on his or her family. Partly for this reason, the relatively high rate of alcoholism found among children of alcoholics has often been interpreted in social-psychological terms (e.g., Blane & Barry, 1973; Burk, 1972; Fox, 1962; Zucker & Gomberg, 1986).

Over the last two decades, however, evidence has mounted to suggest a genetic component in the development of alcoholism (Bohman, 1978; Cadoret & Gath, 1978; Goodwin, 1976, 1981; Goodwin et al., 1974; Kaij & Dock, 1975; McKenna & Pickens, 1981; Schuckit, 1984; Schuckit, Goodwin, & Winokur, 1972; Schuckit & Rayses, 1979; Tarter, Alterman, & Edwards, 1985; Templer, Ruff, & Ayers, 1974).

Whether genetic factors placing a person at risk for alcoholism will produce alcoholism seems to depend, at least in part, on the environment. Various authors suggest that poverty contributes to the risk (El-Guebaly & Offord, 1977), that affection reduces it (Werner, 1986) and that attitudes toward drinking and drunkenness probably have an important impact (Cahalan, 1970; Robins, Bates, & O'Neal, 1962).

A prospective approach is particularly important for studying the impact of childhood on alcoholism. Alcoholics asked to recall their childhood may well have distorting memories that either justify their behavior or serve its self-punitive formula. The present study uses data collected as part of case materials when the subjects of interest were children.

Joan McCord is with the Department of Criminal Justice, Temple University, Philadelphia, Pa. This chapter is reprinted with permission from *Journal of Studies on Alcohol*, 49, pp. 357–362, 1988. An earlier version of this study was presented at the joint meeting of the Research Society on Alcoholism and the Committee on Problems of Drug Dependence, Philadelphia, Pa., June 14–19, 1987.

Method

Subjects for this study were drawn from cases in a project designed as a treatment program to prevent delinquency. The boys, born between 1926 and 1933, typically had received help between their 10th and 16th birthdays. All had lived in congested, urban areas near Boston, Massachusetts. Counselors had visited their homes about twice a month, over a period of more than 5 years, recording what they saw and heard (see Powers & Witmer, 1951).

Intake records had included reports by the children's teachers in elementary school. These, plus the counselors' records, provided data upon which to classify children and their families along many dimensions.

In 1957, records describing the families of those 253 boys who had remained in the program after an initial cut in 1941 were coded (see McCord & McCord, 1960). The codes included ratings of parental alcoholism, family structure and conflict, esteem of each parent for the other, parental supervision and disciplinary characteristics, and parental warmth and aggressiveness. To estimate the reliability of the coding, two raters independently read a 10% random sample of the cases. Agreement for these ratings ranged from 76% to 96% (see Table 1).

The coded records, uncontaminated by retrospective bias, contain the information about family life used in the present study. The predictive validity of the scales has been demonstrated in relation to adult criminal behavior (McCord, 1979).

The 253 boys in the follow-up study included 21 boys whose brothers were also in the study. To avoid counting particular constellations of families more than once, only one child from a family was used in analyses. The follow-up study also had included 29 boys whose acting parents were not their biological parents. Information about their biological parents was unavailable. They, too, were dropped from the present analyses.

Among the remaining 203 families, 65 fathers (32%) had been classified as alcoholics. The designation of alcoholic was given a father if he had lost jobs because of drinking or had marital problems attributed primarily to excessive drinking, if welfare agencies repeatedly noted that his heavy drinking was the source of problems, if he had received treatment for alcoholism, or if he had been convicted at least three times for public drunkenness. Using the same criteria to identify alcoholic mothers, researchers had classified 14 women as alcoholics; 8 were mothers of men whose fathers also were alcoholics.

The alcoholic fathers and their families differed from nonalcoholic fathers and their families in many ways (see Table 2). Alcoholic fathers were more likely to have been convicted for serious crimes ($\chi^2 = 21.334$, 1 *df*, $p = .0001$), to be living apart from the mothers of their children

TABLE 1
Interrater reliability: Dichotomous Variables
(2 Raters on 10% Random Sample)

	Percent agreement	Scott[a] π
Father's alcoholism	96	91
Family structure	96	92
Family conflict	80	55
Father's esteem for mother	84	68
Mother's esteem for father	88	76
Mother's self-confidence	84	60
Mother's control over boy	84	65
Boy's supervision	88	76
Expectations for boy	76	35
Father's discipline	88	52
Mother's discipline	84	62
Mother's leadership	96	91
Mother's "martyrdom"	88	25
Mother's attitude to son	84	68
Father's attitude to son	84	57
Mother's aggressiveness	92	56
Father's aggressiveness	84	41

[a]This measure provides a ratio of actual to possible improvement over chance, where chance equals the square of the proportion of the population with the characteristic plus the square of the proportion without the characteristic (Scott, 1955).

(χ^2 = 12.487, 1 *df,p* = .0004), and to be highly aggressive (χ^2 = 18.476, 1 *df, p* = .0001). They were less likely to be affectionate toward their sons (χ^2 = 4.379, 1 *df, p* = .0364) and less likely to show respect for their wives (χ^2 = 18.739, 1 *df, p* = .0001). Families in which the father was alcoholic were more likely to exhibit considerable conflict (χ^2 = 28.567, 1 *df, p* = .0001) and less likely to provide supervision for the child (χ^2 = 9.524, 1 *df, p* = .0020). Mothers in such families were less likely to be self-confident (χ^2 = 3.985, 1 *df, p* = .0459), to exert control over the child (χ^2 = 5.715, 1 *df, p* = .0168), or to hold the boy's father in high regard (χ^2 = 21.302, 1 *df, p* = .0001).

Between 1975 and 1980, the men were traced through courts, clinics for treatment of alcoholism, and public records. Once found, they were asked to respond to questionnaires and to participate in interviews. Information from these sources as well as from the records helped identify some of the men as alcoholics.

Sons were considered alcoholics on the basis of the follow-up, when they were 45–53 years of age. Both the questionnaire and the interview

TABLE 2
Families of Alcoholic and Nonalcoholic Men (in percent)

	Father not alcoholic (n = 138)	Father alcoholic (n = 65)
Father had criminal record[§]	14	43
Intact family[‡]	72	46
Considerable family conflict[§]	20	57
Father had high esteem for mother[§]	59	26
Mother had high esteem for father[§]	61	26
Mother was self-confident*	32	18
Mother had little control over boy*	30	48
Boy was supervised[†]	66	43
High expectations for boy	28	18
Father was consistently punitive	20	14
Mother was consistently nonpunitive	30	26
Mother was a leader	64	54
Mother acted as a martyr	7	14
Mother was affectionate to son	50	40
Father was affectionate to son*	33	18
Mother was aggressive	9	14
Father was aggressive[§]	9	34

*$p < .05$. [†]$p < .01$. [‡]$p < .001$. [§]$p < .0001$.

included the CAGE test for alcoholism (Ewing & Rouse, 1970). In this test, respondents are asked if they have ever taken a morning eye-opener, felt the need to cut down on drinking, felt annoyed by criticism of their drinking, or felt guilty about drinking. A man was considered an alcoholic if he responded affirmatively, as alcoholics do (Mayfield, McLeod, & Hall, 1974), to at least three of these questions.

There were 38 men who met the criteria for alcoholics used in the CAGE test. A man was also considered an alcoholic if he had received treatment for alcoholism (17 met this criterion), if he had been arrested at least three times for public drunkenness or driving while intoxicated (28 met this criterion), if he described himself as an alcoholic (true for 1 man), or if he had been arrested twice for alcoholism and answered affirmatively to two of the CAGE questions (true for 2). Altogether, 61 men (32%) met at least one of the criteria for alcoholism.

To qualify as nonalcoholic, a man had to have not been arrested for public drunkenness or driving while intoxicated, had to have not been treated for alcoholism, had to have scored less than three on the CAGE test, and had to have not described himself as an alcoholic. There were 132 nonalcoholics.

Ten men could not be classified. Although they met none of the criteria for classification as alcoholics, two had been arrested twice, and eight once, for public drunkenness. Of these 10, two had died in California, five were living in Massachusetts, two were found elsewhere, and one was not found. The analyses used the sample of 193 men who could be classified for alcoholism.

About two thirds of the alcoholics were identified through records and about two thirds through self-descriptions. Sources for the diagnoses of alcoholism among sons of alcoholics ($n = 28$) and sons of nonalcoholics ($n = 33$) differed little: 9 (32%) sons of alcoholics were identified from records only, 11 (39%) from self-description only, and 8 (29%) from records and self-description; 11 (33%) sons of nonalcoholics were identified from records only, 13 (39%) from self-description only, and 9 (27%) from records and self-description.

Criminal records had been searched in 1948, as part of treatment evaluation. In 1975, criminal records were searched again. These records showed the age of conviction as well as the charge for each of the men who had appeared in court. Sons could be classified by their juvenile records as well as by whether they had been convicted of the more serious street crimes (so-called index crimes) that appear on the Federal Bureau of Investigation crime index.

Results

Like other studies, this one shows that sons of alcoholics are at increased risk for alcoholism. Whereas 25% of the 133 men whose fathers were not alcoholic had become alcoholics, 47% of the 60 men with alcoholic fathers had become alcoholics. Additionally, alcoholic fathers and their sons were at increased risk for serious criminal behavior: In the alcoholic father group, 45% of the fathers and 47% of the sons were convicted of an index crime; and in the nonalcoholic father group, 19% of the fathers and 28% of the sons were convicted of an index crime.

Cross-classification of criminal histories for the sons and alcoholism in two generations suggests that both the father's alcoholism and that of the son are related to criminality. The relationships are shown in Table 3.

Juvenile delinquency was tied to paternal alcoholism. Boys whose fathers were alcoholics were more likely than their peers to have a record for juvenile delinquencies. Subsequent serious crimes, however, appeared to be more closely tied to the subject's own alcoholism.

As part of the selection process, teachers had completed "Trait Record Cards," describing the boys. Gathered in 1936 and 1937, these records

TABLE 3
Father's Alcoholism and Son's Criminal History (Percent Convicted, by Record Type)

	FATHER ALCOHOLIC		FATHER NOT ALCOHOLIC	
	Son alcoholic (*n* = 28)	Son not alcoholic (32)	Son alcoholic (33)	Son not alcoholic (100)
Son convicted as juvenile*	46	41	27	22
Son convicted for index crime as a juvenile†	39	31	24	13
Son ever convicted for an index crime†	57	38	45	22

$df = 3$. *$p < .05$. †$p < .01$.

TABLE 4
Boys According to Teachers' Descriptions (1936–37) and Type of Deviance (Percent of Each Category of Boys)

Deviance type	Aggressive (*n* = 33)	Shy (57)	Aggressive and shy (18)	Neither (85)
Neither	45	63	28	49
Alcoholic only	12	19	17	14
Criminal only	30	12	22	15
Alcoholic and criminal	12	5	33	21
Total	99[a]	99[a]	100	99[a]

$\chi^2 = 18.300$, 9 *df*, $p = .032$.
[a]Deviation from 100% due to rounding.

were used to classify the boys as aggressive (if their teachers had checked the description "fights") or as shy (if their teachers had checked "shy," "easily hurt," or "easily moved to tears"). Eighteen boys were classified as both shy and aggressive. These teachers' ratings of the boys in elementary school proved to be predictive of behavior decades later (see Table 4).

Shy children were least likely to become alcoholics or criminals. Boys rated as both aggressive and shy, however, were most likely to become alcoholics or criminals. Of the 8 shy-aggressive boys who had alcoholic fathers, 6 had become alcoholics, as had 3 of the 10 whose fathers were not alcoholic.

TABLE 5
Stepwise Logistic Regression

Term	Coefficient	Standard error	Coeff./SE
Alcoholic father	0.568	0.194	2.929
Mother's esteem	0.669	0.238	2.808
Mother's control	0.336	0.173	1.948
Parental conflict	0.366	0.245	1.495
Constant	−0.406	0.184	−2.201

CORRELATION COEFFICIENTS

	Alcoholic father	Mother's esteem	Mother's control	Conflict	Constant
Alcoholic father	1.000				
Mother's esteem	0.236	1.000			
Mother's control	−0.007	0.178	1.000		
Parental conflict	−0.187	0.612	0.037	1.000	
Constant	0.167	0.273	0.215	0.335	1.000

To identify parental behaviors related to alcoholism of the son, 18 variables were entered into a stepwise logistic regression analysis, with the use of BMDPLR (Dixon, 1985). These variables described the father in terms of alcoholism, criminality, affection for his son, esteem for the boy's mother, disciplinary techniques, and aggressiveness. They described the mother in terms of role, affection for her son, self-confidence, esteem for the boy's father, disciplinary techniques, and aggressiveness. Further description of the families dealt with whether they were intact or broken, whether or not the parents were in considerable conflict, whether anyone supervised the boy after school, and whether the boy was expected to perform well.

Using remove and enter limits of .15 and .10, the analysis identified four variables as predictors of subsequent alcoholism of the sons. The father's alcoholism accounted for the greatest amount of variance ($F = 8.32$, 1/187 *df*, $p = .0044$). The mother's esteem for the boy's father accounted for an additional significant proportion of the variance ($F = 7.64$, 1/187 *df*, $p = .0063$). The mother's control over the boy improved the fit ($F = 3.68$, 1/187 *df*, $p = .0568$). And parental conflict ($F = 2.17$, 1/187 *df*, $p = .1428$) provided the next-best predictor. Regression coefficients, standard errors, their ratios, and associated probabilities are shown in Table 5, along with correlations among the four variables. Together, the linear function based on these four variables provided a logistic model having a reasonable fit to the data ($\chi^2 = 0.327$, 2 *df*, $p = .849$).

TABLE 6
Family Backgrounds for Alcoholism (Log-Linear Analysis)

Source (effect)	*df*	χ^2	Probability
SATURATED MODEL			
Father's alcoholism	1	19.78	.0001
Mother's esteem for father	1	12.58	.0004
Mother's control of boy	1	4.84	.0278
Father's alcoholism × mother's esteem	1	6.37	.0116
Father's alcoholism × mother's control	1	0.06	.8075
Mother's esteem × mother's control	1	0.22	.6395
Father's alcoholism × mother's esteem × control	1	0.71	.3992
Residual	1	3.16	.0755
FULLY SPECIFIED MODEL			
Father's alcoholism	1	20.75	.0001
Mother's esteem for father	1	12.15	.0005
Mother's control of boy	1	5.23	.0222
Father's alcoholism × mother's esteem	1	6.70	.0096
Residual	4	4.38	.3569

Log linear analyses were used to inspect the interaction terms and main effects of the three most important predictors: father's alcoholism, mother's esteem for the father, and mother's control of the boy during childhood (see Table 6). As can be seen, father's alcoholism, mother's esteem for the father, and mother's control of the boy each significantly differentiated alcoholic from nonalcoholic sons. In addition, the interaction between father's alcoholism and mother's esteem contributed reliably to prediction of the son's alcoholism.

Maternal control and the mother's esteem for her husband differed in their effects depending on whether the father was alcoholic (Table 7). Maternal control over the boy had a stronger effect among sons of nonalcoholics than among sons of alcoholics. Maternal esteem for the boy's father, on the other hand, had a particularly strong effect on alcoholism among boys whose fathers were alcoholics. Among the sons of alcoholics, 73% of the 15 whose mothers showed high esteem for their sons' fathers had become alcoholics compared with only 38% of the 45 whose mothers had not shown high esteem for the fathers.

These analyses suggest two paths toward alcoholism. One path appears to be imitative, the other to represent uninhibited behavior. Additional differences between sons of alcoholics and sons of nonalcoholics were sought in responses to interview questions about behavior related to drinking.

TABLE 7
Family Type and Alcoholism

Father's alcoholism?	Mother's high esteem?	Mother's little control?	% Alcoholic	*n*
No	No	No	12	33
No	No	Yes	32	19
No	Yes	No	25	61
No	Yes	Yes	40	20
Yes	No	No	35	23
Yes	No	Yes	41	22
Yes	Yes	No	70	10
Yes	Yes	Yes	80	5

During the interviews, men were asked whether they had ever had an accident because of drinking, hurt a friendship or marriage because of drinking, or missed an appointment or been involved in a fight because of drinking. A majority of the alcoholics (85%) answered affirmatively, whether or not their fathers had been alcoholics: 87% of the alcoholic sons of alcoholics (n = 23), and 83% of the alcoholic sons of nonalcoholics (n = 23). Among nonalcoholics, however, men whose fathers had been alcoholics were more likely to have been in these types of trouble: 47% of the nonalcoholic sons of alcoholics (n = 19) and only 14% of the nonalcoholic sons of nonalcoholics (n = 65) had gotten into trouble through drinking (χ^2 = 9.813, 1 *df*, p = .0017).

During the interviews, the men were asked about their drinking patterns. Alcoholics from families in which the father was also alcoholic were almost twice as likely as other alcoholics to report daily drinking. Whereas 83% of the alcoholic sons of alcoholics reported daily drinking, 48% of the alcoholic sons of nonalcoholic men reported that behavior pattern (χ^2 = 6.133, 1 *df*, p = .0133). The proportions of nonalcoholics from alcoholic families approximated the proportions of those from nonalcoholic families in daily drinking: 21% of the former and 23% of the latter said they drank every day.

Drinking and trouble seemed linked for sons of alcoholics. Drinking and companionship seemed linked for alcoholics from nonalcoholic families. These latter alcoholics tended to be active in social clubs. Among alcoholics whose fathers were not alcoholic, 87% reported being active in clubs, whereas only 41% of the alcoholic sons of alcoholic men reported such activity (χ^2 = 10.405, 1 *df*, p = .0013). The proportions of nonalcoholics from alcoholic families approximated the proportions of those from nonalcoholic families in activity in clubs: 61% of the former and 63% of the latter described at least one club in which they were active.

Discussion

This study has considered development of alcoholism using a prospective design. Subjects were drawn from youths who had participated in a program designed to prevent delinquency. Their case material dated from the 1930s. The case records, uncontaminated by retrospective bias, included descriptions of their behavior in elementary school and observations of family interactions over several years.

The records indicate that shy-aggressive children tended to be at risk for criminal and alcoholic behavior. The records also indicate that alcoholic fathers and their families differed from nonalcoholic fathers and their families in many ways. Although having an alcoholic father increased exposure to conflict and rejection, these conditions appeared not to increase the risk of alcoholism.

Two developmental paradigms emerged from the analyses. In one, men with alcoholic fathers seemed to be imitating a respected father. Sons of alcoholic men were more likely to become alcoholics if their mothers had shown high esteem for their sons' fathers. These alcoholics were likely to drink on a daily basis. Even if not alcoholic themselves, sons of alcoholics tended to report getting into trouble while drinking. Men whose fathers were alcoholics may have been taught that their fathers' behavior was acceptable—or, at least, forgivable. They may have accepted drinking as part of the masculine role. In the other developmental paradigm, lack of maternal control increased the probability that sons would become alcoholics. These alcoholics, especially those whose fathers were not alcoholics, apparently associated drinking with uninhibited, often antisocial behavior.

The models built on paternal alcoholism, mother's approval of her husband, and maternal control clearly provide only a partial explanation for the development of alcoholism. Other variables, both biological and social, could undoubtedly improve both the specificity and the sensitivity of the predictions. It seems reasonable to suggest that better measures of genetic loading might improve predictions, as might the addition of such social variables as approval of heavy drinking.

Acknowledgment

This study was supported in part by U.S. Public Health Service research grant MH26779 (Center for Studies of Crime and Delinquency, National Institute of Mental Health).

References

Blane, H.T., & Barry, H., III. (1973). Birth order and alcoholism: A review. *Quarterly Journal of Studies on Alcohol, 34,* 837–852.

Bohman, M. (1978). Some genetic aspects of alcoholism and criminality: A population of adoptees. *Archives of General Psychiatry, 35,* 269–276.

Burk, E.D. (1972). Some contemporary issues in child development and the children of alcoholic parents. *Annals of the New York Academy of Sciences, 197,* 189–197.

Cadoret, R.J., & Gath, A. (1978). Inheritance of alcoholism in adoptees. *British Journal of Psychiatry, 132,* 252–258.

Cahalan, D. (1970). *Problem drinkers.* San Francisco: Jossey-Bass.

Dixon, W.J. (1983). *BMDP Statistical Software 1985.* Berkeley, CA: University of California Press.

El-Guebaly, N., & Offord, D.R. (1977). The offspring of alcoholics: A critical review. *American Journal of Psychiatry, 34,* 357–365.

Ewing, J.A., & Rouse, B.A. (1970, February). Identifying the hidden alcoholic. Paper presented at the 29th International Congress on Alcohol and Drug Dependence, Sydney, New South Wales, Australia.

Fox, R. (1962). Children in the alcoholic family. In W.C. Bier (Ed.), *Problems in addiction: Alcohol and drug addiction* (pp. 71–96). New York: Fordham University Press.

Goodwin, D.W. (1976). *Is alcoholism hereditary?* New York: Oxford University Press.

Goodwin, D.W. (1981). Family studies of alcoholism. *Journal of Studies on Alcohol, 42,* 156–162.

Goodwin, D.W., Schulsinger, F., Møller, N., Hermansen, L., Winokur, G., & Guze, S. (1974). Drinking problems in adopted and nonadopted sons of alcoholics. *Archives of General Psychiatry, 31,* 164–169.

Kaij, L., & Dock, J. (1975). Grandsons of alcoholics: A test of sex-linked transmission of alcohol abuse. *Archives of General Psychiatry, 32,* 1379–1381.

McCord, J. (1979). Some child-rearing antecedents of criminal behavior in adult men. *Journal of Personality and Social Psychology, 37,* 1477–1486.

McCord, W., & McCord, J. (1960). *Origins of alcoholism.* Stanford, CA.: Stanford University Press.

McKenna, T., & Pickens, R. (1981). Alcoholic children of alcoholics. *Journal of Studies on Alcohol, 42,* 1021–1029.

Mayfield, D., McLeod, G., & Hall, P. (1974). The CAGE questionnaire: Validation of a new alcoholism screening instrument. *American Journal of Psychiatry, 131,* 1121–1123.

Powers, E., & Witmer, H. (1951). *An experiment in the prevention of delinquency: The Cambridge-Somerville Youth Study.* New York: Columbia University Press.

Robins, L.N., Bates, W.M., & O'Neal, P. (1962). Adult drinking patterns of former problem children. In D.J. Pittman & C.R. Snyder (Eds.), *Society, Culture, and Drinking Patterns* (pp. 395–412). New York: Wiley.

Schuckit, M.A. (1984). Relationship between the course of primary alcoholism in men and family history. *Journal of Studies on Alcohol, 45,* 334–338.

Schuckit, M.A., Goodwin, D.W., & Winokur, G. (1972). A study of alcoholism in half siblings. *American Journal of Psychiatry, 128,* 122–126.

Schuckit, M.A., & Rayses, V. (1979). Ethanol ingestion: Differences in blood acetaldehyde concentrations in relatives of alcoholics and controls. *Science, 203,* 54–55.

Scott, W.A. (1955). Reliability of content and analysis: The case of nominal scale coding. *Public Opinion Quarterly, 19,* 321–325.

Tarter, R.E., Alterman, A.I., & Edwards, K.L. (1985). Vulnerability to alcoholism in men: A behavior-genetic perspective. *Journal of Studies on Alcohol, 46,* 329–356.

Templer, D.I., Ruff, C.F., & Ayers, J. (1974). Essential alcoholism and family history of alcoholism [Notes and Comment]. *Quarterly Journal of Studies on Alcohol, 35,* 655–657.

Werner, E.E. (1986). Resilient offspring of alcoholics: A longitudinal study from birth to age 18. *Journal of Studies on Alcohol, 47,* 34–40.

Zucker, R.A., & Gomberg, E.S.L. (1986). Etiology of alcoholism reconsidered: The case for a biopsychosocial process. *American Psychologist, 41,* 783–793.

CHAPTER 25

The Etiology of Alcoholism: A Prospective Viewpoint

George E. Vaillant and Eva S. Milofsky

In retrospect, it has always seemed clear that alcoholism is a symptom of unhappy childhood and of an unstable or dependent personality. In his classic monograph on alcoholism, Jellinek (1960) states, "In spite of a great diversity of personality structures among alcoholics there appears in a large proportion of them a low tolerance for tension coupled with an inability to cope with psychological stresses" (p. 153). In the most recent edition of the *Comprehensive Textbook of Psychiatry,* Selzer (1980) writes: "Alcoholic populations do display significantly more depression, paranoid thinking trends, aggressive feelings and acts, and significantly lower self-esteem, responsibility and self-control than non-alcoholic populations. Despite occasional disclaimers, alcoholics do not resemble a randomly chosen population" (p. 1629). Other retrospective clinical studies of alcohol abuse have equated the antecedents of alcoholism with those of the oral dependent personality (Blane, 1968).

However, since alcoholism profoundly distorts the individual's personality, his or her social stability, and recollection of relevant childhood variables, retrospective impressions are suspect. In recent years six prospective studies (Jones, 1968, 1971; Kammeier, Hoffmann, & Loper, 1973; McCord & McCord, 1960; Robins, Bates, & O'Neal, 1962; Vaillant, 1980) have demonstrated many fallacies in our retrospective conception of the alcoholic's personality. The McCords (1960) observed that, in contradiction to retrospective studies, alcoholics were premorbidly no more likely than controls to manifest phobias, "strong inferiority feelings," more "feminine feelings," "oral tendencies," or "strong encouragement of depen-

George E. Vaillant is Raymond Sobel Professor of Psychiatry, Dartmouth Medical School, Hanover, N.H. Eva S. Milofsky is administrator, Office of the Dean, Harvard University, Cambridge, Mass. This chapter is reprinted with modifications from *American Psychologist, 37,* pp. 494–503, 1982. It was prepared in part while the first author was a fellow at the Center for Advanced Study in the Behavioral Sciences, Stanford, California. A postscript prepared by the first author especially for this book is included at the end of the chapter.

dency" from their mothers. Jones (1968) and Loper, Kammeier, and Hoffmann (1973) observed that contrary to popular beliefs, alcoholics were, if anything, premorbidly outwardly more self-confident, aggressive, and heterosexual than their peers. Alcoholism also appears to be the cause rather than the result of passive-dependent traits (Vaillant, 1980), of elevations on the D and Pd scales of the Minnesota Multiphasic Personality Inventory (MMPI) (Kammeier, Hoffmann, & Loper, 1973), and of passivity, low self-esteem, and introversion (McCord & McCord, 1960). Could alcoholism also be the cause, not the result, of unhappy childhood, broken families, and personality disorder?

Previous work (McCord & McCord, 1960; Robins, Bates, & O'Neal, 1962) identified premorbid antisocial behavior as an important predictor of alcoholism, but these two important prospective studies were derived from samples that premorbidly exhibited a disproportionate amount of antisocial behavior (Powers & Witmer, 1951; Robins, 1966). To evaluate the previous etiological conclusions, we chose to restudy the junior high school boys that the Gluecks (1950) had used as a control group in the monograph *Unraveling Juvenile Delinquency.* As Table 1 suggests, this second follow-up study offered certain methodological advantages. Originally, the sample of inner-city men was selected in early adolescence for nondelinquency. The men were personally reinterviewed at three points in time: ages 25, 31, and 47—well past the peak age of onset of alcoholism. Besides 19 deaths, attrition was held to 7%, and no subject was completely lost (Vaillant & Milofsky, 1980). The present study identified more alcohol abusers than five of the six previous studies cited in Table 1 combined. In contrast to the previous investigators, the Gluecks had prospectively and painstakingly studied the men's relatives over three generations both for cultural background and for alcoholism—variables that recent reviews (Goodwin, 1979; Heath, 1975) suggest are very important in the etiology of alcohol dependence.

The present report will examine the independent contribution of five major variables to the development of alcoholism.

The Sample and Methods of Study

The sample included the 456 boys studied in their early adolescence by the Gluecks between 1940 and 1963 as a control group for their well-known studies of juvenile delinquents (Glueck & Glueck, 1950, 1968). In more recent follow-ups (Vaillant, 1980; Vaillant & Milofsky, 1980; Vaillant & Vaillant, 1981) this group has been called the *core city* sample. This sample had been carefully matched for IQ, ethnicity, and residence in high-crime neighborhoods with 456 Boston youths who had been remanded

TABLE 1
Prospective Studies of the Development of Alcoholism

Study	Age at follow-up	*n* original sample	*n* lost to follow-up	*n* alcohol abusers	*n* not antisocial abusers	% abusers also dependent[b]	Recent interview	Multiple contacts over time
McCord & McCord (1960)	30–35	325	70	29	11	67	No	No
Robins et al. (1962)	c. 43	367	81	78[a]	29	34	Yes	No
Jones (1968)	36–38	106	40	6	6	–	Yes	Yes
Jones (1971)	36–38	106	61	3[d]	3	–	Yes	Yes
Kammeier et al. (1973)	30–35	–[c]	–[c]	38	38	100	No	No
Vaillant (1980)	55	204	2	26	25	34	Yes	Yes
Core city sample	47	456	56	110	90	67	Yes	Yes

[a]This figure includes men who met Robins' diagnostic criteria for schizophrenia and sociopathy and were therefore excluded from her later (1966) report. Women and controls who abused alcohol are also excluded.
[b]These percentages are our rough estimates of the severity of the alcohol abuse in the reviewed studies based on the authors' criteria for inclusion.
[c]This was a follow-back study.
[d]In an unpublished study by Jones this has been increased to 8.

to reform school. Their average IQ was 95 ± 12; eventually 48% graduated from high school or attained high school equivalency certification. Although there were no blacks in the Glueck study, one or both parents of 61% of the boys had been born in a foreign country. Although at age 14 ± 2 years the youths had been chosen for the absence of obvious delinquency, eventually 19% spent time in jail, a datum suggesting a sample only modestly biased toward good behavior. Eventually, 7% (a rather high proportion) met Robins' (1966) criteria for sociopathy.

At the time of original study, the boys, their parents, and their teachers were individually interviewed. Public records were searched for evidence of alcoholism, criminal behavior, and mental illness in all first-degree relatives. Multiple social agency reports were available on most families. Over 90% of the surviving subjects were reinterviewed at ages 25 and 31 (Glueck & Glueck, 1968). At these interviews alcohol abuse or its absence was specifically recorded. When they reached age 47, a two-hour semistructured interview with a detailed 23-item section on problem drinking was used to reinterview 87% of the surviving subjects. The time frame used was the period from age 21 to age 47. Most of these interviews were performed by individuals with two or more years of experience in alcohol clinics; but unfortunately, significant others were not usually interviewed. In addition to the serial interviews, recent psychiatric, medical, and arrest records were also obtained on most subjects. These records helped to confirm that the subjects were reliable informants and to identify additional alcohol-related problems.

On the basis of a search through social service records and interviews with the boy, his parents, and his teacher, clinicians blind to all information about the boy gathered after adolescence rated each of the men on the following 12 premorbid variables:

1. *Parental cultural background.* If a boy's parents had been born in different foreign countries, ethnicity was assigned to the place of birth of the father. If only one parent was foreign born, ethnicity was assigned to the birthplace of that parent. If both parents had been born in the United States, ethnicity was assigned to the birthplace of the parental grandparents.
2. *Parental social class.* This was the 5-point classification devised by Hollingshead and Redlich (1958).
3. *Parental delinquency.* Criteria similar to the parental alcohol abuse category were employed.
4. *Parental alcohol abuse.* A rating of 1 = neither parent alcoholic; 2 = one parent with minor evidence (two convictions for alcoholism, mention of alcoholism in official records, or strong suspicion in the case record that was compiled from family interview and multiple social agency reports); 3 = one parent with major evidence (two or

more of the criteria for minor evidence as well as evidence of chronicity) or two parents with minor evidence; 4 = both parents alcoholic, at least one with major evidence.

5. *Alcoholic abuse in ancestors.* Clinicians rated alcoholism in first- and second-degree relatives who did not live with the subject. If reliable judgment was not possible because of ambiguous or absent familial data, a rating was not made. Otherwise, 1 = no evidence of alcoholism in family; 2 = one relative with minor evidence of alcoholism (defined as in parental alcohol abuse, above); 3 = two relatives with minor evidence of alcoholism or one relative with major evidence (defined as in parental alcohol abuse, above); 4 = three relatives with minor evidence or two relatives with major evidence.
6. *Boyhood competence.* This 8-point scale (Vaillant & Vaillant, 1981) is intended as a crude scale of ego strength. At entrance into the study, the boys received points for doing regular chores, adjusting well to school socially and academically (the ratings controlled for IQ), participating in after-school jobs, and coping with difficulties in their inner-city homes. Interrater reliability ranged from .70 to .91.
7. *Childhood environmental strengths.* This is a 20-point scale described in detail elsewhere (Vaillant, 1974) that rated the men by clinical judgment of childhood environmental strengths. This scale gave points for what went well rather than wrong. Interrater reliabilities among three raters ranged from .70 to .89.
8. *Childhood emotional problems.* A subscale of the above scale was used: 2 = good natured, unusually social; 1 = average, with no known problems; 0 = having emotional problems (e.g., dissociability, feeding problems, pronounced shyness, phobias, enuresis after age 8, stuttering, hyperactivity). Interrater reliability was .56.
9. *Childhood environmental weaknesses.* This is a 50-point scale (Vaillant & Vaillant, 1981) based on 25 concrete criteria that reflected the Gluecks' more clinically defined Delinquency Prediction Scale (Glueck & Glueck, 1950). The scale included 5 items reflecting gross lack of family cohesion (e.g., 9 or more social agency contacts, more than 6 months apart from both parents during childhood). Five items reflected lack of affection and 5 lack of supervision by the mother, and 10 items reflected grossly inadequate supervision and affection by the father. If 10 or more of the 25 items were present, the boy was said to belong to a multiproblem family (n = 62, or 14%). Interrater reliability was .91.
10. *IQ.* Upon entering the study, each subject was given the Wechsler-Bellevue test.
11. *School problems and truancy.* Clinicians recorded the presence or absence of repeated truancy or of disciplinary complaints from teachers, fights with students, and so on.
12. *Infant health problems.* This is the Gluecks' (1950) dichotomous variable regarding whether the mother recollected her son as "cranky, nervous, and fretful as an infant."

On the basis of two-hour interviews at ages 31 and 47 and on the basis of recent mental health and arrest records, clinicians blind to all data gathered before age 31 rated the men at age 47 ± 2 on the following 8 variables:

1. *Health Sickness Rating Scale (HSRS).* This 100-point scale is described in detail elsewhere (Luborsky, 1962; Luborsky & Bachrach, 1974). Interrater reliability was .89.
2. *Social class.* See Hollingshead and Redlich (1958).
3. *Sociopathy scale.* This scale is based on the presence and absence of the 19 criteria used by Robins (1966) for diagnosing sociopathic personality.
4. *Education.* Since this measure incorporated all adult education, it was regarded as an outcome measure. Many of the men attended technical school or achieved high school equivalency in their late 20s.
5. *Alcohol abuse in heredity.* This scale is identical to that measuring alcohol abuse in ancestors but the ratings (1) were made when the men were aged 47 and (2) included available data on both parental and sibling alcohol abuse. Raters were blind to the men's alcohol use.

Because evidence of alcoholism differs depending on the criteria used for its identification and whether it is viewed from a medical or a sociological vantage point, three scales were used for its assessment.

6. *DSM-III scale.* This is a 3-point scale defined by the American Psychiatric Association's (1980) *Diagnostic and Statistical Manual* (3rd ed.) and reflects the medical model for diagnosing alcoholism: 1 = no alcohol abuse; 2 = alcohol abuse; 3 = alcohol dependence (evidence for alcohol abuse and tolerance to, or withdrawal symptoms from, alcohol). Seventy-one men met the criteria for alcohol dependence.
7. *Cahalan scale.* This is the 11-item scale devised by Cahalan (1970) and co-workers to assess the national prevalence of alcohol abuse; it reflects a sociological model.
8. *Problem Drinking Scale (PDS).* The Problem Drinking Scale (Vaillant, 1980) is an equally weighted 16-point scale devised to combine the emphasis of the DSM-III on physiological dependence and the emphasis of the Cahalan (1970) scale on social deviance. Of the 456 men, 442 could be rated with some confidence, and 400 with great confidence, on this scale. In the 400 best-studied cases, men with 0 to 1 symptoms on the scale were, by definition, *social drinkers* (n = 256). Men with 4 or more problems were defined as *alcohol abusers* (n = 110).

Premorbid Predictors of Alcoholism

Our subjects were interviewed at four points over a 33-year span. We were able at least to locate all of our surviving subjects after age 40, and

TABLE 2
Adult Variables That Are More a Result Than a Cause of Alcoholism (in percent)

Variable	Social drinkers (n = 260)	Alcohol dependent (n = 71)
Adult		
Lowest adult social class (V)	4	21
10+ years unemployed	4	24
Never completed high school	56	41
HSRS < 70	24	51
Corresponding childhood		
Lowest parent's social class (V)	32	30
Multiproblem family membership	11	14
IQ < 90	28	30
Childhood emotional problems	32	30

Note: HSRS = Health Sickness Rating Scale (Luborsky, 1962).

we could make a reasonable judgment regarding the presence or absence of alcohol abuse in 442 of the men (Vaillant, Gale & Milofsky, 1982). We obtained essentially complete data sets on 400 surviving men and personal interviews on 367 of our surviving subjects. (Nineteen men died before age 40 and were excluded from most data analyses.) Available evidence (Vaillant, 1982) suggests that attrition was more common among the antisocial and emotionally unstable than among the alcohol abusers.

Table 2 illustrates that in our sample the differences frequently observed between alcoholic and nonalcoholic subjects to which retrospective studies often attribute etiological significance—differences in social class, unemployment, eventual educational achievement, and mental illness—in fact appear after, not before, the development of alcoholism. Although as adults our alcohol-dependent subjects appeared to have personality disorders and to be socially inadequate, as children they were no more underprivileged than their peers who were to drink socially as adults; they were no less intelligent; and they manifested no more evidence of childhood emotional vulnerability. Other childhood variables that did not correlate with subsequent alcoholism included the quality of maternal affection (a subscale of the childhood weakness scale) and the number of mentally ill relatives.

Table 3 puts into perspective the relationship between childhood variables that did correlate with alcoholism and those that correlated with other facets of adult outcome. The purpose of the table is to tease out differences in the etiology of alcoholism, the etiology of sociopathy (as defined by Robins, 1966), and the etiology of poor adult mental health as

TABLE 3
Childhood Variables Predicting Poor Adult Mental Health, Sociopathy, and Alcoholism

Childhood variable	DSM-III (n = 398)	Cahalan scale[a] (n = 399)	Problem drinking scale[b] (n = 442)	HSRS (n = 378)	Sociopathy[c] (n = 430)
MENTAL HEALTH PREDICTORS					
Childhood environmental strengths	–.18***	–.15**	–.14**	.21***	–.17***
Boyhood competence	–.12**	–.14**	–.11**	.24***	–.22***
Childhood emotional problems	NS	NS	NS	–.19***	.13**
IQ	NS	NS	NS	.15**	–.10*
SOCIOPATHY PREDICTORS					
Childhood environmental weaknesses	.15**	.16***	.14**	–.10*	.18***
Delinquent parents	NS	NS	NS	–.10*	.13**
Truancy/school behavior problems	.20***	.20***	.19***	NS	.35***
Poor infant health	.11**	.10*	.09*	.12**	.17***
ALCOHOLISM PREDICTORS					
Alcoholism in parents (1945)	.20***	.17***	.20***	NS	.10*
Alcoholism in ancestors[d]	.14**	.15**	.10*	NS	NS
Alcoholism in heredity (1978)	.26***	.28***	.23***	NS	.11**
Cultural background[e]	–.27***	–.25***	–.27***	NS	–.12**

Note: Pearson product-moment correlation coefficients were the statistics used; HSRS = Health Sickness Rating Scale (Luborsky, 1962); DSM-III = Alcohol abuse and dependence as defined by the *Diagnostic and Statistical Manual of Mental Disorders* (3rd ed.; American Psychiatric Association, 1980).

[a]See Cahalan (1970).
[b]See Vaillant (1980).
[c]See Robins (1966).
[d]Excluding parents.
[e]For purposes of statistical depiction, Irish ethnicity was assigned a weight of 1; old American, Canadian, or British ethnicity, a weight of 2; other northern European ethnicity, a weight of 3; and southern European ethnicity, a weight of 4 (see Jellinek, 1960). For purposes of multiple regression and statistical analysis, a given ethnicity is treated as a dummy variable.

*$p < .05$. **$p < .01$. ***$p < .001$.

defined by the Health Sickness Rating Scale (Luborsky, 1962). The first four childhood variables—childhood environmental strengths (Vaillant, 1974), boyhood competence (Vaillant & Vaillant, 1981), childhood emotional problems, and IQ—significantly predicted independent ratings of adult mental health made 33 years later. These four variables predicted sociopathy less well, and they were only marginally correlated with adult alcoholism. The next four variables—delinquent parents, multiproblem families (childhood environmental weaknesses), poor infant health, and premorbid antisocial personality—correlated most powerfully with sociopathy, less so with alcoholism, and only marginally with subsequent mental health. Finally, the last four variables seemed significantly associated with adult alcoholism but not with sociopathy or poor mental health. (As can be seen in Table 3, the association between alcoholism and premorbid etiological variables was not really affected by whether alcoholism was defined according to the medical model of the DSM-III or the social deviance model of Cahalan, 1970.)

The next three tables examine some of the variables in Table 3 in greater detail. As Table 4 illustrates, the 71 men with several alcoholic relatives were 3 times more likely than the 178 men with no alcoholic relatives to develop alcohol dependence. This relationship remained significant but not as dramatic if parents and siblings (who could exert environmental effects, genetic effects, or both) were excluded from the comparison and the more conservative scale measuring alcohol abuse in ancestors was substituted.

Alcoholism, especially in young men, may be a cause as well as a symptom of antisocial behavior. Table 5 presents the association between alcohol dependence and premorbid truancy and school behavior problems at age 14 ± 2 (such misbehavior antedated alcohol abuse in all but two of the men). Men with serious premorbid behavior problems were four times as likely to develop alcohol dependence as those without such problems; but since there were only 16 men with severe behavior problems, premorbid antisocial behavior cannot be invoked as playing a major etiological role for most of the 71 alcohol-dependent men in the study.

Although the men themselves were native born and shared a common social environment, the cultural mores of their parents (half of whom were foreign born) differed greatly. Such differences in parental ethnicity were highly correlated with the men's subsequent use of alcohol. Table 6 contrasts the culture in which the men's parents were raised with the likelihood of the development of alcohol dependence among the men themselves. The 75 men of Irish extraction were 7 times more likely to manifest alcohol dependence than the 130 men of Italian, Syrian, Jewish, Greek, or Portuguese extraction.

TABLE 4
Number of Alcohol-Abusing Relatives and Development of Alcohol Dependence among Core City Men (in percent)

Scale	DSM-III ALCOHOL CLASSIFICATION		
	No abuse (n = 267)	Abuse without dependence (n = 60)[a]	Dependence (n = 71)
Alcohol abuse in heredity			
No relatives (n = 178)	78	12	10
1–2 relatives (n = 149)	60	20	20
Several relatives (n = 71)	52	14	34[c]
Alcohol abuse in ancestors[b]			
No relatives (n = 213)	73	31	14
1–2 relatives (n = 133)	63	18	19
Several relatives (n = 52)	58	13	29[c]
Alcohol abuse in parents			
Mother alcoholic (n = 36)	50	14	36[c]
Father alcoholic (n = 149)	58	14	28[c]
Neither (n = 244)	75	15	11
Total sample (n = 398)	67	15	18

Note: Rows, not columns, total 100%. DSM-III = *Diagnostic and Statistical Manual of Mental Disorders* (3rd ed.; American Psychiatric Association, 1980). Values are in percentages.
[a]Alcohol abuse as defined by the DSM-III criteria was a somewhat more inclusive category than alcohol abuse defined by having 4 or more symptoms on the Problem Drinking Scale (Vaillant, 1980).
[b]Excluding parents and siblings.
[c]$p < .01$ by chi-square test.

Admittedly, the relationship between alcohol use and culture is extremely complex (Greely, McCready, & Theisen, 1980; Heath, 1975; Marlatt & Rohsenow, 1980; Stivers, 1976), but in this study the relationship is reduced to a single common denominator—cultural characteristics of alcohol use that empirically affect the likelihood of alcohol dependence. Thus, ethnicity was crudely scaled to contrast cultures that forbid drinking in children but condone drunkenness in adults with those cultures that teach children how to drink responsibly but forbid adult drunkenness (Jellinek, 1960; Pittman & Snyder, 1962). Space does not permit full elaboration of the rationale for this division. The non-Moslem Mediterranean countries have no sanctions against children learning to drink but have strong sanctions against drunkenness in adults. On the basis of Snyder's

TABLE 5
Relationship of Premorbid Antisocial Behavior to Development of Alcoholism

DSM-III classification	TRUANCY OR SCHOOL BEHAVIOR PROBLEMS	
	Absent (n = 381)	Present (n = 16)
No alcohol abuse	69	31
Alcohol abuse	15	13
Alcohol dependent	16	56

Note: DSM-III = *Diagnostic and Statistical Manual of Mental Disorders* (3rd ed.; American Psychiatric Association, 1980). All p's < .001 by chi-square test. Values are in percentages.

TABLE 6
Relationship between Parents' Culture and Development of Alcohol Dependence (in percent)

DSM-III classification	CULTURE OF PARENTS		
	Irish (n = 75)	Other[a] (n = 193)	Mediterranean (n = 130)
No alcohol abuse	59	58	86
Alcohol abuse without dependence	13	19	10
Alcohol dependent	28	23	4

Note: DSM-III = *Diagnostic and Statistical Manual of Mental Disorders* (3rd ed.; American Psychiatric Association, 1980). All p's < .001 by chi-square test.
[a]Canadian, American, northern European.

(Pittman & Snyder, 1962) work, the eight Jews in the study were included with the Mediterraneans, and the one Chinese was excluded. Conversely, the Irish and North American cultures forbid drinking in adolescence, have flirted with total prohibition for a century, but give tacit or explicit approval to drunkenness in male adults. For purposes of statistical depiction, other northern European countries were assigned an intermediate position. Jellinek (1960) points out in some detail that although France teaches children responsible drinking, it also condones drunkenness and alcohol use independent of meals.

The interaction between parental culture and familial alcoholism was interesting. Among men of Irish extraction the presence of alcoholic relatives only slightly increased the risk of alcohol dependence in the subjects: Many Irish subjects with alcoholic relatives, in fact, became lifelong teetotalers. Among other ethnic groups (both among the southern Europeans, in whom alcohol abuse was rare, and among the northern Europeans and old Americans, in whom it was common), alcohol dependence occurred five times as often in men with several alcoholic relatives as it did in men with no alcoholic relatives.

Finally, we do not wish to imply that psychological stress and vulnerability are irrelevant in alcoholism, but only that a prospective design and a multifactorial design greatly diminish their importance.

Analysis of the Findings

The correlations in Table 3 reflect the essential findings of the study: Except for the scale measuring childhood environmental strengths, the premorbid variables that most clearly predict alcoholism do not predict mental health, and vice versa. Although correlations in Table 3 seem low, it must be remembered that 33 years separate the childhood and adult ratings. In an exhaustive review of the prospective literature on childhood-to-adult studies, Kohlberg, LaCrosse, and Ricks (1972) report that if IQ is excluded, correlations of .2 to .4 are the highest observed between ratings of childhood and adult behavior. Wordsworth and Freud not-withstanding, prospective study of life spans has taught us that "the child is father of the man" in a most limited fashion.

To be fully understood, further analysis of the Table 3 variables is necessary. For example, can the increased risk of alcoholism in children from multiproblem families be explained by the fact that many of the problems in such families are the result of alcoholic parents and grandparents? For the purposes of multiple regression analysis, it seemed appropriate to combine the Cahalan scale and the PDS to produce a single scale ranging from 0 to a maximum of 27 alcohol-related problems. Table 7 examines the independent contribution to the total number of alcohol-related problems for each of the potential etiological variables identified in Table 3. Alcoholic heredity, school behavior problems, and not being brought up in a Mediterranean culture (entered into the multiple regression analysis as a dummy variable) each made an important contribution to the explained variance in subsequent alcohol problems. In addition to these three variables, boyhood competence made a small additional independent contribution to alcohol abuse. If the number of alcoholic relatives is controlled for, multiproblem families are not associated with an increased number of alcohol-related problems but are associated with an increased

TABLE 7
Contribution of Selected Premorbid Variables to Number of Alcohol-Related Problems and Sociopathic Behaviors

Variable	ALCOHOL-RELATED PROBLEMS		SOCIOPATHIC BEHAVIORS[a]	
	% explained variance	beta weight	% explained variance	beta weight
Alcoholism in heredity	7.6	.19	1.3	.05
No Mediterranean ethnicity[b]	3.5	.20	1.0	.11
School behavior problems[b]	3.0	.18	8.0	.30
Boyhood competence	1.3	.07	2.3	.14
Childhood environmental weaknesses	.1	.05	2.1	.07
Poor infant health	.1	.10	2.9	.17

[a]See Robins (1966).
[b]To minimize the variance that they could explain, these two variables were entered into the multiple regression after the other four variables.

number of antisocial problems. Similarly, in Table 3, childhood environmental strengths was significantly correlated with both alcoholism and sociopathy, but entering the variable in the regression equation in Table 7 explained less than a 0.1% additional variance.

The present data suggest that having many alcoholic relatives is associated with an increased risk of alcoholism, but the data cannot distinguish the effect of genes (having a blood relative with alcoholism) from that of environment (being raised by an alcoholic parent or surrogate parent). However, review of all available cross-fostering studies that have distinguished heredity from environment (Bohman, 1978; Goodwin, 1979; Goodwin et al., 1974; Schuckit, Goodwin, & Winokur, 1972) suggests that alcoholic biological relatives (even if physically absent) apparently contribute far more to the observed increased risk for alcoholism in children than alcoholics in the environment. These observations would help explain why the association of disorganized families with subsequent alcoholism disappears if one controls for familial alcoholism.

Another variable that contributed significant independent variance to the likelihood of adult alcohol abuse was premorbid antisocial behavior. Certainly, antisocial inner-city youths are at greater risk for developing all types of drug abuse (Chein, 1964; McCord & McCord, 1960; Robins, 1966, 1974), and the rate of alcohol abuse that the Gluecks observed among their 31-year-old delinquents was higher than the rate observed among the controls who make up this study (Glueck & Glueck, 1968). Jones (1968) and unpublished findings from a longitudinal study of college men (Vaillant, 1980) agree that middle-class adolescents who become alcoholics also exhibit more rapid tempo and greater impulsivity than their peers.

However, the relationship between alcoholism and antisocial personality is a two-way street. Antisocial symptoms are often a result rather than a cause of alcohol abuse. Thus, if only 9 of the 110 core city alcohol abusers were antisocial before developing patterns of alcohol abuse, at least 12 of the 32 adults who eventually met five or more of Robins' criteria for sociopathy were alcohol-dependent before meeting the criteria for sociopathy. In the same vein, the differences in the etiology of sociopathy and alcohol abuse that were suggested in Table 3 were further supported by multiple regression analysis. The three variables in Table 7 that account for the greatest independent variance in adult sociopathy are school behavior problems, boyhood competence, and poor infant health. The ethnic and familial factors that were important in alcohol abuse were unimportant in sociopathy.

The possibility that the childhood syndrome of "minimal brain damage," or so-called hyperactivity, could be an etiological link shared by both adult alcoholism and sociopathy is intriguing (Cantwell, 1972; Tarter, McBride, Buonpane, & Schneider, 1977); but in presenting their evidence for the possibility, Goodwin, Schulsinger, Hermansen, Guze, and Winokur (1975) admit that their hypothesis is still "based on retrospective information and small numbers" (p. 349). Unfortunately, when they were first studied, not only were the core city subjects already past the age of maximum risk for hyperactivity, but the syndrome was not to be delineated for another 20 years.

As an independent check on the conclusions of Table 7, the contributions of the six variables to alcohol abuse as defined by the DSM-III, by the PDS, and by the Cahalan scale were separately examined. The multiple regression results did not differ appreciably from those in Table 7 either in percentage of explained variance or in beta weights.

As a second independent check, all the 113 psychosocial variables originally coded dichotomously by the Gluecks (1950) were assessed by their value in predicting subsequent alcohol abuse. Six Glueck items—father's alcoholism, marital conflict, lax maternal supervision, many moves, no attachment to father, and no family cohesiveness—were very significantly ($p < .001$) correlated with the subsequent development of alcohol abuse in study subjects. These six dichotomously rated Glueck items—almost identical to the predictor variables identified by the McCords (1960) as having the greatest etiological impact in alcoholism—were used to create a 7-point composite predictor variable. If this composite variable were entered first into the multiple regression analysis shown in Table 7, it could explain 7% of the observed variance in subsequent alcohol-related problems. However, if this were done, the number of alcoholic relatives and the dummy variable, Mediterranean culture background, could still explain an additional 8% of variance. Conversely, if alcoholic heredity, truancy, and ethnicity were entered first into the regression equation, then the six Glueck variables explained only 1% of further independent variance.

As a check to our conclusions that Syme (1957) and Cahalan and Room (1972) are correct in asserting that specific personality facets are not major etiological factors in alcoholism, we compared remitted core city alcoholics with nonalcoholics (Vaillant, 1982). Former alcoholics with current stable abstinence of more than three years duration did not significantly differ from lifelong social drinkers in terms of current mental health. However, of equal importance, the remitted alcoholics did not differ from the unremitted alcoholics in terms of premorbid risk factors.

Conclusions

Obviously, a life-span perspective is only one of a myriad of techniques required to elucidate a subject as complex as alcohol abuse. To achieve the advantages of a 3-decade prospective study, methodological compromises were necessary that may render our conclusions inapplicable to other populations. First, although very similar findings to those in this chapter were derived from a parallel 40-year prospective study of 204 highly educated upper-middle-class white males (Vaillant, 1980), the present inner-city, poorly educated sample includes no women, blacks, Native Americans, or rural males. Second, by excluding the most severe delinquents, the Glueck sample minimizes the contribution of premorbid social deviance to alcoholism. However, it is not news that sociopaths have high rates of psychoactive substance abuse. Studies to complement the follow-ups of disproportionately antisocial youth by Robins and the McCords are needed. Third, the enormous number of data available on each subject and the long duration of the study necessitated both composite and simplified goal assessment and dependence on redundant, naturalistic (rather than specific) experimentally controlled observations. Examination of discrete items possible in a monograph (Vaillant, 1983) or in a short-term study is not possible in this chapter. Thus, ethnicity, family history of alcoholism, and childhood weaknesses reflect simplified approximations based on available data. If such methods fail to do full justice to the complexity of the variables in question, nevertheless, like aerial photography, they add fresh perspective to our understanding of the alcoholic's landscape.

We do not wish to imply that in adult life other variables do not possess etiological importance. Demographic variables (Cahalan, 1970), occupation (Plant, 1979), social peer groups (Jessor & Jessor, 1975), legal availability (Terris, 1967), societal instability (Pittman & Snyder, 1962), affective illness (Winokur, Clayton, & Reich, 1969), attribution and expectancy effects (Marlatt & Rohsenow, 1980), and other, as yet unidentified, factors certainly contribute independent variance to the development of alcohol dependence.

Nevertheless, if we wish to understand alcoholism, a prospective vantage point is essential. Available prospective studies suggest that if one controls for antisocial childhood, for cultural attitudes toward alcohol use and abuse, for alcoholic heredity, and most especially for the effects of alcohol abuse then many of the childhood and adult personality variables to which adult alcoholism has traditionally been attributed will appear as carts and not horses.

Indeed, our findings suggest that even when they employed prospective design, previous investigators may have sometimes erroneously interpreted their data to support the retrospective illusion that alcoholism must be a symptom of personality disorder (Jones, 1971; Gomberg, 1968; McCord & McCord, 1960; Robins, 1966). For example, in concluding that dominant father, "immigrant" Catholicism, and low social class protected against the development of alcoholism, the McCords failed to note that in Cambridge and Somerville in 1940 such individuals were predominantly first- and second-generation Italians. Italian-American drinking practices, rather than other characteristics, may have protected their children from later alcohol dependence. Similarly, although the 51 alcoholic fathers and the 15 alcoholic mothers in the McCords' study parented a disproportionate number of alcoholic subjects, the McCords wrote that "evidence for an hereditary explanation of the disorder was unlikely" (p. 28). Until more definitive cross-fostering studies (Goodwin, 1979) were available, few investigators could believe that the association between parental and child alcoholism might be more hereditary than environmental. When we are on land and view fish under water, we tend to trust our eyes and not the student of parallax who tells us that the location of the fish is an illusion. Similarly, the parallax of retrospective vision distorts what our "eyes" tell us about the etiology of alcoholism.

Postscript

In 1986 Zucker and Gomberg questioned the preceding article, suggesting that it "dismisses childhood effects out of hand." They reexamined the data and suggested that the etiology of alcoholism would be best considered "in the context of a biopsychosocial process"—a suggestion with which I heartily agree.

Zucker and Gomberg correctly pointed out what I also acknowledged in my discussion above and elsewhere (Vaillant, 1983): that the Glueck sample of nondelinquents minimized the contribution of antisocial "personality" to alcohol abuse. However, in reviewing their criticisms, I conclude that in part they simply could not believe the paradigm shift required by prospective data; the illusion of cross-sectional studies was too compelling. For example, it is my belief that Zucker and Gomberg (1986) failed to appreciate what the studies of Kammeier, Hoffmann and Loper (1973), and the more recent data of Pettinatti, Superman, and Maurer (1982), illustrate, namely, that the MMPIS of alcoholics referred to private clinics are not different from normals before they lose control of their drinking and that the MMPIS return to normal after such alcoholics have maintained abstinence for several years.

In addition, studies to which Zucker and Gomberg refer place the blame of subsequent alcoholism on distance from the father and disrupted families in childhood; however, such studies—unlike the study above—fail to control for the genetic contribution that parents make to their children's alcoholism. By this I mean that children who grew up in tranquil families but with alcoholic biological parents were at 4 or 5 times the risk of developing alcoholism as the children in our sample who did not have an alcoholic parent but who grew up in severely disrupted families in which they were not close to their fathers. Zucker and Gomberg (1986) also fail to cite work by William Beardslee (a dynamically oriented child psychiatrist), who conducted a careful and blind review of the childhood data gathered on the Harvard sample from the Study of Adult Development (Beardslee & Vaillant, 1984). Although Beardslee focused on putative environmental familial contributions to alcoholism, such as those suggested by Zucker and Lisansky Gomberg, he was unable to distinguish future alcoholics from matched controls at better than chance levels. However, Beardslee was, in contrast, able to identify on the basis of disrupted childhoods those subjects who would encounter future psychological difficulties.

In emphasizing genetic contributions over familial contributions to alcoholism, I feel it is important to underscore that follow-up studies of the Glueck subjects did not, in fact, confirm the distinction between Type 1 and Type 2 alcoholism. Work by Cloninger and colleagues and by other researchers (Buydens-Branchey, Branchey, & Noumair, 1989; Cloninger, Bohman, & Sigvardsson, 1981; Goodwin, 1979) has suggested that alcoholics with heavy genetic loading may reflect "primary," or Type 2, alcoholism. This hypothesis is based on the clinical observation that in alcoholics with many alcoholic relatives, alcoholism begins earlier, has a worse prognosis, and is more severe. In contrast, late-onset middle-class alcoholics (Type 1) often have a better social adjustment, have fewer alcoholic relatives, and develop "reactive" alcoholism. However, the difficulty is that most of these studies have been drawn from cross-sectional data or from data where only the hereditary and not the environmental contribution was studied. In the present sample, when Beardslee and I (Beardslee, Son, & Vaillant, 1986; Vaillant, 1988) reanalyzed the contribution of genetic loading and the environmental disruption produced by having an alcoholic parent, we could discern no effect of the number of blood relatives on the age of onset of alcoholism or on its severity. In contrast, men from severely disruptive childhoods, whether or not these were caused by alcoholic blood relatives, developed more symtomatic alcoholism and at an early age. In short, if (contrary to Zucker and Gomberg's position) unstable familial environment per se did not predict whether an individual lost control of alcohol, family instability did predict whether he lost control of alcohol *at an early age* and whether he was severely symptomatic.

Acknowledgments

This study was supported by the Grant Foundation, Inc., the Spencer Foundation, and research grant AA-01372 from the National Institute on Alcohol Abuse and Alcoholism.

References

American Psychiatric Association. (1980). *Diagnostic and statistical manual of mental disorders* (3rd ed.). Washington, DC: Author.

Beardslee, W.R., Son, L., & Vaillant, G.E. (1986). Exposure to parental alcoholism during childhood and outcome in adulthood: A prospective longitudinal study. *British Journal of Psychiatry, 149,* 584–591.

Beardslee, W.R., & Vaillant, G.E. (1984). Prospective prediction of alcoholism and psychopathology. *Journal of Studies on Alcohol, 45,* 500–503.

Blane, H.T. (1968). *The personality of the alcoholic: Guises of dependency.* New York: Harper & Row.

Bohman, M. (1978). Some genetic aspects of alcoholism and criminality. *Archives of General Psychiatry, 35,* 269–276.

Buydens-Branchey, L., Branchey, M.H., & Noumair, D. (1989). Age of alcoholism onset: I. Relationship to psychopathology. *Archives of General Psychiatry, 46,* 225–230.

Cahalan, D. (1970). *Problem drinkers: A national survey.* San Francisco: Jossey-Bass.

Cahalan, D., & Room, R. (1972). Problem drinking among American men aged 21–59. *American Journal of Public Health, 62,* 1473–1482.

Cantwell, D.P. (1972). Psychiatric illness in the families of hyperactive children. *Archives of General Psychiatry, 70,* 414–417.

Chein, I. (1964). *The road to h.* New York: Basic Books.

Cloninger, C.R., Bohman, M., & Sigvardsson, S. (1981). Inheritance of alcohol abuse: Cross-fostering analysis of adopted men. *Archives of General Psychiatry, 38,* 861–868.

Glueck, S., & Glueck, E. (1950). *Unraveling juvenile delinquency.* New York: The Commonwealth Fund.

Glueck, S., & Glueck, E. (1968). *Delinquents and non-delinquents in perspective.* Cambridge: Harvard University Press.

Gomberg, E.S.L. (1968). Etiology of alcoholism. *Journal of Consulting and Clinical Psychology, 32,* 18–20.

Goodwin, D.W. (1979). Alcoholism and heredity. *Archives of General Psychiatry, 36,* 57–61.

Goodwin, D.W., Schulsinger, F., Hermansen, L., Guze, S.B., & Winokur, G. (1975). Alcoholism and the hyperactive child syndrome. *Journal of Nervous and Mental Disease, 160,* 349–352.

Goodwin, D.W., Schulsinger, F., Møller, N., Hermansen, L., Winokur, G., & Guze, S.B. (1974). Drinking problems in adopted and non-adopted sons of alcoholics. *Archives of General Psychiatry, 31,* 164–169.

Greely, A., McCready, W.C., & Theisen, G. (1980). *Ethnic drinking subcultures.* New York: Praeger.

Heath, D.B. (1975). A critical review of ethnographic studies of alcohol use. In R.J. Gibbons, Y. Israel, H. Kalant, R.E. Popham, W. Schmidt, & R.G. Smart (Eds.), *Research advances in alcohol and drug problems* (Vol. 2, pp. 1–92). New York: Wiley.

Hollingshead, A.B., & Redlich, F.E. (1958). *Social class and mental illness.* New York: Wiley.

Jellinek, E.M. (1960). *The disease concept of alcoholism.* New Brunswick, NJ: Rutgers Center of Alcohol Studies.

Jessor, R., & Jessor, S.L. (1975). Adolescent development and the onset of drinking. *Quarterly Journal of Studies on Alcohol, 36,* 27–51.

Jones, M.C. (1968). Personality correlates and antecedents of drinking patterns in adult males. *Journal of Consulting and Clinical Psychology, 32,* 2–12.

Jones, M.C. (1971). Personality antecedents and correlates of drinking patterns found in women. *Journal of Consulting and Clinical Psychology, 36,* 61–69.

Kammeier, M.L., Hoffmann, H., & Loper, R.G. (1973). Personality characteristics of alcoholics as college freshmen and at time of treatment. *Quarterly Journal of Studies on Alcohol, 34,* 390–399.

Kohlberg, L., LaCrosse, J., & Ricks, D. (1972). The predictability of adult mental health from childhood behavior. In B.B. Wolman (Ed.), *Manual of childhood psychopathology.* New York: McGraw-Hill.

Loper, R.G., Kammeier, M.L., & Hoffmann, H. (1973). MMPI characteristics of college freshmen males who later became alcoholics. *Journal of Abnormal Psychology, 82,* 159–162.

Luborsky, L. (1962). Clinicians' judgments of mental health. *Archives of General Psychology, 7,* 407–417.

Luborsky, L., & Bachrach, H. (1974). Factors influencing clinicians' judgments of mental health. *Archives of General Psychiatry, 31,* 292–299.

Marlatt, G.A., & Rohsenow, D.J. (1980). Cognitive processes in alcoholic use: Expectancy and the balanced placebo design. In N.K. Marlow (Ed.), *Advances in substance abuse: Behavioral and biological research* (Vol. 1 pp. 159–199). Greenwich, CT: Jai Press.

McCord, W., & McCord, J. (1960). *Origins of alcoholism.* Stanford, CA: Stanford University Press.

Pettinatti, H.M., Superman, H., & Maurer, H.S. (1982). Four-year MMPI changes in abstinent and drinking alcoholics. *Alcoholism: Clinical and Experimental Research, 6,* 487–494.

Pittman, D.J., & Snyder, C.R. (1962). *Society, culture, and drinking patterns.* New York: Wiley.

Plant, M.L. (1979). *Drinking careers: Occupations, drinking habits and drinking problems.* London: Tavistock.

Powers, E., & Witmer, H. (1951). *An experiment in the prevention of delinquency: The Cambridge Somerville Youth Study.* New York: Columbia University Press.

Robins, L.N. (1966). *Deviant children grown up: A sociological and psychiatric study of sociopathic personality.* Baltimore, MD: Williams and Wilkins.

Robins, L.N. (1974). *The Vietnam drug user returns* (Special Action Office for Drug Abuse Prevention Monograph, Series A, No. 2). Washington, DC: Government Printing Office.

Robins, L.N., Bates, W.N., & O'Neal, P. (1962). Adult drinking patterns of former problem children. In D.J. Pittman & C.R. Snyder (Eds.), *Society, culture, and drinking patterns* (pp. 395–412). New York: Wiley.

Schuckit, M.A., Goodwin, D.W., & Winokur, G. (1972). Study of alcoholism in half-siblings. *American Journal of Psychiatry, 128,* 1132–1136.

Selzer, M.L. (1980). Alcoholism and alcoholic psychoses. In H.I. Kaplan, A.M. Freedman, & B.J. Sadock (Eds.), *Comprehensive textbook of psychiatry.* Baltimore, MD: Williams and Wilkins.

Stivers, R. (1976). *A hair of the dog.* University Park, PA: Pennsylvania State University Press.

Syme, L. (1957). Personality characteristics of the alcoholic: A critique of recent studies. *Quarterly Journal of Studies on Alcohol, 18,* 288–301.

Tarter, R.E., McBride, H., Buonpane, N., & Schneider, D.V. (1977). Differentiation of alcoholics. *Archives of General Psychiatry, 34,* 761–768.

Terris, M.A. (1967). Epidemiology of cirrhosis of the liver: National mortality data. *American Journal of Public Health, 57,* 2076–2088.

Vaillant, G.E. (1974). Natural history of male psychological health: II. Some antecedents of healthy adult adjustment. *Archives of General Psychiatry, 31,* 15–22.

Vaillant, G.E. (1980). Natural history of male psychological health: VIII. Antecedents of alcoholism and "orality." *American Journal of Psychiatry, 137,* 181–186.

Vaillant, G.E. (1983). *The natural history of alcoholism.* Cambridge: Harvard University Press.

Vaillant, G.E. (1983). Natural history of male alcoholism: V. Is alcoholism the cart or the horse to sociopathy? *British Journal of Addiction, 78,* 317–326.

Vaillant, G.E. (1988). Some differential effects of genes and environment on alcoholism. In R. Rose & J.E. Barrett (Eds.), *Alcoholism: Origins and etiology* (pp. 75–82). New York: Raven Press.

Vaillant, G.E., Gale, L., & Milofsky, E.S. (1982). Natural history of male alcoholism: II. The relationship between different diagnostic dimensions. *Journal of Studies on Alcohol, 43,* 216–232.

Vaillant, G.E., & Milofsky, E.S. (1980). Natural history of male psychological health: IX. Empirical evidence for Erikson's model of the life cycle. *American Journal of Psychiatry, 137,* 1348–1359.

Vaillant, G.E., & Vaillant, C.O. (1981). Natural history of male psychological health: X. Work as a predictor of positive mental health. *American Journal of Psychiatry, 138,* 1433–1440.

Winokur, G., Clayton, D.J., & Reich, T. (1969). *Manic depression illness.* St. Louis, MO: Mosby.

Zucker, R.A., & Gomberg, E.S.L. (1986). Etiology of alcoholism reconsidered. *American Psychologist, 41,* 783–793.

CHAPTER 26

The Concept of Risk and the Etiology of Alcoholism: A Probabilistic-Developmental Perspective

ROBERT A. ZUCKER

A man I know is a cook in the State Prison of Southern Michigan, one of the largest walled prisons in the world. When one compares his work to that of most others, it is clear that he is in substantially greater danger of assault, and possibly even death, than is true of others of his occupation. And yet he reports, I believe accurately, that should an uprising come, his life would be protected although he is unarmed, because he has befriended a "natural lifer" who is a man both of large physical stature and of considerable political power in the prison social structure. He notes as well that his life would probably be in less jeopardy than that of one of the armed but disliked guards in the prison.

The relevance of this little vignette to the topic under consideration is that both my story and the problem I address have to do with the anticipation of problematic outcomes in risky circumstances. The story makes the point that even living in risky environments does not automatically preclude successful outcomes and that the possession of protective factors does not insure successful ones. Contextual factors that are themselves both the outgrowth of an interaction and a set of processes that unfolds over some time sequence may make the difference between a safe and a damaged ending. The present review deals with these issues in detail, elaborating on the nature of risk itself, examining the theoretical structure that has guided much of the research relevant to risk for alcohol problems and alcoholism, briefly summarizing the evidence upon which current predictive statements are made, and addressing the question of

Robert A. Zucker is with the Department of Psychology, Michigan State University, East Lansing, Mich. This chapter is an updated and revised version of "Is Risk for Alcoholism Predictable?" which appeared in *Drugs and Society, 4,* pp. 69–93, 1989. Reprinting of the unchanged portions is by permission of Haworth Press.

the extent to which certainty about outcome can be anticipated on the basis of a knowledge of risk factors alone.

The Concept of Risk

In *Webster's New International Dictionary* (Neilson & Carhart, 1956) there are several alternative definitions of *risk:*

1. The chance of loss or peril.
2. The degree of probability of such loss.
3. Abbreviation of *amount at risk,* that is, the amount the (insurance) company may lose.
4. The character of hazard involved (syn.: danger).

These definitions deal with three aspects of the phenomenon: Risk implies an *anticipated outcome* (i.e., the occurrence of a negative event, called a loss). It also implies a *likelihood estimate;* or, to put the matter another way, the concept of risk is a probabilistic statement. It implies the attachment of some kind of weight to the possibility of occurrence of a future event. And last, the concept implies that there are some attributes of the ongoing event structure, before the outcome, that are best characterized as *hazard or danger.* This attribute of risk implies an actuality of the moment as well as the possibility of a later troublesome consequence. As in my story, that which is dangerous at the moment may still be negotiated without jeopardy, but it is usually considered to require greater attentiveness to the process, and to the details of the negotiation, than would be the case were one not in a dangerous situation.[1] Such is the nature of risk.

To pursue this definition a bit further, risk is an attribute of populations rather than individuals, and is typically inferred on the basis of an observation that some proportion of a group with specified characteristics eventually displays some sign or symptom. It is the manifest display, the actuality of risk transformed into bona fide trouble, that then becomes the basis for prospective statements about risk characteristics for a new but similar population.

Elements of risk or danger are inferential in a different sense at the individual level; they are best regarded as signs of covert processes that only some of the time are detectable by their overt display. This is precisely the dilemma in anticipating individual outcome for such a disorder as alcoholism.[2] One needs to infer and understand the interplay of factors that may never come to fruition, but that nonetheless must be regarded as still present and operational. The remainder of the chapter deals with the evidence relating to the presence of such factors and of the processes that undergird them. First along the way, we consider the theory structure that guides much of this research, because it has in some instances helped and in others hindered a comprehensive understanding of these phenomena.

Factors Necessary for the Anticipation of Risk: The Problem of Theory in Etiologic Research

Despite the weights assigned to different factors in a predictive equation, one can never hope to be successful in accounting for individual variation if the variables included are trivially to nonexistently related to the criterion. Similarly, if the measures used to assess a factor are crude or heavily laden with error, the estimates of variance accounted for will be no more than imprecise and low-order bounds for the amount of variation attributable to that factor. When these principles are stated in the abstract form just used, they are truisms that one would expect a first-year graduate student to understand after having completed a course in experimental design or multivariate statistics. Yet when these same principles are applied as judgment criteria, in a practical sense, for some of the most significant research programs attempting to map and understand alcoholic etiology, they are too often violated. The result is that our knowledge of etiologic connections—the basis for prediction of both risk and outcome—continues to be more flawed than it needs to be.

From the perspective of theory, there are five types of conceptual or design problems that have repeatedly invaded major research studies of the past decade, in ways that make it difficult to come to a more definitive understanding of causes and outcomes. Each problem will be discussed in some detail.

1. Failure to sample adequately the domain of known and significant influences. For example, within the realm of psychosocial influences upon behavior, it has repeatedly been demonstrated in research on the development of drinking and problem drinking that the role of peers in introducing, stabilizing, and reinforcing these patterns of drug use is one of the most powerful sources of variation of any in the variable domain (Jessor & Jessor, 1977; Kandel, 1980; Zucker & Noll, 1987). [See also chapter 9, by White, Bates, & Johnson, in this volume.] A like pattern of salience of the peer influencing structure—at the level of the family group—has also more recently been demonstrated for older, alcoholic drinkers (Jacob & Seilhamer, 1987). Despite the major power of this level of influencing structure, it is more often noticeable by its absence than by its presence in all but the studies that explicitly are concerned with charting its power as a main effect.

This problem is not one found only among the more biologically based research community; given the strong evidence for a genetic component operating in the regulation of differences in patterns of alcoholism (Cloninger, Bohman, & Sigvardsson, 1981), it becomes increasingly important for primarily psychosocially focused research to attempt to mark and evaluate this source of contributory variation. [See also chapter 30, by Goodwin, in this volume.] At the moment such efforts most often are confined to high-risk research paradigms (cf. Begleiter, 1988; Goodwin, Van Dusen, & Mednick, 1984) than to the mainstream of etiologically focused studies.

2. Failure to represent adequately the interactional nature of this disorder. At a more general level, what is perhaps most unfortunate about this state of affairs is that the discussion of domain issues in understanding this complex set of phenomena is still quite rare, with some exceptions (e.g., Cloninger, 1987; Tarter, Alterman, & Edwards, 1985; Zucker & Gomberg, 1986). It is noteworthy that the Institute of Medicine (IOM), in its most recent review of the state of research in the field of alcohol studies (1987), acknowledges this as a major research issue for the 1980s, and emphasizes the need for multidisciplinary research to address explicitly the across-domain interrelationships that are so much a part of alcoholic phenomena (cf. Institute of Medicine, 1987, p. 12). The IOM recommendation serves to emphasize this second set of design flaws.

Alcoholism is, after all, a biobehavioral disorder. At the biological level it is most appropriately classified as a pharmacogenetic disorder, one that involves the interaction between a self-administered pharmacologic agent and a neurophysiologic system that—to some degree—has been hard wired by way of the genetic structure. At the same time, it is a distinctly different disorder than schizophrenia, whose primary manifestations are best understood as involving central nervous system deficits. In contrast, the self-administered nature of the agent of damage means that for an adequate model of alcoholic etiology there must be adequate representation of an ecology of contexts of availability and social structures, within which patterns of use are shaped, reinforced, and punished. Along with the biological sequelae of sustained alcohol involvement (those factors that lead to tolerance and dependence), there is a massive set of social sequelae of use that play a role in the cumulation process, a process that eventuates, for some individuals, in the development of an alcoholic career.

This issue of neglected interactional influences, as well as of failure to assess adequately the domains of causal structure, is nowhere better illustrated than in the Cloninger et al. (1981) Stockholm Adoption Study. These investigators conclude that among men there are two kinds of genetic predisposition to alcoholism: one substantially limited by the postnatal environment (called Type I, or milieu-limited), the other, apparently unaffected by environmental variations (called Type II, or male-limited). In an incisive reanalysis of these data, Searles (1988) calls attention to the exceptionally sparse and inadequate set of environmental variables included (age at final adoptive placement, extent of postnatal hospital care, occupational status level of adoptive father, rearing by biological parent for more than six months, adoptive parents' alcohol abuse record and criminality record). He notes that the manner in which the data are presented obscures the heavy contribution of unspecified (and unmeasured) environmental variation to the outcome of alcohol abuse, and he demonstrates that the effect size for these influences is "substantially more important in determining alcohol abuse than are genetic factors" (p. 161).

One last point about the mapping of causal domains as this contributes to the evaluation of risk: The Stockholm study is but one example of an inadequate set of measures with which to chart environmental variation. Given the field's current ability to begin to assess genetic load by way of family history measurement, the lack of studies that adequately assay both sets of domains ends up restricting our ability to capture interactional variations that might contribute to a more precise anticipation of differences in alcohol abuse outcome.

3. Failure to understand the importance of "nonspecific factors" in etiology. Despite the articulate and carefully reasoned assaults on the disease model over this last generation (Cahalan, 1970; Peele, 1986), it is still with us in ways that erode the research models we construct and the studies we carry out to understand alcoholic etiology. [See chapters 21 and 22, by Fingarette and by Hore, respectively, in this volume.] The disease model carries within it the promise that a set of factors can be identified that will be the sine qua non of the disorder (cf. Meehl, 1989). Such factors, particularly when they may involve a heritable element, are most often construed to be alcohol-specific in their mechanism (cf. National Institute on Alcohol Abuse and Alcoholism, 1985). In one instance, nonspecific factors have been rejected outright as inappropriate elements even to include in the etiologic equation (Nathan, 1988).

Such a perspective is limiting in two ways. First, it avoids dealing with the empirical criterion of amount of predictive variance apportionable to particular factors; yet from both predictive and public health vantage points, this is undoubtedly a central issue in any theory and any research program. Ultimately, the isolation of a sine qua non is of little practical use if it does not contain the key to modification of the disorder. Yet this solution is not always in the "specific factor" realm (e.g., draining swamps to cure malaria).

Second, such a perspective currently leads investigators into a research paradigm that ends up avoiding some of the most substantial sources of variation in patterns of alcohol (and other drug) involvement of any now known (see section 1, above; also, Kandel, 1978, Peele, 1988). Severe drug involvement is a human act, involving a biopsychosocial process over long spans of developmental time. As such, it is probably not adequately capturable by an etiologic paradigm that ignores nonalcohol specific, but powerful, ecological factors along the way (Jessor & Jessor, 1973; Moos, 1973).

4. Failure to take account of the developmental nature of alcoholic processes, which involve several alcoholisms as well as the more moder-

ate forms of abuse. If there is one characteristic that is both troublesome and ubiquitous about alcohol involvement at different levels of severity, it is that the phenomenon is a fluid one, shifting over developmental epochs, remaining steady for some drinkers and shifting for others, changing from problem to nonproblem status or vice versa, and so on (Cahalan, 1970; Fillmore, Bacon, & Hyman, 1979; Kandel & Logan, 1984; Zucker, 1979). Current evidence indicates that this statement is as true for the severe and chronic end of the alcohol abuse continuum—what gets called alcoholism—as it is for more moderate and more acute manifestations of drinking difficulty (Goodwin, Crane, & Guze, 1971; Polich, Armor, & Braiker, 1980; Roizen, Cahalan, & Shanks, 1978; Skog & Duckert, in press).

The field is increasingly aware of the evidence for epiphenomenality. However, what is still missing is an awareness that the move into and out of severe drinking difficulties occurs in a developmental, life trajectory context and that the difficulties are likely to have different long-term ramifications depending upon the developmental context in which they appear (Zucker, 1987, 1988). Thus, onset of drinking problems is a different issue when it occurs at age 11 than when it occurs at age 16 and different from either of these two when it occurs at age 65 (Gomberg, 1982; Robins, 1984). The recalcitrance of the problem and its susceptibility to outside modification may be very different at these different developmental epochs, because of the meaning structure that surrounds the abuse at the time it takes place and because the processes that work to create the problematic outcome may, in fact, be different in these different periods (e.g., Kandel & Andrew, 1987; Robins & Przybeck, 1985).

A related developmental issue that is only beginning to receive attention is the need to distinguish between a variety of different "alcoholisms," which are marked by different etiologies and different developmental pathways (Cloninger, Sigvardsson, & Bohman, 1988; McCord, 1988; Zucker, 1987). Thus, the appearance of a symptomatic display that is part of several different alcoholic trajectories will tell us nothing about risk if we cannot simultaneously identify which pathway it maps onto. Accurate assessment of risk needs to take account of these different developmental trajectory and contextual issues in weighting relative impact.

5. Failure to distinguish between maintenance processes and onset processes in the formulation of etiologic paradigms. This problem is related to those covered in sections 3 and 4, above. If one conceives of alcoholism (alcohol dependence) in its full-blown form to be closely tied to, if not synonymous with, the alcohol dependence syndrome (Edwards & Gross, 1976), then it is not surprising that one would heavily emphasize the role of alcohol-specific factors in its etiology and would heavily direct research to those neurobiological factors that control the syndrome. Conversely, if one understands that the onset and early movement into heavy alcohol involvement is largely controlled by contextual, attitudinal, and social influence structures (Jessor & Jessor, 1975; Kandel, Kessler, & Margulies, 1978), then it is not surprising that one would heavily direct research to understand the interplay and timing of these variables. The current evidence implicates both sets of factors; but the timing, and the role that each plays, has not been well enough emphasized (Zucker & Gomberg, 1986). The former appear to be elements concerned more with maintenance of the processes; the latter appear to be concerned more with onset. This distinction rarely is made; more important, studies tracking the relative importance of one versus the other and simultaneously allowing for evaluation of the time-linked nature of the relative contributions have not yet been carried out.

Anticipating Risk for Alcoholism: Evidence for Continuity and Cumulation of Process

The above points need to be taken as a series of lenses through which to scrutinize existing studies. Despite the relative lack of awareness of these issues, some sets of influences have systematically been replicated in longitudinal data sets with fairly large variable domains. Thus, as when one is assembling two identical jigsaw puzzles, both of which are missing pieces (but not the same ones), it is possible to reconstruct more of the shape of the original than would be true were only one of the puzzle copies available. We turn now to the replicated evidence that identifies some common factors indicating continuity of outcome of alcoholic process.

The first work we examine involves six studies, all of which originated in their subjects' childhood and adolescence, followed respondents on into adulthood rather than terminate in adolescence, and established a diagnosis of either alcoholism or problem drinking (what would be called alcohol abuse in this era of DSM-III-R terminology) when respondents reached adulthood. Thus, all of these studies began at a time when, it is reasonable to infer, none of the subjects were yet alcoholic, although it is equally reasonable to expect that drinking had already been in place among a substantial subset of those who were adolescent at the Time-1 data collection. The studies referred to are the Oakland Growth Study, the St. Louis Child Guidance Clinic Study, the Columbia Follow-up Study, the Cambridge-Somerville Youth Study, and the Physique and Delinquency Study. A series of papers over the last decade carefully describes these studies in detail, reviews their commonalities, and also points out correspondences between the longitudinal work and other studies of high-risk but not yet alcoholic samples (Zucker, 1979; Zucker & Gomberg, 1986; Zucker & Noll, 1982). Drawing primarily from the Zucker and Gomberg review, the following points summarize the across-study commonalities:

1. Childhood antisocial behavior is consistently related to later alcoholic outcome (five studies).
2. More childhood difficulty in achievement-related activity is consistently found in those who later become alcoholic (four studies).
3. Males who later become alcoholics are more loosely tied to others interpersonally (four studies).
4. Heightened marital conflict is reported with consistently greater frequency in the prealcoholic homes (four studies).[3]
5. Parent-child interaction in prealcoholic homes is characterized by inadequate parenting and by the child's lack of contact with the parent, or parents (all six studies).
6. Parents of prealcoholics are also more often inadequate role models for later normality; they are more likely to be alcoholic, antisocial, or sexually deviant (four studies).
7. Greater activity level in childhood is identified as a precursor in two studies.
8. Ethnic differences are systematically linked to alcohol outcomes; within the range of ethnic heritages sampled, alcoholics are more likely to come from Irish than Italian backgrounds (three studies).

These across-study correspondences are significant, but from a developmental perspective they still leave several issues untouched. Given that the bulk of this work begins at the end of the subjects' middle childhood, it does not inform us about what might have taken place earlier on. However, there is another set of studies that is useful in filling in some of the gaps; I refer to those several longitudinal projects that have begun to examine what early-childhood characteristics might be antecedent to an alcohol- or, in some cases, drug-involved outcome during the teen years. The special utility of these studies—even though they concern quite diverse populations—is that they examine some of the same attributes that were identified as important in the later-stage longitudinal studies. Thus, they allow us to fill in the developmental flow chart running from childhood to adulthood. Table 1 lists the relevant studies, and Table 2 charts the extent to which they provide evidence for correspondence with the findings of the later-state, high-risk longitudinal studies just summarized.

Given the diversity of populations examined here, and the substantial lack of risk loading in two of the four studies (Block, Block, & Keyes, 1988; Brook, Whiteman, Gordon, & Cohen, 1986), the evidence for continuity of behavior from the preschool years onward is significant, if not impressive. In every area where an association could be tested, the relationships either parallel or do not contradict what has been observed in the studies of older populations. Further, the evidence for a precursive relationship between early-childhood antisocial involvement and alcohol-related problem outcomes in later years seems at this juncture to have been sufficiently replicated among boys to be regarded as an established sequence.

TABLE 1
Prospective Studies of Influences During Early Childhood Related to Alcohol and Drug Involvement in Adolescence: Study Characteristics

Study	Subject characteristics and source	Age at first contact	Age at follow-up
1. Rydelius (1981, 1984); Nylander (1960)	229 Stockholm children with severely alcoholic, clinic-treated fathers (50% male) and 163 matched "social twins" without paternal alcoholism	4–12	24–32
2. Kellam, Ensminger, and Simon (1980); Kellam, Brown, Rubin, and Ensminger (1983); Ensminger, Brown, and Kellam (1982)	Complete population sample; N = 1,242 black 1st graders; approx. 50% male; from Chicago's South Side	6	6–7; 9; 16–17
3. Brook et al. (1986)	N = 356; 94% white; 49% male at T-1; random selection from two New York counties	5–10	13–18
4. Block et al. (1988); Block & Block (1980)	N = 105; 67% white; 49% male; recruited through nursery schools, overrepresents middle class	3	4, 5, 7, 11, 14

Evidence from the Michigan State University Longitudinal Study is consistent with all elements of this picture (Zucker, 1987). Over an anticipated 20-year time span, this project is following families that at the initial stage have male children between 3 and 6 years of age who run a high risk of alcoholism in later life because of their being both males and the children of alcoholic fathers. The work is especially relevant to the evolution of risk because it utilizes a population-based sample of families and is able to identify alcoholism in the parent before any formal diagnosis or treatment of the problem has taken place.

In this data set, as hypothesized, cross-sectional evidence is already accumulating that the familiarity with alcoholic beverages, as well as the development of cognitive structures about them (relating to being able to identify alcohol and have a clearer sense of the appropriate contexts of its

TABLE 2
Correspondences between Earlier-Stage Developmental Studies and Later-Stage, High-Risk-for-Alcoholism Longitudinal Studies

	STUDY			
	Rydelius (1981, 1984); Nylander (1960)	**Kellam et al. (1980, 1983)**	**Brook et al. (1986)**	**Block et al. (1988)**
Characteristics identified in longitudinal studies with older subjects[a]				
1. Childhood antisocial Behavior and aggression	+	+ (Boys) 0 (Girls)	+[a]	+[b]
2. Childhood achievement problems	+	0	0[c]	NA
3. Poorer childhood interpersonal connections	NA	+ (Boys)[d]	0[c]	+
4. Heightened activity level in childhood	+	NA	0	+ (Boys) 0 (Girls)
5. Less parent-child contact and more inadequate parenting in childhood	+	NA	NA	+ (Girls) 0 (Boys)
6. More marital conflict in childhood homes	NA	NA	NA	NA
7. Parents more often inadequate role models with more psychopathology and no alcoholism and other psychopathology	+[e]	NA	NA	NA

Notes: A (+) in the columns indicates a correspondence between the later age longitudinal studies and the earlier develomental stage studies: a (0) indicates no relationship and (NA) indicates no data available on this issue. In all of the above tallies, a correspondence is judged to be present if in the younger sample, the identified characteristics are precursive of later, greater alcohol involvement, more alcohol-related problems, or a later stage of alcohol or drug use (cf. Kandel, 1975).

[a]Predelinquency measure was related; aggression against peers and siblings was not.
[b]Marginally significant in girls.
[c]Earlier childhood factors only influenced by way of their impact upon later (adolescent) factors.
[d]Trend for shyness together with aggressiveness.
[e]True of both parents.

use), differentiates the high-risk children from matched controls as early as 3 years of age (Noll & Zucker, 1983). Other analyses (Reider, Zucker, Noll, Maguin, Fitzgerald, 1988; Reider, Zucker, Maguin, Noll, Fitzgerald, 1989) of the patterns of violence in these very young families are consistent with the longitudinal evidence just reviewed concerning parental behavior and risk in adolescence and thereafter (i.e., the later-stage studies). Quite strong relationships are being observed between extent and severity of parents' alcohol problems and the levels of current violence displayed in these families, both spouse to spouse and also parent to child. In addition, the severity of parents' alcohol problems is positively and strongly related to their own antisocial histories and to their level of depression. These are characteristics not well covered by the studies summarized in Table 2. Thus, the evidence here, albeit cross-sectional, is again consistent with what one would expect on the basis of a continuity-of-process hypothesis going back at least as early as age 3.

Specific Versus Nonspecific Etiology: A Cautionary Note

On all of these grounds the evidence for a steady progression is incomplete but very strong. Both longitudinal data sets and a subset of cross-sectional studies on subjects considered high-risk but not yet alcoholic are in accord on these points. In addition, the individual difference findings among the children suggest that the *problem behavior syndrome*—identified by Donovan and Jessor (1985; Donovan, Jessor, & Costa, 1988) in adolescent and young adult populations as the core of a variety of delinquent, substance-abusing, and precociously unconventional activities—may be found in place as early as the preschool years in these very high risk families; this aggressive-antisocial component appears to provide the core around which other elements of the syndrome may accrete.[4] The evidence is consistent with such a hypothesis, but the longitudinal work following it from early on and tracking it into adulthood needs elaboration in several different directions.

The above review implies that the pathway into problematic alcohol involvement is by way of influences that are nonspecific to alcoholism (also see Werner, 1986). The fact that parallel findings concerning this syndrome are also noted in studies that have tracked alcoholic outcome from homes where there was no alcoholic parent (McCord, 1988) lends additional support to this position. But given the likelihood that some aspects of the ultimate disorder are under genetic control, what these studies do not clarify is the extent to which these manifestations are simple prolegomena to the later alcoholic syndrome rather than core mediational structures that form an intrinsic part of the etiologic chain. Current evidence is more consistent with the second alternative (Cadoret, Troughton, O'Gorman, & Heywood, 1986; Hesselbrock et al., 1984), but without a factoring of some measure of genetic loading, however crude, into the data set, these alternate etiologic pathways cannot be differentiated. This factoring in was actually done in McCord's very elegant data set, but since the focus of her report was upon parental contributions to alcoholism in a child, the regression analysis she reports does not allow us to evaluate comparative effects.

Alcohol-specific factors (as opposed to alcoholism-specific ones) operate both in and outside of alcoholic homes and can be expected to be a significant dimension of exposure around which the learning of drinking takes place (Cahalan, Cisin, & Crossley, 1969).[5] Earlier formulations in this arena have focused almost exclusively on the adolescent years as the developmental time frame within which such exposure takes place and needs to be understood (Kandel, 1978). However, recent studies with children as young as 3 years of age indicate that learning about alcohol takes place far earlier than had previously been suspected. Eight investigations have been carried out with younger groups. These studies, summarized in a recent review paper (Zucker & Noll, 1987), indicate that young children already are on their way to developing internalized schemas about alcohol as a drug and about its effects and its appropriate contexts of use. Significant evidence also exists for the presence of individual differences in the development of this knowledge base, as a function of individual differences in parent use patterns (Noll, Zucker, & Greenberg, 1989).

Given these findings and the strong correlation commonly noted between exposure to these experiences (i.e., exposure to drinking peers) in later childhood and exposure to other aspects of the problem behavior syndrome, the contribution of each of these separate sources of variation to the problem outcome of alcohol abuse is not clear. And insofar as the biological aspects of alcoholism are linked intimately to the antecedents, sequelae, and consequences of consuming alcohol in varying amounts and schedules, it is critial at the behavioral level to distinguish among these different sources of risk variation.

Anticipating Risk for Alcoholism and Alcohol Abuse: Evidence for Discontinuity and for the Operation of Mediational and Ameliorative Processes

The current clinical myth, based upon research with alcoholic individuals drawn from treatment settings, is that the disorder is an irreversible one that may remit but not reverse itself (American Psychiatric Association, 1987). The controversy about controlled drinking (Marlatt, 1983) is precisely about this issue, although it rarely is formulated in this way. Disagreements about whether alcoholism is a disease also are, in part, concerned with this question.

From a risk perspective, those who adhere to an irreversibility position are implicitly making the case that a set of factors have coalesced (or have always been present, but have evolved to an advanced enough stage) in such a way that a process is in place that is no longer capable of being disaggregated. In fact, the evidence reviewed above points in just such a direction, although the data suggest that what is operating is more an accretion process than an irreversible syndrome. However, to make this issue a bit more interesting (and challenging), it is also important to note that the figures on continuity of such processes in risky environments (such as having an alcoholic parent) show that it is in fact the minority of children who move on to develop the problematic, alcohol-abusive outcome (Cotton, 1979).[6]

What makes the difference? Two projects currently have data sets that are far enough along in a developmental sense, encompass measures of alcohol and drug involvement (i.e., the end criterion of risky processes), and also address the question of relative risk in early childhood. One is the previously cited Brook, Whiteman, Gordon, and Cohen (1986) longitudinal study (see Tables 1 and 2), which explicitly examines the issue of risk dilution as well as risk enhancement, albeit at an early stage of the drug involvement career. In a creative set of interactive analyses, these investigators have demonstrated that the presence, in adolescence, of greater tolerance for deviance, impulsivity, aggression toward peers, and a variety of other measures that may be subsumed under the rubric of the Donovan and Jessor problem behavior syndrome (also the Block & Block, 1980, Ego Undercontrol rubric), all operate to vitiate earlier childhood Time-1 factors that, on their own, would moderate their level of drug involvement (and very probably their level of alcohol-related problems). Conversely, a variety of high-risk characteristics of earlier childhood (ages 5 to 10), involving both the problem behavior syndrome and other characteristics (depressive mood, eating problems), were ameliorated by factors in adolescence that indicated a low level of the problem behavior cluster.

It is important in this context to stress that this is an interaction effect, which occurs in addition to a main effect. The main effect is for continuity of behavior, by way of a *mediational sequence* (Baron & Kenny, 1986). This process operates by way of chaining: Childhood personality characteristics shape adolescent personality traits (i.e., a process of continuity in development), and they in turn impact upon the level of adolescent drug involvement. The interaction effect, either to enhance or to dilute adolescent drug involvement, overlays and runs counter to the mediational main effect sequence.

There is another critical question that needs to be posed. Given that the more straightforward, main effect sequence involved continuity, the interaction effect shifts in problem behavior from childhood to adolescence thus involve an invasion upon such continuity of process. How can this fact be accounted for? The Brook, Whiteman, Gordon, and Cohen study (1986) does not speak to this question, but other work by these investigators (Brook, Whiteman, & Gordon, 1983) demonstrates that more than personality factors are in operation here: the level of drug involvement in that study was related to the influence of peer and family factors and to their interaction with personality variables.

The other study is McCord's Cambridge-Somerville project. Her 1988 article does not explicitly focus on issues of risk amelioration, but the study's findings concerning pathways leading into alcoholism are interpretable in the obverse direction as protective factors. Shyness in childhood (c. age 10) is a protective factor; so also is coming from a home without an alcoholic father. But most important, this study shows that pathways into (and out of) risk are different for children who have an alcoholic father and for those who do not. For the former, the mother's low esteem for her alcoholic husband is a protection. For the latter, a different set of issues appears central; that is, the mother's high control over her son during his adolescence appears to be more salient as a protection against the son's later becoming alcoholic.

Such studies, which are able to evaluate interactional processes in the earlier childhood years, which deal with the possibility of multiple pathways into and out of risk, and which are able to spread a net across a sufficient array of the salient predictive domains, are critical if the field is to become better able to specify level of risk. Equally central in this process is the willingness of investigators to formulate models of developmental interplay that take account of mediational effects, chaining sequences, and the vulnerability and opportunity to change that seems to be present at key turning points in people's lives (Elder & Caspi, 1990; Rogosh, Chassin, & Sher, 1989). This work is already taking place in other areas of developmental child psychopathology (e.g., Ensminger, Kellam, & Rubin, 1983; Rutter, 1987; Zubin & Spring, 1977), but it has not yet heavily penetrated the alcohol field.

A Probabilistic Framework for Risk and Outcome

The studies reviewed above provide evidence for both continuity and discontinuity in risk progression over time. Were one not to have a developmental lens through which to view such findings, one would be forced to conclude that the data are conflicting; but that is not at all the case. To understand this apparent contradiction it is necessary to move way from the notion of course as an unvarying coherency of process over developmental time and to view it instead as the outcome of a probabilistic process, the result of multiple factors that independently come into play, sometimes simultaneously and sometimes at different developmental periods (Zucker, 1987). This perspective requires a life course framework, or risk grid, that maps both multiple domains of influence and the succession of developmental periods, within which the extent of risk of alcohol use or alcohol-related problems (or both) may build up, remain steady, or subside. Such a model has been presented in detail elsewhere (Zucker, 1987, 1988).

Within the context of such a model, the rudimentary element of risk becomes the event, the display of an individual or contextual attribute (separately or in interaction), which—if repeated often enough—will eventuate in a troublesome (i.e., symptomatic) outcome. The probability of such a display is lessened when the high-risk attribute driving it is diluted by an insulating attribute; it is also lessened if the contextual attribute necessary for an environmental trigger is not present. Conversely, the probability of repeated display, and damaged outcome, is heightened when contextual risk or trigger factors are densely packed around the individual or when repetition itself allows for the evolution of a hierarchized display structure (Zucker, 1988) wherein the elicitation of one component is more likely to trigger the appearance of a variety of related components. Such a framework is better able to handle the complex nature of the evidence for both stability and reversibility of risk than is true of schemas like the disease model, which postulate irreversibility, of course.

Thus, developmental stage changes in alcohol abuse, noted in the longitudinal studies that have tracked this process from adolescence into adulthood (Fillmore et al., 1979; Yamaguchi & Kandel, 1984), are accounted for by differences in the salience of contextual triggers for alcohol involvement across the stages. Similarly, the substantial evidence for migration into and out of alcohol abuse in adulthood, noted in both alcoholic (Skog & Duckert, in press) and nonalcoholic (Cahalan, 1970) populations is understandable by way of the occurrence of an event structure that has neither been repeatedly internally driven nor externally triggered sufficiently to cumulate. The obverse is also true: In studies of the manifestations of risk from birth to adolescence (Werner, 1986), within adolescence itself (Labouvie & McGee, 1986), and from middle childhood into middle adulthood (McCord, 1988), when risk is cumulated by way of the presence of multiple risky components, the probability of a hazardous outcome is increased to substantially over the 50% mark. The currently available data suggest that this cumulation process is strictly additive, but this issue has not been explored in any systematic way.

Within the context of such a probabilistic model it becomes reasonable to look early on in the life span for precursors of later, especially abusive, alcohol involvement. One would be particularly interested in individuals and environments where there is (1) early appearance of the problem behavior syndrome and the endogenous factors (e.g., temperament, activity level) that may drive it, (2) inheritance of positive alcohol-specific endogenous factors, (3) early exposure to alcohol use and to environments that are favorable in their attitudes about heavy alcohol use, and (4) contextual support for the practice and stimulation of the behaviors that emerge as a result of the first three precursors. One would anticipate that such support and stimulation would be most forthcoming from family and peer social structure. These circumstances, operating in concert, can be expected to enhance hazard and cumulate risk. Conversely, cumulation into abuse should be less likely to occur if exposure and involvement is in ameliorative environments. As a corollary to this, if one's interest is in the prevention of abusive alcohol involvement at the individual level, it is most reasonable to intervene in those environments where a greater number of factors suggest the likelihood of cumulative process.

Acknowledgments

Support for preparation of the chapter was provided in part by National Institute on Alcohol Abuse and Alcoholism Grant AA07065 to R.A. Zucker, R.B. Noll, and H.E. Fitzgerald, and by grants from the Michigan Department of Mental Health, Prevention Services Unit, from the Michigan Office of Substance Abuse Services, and from the Michigan State University Social Science Research Bureau.

Notes

1. In fact, from the perspective of prevention, it is essential that these precursive, inferred processes be attended to if one is concerned that the hazard not manifest itself.
2. *Alcoholism* is the generic term being used in this discussion to refer to the propensity for individuals to have chronic, alcohol-related troubles that operate as a syndrome and have been shown to have elements of a common etiology and course. The terms *alcohol dependence* or *chronic problem drinking* could just as easily be substituted instead.
3. This term refers to the child's status; in fact, the parents frequently are already alcoholic, and this situation itself is a common contributor to risk (see no. 6).
4. Recent work by Cloninger and his colleagues (Cloninger, Sigvardsson, & Bohman, 1988) is also consistent with this conclusion. Although the Washington University group set their categories within the framework of a much more strictly clinical vocabulary, they appear to be replicating and rediscovering a set of phenomena that have repeatedly been implicated in the etiology of the involvement of youth in alcohol-related problems and that have been reported in the literature at least since 1968 (Jessor, Graves, Hanson & Jessor, 1968; Zucker & Fillmore, 1968).
5. Alcohol-specific factors are also termed *drinking specific factors* (Zucker, 1979) and *proximal influences* (Jessor & Jessor, 1977).
6. This statement needs to be tempered by awareness that the damaged alternative is the less probable outcome when this factor is the only one present. As will be discussed below, when several risk factors are present, the probabilities become greater.

References

American Psychiatric Association. (1987). *Diagnostic and statistical manual of mental disorders* (3rd ed., rev.). Washington, DC: Author.

Begleiter, H. (1988). Symposium: The genetics of alcoholism: Introduction to the symposium. *Alcoholism: Clinical and Experimental Research, 12,* 457.

Block, J., Block, J.H., & Keyes, S. (1988). Longitudinally foretelling drug usage in adolescence: Early childhood personality and environmental precursors. *Child Development, 59,* 336–355.

Block, J.H., & Block, J. (1980). The role of ego-control and ego-resiliency in the organization of behavior. In W.A. Collins (Ed.), *Development of cognition, affect, and social relations* (pp. 39–101). Hillsdale, NJ: Lawrence Erlbaum.

Brook, J.S., Whiteman, M., & Gordon, A.S. (1983). Stages of drug use in adolescence: Personality, peer, and family correlates. *Developmental Psychology, 19,* 269–277.

Brook, J.S., Whiteman, M., Gordon, A.S., & Cohen, P. (1986). Dynamics of childhood and adolescent personality traits and adolescent drug use. *Developmental Psychology, 22,* 403–414.

Cadoret, R.J., Troughton, E., O'Gorman, T.W., & Heywood, E. (1986). An adoption study of genetic and environmental factors in drug abuse. *Archives of General Psychiatry, 43,* 1131–1136.

Cahalan, D. (1970). *Problem drinkers.* San Francisco, CA: Jossey-Bass.

Cahalan, D., Cisin, I., & Crossley, H. (1969). *American drinking practices: A national study of drinking behavior and attitudes.* New Brunswick, NJ: Rutgers Center of Alcohol Studies.

Cloninger, C.R. (1987). Neurogenetic adaptive mechanisms in alcoholism. *Science, 236,* 410–416.

Cloninger, C.R., Bohman, M., & Sigvardsson, S. (1981). Inheritance of alcohol abuse: Cross-fostering analysis of adopted men. *Archives of General Psychiatry, 38,* 861–867.

Cloninger, C.R., Sigvardsson, S., & Bohman, M. (1988). Childhood personality predicts alcohol abuse in young adults. *Research Society of Alcoholism, 12,* 494–505.

Cotton, N. (1979). The familial incidence of alcoholism: A review. *Journal of Studies on Alcohol, 49,* 89–116.

Donovan, J.E., & Jessor, R. (1985). Structure of problem behavior in adolescence and young adulthood. *Journal of Consulting and Clinical Psychology, 53,* 890–904.

Donovan, J.E., Jessor, R., & Costa, F.M. (1988). Syndrome of problem behavior in adolescence: A replication. *Journal of Consulting and Clinical Psychology, 56,* 762–765.

Edwards, G., & Gross, M. (1976). Alcohol dependence: Provisional description of a clinical syndrome. *British Medical Journal, 1,* 1058–1061.

Elder, G.H., & Caspi, A. (1990). Studying lives in a changing society: Sociological and personological explorations. In A.I. Rabin, R.A. Zucker, R.E. Emmons, & S.J. Frank (Eds.), *Studying persons and lives.* New York: Springer.

Ensminger, M.E., Brown, C.H., & Kellam, S.G. (1982). Sex differences in antecedents of substance use among adolescents. *Journal of Social Issues, 38* (2), 25–42.

Ensminger, M.E., Kellam, S.G., & Rubin, B.R. (1983). School and family origins of delinquency: Comparison by sex. In S.A. Mednick (Ed.), *Prospective studies of crime and delinquency.* Boston: Kluwer-Nijhoff.

Fillmore, K.M., Bacon, S.D., & Hyman, M. (1979). *The 27-year longitudinal panel study of drinking by students in college, 1949–1976.* Final report to the National Institute on Alcohol Abuse and Alcoholism under contract No. (ADM) 281-76-0015. Berkeley, CA: Social Research Group, University of California.

Gomberg, E.S.L. (1982). Alcohol use and alcohol problems among the elderly. *Alcohol and Health Monograph, 4,* 263–290.

Goodwin, D.W., Crane, J.B., & Guze, S.B. (1971). Felons who drink: An 8-year follow-up. *Quarterly Journal of Studies on Alcohol, 32,* 136–147.

Goodwin, D.W., Van Dusen, K.T., & Mednick, S.A. (Eds.). (1984). *Longitudinal research in alcoholism.* Boston: Kluwer-Nijhoff.

Hesselbrock, M.N., Hesselbrock, V.M., Babor, T.F., Stabenau, J.R., Meyer, R.E., & Weidenman, M. (1984). Antisocial behavior, psychopathology and problem drinking in the natural history of alcoholism. In D.W. Goodwin, K.T. Van Dusen, & S.A. Mednick (Eds.), *Longitudinal research in alcoholism,* Boston: Kluwer-Nijhoff.

Institute of Medicine. (1980). *Alcoholism, alcohol abuse, and related problems: Opportunities for research.* Washington, DC: National Academy Press.

Institute of Medicine. (1987). *Causes and consequences of alcohol problems: An agenda for research.* Washington, DC: National Academy Press.

Jacob, T., & Seilhamer, R.A. (1987). Alcoholism and family interaction. In T. Jacob (Ed.) *Family Interaction and psychopathology: Theories, methods and findings.* New York: Plenum.

Jessor, R., Graves, T.D., Hanson, R.C., & Jessor, S.L. (1968). *Society, personality, and deviant behavior: A study of a tri-ethnic community.* New York: Holt, Rinehart and Winston.

Jessor, R., & Jessor, S.L. (1973). The perceived environment in behavioral science: Some conceptual issues and some illustrative data. *American Behavioral Scientist, 16,* 801–828.

Jessor, R., & Jessor, S.L. (1975). Adolescent development and the onset of drinking: A longitudinal study. *Journal of Studies on Alcohol, 36,* 27–51.

Jessor, R., & Jessor, S.L. (1977). *Problem behavior and psychosocial development: A longitudinal study of youth.* New York: Academic.

Kandel, D.B. (1975). Stages in adolescent involvement in drug use. *Science, 190,* 912–914.

Kandel, D.B. (1978). Convergences in prospective longitudinal surveys of drug use in normal populations. In D.B. Kandel (Ed.), *Longitudinal research on drug use: Empirical findings and methodological issues.* Washington, DC: Hemisphere.

Kandel, D.B. (1980). Drug and drinking behavior among youth. In J. Coleman, A. Inkeles, & N. Smelser (Eds.), *Annual review of sociology* (Vol. 6) (pp. 235–285). Palo Alto, CA: Annual Reviews.

Kandel, D.B., & Andrew, K. (1987). Processes of adolescent socialization by parents and peers. *International Journal of the Addictions, 22,* 319–342.

Kandel, D.B., Kessler, R.C., & Margulies, R.Z. (1978). Antecedents of adolescent initiation into stages of drug use: A developmental analysis. In D.B. Kandel (Ed.), *Longitudinal research on drug use: Empirical findings and methodological issues* (pp. 73–99). Washington, DC: Hemisphere.

Kandel, D.B., & Logan, J.A. (1984). Patterns of drug use from adolescence to young adulthood: I. Periods of risk for initiation, continued use, and discontinuation. *American Journal of Public Health, 74,* 660–666.

Kellam, S.G., Brown, C.H., Rubin, B.R., & Ensminger, M.E. (1983). Paths leading to teenage psychiatric symptoms and substance use: Developmental epidemiological studies in Woodlawn. In S.B. Guze, F.J. Earls, & J.E. Barrett (Eds.), *Childhood psychopathology and development* (pp. 17–47). New York: Plenum Press.

Kellam, S.G., Ensminger, M.E., & Simon, M.B. (1980). Mental health in first grade and teenage drug, alcohol, and cigarette use. *Drug and Alcohol Dependency, 5,* 273–304.

Labouvie, E.W., & McGee, C.R. (1986). Relation of personality to alcohol and drug use in adolescence. *Journal of Consulting and Clinical Psychology, 54,* 289–293.

McCord, (1988). Identifying developmental paradigms leading to alcoholism. *Journal of Studies on Alcohol, 49,* 357–362.

Marlatt, G.A. (1983). The controlled-drinking controversy: A commentary. *American Psychologist, 38,* 1097–1110.

Meehl, P.E. (1989). Schizotaxia as an open concept. In A.I. Rabin, R.A. Zucker, R.E. Emmons, & S.J. Frank (Eds.), *Studying persons and lives.* New York: Springer.

Moos, R. (1973). Conceptualization of human environments. *American Psychologist, 28,* 652–665.

Nathan, P.E. (1988). The addictive personality is the behavior of the addict. *Journal of Consulting and Clinical Psychology, 56,* 183–188.

National Institute on Alcohol Abuse and Alcoholism (1985). *Alcoholism: An Inherited disease* (DHHS Publication No. ADM 85-1426). Washington, DC: U.S. Government Printing Office.

Nielson, W.A., & Carhart, D.W. (Eds.) (1956). *Webster's New International Dictionary* (2nd ed.) Springfield, MA: Merriman.

Noll, R.B., & Zucker, R.A. (1983, August). *Developmental findings from an alcoholic vulnerability study.* Paper presented at the American Psychological Association Meetings, Anaheim, CA.

Noll, R.B., Zucker, R.A., & Greenberg, G.S. (1989). Identification of alcohol by smell among preschoolers: Evidence for early socialization about drugs occurring in the home. Unpublished manuscript, Michigan State University, Department of Psychology, East Lansing, MI.

Nylander, I. (1960). Children of alcoholic fathers. *Acta Paediatrica Scandinavica, 49,* supplement 121.

Peele, S. (1986). The implications and limitations of genetic models of alcoholism and other addictions. *Journal of Studies on Alcohol, 47,* 63–73.

Peele, S. (1988). A moral vision of addiction: How people's values determine whether they become and remain addicts. In S. Peele (Ed.), *Visions of addiction* (pp. 210–233). Lexington, MA: Lexington Books.

Reider, E.E., Zucker, R.A., Maguin, E.T., Noll, R.B., & Fitzgerald, H.E. (1989, August). Alcohol involvement and violence toward children among high risk families. Paper presented at the American Psychological Association Meetings, New Orleans, LA.

Reider, E.E., Zucker, R.A., Noll, R.B., Maguin, E.T., & Fitzgerald, H.E. (1988, August). Alcohol involvement and family violence in a high risk sample: I. Spousal violence. Paper presented at the annual meeting of the American Psychological Association, Atlanta, GA.

Robins, L.N. (1984). The natural history of adolescent drug use. *American Journal of Public Health, 74,* 656–657.

Robins, L.N., & Przybeck, T.R. (1985). Age of onset of drug use as a factor in drug and other disorders. In C.L. Jones & R.J. Battjes (Eds.), *Etiology of Drug Use: Implications for Prevention* (pp. 178–192) (NIDA Research Monograph 56: A RAUS Review Report). (DHHS Publication No. ADM 87-1335). Washington DC: U.S. Government Printing Office.

Rogosh, F., Chassin, L., & Sher, K.J. (1989). Personality variables as mediators and moderators of family history risk for alcoholism: Conceptual and methodological issues. Unpublished manuscript, Arizona State University, Department of Psychology, Tempe, AZ.

Roizen, R., Cahalan, D., & Shanks, P. (1978). "Spontaneous remission" among untreated problem drinkers. In D.B. Kandel (Ed.), *Longitudinal research on drug use* (pp. 197–221). New York: Wiley.

Rutter, M. (1987). Psychological resilience and protective mechanisms. *American Journal of Orthopsychiatry, 57,* 316–331.

Rydelius, P. (1981). Children of alcoholic fathers: Their social adjustment and their health status over 20 years. *Acta Paediatrica Scandinavica, 286,* 1–83.

Rydelius, P.A. (1984). Children of alcoholic fathers: A longitudinal prospective study. In D.W. Goodwin, K.T. Van Dusen, & S.A. Mednick (Eds.), *Longitudinal research in alcoholism.* Boston: Kluwer-Nijhoff.

Searles, J.S. (1988). The role of genetics in the pathogenesis of alcoholism. *Journal of Abnormal Psychology, 97,* 153–167.

Skog, O., & Duckert, F. (in press). The stability of alcoholics' and heavy drinkers' consumption: A longitudinal study. *Journal of Studies on Alcohol.*

Tarter, R.E., Alterman. A.I., & Edwards, K.L. (1985). Vulnerability to alcoholism in men: A behavior-genetic perspective. *Journal of Studies on Alcohol, 46,* 329–356.

Werner, E.E., (1986). Resilient offspring of alcoholics: A longitudinal study from birth to age 18. *Journal of Studies on Alcohol, 47,* 34–40.

Yamaguchi, K., & Kandel, D.B. (1984). Patterns of drug use from adolescence to young adulthood: III. Predictors of progression. *American Journal of Public Health, 74,* 673–681.

Zubin, J., & Spring, B. (1977). Vulnerability: A new view of schizophrenia. *Journal of Abnormal Psychology, 86,* 103–126.

Zucker, R.A. (1979). Developmental aspects of drinking through the young adult years. In H.T. Blane & M.E. Chafetz (Eds.), *Youth, alcohol, and social policy* (pp. 91–146). New York: Plenum Press.

Zucker, R.A. (1987). The four alcoholisms: A developmental account of the etiologic process. In P.C. Rivers (Ed.), *Nebraska symposium on motivation: Vol. 34. Alcohol and addictive behaviors* (pp. 27–83). Lincoln, NE: University of Nebraska Press.

Zucker, R.A. (1988, October). Alcohol involvement over the lifespan: A developmental perspective on theory, course and method. Paper presented at the First Vanderbilt Conference on Lifespan and Developmental Alcohol Research, Vanderbilt University School of Medicine and Institute of Public Policy Studies, Nashville, TN.

Zucker, R.A. (1989). Is risk for alcoholism predictable? A probabilistic approach to a developmental problem. *Drugs and Society,* 3 (Nos. 3/4), 69–93.

Zucker, R.A., & Fillmore, K.M. (1968, September). *Motivational factors and problem drinking among adolescents.* Paper presented at the 28th International Congress on Alcohol and Alcoholism, Wash., DC.

Zucker, R.A., & Gomberg, E.S.L. (1986). Etiology of alcoholism reconsidered: The case for a biopsychosocial process. *American Psychologist, 41,* 783–793.

Zucker, R.A., & Noll, R.B. (1982). Precursors and developmental influences on drinking and alcoholism: Etiology from a longitudinal perspective. In National Institute on Alcohol Abuse and Alcoholism (Ed.), *Alcohol and Health Monograph No. 1: Alcohol consumption and related problems.* (DHHS Publication No. ADM 82-1190, pp. 289–330). Rockville, MD: National Institute on Alcohol Abuse and Alcoholism.

Zucker, R.A., & Noll, R.B. (1987). The interaction of child and environment in the early development of drug involvement: A far-ranging review and a planned very early intervention. *Drugs and Society, 2,* 57–97.

SECTION IV

The Genesis and Patterning of Alcoholism and Alcohol-Related Problems

C. ETIOLOGICAL ASPECTS

Introductory Note

As discussed in the introduction to the last subsection, the search for the causes of alcoholism continues to plague researchers in the field. Whereas the last four chapters focused on early childhood predictors of alcoholism, in this section we present selections that primarily examine envionmental influences on later alcoholism. Sociologists believe that the causes of alcoholism can be found in the culture and social structure, and several sociologists have applied sociocultural models to the study of alcoholism (see White, 1982, for a review). Sociocultural theories of etiology date back to the classic works by Bales (see chapter 27) and Ullman (1958) and have roots in the early anthropological research presented in the first section of this book. These theories assume a relationship between cultural attitudes toward, purposes for, motivations for, or norms of alcohol use and the rate of alcoholism.

Robert Bales' theory (chapter 27) is probably the most comprehensive sociocultural theory. He defines three ways in which culture and social organization can influence the rate of alcoholism and claims that the rate of alcoholism is an interaction of three sets of variables: the dynamic factor, the normative orientation, and the presence of alternatives. In addition, Bales delineates four types of normative orientations that influence drinking problems: forced abstinence and ritual, convivial, and utilitarian drinking. Bales supports his arguments with descriptive and historic data on various cultural groups, and his work has been criticized for the lack of systematic data testing the propositions.

This criticism is met head on in the next chapter (chapter 28) in which Arnold Linsky, John Colby, and Murray Straus empirically test Bales' first two propositions. Using state-level data, these authors rate states in terms of a stress index (e.g., high rates of divorce, unemployment, infant mortality) and a proscriptive norm index (e.g., the number of alcohol outlets, the proportion of dry areas). In actuality their test of proscriptive as compared with prescriptive norms relates more to Ullman's (1958) or Mizruchi and Perrucci's (1962) conceptions of normative orientation than to Bales' four normative orientations. The findings of the study strongly support Bales' first proposition: Those states with higher rates of stress also have the highest levels of per capita consumption and alcohol-related mortality rates. The findings for the second proposition are equivocal and demonstrate that indicators of heavy consumption (i.e., high consumption and mortality) are found more often in states low in proscriptive norms but that alcohol-related disruptions (e.g., drunk driving and alcohol-related arrests) are found more often in states rated high on the proscriptive norm index. The authors demonstrate how this discrepancy can be explained by a social control hypothesis—that is, because there is less tolerance of drinking in proscriptive environments there is stronger policing of behavior. Finally, the data indicate that stress and proscriptive norms interact to increase the rate of heavy drinking, as was originally hypothesized by Bales.

In chapter 29 Ole-Jörgen Skog discusses heavy and problem drinking within a social network theory. He has written this chapter, like its companion piece, chapter 7, especially for this volume. Taking it for granted that biological constitution plays a role in the etiology of alcoholism, in this chapter Skog limits himself to interpersonal and environmental influences. Empirically he demonstrates that as one moves from cultures with low to cultures with high mean consumption, the consumption level of all types of drinkers increases. Finally, Skog suggests that there is a relationship between social isolation and alcoholism. He divides alcoholics into two types: those who drink heavily because they are integrated into a heavy-drinking subculture and those who drink heavily because they are poorly integrated. These types suggest that alcoholism results from an interaction of personal vulnerability and social environmental influences.

It is to the issue of personal vulnerability that the final chapter (chapter 30) in this subsection is addressed. This chapter transcends the discipline of sociology and looks at genetic theories of etiology. Although environmentalists argue that the fact that alcoholism runs in families is due to the social and psychological influences of living with alcoholic parents, Goodwin argues that it is the result of biological vulnerability. He summarizes results from twin and adoption studies and concludes that heredity influences some cases of alcoholism. He then speculates about the possible biological explanations for the familial transmission of alcoholism. It should be noted, however, that the adoption and twin studies that Goodwin reviews have been the brunt of many recent criticisms regarding their methodological soundness (see, for example, Lester, 1988).

References

Lester, D. (1988). Genetic theory: An assessment of the heritability of alcoholism. In C.D. Chaudron & D.A. Wilkinson (Eds.), *Theories on alcoholism* (pp. 1–28). Toronto: Addiction Research Foundation.

Mizruchi, E.H., & Perrucci, R. (1962). Norm qualities and differential effects of deviant behavior: An exploratory analysis. *American Sociological Review, 27,* 391–399.

Ullman, A.D. (1958). Sociocultural backgrounds of alcoholism. *Annals of the American Academy of Political and Social Sciences, 315,* 48–54.

White, H.R. (1982). Sociological theories of the etiology of alcoholism. In E.S. Gomberg, H.R. White, & J.A. Carpenter (Eds.), *Alcohol, Science and Society Revisited* (pp. 205–32). Ann Arbor: University of Michigan Press and Rutgers Center of Alcohol Studies.

CHAPTER 27

Cultural Differences in Rates of Alcoholism

ROBERT F. BALES

There are three general ways in which culture and social organization can influence rates of alcoholism. The first is the degree to which the culture operates to bring about acute needs for adjustment, or inner tensions, in its members. There are many of these; culturally induced anxiety, guilt, conflict, suppressed aggression, and sexual tensions of various sorts may be taken as examples. The second way is the sort of attitudes toward drinking that the culture produces in its members. Four different types of attitudes will be suggested later. The crucial factor seems to be whether a given attitude toward drinking positively suggests drinking to the individual as a means of relieving his inner tensions, or whether such a thought arouses a strong counteranxiety. The third general way is the degree to which the culture provides suitable substitute means of satisfaction. In other words, there is reason to believe that if the inner tensions are sufficiently acute certain individuals will become compulsively habituated in spite of opposed social attitudes unless substitute ways of satisfaction are provided.

These three factors may be used as the outline for the rest of this discussion. Under each heading one or more cultural groups will be discussed whose particular rates seem to reflect the factor in question. It is taken for granted that the three factors work together, and that any given rate depends upon their particular combination. If this is not mentioned explicitly in each case, the omission is only a means of simplifying and saving time. It should also be noted that biological or physiological differences may play a part in some of these differences in rates; but if they do, what they may be or how they may operate is still unknown. Various theories to this effect have been offered in the past, but they are hardly accepted by biologists and physiologists now. The three factors mentioned

Robert F. Bales is Professor Emeritus of Social Relations, Harvard University.
This chapter is reprinted (with minor revisions) with permission from *Quarterly Journal of Studies on Alcohol, 6,* pp. 480–499, 1946.

seem to make sense, both theoretically and practically. However, scientists are still very much in the process of interpreting cultural differences in rates, and in some ways the rates offer more problems than answers.

There is fairly good evidence that inebriety tends to be pronounced where the inner tensions or needs for adjustment of many individuals are high, other things being equal. In a careful statistical study of all the primitive societies for which data were available, Horton (1943, pp. 199–320) found that societies with inadequate techniques or resources for maintaining their physical existence also tended to have "strong degrees of insobriety." This finding does not necessarily mean that the members become compulsive drinkers, but that on the occasions when they drink they do so to the point of unconsciousness and their bouts of drunkenness are likely to last for days. Their drinking, in other words, shows a semicompulsive character once started, although it starts only on socially sanctioned occasions. The direct factor seems to be the anxiety induced by the basic insecurities of their lives, such as the constant danger of drought, insect plagues, floods, crop failures, or other threats to the food supply. Those societies that were being broken up by contact with other more powerful groups invariably had high degrees of insobriety. In most of the societies with high subsistence anxiety there was also a great deal of pent-up aggression, which emerged in the periods of drunkenness, sometimes in very extreme form.

Repressed aggression seems to be a very common maladjustment. It can be created by the way the social organization is set up, as well as by other sorts of deprivation or frustration. The data gathered by Hallowell (1955) on the Northeastern Woodlands Indians relating to the time of their contact with the whites indicate a high degree of inhibition of aggressive impulses. They are forced by their culture to be restrained, stoic, amiable, and mild under all provocations and had to suppress all open criticism of one another. It is assumed that they had a great deal of pent-up aggression, because they had a highly developed system of witchcraft directed against one another. When they were introduced to alcohol, the consequences were disastrous. In their bouts of drunkenness they strangled and beat themselves—a form of aggression directed against the self. They broke up everything in their wigwams and quarreled for hours together. Brothers cut the throats of sisters, husbands attacked their wives, mothers threw their children into the fire or into the river, fathers choked their children, and children attacked their parents. Many others of the Indian tribes of the eastern United States showed similar reactions. Their culture was broken up by the whites. They were crowded off their lands, beaten and cheated in their economic dealings, kidnapped and sent into slavery, and exposed to the ravages of strange diseases. In short, they were subject to the strongest sort of subsistence anxiety. This laid a part of the groundwork for the devastation alcohol worked among them. There was another factor, however, tied up with the fact that they had not known alcohol previously, which will be discussed shortly.

To illustrate more definitely the way in which the social organization can induce inner tensions in its members, the Irish peasantry may be taken as an example.[1] The Irish have been noted for their inebriety during the past several centuries. In statistics of admissions for alcoholic disorders to various hospitals in this country the Irish have consistently had rates 2 to 3 times as high as those of any other ethnic group. In 1840 an Irish priest wrote:

> In truth, not only were our countrymen remarkable for the intemperate use of intoxicating liquors, but intemperance had already entered into, and formed a part of the national character. An Irishman and a drunkard had become synonymous terms. Whenever he was to be introduced in character, either in the theatre or on the pages of the novelist, he should be represented habited in rags, bleeding at the nose, and waving a shillelah. Whiskey was everywhere regarded as our idol. (Birmingham, 1840)

The English at this time wished to keep the Irish an agricultural people, so that the Irish would raise the sort of farm produce needed in England. As a part of this program they hampered the development of adequate means of transportation in order to prevent industrialization in Ireland. They had control of the land, with a complicated system of absentee landlords, and squeezed every last penny they could out of the Irish farmers. The farmers, for the most part, lived on the bare edge of existence. They raised so few different kinds of food that the potato blight resulted in severe periodic famines. Many people died during these famines because food could not be transported to them in time.

The small farms were crowded to capacity. There was no room on them for more than one extended family. The grandparents, who had retired, lived in the West Room of the cottage. The farmer and his wife with their children made up the rest of the family. They all had to work hard for their existence, the girls helping their mother around the cottage and farmyard, the boys helping their father in the fields. There was a strict separation of the sexes, which they managed to maintain in spite of the fact that the whole family usually slept in the same room. The training of the children was apparently a very contradictory affair. The elders teased the little boys unmercifully—"codding" it was called—and sometimes prodded them into "scuffing" with one another. It was not unusual for a child to receive extravagant love and affection at one moment, and at the next moment to be cuffed about or even beaten in a fit of anger. There was also a marked tendency to attempt to control children through an exaggerated fear of the "bogey man," "spooks," and "fairies." Conflict between family members was likely to be frequent and severe. Many children must have grown up in an atmosphere of fear and insecurity, both in the family and outside.

Because the farms were so small, they could not be divided. The boys could not marry until the "old fellow" was ready to give up the farm and retire. Then only one of the boys would get the farm. The other brothers and sisters would have to leave. Some of the brothers went into the priesthood. Others were apprenticed to tradesmen in town, or emigrated. One or two of the girls received a dowry and married. The others became nuns, were apprenticed, or emigrated. Very often the "old fellow" was not willing to give up the farm until his physical powers were spent, and this caused a great deal of resentment. So long as his sons stayed on the farm they had to work for him as "boys" and were treated as boys, even though they might be 45 or 50 years old. By the same token they had to stay away from girls. There were very severe sanctions on premarital sexual activity, which were enforced both by the church and the peasantry. Thus, there were many physically mature men, ready for a life of their own, who were kept under the father's thumb as "boys," dependent upon him even for their spending money, and deprived of sexual contacts.

Even social contacts between the two sexes were at a minimum. It was apparently not considered a good idea to encourage love affairs when the land would not support more families. When the "boys" were not working as laborers for their father, they were expected to spend their time with the other "boys." Small male groups of every age met at various farmhouses or taverns to pass the time. Drinking and aggressive horseplay were major activities. The teetotaler, as a matter of fact, was regarded as a suspicious character, since his abstinence implied he was not likely to be with his peers and might be wandering around with the idea of molesting innocent girls. In short, the culture was such as to create and maintain an immense amount of suppressed aggression and sexuality. Both of these suppressed tensions found their outlet in drinking.

It is not entirely clear just what happened to this family system in the urban United States, but it seems to have broken down in various ways and created still other conflicts.[2] The males came in at the bottom of the U.S. occupational ladder and no longer had the ownership of a farm as the mainstay of their self-respect and prestige. The tenement was not a place where aged parents could easily be kept after their working days were over. It was not easy to provide the money for their support out of small day-wages. There was uncertainty and inner conflict over whether one was obligated to keep parents at all. There was usually nothing a father could pass on to his son; but if he died and did leave a little property, there was likely to be conflict over how it should be divided, since equal inheritance was not the rule in Ireland as it is here. The father in many cases seems to have dropped into a role of impotence and insignificance and the mother to have become the dominant member of the family. She tended to bind her sons to her in the way that was usual and natural in Ireland. In this country, however, a strong attachment between a son and his mother made it very difficult, and in some cases impossible, for the son to make a successful transition to independent adult status. In a survey of some 80 cases of alcoholic patients of Irish descent I found this mother-son dependence and conflict in some 60%. Whether this is a higher percentage than would be found in other ethnic groups it is impossible to say at present, but the mother-son dependence pattern was certainly a prominent factor causing maladjustment in these Irish cases.

Although severe inner tensions or needs for adjustment are nearly always found as a background for compulsive drinking, this factor always works in conjunction with the particular attitudes toward drinking that are structured in the culture. Where these attitudes arouse strong counter-anxiety there may be little compulsive drinking in spite of severe maladjustment. Other outlets will be sought. On the other hand, where the attitudes are such as to permit or positively suggest drinking as a means of satisfying minor tensions, the effects of the drinking itself may generate acute maladjustments that result in its perpetuation. It is possible to distinguish four different types of attitudes that are represented in various cultural groups and that seem to have different effects on the rates of alcoholism.

The first is an attitude that calls for complete abstinence. For one reason or another, usually religious in nature, the use of alcohol as a beverage is not permitted for any purpose. The second might be called a ritual attitude toward drinking. This attitude is also religious in nature, but it requires that alcoholic beverages, sometimes a particular one, should be used in the performance of religious ceremonies. Typically the beverage is regarded as sacred, it is consecrated to that end, and the partaking of it is a ritual act of communion with the sacred. This is a characteristic attitude toward drinking among many aboriginal peoples. The third can be called a convivial attitude toward drinking. Drinking is a social rather than a religious ritual, performed both because it symbolizes social unity or solidarity and because it actually loosens up emotions that make for social ease and good will. This is what is often called social drinking. The fourth type seems best described as a utilitarian attitude toward drinking. This attitude includes medicinal drinking and other types calculated to further self-interest or exclusively personal satisfaction. It is often solitary drinking, but not necessarily so. It is possible to drink for utilitarian purposes in a group and with group approval. The distinction is that the purpose is personal and self-interested rather than social and expressive.

One of the outstanding instances of the adoption of the attitude of complete abstinence, total prohibition on a large scale, is that of the Moslems. The taboo rests on a religious basis, the command given by Mohammed. According to one of the translators of the Koran, during the fourth year of the Hegira, or flight from Mecca, Mohammed and his men were engaged in expeditions against neighboring tribes. In the midst of this some of his leaders quarreled while gambling and drinking and upset the plans of warfare. Mohammed then forbade the use of wine and games of chance forever. He supported his decree by a fable of two angels who were sent to Babylon to teach men righteousness. They disobeyed God's commandment not to drink, got into trouble with a woman, and were severely punished by God, who then forbade the use of wine to His servants forever after. It is evident here how the usual dangers to social order, excess aggression and sexuality, played a part in the prohibition. Another factor, perhaps operating in the beginning and certainly afterward in maintaining the taboo, was the danger that intoxication would profane the performance of religious duties. These duties were very strict and exact. A man might soil himself while intoxicated and say a prayer while in an unclean condition.

There do not seem to be any statistical data that might be used as an index to the actual extent of drinking among the Moslems, but there is a great deal of evidence to show that in the course of history the taboo has been very unevenly observed, and in some cases flagrantly violated. All sorts of expedients and rationalizations have been employed to evade the spirit of the law if not the letter of it. One of these was to mislabel the contents of wine containers. Some Moslems assumed that all other alcoholic beverages except wine were permitted. Some protested that the law referred only to excessive drinking. Smuggling and private use have been common. All in all, the Moslem can hardly be regarded as a model of successful total prohibition. One of the chief defects of total prohibition seems to be the extreme difficulty of getting a genuine acceptance of the attitude. The breaking of this taboo becomes an ideal way of expressing dissent and aggression, especially where the original solidarity of the group is weak and aggression is strong. Thus total prohibition sometimes overshoots the mark and encourages the very thing it is designed to prevent. This situation is frequently found among individual alcoholics whose parents were firm teetotalers and absolutely forbade their sons to drink.

A similar situation arose among many of our East Coast Indian tribes when they were first introduced to alcohol. They made repeated attempts to enforce total prohibition when they saw the effects of the alcoholic beverages brought by the traders and colonists. They produced some famous temperance reformers, but they were unable to stem the tide. The old men were usually the most concerned, but the young men could not be controlled. The Hopi and Zuni Indians in western New Mexico form an interesting exception. Although information about them is somewhat confused and contradictory, it appears that in their aboriginal state they used alcoholic beverages in a ritual manner and often ended up with some rioting and sexual expression. When liquor was brought in by the Spaniards, it was used for a period and then rejected entirely. It is not known just how this reaction came about. It may be that the objectionable effects were accentuated by the new insecurity and stronger beverages. These people now put great emphasis on a quiet, calm, orderly existence. Everybody is expected to cooperate peacefully with his neighbors, and nobody tries to be outstanding. Insanity is greatly feared, and they seem to have identified drunkenness with insanity. Their life is highly and meticulously ritualized. It may be that their ritualism adequately took care of their insecurities and fears. It may be surmised, however, that if they had not previously used alcoholic beverages in a ritual manner, thus accepting its prohibition for all other purposes, they would not have been able to make total prohibition effective. The fact that they drank ritually before contact with the whites is one factor that distinctly sets them off from the East Coast Indians, who had had no alcoholic beverages before and hence had not been able to build up any stable attitudes toward drinking, ritually or otherwise. An attitude capable of restraining strong impulses in all the members of a group cannot be created by fiat or even by rational decision of all the members. It has to be a natural part of the emotional training of the child, repeated and actively practiced throughout the life cycle, in order to be firmly built into the personality.

It would be hard to find a better test of the hypothesis of the ritual attitude's significance than the case of Orthodox Jews. They are not total abstainers, as some people suppose. They drink regularly, mostly in a ritual manner, although to some extent in a social way. Yet they are very seldom apprehended for drunkenness, and their rate of admission with alcoholic psychoses to mental hospitals is remarkably low. In almost any table showing rates of this sort for different ethnic groups in this country the Jews are at the bottom, just as the Irish are at the top. There have been many attempts to explain this situation. Immanuel Kant (1793), who was one of the first to offer an explanation, believed that since the Jews' civic position was weak they had to be very rigidly self-controlled and cautious. They had to avoid the scandal and perhaps persecution that might have resulted from drunkenness. Fishberg (1911) emphasizes that Jews have had to live under persecution and says that the Jews know it does not pay to be drunk. Myerson (1942) emphasized the Jewish tradition itself and the hatred the Jews have developed for the drunkard, along with the factors of danger. These hypotheses are not necessarily contradictory—they simply emphasize different aspects of a ramified pattern. Yet it may be seriously questioned whether any of these factors could have remained truly effective without the ritual use of alcohol. In my opinion it is the ritual activity, repeated and participated in from childhood up, that positively stamps into the personality the sentiments or emotions to back up the rational realization of the dangers. Our East Coast Indians certainly saw the dangers rationally, and repeatedly decided to avoid alcohol; but they had no success. They simply did not have the time or the ritual technique for building into the personality the necessary emotional support for their rational decision. Hence, the rational decision could not hold up in a crisis.

If it should be supposed that Jews simply do not have either acute needs for adjustment or inner tensions, strong evidence may be advanced to the contrary. Besides all of the reasons for maladjustment to be expected from their position and historic role, it can be shown that with one or two minor exceptions they are quite as frequently represented in the major mental disorders as other groups. It is the impression of most psychiatrists that Jews have higher rates of neuroses, that is, the milder mental disturbances, than most other ethnic groups. If it should be imagined that Jews have some kind of mysterious immunity to compulsive habits in general, no supporting evidence can be found. Jews are known to have high rates of drug addiction, at least in certain areas. In a study made in this country of all draftees rejected in World War I for psychiatric reasons, Jews were found to have a higher rate of drug addiction than any other ethnic group. This finding emphasizes the fact that their immunity to alcoholism can hardly be explained in terms of either a general immunity to addiction or a lack of acute inner tensions.

The essential ideas and sentiments that may come to be embodied in the attitude toward drinking as a result of ritual drinking, and that seem to operate as a counterinfluence to the formation of the habit of excessive drinking, may be clearly seen in Jewish culture. In the Jewish family at least one male child is greatly desired, so that he may say prayers for his parents after their death. The continuity of the family is through the male line. Eight days after the birth of the male child a ceremony is held for his circumcision. There should be at least 10 adult males present to make up the "minyan," or legal and ritually sufficient quorum, to represent the community and carry on worship in the synagogue. The circumcision is performed because of God's commandment and signifies the entry of the son into the covenant between Jehovah and the Jewish people. The people say, "Even as he has entered into the Covenant, so may he enter into the Law, the nuptial canopy, and into good deeds." A benediction is then offered over a cup of wine, with the following words: "Blessed art Thou,

O Lord our God, king of the universe, creator of the fruit of the vine." The drinking of the cup of wine is a visible symbol and seal of the completed act of union.

In the Jewish culture wine stands for a whole complex of sacred things. Wine is variously alluded to as "the work of God" and "the commandment of the Lord." Similarly, the Torah (the sacred body of the Law), Jerusalem (the sacred place), Israel (the sacred community), and "the Messiah" (the righteous) are all compared to wine. The wine must be ritually pure, untouched by an idolater, and any vessel in which it is put must be "kosher," or ritually clean. Drinking, like eating, has a sacrificial character and is vested with an element of holiness. The dietary laws are symbolic of the separateness of the Jewish people. They constitute a discipline of all appetites to the end of attaining the all-inclusive state of holiness, which is so much desired in the Jewish religion. Undisciplined appetites are a defilement of the self.

The first religious education of the child is directed to the teaching of the proper benedictions for bread, fruit, milk, and other foods. He is told very early, "Thou shalt not eat any abominable thing." He also learns to observe the prohibition against touching that which is "holy unto the Lord." The "fruit of the vine" in the form of grapes falls in this class of sacred things. Grapes must not be eaten before the fifth year of the life of the vine. In the first three years they are "uncircumcised," and in the fourth year they are "holy, to praise the Lord withal." Jewish parents impress their children with a great awe and reverence for things sacred and divine. Both grapes and wine are referred to as the "fruit of the vine," as in the customary benediction over wine.

There are four rituals each Sabbath in which the drinking of wine is the central act of communion. The first is on Friday evening and is called "Kiddush," that is, the ritual which "sanctifies" the holy day. After a thanksgiving prayer by the master of the house, a cup of wine is blessed. The master first partakes, and the cup is then passed from member to member, in the order of their precedence, down to and including the domestics if they are also Jewish. The males wear their hats, as in the synagogue, for this is a religious as well as a familistic ceremony. Then one of the two special Sabbath loaves is broken and each person is given a portion. After the meal, grace is recited and another cup of wine is blessed and passed around as before. This is called the "cup of benediction." A similar ritual, ironically called "Great Kiddush" because of its lesser importance, precedes the benediction over bread before breakfast on the morning of the Sabbath and other festival days. The ritual called the "Habdalah," on the Sabbath evening, marks the separation of the holy day from the rest of the week. The father chants a prayer of separation, the wine cup is poured to overflowing to symbolize the overflowing of blessing that is hoped for, is blessed in the regular way, and is partaken of first by the master of the house and then by the other males. The females do not partake on this occasion. The wine is again blessed, and this time the father drinks alone. He moistens his eyes with the wine, saying, "The commandment of the Lord is pure, enlightening the eyes." The remaining wine is finally poured upon a plate, and the burning candle is extinguished by being dipped into the wine. It is interesting and important to note that the order in which the family members partake of the wine—first the father, then the lesser males, then the females, and then the domestics—emphasizes their relative status and also their relative closeness to the sacred. The same order in reverse is observed in the Habdalah, where first the females abstain and then the lesser males, so that finally only the father drinks. The various members are separated from the sacred in the order of the lesser first, and finally the most important.

The drinking of wine, or the specific abstention from drinking, figures in the various feasts and fasts of the yearly cycle. In general, where food is forbidden as a sign of mourning or as a sign of guilt and expiation, that is, where the ritual state is one of estrangement from the sacred, wine is forbidden. This is true of the "Black Fast" and of Yom Kippur, the Day of Atonement. After the season of estrangement comes the season of reunion and restored favor with God in the Festival of the Booths. This reunion and restored favor is indicated by the taking of food and wine. On Purim, a more secular holiday, the Jew should drink the Talmud says, until he can no longer distinguish between "cursed be Haman" (the ancient persecutor of the Jews) and "blessed be Mordecai" (their ancient savior). Sometimes the old men pretend to become drunk to amuse the youngsters. Actually, however, this pretense appears to be another one of those ironic inversions, like "Great Kiddush," and only emphasizes the extreme foolishness and danger of drunkenness by linking it with the memory of old persecutions. At the celebration of Passover there are four ritual partakings of wine, each with a blessing. After these no more wine may be tasted that night. A cup of wine, which one of the boys tastes, is left for Elijah, who may come during the night. Finally, in the Rejoicing of the Law, a festival celebrating the time when the Law was given, wine is partaken of in the usual ritual manner.

At the time of marriage, the social union par excellence, the bridal couple and their nearest relatives partake of a consecrated cup of wine, and the glass is broken, apparently with the connotation of the finality and exclusiveness of the union. At the time of death, the final separation, the deep mourning is indicated by abstention from all food and wine. In all of these rituals, in which the child participates from the time he is able to understand, the partaking of the consecrated wine indicates a union with the sacred and the solidarity of the Jewish people in their covenant with

God, whereas abstention from the wine indicates a temporary estrangement from Divine favor, a state of guilt, and the sad dispersion of the Jewish people.

In the Jewish culture the wine is sacred and drinking is an act of communion. The act is repeated again and again, and the attitudes toward drinking are all bound up with attitudes toward the sacred in the mind and emotions of the individual. In my opinion this complex of attitudes is the central reason why drunkenness is regarded as so "indecent"—so unthinkable—for a Jew. Rational precaution also probably plays a part, but the ritual use is the main mechanism that builds in the necessary emotional support for the attitudes. Drunkenness is a profanity, an abomination, a perversion of the sacred use of wine. Hence the idea of drinking "to become drunk" for some individualistic or selfish reason arouses a counteranxiety so strong that very few Jews ever become compulsive drinkers.[3]

We have other evidence that the counteranxiety connected with drinking can act as a factor to prevent drunkenness. Among the Balinese alcoholic beverages are distilled and used ceremonially by the people, but they very seldom become intoxicated. The anxiety here, however, seems not to have been created through specific ceremonial use but because it is so extremely important to these people to maintain their exact spatial and geographic orientation. If a Balinese is put into a car and taken suddenly by a winding way to a place where he loses his directions, he becomes extremely anxious and actually sick. Eastward, inland toward their sacred mountain, and upward toward its summit are sacred directions. Westward, outward to the sea, and downward are profane. Each member of the Balinese culture is very careful to keep other members of higher status than himself to the eastward, inland, or on a higher seat. Drunkenness is likely to make one lose his directions and become confused in these relations. In the weaving and confusion of drunkenness one is likely to trespass on the sacred, put sacred things in profane places, or wander off into the dangerous jungle, which they fear greatly. Hence, the anxiety connected with drunkenness is so strong and immediate that the Balinese avoid it almost entirely, in spite of the fact that they have extremely strong suppressed emotions, as is known from other facts about their culture.

Convivial drinking is a mixed type, tending toward the ritual in its symbolism of solidarity, and toward the utilitarian in the "good feeling" expected. Wherever it is found highly developed, it seems to be in danger of breaking down toward purely utilitarian drinking. This breakdown is to be found in marked form in the Irish culture. A drinking party is in order at all of the principal occasions in the life cycle, in the meeting of friends, in business dealings, at political affairs, on pilgrimages, and on every other occasion when people come together. One writer in the last century says,

"Hallow-E'en, St. Patrick's Day, Easter, and all extra-ordinary days are made apologies for a drinking bout; a week's excess is taken at Christmas." He continues:

> Baptisms are generally debauches; launching a ship; making men pay their footing (on board a crowded ship where the men had to stand on the deck), births, wakes, funerals, marriages, churns in the country, are all jovial and vehement occasions of universal revelry. Pledging, toasting, and offering spirits in courtesy is much in vogue; and if a visitor do not taste at any time a day, he gives offence, as in Scotland. Washerwomen, wet nurses, coach drivers, carmen, porters, and others are all treated by their employers with whiskey. There are no dry bargains; and in provision stores and other places, allowances of whiskey are bound to the workmen in the articles of service. An Irishman is in the last stage when he begins to drink alone; which is the case also with the Scotsman; and numbers treat for the mere purpose of obtaining pleasant company. (Dunlop, c. 1834)

It is important to note that although drinking is a part of gatherings for occasions such as marriages, which have a ritual or ceremonial core, the people never drink as a part of the ritual itself. In the Mass the priest partakes of the wine but not the laity. On the sort of occasions where there is a ritual core, drinking is a purely secular, convivial celebration, before or after. Whiskey is always liberally provided. No "good fellow" would be niggardly in providing whiskey for a celebration. It is thought to be only "decent" to treat a friend. One shows that he regards the other as "a good fellow" by drinking with him. When relatives or "friends," as they are called in Ireland, or acquaintances of the same social standing meet in a public house, it is a matter of strict obligation for one of them to "stand" for all the others. He must order drinks all around and pay for them himself. Each man is then obligated to "stand" in his turn, and so on until all have bought at least one drink around. If there are more than three or four in the party, they are necessarily fairly well intoxicated by the time each has done his duty. It is an unforgivable insult to refuse to take a drink with a man without a long involved explanation and a profuse apology.

The breakdown into a utilitarian attitude can be observed in the use of drinking in economic transactions. At the fairs, where the livestock is sold, there is usually a long, heated argument about the price. When an agreement is finally reached, the bargain is sealed with one or more drinks. Sometimes the seller takes his customer to the public house before the agreement is reached and treats him a few times to "soften him up." In making the bargain for a marriage it is necessary to reach extensive economic agreements in the evaluation of the farm and livestock, since this determines the amount of the dowry the young lady's father will give.

In one of these matchmakings the bargainers treat back and forth until all are well fuddled. One writer says, "To one who has lived for some time in England, the mixture of tippling and business seems like some incredible dream. Little bits of business get in, as if by stealth, between the drinks during the day!" The famer typically comes home from a day at the fair in a very intoxicated condition indeed. His wife does not usually complain. In fact, if she is a very good wife, she may treat her husband the next morning with "a hair of the dog that bit him." Drunken men are usually treated with care and affection in Ireland. To the mother the drunken man is "the poor boy." Laborers seeing an intoxicated man coming home from the fair are prone to regard him with envy, rather than pity, since he is in a much better state than they.

Drinking to get over a hangover, taking "a hair of the dog that bit you," is a pure example of individualistic, utilitarian drinking. Here the alcohol is regarded as medicine. According to Morewood (1824), an Irish historian, whisky, or aqua vitae, as it was called, "was first used in Ireland only as a medicine, considered a panacea for all disorders, and the physicians recommended it to patients indiscriminately for preserving health, dissipating humours, strengthening the heart, curing colic, dropsy, palsy, quartan fever, stone, and even prolonging existence itself beyond the common limits." Aqua vitae was sold only in apothecaries' shops until sometime in the sixteenth or seventeenth century, and it has retained its medicinal virtues in the mind of the people to this day. It was the universal folk remedy to "keep the cold out of the stomach," to produce a feeling of warmth after exposure, to restore consciousness in case of fainting, to cure the stomach ache, to cure insomnia, to reduce fatigue, to whet the appetite, to feel stronger, and to get rid of hangovers. In the cholera plagues of 1831 and 1849, which struck a mortal fear into the hearts of the people, brandy was firmly believed to be a preventive. People even sold their beds in their anxiety to get it. The country doctors prescribed it widely and used it themselves. In fact legends grew up about some of these old topers to the effect that the divine inspiration to cure did not possess them unless they were more or less "under the influence." In many cases a glass of whisky was all the poor peasant had to offer the doctor by way of payment. Whisky was given to children as a reward for good behavior. Drinking was the recommended cure-all for young men in low spirits, for whatever reason, just as prayer was recommended to women and old people.

There is little reason to doubt that the utilitarian attitude toward drinking, if commonly held, is the one of the four types that are most likely to lead to widespread compulsive drinking. There is no counteranxiety attached to the process of drinking in this case, and there is every suggestion for the individual to adopt drinking as the means of dealing with his particular maladjustment. The prevalence of this attitude in the culture of the Irish, along with widespread inner tensions, seems adequately to explain their high rate of alcoholism.

There are certain occupational groups with high rates that seem to trace mainly to an occupationally induced utilitarian attitude toward drinking. It is well known, for example, that rates are particularly high among individuals connected with the manufacture and sale of alcoholic beverages, such as brewery employees, innkeepers and managers of public houses and hotels, barmen, waiters, and traveling salesmen. Certain manual laborers who do heavy, exhausting work, such as longshoremen, drink to reduce their fatigue or to escape chronic unpleasant conditions of work. They, too, are likely to have high rates. Sullivan, in a statistical study of England made in 1905, drew a distinction between "convivial drinking," much as it is defined here, and "industrial drinking," which is one type of utilitarian drinking. He found that the highest rates of alcoholic disorders were associated with "industrial drinking, and not with convivial drunkenness." It is a well-known fact that a very high percentage of prostitutes become inebriates. The use of alcohol, both for themselves and as an attraction to their customers, is an indispensable part of their trade.

Finally, there is the third cultural factor that seems to have a bearing on rates of alcoholism—the degree to which the culture provides suitable substitutes for both the inner tensions and the needs for adjustment that it creates. One of the most common substitutes is the use of other narcotic drugs. The high rate of drug addiction among the Jews has been mentioned. Many Moslems, it is said, are users of hashish and seem to be habituated to very strong tea and coffee. Among the Brahmins, who have a severe prohibition against alcoholic beverages, opium is used to a considerable degree, and this pattern seems to have spread to some of the peoples in the East Indies. The Japanese, who are supposed to have a low rate of alcoholism, are frequently users of opium.

Among the Balinese, who avoid drunkenness because of their anxiety about orientation, there is a peculiar trancelike state which seems to act as an outlet for their tensions. Their childhood training is one of constant stimulation to emotional response, followed by frustration. The mother fondles the child, but as soon as he notices and begins to respond emotionally, she "cuts him dead." Under this treatment the children finally withdraw into themselves and refuse to respond. But in their ceremonial dances, which symbolically recall the childhood situation, they go into trances and express extreme self-aggression with mingled emotions of "agony and ecstasy." Thus the trances seem to provide a way of restoring the balance and to give the emotional purging that, in another culture, alcohol might provide. The complicated ritualism of the Hopi and Zuni

Indians, who were successful in total prohibition, may provide a somewhat similar substitute for reducing anxiety.

It is impossible to name all the sorts of things that might possibly serve as substitute means of adjustment, since the nature of any given substitute would depend upon the maladjustments that are acute in particular cultures, and these are very numerous and complicated. The three general factors discussed—the acute inner tensions or needs for adjustment, the attitudes toward drinking, and the provision of substitutes—seem to be fairly adequate in an overall logical way and give some insight into a few of the outstanding cultural differences in rates of alcoholism. There are many rates, however, that are not yet understood. It is still not possible to formulate precisely all the different types of maladjustment or tension that can be involved in compulsive drinking, to say nothing of our inability to measure just how acute they are in specific cultural settings. As to the types of attitude toward drinking, very little is known concerning just how they come about, what makes them endure or break down, or just how widespread particular attitudes are in particular places and cultures. These problems still require an immense amount of careful theoretical and research work.

It can safely be said, however, that all three factors are important. They all work together, and the rates are complicated end results. With regard to the problem of reducing the rates of alcoholism in a particular place, there is no doubt that it must be attacked from all three angles. Anything that can be done to relieve the acute tensions of people or steer them away from utilitarian attitudes toward drinking or provide them with suitable and effective substitutes for the rewards that drinking brings them may have a preventive effect in the long run. It seems clear, however, that no very conspicuous success can be expected unless all three factors can be modified together, and modified considerably for the whole group.

Notes

1. The material on Irish social structure has been drawn chiefly from Arensberg (1937), *The Irish Countryman;* Arensberg and Kimball (1940), *Family and Community in Ireland;* and McCarthy (1911), *Irish Land and Liberty.*
2. These interpretations are based on a study by the present author of some 80 detailed hospital case records of Irish alcoholic patients.
3. Rosenman (1955) has advanced a theory to account for the relative absence of alcoholism among Jews. In his opinion the particular sin committed by the alcoholic is his symbolic alliance with Satan, a pact with or possession by the Bad Father. He seeks power, protection, and pleasure beyond other people; he fights against the restrictive Good Father, he must renounce the Lord and turn to Satan. After his ordeal with Satan, the alcoholic is left helpless, an inferior sinner with Satan's mark on him. This is why the alcoholics form associations, why they think that a virtuous person, that is, a nonalcoholic, cannot under-

stand their weakness. To society the alcoholic is the Devil's agent and must be punished. The idea of the contest between Satan and God as it pertains to alcoholism may offer an explanation of the low rate of alcoholism among Jews. The Hebrews succeeded in integrating all gods and divinities into a single image of God. Satan emerged as his Satanic majesty only with Christianity. The Jew deals only with his one God; alcohol, like everything else, remains a gift of God. "It is to be used to eliminate that hostile self-centeredness which prevents one from integrating solidary relations with God's creatures" as well as to make, seal, and consecrate a covenant with God (Pittman & Snyder, 1962).

References

Arensberg, C.M. (1937). *The Irish countryman.* New York: MacMillan.

Arensberg, C.M. & Kimball, S.T. (1940). *Family and community in Ireland.* Cambridge, MA: Harvard University Press.

Birmingham, J. (1840). *A memoir of the Very Reverend Theobald Mathew.* Dublin: Milliken & Son.

Fishberg, M. (1911). *The Jews: A study of race and environment.* New York: Scribners.

Hallowell, A.I. (1955). Some psychological characteristics of the Northeastern Indians. In *Culture and experience.* Philadelphia: University of Pennsylvania Press.

Horton, D. (1943). The functions of alcohol in primitive societies: A cross-cultural study. *Quarterly Journal of Studies on Alcohol, 4,* 199–320.

Kant, I. (1793). *Antyropologie in pragmatischer hinsicht.* Koenigsberg, Germany: Nicolovius. (Cf., Jellinek, E.M., 1941). Immanuel Kant on drinking. *Quarterly Journal of Studies on Alcohol, 1,* 777–778.

McCarthy, M.J.F. (1911). *Irish land and Irish liberty.* London: Scott.

Morewood, S. (1824). An Essay on the Inventions and Customs of Both Ancients and Moderns in the Use of Inebriating Liquors. London: Longmans, Hurst, Rees, Orme, Brown and Green.

Myerson, A. (1942). Alcoholism and induction into military service. *Quarterly Journal of Studies on Alcohol, 3,* 204–20. Also see: Social psychology of alcoholism. *Diseases of the Nervous System, 1,* 43–50, 1940.

Pittman, D.J. & Snyder, C.R. (Eds.). (1962). *Society, Culture, and Drinking Patterns,* New York: Wiley.

Rosenman, S. (1955). Pacts, possessions and the alcoholic. *American Imago, 12,* 241–274.

Sullivan, W.C. (1905). Industrial alcoholism. *Economic Review*, London, *15*, 150–163.

Testimony of John Dunlop. (c. 1834). Testimony of John Dunlop. In *Evidence on drunkenness presented to the House of Commons.* London: S. Bagster, Jun., Printer.

CHAPTER 28

Stress, Drinking Culture, and Alcohol Problems: A Partial Test of Bales' Theory

ARNOLD S. LINSKY, JOHN P. COLBY, JR., AND MURRAY A. STRAUS

Robert Bales' theory of alcoholism clearly focuses on causes of alcoholism that are located in the culture and social structure of societies rather than on casual factors within the motivation of the individual alcoholic (Bales, 1946). [See Chapter 27, by Bales, in this volume.] The theory also has the virtue of being well conceived and conducive to research in a field not notably crowded by well-articulated theoretical frameworks. Perhaps that is why the theory continues to occupy a central place in alcoholism literature and why it continues to be attractive to sociologically oriented researchers over 4 decades after its original publication. Unfortunately, despite its centrality and influence on subsequent investigations, its major propositions have rarely been the subject of systematic empirical testing (Room, 1976).

Bales' Theory

Bales asserted that cultures or social structures influence the rate of alcoholism in three ways. First, there are social structural factors that create stress and inner tension for members of a particular group or society. We will refer to this as the *stress hypothesis*. Second, there are culturally supported attitudes toward drinking and intoxication that determine whether alcohol will be used as a means for relieving that stress and ten-

Arnold S. Linsky, John P. Colby, Jr., and Murray A. Straus are with the Department of Sociology and State and Regional Indicators Archive, University of New Hampshire, Durham, N.H. Portions of this chapter are reprinted, with permission, from "Stressful Events, Stressful Conditions, and Alcohol Problems in the United States: A Partial Test of Bales' Theory," by A.S. Linsky, M.A. Straus, and J.P. Colby, Jr., which appeared in *Journal of Studies on Alcohol, 46,* pp. 72–80, 1985; from "Drinking Norms and Alcohol-Related Problems in the United States," by A.S. Linsky, J.P. Colby, Jr., and M.A. Straus, which appeared in *Journal of Studies on Alcohol, 47,* pp. 384–393, 1986; and from "Social Stress, Normative Constraints, and Alcohol Problems in American States," by A.S. Linsky, J.P. Colby, Jr., and M.A. Straus, which appeared in *Social Science and Medicine, 10,* pp. 875–883, 1987.

sion (the "normative hypothesis"). The third factor is whether the culture provides alternative mechanisms for relief of that tension (the *functional alternative hypothesis*).

Bales developed his theory primarily on the basis of his analysis of drinking problems in 19th-century Ireland. He explained the high rates of alcoholism in that society as resulting from a combination of a social structure that blocked young men from attaining the respectability and status of householder and married man and at the same time permitted and encouraged heavy drinking in local taverns as a release for the tensions and frustrations produced by their denial of status and sexual fulfillment.

This chapter does not report a full test of all of the hypotheses implied by Bales' theory. Instead, it tests the stress and the normative components of the theory by investigating whether the level of stress or tension and level of normative constraints on drinking are related to the level of alcohol problems in populations.

One reason why the stress aspect of the theory has not spawned more empirical research is that severe measurement difficulties abound. The research reported in this chapter was made possible by our development of ways to measure key elements in Bales' theory. First, we developed a State Stress Index (SSI) to provide objective data on whether one social group is subjected to more or less stress than another. Second, we developed ways to measure norms concerning drinking.

Conceptualization and Measurement of Social Stress

Linsky and Straus (1986) examined state-to-state differences in the extent to which stressful situations occur. Their State Stress Index is used in this chapter as the primary measure of the extent to which social structures or cultures require changes in life patterns and therefore induce inner tensions or stress for individuals. The rationale of the SSI is based on the "life events" research tradition in the measurement of stress (Holmes & Rahe, 1967). The general strategy in life events research has been to demonstrate associations between the onset of illness and recent increases in the number of important events occurring in the lives of individuals and requiring adaptive response. The more events to which individuals have to adapt, the greater the presumed impact on the onset of illness.

Researchers, beginning with Holmes and Rahe (1967), have developed somewhat similar checklists or inventories of stressful life events which are administered to individuals in questionnaire form (see also Coddington, 1972; Dohrenwend, Kranoff, Askenasy, & Dohrenwend, 1978; Paykel, Prusoff, & Uhlenhuth, 1971). What these diverse events have in common is that they are presumed to require important changes in ongoing adjustment. Individuals are asked to check off those events they experienced in the recent past such as divorce, movement to a new community, and so on. Their total life events are then added, in either weighted or unweighted form. The resulting scores have been found to correlate with the subsequent development of physical illness, psychiatric disorders, depressed mood, imprisonment, and pregnancy among others (Holmes & Masuda, 1974).

The State Stress Index was created by translating the life events approach from the original individual level to the macro level. We used the PERI scale developed by Dohrenwend et al. (1978) as the basis for the current macro-level scale because it is the most extensive of the currently available life event scales. Many of the items in this and other individual-level events scales have direct analogs at the societal level. Examples of these parallels include death of a child and the infant mortality rate from the *Vital Statistics of the United States;* loss of home from fire, flood, or other disaster and *Disaster Assistance per 100,000 Families* from the American Red Cross. For other events, the state level indicators only approximate the events in question.

A total of 15 societal indicators of stressful life events are included in the State Stress Index, as shown in Table 1. The SSI is the sum of the 15 indicators (which are first transformed into *Z* scores). The resulting variable was then transformed into "*ZP*" scores (Straus, 1980) to create scores with a more easily interpreted range of 1 to 100 and a mean of 50.

The first column of Table 2 arrays the states in rank order on the State Stress Index. The states with the consistently highest SSI scores are located in the Pacific, Mountain, and Southern regions of the country. The states lowest in stress are those located in the West North Central region of the country, followed by the New England area. Table 2 also shows the three indicators of alcohol-related problems described in the following section.

Indicators of Alcohol-Related Problems

Data on alcohol consumption, and deaths attributed to it, were classified in four ways:

1. Deaths attributed to cirrhosis of the liver per 100,000 adult population (National Center for Health Statistics, 1975–77).
2. The combined death rate for alcoholism and alcoholic psychosis (National Center for Health Statistics, 1975–77).
3. Alcohol consumption in average gallons per person (Licensed Beverage Institute, 1976).

TABLE 1
Life Event Indicators in the SSI

Variable name[a]	Variable and source of data
Economic stressors	
v382r	Business failures per 1 million population, 1976 (The World Almanac and Book of Facts: 1978, New York: Newspaper Enterprise Association, 1978)
v452r	Initial unemployment claims per 100,000 adults, 1976 (U.S. Bureau of the Census. Statistical Abstract of the United States, Washington, D.C.: U.S. Government Printing Office, 1978, p. 59)
t58r	Workers involved in work stoppages per 100,000 adults, 1975 (Statistical Abstract of the U.S. [see v452r], 1978, p. 432)
t183r	Bankruptcy cases commenced per 100,000 population, 1976 (Administrative Office of the United States Courts. Annual Report of the Director, Washington, D.C.: U.S. Government Printing Office, 1976, pp. 152–153)
t200r	Mortgage loans foreclosed per 100,000 population, 1976 (Mortgage Bankers Association of America. National Delinquency Survey, Washington, D.C., 1976)
Family stressors	
t57	Divorces per 1,000 population, 1976 (Statistical Abstract of the U.S. [see v452r], 1979, p. 84)
z120rl	Abortions per 100,000 population, 1977 (Alan Guttmacher Institute. Personal communication, New York, 1981)
t207r	Illegitimate live births per 1,000 population aged 14+, 1976 (U.S. Public Health Service. National Center for Health Statistics. Vital Statistics of the United States, 1976, Vol. 1, Natality, Hyattsville, Md., 1980, pp. 1–172)
t70r	Infant deaths per 1,000 live births, 1976 (Vital Statistics of the U.S. [see t207r], Vol. 2, Mortality, pp. 2–29)
t64r	Fetal deaths per 1,000 live births, 1976 (Vital Statistics of the U.S. [see t207r], Vol. 2, Mortality, pp. 3–9)
Other stressful events	
t187r2	Disaster assistance to families by Red Cross per 100,000 population, 1976 (American National Red Cross. Summary of Disaster Services Activities by Area and State, 1975–76, Washington, D.C.: American Red Cross Disaster Services, 1977)
p168	% of population aged 14+ residing in state ≤ 5 yr, 1976 (U.S. Bureau of the Census. Demographic, Social, and Economic Profiles of States: Spring, 1976. Current Population Reports, Ser. P-20, No. 334, Washington, D.C.: U.S. Government Printing Office, 1979)
v356r	New housing units authorized per 100,000 population, 1976 (Statistical Abstract of the U.S. [see v452r])
t191r2	New welfare recipients per 100,000 population, 1976 (U.S. Social Security Administration. Public Assistance Statistics, Washington, D.C., 1977)
t182r	High-school dropouts per 100,000 population, 1976 (Foster, B.J. and Carpenter, J.M. Statistics of Public Elementary and Secondary Day Schools: 1977–78 School Year. Final Report, Washington, D.C.: National Center for Education Statistics, 1978)

[a]To seek information from the SRIA, these names are necessary for indentification.

TABLE 2
Rank of the 50 States on the SSI and Alcohol-Related Problems

Rank	SSI		CIRRHOSIS DEATHS PER MILLION		DEATHS FROM ALCOHOLISM AND ALCOHOLIC PSYCHOSIS PER MILLION		PER CAPITA ANNUAL CONSUMPTION OF ABSOLUTE ALCOHOL PER POP. 15+ (U.S. GALLONS)[a]	
	State	Score	State	Score	State	Score	State	Score
1	Nev.	104	Nev.	323.91	N.M.	165.31	Nev.	5.33
2	Alaska	87	N.Y.	270.38	Alaska	143.24	N.H.	4.83
3	Ga.	82	Calif.	255.46	Nev.	86.96	Alaska	3.49
4	Wash.	80	Fla.	248.06	Wyo.	67.36	Wis.	3.15
5	Oreg.	78	Del.	229.31	N.C.	66.04	Calif.	3.13
6	Ala.	77	R.I.	227.53	R.I.	61.33	Vt.	3.07
7	Calif.	76	N.M.	218.47	Del.	59.98	Md.	2.96
8	Miss.	74	Mass.	214.17	Ga.	59.31	R.I.	2.94
9	Ariz.	72	N.J.	213.40	Okla.	58.49	Wyo.	2.91
10	Tenn.	71	Mich.	212.99	Utah	56.84	Colo.	2.86
11	Colo.	65	Ariz.	206.12	Mont.	54.01	Mass.	2.85
12	Okla.	62	Ill.	203.70	Ariz.	48.70	Ill.	2.82
13	S.C.	61	N.H.	197.87	Md.	48.54	Ariz.	2.76
14	Fla.	61	Maine	194.99	Conn.	47.72	Del.	2.75
15	Mich.	59	Oreg.	190.83	S.C.	47.46	Wash.	2.74
16	N.Y.	59	Wash.	190.49	W.V.	43.50	Hawaii	2.73
17	Ill.	59	Md.	190.29	Wash.	41.08	N.Y.	2.70
18	Idaho	57	W.V.	189.60	Fla.	40.81	Mont.	2.70
19	Va.	54	Alaska	188.76	Ky.	39.12	Fla.	2.70
20	Ky.	52	Conn.	186.84	Va.	37.94	Mich.	2.83
21	Ohio	51	Penn.	186.69	Maine	36.95	N.J.	2.61
22	La.	51	N.C.	171.63	Tenn.	35.44	Conn.	2.58
23	Del.	50	Wyo.	169.57	Colo.	34.81	N.M.	2.54

24	Md.	50	Ohio	169.15	N.D.	33.76	Oreg.	2.54
25	N.J.	49	Mont.	164.92	Idaho	33.55	N.D.	2.48
26	N.C.	48	Okla.	162.02	Miss.	32.85	Tex.	2.45
27	Tex.	47	Va.	158.17	S.D.	32.76	Minn.	2.43
28	N.M.	47	Ga.	155.75	N.Y.	32.38	La.	2.42
29	Ark.	47	Vt.	155.06	Ala.	31.81	Maine	2.38
30	Pa.	46	S.D.	152.86	Oreg.	31.40	Neb.	2.37
31	W.V.	44	Tex.	150.10	Wis.	31.30	Idaho	2.29
32	Mo.	43	Colo.	149.52	Vt.	30.64	S.C.	2.23
33	Kans.	42	Wis.	147.44	La.	30.49	Penn.	2.21
34	Hawaii	42	La.	145.20	Calif.	30.32	Va.	2.20
35	Ind.	40	S.C.	143.49	Minn.	29.63	Ga.	2.19
36	R.I.	37	Mo.	140.32	Kans.	28.86	S.D.	2.13
37	Conn.	35	Ky.	135.52	N.H.	27.20	Ohio	2.12
38	Maine	35	Neb.	134.93	Ind.	26.36	Iowa	2.07
39	Wyo.	35	Ind.	134.90	Mo.	23.66	Mo.	2.07
40	Vt.	34	Idaho	128.79	Mich.	22.08	Ind.	1.98
41	Mass.	34	Minn.	126.47	Mass.	21.83	N.C.	1.96
42	Mont.	33	Utah	125.68	Ill.	21.80	Miss.	1.88
43	Minn.	28	Ala.	123.37	Ark.	21.23	Okla.	1.85
44	Utah	26	Iowa	123.35	Pa.	20.28	Ala.	1.83
45	N.H.	26	N.D.	123.07	Ohio	20.13	Kans.	1.83
46	Wis.	20	Tenn.	122.55	Tex.	18.57	Ky.	1.79
47	N.D.	19	Kans.	118.31	N.J.	15.85	Tenn.	1.79
48	S.D.	17	Hawaii	110.76	Iowa	14.85	W.V.	1.73
49	Iowa	17	Miss.	106.86	Neb.	14.52	Ark.	1.51
50	Neb.	16	Ark.	106.17	Hawaii	8.88	Utah	1.48

[a]With adjustment for tourism (Hyman et al., 1980).

4. Alcohol consumption, corrected for tourism: gallons per person with corrections made by the authors following Hyman, Zimmerman, Gurioli, and Helrich (1980).[1]

Drinking Norms

Several studies emphasize the quality or content of norms governing drinking and its relationship to alcoholism (Lafferty, Holden, & Klein, 1980; Larsen & Abu-Laban, 1968; Mizruchi & Perrucci, 1962; Pittman, 1967; Skolnick, 1958; Snyder, 1958; Whitehead & Harvey, 1974). Mizruchi and Perrucci (1962) were among the first to call attention to the importance of proscriptive ("Thou shalt not... ") and prescriptive ("Thou shalt... ") norms in the development of alcohol pathology. In reviewing the literature, they conclude that predominantly proscriptive norms are more likely than predominantly prescriptive norms to be tied to pathological reactions when deviation does occur (Mizruchi & Perrucci, 1962, p. 398).

Some investigators have divided societies according to qualitative differences in drinking norms: proscriptive, prescriptive and nonscriptive (Larsen & Abu-Laban, 1968); abstinent, ambivalent, permissive, and overpermissive (Pittman, 1967); and abstinent, prescribed, convivial, and utilitarian (Bales, 1946). Our research follows the approach suggested by Larsen and Abu-Laban, which treats normative approval as a single, more or less continuous variable.

Although the above studies are conceptually rich, they have inspired only qualitative comparative studies or quantitative studies based on individuals as units for the analyses. Thus, they have not provided an empirical test of the hypothesis at the appropriate level, i.e., the social level.

The few quantitative comparative studies of the impact of norm content on drinking and alcohol-related problems investigated preliterate societies. For example, Whitehead and Harvey (1974) used data on 139 preliterate societies. Approval of drunkenness was correlated with alcohol problems between .32 and .43 (in integrated and nonintegrated societies, respectively). Both approval of drunkenness and approval of drinking were positively related to alcohol-related problems, but not as much as the consumption level was.

Stull (1975) has criticized this and other studies of drinking behavior that depend on a single qualitative summary estimation of the cultural characteristics of entire societies. First, this type of study assumes greater homogeneity within preliterate societies than is warranted. Second, they depend on reports of participant observers, which are difficult to replicate and fail to distinguish the normative structure from the actual behavior.

As will be explained in the procedure section of this chapter, the methods we used to investigate the relation of norm content to alcohol-related problems eliminates several of these problems. Our approach concentrates on a single society (the United States) but views the question of degree of internal variation as an empirical question by examining differences between the states of the United States. All indicators of both alcohol-related problems and other variables are quantified, and norms and behavior outcomes are separately measured.

Much more extensive and complex data than is now available would be necessary to measure fully the entire range of normative issues surrounding alcohol. For example, the degree of normative integration or conflict between norms governing alcohol use is an important issue. Researchers usually hypothesize more integration or consistency (or both) between norms as leading to lower rates of alcoholism (Chafetz, 1971; Room, 1976; Ullman, 1958). Our research does not deal with this issue. We focus instead on the degree to which drinking of alcoholic beverages is disapproved and restricted. Although the resultant scale does not reflect the full conceptual richness of some of the discussions in the literature, it is to our knowledge the first normative scale that both measures drinking sentiment and is applicable to states and regions. It has a further advantage of measuring normative standards independently of alcohol consumption and so avoids potential circularity from that source.

Measure of Proscriptive Norms

States vary on a number of measurable characteristics indicative or reflective of norms surrounding alcohol use. These characteristics include prohibitionist status of the state, degree of legal restrictions placed on drinking behavior, and religious composition. The following indicators were used to determine each state's position on the content dimension of proscriptive versus permissive sentiment.

1. Religion: the percentage of a state's population listed as members of fundamentalist and Mormon churches. These churches are known to be proscriptive (cf. Linsky, 1965).[2]
2. Dry areas: the percentage of a state's population residing in legally dry areas (Distilled Spirits Council of the United States, 1978, as reported in Hyman et al., 1980).
3. Liquor outlets: the rate of on-premise liquor outlets per million population (Distilled Spirits Council, 1978). A study by Hyman and Driver reported in Hyman et al. (1980) concludes that dry sentiment has a greater influence on the number of on-premise liquor licenses but financial considerations have a greater influence on the number of off-premise liquor licenses.

TABLE 3
Alcohol Proscriptive Norm Index

Rank	State	Score	Rank	State	Score
1	Miss.	99	26	N.D.	44
2	Utah	95	27	Ind.	43
3	Ky.	90	28	Oreg.	43
4	Ga.	88	29	Mass.	42
5	Tenn.	87	30	Colo.	42
6	Ala.	81	31	Wash.	41
7	S.C.	81	32	Iowa	41
8	Okla.	79	33	La.	40
9	Ark.	76	34	Nebr.	40
10	Idaho	70	35	Calif.	40
11	Texas	70	36	R.I.	38
12	N.C.	69	37	N.H.	38
13	Kans.	57	38	Hawaii	38
14	Minn.	53	39	Wyo.	37
15	Mo.	52	40	Md.	36
16	Va.	51	41	Ohio	36
17	Fla.	49	42	Ill.	36
18	S.D.	46	43	N.Y.	35
19	N.M.	46	44	N.J.	32
20	W. Va.	45	45	Mont.	30
21	Del.	45	46	Pa.	28
22	Maine	45	47	Alaska	28
23	Ariz.	45	48	Vt.	24
24	Mich.	44	49	Wis.	24
25	Conn.	44	50	Nev.	17

4. Sales restrictions: the degree to which on-premise sale of alcohol was restricted in hours or prohibited on Sundays, and on other days of the week (calculated by the authors from Distilled Spirits Council of the United States, 1983).

The Proscriptive Norm Index was computed by *Z* scoring each indicator and summing the four indicators. The resulting variable was then transformed into a *ZP* score to present scores in a more easily interpreted range of 1 to 100.

The internal consistency reliability of the index was evaluated through the alpha coefficient of reliability. The results showed that the Proscriptive Norm Index has an alpha coefficient of .76, indicating a moderately high level of reliability.

Table 3 reports the rank order of the states according to their score on the Proscriptive Norm Index. It can be clearly seen that there are sub-

TABLE 4
Correlations of the SSI with Measures of Alcohol-Related Problems in the 50 States

Dependent variable	Zero-order correlation	Fifth-order partial correlation[a]
Cirrhosis death rate		
Total	.37†	.45†
White men	.33†	.37†
White women	.41†	.47‡
Non-white men[b]	.02	.25
Non-white women	−.03	.26
Alcoholism and alcoholic psychosis death rate		
Total	.30*	.25*
White men	.19	.17
White women	.32*	.35*
Non-white men	.09	.20
Non-white women	.06	.14
Alcohol consumption		
Total	.22	.34*
Corrected for tourism	.20	.31*

[a]Correlations are fourth-order for non-white men and women because % black is omitted from the controls.
[b]Non-white correlations are based on 38 states because their *N*s in 12 states are too small to yield stable death rates.

*$p \leq .05$. †$p \leq .01$. ‡$2p \leq .001$.

stantial differences between states. The state with the highest (most proscriptive) index score is Mississippi, followed by Utah, Kentucky, and Georgia. In fact most of the highly proscriptive states tend to be Southern, with the exception of Utah, with its large Mormon population. At the other end of the continuum are the most permissive states with regard to drinking norms. Nevada is most permissive, followed in order by Wisconsin, Vermont, Alaska, and Pennsylvania.

Stress and Alcohol Problems

The first column of Table 4 shows that the State Stress Index is correlated with the cirrhosis death rate, the death rate for alcoholism and alcoholic psychoses, and the total rate of alcohol consumption—all in the direction expected by the theory.

Although the overall pattern of correlations appears highly consistent with the hypothesis no matter which alcohol-related variable is used, the zero-order correlations in Table 4 can be misleading. One possibility worth considering is that the correlation with alcoholism is partly or wholly spurious because it is accounted for by factors such as age structure or by other variables correlated with stressor events. We therefore used partial correlation to control for five variables: the percentage of the population aged 55 or over, the percentage with 4 or more years of high school, the percentage of families below the poverty line, the percentage of blacks, and the percentage living in metropolitan areas. The right-hand column of Table 4 reports correlations with these five variables controlled for simultaneously. If the correlations between stressful events and alcohol-related behavior were either wholly or partly spurious, we would expect those correlations either to drop out or to become significantly smaller with these variables controlled for. However, the correlations for total rate of death from alcoholism and alcoholic psychosis become only marginally smaller, and the correlations for the cirrhosis and the alcohol consumption variable actually increase. Thus, if anything, the five variables controlled for were partly suppressing the strength of the relationship between stressor events and the dependent variables.

Norms and Alcohol-Related Problems

The evidence presented up to this point appears to be highly compatible with Bales' first hypothesis. Bales' theory, however, emphasizes that it is not stress alone but the combination of stressful conditions and certain types of culturally approved attitudes toward drinking that result in high rates of alcoholism. Our findings suggest that the stressfulness of life in a society alone accounts for some of the state-to-state variation in levels of alcoholism, without reference to normative controls. We now move to examine the effect of normative controls on alcohol-related problems.

Because of the special interest in the social control over alcohol in this part of the study, we expanded the set of dependent variables to include several behavioral outcomes of drinking in addition to the consumption and disease-related variables considered in the investigation of the stress hypothesis. These variables are:

1. The arrest rate for driving while intoxicated (DWI) (U.S. Department of Justice, 1981).
2. The arrest rate for other alcohol-related offenses: the combined arrest rate for violation of alcohol laws, vagrancy, drunkenness, and disorderly conduct (U.S. Department of Justice, 1981).

TABLE 5
Correlation among Indicators of Alcohol-Related Problems in the 50 States

	CORRELATION (r)				
	2	3	4	5	6
1. Death rate from cirrhosis	.62‡	.64‡	–.12	.02	–.31*
2. Alcohol consumption		.99‡	.03	.07	–.21
3. Alcohol consumption corrected for tourism			–.04	.00	–.25*
4. DWI arrest rate				.48‡	.49‡
5. Other alcohol-related arrests					.68‡
6. Alcohol-related arrests as % of all arrests					

*$p \leq .05$. ‡$p \leq .001$.

3. The percentage of all arrests for alcohol-related offenses: the total alcohol-related arrests divided by the total arrests for all causes.

No measure of alcohol problems is completely free of bias, because all the indicators of alcoholism arise or are mediated through social control processes, which in turn reflect a social response to alcoholism. This situation includes arrest data, attribution of cause for alcohol-related deaths, and self-reports of drinking or drunkenness in community surveys. We employ indicators of alcohol-related problems that depend on diverse sources of data. This method in itself does not protect against bias but ensures, at least, that the measures will not all share the same type of bias.

Correlation between Indicators of Alcohol-Related Problems

Before looking at the impact of proscriptive norms on alcohol-related problems, we first consider how the different types of indicators of these problems relate to one another, as shown in Table 5. As expected from other research, the death rate for cirrhosis of the liver is highly correlated with both measures of per capita consumption of alcohol. In addition, the three measures of disruptive alcohol-related behavior (variables 4, 5, and 6) are also highly intercorrelated. However, since variables 5 and 6 use a

TABLE 6
Correlation of Proscriptive Norm Index with Indicators of Heavy Drinking and Disruptive Alcohol-Related Behavior

Indicator of alcohol-related Problems	Zero-order correlations	5th-order partial correlations
Indicators of heavy drinking		
Death rate from cirrhosis	–.52‡	–.56‡
Alcohol consumption	–.60‡	–.55‡
Alcohol consumption corrected for tourism	–.63‡	–.58‡
Indicators of disruptive alcohol-related behavior		
DWI arrest rate	.37†	.25*
Other alcohol-related arrests	.32†	.17
% of all arrests alcohol-related	.49‡	.36†

*$p \leq .05$. †$p \leq .01$. ‡$p \leq .001$.

partly overlapping count of arrests, the correlations between them may be inflated.

At the same time, the rate of arrests for alcohol offenses and for DWI—both of which are measures of the extent to which alcohol-related behavior becomes socially disruptive—were uncorrelated with cirrhosis deaths and with the total rate of consumption of alcoholic beverages. Correlations in both cases approach zero. This suggests that DWI arrests and arrests for other alcohol-related offenses do not arise as a response or reaction to heavy drinking, because there is apparently no more drinking in states with high arrest rates than in states with low arrest rates. In the next section of this chapter we return to this issue in terms of what social factors are related to DWI and other alcohol-related arrests.

Proscriptive Norms and Alcohol-Related Problems

Table 6 provides data on the question whether there is a relationship between the Proscriptive Norm Index and the indicators of alcohol-related problems. Attention is directed to the difference in the correlations in the upper part of Table 6, which deals with the indicators of heavy drinking, and the lower part, which deals with disruptive alcohol-related problems.

The first three correlations in the upper part of the table show that the Proscriptive Norm Index is strongly and inversely related to the death rate from cirrhosis of liver and also with the two measures of alcohol consumption (–.52, –.60, –.63). All three correlations are highly significant. The right column of the table shows that there are no important changes in the relationship between proscriptive norms and alcohol when controls are introduced for five variables that plausibly could affect the relationships (poverty, educational level, metropolitan population, nonwhite population, and age).

The situation changes radically, however, when alcoholism is measured by disruptive behavior related to alcohol (lower part of Table 6), instead of by disease-related deaths and consumption of alcohol. Here there is an abrupt turnaround in the direction of the relationship. The Proscriptive Norm Index is positively and significantly correlated with the arrest rate for DWI, the arrest rates for other alcohol-related offenses, and the percentage of all arrests that are related to alcohol (.37, .32, .49).

When the five control variables are partialed out, the correlations between the Proscriptive Norm Index and disruptive alcoholic behavior are somewhat smaller, but two of the three (the DWI arrest rate and the percentage of all arrests for alcohol-related offenses) remain statistically significant. Thus, the more proscriptive the norms concerning alcohol consumption, the greater the incidence of behavior that is defined as socially disruptive. Moreover, as we pointed out above, arrests for DWI and arrests for other alcohol-related offenses do not seem to arise as a response to the amount of drinking, because there is apparently no correlation between level of alcohol consumption and arrests for DWI or other alcohol-related offenses. Instead, such alcohol-related problems appear to be a response to the strong cultural disapproval of drinking, with the proscriptively oriented states experiencing the highest rates of arrests related to alcohol.

Social Control versus Psychological Ambivalence

So far we have not explained the somewhat paradoxical finding of why high DWI and other alcohol-related arrest rates exist within proscriptive, or dry, communities. At least two possible intervening processes could account for the observed correlation: the "social control" hypothesis and the "ambivalence," or "inoculation," hypothesis. Each has plausibility based on sociological and alcoholism theory. The social control approach emphasizes the fact that the observed arrest rates are a product of both the true incidence of the behavior in question and the society's reaction to that behavior. According to this reasoning, dry communities may be less tolerant of public drunkenness and may encourage stricter policing of that behavior. Other communities, with more liberal alcohol traditions, may be more forgiving of the same types of public behaviors under influence of alcohol.

The ambivalence hypothesis, in contrast, suggests that the true incidence of disruptive and dangerous drinking is actually higher in proscriptive communities despite the lower overall consumption of alcohol. This situation occurs, according to several authors (Chafetz, 1971; Room, 1976; Ullman, 1958) because those who drink within predominantly dry cultures are especially vulnerable to alcohol-related problems. The normative conflict between acceptance and rejection of alcohol competes within the psyche of those who drink, and they experience anxiety and guilt in connection with drinking, a reaction that in turn leads to loss of control.

A related form of the ambivalence hypothesis is, as noted above, sometimes referred to as the inoculation hypothesis. Individuals reared within normatively dry environments are not properly socialized to maintain control over their drinking and over their behavior when they are under the influence of alcohol (Chafetz, 1971; Room, 1978; Skolnick, 1958). For example, research by Globetti (1978) on drinking by high school students and research by Straus and Bacon (1953) on drinking by college students within strongly proscriptive environments found that fewer students drank but a disproportionate number of those who drank did so excessively and were more disruptive when drinking. There were more reports of students getting into fights, destroying property, and driving while under the influence of alcohol. Thus, teaching abstinence seems inadvertently to encourage the behavior it deplores (Skolnick, 1958).

In this analysis, we group the ambivalence and the inoculation hypotheses together because they both assume a higher true prevalence of unruly drinking behavior among those who drink in proscriptive communities and thus lead to identical predictions in the data. The expectations of the ambivalence-inoculation hypothesis are that there will be less drinking in proscriptive environments but that when drinking occurs it will be more dangerous and disruptive. The finding of higher arrest rates in proscriptive communities according to this hypothesis simply reflects the true prevalence of this dangerous and disruptive drinking.

A critical test of these two alternative hypotheses requires additional data beyond the arrest rates for DWI and other alcohol-related offenses, namely, information on the "true incidence" of alcohol-related disruptive behavior independent of the police response to that behavior. Such data are partially available for the United States through the recently initiated

TABLE 7
Mean Rates for DWI Arrests, DWI Self-Reports, and Consumption of Alcohol, by Proscriptive Norm Index Rank[a]

	Proscriptive (n = 8)	Moderate (8)	Permissive (8)	Total (24)
Alcohol consumption (gallons per capita)	2.06	2.66	3.26	2.66
% DWI self-report	4.51	4.95	6.91	5.46
DWI arrests per 100,000	11.15	6.41	6.19	7.92
DWI arrests per 100,000 self-reports	247.2	129.5	89.6	145.1

[a]Findings are based on the 24 states for which self-report data are available.

Behavioral Risk Factor Telephone Survey conducted by the Centers for Disease Control (1984).

Approximately 21,000 survey respondents in 24 states (selected to represent the adult population of their state) were asked if over the last month they had driven when they might have had too much to drink. This data is of course subject to the same limitations as any self-reported survey data on crimes or misconduct. In addition, respondent reports on this subject could be influenced by community views on drinking. We use this data despite the limitations because it is the most broadly based data set allowing state-by-state comparisons of the prevalence of drinking and driving.

In Table 7 the 24 states are divided into three categories ranging from proscriptive to permissive. The three sets of states are then compared according to their average level of consumption of alcohol per capita, DWI self-reports, and DWI arrests. This table allows a partial test of the ambivalence hypothesis through comparison of consumption of alcohol with amount of drunken driving, as well as a test of the social control hypothesis through the ratios of DWI arrests per 100,000 to DWI self-reports under different normative contexts.

In a finding consistent with the correlation data reported in Table 3, average consumption of alcohol increases markedly as one moves from proscriptive to permissive states, whereas DWI arrest rates decrease. DWI self-reports increase from proscriptive to permissive states, with the increase roughly proportional to the increase in average amounts of alcohol consumed. Thus, there is no indication that drinkers behave more recklessly in proscriptive environments. At least as far as drinking and driving is concerned, the ambivalence hypothesis is not supported by this data.

The relationship of self-reported driving while drinking with DWI arrests reveals another pattern. As we move from permissive to proscriptive states, the self-reported incidence of driving while under the influence of alcohol decreases while arrests for DWI increase. This pattern results in dramatically higher ratios (bottom row) of arrest rates to prevalence rates in proscriptive states (247.2) than in permissive (89.6) states. Higher DWI arrests in proscriptive areas appear to be a reflection of the lower tolerance and tougher law enforcement with regard to drinking and driving. Thus, the higher DWI arrest rates for drivers in proscriptive states appear to be highly consistent with the control hypothesis.

To summarize, two distinct patterns in the data are salient. First, permissive normative systems are significantly correlated with all the indicators of heavy drinking and deaths from cirrhosis. Second, proscriptive normative systems are significantly correlated with all of the indicators of disruptive behavior related to alcohol. States that have the strongest cultural biases against beverage alcohol tend to be the same states that experience the most problems, that is, the highest arrest rates associated with drinking.

The DWI rates and other arrests related to alcohol do not appear to arise as a response to the incidence of heavy drinking, because we also found that alcohol-related arrests were completely uncorrelated with the amount of alcohol consumed, nor did they correspond with the self-reported incidence of driving while under the influence of alcohol. The data suggest that even in respect to behaviors that are seemingly objective and obviously dangerous, such as driving while intoxicated, norms regarding drinking may be as important as the drinking behavior itself in the determination of the extent to which alcoholism is defined as a social problem.

Thus, strong normative proscriptions regarding alcohol seemingly produce results opposite to their intent, not so much by increasing disruptive behavior among these who drink but rather by increasing the law enforcement against such behavior once it occurs.

The somewhat paradoxical pattern discussed above has been observed with regard to other types of deviance. As Erikson (1966, p. 22) concluded from his analysis of deviancy and societal values in the early Puritan colonies of New England, "it is not surprising that deviant behavior should seem to appear in a community at exactly those points where it is most feared. Men who fear witches soon find themselves surrounded by them; men who become jealous of private property soon encounter eager thieves." The results of the present study suggest a parallel conclusion, namely, that societies that fear alcohol soon encounter problems with disruptive alcoholics.

The Interaction of Stress and Drinking Norms

So far we have reported investigation of Bales' two major propositions, the stress hypothesis and the normative hypothesis, and have presented evidence that stress and proscriptive norms are each separately related to alcohol-related problems. Although these findings are consistent with Bales' theory, they fail to provide a full test of it, because that theory clearly implies an interactive relationship between stress, alcohol norms, and alcohol-related problems rather than a simple catalog of the determinants of alcoholism. In his study of alcohol-related problems in Irish culture, Bales (1946) argues that it was the combination of stress produced in young men by a social system that denied them sexual and status fulfillment in the context of a normative system that allowed and encouraged them to release their frustration by drinking in local taverns that led to the especially high rates of alcoholism among that population. Hence, to test the theory fully, both variables must be combined within the same research design.

According to Bales' theory, the highest rates of alcohol-related problems should prevail in situations in which high stress is linked with norms encouraging and allowing alcohol consumption. Consequently, this final section of the chapter considers the question whether differences between states in the normative systems influence the relationship between stress and alcohol-related problems. In Table 8, states are arranged in quartiles, according to their permissiveness on alcohol use. If Bales' theory is correct, the State Stress Index should be linked most strongly to alcohol problems in states that are most permissive on alcohol use. The correlations of Table 8 support Bales' theory. Stress is most strongly linked to the amount of alcohol consumed and to deaths from cirrhosis within the most permissive group of states (see right-hand column of Table 8). For all three indicators of heavy drinking, the correlations with stress are highest, and in each case statistically significant, within the most permissive quartile of states (r- .56, .60, and .75). Stress and the indicators of heavy drinking are correlated in some of the other quartiles as well, but the correlations are not as high for the most part, and only one of the nine other correlations is statistically significant.

Summary

This study represents the first time that the links between social stress, drinking norms, and alcohol-related problems have been investigated with the use of broadly based and systematic measurements of these concepts.

TABLE 8
Correlation of State Stress Index with Heavy Drinking and Alcohol-Related Arrests for Proscriptive, Moderately Proscriptive, Moderately Permissive, and Permissive States (Zero-Order Correlations)

	CORRELATION OF STRESS WITH ALCOHOL-RELATED PROBLEMS FOR STATES IN WHICH NORMATIVE CONSTRAINTS ARE:[1]			
Indicators of Alcohol Problems	**Proscriptive (n = 13)**	**Moderately Proscriptive (n = 13)**	**Moderately Permissive (n = 12)**	**Permissive (n = 12)**
Avg. consumption of alcohol	.26	.38	–.02	.56*
Avg. consumpt. corrected for tourism	.31	.38	–.02	.60*
Death rate from cirrhosis	.04	.69**	.40	.75**

*$p < .05$. **$p < .01$.

1. The highest-ranking 13 states on the Proscriptive Norm Index constitute the proscriptive states, the second 13 are the moderately proscriptive, the next 12 are the moderately permissive, and the last 12 are the permissive. The exact n for the quartiles was governed by the convenience of cutting points.

The stressfulness of the social environment in the 50 states of the United States was consistently correlated with heavy drinking and alcohol-related disease problems. Those correlations were, for the most part, undiminished, and even enhanced in some cases, when we controlled for additional variables. This evidence is highly compatible with Bales' stress hypothesis of alcoholism.

With regard to Bales' normative hypothesis, the findings are more complex. States with more permissive drinking norms do experience greater problems with alcohol-related disease deaths from cirrhosis and higher consumption levels. However, it is the proscriptively oriented states that experience the most problems with disruptive drinking (DWI and other alcohol-related arrests), despite the fact that disruptive behavior problems related to alcohol were not correlated with the consumption of alcohol. Disruptive behavior problems related to alcohol apparently arise because of the lower tolerance of public drunkenness and the stricter policing of that behavior within proscriptive communities.

Finally, we found strong support for Bales' original theory because when both the stressfulness of life and normative approval of alcohol are combined within the analysis it appears that the stress is most clearly linked to deaths from cirrhosis and to heavy drinking within the context of strong cultural support for the use of alcohol.

Notes

1. We did this correcting by multiplying the total apparent alcohol consumption by a correction factor formed by the ratio of out-of-state tourist expenditures to total personal income for states. This correction factor probably undercorrects for states that have notably lower beverage prices, because nontourist residents of neighboring states may enter the state specifically to purchase alcoholic beverages.
2. The fundamentalist churches, in order of size, include: churches affiliated with the Southern Baptist Convention, the Lutheran Church (Missouri Synod), Christian Churches (formerly Disciples of Christ), Churches of Christ, the Church of the Nazarene, and the Seventh-Day Adventist Church. The figures are from a study of church membership by Johnson, Picard, and Quinn (1974), as summarized by state in Hyman et al., (1980).

References

Bales, R.F. (1946). Cultural differences in rates of alcoholism. *Quarterly Journal of Studies on Alcohol, 6,* 480–499.

Centers For Disease Control (1984). Behavioral risk factor surveillance, 1981–1983. *Mortality and Morbidity Weekly Report, 33,* 1.

Chafetz, M. (1971). Introduction. In *First special report to the U.S. Congress on alcohol and health* (pp. 1–4). National Institute on Alcohol Abuse and Alcoholism, Washington, DC.

Coddington, R.D. (1972). The significance of life events as etiologic factors in the diseases of children: II. A study of a normal population. *Journal of Psychosomatic Research, 16,* 205–213.

Distilled Spirits Council. (1977). Summary of state laws and regulations relating to distilled spirits. Washington, DC: Author.

Distilled Spirits Council. (1983). Summary of state laws and regulations relating to distilled spirits. Washington, DC: Author.

Dohrenwend, B.S., Kranoff, L., Askenasy, A.R., & Dohrenwend, B.P. (1978). Exemplification of a method for scaling life events: The PERI life events scale. *Journal of Health and Social Behavior 19,* 205–229.

Erikson, K.T. (1966). *Wayward puritans: A study in the sociology of deviance.* New York: Wiley.

Globetti, G. (1978). Prohibition norms and teenage drinking. In J. Ewing & B. Rouse (Eds.), *Drinking; Alcohol in American society* (pp. 159–170). Chicago: Nelson-Hall.

Holmes, T.H., & Masuda, M. (1974). Life change and illness susceptibility. In B.S. Dohrenwend & B.P. Dohrenwend (Eds.), *Stressful life events; Their nature and effects* (pp. 140. 45–71). New York: Wiley.

Holmes, T.H., & Rahe, R.H. (1967). The social readjustment rating scale. *Journal of Psychosomatic Research, 11,* 213–218.

Hyman, M.M., Zimmermann, M.A., Gurioli, C., & Helrich, A. (1980). *Drinkers, drinking and alcohol-related mortality and hospitalizations.* New Brunswick, NJ: Rutgers Center of Alcohol Studies.

Johnson, D., Picard, P., & Quinn, B. (Eds.). (1974). *Churches and church membership in the United States.* Washington, DC: Glenmary Research Center.

Lafferty, N.A., Holden, J., & Klein, E. (1980). Norm qualities and alcoholism. *International Journal of Social Psychiatry 26.*

Larsen, D., & Abu-Laban, B. (1968). Norm qualities and deviant drinking behavior. *Social Problems, 15,* 41–45.

Linsky, A.S. (1965). Religious differences in lay attitudes and knowledge on alcoholism and its treatment. *Journal for the Scientific Study of Religion, 5,* 41–50.

Linsky, A.S., Colby, J.P., Jr., & Straus, M.A. (1986). Drinking norms and alcohol-related problems in the United States. *Journal of Studies on Alcohol, 47,* 384–393.

Linsky, A.S., Colby, J.P., Jr., & Straus, M.A. (1987). Social stress, normative constraints and alcohol problems in American states. *Social Science and Medicine, 24,* 875–883.

Linsky, A.S., & Straus, M.A. (1986). *Social Stress in the United States: Links to regional patterns in crime and illness.* Dover, MA: Auburn House.

Linsky, A.S., Straus, M.A., & Colby, J.P., Jr. (1985). Stressful events, stressful conditions and alcohol problems in the United States: A partial test of Bale's theory. *Journal of Studies on Alcohol, 46,* 72–80.

Mizruchi, E.H., & Perrucci, R. (1962). Norm qualities and differential effects of deviant behavior: An exploratory analysis. *American Social Review, 27,* 391–399.

National Center for Health Statistics: Vital Statistics of the United States (1975–1977). Vol. II, Mortality, Part B. Public Health Service. Washington, DC: U.S. Government Printing Office.

Paykel, E.S., Prusoff, B.A., & Uhlenhuth, E.H. (1971). Scaling of life events. *Archives of General Psychiatry, 25,* 340–347.

Pittman, D.J. (1967). International overtones: Social and cultural factors in drinking patterns, pathological and nonpathological. In D.J. Pittman (Ed.), *Alcoholism* (pp. 3–20). New York: Harper & Row.

Room, R. (1976). Ambivalence as a sociological explanation: The case of cultural explanations of alcohol problems. *American Sociological Review, 41,* 1047–1065.

Room, R. (1978). Evaluating the effects of drinking laws on drinking. In J.A. Ewing & B.A. Rouse (Eds.), *Drinking: Alcohol in American society* (pp. 267–289). Chicago: Nelson Hall.

Skolnick, J.H. (1958). Religious affiliations and drinking behavior. *Quarterly Journal of Studies on Alcoholism, 19,* 452–470.

Snyder, C.R. (1958). *Alcohol and the Jews.* New Brunswick, NJ: Rutgers Center of Alcohol Studies Monograph No. 1.

Straus, M.A. (1980). The zp scale: A percentage z score scale. In M.A. Straus (Ed.), *Indexing and scaling for social science research with SPSS. Available on request.*

Straus, R., & Bacon, S. (1953). *Drinking in college.* New Haven: Yale University Press.

Stull, D. (1975). Hologeistic studies of drinking: A critique. *Drinking and Drug Practices Surveyor, 10,* 4–10.

Ullman, A.D. (1958). Sociocultural backgrounds of alcoholism. *Annals of the American Academy of Political and Social Science, 315,* 48–54.

U.S. Department of Justice. (1981). Sourcebook of Criminal Justice Statistics. Washington, DC.: Bureau of Justice Statistics.

Whitehead, P.C., & Harvey, C. (1974). Explaining alcoholism: An empirical test and reformulation. *Journal of Health and Social Behavior, 15,* 57–65.

CHAPTER 29

Implications of the Distribution Theory for Drinking and Alcoholism

Ole-Jörgen Skog

In chapter 7 the theoretical arguments for the existence of regularities in the distribution of alcohol consumption were outlined. These arguments suggest that the distribution is highly skewed and may resemble a class of theoretical distributions called the lognormal family. Further, they suggest that individual drinking is affected by many different factors and that these factors tend to combine multiplicatively. Many of these factors are social and, thus, suggest that there may be a cultural foundation for drinking patterns and drinking problems. In this chapter the distribution problem is placed in the context of social network theory.

Drinking in a Social Network Perspective

Factors Affecting Drinking Behavior

Like most behaviors, drinking is influenced by both nature and nurture. It is well documented that genetic factors "explain" a significant fraction of the variance of the distribution (Partanen, Bruun, & Markkanen, 1966) and that there are genetic factors that strongly increase the risk for alcoholism in some individuals (Cloninger, Bohman, & Sigvardsson, 1981; Goodwin, 1976). However, there is also much variation that cannot be explained by genetic factors—for instance, differences between identical twins or the increase in alcohol consumption in most of the Western world after World War II. In the argument below I take for granted the fact that biological constitution plays a certain role, but I will restrict myself to discussing environmental factors.

Ole-Jörgen Skog is with the National Institute for Alcohol and Drug Research, Oslo, Norway. This chapter is the second of two chapters prepared especially for this volume. The reader is referred to chapter 7, which should be read first.

We take as point of departure the fact that drinking behavior is a social behavior—something we learn from and practice together with other people. It certainly makes a difference whether you were born in France, the United States, Japan, or Iran. And it makes a difference if your family and friends are teetotallers, moderate drinkers, or heavy drinkers.

A corollary of these facts is that individual drinkers tend to model and modify each others' drinking and, hence, that there is a strong interdependence between the drinking habits of individuals who interact. This interdependence suggests a social network perspective on drinking. The fact that people influence each other in many different ways suggests that an individual's behavior can only be fully understood as a part of a larger entity. First, his or her drinking must be understood in relation to his or her personal social network, that is, the people with whom he or she interacts. Second, since the individuals in his or her network are influenced by still other individuals, these indirect secondary contacts may also have an impact on ego. And so it continues. Potentially, each individual is linked, directly or indirectly, to all members of his or her culture, and perhaps even beyond that. Since the same argument works for every member—or practically every member—of a culture, we must conceive of a population as an integrated structure of individual drinkers.

Before outlining some implications of this perspective for the distribution of alcohol consumption, we shall review some empirical evidence on this issue.

Modeling and Informal Social Control

In several empirical studies the drinking behavior of the subjects has been related to the drinking behavior of parents and friends, as perceived by the subjects, and substantial positive correlation has been reported (Alexander & Campbell, 1967; Dight, 1976; Gusfield, 1961; Haer, 1955; Sorosiak, Thomas, & Balet, 1976; Wieser, 1968). [See also chapter 9, by White, Bates, & Johnson, in this volume.] This kind of evidence for interdependence should not be accepted uncritically, however, since ego's perception of alter's drinking behavior may very well be biased. This kind of perceptual bias may easily produce spuriously high correlations (Vogt, 1976).

This problem is avoided in other studies, where self-reported consumption patterns of identified friends have been related. These studies demonstrate that natural cliques are more homogeneous with respect to drinking behavior than random groups are (Alexander, 1964; Alexander & Campbell, 1967). Even though these studies clearly support the interdependence hypothesis, the problem of cause and effect remains to be solved: Do people choose as friends those who have similar drinking habits, or do they become similar because they interact? To put it differently: Does friendship cause similarity or does similarity cause friendship?

Both of these alternatives are likely to be correct, and the relationship between friendship and behavioral similarity is likely to be two-way rather than one-way. Some degree of similarity may facilitate the establishment of friendship (see Britt & Campbell, 1977), but it is probably also true that friendship in general will bring about further harmonization of behavior.

The usefulness of the social network perspective obviously requires that one accept the idea that "interaction causes similarity," and hence that the observed within-clique homogeneity of drinking habits is not produced only by the mechanism of peer selection. A number of experimental studies have in fact demonstrated that interaction causes similarity in drinking behavior. Bruun (1959), in his classical study of drinking behavior in small groups, was able to demonstrate that interaction produces homogeneity of drinking within the groups, with respect to both amounts consumed and beverage preferences.

Caudill and Marlatt (1975) conducted a laboratory experiment designed to study the effects of modeling influences upon social drinking behavior. The subjects were assigned at random to a high-consumption model, a low-consumption model, or no model at all (the controls). It turned out that those drinking with a high-consumption model drank significantly more than the control subjects, and those with a low-consumption model drank less than the controls.

Reid (1978) conducted a similar experiment in a natural setting. Trained models took a seat beside single patrons in an ordinary bar, and the patron's drinking behavior was recorded by trained observers. The models were either high consumers (5 drinks per hour) or low consumers (1 drink per hour). Single patrons, not being exposed to models, served as controls, and the latter had a consumption rate of 2.76 drinks per hour. Reid was able to demonstrate strong modeling effects. Those who were exposed to high-consuming models drank more than the controls, who again drank more than those exposed to low-consuming models. Hence, it appears that models may modify patrons' drinking behavior in both directions.

DeRicco and Garlington conducted a series of experiments that further corroborate the hypothesis that interaction produces similarity. In the first study (Garlington & DeRicco, 1977), each subject, believing he or she was participating in a study of normal drinking and receiving the instruction to drink at his or her normal rate, drank together with one model in an experimental room resembling a tavern. In Phase 1 of the drinking session the models matched the drinking rate of the experimental subjects. In Phase 2, the model increased (or decreased) his or her rate by a third and remained at this level for a while. Then (Phase 3) the model returned to a baseline rate and stayed there for a while, before decreasing (or increasing) his or her consumption by a third (Phase 4). The session was ended by return to baseline (Phase 5). Garlington and DeRicco (1977) demonstrated that the three subjects followed their respective models quite closely in both directions. This result was replicated in their next study, where they also demonstrated that unveiling the real purpose of the experiment did not affect the result (DeRicco & Garlington, 1977).

It is remarkable how closely the drinking rate of the subjects matched that of the models in DeRicco and Garlington's experiments; the matching was in fact nearly perfect. The same applied to Reid's models. Those exposed to models with a rate of 5 drinks per hour consumed 5.56 drinks per hour, and those exposed to models with a rate of 1 drink per hour drank 1.57 drinks per hour.

In Caudill and Marlatt's experiment the modeling effect was somewhat weaker, however. Those exposed to models drinking 700 mL of wine during the experimental session drank 364.1 mL, and those exposed to models drinking 100 mL consumed 141.9 mL. In view of the fact that the consumption level of the control subjects was 180.8 mL, the experimental subjects appear to have drunk at a level approximately halfway between their "natural" level and the level of the models. This somewhat weaker modeling effect is probably due to the fact that interaction was less intense in this experiment than in Reid's and DeRicco and Garlington's. Verbal interaction was not allowed during Caudill and Marlatt's experiment.

Taken together, these studies clearly demonstrate that interaction causes similarity in drinking behavior and, hence, that the interdependence observed in the nonexperimental studies reviewed above is not solely caused by the reverse mechanism, namely, peer selection. The studies in question also suggest that the interactive modification of drinking behavior in social groups may be quite strong. Since most natural drinking occasions are group occasions, it becomes evident that social interaction is a very important factor in the process of formation and change of drinking habits.

The evidence reviewed is of limited scope, however, because it focuses only on the direct effects of drinking together. The total effects of interacting with other persons—that is, the effects of being raised in and participating in a particular culture—are much more profound than this. Via his or her social network, the individual learns about drinking as a culturally meaningful activity, and this practical, everyday knowledge affects his or her way of acting.

Implications for the Distribution of Consumption

The basic presumption of the social network model is that an individual's drinking habits are the product of an interplay between three types of factors: (1) individual characteristics (endogenous factors), including biological constitution, psychological factors, and the like; (2) his or her social network, that is, the persons he or she has ties to, in one way or another; and (3) general and specific material (including availability) and sociocultural (including differential norms) conditions in his or her environment (exogenous factors).

If mutual influence within a network were very weak, there would be no tendency for clustering of individuals with similar drinking habits. High and low consumption levels would be randomly distributed in the network. However, if there were a significant interaction between network members' consumption levels, it would produce a pattern where high and low consumption levels tend to form clusters. Under these circumstances potentially heavy drinkers may in some cases be effectively controlled by their network. In other cases, potentially heavy drinkers may increase the drinking in their network, thus producing a subculture of higher-than-average drinking habits. In still other cases, neither of these modifications needs to occur, and we may find a single heavy drinker in the middle of a group of very light drinkers.

In a network with strong mutual influences, changes in some individuals' drinking would have several effects. First, increases in their drinking would be partly restrained by the informal social control exerted by their friends. Second, any change that does occur would induce some degree of change in the drinking of their friends, and the inhibitory effect of friends might thus be slightly reduced. Third, the changes that are induced among the "friends" of those who initiated the process will typically spread to friends of friends, and so on. A rather complex process is therefore to be expected. By and large a displacement of the whole population along the consumption scale would take place. Hence, it appears that interaction, provided that it is not too weak, may strongly regulate how the distribution changes when the general consumption level of the population changes. (These changes can be demonstrated with formal models that are available from the author upon request. See also Skog, 1977, 1979.)

The Collectivity of Drinking Culture: Empirical Results

Collective Differences and Patterns of Change

The social network perspective thus suggests that changes in population drinking typically may be a collective phenomenon, and one would

FIGURE 1
Empirical Relationship between Mean Consumption Level and Selected Percentiles (prototype drinkers) in Consumption Surveys from Different Countries. The points are observations, and the lines are the least squares regressions (source: Skog, 1985).

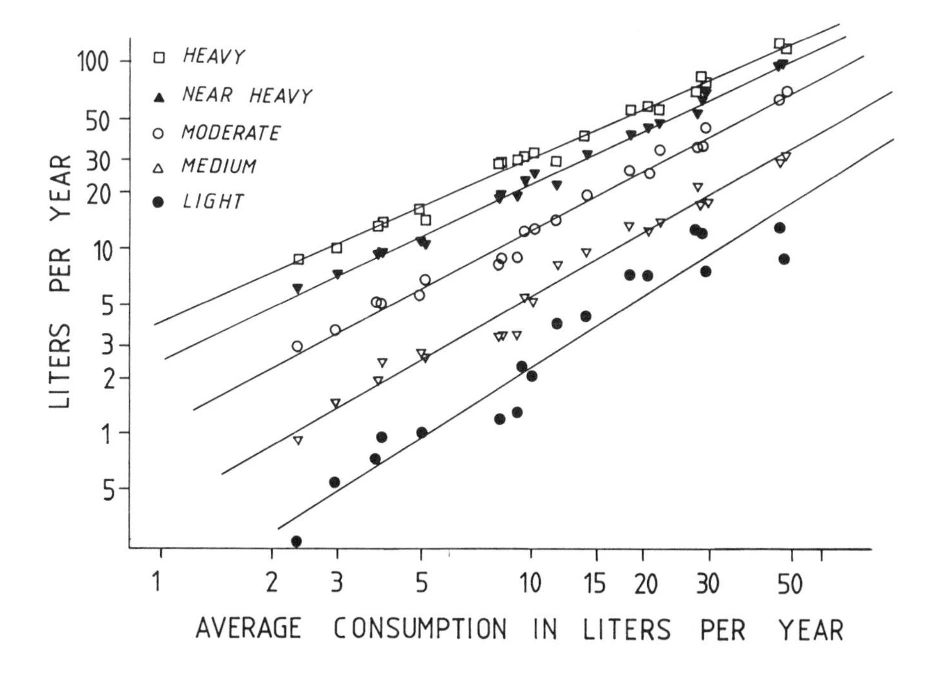

expect that the individual drinker will adjust his or her drinking according to the changes in the overall level of consumption in his or her culture. A similar argument should apply for cross-cultural comparisons.

This prediction can be tested by an investigation of how the consumption level of different prototype drinkers changes in relation to changes in the mean consumption level of the population Operationally we let the 25th percentile represent the prototypical "light drinkers," and let the 50th, 75th, 90th, and 95th percentiles represent prototypical "medium drinkers," "moderate drinkers," "near heavy drinkers," and "heavy drinkers," respectively. In Figure 1 one sees the consumption level of these prototypical drinkers plotted in relation to the mean consumption level in their culture. The data derive from 21 different population surveys in nine different countries (cf. Skog, 1985).

The diagram clearly confirms that the consumption level of all types of drinkers increases as we go from cultures with low to cultures with high mean consumption levels. Thus, the population moves in concert upwards along the consumption scale, as the social interaction argument suggests. Moreover, since the relationships are linear on logarithmic scales, the result is yet another example of multiplicative, or proportional, effects.

It is worth noting that the rate of increase in consumption as the mean increases is not exactly the same across consumption groups (or prototypes). The lines of regression in Figure 1 are steeper for light and intermediate drinkers than for the near heavy and heavy drinkers. Thus, a 10% increase in mean consumption would be expected to lead to a slightly larger rate of increase in the consumption of light and medium drinkers than among heavy drinkers. As a result of this, the relative dispersion of the distribution, and hence the concentration of drinking as measured by the Lorentz diagram, decreases slightly as we go from low-consuming countries to high-consuming countries.

Individual Changes in a Changing Culture

In the preceding section, aggregate level changes and differences are described in terms of prototypes. The data demonstrate that an increase in the overall consumption level is the result of a parallel increase among all types of drinkers. However, it should be made clear that the preceding does not imply that each single drinker increases his or her consumption level.

As was noted earlier, individuals' drinking patterns are in a constant flux: Some increase and some decrease. Consider first a culturally stable situation, in which the mean consumption does not change and the distribution is in a steady state. In this situation it must necessarily be the case that—on the average—heavy drinkers decrease, whereas light drinkers increase, their consumption. If this were not the case, the variance of the distribution would increase and the situation would not be a steady state. This regression effect was shown in both Table 2 and Figure 3 in chapter 7.

In a situation in which the overall level of drinking increases, the pattern of individual changes will be a combination of this regression effect and the collective increase observed in the preceding section. One should expect that very heavy drinkers will—on the average—tend to decrease their intake from Time 1 to Time 2. This decrease would occur in spite of the fact that the prevalence of heavy drinkers would increase markedly when mean consumption increased. This apparent paradox is resolved once one realizes (1) that those heavy drinkers who decrease their intake are replaced by persons who drink slightly less heavily and who have increased their intake and (2) that the latter group is larger.

FIGURE 2
Empirical Relationship between Mean Consumption Level and Prevalence of Drinkers Whose Annual Intake Exceeds 18, 25, and 36.5 L of Pure Alcohol per Year in Consumption Surveys from Different Countries. The points are observations, and the curves are the least squares regressions (source: Skog, 1985).

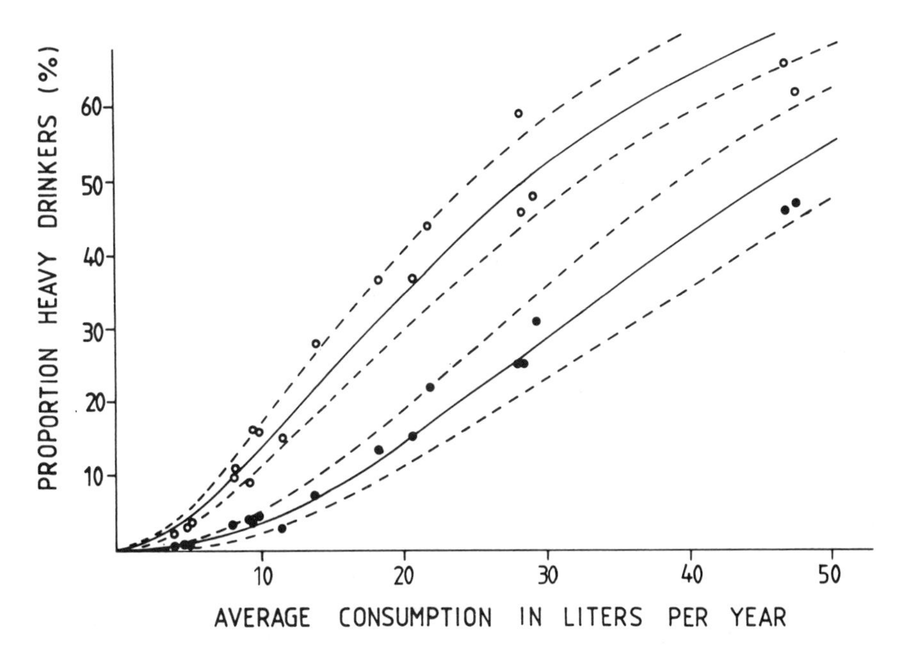

Since, as a main rule, the whole distribution moves upwards along the consumption scale when the mean increases, the number of persons above a fixed consumption limit ought to be increasing as well. In effect, there should be a close relationship between the mean consumption level in a population and the prevalence of heavy drinking.

This supposition is confirmed in Figure 2, where the proportion of the samples with annual intake exceeding 18.25 L and 36.5 L (or 40 g and 80 g daily) are plotted as a function of the mean consumption level. In low-consumption populations the prevalence of heavy drinking is very low, whereas it increases rapidly as mean consumption increases.

Discussion

It should be clear from the data reviewed above that there is substantial regularity in the distribution of alcohol consumption. Not only is the distribution genuinely skewed and roughly of lognormal shape, but the parameters of the distribution covary in a predictable way.

However, as was mentioned in chapter 7, it is well known that self-reported consumption data have limited validity and that surveys seldom cover more than about half of all the alcohol known to be consumed. Not much is known about the true relationships between self-reported and actual intake. It is therefore difficult to know exactly how measurement errors affect the results presented above. Is it possible that the regularities observed in the distribution are only an illusion?

Probably not. Independent evidence corroborating some of the results is available. For instance, it is well documented that liver cirrhosis mortality rates covary with per capita alcohol consumption both cross-culturally and within countries over time (Popham, 1970; Skog, 1986). Liver cirrhosis is a good indicator of heavy alcohol consumption because the risk for this disease is very small at low-to-moderate consumption levels, whereas it increases nearly exponentially at higher levels (Pequignot, Tuyns, & Berta, 1978). It should be noted, however, that since cirrhosis develops only after prolonged abuse, the relationship between aggregate consumption and mortality rates is strongly lagged (Skog, 1980a). This fact has to be taken into consideration when mortality rates are used as indicators of heavy alcohol use.

Although the great majority of the cases that have been reported in the literature confirm the relationship between mean consumption and prevalence of heavy drinking, a few exceptions are known to exist. This fact is hardly surprising. After all, the regularities observed in the distribution of alcohol consumption are not a law of nature. They are a product of social forces and mechanisms, and under certain circumstances deviating patterns may obtain.

A well-documented example of this derives from Sweden, where a rationing system of alcoholic beverages was brought to an end in 1955. Norström (1987) has shown that this change had a dramatic effect on the distribution pattern. Before 1955 the distribution was much less skewed than in other countries, having a much thinner tail. After 1955 the distribution came to resemble the pattern found in other countries. The mean consumption level did not change very much during this period, but the prevalence of heavy drinkers increased strongly, and so did liver cirrhosis mortality rates over the next decades.

Less dramatic deviations from the "rule" may follow in the footsteps of the rapid expansion of the treatment sector for alcoholism. Because of the exponential form of the risk function for cirrhosis, even moderate reductions in the long-term consumption of alcoholics could have significant effects on their health, and a decrease in cirrhosis mortality could occur even in the absence of a noticeable decrease in the mean consumption of the population. Some evidence to this effect has recently been published (Mann, Smart, Anglin, & Rush, 1988). Although the issue is far from settled, it could explain the divergence observed in recent years in consumption trends and liver cirrhosis mortality in some countries (e.g., Grant, Noble, & Malin, 1986).

Alcoholism and the Tail of the Distribution

Implications of the Distribution Theory for Alcoholism

During the last decade the rigid conception of alcoholism as a distinct entity, qualitatively different from normal drinking, has been replaced by the more flexible concept *alcohol dependence syndrome* (Edwards, Gross, Keller, Moser, & Room, 1977). This conceptual change acknowledges the idea that the condition normally called alcoholism is a many-faceted one, with a multiplicity of different symptoms. Furthermore, it is acknowledged that abuse and dependence come in many different degrees and that the transition between so-called normal drinking and abnormal drinking is gradual, rather than abrupt.

This new conception of abnormal drinking is consistent with the perspective outlined previously in chapter 7. There I argued that drinking behavior has a pluralistic anchorage—being influenced by a host of different factors. Hence, heavy drinking has a multiplicity of causes, and a heavy drinker is a person who combines a large number of constitutional and environmental "predisposing" factors that tend to increase drinking. Each separate factor may not in itself induce heavy drinking. However, the existence of many factors operating in the same direction and amplifying each other by multiplicativity of effects produces heavy drinking as a result. Clearly, this implies a syndrome with multiple etiologies and the absence of a clear-cut distinction between normal and abnormal.

The data that have been presented in chapter 7 and above clearly underline the importance of the environment as a determinant of risk for alcohol abuse and alcoholism. This environmental influence is particularly clear from the fact that heavy drinking is closely related to drinking in general. The collective movements of the whole population up and down the consumption scale obviously suggest that normal drinking represents the cultural foundation of heavy drinking and that the consumption level—and hence the risk of drinking problems—for a drinker with a given constitutional disposition is directly related to the amount of alcohol he or she is exposed to in his or her cultural environment.

This suggestion is not a renunciation of the importance of constitutional disposition as a determinant of drinking. Whether an individual will become a so-called chronic alcoholic depends on his or her disposition, which is in turn dependent on constitutional factors—both biological and psychological—as well as on the extent to which he or she is exposed to the agent alcohol. Even individuals who have a very strong constitutional disposition toward alcoholism will not become alcoholics in the absence of alcohol or if alcohol is a very rare commodity in their environment. On the other hand, even individuals with a limited constitutional disposition toward alcoholism may develop a drinking problem when exposed to a particularly wet environment where alcohol is cheap, easy to come by, and frequently used by everybody. Hence, there is no contradiction in saying that both biological constitution and the environment are important.

Consider the following example. If you drop a dozen eggs on the floor, some of them will break, others will not. Which will break is probably to a very large extent determined by each egg's "constitution"—the thickness and microscopic structure of the eggshell, and so forth. One would, in fact, expect a very high correlation between the robustness of the eggshell and its breaking. Consequently, each egg's biophysical disposition is an important predictor. However, the fate of the eggs will also critically depend on the shock they are exposed to when hitting the floor. It certainly makes a difference whether you dropped the eggs from a height of only 5 cm or dropped them from 50 cm. In the first case most of the eggs may hold; in the latter, most will certainly break. Seeing the height in this example as an analogue to the drinking culture and per capita alcohol consumption, one can understand that there is no contradiction between biological and sociological research findings on alcoholism.

Types of Alcoholism, Social Integration, and Loneliness

Almost any study of representative groups of alcoholics demonstrates an enormous heterogeneity in areas such as drinking patterns, consequences of drinking, and psychological and psychiatric profiles before and after the onset of alcohol abuse. This fact has inspired students of alcoholism to develop typologies, the most well known being Jellinek's (e.g., gamma and delta alcoholics). Recently, Cloninger et al. (1981) have suggested another, but clearly related, typology (so-called Type 1 and Type 2 alcoholics).

The social network theory of the distribution of alcohol consumption also predicts different types of abusers. As we have seen in this theory, informal social control is a basic dimension for the understanding of drinking in general, as well as heavy drinking. People who are poorly socially integrated are also weakly controlled by their environment. The implication is, therefore, that they will be "spread out" more strongly along the consumption scale than those who are more effectively controlled by their network (i.e., the variances will be different in the two groups). Hence, the poorly socially integrated will be overrepresented both at high and low consumption levels, and a curvilinear relationship between sociability and consumption level is to be expected (Skog, 1979, 1980b).

This prediction finds some support in the literature. First, it is a well-documented fact that alcoholics and excessive users of alcohol tend to be more socially isolated than others (Bacon, 1945; Mowrer & Mowrer, 1945; Singer, Blane, & Kasschau, 1964). To some extent, their isolation may be a consequence of their deviant drinking, but this possibility does not rule out another—that isolation also existed before the drinking problem. Conceivably, isolation may sometimes lead to heavy drinking, which in due course may increase isolation even further. Second, some studies have found that among normal drinkers a positive correlation between consumption and different measures of sociability obtains (Irgens-Jenssen, 1965; Kuusi, 1957). Hence, those who occupy the lower part of the consumption scale are more socially isolated than others. In effect, social isolation is prevalent at both extremes of the consumption scale, as is expected.

Even though the socially isolated are overrepresented among heavy drinkers, one would also expect a large number of heavy drinkers with normal social relations. The social network model predicts that such persons will be heavy drinkers if they have many heavy-drinking friends. In effect, the model predicts that those who occupy the upper tail of the distribution may form a fairly heterogeneous group, and we can distinguish two main "types" of alcoholics: Type A, those who drink heavily because they are integrated into a heavy-drinking subculture, and Type B, those who are heavy drinkers because they are poorly integrated.

Type A become heavy drinkers not primarily because they have a high psychobiological disposition but because they live in a wet environment. They would be most numerous in countries with a high per capita consumption level. In fact, the empirical relationship between per capita consumption and prevalence of heavy drinking can be seen as a direct manifestation of this type. Type B—the poorly socially integrated—would typically be persons with a high disposition toward alcoholism. Their disposition is allowed to manifest itself because of lack of informal social control to restrain their drinking. Whereas the strongly socially integrated person is effectively controlled by his or her environment, those who are weakly integrated are less effectively controlled. This type would be expected to be most numerous, relatively speaking, in low-consumption countries, since Type A would be rare in these countries. This line of reasoning suggests that the social network of drinkers may prove to be an important predictor of who develops a drinking problem and who does

FIGURE 3
Consumption Level (liters of pure alcohol per year) before Admission to Treatment and at Four Annual Follow-ups in Four Groups of Alcoholics Who Were Treated in an Inpatient Alcoholism Clinic in Oslo. The four groups were defined in terms of their response to a question about drinking habits among friends at admission to treatment (source: Skog, 1989).

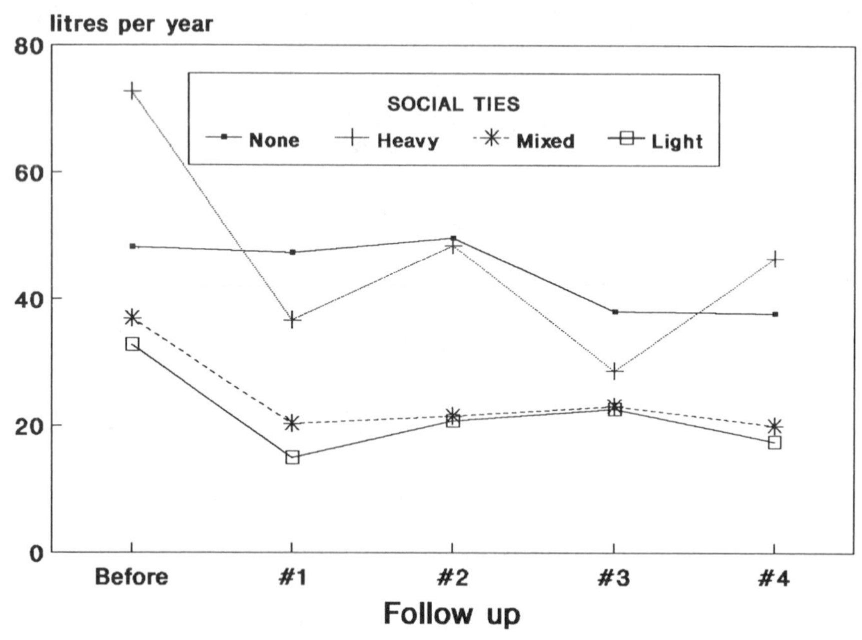

not. And it may also, by a similar argument, predict who recovers from a drinking problem and who does not.

Data from a longitudinal, prospective study of alcoholics in a Norwegian inpatient treatment center (cf. Duckert, 1988; Skog, 1989), may illustrate this point. When admitted to treatment, as well as at four annual follow-ups, the subjects were interviewed by a clinical psychologist about, among other things, their drinking pattern. The subjects were also asked at the first interview about their friends' drinking. In Figure 3 the average annual intake of subjects who initially said they had no friends, mainly heavy-drinking friends, both heavy and moderately drinking friends, and mainly moderate-drinking friends are displayed.

Subjects who initially reported that their friends were mostly heavy drinkers consistently drank substantially more than those whose friends were drawn from a network of moderate or mixed drinkers. This finding remained so during all follow-ups except the third, that is, across a time span of nearly 5 years. Apparently, an alcoholic's social network has a significant predictive value in relation to his or her future drinking career.

It is also of interest to note that those who said they had no friends, as opposed to all other groups, failed to reduce their intake after treatment. At both the first and the second follow-up the socially isolated alcoholics drank at the same level as before treatment, whereas the remaining subjects typically drank 40% to 50% less at those occasions. Furthermore, the socially isolated alcoholics had the highest consumption level of all groups at all but the last follow-up, in spite of the fact that their consumption level initially, upon admission to treatment, was not above the average level.

In short, the two groups with the most unfavorable prognosis in terms of consumption level were those with a poor social integration and those integrated into a wet environment. Being poorly socially integrated is likely to be a risk factor, not only because of weak informal social control but also because the condition often implies loneliness. As many studies have shown, a considerable fraction of alcoholics report feelings of loneliness, and it is tempting to suggest that excessive drinking may, in many cases, be a way of "coping" with feelings of loneliness and inadequate social relations. As Edwards (1987) has pointed out, one should not exaggerate the importance of loneliness. But it is my impression that the modern research literature on alcoholism tends to do the opposite. As Hörnqvist, Hansson, and Åkerlind (1988) have pointed out, the research literature on alcoholism and loneliness is rather limited. It is of interest to note that the prospective, longitudinal study of Swedish alcoholics by Hornqvist et al. does in fact suggest that the two most important prognostic factors are an alcoholic's feeling of loneliness and whether he or she has a "drinking buddy." In our terminology, the factors would be called poor social integration and integration into a wet environment, respectively.

The social typology that has been briefly outlined here obviously has something in common with Jellinek's types (Jellinek, 1960). Type A—the socially well integrated—may remind you of Jellinek's beta and delta alcoholics. In fact, Jellinek (1960, p. 38) described his delta (or French) type by saying that "the incentive to high intake may be found in the general acceptance of the society to which the drinker belongs, while prealcoholic psychological vulnerability, more often than not, may be of a low degree." Some similarities may also exist between our Type B—those with poor social integration and a high psychobiological disposition—and Jellinek's alpha and gamma alcoholics.

There are also similarities with Cloninger, Sigvardsson, and Bohman's (1988) Type 1, Type 2 scheme. Type 1 alcoholics are characterized by a late onset of drinking problems, which are reinforced by external circumstances, and they have few social complications. Type 2 alcoholics have an early onset of drinking problems, regardless of external circumstances, and they have multiple social problems. According to the authors, Type 2 subjects are typically aggressive, and high on novelty seeking, but they are low on harm avoidance and reward dependence and frequently show signs of antisocial behavior, whereas Type 1 subjects typically have opposite characteristics (Cloninger et al., 1988). Hence, Type 1 may resemble Type A, above, and Type 2 resembles Type B.

Drinking Careers and Return to Normal Drinking

In recent years, social learning theory (Bandura, 1977) has attracted much interest in alcoholism research (e.g., Heather & Robertson, 1981). Alcoholism is conceived as learned behavior, and social modeling and interpersonal relations have come into focus. One of the basic dogmas of this school of thought has been that what is learned can be relearned. Hence, the idea of an alcoholic's possibly being able to return to controlled or normal drinking has gained popularity. This issue has caused heated controversy for many years. Some argue that alcoholics can never return to normal drinking, whereas others say that just such a thing occurs in quite a number of cases.

No doubt, the answer is to a large extent dependent on how the basic concepts involved are defined: what exactly you mean by alcoholism, alcohol dependence, addiction, or whatever term you prefer, and what criteria are used for controlled or normal drinking. Using strict definitions of both alcoholism (or alcohol dependence) and controlled (or normal) drinking may imply that such changes are rare (Edwards, 1985; Vaillant, 1983). Less strict criteria will no doubt produce positive findings in a substantial number of cases (Cahalan & Room, 1974; Clark, 1976; Clark & Cahalan, 1976; Nordström, 1987).

It is a problem that evaluation studies often lump together many different types of criteria of progress into summary measures (e.g., successful or not). Hence, substantial changes may pass unnoticed. For instance, an abuser may have reduced his or her intake substantially, but not enough to deserve the social label *successful outcome*. Still, the reduction may be large enough to reduce substantially his or her risk of liver cirrhosis or other complications. Therefore, in order to obtain a reliable picture of drinking careers, it is necessary to study individual developments along a number of different dimensions in more detail.

The transition matrix in Table 2 in chapter 7 focused exclusively on changes in alcoholics' intake. Smooth and gradual change was a more typical feature than dramatic shifts and "loss of control." The time span was only one year in this particular table, but the picture did not change much in a more long-term perspective (Skog & Duckert, forthcoming). The data obviously suggest that some alcoholics may drink at a fairly low level for extended periods of time. Whether this is in fact a return to normal drinking is an open question, partly depending on definitions. Whatever you choose to call it, the finding is in any case of considerable importance.

In one sense, the whole debate on lifelong abstinence versus a return to normal drinking is hopelessly static. An alcoholic may become abstinent or may start to drink in a socially acceptable manner. Although this behavior may last for a long period of time, there is no guarantee that he or she will remain in that state forever. Life is a process, not a state. A more adequate perspective for describing drinking careers is one of a partly structured, "never" ending process of change. As long as they are still alive, most individuals have the possibility of changing. The behavioral states that an individual may occupy are typically unstable, and changes are likely to occur, sooner or later—particularly when the behavioral dimension in question is one that exists in degrees.

Many studies have demonstrated that the direction in which any particular individual will change is difficult to predict (e.g., Polich, Armor, & Braiker, 1981). Changes in both directions are very likely at all levels of consumption (cf. chapter 7, Table 2), and subsequent changes tend to accumulate over time and to amplify diversity (Skog & Duckert, forthcoming). Therefore, a typical developmental pattern or "natural history" of alcoholism may simply not exist. In general, individual life histories are complex processes, and there is no reason to expect that drinking careers should be less complex, because they are influenced by numerous life events of different types.

A study by Drew (1968) is relevant in this context. He studied incidence and prevalence rates of alcoholism as a function of age and demonstrated an increasing divergence between prevalence rates and cumulative incidence rates with increasing age. Apparently, many former alcoholics disappear, and the rate at which they disappear is much too high to be explained by excessive mortality. Drew argued that the rate is most likely a sign of considerable spontaneous remission. Whether this is a remission to abstinence, "normal" drinking, or only "less problematic" drinking is an open question.

Besides depending on definitions and criteria, the possibility of a "return to normal drinking" obviously will depend on the material and social context of the drinker. Experimental studies clearly demonstrate that even chronic alcoholics may control their drinking under certain circumstances (Mello & Mendelson, 1965, 1966, 1972; Pattison, Sobell, & Sobell, 1977). Thus, when excessive drinking has undesirable consequences, or when necessary efforts and costs are large enough, an alcoholic may prefer to reduce his or her drinking. Hence, the possibility of "regaining control" should not be abstracted from the environmental situation representing the life space of the drinker.

When we recognize the potential importance of the drinkers' sociocultural situation, it becomes of interest to look at how heavy drinkers and alcoholics change their drinking when their sociocultural context changes. In the light of the results presented both above and in chapter 7 it is of particular interest to investigate this question in relation to changes in the overall level of consumption in the culture.

The return-to-normal-drinking battles have mainly been fought with individual-level data, that is, recordings of individual drinking careers. However, aggregate level data may also shed some light on the issue. In particular, evidence demonstrating that long-term heavy drinkers with severe, alcohol-induced liver problems may—under certain circumstances—dramatically reduce their drinking is clearly relevant. Several cases have been described in the literature. Perhaps the best-known case derives from Paris during World War II (Ledermann, 1964), where cirrhosis mortality rates dropped by about 50% during a single year, namely from 1941 to 1942, and were reduced by about 80% after four years (1945). Obviously, this reduction took place because drinkers of cirrhogenic quantities—many of whom would qualify for a diagnosis of alcoholism or severe alcohol dependence—dramatically reduced their intake. Whether their new drinking would qualify for the social label *normal* is of course an open question, but the change in any case had dramatic consequences for their health, and this fact is probably more important.

Of course, the situation in Paris in 1942 was special, and the decrease in liver cirrhosis mortality was due to strict rationing of alcoholic beverages. The control of the alcoholics' drinking was therefore not a voluntary, cognitive control but a strict environmental control. This fact reminds us that the issue of return-to-normal drinking must be seen in a sociocultural perspective. It is not entirely a psychological and biological issue. However, examples of substantial reductions in alcoholics' drinking under less dramatic circumstances are also available. Many countries experienced a long-term downward trend in alcohol consumption around the turn of the century, and these changes in drinking practices were accompanied by parallel trends in liver cirrhosis mortality. In England, for instance, cirrhosis mortality rates declined by 50% from 1920 to 1940. A close analysis of the British experience, taking into consideration the fact that cirrhosis of the liver develops only after prolonged abuse, shows that this trend is only explainable if heavy drinkers systematically reduced their drinking during this period (Skog, 1980a).

What is observed in many countries in this era is a cultural transformation process in which the drinking habits of the population at large change and the drinking of alcoholics and severely dependent persons changes in response. The drinking habits of alcoholics thus appear to be controlled by the drinking habits in their culture. When everybody else uses alcohol somewhat less freely and less often, so does the "drunkard."

Although these and similar experiences do not directly demonstrate a return to controlled drinking in the strict sense, they show that alcoholics can "control" their intake when alcohol becomes extremely difficult to come by, or when the culture "demands" a reduction of intake. Like everybody else, alcoholics are cultural products. They are not automata whose drinking is unaffected by their culture.

Summary and Conclusions

The data presented above have examined the distribution of consumption within a social network model. In summary, these data suggest the following. Most individuals' drinking habits are strongly influenced by the drinking habits in their environment. Informal social control is therefore very important in relation to drinking. This influence produces a strong collective component in drinking behavior, with the result that the whole population typically moves in concert up and down the consumption scale. This mass fluctuation, again, has the effect of making the prevalence of heavy drinking closely related to drinking in general—as measured by the mean consumption level in the population. In relative terms, the prevalence rate typically increases more that the mean. This fact implies that an individual's risk of becoming a heavy drinker or alcoholic strongly depends on environmental factors, particularly the wetness of his or her environment. However, this dependence on environmental factors does not mean that constitutional disposition is unimportant. The importance of informal social control suggests that heavy drinkers and alcoholics may be of at least two types: socially isolated drinkers with a high constitutional disposition, and socially integrated drinkers without a strong constitutional disposition. These types may correspond to similar types in Jellinek's and Cloninger's typologies.

These results strongly underline the cultural foundation of drinking problems. Abuse and alcoholism are not only individual problems but also sociocultural problems. An individual's fate vis-à-vis alcohol strongly depends on his or her environment, and in particular on the wetness of this environment. And the number of persons in a culture who develop a drinking pattern that represents a health hazard depends on the general level of consumption in the culture.

This finding has important implications for the prevention of alcohol-related problems—particularly chronic health problems. Since abuse is so closely tied to normal use, it may be quite difficult to reduce drinking problems to any significant extent without at the same time reducing normal use. Therefore, selective preventive measures focused on abuse and heavy drinking should—in order to be effective—be combined with measures that are aimed at drinking in general. If not, alcoholics and abusers will continue to be exposed to the environment that was at least partly responsible for their problem in the first place.

It is a well-documented fact that the volume of chronic alcohol-related diseases in a society are related to the general level of consumption (Bruun et al., 1975). For acute alcohol problems the relationship is more complex. In cross-cultural comparisons, only very weak relationships between rates of acute problems and the general consumption level have been found (Mäkelä, 1978). The reason is that acute problems are more dependent on drinking patterns than on the amounts consumed. However, in studies of the development of each culture over time, rates of acute problems often seem to follow the trends in the general consumption level (Mäkelä, Room, Single, Sulkunen, & Walsh, 1981; Skog, 1989). This correlation is probably explained by cultural inertia in drinking patterns. The per capita consumption level apparently changes much more rapidly than the cultural rules of drunken comportment, and therefore changes in consumption levels typically also mean parallel but delayed changes in drunkenness.

References

Alexander, C.N. (1964). Consensus and mutual attraction in natural cliques: A study of adolescent drinkers. *American Journal of Sociology, 69,* 395–403.

Alexander, C.H., & Campbell, E.Q. (1967). Peer influences on adolescent drinking. *Quarterly Journal of Studies on Alcohol, 28,* 444–453.

Bacon, S.D. (1945). Alcoholism and social isolation. In M. Bell (Ed.), *Cooperation in crime control* (pp. 209–234). New York: National Probation Association.

Bandura, A. (1977). *Social learning theory.* Englewood Cliffs, NJ: Prentice-Hall.

Britt, D.W., & Campbell, E.Q. (1977). A longitudinal analysis of alcohol use, environmental conduciveness and normative structure. *Journal of Studies on Alcohol, 38,* 1640–1647.

Bruun, K. (1959). *Drinking behavior in small groups.* Helsinki: Finnish Foundation for Alcohol Studies.

Bruun, K., Edwards, G., Lumio, M., Mäkelä, K., Pan, L., Popham, R.E., Room, R., Schmidt W., Skog, O-J., Sulkunen, P., & Österberg, E. (1975). *Alcohol control policies in public health perspective.* Helsinki: Finnish Foundation for Alcohol Studies.

Cahalan, D., & Room, R. (1974). *Problem drinking among American men.* New Brunswick, NJ: Rutgers Center of Alcohol Studies.

Caudill, B.D., & Marlatt, G.A. (1975). Modelling influences in social drinking: An experimental analogue. *Journal of Counselling and Clinical Psychology, 43,* 405–415.

Clark, W. (1976). Loss of control, heavy drinking, and drinking problems in a longitudinal setting. *Journal of Studies on Alcohol, 37,* 1256–1290.

Clark, W., & Cahalan, D. (1976). Changes in problem drinking over a four-year span. *Addictive Behaviors, 1,* 251–259.

Cloninger, C.R., Bohman, M., & Sigvardsson, S. (1981). Inheritance of alcohol abuse: Cross-fostering analysis of adopted men. *Archives of General Psychiatry, 38,* 861–868.

Cloninger, C.R., Sigvardsson, S., & Bohman, M. (1988). Childhood personality predicts alcohol abuse in young adults. *Alcoholism: Clinical and Experimental Research, 12,* 494–505.

DeRicco, D.A., & Garlington, W.K. (1977). The effect of modelling and disclosure of experimenter's intent on drinking rate of college students. *Addictive Behaviors, 2,* 135–139.

Dight, S. (1976). *Scottish drinking habits.* London: Her Majesty's Stationery Office.

Drew, L.R.H. (1968). Alcoholism as a self-limiting disease. *Quarterly Journal of Studies on Alcohol, 29,* 956–967.

Duckert, F. (1988). Recruitment to alcohol treatment: A comparison between male and female problem drinkers recruited to treatment in two different ways. *British Journal of Addiction, 83,* 285–293.

Edwards, G. (1985). A later follow-up of a classical case series: D.L. Davies' 1962 report and its significance for the present. *Journal of Studies on Alcohol, 36,* 181–190.

Edwards, G. (1987). *The treatment of drinking problems* (2nd ed.). Oxford: Blackwell Scientific Publications.

Edwards, G., Gross, M.M., Keller, M., Moser, J., & Room, R. (1977). *Alcohol-related disabilities.* Geneva: World Health Organization.

Garlington, W.K., & DeRicco, D.A. (1977). The effect of modeling on drinking rate. *Journal of Applied Behavioral Analysis, 10,* 207–211.

Goodwin, D. (1976). *Is alcoholism hereditary?* New York: Oxford University Press.

Grant, B., Noble, J., & Malin, H. (1986). Decline in liver cirrhosis mortality and components of change: United States, 1973–1983. *Alcohol Health and Research World, 3* (10), 66–69.

Gusfield, J.R. (1961). The structural context of college drinking. *Quarterly Journal of Studies on Alcohol, 22,* 428–443.

Haer, J.L. (1955). Drinking pattern and the influence of friends and family. *Quarterly Journal of Studies on Alcohol, 16,* 178–185.

Heather, N., & Robertson, I. (1981). *Controlled drinking.* London: Methuen.

Hörnqvist, J.O., Hansson, B., & Åkerlind, I. (1988). The working capacity of the alcohol abuser. *Scandinavian Journal of Social Medicine, 16,* 27–33.

Irgens-Jenssen, O. (1965). *Alkoholvaner i en Utkantkommune.* Oslo: Universitetsforlaget.

Jellinek, E.M. (1960). *The disease concept of alcoholism.* New Brunswick, NJ: Rutgers Center of Alcohol Studies.

Kuusi, P. (1957). *Alcohol sales experiment in rural Finland.* Helsinki: Finnish Foundation for Alcohol Studies.

Ledermann, S. (1964). *Alcool, alcoolism, alcoolisation* (Vol. 2). Paris: Presses Universitaires de France.

Mäkelä, K. (1978). Levels of consumption and social consequences of drinking. In Y. Israel, F.B. Glaser, H. Kalant, R.E. Popham, W. Schmidt, & R.G. Smart (Eds.), *Research advances in alcohol and drug problems* (Vol. 4, pp. 303–348). New York: Plenum Press.

Mäkelä, K., Room, R., Single, E., Sulkunen, P., & Walsh, B. (1981). *Alcohol, society and the state.* Toronto: Addiction Research Foundation.

Mann, R.E., Smart, R., Anglin, L., & Rush, B.R. (1988). Are decreases in liver cirrhosis rates a result of increased treatment for alcoholism? *British Journal of Addiction, 83,* 683–688.

Mello, N.K., & Mendelson, J.H. (1965). Operant analysis of drinking patterns of chronic alcoholics. *Nature, 206,* 43–46.

Mello, N.K., & Mendelson, J.H. (1966). Experimental analysis of drinking behavior of chronic alcoholics. *Annals of New York Academy of Sciences, 133,* 828–856.

Mello, N.K., & Mendelson, J.H. (1972). Drinking pattern during work contingent and non-contingent alcohol acquisition. *Psychosomatic Medicine, 34,* 139–164.

Mowrer, H.R., & Mowrer, E.R. (1945). Ecological and familial factors associated with inebriety. *Quarterly Journal of Studies on Alcohol, 6,* 36–44.

Nordström, G. (1987). *Successful outcome in alcoholism: A prospective long-term follow-up.* Lund: University of Lund.

Norström, T. (1987). The abolition of the Swedish alcohol rationing system: Effects on consumption distribution and cirrhosis mortality. *British Journal of Addiction, 82,* 633–642.

Partanen, J., Bruun, K., & Markkanen, T. (1966). *Inheritance of drinking behavior.* Helsinki: Finnish Foundation for Alcohol Studies.

Pattison, M., Sobell, M., & Sobell, L. (1977). *Emerging concepts of alcohol dependence.* New York: Springer.

Pequignot, G., Tuyns, A., & Berta, J.L. (1978). Ascitic cirrhosis in relation to alcohol consumption. *International Journal of Epidemiology, 7,* 113–120.

Polich, J.M., Armor, D.J., & Braiker, H.B. (1981). *The course of alcoholism: Four years after treatment.* New York: Wiley.

Popham, R.E. (1970). Indirect methods of alcoholism prevalence estimation: A critical evaluation. In R.E. Popham (Ed.), *Alcohol and alcoholism* (pp. 294–306). Toronto: University of Toronto Press.

Reid, J.B. (1978). Study of drinking in natural settings. In G.A. Marlatt and P.E. Nathan (Eds.), *Behavioral approaches to alcoholism* (pp. 58–74). New Brunswick, NJ: Rutgers Center of Alcohol Studies.

Singer, E., Blane, H.T., & Kasschau, T. (1964). Alcoholism and social isolation. *Journal of Abnormal and Social Psychology, 69,* 681–685.

Skog, O.-J. (1977). On the distribution of alcohol consumption. In D.L. Davies (Ed.), *The Ledermann curve: Report of a symposium.* London: Alcohol Education Center.

Skog, O.-J. (1979). *Modeller for drikkeatferd.* Oslo: National Institute for Alcohol Research.

Skog, O.-J. (1980a). Liver cirrhosis epidemiology: Some methodological problems. *British Journal of Addition, 75,* 227–243.

Skog, O.-J. (1980b). Social interaction and the distribution of alcohol consumption. *Journal of Drug Issues, 10,* 71–92.

Skog, O.-J. (1985). The collectivity of drinking cultures. A theory of the distribution of alcohol consumption. *British Journal of Addiction, 80,* 83–99.

Skog, O.-J. (1986). The long waves of alcohol consumption: A social network perspective on cultural change. *Social Networks, 8,* 1–32.

Skog, O.-J. (1989). The socio-cultural foundation of drinking problems. In R.B. Waahlberg (Ed.), *Proceedings of the 35th International Congress on Alcoholism and Drug Dependence* (pp. 317–348). Oslo: National Directorate for the Prevention of Alcohol and Drug Problems.

Skog, O.-J., & Duckert, F. (forthcoming). The development of alcoholics' and heavy drinkers' consumption: A longitudinal study. *Journal of Studies on Alcohol.*

Sorosiak, F.M., Thomas, L.E., & Balet, F.N. (1976). Adolescent drug use: An analysis. *Psychological Reports, 38,* 211–221.

Vaillant, G. (1983). *The natural history of alcoholism.* Cambridge: Harvard University Press.

Vogt, I. (1976). *Mother-child interaction and patterns of drug consumption.* Unpublished manuscript.

Wieser, S. (1968). Über das Trinkverhalten der allgemeinen Bevölkerung und Stereotype des Abstinenten und Trinkers; eine empirische soziopsychiatrische Studie im Bundesland Bremen. (On the drinking patterns of the general population and the stereotype of the abstainer and the drinker: An empirical social-psychiatric study in the district of Bremen.) *Fortschritte der Neurologie, Psychiatrie, und Ihrer Grenzgebiete 36,* 485–509.

CHAPTER 30

The Etiology of Alcoholism

DONALD W. GOODWIN

If the title seems grandiose, it is. In fact, the etiology of alcoholism is unknown. But my comments in this brief chapter bear on etiology, which, after all, is the central mystery of alcoholism: Why do most people drink and a rather small minority become alcohol-dependent?

The chapter has four sections: (1) comments on the definition of alcoholism, (2) a discussion of risk factors, (3) evidence that heredity influences the development of alcoholism, and (4) speculation about mechanisms of inheritance.

Definition

Charles Jackson (1944), author of *The Lost Weekend,* defined the alcoholic as a person who can take it or leave it, so he takes it. This definition may sound frivolous, but in fact it captures a central feature of the illness: self-deception. "We are all victims of systematic self-deception," Santayana said, and the alcoholic is a victim par excellence. People are victims of many things—passion, greed, pollution—and can accept it. But, deep down, the alcoholic believes he is doing it to himself; he is the perpetrator, not the victim. This he cannot accept, so he lies to himself: "I can stop drinking anytime. Important people drink. Churchill drank. Today is special—a friend is in town. Nothing is going on—why not? Life is tragic—why not? Tomorrow we die—why not?" As he lies to himself, he lies to others. In the professional jargon, this lying is called denial.

The Diagnostic and Statistical Manual of Mental Disorders (DSM-III-R), the official classification system of the American Psychiatric Association (1987), has a more cumbersome definition (Table 1). If a person meets three or more of the criteria, he or she is called alcohol-dependent. (Official organizations, including the World Health Organization, encourage the term *alcohol dependence* rather than *alcoholism* on the grounds that

Donald W. Goodwin is with the Department of Psychiatry, University of Kansas Medical Center, Kansas City, Kans. This chapter was written especially for this book.

TABLE 1
Diagnostic Criteria for Alcohol Dependence
(DSM-III-R, 1987)

Three or more manifestations required for diagnosis:

1. Alcohol often taken in larger amounts or over a longer period than the person intended.
2. Persistent desire or one or more unsuccessful efforts to cut down or control alcohol use.
3. A great deal of time spent in activities necessary to get alcohol, taking the substance, or recovering from its effects.
4. Frequent intoxication or withdrawal symptoms when expected to fulfill major role obligations at work, school, or home (e.g., does not go to work because hung over, goes to school or work intoxicated, or intoxicated while taking care of his or her children), or when alcohol use is physically hazardous (e.g., drives when intoxicated).
5. Important social, occupational, or recreational activities given up or reduced because of alcohol.
6. Continued alcohol use despite knowledge of having a persistent or recurrent social, psychological, or physical problem that is caused or exacerbated by the use of the substance.
7. Marked tolerance: need for markedly increased amounts of the substance (i.e., at least a 50% increase) in order to achieve intoxication or desired effect, or markedly diminished effect with continued use of the same amount.
8. Characteristic withdrawal symptoms.
9. Alcohol often taken to relieve or avoid withdrawal symptoms.

isms imply moral judgments. However, *alcoholism* was introduced 150 years ago by a Swedish public health official and is not likely to disappear. It is the word used in this chapter.)

The elaborate definition of alcohol dependence proposed by the American Psychiatric Association can be collapsed more or less to a single sentence: Alcoholism involves a compulsion to drink, causing damage to self and others.

Risk Factors

Although the cause of alcoholism is unknown, clearly some people are more susceptible to the illness than others. Known risk factors for alcoholism include (1) gender, (2) family history, (3) ethnic background, and (4) vocation. Most of this chapter focuses on family history, but other risk factors will be discussed.

More men than women are alcoholic (Goodwin, 1988b). For every female alcoholic there are three to five male alcoholics. This imbalance appears to have been in effect throughout history and in all cultures. The sex difference is a real one, but it has received relatively little attention. Cultural factors seem obviously important. Men, especially in the United States, have traditionally gone to bars not only to drink but to avoid women (London, 1982). Until recently saloons were occupied mainly by men while women stayed at home, raised children, and worried about their drinking spouses. The feminist movement changed all that. Now women can go to bars as freely as men, but most do not or feel uncomfortable when they do, unless accompanied by others, preferably males.

Hormones may play a role in the sex difference. There is a small body of literature indicating that female animals drink more or less, depending on the phase of their estrous cycle (Goodwin, 1988b). Some anecdotal evidence suggests that women drink less (or more) before the beginning of each menstrual cycle (Mello, 1980).[1] They certainly drink less during pregnancy, a time of great hormonal change. This lessening of intake is probably fortunate, if drinking is deleterious to the fetus (Little & Ervin, 1984).

Alcohol use and abuse are also associated with ethnic background. [See chapters 15, 16, & 17, by Glassner, Harper & Saifnoorian, and Neff, respectively, in this volume.] Despite intermarriage and cultural assimilation, Jews still have a low rate of alcoholism compared with others in the Judeo-Christian culture (Goodwin, 1988b). The Irish have a relatively high rate, particularly Irish who immigrate to England and the United States. (This was true of those who immigrated in the 19th and early 20th centuries; fewer immigrate today, and it is not clear how current immigration affects drinking by the Irish.)

People who live in northern climates apparently have higher alcoholism rates than those who live in warmer southern regions. French authorities maintain that there is more alcoholism in northern France than in the south, and cirrhosis data in the country appear to support this claim. Since geography is related to ethnicity, it is not clear whether weather or genes are responsible for the geographic differences.

Vocation also influences alcoholism rates. Cirrhosis data show that there is a relatively high rate of alcoholism among bartenders and newspaper reporters and a low rate among postmen. The group with the highest cirrhosis rate ever reported consists of U.S. citizens who won the Nobel prize for literature (Goodwin, 1988a). Of the seven Americans who have won the prize, five were clearly alcoholic (Sinclair Lewis, Eugene O'Neill, Ernest Hemingway, William Faulkner, and John Steinbeck), and two were nonalcoholic (Pearl Buck and Saul Bellow). There is a high rate of alcoholism among American writers in general (Goodwin, 1988a). The explanation is unknown.

Hereditary Influences

The possibility that alcoholism may be influenced by heredity comes from three sources: (1) family studies, (2) twin studies, and (3) adoption studies. Each will be discussed briefly.

Family Sudies

Alcoholism runs in families. This fact has been known for centuries. Plutarch, 2,000 years ago, said that "drunkards beget drunkards," and numerous studies in the 20th century support this observation. In general, about 25% of sons of alcoholics become alcoholic and about 5% of daughters (Cotton, 1979). There is controversy about rates of alcoholism in the general population, but a reasonable figure for men is 5% and 1% for women. Thus, having an alcoholic parent or sibling increases the risk of a person's becoming alcoholic by a factor of 4 or 5.

Many things run in families that are not hereditary. Speaking French is an example. The problem with family studies is that most individuals, including alcoholics, are raised by the progenitors; thus, nature and nurture arise from the same source and are, in usual circumstances, inseparable. There are two methods for partly separating nature from nurture under unusual circumstances. One is to compare identical with fraternal twins, where environmental factors presumably are held constant but genetic make-ups differ. (Identical twins have identical genes; fraternal twins are genetically related only to the extent that siblings in general are related.) The other method involves studying individuals separated from their biological parents soon after birth and raised by nonrelatives. If the biological parent or parents have an illness that runs in families and it still does so even when the offspring are separated from the ill biological parent (or parents), then the indication is that environmental factors are not solely responsible for the transmission of the illness in the families. The explanation may not be genetic in the sense that specific genes in the chromosomes are responsible for the illness, but the predisposition might be assumed to involve "constitutional" factors; that is, the individual might have been born with a susceptibility to the illness in question. The cause for the susceptibility may involve intrauterine exposure.

Both the twin and adoption approaches have inherent weaknesses. For example, the assumption that identical and fraternal twins have equally similar environments is dubious. In Partanen's study (Partanen, Bruun, & Markkanen, 1966), identical twins differed from fraternal twins in that they lived longer together, were more concordant with respect to marital status, and were more equal in "social, intellectual and physical dominance relationships." Even in rare instances in which monozygotic twins are reared apart, zygosity may influence environmental effects. A person's appearance, for example, influences people's behavior toward him; individuals who look alike may be treated alike. In this and in other ways, the interaction between physical characteristics and the environment tends to reduce intrapair differences in identical twins and to increase differences in fraternal twins.

In addition, generalizing from twin data is hazardous. Twins obviously represent a genetically selected population; they have higher infant mortality, lower birth weight, and slightly lower intelligence; and their mothers' ages are on the average higher (Partanen, Bruun, & Markkanen, 1966).

Adoption studies pose other difficulties. For the results to be meaningful, the following requirements must be met: (1) Reliable information must be available about the biological and adoptive parents of two groups of adoptees—those of alcoholic parentage and those of normal parentage—as well as detailed information about the adoptees themselves. (2) The adoptive parents of both groups should be comparable with regard to drinking habits, psychopathology, and socioeconomic and other demographic variables. (3) The adoptees should have been separated from their biological parents within a few months, at most, after birth, should have had no subsequent contact with them, and should be matched for age and sex. (4) The adoptees should have transversed the age of risk for the illness being studied, which, in the case of alcoholism, is estimated to be between 20 and 40 years of age (Goodwin, 1988b).

Nevertheless, twin and adoption studies have been conducted in regard to alcoholism, and their conclusions, briefly reviewed here, are interesting if not conclusive.

Twin Studies

Kaij (1960), in Sweden, found that identical twins were more concordant for alcoholism than fraternal twins; the more severe the alcoholism, the greater the difference. The previously mentioned Finnish study (Partanen, Bruun, & Markkanen, 1966) found that younger identical and fraternal twins differed with regard to alcohol-related problems—identical twins being more often concordant—but that there was no difference in the total sample. An English study (Murray, Clifford, & Gurlin, 1983) showed no difference between identical and fraternal twins. Still another study (Hrubec & Omenn, 1981), analyzing Veterans Administration records, found identical twins more often concordant for alcoholism than fraternal twins.

In summary, two studies of twins produced results indicative of a genetic influence, one did not, and a fourth was equivocal.

Adoption Studies

In 1944, Roe found no difference in drinking behavior between children of alcoholics and children of nonalcoholics, both in their early twenties, all adopted to presumed nonalcoholics. The sample size was small; no criteria were given for alcoholism. The findings thus tend to be discounted. One should be cautious about discounting findings discordant with one's bias. If the 1944 study supported a genetic hypothesis, it would probably be accepted uncritically by many today.

A rash of adoption studies erupted in the 1970s. The first compared half siblings of alcoholics with full siblings (Schuckit, Goodwin, & Winokur, 1972). Behind the study was an assumption that if genetic factors were important full siblings more often would be alcoholic than half siblings. This assumption did not prove true. The study did find that the half siblings were from broken families. This finding created an opportunity to compare alcoholism in offspring with alcoholism in biological parents as well as in parents of upbringing. Having a biologic father who was alcoholic was highly correlated with alcoholism in the sons. There was no correlation with alcoholism in the surrogate fathers. The study suggested that biological factors were more important than environmental factors in producing alcohol-related problems.

Next came three adoption studies (Bohman, 1978; Cadoret, O'Gorman, Troughton, & Heywood, 1984; Goodwin, Schulsinger, Hermansen, Guze, & Winokur, 1973) from three countries: Denmark, Sweden, and the United States (Iowa). The findings were extraordinarily similar: (1) Sons of alcoholics were 3 or 4 times more likely to be alcoholic than sons of nonalcoholics, whether raised by their alcoholic biological parents or by nonalcoholic adoptive parents. (2) Sons of alcoholics were no more susceptible to other adult psychiatric disturbances, such as depression or sociopathy, than sons of nonalcoholics, when both groups were raised by adoptive parents. The Iowa study found a higher rate of childhood conduct disorder in the sons of alcoholics and was supported by studies reporting increased rates of hyperactive syndrome and conduct disorder in the childhood of alcoholics (Tarter, 1981).

Data regarding female alcoholism were more equivocal. One Danish study found that 4% of the daughters of alcoholics were alcoholics, but 4% of the control women also were alcoholic, a finding introducing the possibility that adoption may contribute to alcoholism (Goodwin, Schulsinger, Knop, Mednick, & Guze, 1977). The Swedish study (Bohman, 1978) found a low rate of alcoholism in the adopted-out daughters of alcoholics, although a subsequent analysis found a correspondence between alcoholism in the biological mothers and alcoholism in the adopted-out daughters (Cloninger, Bohman, & Sigvardsson, 1981).

A comment about methods: The Danish and Iowa studies involved personal (blind) interviews with the adopted children and the application of DSM-III-type criteria for the diagnosis of alcoholism. The Swedish study was based on records listing alcohol-related arrests, hospitalizations and clinic visits for alcoholism, and demographic information. The records did not lend themselves to diagnostic criteria, and biases may have occurred because of the heavy emphasis on alcohol-related arrests; men are arrested more than women for almost all causes. The latter may explain the original report of a low alcoholism rate in the adopted-out daughters of alcoholics. The Swedish study has undergone a number of statistical refinements conducted by Cloninger, Bohman, and Sigvardsson (1981) in the United States, which have generated some tentative, but highly testable, hypotheses relating to genetic-environmental interaction.

On balance, the evidence that heredity influences some cases of alcoholism seems fairly good. Rather than use the word *heredity,* some would prefer *biological vulnerability.* It is true that the transmission of alcoholism does not follow a predictable Mendelian pattern and that concepts such as polygenesis and incomplete penetrance must be invoked to introduce any element of genetic control. It should also be realized that many alcoholics do not, in fact, have alcoholism in the family. In other family illnesses these are called sporadic cases. Considerable evidence suggests that alcoholism that runs in families (familial alcoholism) has an earlier onset than alcoholism that appears *de novo.* Familial alcoholism may also be more severe and have a worse prognosis (Goodwin, 1983).

To summarize: Alcoholism runs in families. It runs in families even when the children are separated from their alcoholic biological parents and raised by nonalcoholic adoptive parents. Twin data, in general, supports a genetic influence. Thus—to repeat—the evidence is fairly good that heredity influences at least some cases of alcoholism.

What Is Inherited?

If alcoholism is transmitted by heredity—even partly—how does the process work?

Whatever is transmitted is clearly chemical. The chemistry of the brain is still largely a mystery. Even less is known about the effects of alcohol on brain chemistry. Still, several theories are being pursued.

First, persons with a genetic propensity for alcoholism may be deficient in certain forms of biochemical activity required for optimal well-being. These persons, given available alcohol, a suitable culture, and an absence of countervailing traits, might discover that alcohol temporarily corrects this hypothetical deficiency, producing an intensity of mood change foreign to those without the deficiency. The model then requires that alcohol would have a biphasic effect, causing subsequent underactivity of the reward system.

Serotonin might fit the deficiency model. Alcohol has a biphasic effect on serotonin metabolism (Kent et al., 1985). Alcohol increases serotonergic activity during acute intoxication and then subsequently reduces seroton activity to subnormal levels. The "deficient" person then would have two reasons for drinking. First he would drink to correct the deficiency and then continue to drink to correct an even greater deficiency resulting from the biphasic effect of alcohol on serotonin. Chemically, this pattern might explain the "addictive cycle," in which a person initially drinks to feel good and then later drinks to stop feeling bad from the substance that originally made him feel good.

Support for the serotonin theory has recently come from several sources. Li and associates (McBride, Murphy, Lumeng, & Li, 1989) have developed an animal model for alcoholism by inbreeding rats. The "alcoholic" or "high-preference" rats have lower levels of serotonin in regions ordinarily rich in serotonin in the brain (such as the hypothalmus and the septohippocampi).

Early-onset alcoholics with a family history of alcoholism have lower serotonergic activity—inferred from tryphophon levels in blood—than late-onset alcoholics with negative family histories (Buydens-Branchey, Branchey, Noumair, & Lieber, 1989). Finally, a new family of drugs called serotonin reuptake blockers is just being released and has been reported to suppress drinking in animals (Zabik, Blinkerd, & Roache, 1985) as well as human subjects (Naranjo et al., 1987). These drugs increase serotonin activity while leaving other central neurotransmitters relatively unaffected.

At the moment, the serotonin deficiency hypothesis probably has more evidence to support it than any other going theory. However, much confirmatory and supportive work must be done.

There is also the overproduction model. This theory holds that a genetic propensity to alcoholism involves the overproduction of substances that in some way facilitate addiction. For example, alcohol produces minute amounts of morphinelike compounds in the brain (Davis & Walsh, 1970). One study found these compounds in the spinal fluid of alcoholics in greater quantities after alcohol ingestion than in nonalcoholics (Borg, Kvande, Magnusson, & Sjoquist, 1980). Rats and monkeys drink increased amounts of alcohol when these substances are injected into the brain (Myers, McCaleb, & Ruwe, 1982). Thus, the possibility exists that the genetic defect operating in alcoholism involves a reduced capacity of the brain to oxidize aldehydes, resulting in the overproduction of morphine-like alkaloid compounds (aldehyde condensation products), which may facilitate alcohol addiction.

One more possibility will be mentioned. Data from a Danish study (Pollock et al., 1983) indicate that sons of alcoholics generate more slow-wave alpha activity on the EEG after drinking alcohol than sons of nonalcoholics. Earlier studies found reduced alpha activity in alcoholics, usually interpreted as representing damage from heavy drinking. The finding that sons of alcoholics (not heavy drinkers) have increased amount of alpha activity after drinking suggests that a risk factor in alcoholism may be a genetically determined increase in alpha activity from alcohol. The finding needs much replication before the theory can be accepted.

Speculation about genetic factors in alcoholism is not complete without comment on physiologic reactions that deter, or protect, large numbers of people from becoming alcoholic.

Many women are intolerant of alcohol. After drinking small amounts they experience unpleasant physiologic reactions, ranging from mild nausea to dizziness or somnolence. This intolerance may partly explain the fact that fewer women than men are alcoholic. However, the role of physiologic protective factors in deterring heavy drinking is best documented by studies of the so-called Oriental flushing phenomenon.

Roughly two thirds to three quarters of Asians experience a cutaneous flush, mainly of the face and upper part of the body, after drinking a small amount of alcohol (fewer than one or two beers) (Wolff, 1972). The flush is usually accompanied by unpleasant subjective reactions, including feelings of warmth and queasiness, as well as an increase in heart rate and a decrease in blood pressure. The Oriental flushing phenomenon is undoubtedly genetic in origin. Asian infants have been given tiny amounts of alcohol, and they too have flushed. Some studies—not all—find elevated levels of acetaldehyde correlated with the flush, and atypical forms of both alcohol dehydrogenase and aldehyde dehydrogenase have been reported in a high proportion of Japanese men and may presumably be related to the flush (Harada, Agarwah, & Goedde, 1981). Whatever the biochemical mechanism, the flushing reaction undoubtedly constitutes a powerful deterrent to heavy drinking for many Asians. The reported low alcoholism rate in much of Asia thus may have a genetic explanation, reinforced perhaps by cultural factors.

Note

1. "The investigation of any trait in alcoholics will show that they have either more or less of it."—Mark Keller (Keller's Law).

References

American Psychiatric Association. (1987). *Diagnostic and Statistical Manual of Mental Disorders* (DSM-III-R). (3rd ed., rev.) Washington, DC: Author.

Bohman, M. (1978). Some genetic aspects of alcoholism and criminality: A population of adoptees. *Archives of General Psychiatry, 35,* 269–276.

Borg, S., Kvande, H., Magnusson, E., & Sjoquist, B. (1980). Salsolinal and salsoline in cerebrospinal lumbar fluid of alcoholic patients. *Acta Psychiatry Scandinavian Supplement, 286,* 171–177.

Buydens-Branchey, L., Branchey, M.H., Noumair, D., & Lieber, C.S. (1989). Age of alcoholism onset: II. Relationship to susceptibility to serotonin precursor availability. *Archives of General Psychiatry, 46,* 231–236.

Cadoret, R.J., O'Gorman, T.W., Troughton, E., & Heywood, E. (1984). Alcoholism and antisocial personality: Interrelationships, genetic and environmental factors. *Archives of General Psychiatry, 42,* 161–167.

Cloninger, C.R., Bohman, M., & Sigvardsson, S. (1981). Cross-fostering analysis of adopted men. *Archives of General Psychiatry, 36,* 861–868.

Cotton, N.S. (1979). The familial incidence of alcoholism: A review. *Journal of Studies on Alcohol, 40,* 89–116.

Davis, V.E., & Walsh, M.J. (1970). Alcohol, amines and alkaloids: A possible biochemical basis for alcohol addiction. *Science, 167,* 1005–1007.

Goodwin, D.W. (1983). Familial alcoholism. *Substance and Alcohol Actions/Misuse, 4,* 129–136.

Goodwin, D.W. (1988a). *Alcohol and the writer.* New York: Andrews & McMeel.

Goodwin, D.W. (1988b). *Is alcoholism hereditary?* New York: Ballantine Books.

Goodwin, D.W., Schulsinger, F., Hermansen, L., Guze, S.B., & Winokur, G. (1973). Alcohol problems in adoptees raised apart from alcoholic biological parents. *Archives of General Psychiatry, 28,* 238–242.

Goodwin, D.W., Schulsinger, F., Knop, J., Mednick, S., & Guze, S.B. (1977). Psychopathology in adopted and nonadopted daughters of alcoholics. *Archives of General Psychiatry, 34,* 1005–1009.

Harada, S., Agarwah, D.P., & Goedde, H.W. (1981). Aldehyde dehydrogenase deficiency as cause of facial flushing reaction to alcohol in Japanese. *Lancet, 1,* 982.

Hrubec, Z., & Omenn, G.S. (1981). Evidence of genetic predisposition to alcoholic cirrhosis and psychosis: Twin concordances for alcoholism and its biological end points by zygosity among male veterans. *Alcoholism: Clinical and Experimental Research, 5,* 207–215.

Jackson, C. (1944). *The lost weekend.* New York: Modern Library.

Kaij, L. (1960). *Studies on the etiology and sequels of abuse of alcohol.* Lund, Sweden: University of Lund.

Kent, T.A., Campbell, J.L., Pazdernik, T.L., Hunter, R., Gunn, W.H., & Goodwin, D.W. (1985). Blood platelet uptake of serotonin in men alcoholics. *Journal of Studies on Alcohol, 46,* 357–359.

Little, R.E., & Ervin, C.H. (1984). Alcohol use and reproduction. In S.C. Wilsnack and L.J. Beckman (Eds.), *Alcohol problems in women* (pp. 155–188). New York: Guilford Press.

London, J. (1982). *Novels and social writings.* New York: Literary Classics of the United States.

McBride, W.J., Murphy, J.M., Lumeng, L., & Li, T.K. (1989). Serotonin and ethanol preference. In M. Galanter (Ed.), *Recent developments in alcoholism* (Vol. 7, pp. 187–209). New York: Plenum Press.

Mello, N.K. (1980). Some behavioral and biological aspects of alcohol problems in women. In O.J. Kalant (Ed.), *Research advances in alcohol and drug problems* (Vol. 5, pp. 263–298). New York: Plenum Press.

Murray, R.M., Clifford, C., & Gurlin, H.M. (1983). Twin and alcoholism studies. In M. Galanter (Ed.), *Recent developments in alcoholism* (Vol. 1, pp. 25–48). New York: Gardner Press.

Myers, R.D., McCaleb, M.L., & Ruwe, W.D. (1982). Alcohol drinking induced in the monkey by tetrahydropapaveroline (THP) infused into the cerebral ventricle. *Pharmacological Biochemical Behavior, 16,* 995–1000.

Naranjo, C.A., Sellers, E.M., Sullivan, J.T., Woodley, D.V., Kadlec, K., & Sykora, K. (1987). The serotonin uptake inhibitor citalopram attenuates ethanol intake. *Clinical Pharmacology Therapy, 4,* 226–274.

Partanen, J., Bruun, K., & Markkanen, T. (1966). Inheritance of drinking behavior: A study on intelligence, personality, and use of alcohol of adult twins. Helsinki: Finnish Foundation for Alcohol Studies.

Pollock, V.E., Volavka, J., Goodwin, D.W., Mednick, S.A., Gabrielli, W.F., Knop, J., & Schulsinger, F. (1983). The EEG after alcohol in men at risk for alcoholism. *Archives of General Psychiatry, 40,* 857–861.

Roe, A. (1944). The adult adjustment of children of alcoholic parents raised in foster homes. *Quarterly Journal of Studies on Alcohol, 5,* 378–393.

Schuckit, M.A., Goodwin, D.W., & Winokur, G. (1972). A half-sibling study of alcoholism. *American Journal of Psychiatry, 128,* 1132–1136.

Tarter, R.E. (1981). Minimal brain dysfunction as an etiological predisposition in alcoholism. In R.E. Meyer, B.C. Glueck, J.E. O'Brien, T. Babor, J. Jaffe, & J. Stabenau (Eds.), *Evaluation of the alcoholic: Implications for research, theory, and treatment* (Research Monograph 5, pp. 167–191). (DHHS Publication No. ADM 81–1033). Washington, DC: U.S. Government Printing Office.

Wolff, P.H. (1972). Ethnic differences in alcohol sensitivity. *Science, 175,* 449–450.

Zabik, J.E., Blinkerd, K., & Roache, J.D. (1985). Serotonin and ethanol aversion in the rat. In C.A. Naranjo & E.M. Sellers (Eds.), *Research advances in new psychopharmacological treatment of alcoholism* (pp. 87–101). New York: Excerpta Medica.

SECTION IV

The Genesis and Patterning of Alcoholism and Alcohol-Related Problems

D. RELATION TO INSTITUTIONS: FAMILY, ECONOMIC, AND LEGAL

Introductory Note

Although implying a good deal about such matters, the studies just presented do not give full consideration to the impact of alcoholism on those social contexts in which most persons in our society experience the bulk of their face-to-face interaction, namely, in the family and at work. Actually, this is a two-sided problem. There is, on the one hand, the question of the impact of alcoholism on the familial and occupational systems. On the other hand, there is the question of the effects these systems themselves have on the development and patterning of alcoholism. The first two chapters in this subsection address these questions.

Picking up from where Joan Jackson (1962) left off in the first volume, Theodore Jacob and Ruth Ann Seilhamer (chapter 31) present a review of recent research on the behavior of family members in relation to alcoholism. They suggest that most of the literature has focused on the alcoholic or the spouse of an alcoholic as individuals within the family, but what is needed are more studies that examine the family as a social system. Jacob and Seilhamer describe recent promising research that studies family interaction with and without alcohol consumption in laboratory situations. They argue, however, that there are still significant gaps in the research literature on alcoholism and family interaction, such as assessments of the acute effects of alcohol on the family, examinations of the families of female alcoholics (see chapter 13, by Gomberg, in this volume), and examinations of the children of alcoholics (see chapter 37, by Rudy, in this volume).

In chapter 32 we turn to the question of the relations of drinking and occupational patterns. This study by Harrison Trice and William Sonnenstuhl takes up one particular facet of this whole question, approaching the matter initially from the point of view of the effects of drinking upon the job, rather than vice versa. Trice and Sonnenstuhl's main concern is with alcoholism and the possible impact of alcoholism on certain key aspects of job performance, such as productivity, absenteeism, and accidents. The findings reported here are remarkably consistent with those reported by Trice (1962) in the original volume.

In the course of their analysis, however, Trice and Sonnenstuhl effectively show how differences in the structure of the work situation that are related to occupational type and status influence the form in which alcoholism is expressed. They also analyze how the denial syndrome expresses itself in work environments and how adhering to a performance standard in a constructive confrontation strategy counteracts denial.

As the last two chapters of this subsection we have chosen to present reports that explore the relationships of alcoholism to phenomena of deviance that have sustained the interest of social scientists for generations, namely, crime and suicide. In chapter 33 James Collins reviews the relationship between alcohol use and crime, especially public intoxication, drunk driving, violent crime, and property crime. He explains why there has been a decrease in law enforcement against public intoxication (see chapter 38, by Rubington, in this volume) and an increase in enforcement of laws against driving while intoxicated. Collins also refutes causal assumptions about the alcohol-crime connection. He argues that although the pharmacological effects of alcohol can account for some violence, the relationship between alcohol use and violence is overstated and that drinking is not etiologically important for property crime. This chapter refutes some commonly held beliefs and raises important questions about interventions for offenders with alcohol-related problems.

Chapter 34, the last of the subsection, is a reprint of a British study conducted by Keith Hawton, Joan Fagg, and Stephen McKeown of the relationship between suicide attempts and the use of alcohol. Rather than examine the extent of suicidal behavior among alcoholics, these researchers studied the rate of alcoholism among suicide attempters and found it to be about 15% and increasing over time. The rate of alcoholism was especially high among subjects who were unemployed or were housewives. It appears also that alcoholic suicide attempters are more likely to repeat attempts. In a new twist, the researchers also assessed the extent of alcohol use before and during suicide attempts and found that such use is very common, especially among alcoholics. Thus, they warn that alcohol use not only increases the likelihood of an attempt but adds to the dangers of an overdose.

In this subsection we present chapters dealing with the problems of alcoholism in relation to major societal institutions. In the final section of the book we examine the social movements that have arisen in response to these problems.

References

Jackson, J. (1962). Alcoholism and the family. In D.J. Pittman & C.R. Snyder (Eds.), *Society, culture, and drinking patterns* (pp. 472–92). New York: Wiley.

Trice, H.M. (1962). The job behavior of problem drinkers. In D.J. Pittman & C.R. Snyder (Eds.), *Society, culture, and drinking patterns* (pp. 493–510). New York: Wiley.

CHAPTER 31

Alcoholism and the Family

THEODORE JACOB AND RUTH ANN SEILHAMER

Twenty-five years ago Joan Jackson reviewed the existing literature on families of alcoholics and reported major limitations regarding well-articulated theories and systematic research (Jackson, 1962). The few studies that existed were largely helter-skelter reports; there were virtually no programmatic efforts to build an empirical or theoretical base upon which to structure future work. Jackson criticized research in this area for its use of small, unrepresentative samples, its lack of control groups, and its assessments gathered when families presented at social or legal agencies in times of crisis. Moreover, approaches that focused on the alcoholic or the spouse or the children as separate entities dominated the literature at that time. With unusual foresight, Jackson identified gaps in these individually focused literatures that continue today. For example, she pointed to the need for research regarding female alcoholism and its impact on spousal and maternal roles and to the need for longitudinal studies aimed at identifying factors that mitigate adverse outcomes for children of alcoholics. Most important, Jackson stressed the need for studies of the alcoholic's family as a social and behavioral unit.

In addition to the investigation of molecular family interactions, Jackson encouraged molar conceptualizations of the alcoholic family in the context of its cultural milieu and stages of adaptations to parental alcoholism. Using the reports of Al-Anon wives, she developed a seven-stage model of family disorganization and role realignments that occur as the family adjusts to living with an alcoholic (Jackson, 1954). Although not empirically substantiated, this model emphasizes the need to consider stage-related variables such as the duration of drinking, the child's age

Theodore Jacob is with the Department of Family Studies and Psychology, University of Arizona, Tucson, Ariz. Ruth Ann Seilhamer is with Western Psychiatric Institute and Clinic, University of Pittsburgh, Pittsburgh, Pa. This chapter is a revision of "Alcoholism and Family Interaction," which appeared in *Recent Developments in Alcoholism, Vol. 7: Treatment Issues,* M. Galanter, Ed. (New York: Plenum, 1989), pp. 129–145, and appears here by permission.

when problem drinking began, and progressive behavioral adaptations of the family to the course of alcoholism.

The past 20 years have witnessed a vast expansion of studies addressing the issues that Jackson provocatively presented. Although much of this literature continues to be subject to the same criticisms, there have been many theoretical and methodological advances in several areas of endeavor. This chapter will review the development of the family interaction literature related to alcoholism, beginning with early individually oriented reports and progressing to more recent approaches that study the family as an interacting unit.

Individual Approaches

As mentioned, much of the literature relevant to families of alcoholics focuses on individuals within the family matrix. By comparison, studies that deal with the family as a unit—either as an influence in the onset and maintenance of abusive drinking or as a system disturbed by the effects of alcoholism—are quite limited (Jacob & Seilhamer, 1987). This individual emphasis can be traced to traditions that defined alcoholism as a personal moral deficit or medical problem, as well as to the predominance of psychodynamic perspectives in early clinical and empirical reports.

This latter influence is most apparent in descriptions of alcoholics' wives as "disturbed personalities" who sought to dominate or punish (or do both to) their spouses (Futterman, 1953; Kalashian, 1959; Lewis, 1937). Later, researchers with environmental perspectives recast wives as victims rather than villains, claiming that their disturbances were due to the accumulative stress of living with an alcoholic (Jackson, 1954; Jacob & Seilhamer, 1982). More recently, theoretical and empirical efforts have focused on the association of individual and environmental factors, with attempts to describe and categorize the coping strategies used by wives of alcoholics (James & Goldman, 1971; Orford et al., 1975; Schaffer & Tyler, 1979; Wiseman, 1980b). Notwithstanding these contributions, information that clarifies marital interactions that potentiate or reinforce abusive drinking is largely absent. However, the literature on spouses of alcoholics clearly implies that their marriages are characterized by identifiable and repetitive relationship patterns that influence the onset and course of alcohol abuse.

Similarly, deficits in personality and in psychosocial and psychiatric status have been major themes in the diverse literature concerned with children of alcoholics. However, the general assumption that adverse outcomes are inevitable for children of alcoholics has been weakened by observations of wide variability among these offspring (Adler & Raphael,

1983; el-Guebaly & Offord, 1977; Jacob, Favorini, Meisel, & Anderson, 1978; Wilson, 1982). Additionally, empirical studies have been marked by a lack of consistent findings, sound research methods, and coherent conceptualizations (Russell, Henderson, & Blume, 1984). Nevertheless, the collective literature has well established that children of alcoholics are a population at risk for subsequent biopsychosocial difficulties, and currently researchers are stressing an interplay of multiple factors that affect child outcome. Again, the relevant literature implies that such outcomes are associated with disturbed marital and parent-child interactions that evolve in family structures with an alcoholic parent. These family interactions, however, are not well described, particularly with reference to how family processes mediate negative child outcome.

Thus, the study of family interaction in the alcoholism literature has been relatively neglected, primarily because of the historical emphasis on alcoholism as an individual problem. Despite recurrent references to the interpersonal aspects associated with alcohol abuse, it was not until the late 1960s and early 1970s that observational studies of family interactions began to appear in the literature. These studies are based primarily on observations of small samples of subjects performing structured tasks in laboratory settings; there are less than a such dozen empirical reports in the literature to date. Although these efforts are preliminary, they have introduced innovative methods and provocative conceptualizations for exploring the interaction patterns of families of alcoholics.

The application of an interaction perspective to studies of alcoholism offers several possibilities for clarifying the association of family factors and alcohol abuse. First, descriptions based on empirical data can dispel myths and misinformation generated by reports gathered within clinical contexts and based on small, unrepresentative samples. Second, clearly identified and empirically validated patterns of interaction can be related to current and future states of the family and individual members. In this context, family interaction patterns can be either a dependent or an independent variable in broader levels of analysis, depending on the research question addressed. Third, to the extent that interaction research can provide further understanding of affective interchanges, problem-solving style and effectiveness, dominance patterns, and parent-child socialization practices that are associated with alcohol and nonalcohol situations, treatment and prevention programs will be founded on greater substance and less supposition.

Transitional Studies

During the past 30 years several studies have attempted to move beyond individual assessments to descriptions of relationships in families of alcoholics. The majority of the early investigations were based on the theories of interpersonal perception and attempted to evaluate how alcoholics and their spouses appraised themselves and each other. The basic premise was that perceptions of oneself and one's partner are a better gauge of the way a relationship is functioning than individually derived trait descriptions. These early reports served as an impetus for subsequent efforts that actually observed and recorded interchanges between alcoholics and their spouses and, as such, may be considered "transitional" studies.

The earliest of these efforts was reported by Mitchell (1959) and required subjects to rate 17 traits with regard to their applicability to self and to spouse. The subjects included 28 couples with alcoholic husbands and 28 maritally conflicted but nonalcoholic couples. Although the experimental and the control couples were similar on most measures, there were important group differences on dimensions concerned with control, dominance, and sensitivity. Continuing this line of research, Drewery and Rae (1969) conducted analyses of perceptual and attributional differences between husbands and wives in 22 alcoholic and 26 control marriages. Among the many comparisons, the most interesting results were found in differences between alcoholic and control groups regarding wives' descriptions of their husbands. Essentially, wives in the control group were more in accord with their husbands' self-descriptions than the wives of alcoholics.

Two later studies focused on differences in perception of the alcoholic during sober and intoxicated states and on the congruence of spouses regarding these differences. Tamerin, Tolar, DeWolfe, Packer, and Neuman (1973) asked 20 alcoholic inpatients and their spouses to provide retrospective descriptions of moods and behaviors that characterized the alcoholic when sober and when intoxicated. Major findings were as follows: (1) the alcoholic and spouse described the alcoholic in generally positive terms when the alcoholic was sober; (2) the alcoholic and spouse reported an increase in negativism and depression from sober to intoxicated states; (3) the increased negativity from sober to intoxicated states was reported more often by the spouse than by the alcoholic.

Using the Tamerin et al. (1973) report as a guide, Davis (1976) conducted a study of interpersonal perceptions. Data were obtained both while the alcoholic husband was sober and while he was intoxicated. In a third condition, spouses were asked to predict their responses if the alcoholic mates were intoxicated. Several discrepancies between these types of data were found, indicating the need to corroborate "as if" data with actual observations within the context of interest (e.g., during periods of actual drinking).

In a transitional study by Hanson, Sands, and Sheldon (1968), 19 alcoholic couples completed a behavior questionnaire on themselves and then completed it again as they predicted their spouse would. Analyses supported the primary hypothesis: Alcoholic husbands predicted their wives' responses more accurately than the spouses predicted their husbands' responses. The finding suggests that communication is more often unidirectional than bidirectional and thus that the alcoholic knows more about the thoughts, feelings, and beliefs of the spouse than the spouse knows about the alcoholic's.

Confidence in the findings of these transitional studies is undermined by a variety of methodological flaws, namely, absent or inappropriate comparison groups; small, unrepresentative samples; and measures of unreported validity and reliability. From a substantive viewpoint, these studies stopped short of moving into interactional issues or behaviors that could be implied from perceptual data. However, these reports focused attention on relationships rather than individuals and suggested that spouses' perceptions differed according to intoxicated and sober states of the alcoholic partner. Given the difficulty in bridging the gap between perceptions and actual behavior, these studies served to stimulate interest in observed interactions as a more direct avenue to the understanding of the complex association of alcoholism and family processes.

Outcome-Oriented Studies

Although a small body of literature, family interaction studies of alcoholics have introduced important concepts and models, most notably the direct observation of families and marital dyads in laboratory and natural settings.

Three studies have focused on outcomes of laboratory games and have sought to clarify role relationships, power structures, and communication styles of alcoholics and their spouses. The most important contribution within this group comes from the work of Gorad (Gorad, 1971; Gorad, McCourt, & Cobb, 1971), who introduced a communication systems framework to the study of families of alcoholics. Specifically, Gorad hypothesized that alcoholics deny responsibility for their messages by means of metacommunications (verbal or nonverbal messages about a message) such as: "I am under the control of alcohol. Therefore, I am not responsible for what I say or do."

Gorad tested this hypothesis in a laboratory game that enabled participants to select options for sharing or monopolizing payoffs. The game was designed so that one could choose to dominate the situation while avoiding disclosure of this choice to other participants. As predicted, alcoholics

(n = 20) selected the responsibility-avoiding option significantly more often than their spouses and the matched normal controls (n = 20). In addition to its theoretical contributions, this study introduced an objective, standard procedure whereby key constructs could be operationalized and objectively measured.

In an effort to assess the problem-solving competencies of alcoholic couples, Cobb and McCourt (1979) administered a variety of laboratory tasks to 36 alcoholic couples and 33 normal controls. Generally, there were few differences, although alcoholic couples were less cooperative and more competitive than controls. Gorad's laboratory game was used in this study, but no differences were found in responsibility-avoiding behaviors. However, this unpublished report lacks sufficient description of procedures and analyses for a thorough evaluation.

A third study, by Kennedy (1976), focused on cooperative and competitive interactions during a simulation game administered to three groups: 11 couples with alcoholic husbands, 11 nondistressed couples, and 6 couples with psychiatrically disturbed wives. Significant group differences were minimal, although qualitative analyses suggested greater heterogeneity within the alcoholic group and differences related to inpatient versus outpatient subgroups. Again, this study failed to replicate group differences in the interaction styles that were exhibited by Gorad's subjects. However, comparisons across studies are hampered by differences in sample characteristics, procedures, and conceptual dimensions.

In summary, these outcome studies contributed intriguing concepts from the communication systems literature, suggesting that issues of control and influence characterize interchanges between alcoholics and their spouses. Moreover, these reports implemented precise, objective methods of small-group research, allowing for the control of interactional directions and the elicitation of specific, relevant dimensions of communication behavior.

Process-Oriented Studies

Hersen, Miller, and Eisler (1973) reported the first empirical study that focused on the exchange of verbal and nonverbal behavior during an observed marital interaction. These researchers videotaped four couples during discussions of drinking-related and drinking-unrelated topics that alternated every 6 minutes over a 24-minute period. Results showed that wives looked at their husbands more during drink-related than drink-unrelated discussions, and the authors suggested that the spouse's attention may reinforce abusive drinking. An expansion of this design by Becker and Miller (1976) included a larger sample of alcoholics (n = 6), a nonalcoholic psychiatric control group (n = 6), and a wider range of verbal and nonverbal measures. Analyses of variance detected one significant main effect: alcoholic couples interrupted more frequently than control couples, regardless of the discussion topic. Again, the findings reported in these two studies are tentative; small samples and the lack of a normal comparison group limit generalizability.

A major advance in this research area was initiated with a programmatic series of studies by Steinglass and his colleagues. This research direction began during the late 1960s with a series of experimental studies at the National Institutes of Mental Health (NIMH) that included observations of inpatient alcoholics during both an experimental drinking phase and a subsequent withdrawal period (Steinglass, 1975). One of these drinking experiments involved related pairs of alcoholics and represented the first assessment of family units during intoxicated and sober states (Steinglass, Weiner, & Mendelson, 1971). The affective and structural characteristics of the relationship were dramatically altered during the drinking period; role reversals, intense expression of affect, and changes in the patterns and organization of interactions occurred.

For Steinglass, these observations confirmed the significance of reciprocal effects involving alcohol and interpersonal interaction, and they led to a preliminary model of alcoholism based on family systems theory. Specifically, Steinglass (Steinglass et al., 1971) suggested that abusive drinking can serve two different functions. In one case, drinking is a signal that individuals and relationships within the system are experiencing significant stress. In contrast with the "signal" function, which is most likely in families where problem drinking has not yet become an ongoing process, drinking can maintain and stabilize the family unit. Steinglass observed that alcohol abuse appeared to effect very different behaviors in different family dyads; that is, alcohol abuse was associated with the controlled release of aggression in one pair and with the clarification of dominance patterns in another pair. "Although the style was different, in each instance the end result was the stabilization of a dyadic system which might otherwise be expected to have been characterized by chaos" (p. 408).

Steinglass's research then moved into assessments of marital pairs in an experimental treatment program during which couples were housed in apartmentlike settings and were encouraged to behave as normally as possible (Steinglass, Davis, & Berenson, 1977). Alcohol was freely available for 7 days of this period, and the couples' activities were videotaped. The participants were 10 couples in which the husband, the wife, or both were alcoholic. Conclusions were derived from interaction summaries that were based on the clinical observations of the program staff. Generally, intoxicated behaviors were more exaggerated and more restricted in range than interactions displayed in sober states. These findings confirmed Steinglass's earlier impressions that intoxication enhanced the regularity and rigidity of interactions. Furthermore, striking differences in patterns of interaction were observed between sober and intoxicated periods—differences that appeared to serve important "adaptive" functions for the couple. That is, the behavior that emerged during intoxicated periods appeared to potentiate or inhibit certain aspects of the relationship, thus in effect reducing tensions by providing a temporary solution to a conflictual or stressful process.

These observations suggested a second-generation theory, the "alcohol maintenance model." As conceptualized by Steinglass (Steinglass et al., 1977), the theory held that if alcohol effectively reduced tension or temporarily solved a problem (or did both), short-term family stability could be achieved and thus that the change from sober to intoxicated interactional states served to stabilize an unstable system.

The third phase of Steinglass's work involved a more systematic and empirical effort that emphasized a macroscopic, longitudinal view of drinking patterns. The life history model of alcoholism (Steinglass, 1980) suggested that periods of sobriety and active drinking tend to cycle over long time periods in the lives of most alcoholics. In contrast with the maintenance model, which spoke of rapid changes from sober to intoxicated states with associated changes in patterns of interaction, the life history model suggested that three important phases—*dry, wet,* and *transitional*—appear and reappear many times over a 20-to-30-year period. For some families, there may be many occurrences of dry, wet, and intervening transition phases, whereas for other families there may be only one stable wet phase that is ultimately resolved into a stable dry phase or simply continues until death or divorce changes the structure of the family system.

The life history model incorporated the earlier maintenance model; that is, within the wet phase there are alternating sober and intoxicated states, each having a distinct pattern of interaction and perpetuated because of the short-term solution provided by the intoxicated interactions. A second, more overarching pattern of change operated over longer time periods, with wet, dry, and transitional phases extending over weeks, months, and years. A key implication of this broader framework was that the family failed to progress along a normal developmental course characterized by greater complexity and differentiation. For Steinglass, it was this distortion and constriction of the family's natural growth and development that came to be a major adverse outcome of alcoholism.

In an effort to validate and explore the implications of the life history model, Steinglass undertook a multifaceted study of 31 families of alcoholics. The basic research strategy was to categorize families into *stable wet* (SW), *stable dry* (SD), or *transitional* (TR) phases and then to assess for group differences in family interaction in three contexts: the home, the interaction laboratory, and multiple-family group meetings.

Results of home observations (Steinglass, 1981) revealed that SW families were characterized by "a tendency of family members to disperse in the house, physically interacting only when they intend[ed] to talk with one another for some purposeful reason" (p. 581). In contrast, SD families exhibited "relatively high rates of decision-making behavior and greater affective display, especially in the direction of allowing disagreements to be expressed" (pp. 581–582). TR families, manifesting extreme scores on both dimensions, were viewed as physically close "to a degree that [gave] them the appearance of huddling together for warmth and protection" (p. 582) and as manifesting a very narrow range of tasks, affect, and outcome.

Analyses of laboratory interactions (Steinglass, 1979) revealed that dry families ($n = 2$) functioned as a cohesive group significantly better than wet families ($n = 5$) on a structured card sort procedure. In discussing these findings, Steinglass suggested that members of wet families tended to behave in a relatively independent manner (low coordination) whereas individuals from dry families appeared to emphasize togetherness, agreement, and family solidarity (high coordination). Steinglass also suggested that the findings appeared consistent with earlier formulations of the alcoholic family's "biphasic" nature involving different patterns of interaction associated with intoxicated and sober states.

The innovative work of Steinglass has guided and shaped theory development and empirical methods in family interaction research for 15 years. There has been a major limitation of his efforts, however: insufficient empirical support for his proposed models. Specifically, his work was based on small, highly selective samples that lacked normal comparison groups and were uncontrolled for possible confounds such as the co-occurrence of other psychiatric disorders in the alcoholic proband or spouse. Moreover, the data used to generate early conceptual models were based on impressionistic clinical summaries. Although later reports were founded on an objectively defined recording system (the Home Observation Assessment Method) (Steinglass, 1979), this procedure needs further elaboration and validation. Taken together, these limitations raise questions about the stability and interpretability of obtained results.

Drinking Experiments

The basic tenet of Steinglass's work is that interaction differs in intoxicated versus sober states and that interactions during intoxication provide sufficient reinforcement for the maintenance of abusive drinking. As mentioned, these assumptions were based on clinical impressions of very small samples, and Steinglass's later efforts did not directly assess the acute effects of alcohol on family interchanges. During the past several years, a number of studies have emerged that include experimental drinking procedures with families of alcoholics.

Jacob, Ritchey, Cvitkovic, and Blane (1981) videotaped the laboratory interaction of eight families with an alcoholic father and eight normal control families. Both parents and two of their children aged 10–18 engaged in a structured discussion task, the Revealed Difference Questionnaire (RDQ) (Jacob, 1974), and in discussions of personally relevant topics gathered from the Areas of Change Questionnaire (ACQ) (Weiss, Hops, & Patterson, 1973). On separate occasions, these procedures were conducted under two different conditions: the presence versus the absence of alcoholic beverages.

The primary analysis focused on group differences in affective and problem-solving communications that were coded according to the Marital Interaction Coding System (MICS) (Weiss, 1976). Analyses of the ACQ interactions of the marital dyads showed that: (1) the couples with an alcoholic husband expressed more negative and less positive affect than the controls; (2) the couples with an alcoholic husband were more negative in drink versus nondrink sessions, whereas the normal couples did not differ between these two conditions; (3) the wives of alcoholics were more disagreeable in the drink versus the nondrink session, whereas the opposite effect was found in wives from normal couples; and (4) the husbands from normal couples exhibited more problem-solving behaviors than their wives, whereas alcoholics and their wives engaged in similar levels of problem solving. Analyses of RDQ discussions showed that husbands in the normal group displayed more problem solving than their wives; an opposite effect was found in the couples with an alcoholic husband.

RDQ parent-child discussions showed interesting patterns in instrumental behavior: (1) normal fathers were more problem solving than their children; (2) alcoholic fathers and their children showed similar levels of problem solving; and (3) mothers in the alcoholic group were more problem solving than their children. These findings support the clinical-theoretical accounts of structural changes in alcoholic families (Jackson, 1954) as well as other family interaction research, which suggests that fathers in normal families are more influential in parent-child relationships than fathers in disturbed families (Jacob, 1975). Although the Jacob et al. (1981) study produced some provocative results, confidence in these findings is limited by design features, namely, a small sample, absence of a psychiatric or distressed control group for examination of the specificity of alcoholism's impact, and failure to include discussions with both parents present.

Both distressed and nondistressed control groups were included in a study of videotaped marital interactions that was conducted by Billings,

Kessler, Gomberg, & Weiner (1979). Twelve couples were in each of the demographically matched groups (alcoholic, distressed, nondistressed). As in the Jacob et al. study (1981), alcoholic beverages were served one evening, whereas only soft drinks were available during a second evening. The principal finding was that alcoholic and distressed couples engaged in significantly less problem solving and in more negative and hostile behaviors than nondistressed controls. The only major difference to emerge in comparisons of drink versus nondrink situations was that alcoholic and distressed couples spoke more during drinking sessions, whereas normal controls were relatively unaffected. Again, methodological shortcomings limit conclusive interpretation. Of particular relevance to the central hypothesis of this study is the fact that almost half of the couples in each group did not consume alcohol during the drinking session and those who drank consumed only minimal amounts (blood alcohol concentration [BAC] = 0.019-0.026%). Moreover, the failure to define subjects carefully according to their psychiatric status presents a potentially confounding influence.

An experimental drinking study by Frankenstein, Hay, and Nathan (1985) controlled the alcohol consumption of subjects by administering fixed doses of alcohol before videotaped interaction sessions. Analyses of six summary MICS codes showed that the eight alcoholic couples demonstrated more positive interactions in the drink versus the nondrink situation, a finding that resulted primarily from the spouses' change between situations. Additionally, the alcoholics talked more than their spouses when drinking. They also tended to express more problem-solving statements when drinking, whereas this effect was not found for spouses. The authors proposed that these results support theories that alcohol consumption is reinforced and maintained by its beneficial effect on family relationships.

The discrepancies in findings between the Jacob et al. (1981) and Frankenstein et al. (1985) studies may be due to design characteristics. Specifically, critical differences are apparent in the amount and administration of alcohol, the presence versus absence of alcohol during ongoing interactions, and the application of MICS codes (see Jacob & Seilhamer, 1987, for further elaboration of these issues).

Experimental drinking was not part of an interaction study by O'Farrell and Birchler (1985), but improvements in design make this effort particularly noteworthy. Specifically, these researchers used an increased sample size (n = 26 couples); both alcoholic, maritally distressed, and nondistressed, nonalcoholic groups; MICS-coded interactions that involved personally relevant, conflict-oriented topics; and multimethod assessment (i.e., self-reports and direct observation). As expected, alcoholic and maritally conflicted couples manifested more disturbance; but predictions that alcoholic couples would display unique structural and qualitative differences in interaction were not supported. Again, cross-study comparisons are hampered by several methodological deficiencies, namely, the use of treatment-seeking couples in the maritally distressed group, the recent treatment and current abstinence of the alcoholic probands, and the failure to assess the subjects' other possible psychopathology. Most critical, however, is that the lack of a drinking manipulation prohibits any generalizations of nondrinking interactions to drinking situations.

Clearly, these initial drinking studies leave several substantive and methodological issues unaddressed. It is of critical importance that samples be more carefully defined, especially with regard to the psychiatric status of all family members. With reference to index cases, more detailed descriptions of drinking history and patterns of consumption are needed to identify subgroups of drinkers. Also, appropriate control groups are essential to determine what, if any, interactive characteristics are unique to alcoholic family systems. Samples must be representative of the groups to which findings are to generalize, and they must be large enough to allow for cross-validation and the application of multivariate statistics.

In an effort to address these limitations, Jacob and his colleagues (Jacob, Seilhamer, & Rushe, 1989) have been conducting a multifaceted program of interaction studies focused on families of alcoholics. This research design confronts several key deficiencies in the literature by including: (1) diagnostically homogeneous alcoholics who exhibit no additional psychiatric diagnoses and are not married to alcohol-abusing wives; (2) a normal and a psychiatrically disturbed group composed of depressed males; (3) an experimental drinking design in which couples self-administer significant amounts of alcohol during videotaped laboratory interactions; (4) parent-child interactions; (5) theoretically relevant, empirically derived coding systems; and (6) large samples in order to maximize statistical power, explore family typologies, and allow for adequate follow-up of probands and their offspring.

Jacob and Krahn (1988) analyzed MICS ratings of the marital interactions of the 38 families of male alcoholics, 35 families of male depressives, and 34 families of normals (social drinkers with no diagnosed psychopathology) who participated in this investigation. Drinking and nondrinking sessions were identically formatted, except that alcoholic beverages were available during the drinking session, coffee and soft drinks during the nondrinking session. As expected, normal couples exhibited higher rates of congeniality and positivity than disturbed couples. Also, alcoholic couples displayed more negative behavior than normals—a finding consistent with previous results (Jacob et al., 1981). Of particular interest was that all couples displayed more positivity and more negativity in the drinking session—a finding that was more salient with alcoholic couples, who were more negative than depressed or normal couples when drinking, but who were similar to both control groups when not drinking. Jacob and Krahn acknowledge that drinking per se, as well as the presence of alcohol (by virtue of association with disruptive, intoxicated behavior), may act as discriminative stimuli for negative interactions. However, these authors note that higher negativity in the alcoholic group is consistent with Gorad's "responsibility-avoiding" communication theory, in which drinking allows for the expression of unaccountable, and therefore uninhibited, negative behavior and affect (Gorad, 1971; Gorad et al., 1971).

In an exploration of the episodic-steady dimension of alcoholism, other studies of alcoholic couples by the Jacob group (Dunn, Jacob, Hummon, & Seilhamer, 1987; Jacob, Dunn, & Leonard, 1983) revealed important distinctions that were associated with husbands' episodic or steady drinking patterns. Using these findings, Jacob and Leonard (1988) reanalyzed the marital interactions of the alcoholic group with an expanded sample (n = 49). Overall, steady and episodic drinkers displayed significantly different patterns of marital interaction, particularly when drinking. In the nondrinking condition, there was relatively little difference in rates of negativity and problem solving. However, in the drinking condition there was (1) more expressed negativity, except for spouses of episodic drinkers, who showed a decrease in this dimension, and (2) a difference in problem solving, with tendencies of steady drinkers and their wives to show more effectiveness and for episodic drinkers and their wives to decrease problem-solving behaviors.

Jacob and Leonard suggested that the marital interactions of the steady drinkers are consistent with the "adaptive consequences" model of Steinglass in that alcohol consumption facilitated their problem solving, a function vital to the preservation of the family system. In contrast, the marital interactions of episodic drinkers appear to represent a "coercive control" mechanism by which the alcoholic avoids dealing with conflicting issues by manifesting high levels of hostile behaviors while drinking. The authors elaborate further on how the steady-episodic drinking dichotomy dovetails with other typologies suggested in the extant literature on alcoholic subtypes.

In summary, the recent reports of Jacob and his colleagues have demonstrated that the family interactions of alcoholics—even within an intact, selective group of families—can be differentiated from those of both normal families and other distressed families. Moreover, unique effects of alcohol consumption on family interactions have been demonstrated for both marital and parent-child subgroups (Krahn & Jacob, in preparation). Additionally, differentiation of alcoholics according to patterns of consumption has been suggested as an important issue for further exploration.

Future Directions

The present understanding of families of alcoholics has been greatly enhanced by more sophisticated theories and more precise methods that have been developed over the past 10–15 years. Nevertheless, family interaction studies of alcoholics are still at an early stage of development. Beyond the need for more adequate experimental designs, the generation and empirical testing of conceptual models are of major importance to the future credibility of the field. Given the scant number of studies that assess the acute effects of alcohol on family interaction, the theories proposed to date can be considered little more than preliminary hypotheses. Although the recent reports of our research group have provided support for Steinglass's alcohol maintenance model, closer examination of alcoholic subgroups has revealed the necessity for greater refinement of existing paradigms.

In addition, rather than rely on a collection of "one-shot" studies, research in this domain must be grounded in more systematic efforts. Programmatic investigations will provide a broad and sure foundation on which to structure future efforts and will allow for more clarity of purpose and sense of direction. Moreover, multidimensional research probes are important for substantiating findings, not only across methods (e.g., questionnaires, interviews, and direct observations) but also across time frames. Detailed assessments of interaction processes over relatively brief periods as well as macroscopic views (such as drinking stages and individual and family developmental stages) must be considered. Additionally, other major determinants, such as extrafamilial experiences and genetic and environmental influences, may be crucial issues for various research designs.

Another unexplored area in interaction studies involves the families of female alcoholics. The literature strongly suggests that family processes related to the transmission and perpetuation of alcoholism are likely to be markedly different for female versus male alcoholics. Moreover, the impact on the family is likely to be much different in light of the following considerations: (1) the drinking patterns of the alcoholic wife-mother are more closely tied to family dynamics, crises, and developmental transitions than those of the alcoholic husband-father (Beckman, 1975; Beckman, 1976; Blume, 1982; Gomberg, 1981; Knupfer, 1982; Schuckit & Duby, 1983; Williams & Klerman, 1983); the wife's drinking pattern is more influenced by the spouse's drinking pattern, and he is more likely to be an abuser also (Schuckit & Morrisey, 1976; Wilsnack, Wilsnack, & Klassen, 1982); (3) wives and husbands respond differently to their spouse's drinking (Wiseman, 1980a; Wiseman, 1981); (4) given that mothers are primary caretakers, the impact on children is likely to be different; and (5) inheritance patterns may be different in maternal versus paternal alcoholism (Bohman, Sigvardsson, & Cloninger, 1981; Cloninger, Bohman, & Sigvardsson, 1981).

Another major area in need of clarification involves the assessment of offspring of alcoholics. That children of alcoholics are at high risk for subsequent alcohol abuse has been extensively discussed and documented in the extant literature (Jacob et al., 1978; Wilson, 1982). At the same time there remains considerable uncertainty about the influence of various forms and degrees of alcohol abuse, the relative influence of and interaction between family genetic and family environmental effects, the developmental patterns that characterize the emergence and expression of these outcomes, and the nature of familial and nonfamilial variables that serve to protect a sizable portion of these high-risk children from such adverse outcomes. In light of these considerations, it seems necessary for investigators to consider longitudinal studies of these high-risk offspring in order to provide further clarification of these issues.

In summary, there is a vigorous and encouraging movement in the area of family interaction studies of alcoholic families. Future directions encompass the formulation and testing of more comprehensive theories, the application of more exacting methods, and the investigation of relatively untapped domains related to female alcoholism and offspring outcome. These efforts hold much promise for a broader understanding of the myriad ways in which alcoholism and family processes are interwoven.

References

Adler, R., & Raphael, B. (1983). Children of alcoholics. *Australian and New Zealand Journal of Psychiatry, 17,* 3–8.

Becker, J.V., & Miller, P.M. (1976). Verbal and nonverbal marital behavior interaction patterns of alcoholics and nonalcoholics. *Journal of Studies on Alcohol, 37,* 1616–1624.

Beckman, L. (1975). Women alcoholics: A review of social and psychological studies. *Journal of Studies on Alcohol, 36,* 797–824.

Beckman, L.J. (1976). Alcoholism problems and women: An overview. In M. Greenblatt & M.A. Schuckit (Eds.), *Alcohol problems in women and children* (pp. 65–96). New York: Grune and Stratton.

Billings, A., Kessler, M., Gomberg, C., & Weiner, S. (1979). Marital conflict: Resolution of alcoholic and nonalcoholic couples during sobriety and experimental drinking. *Journal of Studies on Alcohol, 3,* 183–195.

Blume, S.B. (1982). Alcohol problems in women. *New York State Journal of Medicine, 82,* 1222–1224.

Bohman, M., Sigvardsson, S., & Cloninger, R. (1981). Maternal inheritance of alcohol abuse: Cross-fostering analysis of adopted women. *Archives of General Psychiatry, 38,* 965–969.

Cobb, J.C., & McCourt, W.F. (1979). *Problem solving by alcoholics and their families: A laboratory study.* Paper presented at American Psychological Association annual meeting.

Cloninger, R., Bohman, M., & Sigvardsson, S. (1981). Inheritance of alcohol abuse: Cross-fostering analysis of adopted men. *Archives of General Psychiatry, 38,* 861–868.

Davis, D.I. (1976). Changing perception of self and spouse from sober to intoxicated state: Implications for research into family factors that maintain alcohol abuse. *Annuals of the New York Academy of Science, 273,* 497–506.

Drewery, J., & Rae, J.B. (1969). A group comparison of alcoholic and nonalcoholic marriages using the Interpersonal Perception Technique. *British Journal of Psychiatry, 115,* 287–300.

Dunn, N.J., Jacob, T., Hummon, N., & Seilhamer, R.A. (1987). Marital stability in alcoholic-spouse relationships as a function of drinking pattern and location. *Journal of Abnormal Psychology, 96,* 99–107.

el-Guebaly, N., & Offord, D.R. (1977). The offspring of alcoholics: A critical review. *American Journal of Psychiatry, 134,* 357–365.

Frankenstein, W., Hay, W.M., & Nathan, P.E. (1985). Effects of intoxication on alcoholics' marital communication and problem solving. *Journal of Studies on Alcohol, 46,* 1–6.

Futterman, S. (1953). Personality trends in wives of alcoholics. *Journal of Psychiatric Social Work, 23,* 37–41.

Gomberg, E. S. (1981). Women, sex roles, and alcohol problems. *Professional Psychology, 12,* (1), 146–155.

Gorad, S. (1971). Communicational styles and interaction of alcoholics and their wives. *Family Process, 10,* 475–489.

Gorad, S., McCourt, W., & Cobb, J. (1971). A communications approach to alcoholism. *Quarterly Journal of Studies on Alcohol, 32,* 651–668.

Hanson, P.G., Sands, P.M., & Sheldon, R.B. (1968). Patterns of communication in alcoholic married couples. *Psychiatric Quarterly, 42,* 538–547.

Hersen, M., Miller, P., & Eisler, R. (1973). Interaction between alcoholics and their wives: A descriptive analysis of verbal and nonverbal behavior. *Quarterly Journal of Studies on Alcohol, 34,* 516–520.

Jackson, J.K. (1954). The adjustment of the family to the crisis of alcoholism. *Quarterly Journal of Studies on Alcohol, 15,* 562–586.

Jackson, J.K. (1962). Alcoholism in the family. In D.J. Pittman & C.R. Snyder (Eds.), *Society, culture, and drinking patterns* (pp. 472–492). New York: Wiley.

Jacob, T. (1974). Patterns of family dominance and conflict as a function of child age and social class. *Developmental Psychology, 10,* 1–2.

Jacob, T. (1975). Family interaction in disturbed and normal families: A methodological and substantive review. *Psychological Bulletin, 82,* 33–65.

Jacob, T., Dunn, N.J., & Leonard, K. (1983). Patterns of alcohol abuse and family stability. *Alcohol Clinical and Experimental Research, 7,* 382–385.

Jacob, T., Favorini, A., Meisel, S., & Anderson, C. (1978). The spouse, children, and family interactions of the alcoholic: Substantive findings and methodological issues. *Journal of Studies on Alcohol, 39,* 1231–1251.

Jacob, T., & Krahn, G.L. (1988). Marital interactions of alcoholic couples: Comparison with depressed and nondistressed couples. *Journal of Consulting and Clinical Psychology, 56,* 73–79.

Jacob, T., & Leonard, K. (1988). Alcoholic-spouse interaction as a function of alcoholism subtype and alcohol consumption. *Journal of Abnormal Psychology, 97,* 231–237.

Jacob, T., Ritchey, D., Cvitkovic, J., & Blane, H. (1981). Communication styles of alcoholic and nonalcoholic families when drinking and not drinking. *Journal of Studies on Alcohol, 43,* 466–482.

Jacob, T., & Seilhamer, R.A. (1982). The impact on spouses and how they cope. In J. Orford & J. Harwin (Eds.), *Alcohol and the family* (pp. 114–126). London: Crown Helm.

Jacob, T., & Seilhamer, R.A. (1987). Alcoholism and family interaction. In T. Jacob (Ed.), *Family interaction and psychopathology: Theories, methods, and findings* (pp. 535–580). New York: Plenum Press.

Jacob, T., Seilhamer, R.A., & Rushe, R. (1989). Alcoholism and family interaction: An experimental paradigm. *The American Journal of Drug and Alcohol Abuse, 15* (1), 73–91.

James, J.E., & Goldman, M. (1971). Behavior trends of wives of alcoholics. *Quarterly Journal of Studies on Alcohol, 32,* 373–381.

Kalashian, M. (1959). Working with wives of alcoholics in an outpatient clinic setting. *Marriage and Family, 21,* 130–133.

Kennedy, D.L. (1976). Behavior of alcoholics and spouses in a simulation game situation. *Journal of Nervous and Mental Disorders, 162,* 23–24.

Knupfer, G. (1982). Problems associated with drunkenness in women: Some research issues. In *Special population issues (NIAAA* Alcohol and Health Monograph No. 4, pp. 3–39). (*DHHS* Publication No. ADM 82–1193). Washington, DC: U.S. Government Printing Office.

Krahn, G.L., & Jacob, T. (in preparation). *Parent-child interactions in families with alcoholic fathers.*

Lewis, M. (1937). Alcoholism and family casework. *Family, 18,* 39–44.

Mitchell, H.E. (1959). The interrelatedness of alcoholism and marital conflict. *American Journal of Orthopsychiatry, 29,* 547–559.

O'Farrell, T.J., & Birchler, G.R. (1985). *Marital relationships of alcoholic, conflicted, and nonconflicted couples.* Paper presented at the American Psychological Association annual meeting, Los Angeles.

Orford, J., Guthrie, S., Nicholls, P., Oppenheimer, E., Egert, S., & Hensman, C. (1975). Self-reported coping behavior of wives of alcoholics and its association with drinking outcome. *Journal of Studies on Alcohol, 9,* 1254–1267.

Russell, M., Henderson, C., & Blume, S.B. (1984). *Children of alcoholics: A review of the literature.* New York: Children of Alcoholics Foundation.

Schaffer, J.B., & Tyler, J.D. (1979). Degree of sobriety in male alcoholics and coping styles used by their wives. *British Journal of Psychiatry, 135,* 431–437.

Schuckit, M.A., & Duby, J. (1983). Alcoholism in women. In B. Kissin & H. Begleiter (Eds.), *The biology of alcoholism: The pathogenesis of alcoholism: Psychological factors* (pp. 215–237). New York: Plenum Press.

Schuckit, M.A., & Morrisey, R.R. (1976). Alcoholism in women: Some clinical and social perspectives with an emphasis on possible subtypes. In M. Greenblatt & M.A. Schuckit (Eds.), *Alcoholism problems in women and children* (pp. 5–35). New York: Grune and Stratton.

Steinglass, P. (1975). The simulated drinking gang: An experimental model for the study of a systems approach to alcoholism: I. Description of the model. II. Findings and implications. *Journal of Nervous and Mental Disorders, 161,* 101–122.

Steinglass, P. (1979). The Home Observation Assessment Method (HOAM): Real-time observations of families in their homes. *Family Process, 18,* 337–354.

Steinglass, P. (1980). A life history model of the alcoholic family. *Family Process, 19,* 211–226.

Steinglass, P. (1981). The alcoholic family at home: Patterns of interaction in dry, wet, and transitional stages of alcoholism. *Archives of General Psychiatry, 38,* 578–584.

Steinglass, P., Davis, D., & Berenson, S. (1977). Observations of conjointly hospitalized "alcoholic couples" during sobriety and intoxication: Implications for theory and therapy. *Family Process, 16,* 1–16.

Steinglass, P., Weiner, S., & Mendelson, J.H. (1971). A systems approach to alcoholism: A model and its clinical application. *Archives of General Psychiatry, 24,* 401–408.

Tamerin, J.S., Tolar, A., DeWolfe, J., Packer, L., & Neuman, C.P. (1974). Spouses' perception of their alcoholic partners: A retrospective view of alcoholics by themselves and their spouses. *Proceedings of the Third Annual Alcoholism Conference of the National Institute on Alcohol Abuse and Alcoholism* (pp. 33–49). Washington, DC.

Weiss, R.L. (1976). *Marital Coding System (MICS): Training and reference manual for coders.* Unpublished manuscript, University of Oregon Marital Studies Program, Eugene.

Weiss, R.L., Hops. H., & Patterson, G.R. (1973). A framework for conceptualizing marital conflict: A technology for altering it, some data for evaluating it. In R. W. Clark & L. Hamerlynck (Eds.), *Critical issues in research and practice: Proceedings of the fourth BANIF international conference.* Champaign, IL: Research Press.

Williams, C.N., & Klerman, L.V. (1983). Female alcohol abuse: Its effects on the family. In S. Wilsnack & L.J. Beckman (Eds.), *Alcohol problems in women* (pp. 280–312). New York: Guilford Press.

Wilsnack, R.W., Wilsnack, S.C., & Klassen, A.D., Jr. (1982). *Women's drinking and drinking problems: Patterns from a 1981 survey.* Paper presented at the annual meeting of the Society for the Study of Social Problems, San Francisco.

Wilson, C. (1982). The impact of children. In J. Orford & J. Harwin (Eds.), *Alcohol and the family* (pp. 151–166). London: Crown Helm.

Wiseman, J.P. (1980a). Discussion summary. Presented at NIAAA Workshop, Jekyll Island, Georgia, April 2–5, 1978. In *Alcohol and women* (NIAAA Research Monograph No. 1, pp. 107–114). (DHEW Publication No. ADM 80–835). Washington, DC: U.S. Government Printing Office.

Wiseman, J.P. (1980b). The "home treatment": The first steps in trying to cope with an alcoholic husband. *Family Relations, 29,* 541–549.

Wiseman, J. (1981). Sober comportment: Patterns and perspectives on alcohol addiction. *Journal of Studies on Alcohol, 42,* 106–126.

CHAPTER 32

Job Behaviors and the Denial Syndrome

HARRISON M. TRICE AND WILLIAM J. SONNENSTUHL

Since the 1960s, the idea that drinking alcohol adversely affects job behavior has become accepted as gospel by employers and workers alike. Consequently, the job performance standard has become an integral part of today's employee (or member) assistance programs for identifying problem drinkers and motivating them to change their behavior. This chapter: (1) reviews the existing evidence on drinking and job behaviors, (2) examines how the job performance standard is used to identify and motivate problem drinkers to change their behavior, and (3) analyzes how the denial syndrome expresses itself in work environments and how the performance standard counteracts it.

Job Performance

In most instances, job performance is clearly and adversely affected by alcohol abuse. Even in cases where alcoholism does not appear to affect performance, alcoholics work extra hard on some days in order to compensate for their poor performance on others (Trice, 1962). Although these findings do not vary by occupational status, amount of job freedom, or off-the-job drinking with fellow workers, the manner in which work performance declines does vary by job status. Professional, managerial, and other white-collar alcoholics are likely to go to work even when feeling incapable of doing a good job, and lower-status employees resort either to absenteeism or to further drinking off the job when they feel incapable of performing.

Impaired performance is not one of the earliest job indicators of alcoholism (Maxwell, 1960), but it consistently shows up as early-stage alcoholism develops into early-middle-stage and middle-stage alcoholism. Recovering alcoholics report as moderately early signs lower quantity of work, lower quality of work, "putting things off," "neglecting details usu-

Harrison M. Trice and William J. Sonnenstuhl are with the School of Industrial and Labor Relations, Cornell University, Ithaca, N.Y. This chapter was written especially for this book.

ally done," "less even, more sporadic work pace," and "making mistakes or errors of judgment." Out of 44 such possible signs these were ranked, respectively, 15, 18, 4, 10, 6, and 17 by Maxwell's respondents. Trice (1964) submitted Maxwell's 44 signs to a sample of supervisors who, in total, had contact with 72 employees who had been diagnosed as alcoholics by a company's medical department. He also submitted them to the alcoholic employees they had supervised. Both groups were asked to classify each sign as (1) noticed early and frequently thereafter, (2) noticed later but frequently thereafter, (3) noticed fairly early but infrequently thereafter, and (4) noticed late and infrequently thereafter.

The supervisors put "lower quality of work" in the "noticed early and frequently thereafter" category, and "lower quantity of work" and "less even, more sporadic work pace," in the "noticed later but frequently thereafter" category. Their problem-drinking subordinates, however, put these signs in the "noticed late and infrequently thereafter" category. The discrepancy in findings suggests that supervisors do recognize these signs relatively early in the development of an employee's alcoholism and, because they may not know how to react to them, delay and vacillate. Consequently, the alcoholic employee tends to believe that he or she has successfully covered up his or her developing alcoholism.

Subsequent studies (e.g., Asma, 1975; Edwards, 1975; Mannello & Seaman, 1979; Pearson, 1982) support Trice and Maxwell's findings. For instance, Warkov and Bacon (1965, p. 65) report that "supervisors consistently rate employees drawn at random as more acceptable in work performance, . . . than the group of problem drinkers." One interesting feature of this study was the finding of "differential tolerance," that is, the finding that managers were significantly more hesitant to link problem drinking to poor performance among higher-status workers, but readily did so among lower-status workers. However, when the linkage was made, higher-status problem drinkers were twice as likely to receive unsatisfactory performance ratings as lower-status workers.

Likewise, Trice and Beyer (1984) found that among a sample of 574 managers in a large corporation who had had a subordinate with a drinking problem, or with any of a number of other forms of behavioral disorders, their alcoholic employees (both male and female) typically ranked slightly above the worst employee they could recall. In some rare instances they estimated that a specific problem-drinking employee approached the best employee they could recall, but the drinker never was selected as the best. Also, in their study of the implementation of the alcoholism policy of the federal government they reported that "about 11% of the supervisors (in a sample of supervisors in 71 federal installations) reported an opportunity to use the alcoholism policy procedures with employees they supervise" (Beyer & Trice, 1978, p. 77). The most frequent triggering events for these instances of use were "poor work performance, excessive absenteeism and visible signs of drinking on the job—sometimes in combination."

Absenteeism

Absenteeism is closely associated with performance because in most cases one must be physically present in order to perform one's job. Maxwell (1960) found that "absenteeism, half-day or day" ranked 24th out of the 44 drinking signs to appear on the job. Thus, the 406 problem drinkers in his sample did not perceive absenteeism to be an early, on-the-job sign. In contrast, Trice's (1964) sample of managers who had experienced an alcoholic employee placed "absenteeism half or full day" in the earliest and most frequent category. In addition, 70% of the Alcoholics Anonymous (AA) respondents in Trice's (1962) study indicated that their absences from work increased as their drinking problems escalated. They estimated their increase ranged from ¼ to 1 day per week over what it had been before their alcoholism developed. Although developing problem drinkers may perceive their absences as being largely early-middle stage, the data suggest that relatively high amounts of absenteeism are, in contrast, noticed early on by immediate supervisors (Trice, 1964).

In addition, Trice's (1962) study of work histories of AA members found that relatively high-status problem drinkers had significantly less off-the-job absenteeism than those in lower-status occupations. His interview data revealed that a heightened sense of guilt and self-hate, along with a tendency to view physical presence at work as a symbol of self-control and responsibility, characterized higher-status respondents, causing them to have less "off-the-job" absenteeism, but more "on-the-job" absenteeism (Trice, 1959). In essence, for high-status problem drinkers—in contrast with those of lower status—going to work was a sign of "managing" their drinking problem. Professional, managerial, and other white-collar personnel tended to go to work even when intoxicated or hung over but accomplished practically nothing of substance once there.

Other studies confirm these differences in absenteeism rates. Pell and D'Alonzo (1970), Stamps (1965), and Trice (1965), for instance, found that the absenteeism frequency rates among unskilled workers were considerably higher than those of workers in other occupational groups. Similarly, Schramm, Mandell, & Archer (1978, p. 74) reported a research population of largely blue-collar workers that "evinced an excessive level of absenteeism," making it the chief factor in referral to treatment. He found absenteeism rates between 3.8 and 8.3 times greater than "normal experience." Likewise, Warkov and Bacon (1965, p. 68) state, "No case was found among lower-status employees where problem drinking occurred unaccompanied by absenteeism—high absence is the crucial index employed by supervisors in identifying problem drinkers among lower-status employees." Mannello and Seaman (1979) found that supervisors in railroad companies estimated that problem drinkers missed four times as much work as other employees, but they did not report on occupational differences in these rates.

Drinking before Work and on the Job, and Hangovers

Nineteen percent of the alcoholics in Maxwell's (1960) study reported that drinking during working hours was among the first 5 on-the-job signs. This signal ranked 25th among the possible 44 signs. Morning drinking before going to work and drinking at lunch were reported more frequently than actual on-the-job drinking (36% among the first 5 signs, with a rank of 13, and 24% among the first 5 signs, with a rank of 12, respectively). Of these 3 signs, the supervisors in Trice's (1964) study selected only "drinking at lunch time" as a clue; and it was noticed infrequently, but early on. In Trice's other study (1962) one of the major reasons reported by a substantial number for avoiding accidents was on-the-job drinking to "steady nerves" and calm hangover distress. On-the-job drinking is apparently not universal; however, there are enough instances to regard it as a distinct job behavior possibility.

Hangovers or withdrawal distress may impair job effectiveness more than actual drinking on the job, or before work, or at lunch time. Maxwell's (1960) respondents collectively gave hangovers a rank of 1 among the 44 on-the-job signs, with 66% reporting it among the first 5 signs; and Trice's (1964) supervisors noticed hangovers somewhat later than reported by alcoholics themselves, but frequently thereafter. In addition, in Maxwell's (1960) study the responses of alcoholics formed a sequence between hangover, drinking before going to work, and noontime drinking. Also, other prominent signs in his study suggested hangover symptoms: increased nervousness or jitters, hand tremors, more edginess or irritability, red or bleary eyes, putting things off, neglecting details, making mistakes or errors of judgment, and a less even, more spasmodic work pace.

Recent studies of the simulated flight performance of 10 airplane pilots tend to support these observations. Simulated flight under nonhangover conditions were contrasted with similar simulated flights 14 hours after the pilots had been drinking enough to attain a high (0.10%) blood alcohol concentration (resulting in a hangover). The pilots' performance under hangover conditions was worse in practically all measures of performance and significantly worse on four of the 12 indexes (Yesavage & Leirer, 1986). During the hangover flights in a similar study of experienced U.S. Navy pilots, the pilots performed significantly worse on a large majority of measures of performance (McConnell, 1987).

On-the-Job Accidents

Whereas Maxwell's (1960) respondents had ranked hangovers 1st, they placed lost-time accidents on the job at the lowest possible rank—44th—and minor accidents on the job as 43rd. Similarly, none of the supervisors in Trice's study (1964) selected any type of accident as a sign of alcoholism. Because the degree of exposure to health and safety risks was not estimated in these two studies, Trice (1965) asked personnel representatives in a large utility to estimate on a 4-point scale the extent of accident risk they believed to be present in the jobs of samples of troubled employees (alcoholics, neurotics, psychotics) compared with "normals." The alcoholics turned out to be in jobs that were significantly more exposed to accidents than any of the other categories, including "normals."

Despite this differential exposure, neither the alcoholics nor any other diagnosed group had on-the-job accident experiences that differed significantly from each other's or from those of "normals," although over a 5-year period the alcoholics were somewhat higher (9% for the alcoholics, 6% for the "normals," 4% for the neurotics, and 5% for the psychotics). This mildly heightened rate could easily be explained by a higher exposure rate. Similarly, Asma (1975) reported that during the 5-year period before entering an alcoholism program 402 problem drinkers experienced 57 accidents.

AA members report that when they were drinking and in poor physical condition they took special precautions to prevent accidents (Trice, 1962). They frequently resorted to absenteeism whenever they were more afraid of accidents than usual, were often transferred from dangerous jobs by supervisors who were concerned about them, and on occasion moderately used alcohol on the job to reduce hand tremors and mistakes. These findings vary by age (Miner & Brewer, 1976; Observer & Maxwell, 1959). Early-stage, younger problem drinkers have higher rates of on-the-job accidents (twice as many as nonalcoholic workers do), but older problem drinkers—age 40 or above—do not show these high rates (their rate was identical with that of nonalcoholic workers). Thus, it may be that age, not problem drinking, tends to explain the difference in accident rates. Although a well-designed study of on-the-job accidents among problem drinkers has yet to be done, the simple notion that they as a distinct group are automatically destined to have more accidents is questionable.

Alcoholics are also believed to have higher rates of off-the-job accidents than nonalcoholics have (Brenner, 1967; Observer & Maxwell, 1959). Despite this belief, Maxwell's (1960) sample ranked "lost time from accidents off-the-job" 41st out of 44 signs. Likewise, Asma (1975) found that her cases had 75 off-the-job accidents (disability lasted an average of 7 days) in a 5-year period, but, in contrast, 57 on-the-job accidents. These rates are not unusually high.

Job Turnover

Somewhat like accidents, the turnover rates of problem drinkers seem to be rather ordinary (e.g., Straus & Bacon, 1951). Trice (1962, p. 502) concluded that among his AA respondents the relevant data "tend to confirm... that many problem drinkers are more stable occupationally than is generally believed." Even among blue-collar populations, Archer (1977) reports that no pattern of frequent turnover was found in her Baltimore sample of male alcoholics. Schramm, Mandell, and Archer (1978, p. 84) interpreted the same data to show "little evidence of a pattern of downward mobility when viewed against the experience of non-drinking counterparts in the same work force." Smart (1974) found that workers whom supervisors identified as problem drinkers actually experienced longer job tenure than their non-problem-drinking counterparts in the same labor force did. In this case, there is evidence that problem drinkers have less turnover than their counterparts.

Cover-Up Experiences

Only Trice's (1962) study of work histories of AA members contains data on cover-up at work. Just over 40% of the AA members reported that they, themselves, were the primary source of cover-up; approximately 20% indicated little, if any, cover-up experiences, and 36% reported cover-up by their work associates. In this latter type, fellow workers reportedly were the agents of cover-up for 17%, the immediate supervisor for 12%, and subordinates for 7%. Of considerable interest were the correlations that emerged in this study between the amount of job freedom and off-the-job drinking experiences with fellow workers. Self cover-up concentrated in jobs with a maximum of freedom from imposed work schedules and close supervision. Apparently, self cover-up occurred largely in managerial and professional positions.

No cover-up, or work associate cover-up, in contrast, clearly tended to be associated with lower-status occupations and with openly becoming intoxicated with fellow workers in off-the-job drinking experiences. Thus, higher-status alcoholics participated in such happenings, but drank "carefully," engaging in isolated heavier drinking later. Lower-status alcoholics, however, drank openly with fellow workers off the job in a relatively uninhibited manner that offered little, if any, cover-up. As a result they were more visible, a factor that may well contribute to reported higher rates of problem drinking in lower-status occupational strata.

Alcohol Problems and Denial

Alcohol problems are characterized by denial (Gallanter, 1979), a "shorthand term for a wide repertoire of psychological defenses and maneuvers that alcoholic persons unwillingly set up to protect themselves from the realization that they do in fact have a drinking problem" (Anderson, 1981, p. 11). Denial may express itself as: (1) simple denial: maintaining that a problem does not exist when, in fact, it does; (2) minimizing: admitting a problem but doing so in a way that it seems less serious than it actually is; (3) blaming: shifting of blame from oneself to someone or something else; (4) rationalizing: providing reasons other than alcoholism for one's behavior; (5) intellectualizing: avoiding personal awareness by dealing with the problem intellectually; (6) diversion: changing the subject in order to avoid discussion of or attention to alcohol-related problems; and (7) hostility: using anger and hostility when reference is made to a person's problems with alcohol.

Although there is little doubt that alcoholics and problem drinkers deny their problems, there is considerable doubt and skepticism concerning the extent to which it is unique to alcoholics and intrapersonal in origin (Brissett, 1988). Rather, denial is a common, interactional phenomenon in everyday life (e.g., Hewitt & Stokes, 1975; Scott & Lyman, 1968; Sykes & Matza, 1957) that influences people's efforts to cope with threats both to their self-esteem and to acceptance in friendship groups. According to Homans (1987, p. 72), "Friendship with others is rewarding to many people and its loss is a punishment." People in general strive by denial to maintain a positive self-concept; consequently, it is no surprise that alcoholics will also resist the label of moral failure—probably much more so than nonalcoholics.

Alcoholics are inclined to disavow their addiction (Pennock & Poudrier, 1978) because, as Osagie (1988, p.23) writes: "Human beings relish a continuous high esteem, and the need to maintain a positive self-image often predominates over all other social considerations." For instance, Forchuk's (1984) study of patients in five alcoholism programs in Ontario, Canada, found that ratings about self-esteem among alcoholics rose correspondingly with the strength of their denial syndrome.

Consequently, the denial is a process between the developing alcoholic and the expectations and actions of people that surround her or him (Davis, 1961). Although the denial behaviors are expressed by discrete individuals, they have meaning primarily in an interactional context. That is, denials "are always negotiated in social interaction" (Brissett, 1988, p. 393). Denzin (1987), for example, provides a vivid description of how the alcoholic believes he or she has personal power and control of his world and sees alcohol as a source of strength for the performances of his tasks. The alcoholic is convinced that abstinence would lead to powerlessness—a premise that, in turn, justifies continued drinking.

Maxwell's (1960) study, cited earlier, included numerous signs of on-the-job behaviors directed at denying the existence of drinking problems. Two of these were ranked quite high: "sensitive to opinions about your drinking" and "indignant when your drinking was mentioned." These two items ranked 7 and 11 out of 44, respectively. Indignation clearly implies a fervent denial that a drinking problem should be even considered, much less assessed. Sensitivity to the opinion of others on the job about one's drinking is an emotional concomitant to indignation. If denial is to be made effective, the developing problem drinker must be on the alert to deny it in some fashion on any and all possible occasions when opinions about his or her drinking might be voiced. The emotional strategy apparently was to attack and undermine the development among workplace associates of the notion that there was any abuse of alcohol on his or her part.

In addition, Maxwell's respondents ranked "avoiding boss or associates" 9th and "more unusual excuses for absences" 22nd. These justifications are not mere everyday excuses, but "unusual" ones. Apparently, as problem drinking becomes more developed, indignation, sensitivity, and avoidance are joined by an active, explicit effort to deny the most obvious feature of job behavior for the majority—absenteeism. Since occupations and jobs play a prominent role in the lives of U.S. workers of both high and low status, these denials are of considerable importance; and in order to overcome them at work it is necessary to understand denial as an interactional process. The respondents also tended to add to their denial mechanisms at work "more tendency to blame other workers" and a tendency to be "more resentful of fellow workers," items they ranked 21 and 27, respectively. Finally, if these various denial tactics are successful, and they often are, they act to delay recognition of a problem and referral to treatment. Those who vigorously deny a problem and adamantly refuse to accept a program's help are apt to be screened out, whereas those who more readily admit a problem are more likely to be included in a program.

Off-the-job drinking behavior in the context of the presence of informal drinking groups and occupational drinking subcultures on the job can also serve to deny and camouflage drinking problems. In order to preserve their self-image, alcoholics tend to avoid drinking situations in which their behavior might be questioned and to gravitate to drinking groups in which their behavior will be accepted (Trice, 1966). It is instructive to view the denial syndrome among alcoholics from the vantage point of the dynamics of the drinking group relocation process. During the early stages of their drinking, they do not deny their behavior, because the vulnerable drinker gets considerable recognition and emotional satisfaction in the drinking context from various "enablers." As this reward turns to rejection, however, the drinker finds it increasingly necessary to resort to denial mechanisms learned previously in everyday life. Because rejection generates defenses to salvage self-esteem, the rejection experience makes for considerable exercise in denial mechanisms. Moreover, according to Connor (1962), alcoholism-prone drinkers tend to manifest sentiments of longing for the warmth and acceptance of primary-type group life. In essence, there is a strong yearning for the comfort and protection of an uncritical and loving set of intimates. Such feelings press for adequate denials through which friendships can be retained and rejection countered.

Lastly, one of the alcoholic's most common proclivities—seeking out other drinking contexts that positively enforce his or her drinking style (Denzin, 1987, p. 123–124)—also reinforces the use of those denial mechanisms already in place and the inclusion of new ones in the syndrome. These observations are consistent with Vaillant's (1983, p. 173) observations that the degree to which alcohol abuse is tolerated or disapproved in the alcoholic's social environment influences his or her choice of accepting or rejecting his or her addiction; that is, the strength of the denial. Relocation into new and tolerant drinking contexts probably reinforces denials, giving them an emotional support network that reinforces and sustains them.

Job Performance and the Denial Syndrome

All societies attempt to control the use of alcohol (McAndrews & Edgerton, 1969; Trice & Pittman, 1958). In the United States, at one extreme, alcohol has been tightly controlled and sanctioned as among the 17th century New England Puritans (Lender & Martin, 1982); and at the other extreme, its use is characterized by a deep-seated collective ambivalence, which approaches anarchy, about what constitutes appropriate drinking (Gusfield, 1963). In the highly pluralistic United States there are few agreed-upon standards applied to that subject (Akers, 1977). In this competitive environment, the pragmatic definition of appropriate drinking employed in many workplaces—drinking that does not disrupt job performance—provides an anchor for defining what is appropriate and inappropriate (Trice & Sonnenstuhl, 1990).

Both management and labor are aware of drinking's effects upon job behavior and have used these insights to develop the job-based strategy of constructive confrontation for helping problem drinkers. Because job performance is the standard of industrial jurisprudence (Sonnenstuhl, 1989), the emergence of impaired performance for most problem drinkers provides an explicit, legitimate, and realistic opportunity for management to confront the problem drinkers and to undermine denial. The inescapable reality present in the tangible and objective fact of impaired performance provides a formidable challenge for the most practiced denial specialist. The legitimacy of such a challenge to the denial syndrome is obvious. Unlike other social groups, work organizations review their own performance regularly and have the legal right to expect unimpaired performance as a basic role obligation. In this regard workplace interventions enjoy a strategic advantage over most other interventions for problem drinkers. In essence, they have a legitimate and relatively powerful reason for challenging the denial syndrome.

Historically, Employee Assistance Programs (EAP) have used deteriorating job performance as the starting point and justification for intervening with problem drinkers. In the early 1970s the Occupational Program Branch of the National Institute on Alcohol and Alcohol Abuse (NIAAA) encouraged the development of job-based programs emphasizing the identification of problem drinkers through job performance indicators rather than symptoms of alcoholism (Steele, 1989). This emphasis remains prominent today despite the weakening effect produced by the enormous diversity of current EAPs.

The constructive confrontation, practiced in many workplaces since the 1960s (Dancey, 1973; Roman, 1981), prescribes a combination of progressive, positive discipline (Beyer & Trice, 1984; Miller & Oliver, 1988) with provisions designed to help problem-drinking employees to rehabilitate themselves. Sonnenstuhl and Trice (1986) describe the strategy as follows:

> For constructive confrontation to be effective, a supervisor may need to hold a number of discussions with an employee whose performance is unacceptable. In the confrontational part of the initial discussion, the employee is given the specifics of unacceptable work performance and warned that continued unacceptable performance is likely to lead to formal discipline. In the constructive part, supervisors remind employees that practical assistance is available through the EAP. Subsequent steps in the process depend on the response of the employee. If performance improves, nothing happens; if unacceptable performance continues, several more informal discussions may follow. (p. 25)

Several studies examine the job performance of alcoholics who were referred to treatment facilities by supervisory use of a constructive confrontation strategy (see, for example, Franco, 1960; Hilker, Asma, & Eggert, 1972). These studies use a variety of criteria to measure job performance that include indices of job maintenance, supervisory ratings, absenteeism, use of sick benefits, and visits to medical departments. On the average, these studies also report that 70% of the employees involved returned to acceptable performance. For example, Heyman (1978) interviewed alcoholic employees who had experienced constructive confrontation. Her data show a significant relationship between constructive confrontation and improved work performance. Griffith Edwards and his associates (1977) reached a somewhat similar conclusion in their studies of alcoholism treatment. They compared a group that received "advice" with one that received intensive alcoholism treatment. The method used in advising the first group was remarkably similar to the constructive confrontation strategy of job-based programs. The authors concluded that "advice," with its minimal intervention, had been as effective as a more costly, elaborate, and intense treatment program.

Likewise, Trice and Beyer (1984), using a multiple regression format, contrasted job performance before and after constructive confrontation and concluded that the policy and program produced impressive success rates—80% of the problem-drinking employees improved in general conduct, 74% in work performance. A second conclusion was that the use of the constructive confrontation strategy was closely related with improvements in performance but that the use of formal, punitive discipline was counterproductive. Finally, they concluded that "it was not just the actions taken with problem drinking employees that produced desired outcomes, but also the legitimacy and predictability accorded these actions by the presence of a formal policy" (Trice & Beyer, 1984, p. 404).

Supervisory Use of Constructive Confrontation

Although the data suggest that constructive confrontation is an effective strategy when implemented, some observers question supervisors' willingness to use it (e.g., Clyne, 1965; Googins & Kurtz, 1980; Shain & Groeneveld, 1980; Shain, Suurvali & Boutilier, 1986). For instance, Googins and Kurtz (1980, p. 1201) state: "Performance criteria as applied in most work settings [may not be] sensitive enough to allow supervisors to identify a drinking problem at an early stage." Available data highlight the idea that performance indicators emerge in the presumed early-middle or the indefinite middle phases of alcoholism but rarely in the very early period. As we have noted, job performance is not an early symptom of alcoholism. The best evidence suggests that the early stages are personally so rewarding and satisfying that few workplace significant others, or developing alcoholics themselves, notice more than occasional impairments.

Trice and Roman (1978) have designated this the "disrupted but normal" stage for supervisors.

On the other hand, performance indicators are unlikely to occur during the advanced late stages, because these are dominated by such severe symptoms as convulsions, uncontrollable tremors, and hiding of supply. "Only the earliest and most advanced stages of the [alcoholism] process" were reliably predicated by a study of the sequencing of symptoms (Trice & Wahl, 1958, p. 648).

Additionally, assumptions that early-stage identification leads to more effective rehabilitation have been sharply questioned in research data. For instance, Moberg (1976) found, in a study of employed alcoholics, that early-phase cases responded significantly less well to treatment than late-phase cases. Similarly, Schramm, Mandell, and Archer (1978) reported that those alcoholics who remained longer in treatment, and responded better, were those who had had a longer period of heavy drinking (see also Jutcovich & Calderone, 1984). These findings are consistent with earlier findings about affiliation with Alcoholics Anonymous (Trice & Wahl, 1958).

Overall, there is evidence to suggest that job impairment, rather than coinciding with "later phases," occurs first in the earlier periods of the disorder as the problem drinker begins to lose control of drinking.

Likewise, there is little evidence to suggest that alcoholics decide to seek help without being confronted by family members, friends, co-workers, or supervisors (Foote & Erfurt, 1981; Hyman, 1978; Schramm & DeFillippi, 1975; Smart, 1974; Sonnenstuhl, 1986). Rather, the concept of "self-referral" appears to be a polite fiction, designed to let those alcoholics who seek help to think better of themselves—they did it on their own—and to facilitate the provision of treatment (e.g., Sonnenstuhl, 1982; Sonnenstuhl, Staudermeier, & Trice, 1988).

Research also demonstrates that supervisors who are familiar with the performance-based policy are likely to use it. Research by Beyer and Trice (1978) revealed a positive relationship between familiarity with such a policy and its actual use. Indeed, familiarity was the most effective predictor of supervisors' use—more influential than their general agreement with it or the assessed need for it or its perceived ease of administration. In a practical sense, then, both training time invested by the organization in supervisory training and the number of training topics covered are of paramount importance in generating familiarity and actual use; that is, the more of each, the most likely supervisors were to use the policy. Also, a prominent part of familiarity involved an accurate knowledge of the supervisor's role in the use of the policy. In essence, supervisory familiarity played a large role in actual policy and program use.

In addition, considerable evidence exists regarding the possible fears and reluctance of supervisors to use a performance-oriented policy. To what extent do managers expect to experience trouble if they use a policy, or do they expect to please their management and achieve a favorable outcome? Data on these questions were collected from a stratified random sample of 19 different locations of a large corporation that since the 1960s had had a performance-based EAP geared toward alcoholism. The study focused upon two types of troubled employees: alcoholics and other problem employees. Fifty percent of the managers of alcoholic employees felt that it was very unlikely they would have trouble, 22% thought it unlikely they would have trouble, 14% were ambivalent about whether they would have trouble, and 7% felt that they were likely to have trouble. Among managers of other types of troubled employees the percentage distribution was practically identical. In sum, some supervisors in this study probably were and still are troubled by the nature of such a policy (Sonnenstuhl & Trice, 1989).

Supervisory vacillations may well go through stages that tend to lead to a resolution in favor of taking action in the so-called middle period. Trice and Roman (1978) describe four such stages as follows. In the disrupted but normal stage, the job performance of employees is intermittently disrupted, but the disruptions are neither frequent enough nor serious enough to indicate abnormality. In the blocked-awareness stage, supervisors and co-workers link the employees' behavior with alcohol use, but the recognition of a "drinking problem" strikes numerous barriers (i.e., low visibility of some workers, the status level of workers, the tolerance and encouragement shown by co-workers and even supervisors).

In the see-saw stage, employees' unsatisfactory performance increases, but supervisors are indecisive about whether to define the behavior as a deviant-drinking problem or to consider it within the "normal" range of drinking behavior. In the decision-to-recognize stage, supervisors, because of accumulated evidence on job performance, recognize that employees' behavior is abnormal and that referral to treatment is appropriate. Supervisors progress through these four stages in reacting to other troubled employees as well (Trice & Roman, 1978).

There is also considerable evidence that superiors may be anxious about union reactions to a performance-oriented policy; this concern, however, may produce program use, not avoidance. Data from the study of the federal alcoholism policy described earlier showed that supervisory awareness of a federal union's position on the policy, whether supportive or opposed, was significantly associated with actual supervisory use (Beyer, Trice, & Hunt, 1980). The mere presence of a union was associated with a larger proportion of the supervisors having used the performance-oriented policy.

Finally, the make-up, structure, and environment of the work organization indeed seems to hinder or encourage supervisory use. Despite supervisory attitudes toward the policy, the size of a given installation in the federal study was positively associated with use (Beyer & Trice, 1978). Organizational centralization, however, had negative effects on supervisory use and policy implementation in general. In contrast, the amount of decentralization in an installation was positively associated with supervisory use. Similarly, installations that have more formalization in their structure were less likely to contain supervisors who used the policy.

Peer Use of Job Performance and Constructive Confrontation

Although the majority of research on the constructive confrontation strategy has focused on supervisors, there is a growing body of literature on its use by peers (Denzin, 1989; Johnson, 1981; McCrody, 1989; Molloy, 1989; Perlis, 1980; Reichman & Beidel, 1989; Weed, 1980). For instance, Sonnenstuhl and Trice (1986) conducted an ethnographic study of workers (affectionately known as "sandhogs") in the Tunnel and Construction Workers' Member Assistance Program (MAP) in order to describe and understand the dynamics of peer referral in a craft union of miners characterized by a heavy-drinking culture. The program consisted of two alcoholism counselors and a peer-referral network consisting of approximately 125 members of Alcoholics Anonymous. Network members identified problem drinkers on the basis of gossip about, and direct observations of, drinking behavior, job performance, and family difficulties. Network members confronted the problem drinkers, emphasizing that if they did not stop drinking and join the network they would ultimately either die or lose their union books. After repeated confrontations, some problem drinkers returned to "normal drinking"; some recognized their alcoholism and joined the network; some died. Sonnenstuhl and Trice concluded that although there are significant differences between the way joint management–labor EAPs and MAPs operate both use job performance to identify problem drinkers and employ a combination of constructive and confrontative elements to break through their denial and motivate them to seek help.

Similarly, the Air Line Pilots Association (ALPA) has established a peer referral process for rehabilitating alcoholic members (Weed, 1980). Fellow pilots confront problem drinkers about job performance and urge them to seek help from the program. Sometimes confrontations are carried out by the government agencies in formal hearings on specific incidents and behaviors deemed in possible violation of accepted safety procedures. This is a very powerful form of confrontation, because both employer and ALPA generally support the government position.

Likewise, professional associations, building upon their authority to prescribe appropriate codes of behavior and discipline recalcitrant members, are establishing programs (Bissell & Haberman, 1984). For instance, in 1973 the American Medical Association's (AMA) Council on Mental Health published a report recognizing a colleague's impairment because of physical or mental illness, including alcoholism or drug dependencies. In an effort to implement the policy, the AMA's legislative department designed a model law expressing the profession's policy in action:

> First enacted in Florida, various sick physician statutes, many based on the AMA model, had by 1980 been enacted by nearly 40 states. At present, all 50 states have in place, or at least are in the process of forming, some type of impaired physician committee. (Bissell & Haberman, 1984, p. 139)

In addition, some medical doctors have been working with state medical societies and local AMA chapters to develop programs (John, 1978).

The California Bar Association program applies constructive confrontation to its alcoholic members through peer pressure. Through its disciplinary committee and its procedures, the association suggests and offers treatment opportunities while insisting on conformity to ethical and competency standards.

Summary

On-the-job behaviors of problem drinkers, especially impaired performance, are potent mechanisms for breaking down denial, motivating them to seek help, and increasing their readiness to respond to treatment. Constructive confrontation, which builds upon the relationship between deteriorating performance and drinking, seems to be an effective strategy for reaching employed problem drinkers. Alternatives to this strategy, such as early referrals and self-referrals, however, appear to be unrealistic. Finally, programs of intervention based in occupational cultures, in contrast to those in hierarchial work organizations, appear also to be fruitful locations for the use of the strategy of constructive confrontation.

References

Akers, R.L. (1977). *Deviant behavior: A social learning approach.* Belmont, CA: Wadsworth.

Anderson, D.J. (1981). *Psychopathology of denial.* Center City, MN: Hazelden Professional Education Series.

Archer, J. (1977). Social stability, work force behavior, and job satisfaction of alcoholic and non-alcoholic blue-collar workers. In C.J. Schramm (Ed.), *Alcoholism and its treatment in industry.* Baltimore: The Johns Hopkins University Press.

Arnoff, M. (1980). A content analysis of formal company policy statements on employee alcoholism and employee assistance programs. Unpublished seminar report ILR-677, Cornell University, School of Industrial and Labor Relations, Ithaca, NY.

Asma, F.E. (1975). Long-term experience with rehabilitation of alcoholic employees. In R. Williams and G. Moffat (Eds.), *Occupational alcoholism programs.* Springfield, IL: Charles C. Thomas.

Bayer, G., & Gerstein, L. (1988). Supervisory attitudes toward impaired workers: A factor analytic study of the behavioral index of troubled employees (BITE). *Journal of Applied Behavioral Science, 24,* 413–422.

Beyer, J., & Trice, H.M. (1984). The best/worst technique for measuring work performance in organizational research. *Journal of Organizational Behavior and Statistics, 1,* 95–115.

Beyer, J., & Trice, H.M. (1978). *Implementing change: Alcoholism policies in work organizations.* New York: Free Press.

Beyer, J., & Trice, H. (1984). A field study of the use of perceived effects of discipline in controlling work performance. *Academy of Management Journal, 27,* 743–764.

Beyer, J., Trice, H.M., & Hunt, R.E. (1980). The impact of federal sector unions on supervisor's use of personnel policies. *Industrial and Labor Relations Review, 33,* 212–231.

Bissell, L., & Haberman, P. (1984). *Alcoholism and the professions.* New York: Oxford University Press.

Brenner, B. (1967). Alcoholism and fatal accidents. *Quarterly Journal of Studies on Alcohol, 28,* 517–527.

Brissett, D. (1988). Denial in alcoholism: A sociological interpretation. *Journal of Drug Issues, 18,* 385–402.

Clyne, R.M. (1965). Detection and rehabilitation of the problem drinker in industry. *Journal of Occupational Medicine, 7,* 265–268.

Cohen, A.K. (1959). The study of social disorganization and deviant behavior. In R.K. Merton, L. Broom, & L.S. Cottrell (Eds.), *Sociology today.* New York: Basic Books.

Connor, R. (1962). The self-concepts of alcoholics. In D.J. Pittman and C.R. Snyder (Eds.), *Society, culture, and drinking patterns* (pp. 455–467). New York: Wiley.

Dancey, T.E. (1973). The constructive coercion technique in alcoholism and drug dependency programs. In J.M. Scher (Ed.), *Drug abuse in industry: Growing corporate dilemma.* Springfield, IL: Charles C. Thomas.

Davis, F. (1961). Deviance disavowal. *Social Problems, 9,* 120–132.

Dentler, R.A., & Erikson, K.T. (1959). The function of deviance in groups. *Social Problems, 7,* 98–108.

Denzin, N.K. (1986). *Treating alcoholism.* Beverly Hills, CA: Sage Publications.

Denzin, N.K. (1987). *The alcoholic self.* Beverly Hills, CA: Sage Publications.

Denzin, N.K. (1989). The natural history of a university employee assistance program. *Journal of Drug Issues, 19,* 385–401.

Edwards, G. (1975). The alcoholic doctor: A case of neglect. *Lancet, 2,* 1297–98.

Edwards, G., Orford, J., Egert, S., Guthrie, S., Hawkins, A., Hensman, C., & Mitcheson, M. (1977). Alcoholism: A controlled trial of treatment and advice. *Journal of Studies on Alcohol, 38,* 1004–1029.

Etizioni, A. (1968). Social control: Organizational aspects. In D.L. Sills (Ed.), *International encyclopedia of the social sciences.* New York: Macmillan and Free Press.

Foote, A., & Erfurt, J.C. (1981). Effectiveness of comprehensive employee assistance programs at reaching alcoholics. *Journal of Drug Issues, 11,* 217–232.

Forchuk, C. (1984). A company program for problem drinkers: Ten year follow-up. *Journal of Occupational Medicine, 2,* 157–162.

Franco, S.C. (1960). A company program for problem drinking: Ten years' follow-up. *Journal of Occupational Medicine, 2,* 157–162

Gallanter, M. (1979). Religious conversation: An experimental mode for affecting alcoholic denial. *Currents in Alcoholism, 6,* 69–78.

Goffman, E. (1959). *The presentation of self in everyday life.* New York: Doubleday.

Googins, B., & Kurtz, N.R. (1980). Factors inhibiting supervisory referrals to occupational alcoholism intervention programs. *Journal of Studies on Alcohol, 41,* 1196–1208.

Gusfield, J. (1963). *Symbolic crusade: Status politics and the American temperance movement.* Urbana: University of Illinois Press.

Hewitt, J.P., & Stokes, R. (1975). Disclaimers. *American Sociological Review, 40,* 1–11.

Heyman, M. (1978). *Alcoholism programs in industry.* New Brunswick, NJ: Rutgers Center of Alcohol Studies.

Hilker, R., Asma, F.E., & Eggert, R. (1972). A company sponsored alcoholic rehabilitation program: Ten years evaluation. *Journal of Occupational Medicine, 14,* 769–771.

Homans, G.C. (1987). Behaviorism and after. In A. Giddens & J.A. Turner (Eds.), *Social theory today.* Stanford: Stanford University Press.

John, H. (1978). Alcohol and the impaired physician. *Alcohol Health and Research World,* Winter, 2–9.

Johnson, L. (1981). Union responses to alcoholism. *Journal of Drug Issues, 11,* 263–277.

Jutcovich, J., & Calderone, J. (1984). Evaluation of the treatment process for employee alcohol abusers: Structure and strategies for successful programming. *EAP Research, 1,* 48–62.

Kelley, M. (1977). Subjective performance evaluation and person-role conflict under conditions of uncertainty. *Academy of Management Review, 20,* 301–314.

Lender, M.E., & Martin, J.M. (1982). *Drinking in America.* New York: Free Press.

Mannello, T.A., & Seaman, F.J. (1979). Prevalence, costs, and handling of drinking problems on seven railroads. *Final Report.* Washington, DC: University Research Corporation.

Maxwell, M.A. (1960). Early identification of problem drinkers in industry. *Quarterly Journal of Studies on Alcohol, 21,* 655–678.

MacAndrew, C., & Edgerton, R.B. (1969). *Drunken comportment: A social explanation.* Chicago: Aldine.

McConnell, H. (1987). Hungover pilots miscue on maneuvers. *Journal of the Addiction Research Foundation, 16,* 3–4.

McCrody, B.S. (1989). The distressed or impaired professional: From retribution to rehabilitation. *Journal of Drug Issues, 19,* 337–349.

Miller, T., & Oliver, S. (1988). Alcohol and drugs in airline operations. In J. McKelvey (Ed.), *Cleared for takeoff: Airline labor relations since deregulation.* Ithaca, NY: ILR Press.

Miner, J.B., & Brewer, J.F. (1976). The management of ineffective performance. In M. Dunnette (Ed.), *Handbook of industrial and organizational psychology.* Chicago: Rand-McNally.

Moberg, P. (1976). Treatment outcomes for earlier-phase alcoholics. *Annals of the New York Academy of Sciences, 273,* 543–552.

Molloy, D.J. (1989). Peer intervention: An exploratory study. *Journal of Drug Issues, 19,* 319–336.

Observer (Pseud.) & Maxwell, M.A. (1959). A study in absenteeism, accidents, and sickness payments among problem drinkers in one industry. *Quarterly Journal of Studies on Alcohol, 20,* 301–312.

Osagie, A. (1988). *Labeling processes in alcoholism: Alcoholic denial and identity negotiation* (Working Paper No. 4). Ithaca, NY: Cornell University, Program on Alcoholism and Occupational Health. School of Industrial and Labor Relations.

Pearson, M.M. (1982). Psychiatric treatment of 250 physicians. *Psychiatric Annals, 251,* 743–746.

Pell, S., & D'Alonzo, C.A. (1970). Sickness absenteeism of alcoholics. *Journal of Occupational Medicine, 12,* 198–210.

Pennock, M., & Poudrier, L.M. (1978). Overcoming denial: Changing the self-concepts of drunken drivers. *Journal of Studies on Alcohol, 39,* 918–921.

Perlis, L. (1980). Labor and employee assistance programs. In R.H. Egdahl & D.C. Walsh (Eds.), *Mental wellness programs for employees.* New York: Springer-Verlag.

Reichman, W., & Beidel, B.E. (1989). Implementation of state police EAP. *Journal of Drug Issues, 19,* 369–383.

Roman, P. (1981). From employee alcoholism to employee assistance: Deemphases on prevention and alcohol problems in work-based programs. *Journal of Studies on Alcohol, 42,* 244–272.

Schramm, C.J., & DeFillippi, R.J. (1975). Characteristics of successful alcoholism treatment programs for American workers. *British Journal of Addiction, 70,* 271–275.

Schramm, C.J., Mandell, W., & Archer, J. (1978). *Workers who drink: Their treatment in an industrial setting.* Lexington, MA: Lexington Books.

Scott, M.B., & Lyman, S.M. (1968). Accounts. *American Sociological Review, 33,* 46–62.

Shain, M., & Groeneveld, J. (1980). *Employee assistance programs.* Lexington, MA: D.C. Heath.

Shain, M., Suurvali, H., & Boutilier, M. (1986). *Healthier workers: Health promotion and employee assistance programs.* Lexington, MA: Lexington Books; D.C. Heath.

Smart, R. (1974). Employed alcoholics treated voluntarily and under constructive coercion. *Quarterly Journal of Studies on Alcohol, 35,* 196–209.

Sonnenstuhl, W.J. (1982). Understanding EAP self-referral: Toward a social network approach. *Contemporary Drug Problems, 11,* 269–293.

Sonnenstuhl, W.J. (1986). *Inside an emotional health program: A field study of workplace assistance for troubled employees.* Ithaca, NY: ILR Press.

Sonnenstuhl, W.J., Staudenmeier, W., & Trice, H.M. (1988). Ideologies and referral categories in EAP research. *Journal of Applied Behavioral Science, 24,* 383–396.

Sonnenstuhl, W.J., & Trice, H.M. (1989). *The myth of the troubled supervisor.* Paper presented at the 59th annual meeting of the Eastern Sociological Society, Baltimore, MD.

Sonnenstuhl, W.J., & Trice, H.M. (1986). *Strategies for employee assistance programs: The crucial balance* (Key Issues Series No. 30). Ithaca, NY: Cornell University, New York State School of Industrial and Labor Relations.

Stamps, R. (1965). *Alcoholic employees and problem concealment.* Unpublished master's thesis, Washington State University.

Steele, P. (1989). A history of job-based alcoholism programs: 1955–1972. *Journal of Drug Issues, 19,* 511–532.

Straus, R., & Bacon, S. (1951). Alcoholism and social stability: A study of occupational integration in 2,023 male clinic patients. *Quarterly Journal of Studies on Alcohol, 12,* 231–260.

Sykes, G.M., & Matza, R. (1957). Techniques of neutralization. *American Sociological Review, 22,* 664–670.

Trice, H.M. (1959). *The problem drinker on the job* (Bulletin No. 40). Ithaca, NY: Cornell University, New York State School of Industrial and Labor Relations.

Trice, H.M. (1962). The job behavior of problem drinkers. In D.J. Pittman and C.R. Snyder (Eds.), *Society, culture, and drinking patterns* (pp. 493–510). New York: Wiley.

Trice, H.M. (1964). New light on identifying the alcoholic employee. *Personnel, 41,* 18–25.

Trice, H.M. (1965). Alcoholic employees: A comparison of psychotic, neurotic, and "normal" personnel. *Journal of Occupational Medicine, 7,* 94–99.

Trice, H.M. (1966). *Alcoholism in America.* New York: McGraw-Hill.

Trice, H.M. (1984). Alcoholism in America revisited. *Journal of Drug Issues, 14,* 109–123.

Trice, H.M., & Beyer, J.M. (1977). A sociological property of drugs: Acceptance of users of alcohol and other drugs among university undergraduates. *Journal of Studies on Alcohol, 38,* 58–74.

Trice, H.M., & Beyer, J.M. (1984). Work-related outcomes of the constructive-confrontation strategy in a job-based alcoholism program. *Journal of Studies on Alcohol, 45,* 393–405.

Trice, H.M., & Pittman, D.J. (1958). Social organization and alcoholism. *Social Problems, 5,* 294–307.

Trice, H.M., & Roman, P. (1978). *Spirits and demons at work: Alcohol and other drugs on the job* (ILR paperback, No. 11). Ithaca, NY: Cornell University, New York State School of Industrial and Labor Relations.

Trice, H.M., & Sonnenstuhl, W.J. (1990). On the construction of drinking norms in work organizations. *Journal of Studies on Alcohol, 51,* 201–220.

Trice, H.M., & Wahl, J.R. (1958). A rank order analysis of the symptoms of alcoholism. *Quarterly Journal of Studies on Alcohol, 19,* 636–648.

Vaillant, G.E. (1983). *The natural history of alcoholism.* Cambridge: Harvard University Press.

Warkov, S., Bacon, S., & Hawkins, A.C. (1965). Social correlates of industrial problem drinking. *Quarterly Journal of Studies on Alcohol, 26,* 58–71.

Weed, E.D. (1980). Pilots who drink: FAA regulations and policy, an airline pilots' association treatment program. *Journal of Air Law and Commerce, 45,* 1089–1114.

Weick, K. (1969). *The social psychology of organizing.* Reading, MA: Addison-Wesley.

Wrich, J.T. (1980). *The employee assistance program.* Minneapolis, MN: Hazelden Foundation.

Yesavage, J.A., & Leirer, V.O. (1986). Hangover effects on aircraft pilots 14 hours after alcohol ingestation: A preliminary report. *American Journal of Psychiatry, 143,* 1546–1550.

CHAPTER 33

Drinking and Violations of the Criminal Law

JAMES J. COLLINS

The use of alcohol is involved in violations of the criminal law in two major ways: when laws regulating its use or distribution are violated and when the effects of drinking generate behavior that violates the law. Most governments regulate alcohol use and distribution. In the United States persons under age 21 may not purchase alcohol. Most jurisdictions control the sale of alcohol by operating or licensing outlets and by regulating hours of sales. Alcohol use in some public places is not allowed. Public intoxication is often a violation of the criminal law. Operating a motor vehicle with a blood alcohol concentration (BAC) above some threshold is prohibited. These last two categories, public drunkenness and driving while intoxicated (DWI), have been important issues for the criminal law and the criminal justice system for at least the last quarter century. Public drunkenness has been decriminalized in many places, and the criminal justice system is less likely to have sole responsibility for dealing with the problem. The trend has been in the opposite direction for DWI. Both laws and their enforcement have become more strict. Public drunkenness and DWI are discussed below.

Drinking is associated with serious criminal behavior and has been shown to be an important factor in the occurrence of some violence. Its role is overstated, however. The role of alcohol in interpersonal violence is summarized in a separate section of this chapter. Drinking is also related to some property crime; this subject, too, is discussed later.

Public Drunkenness

Arrests for public drunkenness totaled more than 822,500 in the United States in 1989 (Federal Bureau of Investigation, 1989). These arrests were

James J. Collins is with the Research Triangle Institute, Research Triangle Park, N.C. This chapter was written especially for this book.

5.7% of total arrests in the country. By comparison, there were approximately 2 million drunkenness arrests in 1965—about one third of the total (President's Commission on Law Enforcement and Administration of Justice, 1967, hereafter referred to as the President's Crime Commission). In that 24-year period, the percentage of recorded arrests represented by drunkenness was reduced by more than 80%. The decrease in arrests over that time period reflects major changes in laws and law enforcement emphasis as well as other changes in attitudes and public response to drunkenness. [See chapter 38, by Rubington, in this volume.]

Drunkenness offenders had become a major problem for police and local jails in the 1960s. The criminal justice system resources that were used to process these cases were out of proportion to the public safety threat these cases posed. The system's capacity to respond to serious street crime was being impaired. Moreover, many drunkenness offenders were arrested repeatedly, incarcerated briefly, and released, creating a "revolving door." It was becoming clear that the criminal justice system should not have primary responsibility for dealing with public drunkenness.

Attitudes about problem drinking were also changing. The disease concept of alcoholism had gained wide acceptance. The view of an alcoholic as an individual who is sick and whose drinking is not fully voluntary was and is somewhat inconsistent with the major goals of the criminal justice system—that is, apprehending and punishing offenders for their volitional acts. Court decisions also fostered change. In the *Easter v District of Columbia* (1966) case, for example, it was held that alcoholics could not be convicted for drunkenness. Finally, death is a real risk when acutely intoxicated individuals are held in police custody with no medical supervision.

In 1967, the President's Crime Commission made three recommendations regarding public drunkenness:

> Drunkenness should not in itself be a criminal offense. Disorderly and other criminal conduct accompanied by drunkenness should remain punishable as separate crimes. The implementation of this recommendation requires the development of adequate civil detoxification procedures.
>
> Communities should establish detoxification units as part of comprehensive treatment programs.
>
> Communities should coordinate and extend aftercare resources, including supportive residential housing. (pp. 236–237)

Many communities implemented some or all of these recommendations, although the detoxification and treatment responses to the problem have not been adequate in many places. The police continue to respond to public drunkenness in many cases. Sometimes their response is less for a "law enforcement" purpose than for the protection of the individual who may be at risk of the elements or victimization. The President's Crime Commission recommendations have had a major influence on law enforcement responses to public drunkenness.

Public drunkenness, however, continues to be a significant problem for law enforcement; and the 822,500 arrests for drunkenness in 1989 probably understate the enforcement resources expended on the problem. Because detoxification facilities exist in many places, and because drunkenness is less likely to be viewed as a crime, police often transport intoxicated persons to a detoxification or other facility instead of to the local jail. These activities are not reflected in arrest statistics.

The criminal justice system must also deal with various other alcohol control violations, such as the illegal sale of alcohol. Classified as "liquor law" violations by the Federal Bureau of Investigation's Uniform Crime Reporting System, they were 4.6% of the total arrests in 1989 (FBI, 1990). When drunkenness and liquor law arrests are combined, they represent 10% of the total arrests. "Driving under the influence" arrests were 12% of the total in 1989. Thus, more than 22% of all arrests in 1989 were for an alcohol-related offense. Drug abuse violations made up an additional 9% of all 1989 arrests. If all alcohol and drug offense arrests are combined, they constitute almost one third of all arrests. Clearly, although change has occurred, substance use and abuse violations consume very substantial proportions of U.S. criminal justice resources.

Driving While Intoxicated (DWI)

In some ways criminal justice system reactions to DWI in recent years are in contrast to the system's reaction to public drunkenness. Arrests have increased, and system reactions have become more, not less, punitive. The arrest rate for "driving under the influence" was 507 per 100,000 population in 1975. The rate had increased to 626 per 100,000 in 1980 and to 685 in 1988—a 35% increase between 1975 and 1988. Clearly, different forces were operating for public drunkenness and DWI.

Probably one of the factors influencing law enforcement responses to DWI was an increase in alcohol-related traffic casualties. The proporion of traffic crash deaths that were alcohol related increased by 6 percentage points between 1977 and 1987—from 37 to 43% (Zobeck, Grant, Williams, & Bertolucci, 1989). This represents a loss of nearly 20,000 lives in 1987 alone, and the increasing toll was a factor in stepped-up enforcement. Also, in recent years more and stricter laws against DWI have been enacted. One major change has been the lowering of the BAC threshold for impairment. Many jurisdictions have reduced this level from .15% to .10% or lower, and there are continuing recommendations to move the threshold to .05% or even to zero.

Another force encouraging stricter laws and law enforcement against DWI is the belief that criminal sanctions against those who drive while impaired might deter offenders and potential offenders from driving drunk. Some research supports this view. The most prominent researcher in this area, after a review of the entire research literature, concluded that "changes in the law promising increased certainty or combined certainty and severity of punishment reduce the amount of drinking and driving" (Ross, 1982, pp. 102–103). Apparently, however, these deterrent effects quickly deteriorate. Unless enforcement activity continues visibly, deterrence effects diminish.

Ross (1982) notes that the deterrent effect of the law and enforcement depends largely on the subjective perceptions of potential violators. Some offenders will be deterred if they think their chance of being detected is high. Ross suggests some ways of elevating risk perception, such as (1) visible police patrols in areas where, and at times when, the offense is most likely to occur and (2) roadblock breath-analyzer testing.

The Mothers Against Drunk Driving (MADD) movement has influenced the elevation in legal sanction levels against DWI. This grass-roots organization has been an effective force in calling attention to the human suffering caused by the injury and deaths associated with drunk driving and in influencing public policy and legislation. President Ronald Reagan appointed Candy Lightner, founder and president of MADD, to the President's Commission on Drunk Driving (1983). Local MADD chapters have proliferated across the country, and many have lobbied effectively for increased enforcement of DWI laws.

The elevated level of legal sanctions against DWI has also been incorporated into the U.S. concept of appropriate social control. Five different federal departments (Defense, Education, Justice, Health and Human Services, and Transportation) sponsored the *Surgeon General's Workshop on Drunk Driving* (U.S. Surgeon General, 1988). The workshop proceedings, published in December 1988, make wide-ranging recommendations and include educational, health, and prevention messages. The campaign against drunk driving has moved from the law and criminal justice arenas to encompass broad public sponsorship.

Drinking and Interpersonal Violence

The relationship of drinking to violent behavior has received much attention over the years. Studies of violent events, violent offenders, and victims of violent crime all suggest that the pharmacological effects of alcohol precipitate some violence. Alcohol has been found in high percentages of homicide and assault offenders (Mayfield, 1976; Voss & Hepburn, 1968; Welte & Abel, 1989; Wolfgang, 1958). Arrested violent offenders have often been drinking—approximately three quarters in a Columbus, Ohio, study, for example (Shupe, 1954). Prisoners with drinking problems have higher assault rates than prisoners without drinking problems (Barnard, Holzer, & Vera, 1979; Chaiken & Chaiken, 1982; Edwards, Hensman, & Petro, 1971; Gibbens & Silberman, 1970; Institute for Scientific Analysis, 1978; Mayfield, 1976; Myers, 1982). In one study about half of the rape offenders were found to have been drinking at the time of the offense (Johnson, Gibson, & Linden, 1978; Rada, 1975; Shupe, 1954). This literature has been reviewed several times over the last 12 years (Collins, 1981, 1986, 1989; Roizen & Schneberk, 1977). Another detailed review will not be repeated here. Instead, selected findings and their limitations will be discussed. The limitations are emphasized because the importance of alcohol to the occurrence of violence is exaggerated.

The ubiquitous presence of drinking in violent events and the disproportionate problem drinking in violent offenders almost certainly overstate alcohol's importance in the occurrence of violence. There are several reasons why it is logical to conclude that drinking is less relevant to violent behavior than a noncritical assessment of the literature suggests:

- Most drinking occasions are not followed by violence.
- Violent problem drinkers act violently only occasionally.
- The capacity of drinking to account for violence is slight in studies that control for the effects of multiple correlates of violence.
- There is a tendency on the part of violent offenders and others to blame drinking for violence as a way to excuse or mitigate responsibility.
- Serious methodological problems characterize most research on alcohol-related violence; for example, study populations are usually prisoners or some other sample of convenience, and measurement of the drinking variable is usually crude.
- The mechanisms by which drinking precipitates violence are poorly understood.

The last point requires further discussion.

On those occasions when drinking does have a role in the occurrence of violence, it very probably (1) does so through the effects that it has on individual's cognitive process, and (2) acts in conjunction with other factors as well. Pernanen (1976, 1981) has done the most careful and systematic thinking about these matters. He noted that alcohol distorts cognition and judgment. Drinking increases the likelihood that behavioral cues will be misinterpreted; miscommunication between individuals may result. The cognitive impairment associated with drinking may also narrow one's behavioral repertoire. These effects probably increase the likelihood that interpersonal interactions will escalate to violence. Other factors may promote violence. A drinking individual whose cultural roots include an ethos of violence, for example, may more readily resort to violence in circumstances in which his or her cognition has been impaired after drinking.

It is helpful to distinguish alcohol's immediate pharmacological (acute) effects from its longer-term (chronic) ones. Drinking has various acute effects on mood, cognition, reaction time, and so on. Over a period of months and years, heavy drinkers can experience a variety of mental, psychological, and physical changes, that is, chronic effects. There is some evidence that the acute effects of alcohol are associated with violence but little, if any, evidence that the same is true of chronic effects (Collins & Schlenger, 1988; Welte & Miller, 1987). This evidence is consistent with the view that alcohol's violence-producing effects are the result of its effects on cognition and judgment.

A few community studies have found evidence of a causal role for alcohol in the occurrence of interpersonal violence. Room (1983) reviewed a number of studies of the effects of temporary reductions in alcohol availability as a result of strikes and other factors. Reductions in violence levels were sometimes observed when alcohol supplies were interrupted. A Swedish experiment studied the effects of closing state-owned retail liquor stores on Saturdays for several months (Olsson & Wikstrom, 1982). Assault rates declined for all days of the week, but especially for Saturdays.

In summary, it is very likely that drinking accounts for some incidents of interpersonal violence, but it is difficult to assess the strength of this relationship for several reasons:

- Drinking is a very common activity, and it is difficult to distinguish its mere presence in violent events and offenders from its etiological relevance.
- Drinking is sometimes used as an excuse for acting violently; this tendency confounds attempts to understand when it is pharmacologically important.
- The mechanisms by which drinking precipitates violence are poorly understood; this fact inhibits etiological analyses.

It is reasonable to conclude that drinking does not by itself explain a large proportion of interpersonal violence and, further, that the drinking-to-violence relationship is a complex one, in which multiple factors are likely to act in combination to explain most incidents.

Property Crime

Property offenders drink more, more often, and have higher rates of problem drinking than nonoffenders (Bureau of Justice Statistics, 1983; Neighbors et al., 1987; Roizen & Schneberk, 1977). One is thus led to ask whether the empirical relationship between drinking and property offense is simply happenstance—spurious, in the language of science—or whether it indicates that drinking has a role in property offense. A number of points are relevant in this regard.

Some individuals report drinking to steady their nerves in preparation for committing a crime (Cordelia, 1985; Strug et al., 1984). In such cases, the motivation to offend is independent of alcohol. This finding suggests a very different relationship between drinking and property crime than one in which alcohol is a behavioral precipitant.

There is also evidence that a drinking offender is less able to avoid detection than one who is not drinking. Petersilia, Greenwood, and Lavin (1978), for example, found that alcohol-abusing offenders were more likely to be arrested for offenses committed than drug-abusing or non-substance-abusing offenders. This finding is not surprising. It seems likely that the adverse effects of drinking on cognition and judgment make it more difficult for a drinking offender to avoid detection and apprehension. In any case, the alcohol-abusing offender's inability to avoid detection may help to account for the higher-than-expected rates of drinking among property offenders. Drinking may influence not the property-offending behavior but, simply, the likelihood of getting caught for such offenses.

There is substantial evidence that many frequent users of expensive drugs such as heroin and cocaine commit property offenses to get cash to support their drug habits. Alcohol abusers rarely support their drinking by stealing, because alcohol is relatively cheap. A relatively few dollars a day can support heavy daily drinking.

The correlation between drinking and property crime is sometimes explained by "lifestyle" (Peterson & Braiker, 1980). Property offenders are characterized as engaging in various deviant behaviors: having marginal family and vocational lives, drinking heavily, using drugs, and so forth. This view suggests that the association of drinking and property crime is simply correlational, not causal.

In summary, there is little reason to think that drinking is etiologically important to involvement in property crime in the sense that alcohol has effects that precipitate such offending. The often-observed association between drinking and involvement in property crime is probably a result of lifestyle factors, the use of alcohol to steady nerves before offending, and the higher probability of detection for drinking offenders.

Summary

Control of the use and distribution of alcohol is a major focus of the criminal law and the criminal justice system. Major aspects of this control are laws against public drunkenness and enforcement of DWI statutes. Recent history has seen major changes in the way that the United States deals with these problems. Beginning about the mid-1960s, attempts were made to decriminalize public drunkenness and to develop an alternative system for dealing with the acute and chronic effects of drunkenness. Arrests for public drunkenness and incarceration of offenders have been replaced to some extent by a quasi–health care system that emphasizes detoxification and treatment referral, but the police and criminal justice retain an important role in dealing with public intoxication. This role continues in part because it is a public safety issue in some ways and in part because the system of alternative responses recommended by the President's Crime Commission in 1967 was never fully developed.

Criminal justice system responses to DWI have escalated in recent years. Several factors account for this escalation. First, the problem appears to have worsened, thus raising public health, mortality, and other costs. Social disapproval of driving while intoxicated has increased in response to public pronouncements and the effectiveness of organizations such as Mothers Against Drunk Driving. Finally, the evidence that visible enforcement does deter drunk driving has further encouraged enforcement.

Very substantial public resources are expended by the criminal justice system to enforce laws regulating the use and distribution of alcohol. In spite of decriminalization and reduced enforcement of public drunkenness laws, more than 22% of all arrests in the United States in 1989 were for public drunkenness, liquor law violations, or driving while intoxicated.

The pharmacological effects of drinking account for some interpersonal violence. The violence-producing capacity of alcohol probably operates through the drug's detrimental effects on cognitive function and judgment—in effect elevating the likelihood that interpersonal interactions will deteriorate to violence. On the other hand, the very high empirical correlation between drinking and violence overstates the causal relevance of drinking to violence. Drinking is sometimes present but not etiologically relevant and is sometimes unjustifiably blamed for causing violence.

Property offenders have high rates of drinking and problem drinking, but drinking is probably not etiologically important to the commission of acquisitive crime. The empirical relationship of drinking to property crime is likely to be a function of drinking to steady one's nerves before offending, a higher likelihood that a drinking property offender will be caught, and the fact that both drinking and stealing are common features of what might be called deviant lifestyles.

Alcohol control activities and the behavioral effects of drinking have important implications for the criminal law and the criminal justice system. Major resources are expended to deal with violations of laws that control alcohol use and distribution. Some violent behavior is precipitated by drinking. Substantial percentages of offenders have drinking problems. This chapter has not discussed the last point—the treatment of offenders who have drinking problems—but this is another aspect of the alcohol–law–justice system complex. It is well known that many convicted offenders who are on probation, on parole, or incarcerated have drinking problems and that for some offenders these problems are closely connected with their violations of law and marginal participation in society.

Therapeutic interventions to address the alcohol-related problems of offenders generally have not been implemented or are of unknown or marginal effectiveness. There are a number of reasons for this situation:

- Financial and human resources are scarce.
- Compulsory or involuntary treatment is thought inappropriate by many because successful treatment requires a client interested in dealing with his or her problem.
- Alcohol-related problems are difficult to treat, even with adequate resources and motivated subjects.
- Some argue that the criminal justice system is not an appropriate vehicle for pursuing therapeutic goals.
- Even where alcohol treatment has been implemented for offenders, systematic evaluations have not been done.

Because drinking problems are so prevalent among offenders, the potential economic and health benefits of successful interventions are great. This is now an undeveloped frontier worthy of systematic inquiry. The drinking problems among offenders may also be an appropriate focus of additional public support.

Acknowledgment

The author appreciates the editorial assistance of Ms. Elizabeth Cavanaugh of the Research Triangle Institute.

References

Barnard, G., Holzer, C., & Vera, H. (1979). A comparison of alcoholics and nonalcoholics charged with rape. *Bulletin of the American Academy of Psychiatry and the Law, 7,* 432–440.

Bureau of Justice Statistics. (1983). *Prisoners and alcohol.* Washington, DC: Author.

Chaiken, J., & Chaiken, M. (1982). *Varieties of criminal behavior.* Santa Monica, CA: Rand.

Collins, J.J. (Ed.). (1981). *Drinking and crime: Perspectives on the relationship between alcohol consumption and criminal behavior.* New York: Guilford.

Collins, J.J. (1986). The relationship of problem drinking to individual offending sequences. In A. Blumstein, J. Cohen, J. Roth, & C.A. Visher (Eds.), *Criminal careers and "career criminals"* (Vol. 2, pp. 89–120). Washington, DC: National Academy Press.

Collins, J.J. (1989). Alcohol and interpersonal violence: Less than meets the eye. In N.A. Weiner & M.E. Wolfgang (Eds.), *Pathways to criminal violence* (pp. 49–67). Beverly Hills, CA: Sage.

Collins, J.J., & Schlenger, W.E. (1988). Acute and chronic effects of alcohol use on violence. *Journal of Studies on Alcohol, 49,* 516–521.

Cordelia, A. (1985). Alcohol and property crime: Explaining the causal nexus. *Journal of Studies on Alcohol, 46,* 161–171.

Easter v. District of Columbia, 361 F. Supp. 21.50 (D.C. Cir. 1966).

Edwards, G., Hensman, C., & Peto, J. (1971). Drinking problems among recidivist prisoners. *Psychological Medicine, 1,* 388–399.

Federal Bureau of Investigation. (1990). *Crime in the United States, 1989.* Washington, DC: Department of Justice.

Gibbens, T., & Silberman, M. (1970). Alcoholism among prisoners. *Psychological Medicine, 1,* 73–78.

Institute for Scientific Analysis. (1978). *Drinking patterns and criminal careers: A study of 310 imprisoned male felons* (Final report prepared for the National Institute on Alcohol Abuse and Alcoholism). San Francisco: Author.

Johnson, S.D., Gibson, L., & Linden, R. (1978). Alcohol and rape in Winnipeg, 1966–1975. *Journal of Studies on Alcohol, 39,* 1887–1894.

Mayfield, D. (1976). Alcoholism, alcohol intoxication, and assaultive behavior. *Diseases of the Nervous System, 37,* 288–291.

Myers, T. (1982). Alcohol and violent crime re-examined: Self-reports from two subgroups of Scottish male prisoners. *British Journal of Addiction, 77,* 399–413.

Neighbors, H.W., Williams, D.H., Gunnings, T.S., Lipscomb, W.D., Broman, C., & Lepkowski, J. (1987). *The prevalence of mental disorder in Michigan prisons.* Lansing: Michigan Department of Corrections.

Olsson, O., & Wikstrom, P.O.H. (1982). Effects of the experimental Saturday closing of liquor retail stores in Sweden. *Contemporary Drug Problems, 11* (3) 325–353.

Pernanen, K. (1976). Alcohol and crimes of violence. In B. Kissin & H. Begleiter (Eds.), *The biology of alcoholism: Vol. 4. Social aspects* (pp. 351–444). New York: Plenum.

Pernanen, K. (1981). Theoretical aspects of the relationship between alcohol and crime. In J.J. Collins (Ed.), *Drinking and crime: Perspectives on the relationship between alcohol consumption and criminal behavior* (pp. 1–69). New York: Guilford.

Petersilia, J., Greenwood, P.W., & Lavin, M. (1978). *Criminal careers of habitual felons.* Washington, DC: National Institute of Law Enforcement and Criminal Justice.

Peterson, M., & Braiker, H. (1980). *Doing crime: A survey of California prison inmates.* Santa Monica, CA: Rand.

President's Commission on Law Enforcement and the Administration of Justice. (1967). *The challenge of crime in a free society.* Washington, DC: Author.

President's Commission on Drunk Driving. (1983). *Final report.* Washington, DC: Author.

Rada, T. (1975). Alcoholism and forcible rape. *American Journal of Psychiatry, 132,* 444–446.

Roizen, J., & Schneberk, D. (1977). Alcohol and crime. In M. Aren, T. Cameron, J. Roizen, R. Roizen, R. Room, D. Schneberk, & D. Wingard (Eds.), *Alcohol, casualties and crime* (pp. 286–465). Berkeley, CA: Social Research Group.

Room, R. (1983). Alcohol and crime: Behavioral aspects. In S. Kadish (Ed.), *Encyclopedia of crime and justice* (Vol. 1, pp. 35–44). New York: Macmillan.

Ross, H.L. (1982). *Deterring the drinking driver. Legal policy and social control.* Lexington, MA: Lexington Books.

Shupe, L.M. (1954). Alcohol and crime. *Journal of Criminal Law, Criminology, and Police Science, 44,* 661–664.

Strug, D., Wish, E., Johnson, B., Anderson, K., Miller, T., & Sears, A. (1984). The role of alcohol in the crimes of active heroin users. *Crime and Delinquency, 30,* 551–567.

U.S. Surgeon General. (1988). *Surgeon General's Workshop on Drunk Driving: Proceedings.* Rockville, MD: Office of the Surgeon General, Public Health Service.

Voss, H.L., & Hepburn, J.R. (1968). Patterns in criminal homicide in Chicago. *Journal of Criminal Law, Criminology and Police Science, 59,* 499–508.

Welte, J.W., & Abel, E.L. (1989). Homicide: Drinking by the victim. *Journal of Studies on Alcohol, 50,* 107–201.

Welte, J.W., & Miller, B.A. (1987). Alcohol use by violent and property offenders. *Drug and Alcohol Dependence, 19,* 313–324.

Wolfgang, M.E. (1958). *Patterns in criminal homicide.* New York: Wiley.

Zobeck, T.S., Grant, B.F., Williams, G.D., & Bertolucci, D. (1989). *Trends in alcohol-related fatal traffic crashes, United States: 1977–1987.* Rockville, MD: Alcohol, Drug Abuse, and Mental Health Administration.

CHAPTER 34

Alcoholism, Alcohol, and Attempted Suicide

KEITH HAWTON, JOAN FAGG, AND STEPHEN P. MCKEOWN

The association between alcohol abuse and suicidal behaviour has long been recognized. The relationship between alcoholism and completed suicide has been particularly well documented (Roy & Linnoila, 1986). In three major studies of suicides a diagnosis of alcoholism was made in 15 to 26% of cases (Barraclough, Bunch, Nelson, & Sainsbury, 1974; Dorpat & Ripley, 1960; Robins, Murphy, Wilkinson, Gassner, & Kayes, 1959). As many as 15% of alcoholics may eventually commit suicide (Miles, 1977), their risk of suicide being at least 10 times that of the general population (Lemere, 1953), with this figure probably being considerably higher among male alcoholics (Kessel & Grossman, 1961).

Although it is also recognized that alcohol dependence is common among suicide attempters (Holding, Buglass, Duffy, & Kreitman, 1977; Morgan, Burns-Cox, Pocock, & Pottle, 1975; Smith & Davison, 1971; Urwin & Gibbons, 1979), rather less attention has been paid to the nature of the association between attempted suicide and alcohol use and abuse. This state of affairs is unfortunate because a suicide attempter's admission to a general hospital might provide a special opportunity for assessment and treatment of alcohol-related problems.

This chapter presents the findings of a large-scale survey of alcoholism among suicide attempters and the use of alcohol in association with attempts. The specific aims of the study were: (1) to determine the extent of the association between alcoholism and attempted suicide; (2) to investigate the characteristics of alcoholics who make suicide attempts, especially in relation to sex, age, unemployment, and nature and repetition of attempts; and (3) to determine how frequently alcohol is consumed shortly before suicide attempts, and as part of the act itself, and by what types of patients.

Keith Hawton, Joan Fagg, and Stephen P. McKeown are with the University Department of Psychiatry, Warneford Hospital, Oxford, U.K. This chapter is reprinted with permission from *Alcohol & Alcoholism, 24,* pp. 3–9, 1989.

Subject and Methods

The study population consisted of all patients referred to the Emergency Psychiatric Service in the general hospital in Oxford between 1976 and 1985 because of self-poisoning or self-injury. The general hospital receives all hospital-referred cases from Oxford and the surrounding area.

Alcoholism and Alcohol Use in Association with Suicide Attempts among Attempted Suicide Patients

Patients referred to the general hospital in Oxford following suicide attempts are identified by a monitoring service maintained by the University Department of Psychiatry (Hawton, Gath, & Smith, 1979). The majority of attempts involve self-poisoning, the rest either self-injury or both methods.

Self-poisoning. This term is defined as the intentional self-administration of more than the prescribed dose of any drug whether or not there is evidence that the act was intended to cause self-harm. This category also includes overdoses of "drugs for kicks" and poisoning by non-ingestible substances and gas, provided the hospital staff members consider these to be cases of deliberate self-harm. Alcohol intoxication is not included unless accompanied by other types of self-poisoning or self-injury. *Self-injury* is defined as any injury recognized by the hospital staff as having been deliberately self-inflicted. Throughout the paper the conventional term *attempted suicide* is used to denote both groups of patients.

Approximately 85% of the attempted suicide patients are routinely referred to the Emergency Psychiatric Service in the hospital (Hawton, Fagg, Marsack, & Wells, 1982). All patients referred to the service receive a detailed psychosocial assessment by a specially trained psychiatrist, psychiatric nurse, or social worker. Each assessment is discussed in detail with a senior psychiatrist (Hawton et al., 1979). Several specific items (see below) are recorded by the assessors on data sheets; the sheets are then coded and the data entered into a computerized data file.

Alcoholism. Questions about drinking habits are asked during the assessment. In addition to asking the patient, the assessor obtains information from any other available informants and the patient's general practitioner. On the basis of all this information the assessor decides on clinical grounds whether the patient suffers from alcoholism. Although no specific criteria are laid down for the threshold of diagnosis of alcoholism, particular attention is paid to symptoms of dependence and the physical consequences of chronic alcohol abuse. The diagnosis would not be made solely on a history of heavy drinking. Where uncertainty exists, it is recorded. In addition to demographic data, the assessor records information on the data sheets about many other factors, including the nature of the attempt, alcohol use both shortly before (within 6 hours) and during the attempt, employment status, and any previous attempts. Findings with regard to alcoholism were available for the period 1976–1985, and for alcohol use in association with attempts from 1978 to 1985. For some items the results are based only on information obtained at the first referral of each person during the study period (the "index referral").

Repetition of attempts by alcoholics and nonalcoholics has been studied according to three factors: (1) previous attempts leading to hospital referral before the index referral; (2) repeat attempts resulting in re-referral to the general hospital in Oxford during the year following the index referral; and (3) chronic repetition, defined as five or more further attempts, each resulting in re-referral to the general hospital, during the whole study period. The period of follow-up for this last item varied according to the date of the index referral; the average period was 5 years.

Results

Alcoholism among Attempted Suicide Patients

During the 10-year study period, 1976–85, 5,269 attempted suicide patients aged 16 years and over were assessed by the staff of the general hospital Emergency Service. Whether or not the patients suffered from alcoholism at the time of referral was known definitely in 4,756 cases (90.3%). During the study period the Emergency Service staff assessed a total of 7,006 attempts.

Overall, 7.9% of the patients received a diagnosis of alcoholism on their first referral during the study period (Table 1), this diagnosis being made far more often for male than female attempters ($\chi^2 = 161.79, p < .001$).

The percentage of patients receiving a diagnosis of alcoholism on their first referral in each year increased during the study period (Table 2) from a mean annual percentage of 7.1% during 1976–79 to 10.9% during 1982–85. Although there was considerable annual variation, especially among the male attempters, the general pattern was for this increase to occur in both sexes (from 12.8% to 19.9% in males, and from 4.4% to 5.6% in females). However, in the males the increase in the percentage of alcoholics partly reflected an increase in the numbers of persons diagnosed as alcoholic each year (from a total of 101 in 1976–79 to 145 in 1982–85) and partly a fall in the total number of persons referred follow-

TABLE 1
Alcoholism in Attempted Suicide Patients
(Diagnosis at first referral to general hospital Emergency Service)

	Alcoholic N (%)		Nonalcoholic N (%)	
Males (n = 1675)	245	(14.6)	1430	(85.4)
Females (n = 3081)	129	(4.2)	2952	(95.8)
Both sexes (n = 4756)	374	(7.9)	4382	(92.1)

TABLE 2
Attempted Suicide Patients Receiving a Diagnosis of Alcoholism Each Year, 1976–85 (only the first attempt in each calendar year was included—patients could appear in the figures for more than one year)

	MALES		FEMALES		BOTH SEXES	
Year	Total no. of patients	% alcoholic	Total no. of patients	% alcoholic	Total no. of patients	% alcoholic
1976	(196)	18.9	(437)	3.4	(633)	8.2
1977	(220)	14.5	(445)	4.5	(665)	7.8
1978	(184)	7.6	(424)	4.5	(608)	5.4
1979	(188)	9.6	(358)	5.3	(546)	6.8
1980	(180)	16.1	(326)	6.4	(506)	9.9
1981	(210)	17.6	(384)	6.0	(594)	10.1
1982	(177)	24.3	(340)	4.4	(517)	11.2
1983	(194)	20.1	(285)	4.2	(479)	10.6
1984	(173)	20.2	(319)	6.0	(492)	11.0
1985	(183)	15.3	(288)	8.0	(471)	10.8

ing attempts (from 788 to 727). In the females, by contrast, the increased percentage was entirely due to the reduced overall numbers of persons referred each year (from a total of 1,664 in 1976–79 to 1,232 in 1982–85), the number of alcoholic attempters in the two periods having been very similar (73 and 69).

Age and alcoholism. There was an increase in the percentage of attempted suicide patients diagnosed as alcoholic in each age group, up until age 51–65 (Figure 1) (χ^2 = 191.32, 4 df, $p < .001$). This age pattern was found among both the male (χ^2 = 13.62, 4 df, $p < .01$) and the female attempters (χ^2 = 93.14, 4 df, $p = < .001$), although the peak prevalence for alcoholism among the male attempters was reached by age 36–

FIGURE 1
Percentage of Alcoholic Attempters in Each Age Group, According to Sex

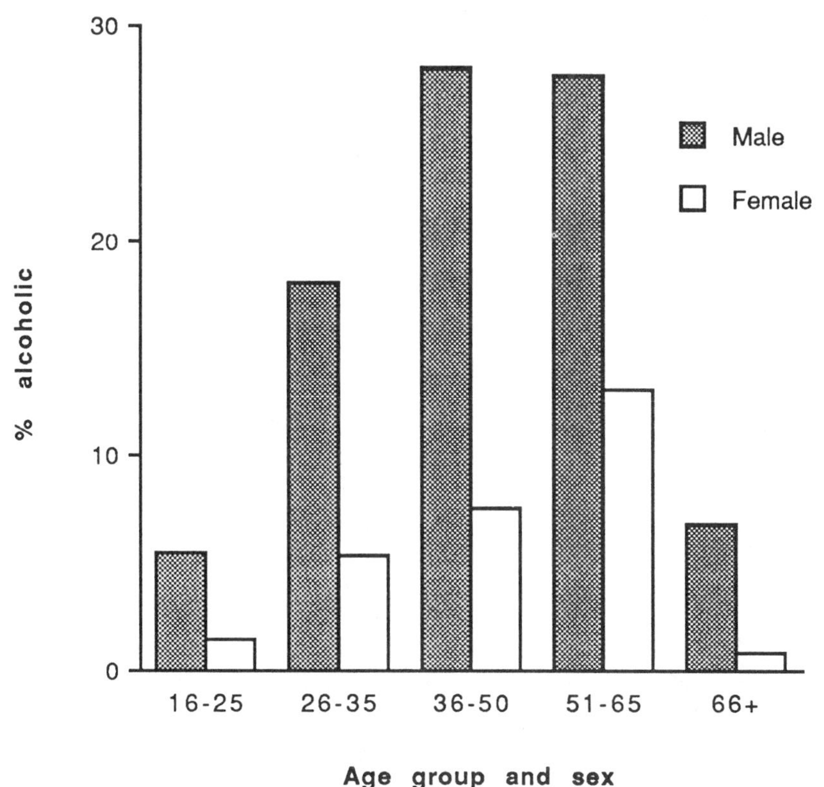

50. Within every age group the proportion of male alcoholics was significantly greater than the proportion of female alcoholics.

Unemployment. Alcoholism was considerably more common among the unemployed than the employed of both sexes (Table 3) (males: $\chi^2 = 60.3$, $p < .001$; females: $\chi^2 = 17.86$, $p < .001$). Nearly two thirds (61.8%) of the economically active male alcoholic attempters were unemployed. In female attempters, alcoholism was almost as common among housewives as among the unemployed.

Methods Used in Attempts

The majority of attempts involved self-poisoning (88.2%). However, alcoholics were more likely than nonalcoholics to injure themselves deliberately (10.3% vs. 5.5%; $\chi^2 = 12.63$, $p < .001$). The overdoses of the alcoholics more often involved minor tranquilizers and sedatives (49.6%

TABLE 3
Alcoholism and Employment Status among Attempters

	% ALCOHOLIC					
	Employed		Unemployed		Housewife	
Males (N = 1,357)	10.0%	(84/837)	26.2%	(136/520)	–	
Females (N = 2,504)	2.9%	(37/1257)	7.8%	(34/437)	6.0%	(49/810)

Note: Patients in other economic groups are omitted from these figures.

vs. 36.2%; $\chi^2 = 23.21$, $p < .001$), and less often nonopiate analgesics (33.5% vs. 43.4%; $\chi^2 = 12.10$, $p < .001$), especially salicylates (15.7% vs. 23.2%; $\chi^2 = 9.64$, $p < .01$).

Repetition of attempts. Alcoholism was significantly associated with increased risk of repeated suicidal behavior on all three measures of repetition (attempts prior to index attempt, repetition within year following index attempt, and five or more attempts during the 10-year study period [Table 4]).

Alcohol Consumption in Association with Suicide Attempts

Alcohol had been consumed during the 6 hours before the attempt in 40.8% of the cases (Table 5), the proportion of attempts by males (53.4%) in which this was so being significantly greater than that for females (33.4%; $\chi^2 = 190.18$, $p < .001$). There was little variation in these figures during the study period. There were marked differences between the age groups ($\chi^2 = 125.05$, 4 *df*, $p < .001$), with alcohol consumption preceding attempts being especially common in age groups 26–35 and 36–50 for both sexes. Alcoholic attempters had far more often consumed alcohol before their attempts than had nonalcoholic attempters (83.4% vs. 30.8%; $\chi^2 = 321.62$, $p < .001$).

Alcohol had been consumed as part of the attempt in over a quarter of cases (Table 5), and more commonly in attempts by males than by females ($\chi^2 = 43.32$, $p < .001$). There was little variation in the figures during the study period. Again there were differences between the age groups ($\chi^2 = 99.58$, 4 *df*, $p < .001$), alcohol consumption being most common in attempts by those in the age range 36–50. Almost two thirds of the attempts by alcoholics (63.6%) involved alcohol consumption, whereas only 18.6% of the attempts by nonalcoholics did ($\chi^2 = 296.21$, $p < .001$).

TABLE 4
Repetition of Attempted Suicide: Comparison of Alcoholic and Nonalcoholic Attempted Suicide Patients

	MALES		FEMALES		BOTH SEXES	
	Alcoholic %	Nonalcoholic %	Alcoholic %	Nonalcoholic %	Alcoholic %	Nonalcoholic %
History of attempts prior to index attempt	45.0* (104/231)	24.8 (342/1381)	56.2* (68/121)	31.9 (911/2858)	48.9* (172/352)	29.6 (1253/4239)
Repetition within 1 year of index referral (only patients referred between 1976 and 1984)	18.1* (41/226)	9.0 (116/1296)	17.7‡ (20/113)	9.9 (270/2736)	18.0* (61/339)	9.6 (386/4032)
Five or more attempts during study period	4.9† (12/245)	1.7 (24/1429)	5.4‡ (7/129)	1.9 (56/2953)	5.1* (19/374)	1.8 (80/4382)

*$p < .001$. †$p < .01$. ‡$p < .05$.

TABLE 5

Consumption of Alcohol by Attempted Suicide Patients, 1978–85: (a) During the 6 hours Preceding the Attempt; and (b) as Part of the Attempt (based on all episodes)

Consumption of alcohol		AGE GROUP (YEARS)					
		16–25	26–35	36–50	51–65	66+	All ages
(a) During 6 hours preceding the attempt	Males	50.6% (341/674)	61.4% (347/565)	57.5% (214/372)	39.9% (63/158)	24.7% (18/73)	53.4% (983/1842)
	Females	31.5% (409/1299)	38.3% (283/738)	38.6% (251/651)	31.8% (94/296)	7.8% (12/154)	33.4% (1049/3138)
	Both Sexes	38.0% (750/1973)	48.3% (630/1303)	45.5% (465/1023)	34.6% (157/454)	13.2% (30/227)	40.8% (2032/4980)
(b) As part of the attempt	Males	24.3% (163/670)	35.8% (200/559)	38.9% (140/360)	37.2% (58/156)	20.8% (15/72)	31.7% (576/1817)
	Females	17.6% (228/1293)	27.0% (198/732)	30.2% (197/652)	27.2% (79/290)	12.4% (19/153)	23.1% (721/3120)
	Both sexes	19.9% (391/1963)	30.8% (398/1291)	33.3% (337/1012)	30.7% (137/446)	15.1% (34/225)	26.3% (1297/4937)

Discussion

This study has confirmed the very important association between alcoholism and attempted suicide, with 14.6% of male attempters and 4.2% of female attempters being diagnosed as alcoholic. Although not directly comparable, these figures are considerably greater than the general population's prevalence rates of alcohol-related problems, which in one investigation were reported as 5.3% for males and 1.6% for females (Wilson, 1980). The results of this study are in keeping with findings from elsewhere in the United Kingdom, in that several reports have commented on relatively high rates of alcoholism in suicide attempters, especially males (Holding et al., 1977; Morgan et al., 1975; Smith & Davison, 1971; Urwin & Gibbons, 1979). Although findings for Oxford cannot necessarily be generalized to elsewhere, it is worth noting that similar findings were reported for Edinburgh both for attempted suicide in general and for alcoholism, which in 1983–84 was diagnosed in 14.3% of male attempters and 5.5% of female attempters (Platt, Hawton, Kreitman, Fagg, & Foster, 1988).

This study has also indicated a rise in the proportion of attempted suicide patients receiving a diagnosis of alcoholism between the late 1970s and the early-to-mid-1980s. However, it should be noted that there was a reduction in the rate of attempted suicide after 1978, largely among females (Hawton & Catalan, 1987; Platt et al., 1988). This reduction entirely explained the changes in the prevalence of alcoholism among female attempters, a decline having occurred in numbers of nonalcoholics but not alcoholic attempters. By contrast, the rise in the proportion of male alcoholic attempters was due to both a decline in the number of nonalcoholic attempters and a substantial increase in the number of alcoholic attempters. This pattern might reflect the recent increase in alcohol abuse by young men (Plant, 1987), although worsening unemployment rates might be another factor.

Changes in diagnostic practice among the staff in the Emergency Psychiatric Service could be a further explanation for the alterations in the proportions of attempters diagnosed as alcoholic during the study period. Although this explanation cannot be ruled out entirely, it seems unlikely in view of the relatively consistent staffing of the service during the study period, the careful training and supervision of the staff (Catalan et al., 1980; Hawton et al., 1979), and the sex differences in the findings noted above.

The marked association between unemployment and alcoholism among suicide attempters is of particular interest in view of the recent increase in general unemployment rates. However, as has been discussed elsewhere (Hawton & Rose, 1986; Platt, 1986; Smith, 1985), one must be cautious in interpreting findings such as this that are based on cross-sectional data. Thus, it is not possible to say, for example, whether unemployment increases the risk of alcoholism, and hence of suicide attempts, or whether high rates of alcoholism and attempted suicide among the unemployed are a reflection of the fact that people with emotional and psychiatric problems (including alcohol abuse) are more likely to become unemployed. Nevertheless, it is noteworthy that the higher rates of alcoholism among attempters during the latter part of the study period coincided with particularly high general population rates of unemployment.

The relatively high rate of alcoholism among housewives in this study is a surprising finding and warrants further investigation, especially in the light of increasing concern about female alcoholism (Royal College of Psychiatrists, 1979).

The higher incidence of self-injury in attempts by alcoholics than nonalcoholics reflects the known association between self-cutting (the majority of self-injuries) and alcoholism (Simpson, 1975). The greater use of minor tranquilizers and sedatives in the overdoses of alcoholics is probably associated with more extensive psychiatric problems and treatment in this group and hence availability of such medication, although it may also reflect the tendency of alcoholics to use mood-altering substances. Nevertheless, it highlights the need to observe particular caution when considering prescribing psychotropic medication to individuals known to abuse alcohol.

The importance of the association between alcoholism and attempted suicide is emphasized by the far higher rates of repetition of attempts among alcoholics than nonalcoholics. This finding, which is in accord with the findings of specific studies of repetition of attempted suicide (Barnes, 1986; Buglass & Horton, 1974), is particularly significant because there is increased risk of eventual suicide in those who make repeat attempts (Hawton, 1987).

There was extensive use of alcohol, especially by alcoholics, both immediately before and during suicide attempts. The disinhibition produced by intoxication probably facilitates suicidal ideas and increases the likelihood of suicidal thoughts being put into action, often on impulse. This facilitating factor further emphasizes the need to try to help suicide attempters gain control over their drinking, since increased control might reduce the likelihood of repeat attempts. The use of alcohol as part of the attempt itself adds to the potential danger of an overdose, both because alcohol increases the toxicity of drugs commonly used in self-poisoning, especially psychotropic agents, and because it is not unusual for collapse or unconsciousness to be attributed to alcohol alone (particularly in known alcoholics) and hence for medical treatment for drug overdose to be substantially delayed.

Another implication of the high prevalence of alcoholism among attempted suicide patients is the risk of their developing the alcohol withdrawal syndrome during general hospital admission because of the sudden cessation of drinking. Medical staff should be vigilant for any early signs of withdrawal (e.g., anxious arousal, restlessness, insomnia, and confusion), and instigate appropriate treatment or referral for specialist care if that is indicated.

It is abundantly clear that all suicide attempters should be thoroughly assessed with regard to their alcohol use and that the hospital referral resulting from an attempt can provide a unique opportunity for such assessment. There should be a close liaison between all general hospital services for attempted suicide patients and local alcoholism treatment units, with ready availability of staff from the latter to provide clinical opinions and accept referrals for treatment. This liaison should be part of the current policy towards increasing the detection and treatment of problem drinkers.

Acknowledgments

We thank the staff of the Barnes Unit, John Radcliffe II Hospital, in Oxford, whose collaboration with the monitoring service for attempted suicide patients made this study possible, Pat Wells for her assistance with data collection, and Leicester Gill for his help in maintaining our computerized record system. This study was partly supported by a grant from Oxford Regional Health Authority Research Committee.

References

Barnes, R.A. (1986). The recurrent self-harm patient. *Suicide and Life-Threatening Behavior, 16,* 399–408.

Barraclough, B., Bunch, J., Nelson, B., & Sainsbury, P. (1974). A hundred cases of suicide: Clinical aspects. *British Journal of Psychiatry, 125,* 355–373.

Buglass, D., & Horton, J. (1974). A scale for predicting subsequent suicidal behaviour. *British Journal of Psychiatry, 124,* 573–578.

Catalan, J., Marsack, P., Hawton, K.E., Whitwell, D., Fagg, J., & Bancroft, J.H.J. (1980). Comparison of doctors and nurses in the assessment of deliberate self-poisoning patients. *Psychological Medicine, 10,* 483–491.

Dorpat, T.L. & Ripley, H.S. (1960). A study of suicide in the Seattle area. *Comprehensive Psychiatry, 1,* 349–359.

Hawton, K. (1987). Assessment of suicide risk. *British Journal of Psychiatry, 150,* 145–153.

Hawton, K., & Catalan, J. (1987). *Attempted suicide: A practical guide to its nature and management* (2nd Ed). Oxford: Oxford University Press.

Hawton, K., Fagg, J., Marsack, P., & Wells, P. (1982). Deliberate self-poisoning and self-injury in the Oxford area, 1976–80. *Social Psychiatry, 17,* 175–179.

Hawton, K., Gath, D.H., & Smith, E.B.O. (1979). Management of attempted suicide in Oxford. *British Medical Journal, 2,* 1040–1042.

Hawton, K., & Rose, N. (1986). Unemployment and attempted suicide among men in Oxford. *Health Trends, 18,* 29–32.

Holding, T., Buglass, D., Duffy, J.C., & Kreitman, N. (1977). Parasuicide in Edinburgh: A seven year review, 1968–74. *British Journal of Psychiatry, 130,* 534–543.

Kessel, N., & Grossman, G. (1961). Suicide in alcoholics. *British Medical Journal, 2,* 1671–1672.

Lemere, F. (1953). What happens to alcoholics. *American Journal of Psychiatry, 109,* 674–676.

Miles, C.P. (1977). Conditions predisposing to suicide: A review. *Journal of Nervous and Mental Disease, 164,* 231–246.

Morgan, H.G., Burns-Cox, C.J., Pocock, H., & Pottle, S. (1975). Deliberate self-harm: Clinical and socioeconomic characteristics of 368 patients. *British Journal of Psychiatry, 126,* 319–328.

Plant, M. (1987). *Drugs in perspective.* London: Hodder and Stoughton.

Platt, S. (1986). Parasuicide and unemployment. *British Journal of Psychiatry, 149,* 401–405.

Platt, S., Hawton, K., Kreitman, N., Fagg, J., & Foster, J. (1988). Recent clinical and epidemiological trends in parasuicide in Oxford and Edinburgh: A tale of two cities. *Psychological Medicine, 18,* 405–418.

Robins, E., Murphy, G.E., Wilkinson, R.H., Gassner, S., & Kayes, J. (1959). Some clinical considerations in the prevention of suicide based on a study of 134 successful suicides. *American Journal of Public Health, 49,* 888–899.

Roy, A., & Linnoila, M. (1986). Alcoholism and suicide. *Suicide and Life-Threatening Behavior, 16,* 244–273.

Royal College of Psychiatrists (1979). *Alcohol and alcoholism.* London: Tavistock.

Simpson, M.A. (1975). The phenomenology of self-mutilation in a general hospital setting. *Canadian Psychiatric Association Journal, 20,* 429–433.

Smith, R. (1985). Occupationless health. "I couldn't stand it any more": Suicide and unemployment. *British Medical Journal, 291,* 1563–1566.

Smith, J.S. & Davison, K. (1971). Changes in the pattern of admission for attempted suicide in Newcastle-upon-Tyne during the 1960s. *British Medical Journal, 4,* 157–159.

Urwin, P., & Gibbons, J.L. (1979). Psychiatric diagnosis in self-poisoning patients. *Psychological Medicine, 9,* 501–507.

Wilson, P. (1980). *Drinking in England and Wales.* London: HMSO.

SECTION V

Responsive Movements and Systems of Control

Introductory Note

Broadly speaking, one could say that the social control of drinking of alcoholic beverages and related behaviors is a theme that runs throughout this book. Yet in another and narrower sense there are special phenomena and problems relating to the social control of alcohol use in contemporary society that merit particular consideration. There are general conditions for the emergence of those relatively differentiated movements, specialized agencies, and activities with which our past history (pre–World War II) is replete and which aim variously at the social regulation of alcohol, consumption, drunkenness, alcoholism, and related problems. Although it is not within the province of this book to review these developments, such as the U.S. temperance movements in the 19th and early 20th centuries, in historical perspective, we are concerned to call attention in this final section to some of the currently important alcohol-centered movements and control systems that have been, or hold promise of being, excellent subjects for sociological investigation.

First we have chosen to reprint Edwin Lemert's classic paper from the original volume, in which he analyzes four "models" of the social control of alcohol use—models derived from the present and past experiences of societies. These models of control, though not applicable to alcoholism, have as their goal minimizing the costs of intoxication and drunkenness (chapter 35). However, as Lemert indicates, none of them has been particularly successful. In his discussion, Lemert cogently presents the dilemma facing the state in regard to the control of alcohol use, namely, the fact that those groups that place a high premium on sobriety and a low value on intoxication have little need for governmental regulation of their drinking behavior. (The classic illustration here is that of the Jews, discussed by Barry Glassner in chapter 15.) Conversely, those groups or societies that at times value intoxication more than sobriety frequently sabotage or reject governmental controls.

Despite some success of state control plans for alcohol use and alcohol-related problems, Western societies in particular have a vast but realistic problem in bringing controls to bear on alcoholism and alcohol-related behavioral problems. In the last half of the 20th century, a variety of social movements have arisen throughout the Western world to cope with the economic, social, psychological, and physical costs of alcoholism. These movements and activities run the gamut from massive governmental programs in Scandinavia, Canada, and the United States through the activities of professional and voluntary associations such as the American Medical Association, the American Bar Association, the American Council on Alcoholism, and the American Public Health Association, to the organizations run by and for both alcoholics themselves and those close to them: Alcoholics Anonymous (AA), Al-Anon, and Adult Children of Alcoholics. Although precise data are not available, it is perhaps the case that more alcoholics have found sobriety through the vehicle of Alcoholics Anonymous than by means of all other agencies combined. Again, the striking feature of this movement is that its program was developed and continues to operate not through professional therapists or governmental supports but solely through the activities of alcoholics themselves.

In Norman Denzin's chapter 36, the first of two chapters devoted to self-help groups, we present a selection on the AA group, which is partially based on his book *The Recovering Alcoholic.* Using the theoretical framework of symbolic interaction, Denzin uses as his unit of analysis the AA group, which he places within the historical context of the first meeting between the founders of the organization, Bill Wilson and Robert Smith. Alcoholics Anonymous exists only through group meetings, in which the members share their past experiences with each other and interpret them in terms of the AA traditions and philosophy. Despite the concerns of many scholars (Bales, 1962; Tournier, 1979; Trice, 1957) about potential problems in AA related to crises in sponsor-protege relationships, political conflicts among older members, tendencies to schism, and pressures to define future activities within a bureaucratic and hierarchal structure, AA has been the preeminent tool for aiding alcoholics.

Because of the dramatic therapeutic results obtained by many alcoholics in their affiliations with Alcoholics Anonymous, there has been an understandable tendency to present the organization as the ultimate answer to the alcoholism problem and to eulogize its therapeutic features. However, AA does possess certain dysfunctionalities from a therapeutic point of view. Trice (1957) for example, has pointed out in his study of affiliation with AA that certain local groups do not provide adequate sponsorship for new members. Other researchers (Lofland & Lejeune, 1960; Rudy, 1986) have charted the effects of differences in social class upon the readiness of local groups to incorporate the potential member. Some AA groups, although accepting the view that alcoholism is a chronic disease, apparently have a low tolerance for those members who have "slips" or relapses. In these circumstances, the centering of the member's whole life around the AA groups makes it virtually impossible for the person to admit that he or she has had a slip, and thus the stage is set for the member's dissociation from the group and new cycles of uncontrolled drinking.

Moreover, if a relative autonomy from special external therapeutic supports is to be taken as a goal of therapy, then Alcoholics Anonymous tends to fall short of this objective. It is evident the movement fails to reach many alcoholics, some of whom, at least, may be presumed to be amenable to other types of therapy. Also, the movement has clearly had its greatest impact among cultures of Anglo-Saxon derivation and has very probably appealed especially to the type of alcoholic predominant in such cultures (see chapter 20, by Jellinek, in this volume). These limitations need not detract, however, from the great importance that this social invention, Alcoholics Anonymous, has had in helping the alcoholic; they may be understood simply as posing problems for future inquiry.

However, Alcoholics Anonymous is only one of a variety of movements, agencies, and activities that have evolved in complex society to cope with alcoholism, alcohol-related problems, and individuals who have been affected by them. Particularly significant in U.S. society has been the development and diffusion of self-help or 12-step organizations modeled after AA, whose goal is to provide social and psychological support for the victims of alcoholics—the members of their families—in, for example, Al-Anon, Alateen, and Adult Children of Alcoholics (ACOA) groups. The latter groups are the focus of David Rudy's ongoing research, presented as chapter 37, which places ACOA groups within the context of a social movement. That some children reared by an alcoholic parent or parents have been psychologically scarred is an observation made by many clinicians (Black, 1982; Wegscheider-Cruse, 1985). However, systematic longitudinal investigations of these children with matched controls are extremely rare in the field of alcohol studies.

With the reemergence and widespread acceptance of the disease concept of alcoholism in American society in the mid–20th century, increased attention was given to individuals (mainly males) who were arrested for being intoxicated in public. It should be noted that laws have been enacted since the settlement of the United States to control drunkenness, included in which have been prohibitions on public displays thereof. Although public drunkenness remains a criminal offense in certain U.S. legal jurisdictions, there was a social movement in the United States, Canada, Great Britain, and Scandinavia in the 1960s and 1970s to decriminalize it and a concomitant development of detoxification centers to provide treatment. In chapter 38 Earl Rubington presents a view of the life situation of public inebriates, before and after decriminalization, from the perspective of the social institutions responsible for their management—specifically, the inebriate's interaction with jail personnel before decriminalization and with a detoxification center staff afterwards. Rubington's empirical research, both in jails and detoxification centers, focuses on the microstructural factors operating within these contexts that help explain why decriminalization has never achieved its idealistic goals of providing these men not only treatment but rehabilitation. However, we would suggest that macrostructural factors in the 1980s, when attention was directed toward individual entrepreneurship goals that emphasized personal wealth with minimal attention to community needs, played a major role in the demise of the decriminalization movement and in the fact that detoxification centers for indigents were allowed to expire in many communities (Philips, 1990).

Alcoholics Anonymous, the adult children of alcoholics, and decriminalization of public drunkenness represent only some of the movements, agencies, and activities that have evolved in a pluralistic society, manifestly to cope with alcoholism and alcohol-related problems. These movements stand in reciprocal relation to the societal definition of alcoholism as an illness or disease, having facilitated the definition and having received ideological support therefrom. The medicalization of deviance (Conrad & Schneider, 1980), including that of alcoholism, drug abuse, behavioral problems, and so on, has been built upon the ascendancy of medicine and allied fields. The redefinition of alcoholism as a medical problem rather than a legal or moral one (or both) has had significant implications for social policy toward alcohol-related transgressions; these implications are addressed by Paul Roman and Terry Blum in chapter 39. They analyze some of the social consequences for both society and the individual of conceptualizing alcoholism and alcohol problems within an illness framework and put special emphasis on the treatment of alcoholism and alcohol-related problems in medical settings. As these authors note, despite the constant survey research findings that the U.S. population simultaneously holds both a disease and a moral weakness orientation to alcoholism, there does exist a convergence between U.S. cultural values and the medicalized concept of alcoholism. Thus, the social movement to redefine the alcohol-dependent individual in U.S. society as a sick individual has on the whole been successful.

The disease concept of alcoholism established the basis on which alcoholics, as sick individuals, became legitimate concerns of governmental health programs. Thus, the stage was set for the intervention of government into all areas upon which alcoholism and alcohol-related problems impinge, including the conditions that are predicated as significant in the disease's etiology. This tendency can be documented in the vast proliferation of international, national, state, and local programs on alcohol and alcoholism over the last four decades (Wiener, 1981). In the concluding chapter, "The New Temperance Movement," David Pittman analyzes the convergence of governmental (including public health officials) and consumer groups who emphasize the locus for alcohol problems in environmental factors such as the extensiveness of the availability of alcoholic beverages and the highly effective marketing techniques of alcoholic beverage producers. Central to an understanding of the more restrictive climate toward alcoholic beverages in the 1990s is the increased role of governmental regulation in all areas of human conduct in modern society, with the concomitant development of bureaucratic structures to implement these rules.

Pittman and Snyder (1962) warned that the proliferation of alcohol-driven agencies and activities could lead to a situation in which these groups increase

> their dependence at various points upon scientific research for legitimation and direction. In these circumstances, there are at once dangers and challenges to social scientists interested in gaining support for research on alcohol problems. A first danger, as we see it, is that of being drawn into pseudo-research activities where the explicit or implicit concerns for and defense of these programs and their bureaucratic structure limit the selection of problems and the vision of research to administratively safe questions of a pedestrian nature. This would preclude, among other things, realistic appraisal of the programs themselves. (pp. 551–552)

To a certain extent this scenario has occurred in the growth of pseudoscientific studies that have not been peer-reviewed and are fed by advocacy organizations to mass media as startling new discoveries—for example, the claim that the average teenager has seen 100,000 beer commercials by the age of 18. A second danger, given the drift of attention in various historical periods to the primacy of various disciplines, is in the temptation by responsible persons in various alcoholism agencies to oversell a particular point of view—for example, the genetic basis of alcoholism or health warning labels on containers of alcoholic beverages—as panaceas for problems related to alcohol use.

References

Bales, R.F. (1962). The therapeutic role of Alcoholics Anonymous as seen by a sociologist. In D.J. Pittman & C.R. Snyder (Eds.), *Society, culture, and drinking patterns.* New York: Wiley.

Black, C. (1982). *It will never happen to me.* Denver: MAC Printing.

Conrad, P., & Schneider, J. (1980). *Deviance and medicalization: From badness to sickness*. St. Louis: Mosby.

Lofland, J.F., & Lejeune, R.A. (1960). Initial interaction of newcomers in Alcoholics Anonymous: A field experiment in class symbols and socialization. *Social Problems, 8,* 102–111.

Philips, K. (1990). *The politics of rich and poor: Wealth and the American electorate in the Reagan aftermath.* New York: Random House.

Pittman, D.J., & Snyder, C.R. (Eds.) (1962). *Society, culture, and drinking patterns.* New York: Wiley.

Rudy, D.R. (1986). *Becoming alcoholic: Alcoholics Anonymous and the reality of alcoholism.* Carbondale: Southern Illinois University Press.

Tournier, R. (1979). Alcoholics Anonymous as treatment and as ideology. *Journal of Studies on Alcohol, 40,* 230–238.
Trice, H.M. (1957). A study of the process of affiliation with Alcoholics Anonymous. *Quarterly Journal of Studies on Alcohol, 18,* 39–43.
Wegscheider-Cruse, S. (1985). *Alcoholism and the family: A book of readings.* Wernersville, PA: Caron Institute.
Wiener, C. (1981). *The politics of alcoholism: Building an arena around a social problem.* New Brunswick, NJ: Transaction Books.

CHAPTER 35

Alcohol, Values, and Social Control

EDWIN M. LEMERT

A general analysis of alcohol and social control deals with the universal qualities of alcohol, the values and costs of its use, the distribution of power in social structures, the available means of control, the probabilities of resistance, and patterns of values in cultures and individuals. These concepts are designed for a kind of "action analysis," being oriented to research into problems of alcohol use that have emerged from the impact of rapid technological change in Western societies and from the spread of modern technologies and ideologies to nonliterate and "underdeveloped" societies of the world. Central to this analysis is the hypothesis, well grounded in empirical findings of alcohol research, that values are crucial factors in the social control of alcohol use.

The Attributes of Alcohol and Values

The values that have been assigned to alcoholic beverages throughout the world partially rest upon their physical qualities and certain of their recurrent or universal physiological effects. It has been pointed out that alcohol is distinctive for the ease and cheapness of its preparation. Other physical properties permit it to be stored for long periods and transported with facility (Haggard & Jellinek, 1942; Horton, 1943). In many societies alcohol is valued as a food, as a promoter of digestion and sleep, as a protection against cold and fatigue, and as a medicine to relieve pain or to treat specific illnesses. This is not to say that alcohol metabolism determines symbolic associations among human beings. It should be emphasized that the ascription of values to alcoholic beverages diverges from and transcends their demonstrable physiological functions.[1]

The more important symbolic associations of alcohol derive from its function as a behavior modifier. Mild to severe intoxication promotes the

Edwin M. Lemert is with the Department of Sociology, University of California at Davis. This chapter is reprinted from the original D.J. Pittmen & C.R. Snyder (Eds.), *Society, Culture, and Drinking Patterns* (New York: Wiley, 1962).

expression of a variety of idiosyncratic values in the individual and a large measure of socially shared and communicable values. Perhaps the most important of the former is the relief or relaxation from fatigue, tension, apathy, and the sense of isolation.

The social values facilitated by the consumption of alcoholic drinks spring from a recognition of their function in diminishing social distance and strengthening group bonds. These values revolve around fellowship, social amity, and group morale. Often these values are expressed through rituals that symbolize the solidarity of kin groups and work groups or the collective willingness of warriors and soldiers to die for a leader or a cause. Alcohol is further valued for its ritual functions in symbolizing status changes at birth, marriage, coming of age, and bereavement. Alcohol also finds an important place in some societies in the specialized culture of ecstasy and in Dionysian communion with gods. A number of societies have institutionalized groups in which drinking and drunkenness are terminal rather than mediating values.

There are other positive values of alcohol use, less concerned with interpersonal interaction, which come to the fore in larger, nation-state-type societies. One of these is the recognized ease with which revenue for state purposes can be raised through taxing alcohol production. This, of course, is a more specialized valuation of alcohol held by political or administrative elites. Closely related to this valuation is the recognition of the value of induced dependence upon alcohol as a means of social control. This second valuation can be seen in connection with contract labor and peonage, in connection with sex behavior, and in connection with diplomacy and power struggles between ruling elites. Finally, there is a set of sharply defined values attached to alcohol by those whose economic livelihood and occupational status rest upon the production and distribution of alcoholic beverages.

There is a tendency, perhaps universal, for valuations of alcoholic beverages to become polarized. At one extreme, liquors, wine, and beer are glorified in song, poetry, and drama as keys to ecstasy and sublimity; at the other extreme, they are viewed as perverters of human morality and the chief causes of the ills of society as well as of the sorrows of individuals. This polarization imparts a marked ambivalence to attitude and opinion concerning the proper place of alcohol in social life (Myerson, 1940). In part this ambivalence stems from an awareness that satisfactions brought by imbibing alcohol not infrequently have a spurious quality. What seemed to be another's expression of love to the intoxicated maiden turns out in sober retrospect to have been sex exploitation. The comradeship of the college reunion in afterthought is seen realistically as largely inspired by the martinis rather than by common interest long since gone.

A more important ingredient of this ambivalence towards alcohol comes from the perception of its previously mentioned function as a behavior modifier. Modifications in human behavior brought about by intoxication are socially and personally destructive as well as socially integrative. The same object that makes human pleasures makes human pain and unhappiness. Although this statement can be made about many other objects used by human beings, alcohol is distinctive in that the consequences of its consumption are difficult to predict.

The Costs of Alcohol Use

It is a reasonable assumption that there is some kind of hierarchy in the values held by human beings. This idea signifies that in a hypothetically free situation with unlimited means at hand there is an order in which values are satisfied. However, this order is imperfectly manifested in actual choice making by individuals because the environmental situation limits choices, often excluding opportunities for satisfying some values and allowing only a partial or compromised satisfaction of others. The cost of a value, stated most simply, is the degree to which other values must be sacrificed in the process of its satisfaction (Cottrell, 1954). Generally speaking, the cost of a value is estimable in terms of time, energy, and the amount of discomfort expended to satisfy it.

Indulgence in alcohol, though it promotes the fulfillment of previously mentioned values, frequently does so at the cost of others. This pattern is most apparent in the tendency for intoxication to encourage or "release" aggression and deviant sex behavior (Horton, 1943). Intoxication also impairs physiological functions, giving rise to the neglect or inadequate performance of roles as well as to accidents. The values commonly sacrificed by drunkenness are respect for person, life, property, health, longevity, family integrity, parental responsibility, regularity of work, and financial dependability. Apart from these general or universal costs chargeable against intoxication are countless others that are more variable and understandable only in relation to a particular sociocultural system.

The strictly economic costs of drunkenness and alcoholism in our society have been subject to estimates and can be considerable (Landis, 1945). The theoretical limits to such costs can be only speculative. Adam Smith (1950, pp. 456–457) stated that "there is no risk that nations will destroy their fortunes through excessive consumption of fermented liquors." However, the production of alcohol may take place at the cost of necessary commodities and services, as happened in the Hawaiian Islands at one time (Translation of the Laws of the Hawaiian Islands, 1842).

The mutineer colony of Pitcairn Island came perilously close to annihilating itself through drunkenness and conflict over women (Nordhoff & Hall, 1934).

It is possible that rapid social change in the last 150 years, which has been strongly felt in Western societies and is emergent throughout the world, has enhanced the values of alcoholic intoxication. Such things as culture conflict, stress, and anomie may have grown to such proportions that alcohol in many societies is increasingly valuable as a social reagent and as a sedative for personality conflicts. This hypothesis is a large one and, with the present state of sociology, a difficult one to test (Bacon, 1945).

Modern technology changes the costs of satisfying values even though the order of the values may remain unchanged. This is to say that technological and related cultural changes very definitely affect choice-making behavior in relation to social control. As applied to alcohol consumption, this idea means that intoxication and drunkenness levy critically higher costs in certain contexts of technologically mediated, interdependent, high-speed, high-productivity, health-oriented societies. Thus, to older, more universal, costs of drunkenness are added death and injury from traffic accidents, lowered productivity and absenteeism in industry, disease, disturbances of public order, crime, and weakening of military discipline. Furthermore, with the growth of statism and welfare values, costs of policing and treating the chronically inebriate population must be reckoned in the overall economy of alcohol use.

Values and Power Elites

It should not be concluded that policies or lines of action or inaction followed by societies in relation to the control of drinking result from a simple summative economy of values and costs. For this reason serious questions have to be raised about the sufficiency of research into such things as basic personality, themes, cultural patterns, and national character in order to predict the course and results of social control. Social action and control usually emanate from elite power groups who have their own systems of values, which differ from those of the general population, from those of other groups, and even from those of individual members of the elites. The organizational values of such elites and their rules of procedure also have a strong bearing on controlling events. Furthermore, elites are limited by the amount of power they exercise, the kinds of alliances they make, and the means of control available to them to reach their goals.

The position of groups and individuals at the point of their interaction in a social structure is of great significance in predicting the resultant action taken by a society or government to control or decontrol alcohol consumption. Groups and individuals whose values are being sacrificed by intoxication and drunkenness may have no structure to formulate their vaguely felt dissatisfactions. On the other hand, minorities, because their programs are defined and their power is organized and well-timed, more readily have their values cast into the emergent pattern of social action (Cottrell, 1954).

Resistance

Action by groups or whole societies to change the drinking behavior of a population necessarily alters the costs of satisfying values that have been sought directly through the medium of alcoholic beverages. At the same time, other costs may be assessed through modification of laws and reorganization of the political and socioeconomic structures required to institute the new controls. Many of these costs are likely to be unanticipated and to carry the risk that resistance will follow. A consideration of the amount, the duration, and the form of the resistance must be a part of the study of social action to control drinking.

If resistance arises, decisions must be made about how it will be met and dealt with by control agencies. Such choices may be constricted by social structures and generalized values that make some means of control available but exclude others. For example, coercion may be an acceptable sanction in some societies but not in others. The organization of power in space and time may prevent the application of coercive controls to resisting populations, even where they are morally and legally permissible.

A special factor complicating the choice of controls over drinking is the irrationality of the intoxicated person and his unresponsiveness to symbols that limit the responses of sober persons. Added to this complication is the fact that socially deleterious behavior of the drunken person is often followed by acceptable or even praiseworthy behavior. Although the inebriated person is nowhere regarded as mentally disordered, some societies have sought to solve this problem by treating him or her as irresponsible. However, this strategy is not a solution in the sense of control.

The Absence of Social Control—Laissez Faire

A substantial number of societies appear to exercise little or no social control over consumption of alcoholic beverages; they approximate a condition of laissez faire. In these societies there is no organized public opinion unfavorable to drinking or drunkenness. With respect to the drunken person the attitude seems to be *caveat socius* rather than *caveat potor*—let society beware of the drinker rather than the reverse. Efforts at control in such societies are directed toward avoiding the costly consequences of drunkenness rather than toward controlling it. Speeches are made before feasts or drinking sprees, or rituals performed urging drinkers to avoid quarreling and fighting or to "have a good time." Sober persons are told to stay away from quarrelsome drunks and are even blamed if their not doing so results in their being injured. In extreme cases, fighting drunks may be separated and passively restrained. They may be tied to trees or, as in a case familiar to us among the Salish Indians of British Columbia, placed in cooking retorts ordinarily used to pack canned salmon. When the drunken person passes out, he may have a pillow placed under his head or be covered with a blanket or wrapped in a hammock pending his return to sobriety (Horton, 1943).

The apparent tolerance and lack of direct control of the inebriated persons in such societies are attributable in part to the high value placed upon intoxication by individual members. The attenuation of control is also consistent with the fact that the integrating functions of drinking for society are perceived and collectively valued. When urgent and persistent needs are met thereby, drunkenness may assume institutionalized form, as it did among Northwest Coast Indians (Lemert, 1954a, 1958a). Correspondingly, losses from drunkenness in these societies may be relatively low. Consuming beverages of low alcoholic content, such as beers, may mitigate the drunkenness or shorten its duration. The presence or absence of toxic congeners in the beverages and adulteration are additional variables predictive of unwanted consequences of intoxication.

Even when drunkenness is widespread and destructive at the time, its occurrence in the form of sprees or festivals following or preceding planting, harvest, hunting, and fishing seasons minimizes interference with economic activities.[2] If the production of food is carried out by isolated families, the costs of drunkenness will not be felt by the whole community. Finally, in societies living close to a subsistence level, the withdrawal of labor from the food quest tends to be a self-limiting phenomenon. With all of these facts to consider, it is probably correct to say that many primitive and rural societies adjust to, rather than control, intoxication and drunkenness. In all cases there are costs, but they seem to be written off or absorbed.

The *gemeinschaft* qualities of many primitive and isolated rural societies make them ill adapted to take action towards drunkenness even though its costs are substantial and are realistically perceived. Their dependence upon locality and kinship groups as units of control does not seem to work well, because they are bound together on an intensely personal basis. This happens because the aggression and aberrant sex behavior released by intoxication come from impulses in the drunken person that were never fully integrated in social interaction with family members, companions, and neighbors (Kluckhohn, 1944). Close associates seem to sense this, and furthermore they realize that they must "live with" the drunken person for years to come. This realization fosters ambivalence and deters forceful action, a condition that exists in families and neighbors of drunken persons even in modern urban societies.

When societies are organized into clans, a drunken person may be an especially vexatious problem because attempts to restrain him may backfire and lead to demands for damages or may even trigger warfare. When this potential is combined with a diffuse kind of authority in leaders and the absence of supraclan or "superordinate" organization, control over the drunken person may be at a minimum (Horton, 1943). Wives, who often bear the heaviest costs of drunkenness, have low status in most primitive and rural societies and often are products of out-group marriages. Hence, the values they hold are more easily sacrificed.

Ritual and Drunkenness

Evidence from several societies indicates that notwithstanding a high tolerance for drunkenness, serious difficulties are created when drinking coincides with ritual performance. Efforts to integrate the two may be made, but they are seldom successful. The Snohomish of Puget Sound say that intoxication spoils their spirit dancing. Likewise the Kwakiutl of Kingcome Inlet in British Columbia and the Bella Coola farther north say that drinking is contrary to the sacred value of their potlatch dances (Lemert, 1954b; McIlwraith, 1948). Yet attempts at control on these ritual occasions more often than not were completely absent or at best were token efforts. The Pomo Indians of California on one occasion decided that a 4-day winter ceremonial dance had been spoiled by drinking. No one, however, was reprimanded or punished. Instead, the entire ritual was repeated (C. Meighan, personal communication, 1952).

To the degree that basic values of a society are widely expressed in ritual, drunkenness becomes more costly to the community. It is quite possible that the detailed, complicated, and interlocking ceremonial life of the Zuni Indians, in which all feel a sense of participation, may explain why they went through the experience of first accepting and then rejecting alcoholic beverages (Benedict, 1958). The high degree of organization of Pueblo societies seems to have prepared them better to establish policing agencies and to use authoritative controls than most other American Indians. Their response to temperance propaganda was more broadly based, and early 19th-century temperance societies among them were less a missionary implant and more an indigenous growth (Cherrington, 1925). Yet some accounts suggest that the problem of drunkenness was not completely solved by the Pueblo peoples (White, 1942).

Contact and Interaction with White Society

The relation of social structure to the deficiency of social control over intoxicating drinks comes into clearest focus in American Indian societies that experienced direct and continuing contact with white explorers, traders, missionaries, settlers, and soldiers. The introduction of strong liquor into these societies was accompanied by a great deal of social disruption, destructive behavior, and demoralization (Howay, 1942; MacLeod, 1928; Salone, 1904). None of these societies was able to take effective action in the matter. Chiefs and high-ranking Indians in a number of tribes clearly appreciated the social costliness of immoderate liquor consumption, and they themselves sometimes showed restraint. In several instances, in the territories of the Iroquois and the Shawnee, efforts were made to install prohibition but were without success (MacLeod, 1928; Wraxall, 1915).

The problem, while obviously concerned with values, hinged also upon the importance of clans and kin groups as agencies of control in these societies and upon the need for complete unanimity of public opinion before action could be taken (Mead, 1937). The typical way in which nobles and chiefs sought to influence the behavior of others—through precept and example—also proved to be a poor technique for restraining those—often younger men—who chose to get drunk or enter into dealings with liquor sellers.

Looming over the immediate situation was a larger context in which American Indian societies were caught in conflicts of power between imperialistic nations of Europe, especially France and England. Jealous colonial governments and rival fur trade companies gave added dimensions to these struggles. Later the United States became a party in the conflicts. In some cases warring nations deliberately used rum or brandy as a means of weakening an Indian tribe allied with their enemies (MacLeod, 1928). The few agreements reached by competing fur companies to curtail the sale or trading of liquor to Indians quickly fell into disuse (Chittenden, 1902). Many colonial governments and, later, states and Canadian provinces passed prohibition laws, but their history has been one of evasions and unenforceability.

Although missionaries tried to aid the American Indians in their struggles with liquor, their policies and methods were not readily accepted. The price they demanded in return for the Christian sobriety they promised was one the Indians could not afford to pay, being more or less a complete repudiation of deeply cherished values as well as their culture-sustaining rituals. Thus, drunkenness often took on a reactionary virtue—becoming a means of resistance and of discrediting the "missionary way" and at the same time reaffirming older values (Lemert, 1958b).

Although control and, indeed, the idea of control over drinking, as we have shown, have been weakly developed or absent in many societies,[3] other societies, particularly those of the Western world, have followed a much more positive course of action. They have recognized drinking and its consequences as a problem to be solved or as a threat to the integrity of society itself that calls for removal. This task has been undertaken through decrees, passage of laws, and the indoctrination of special ideologies. A variety of control systems and methods have been devised; specialized agencies and organizations have been instituted and given responsibility for control over alcohol consumption. To discuss all of these in the detail they merit is impossible here; at best we can examine a few of the salient ideas around which they have been built.

Four Models of Social Control

A reasonable working assumption is that the objective of social control over alcohol use generally is to minimize the costs of intoxication and drunkenness. For purposes of determining whether and how such an objective can be reached, it is helpful to formulate a number of hypothetical models of social control. These are drawn from the experiences of whole societies and also from the ideologies and programs of power elites that at different times and places have actively sought to bring alcohol use under control.

Model I: The Costs of Intoxication and Drunkenness Can Be Reduced by a System of Laws and Coercive Controls Making It Illegal to Manufacture, Distribute, or Consume Alcoholic Beverages

This, of course, is a familiar model—prohibition. It has been tried in several forms, for long and short periods of time, in Aztec society, ancient China, feudal Japan, the Polynesian Islands, Iceland, Finland, Norway, Sweden, Russia, Canada, and the United States. The well-documented failures of the model can be attributed to its high costs, the instability of power elites favorable to prohibition, the limitations of power and available means of control, and the growth of resistance unresponsive to coercion.

The prohibition model in effect sacrifices all of the values of moderate drinking as well as those associated with intoxication, as well as the vested values of those who earn a livelihood or receive investment returns from the production and distribution of fermented drinks and liquor. In order for such high costs to be willingly paid, a large number of power elites must either see positive value gains in prohibition laws or see them as a means of protection against threatened value losses (Lee, 1944). Consequently the prohibition model most likely can be established only through a social movement during periods of rapid social change, culture conflicts, conquest, or nationalistic movements.

Reform movements, from which prohibition springs, tend to be ephemeral in nature because new issues supplant the old, undermining or destroying the alignments of elites supporting them. Disillusionment of their individual adherents and defection of groups through changing policies are speeded by difficulties of enforcement and consequences of resistance.

Resistance can be predicted for any model of control requiring the abandonment of values deeply held by large segments of a population. In the case of prohibition the problem of enforcement is augmented because it is a form of sumptuary legislation that affects the more personal aspects of human behavior and individual choice making in intimate or private behavior. In the absence of a reinforcing public opinion the application of coercive controls seldom has succeeded. Even the most severe punishments, such as death among the Aztecs and exile to Siberia in Russia, failed to abolish bootlegging in those societies (Lemert, 1958a).

In large, complex societies a control model that prohibits a highly desired item increases its scarcity value and also the probability that collective enterprise will grow up to supply it. The large variety of foods that can be converted into alcohol and the ease and cheapness of its production and its movement make bootlegging and smuggling inevitable. The costs to the state of discovering and stamping out such illegal industry can reach a point where governing elites are unable or unwilling to pay them. The organization of evasion poses threats to other values, such as respect for property and life and even the value of government under law. This contributes further to resistance and reaction against prohibition.

The power of a given state to enforce prohibition may be insufficient if the economic and political values of other states are threatened thereby. In 1841, France intervened under threats of bombardment and compelled the reigning monarch of the Hawaiian Islands to end prohibition there (Thursten, no date). France, Germany, Russia, and Spain at different times variously applied pressures to Finland, Norway, Iceland, and Turkey in order to prevent interference with liquor imports and smuggling (Catlin, 1931).

It may be more profitable to speculate on the conditions under which the prohibition model can succeed rather than to inventory the reasons for its failure. Probably it would require conspicuously high costs of drunkenness on one hand and on the other a positive replacement of the drinking values or substitution of new means of achieving the old values. A precondition of this scenario would be relatively complete geographic isolation, similar to that found on islands, where behavior deviations have a high visibility. A social structure in which power is concentrated and little affected by public opinion or is upheld by supernatural sanctions perhaps would make for successful prohibition. Needless to say, conditions such as these are increasingly anomalous in the world of the present.

Model II: The Costs of Intoxication and Drunkenness Can Be Reduced by a System of Indoctrinating Information about the Consequences of Using Alcohol, Thus Leading to Moderate Drinking or Abstinence

The assumption behind this control model is that a causative relationship holds between controlled presentation of information and change in attitudes and values. The general idea is favored by some research on attitudes in specified areas of behavior. The findings, however, are not altogether consistent; and where giving information has been found to modify attitudes and values, the change is not always in the anticipated direction (Bonner, 1953). In general it would be hard to say whether exposure to information leads people to change specific values or whether the reverse is true—that is, whether the adherence to certain values stimulates people to inform themselves of pertinent facts. Currently there is no conclusive research from which judgments can be made about the kinds of information or educational content best calculated to achieve the goals of abstinence or moderation in drinking.

In the absence of data, the content of alcohol education has tended to be influenced by values and policies of temperance groups. This has been especially the case of alcohol education in the schools of countries where forms of this model have been tried. Those whose values are more directly involved—parents—are not inclined to resist the inclusion of special curricula on alcohol education in the schools because many believe that some guidance is needed, especially in the United States, where education is apotheosized as a solution for social problems. On the other hand, some evidence from England and the United States speaks of indifference to or lack of sympathy with the alcohol curriculum on the part of teachers who do not subscribe to temperance values. This attitude may reflect a failure to reconcile religious and scientific orientations in the recommended curricula (Levy, 1951). Resistance from these sources, of course, need not necessarily be fatal to the working of this model.

A more serious flaw in the educational model of alcohol control lies in the probability that values surrounding drinking and embedded in drinking patterns are primarily shaped by experiences in the family and in peer groups rather than by formal educational agencies (Levy, 1951; Sariola, 1956; Straus & Bacon, 1953). No problem is created for the society relying upon the educational model if it is homogeneous in its drinking or abstinence habits. In the absence of such homogeneity, educators are faced with discontinuity between the learning process in the schools and what goes on in primary groups outside. Examples set by parents at home or pressures to conform in friendship groups easily cancel out the abstinence or moderation precepts of the school.

Education for restraint in drinking might be directed to parents instead of youth, through agencies outside, or peripheral to, the school. Yet prospects for this kind of enlightenment are not promising. Family-life education is generally considered not yet to have proved its worth (Goode, 1950). Other programs, such as those promoted by state mental hygiene departments or by local alcoholism committees—or by citizens' committees, as in Russia—are still largely untested by research or extensive empirical trial. In some places it has been shown that mental hygiene education can have effects opposite to those intended (Cumming, 1957; Davis, 1938).

The means that can be made to work for alcohol education have not been well adapted to the ends sought. Pamphlets, charts, movies, and lectures that dwell upon the results of excessive drinking often seem to run afoul of the ambivalence underlying popular reactions to alcohol. The arousal of fears, the implied warnings, and the threats about what will happen if one drinks too much have been noted to provoke avoidance reactions toward further propaganda. There also seems to be an unwillingness of audiences to particularize such propaganda. This reaction is very apt to occur if the educational materials point to alcoholism as an end result of drinking.

Research into the factors that account for the long history of Jewish sobriety makes it fairly certain that the indoctrination of values is significant. It has not yet been settled what the values are, nor whether they are generalized or specific in nature. Furthermore there is a good possibility that such values may be functional only in a special context of ongoing social control represented by the Jewish community (Glad, 1947; Snyder, 1958). Comparable research into drinking by Italian Americans and Italians concludes that valuation of wine as a food is an important part of their relative sobriety (Lolli et al., 1958). Yet this scarcely seems to be a complete explanation, and even if it were, there are no investigations to clarify the process by which such a value is inculcated or maintained in the ethnic population.

A final necessary comment on the educational model of control is that in a mobile, culturally diversified society that changes rapidly it becomes difficult to predict what pattern of values indoctrinated in children will best serve to adjust them as adults. To the extent that drinking is a response to situations, to adventitious groupings, and to stresses generated by role conflicts or social isolation, reliance upon a preconceived pattern of drinking values to control excessive use of alcohol may fail. Attention must be directed to the controls functioning in the drinking situation.

Model III: The Costs of Intoxication and Drunkenness Can Be Reduced by Legal Regulation of the Kinds of Liquor Consumed, Its Pecuniary Cost, Methods of Distribution, the Time and Place of Drinking, and Its Availability to Consumers According to Age, Sex, and Other Socioeconomic Characteristics

This model rests upon the conviction that the state or its agencies can determine what amounts and what forms of drinking have costly consequences. In its most comprehensive form the model defines drinking as a privilege that, if abused, can be withdrawn from the individual; corollary to this definition is withdrawal of privileges, such as that of driving an automobile, that are affected by drinking. Archetypes of this model are found in the history of Scandinavian countries—particularly Sweden, home of the Gothenberg and Bratt systems of liquor control. Examples also come from the temperance orders of medieval Germany, possibly including the apocryphal *jus potandi,* the drinking code of orders given to heavy drinking (Samuelson, 1878).

Government regulation of alcohol consumption grew up historically largely from nonmoral considerations. Among these were popular agitation that governments make fermented beverages equally available to localities and demands that the quality of such beverages be insured against fraud. The willingness of populations to accept taxes on alcohol production as a revenue measure has been an enticing path to regulation for financially hard-pressed governments (Catlin, 1931). This last possibility has caused a persistent dilemma for governing officials who have to choose between raising money the "easy way" and at the same time taking steps to diminish heavy drinking, the latter course of action inevitably having the effect of decreasing revenues. The dilemma has been conspicuous in the history of Russia, where, under the tsar's *kabak* system, and subsequently under the "farming out" system of vodka monopoly, the government abetted or encouraged widespread drunkenness (Efron, 1955; Johnson, 1915). The dilemma still lives today in muted controversies over the respective merits of licensing versus monopoly and trust systems of alcohol distribution.

Effective alcoholic beverage control may strike heavily at the economic values of producers and distributors even when it does not threaten drinking values of the population. If these elites are numerous and well organized, their resistance may well nullify efforts at regulation. A well-documented case in point is France, whose parliaments passed a series of regulatory laws after the first and second world wars. Yet the great power of the wine industry there has prevented anything beyond token enforcement of the laws. The presence of approximately a third of the electorate who are either workers in the wine industry or their family members does much to explain this phenomenon (Mignot, 1955). The so-called *bouillers de cru* (home distillers), who are entrenched in certain areas of France, seem to openly defy regulation, protecting their traditional privileges largely through sheer power of numbers at the polls.

In the latter part of the 19th century, in some countries pressures from temperance groups exerted a mounting influence on government regulation. The result has been that regulation more and more has reflected a power conflict between temperance organizations and those of the liquor industry. Regulations formulated by legislative bodies have been the incorporation less of consistent policy and designed control than of compromises, special concessions, exceptions, and arbitrary requirements. The atmosphere of mutual distrust between the two power alignments, of which these have often been the products, is not conducive to enforcement.

As other power elites—such as health, welfare, and law enforcement agents; researchers; and tax officials—become more professionalized and articulate their values, regulatory laws can become more symmetrical. Administrative rule making also permits a more rational adaptation of the means of control to the ends. If the regulatory model is to work efficiently, however, conflicting elites must be able to believe that their values can be realized or preserved through regulation. Those persons and groups who have to bear the heaviest sacrifice of values must be able to find alternatives.

The organization and the jurisdiction of regulatory agencies present as yet unsolved problems of this control model. In large, heterogeneous societies like our own and perhaps that of the Soviet Union, a high level of uniformity in regulations coupled with centralized control over alcohol use carries a strong probability of resistance. On the other hand, extreme decentralization of control and dependence upon purely local agencies invite connivance and circumvention of rules where they deviate from local drinking customs or run counter to interests of local power elites.

Although this model is not designed to liquidate values associated with drinking or even drunkenness, it may nevertheless have this effect through regulations that significantly alter the form or pattern of drinking. In areas where public opinion does not support such regulations or the actions of the enforcing agencies, the result often is simply the addition of extraneous behavior to the form of drinking without appreciably modifying it. Thus, a rule specifying that children and youth may not enter a liquor establishment unless food is served there may do nothing more than cause proprietors to install a bare minimum of restaurant equipment.

Requirements that wine and liquor can be served only with meals may simply have the effect of adding the cost of a meal to the liquor bill.

Where lawyers or legal-minded elites set the policies of regulatory agencies, there may be little understanding of the functions and values of drinking groups and related institutions. Historical studies and research reports both have shown a significant relationship between drinking groups and the persistence of primary groups' values (Kolb, 1959; Levy, 1951). Ignorance or disregard of such facts easily vitiates regulation. In this country many states have regulations against extending credit in bars and taverns. Yet frequently the success of such a place depends upon personal ties between customers and barkeepers. Hence, many of the latter put drinks "on the tab" or "hold" personal checks as a means of giving credit. Not only does the regulation often fail in its purpose, but it also plays a part in many offenses involving bad checks.

The chief means for implementing regulatory rules are suspensions and withdrawal of distributors' licenses or, in the case of monopoly systems, manipulating the number of outlets and their hours of sale. Along with these are the withdrawal of ration cards or "buyer surveillance" (Lanu, 1956), techniques aimed at errant individual drinkers. Behind all of these is the possibility of police action for persistent and flagrant violators.

Distributors and retailers unquestionably can be hurt economically and made more receptive to rules by suspension or revocation of licenses. Furthermore, experiments in rural Finland have shown that placement of sales outlets does have some limited effects upon the kinds of alcoholic beverages consumed in certain population categories (Kuusi, 1957). No workable methods yet have been invented to control the individual drinker or drinkers who want to "beat the system." If their motivation is strong enough, such persons will find ways of circumventing the regulations. Furthermore, controls focused upon the excessive drinker often inconvenience or alienate persons who comply with the form and meaning of the regulations. Even in Scandinavian countries, where a strong tradition of government paternalism prevails, ration cards and "buyer surveillance" have not been popular (Lanu, 1956).

The preceding discussion underscores the importance of public opinion in securing cooperation necessary for regulatory control. The cooperation of local law enforcement officials is equally important. A broad area of responsibility for dealing with the consequences of drinking must always remain with the local community because it must deal with disturbances of public order, offenses against the family, and juvenile offenses, which in varying degrees involve drinking and drunkenness. Legal procedures and coercive controls, such as fines and jail sentences, are poor methods for handling such cases because the consequences of punishment often are worse than the offense.

Model IV: The Costs of Intoxication and Drunkenness Can Be Reduced by Substitution of Functional Equivalents of Drinking

This model has received most attention in England, where it has been the subject of investigation under the headings of "moderating influences" and "counter attractions" (Levy, 1951). It has interested those who see excessive drinking as a symptom of some kind of "deprivation" of human beings that is essentially due to defects or omissions in social structures (Bales, 1946; Poirier, 1956). It carries the assumption that values satisfied through drinking or drunkenness can be fulfilled through other activities. It calls for an engineering-type reorganization of community life so that time, money, and interests devoted to drinking will be redirected into sports, games, gardens, radio and television programs, motion pictures, travel, and similar diversions. Improved housing to make family life more attractive and building of community centers also are envisioned as part of this model.

In certain kinds of internally controlled or isolated community situations in which boredom and apathy or social isolation have reached critical proportions, diversionary activities may very well decrease the extent of drunkenness. This pattern has been observed in military encampments and isolated outposts (Moore, 1942). Comparable data are also at hand in the history of a missionary system of control among the Salish Indians of British Columbia; the religious pageantry organized by Catholic Oblates for a time, at least, successfully replaced whiskey feasts and decreased other forms of drunkenness (Lemert, 1958a).

It is, of course, naive to expect to convert urban communities into analogues of military camps or missionary societies. Short of this tactic the best that can be done is to introduce new programs, such as recreation, into situations in which many other variables cannot be controlled and to look for changes in the amount and forms of drinking. This is a crudely empirical procedure and can be very costly. Where heavy economic costs must be met, as in housing development, business and governing elites will not easily support the programs.

Despite these reservations this model may have usefulness in many situations that occur in contemporary societies. Here we think of logging camps, long-term construction projects in sparsely settled areas, technical research teams, and diplomatic corps in foreign countries, as well as of military installations throughout the world. Wherever there are situations in which centripetal social integration operates, manipulation of social participation may be significant in reducing drunkenness.

This model has an appeal to the researcher because in the kinds of situations that have been specified it may lend itself to rigorous testing. An important task in such research would be to ascertain whether in given cultural contexts drinking is inescapably associated with attaining certain value satisfactions. The obverse of the question is whether drinking or drunkenness is symptomatic for societies in the same way that some psychiatrists hold it to be for individuals. This query also merges into one about the influence of values on the selection of narcotics, stimulants, and sedatives by societies—a problem similar to that of "symptom choice" in psychiatry.

Conclusion

As yet, no model of social control has been evolved that has been greatly effective in diminishing the costs of excessive drinking. Research is complicated by the fact that the adoption of control programs and a decline in drunkenness both may be functions of changes in larger value systems. In general, those societies and groups that place a high value on sobriety and a low value on intoxication do not have a need for extensive social control. This generalization is subject, however, to the qualification that drunkenness among a small number of persons whose roles express basic values of the society or drunkenness at vulnerable junctures in an industrial system magnifies the need for control. Presumably the necessary controls are more easily established under such conditions.

Societies that place a high premium upon the pleasures of drink and that have the greatest need for control are inclined to reject programs of control or to sabotage them if they are established. Members of these societies who do not share the drinking values or who perceive their high costs may be unable to make their voices heard in the arena of government or in community councils. If they do, they risk unpopularity and ostracism. France is an almost classic example of this situation.

Large societies with mixtures of ethnic minorities, diverse locality, and occupational groups make it unlikely that any one model will suffice to eliminate socially harmful drinking. The problem of choice of a model is complicated by the fact that drinking may be, in turn, a culture pattern, a symptom of psychic stress, a symbolic protest, or a form of collective behavior. Yet a technologically oriented society inexorably demands that drinking, whatever its form, not be permitted to disrupt crucial social integrations that cut across many groups. Formulation of these requirements in the areas of industry, communications, health, and family life probably can be accomplished only by controlling elites "from above." Achievement of these minimum conformities in drinking behavior can best be implemented at the grass-roots level in particular groups in which resources for control at the level of informal interaction can be tapped and brought into play (Lanu, 1956).

An example of the possibilities in such a process was the decision of English labor groups and associations in the 19th century to remove their meetings from public houses. According to some writers (Samuelson, 1878) this action was important among other influences bringing about a decline in drunkenness during the period. In our present-day situation, there is little hope that state officials or police can directly control such indigenous cultural growths as office parties at Christmastime, New Year's Eve celebrations, "martini luncheons," "beer-bust" picnics, and general weekend and holiday drinking behavior. It does seem possible, however, that employers' groups, professional groups, unions, clubs, and civic associations that are close to the drinking phenomena can assume a larger share of responsibility for such control.

The problem of establishing communication between these groups and responsible control agencies is formidable, and it requires more than legal instrumentation. When drinking takes place adventitiously or when it is a form of protest or alienation from society—as with much teenage drinking and with "bottle gangs" in Skid Row—control through the co-optation of groups is difficult, if not impossible. It is here that direct regulation, unsatisfactory though it may be, must be applied.

Control of any kind is a marginal influence in social and cultural change—a consideration no less true of action to reduce the costs of intoxication and drunkenness. Control cannot create behavior *de novo,* but it can strengthen existing tendencies by articulating unspoken values and by organizing the unorganized dissidents in a population. Further, it can define programs of action in a way to minimize resistance or gain the support of otherwise opposed or indifferent groups. Whether this reorganization can be done best at the local, regional, or national level is a question best left to research and open-minded experimentation.

No effort has been made here to devise and discuss a model of control for the addictive drinker. Models suitable for limiting drunkenness in whole societies most assuredly will not apply to the alcoholic. The nature and ordering of values in these persons is such that they usually are unmoved or even made hostile by symbols of control that have an effect upon other drinkers. The extreme tensions under which they labor and their disturbed social interaction distort their value systems in complex ways that preclude the communication necessary for control. Such pathological drinkers presumably can be made responsive only through specially invented therapeutic models of control.

Notes

1. *Function* in this chapter means operation or process; *value* is an object or a state that is desired. Functions may or may not be valued.

2. When hunting is a continuous necessity rather than a seasonal activity, drunkenness may be more disruptive in a hunting society than in an agricultural society. This state of affairs seems to have been the case with the Eastern Woodlands Indians (Kelbert & Hale, no date).
3. It is not intended to claim here that lack of control over drinking is a characteristic of primitive societies in general. Indeed, recent research has revealed the existence of reasonably successful "disciplined" drinking in certain Polynesian societies (Lemert, 1962).

References

Bacon, S.D. (1945). Alcohol and complex society. In *Alcohol, science and society* (pp. 190–193). New Haven, CT: Journal of Studies on Alcohol.

Bales, R.F. (1946). Cultural differences in rates of alcoholism. *Quarterly Journal of Studies on Alcohol, 6,* 482–498.

Benedict, R. (1958). *Patterns of culture* (pp. 78–82). New York: New American Library (Mentor Books).

Bonner, H. (1953). *Social psychology* (pp. 185–192). New York: American Book Company.

Catlin, G. (1931). *Liquor control.* New York: Holt, Rinehart and Winston.

Cherrington, E.H. (1925). *Standard encyclopedia of the alcohol problem* (pp. 5–7, 37–39). Westerville, OH: American Issue.

Chittenden, H.M. (1902). *The American fur trade of the Far West* (Vol. 1, chap. 4). New York: Francis P. Harper.

Cottrell, W.F. (1954). *Research for peace.* Amsterdam: North Holland Publishing Company.

Cumming, E., & Cumming, J. (1957). *Closed ranks: An experiment in mental health education.* Cambridge: Harvard University Press.

Davis, K. (1938). Mental hygiene and class structure. *Psychiatry, 1,* 55–65.

Efron, V. (1955). The tavern and saloon in Old Russia: An analysis of I.G. Pryshov's historical sketch. *Quarterly Journal of Studies on Alcohol, 16,* 484–505.

Glad, D.D. (1947). Attitudes and experiences of American-Jewish and American-Irish male youths as related to differences in adult rates of inebriety. *Quarterly Journal of Studies on Alcohol, 8,* 406–472.

Goode, W.J. (1950). Social engineering and the divorce problem. *Annals of American Academy of Political and Social Sciences, 272,* 86–94.

Haggard, H.W., & Jellinek, E.M. (1942). *Alcohol explored.* Garden City, New York: Doubleday, Doran and Company.

Horton, D. (1943). The functions of alcohol in primitive societies: A cross-cultural study. *Quarterly Journal of Studies on Alcohol, 4,* 199–319.

Howay, F.W. (1942). The introduction of intoxicating liquors amongst Indians of the northwest coast. *British Columbia Historical Review, 6,* 157–169.

Johnson, W.E. (1915). *The liquor problem in Russia.* Westerville, OH: American Issue.

Kelbert, M., & Hale, L. (no date). The introduction of alcohol into Iroquois society. Unpublished manuscript, University of Toronto, Department of Anthropology.

Kluckhohn, C. (1944). *Navaho witchcraft. Papers of the Peabody Museum of Archaeology and Ethnology, 22* (2), 52–54.

Kolb, J.H. (1959). *Emerging rural communities* (pp. 60f). Madison, WI: University of Wisconsin Press.

Kuusi, P. (1957). *Alcohol sales experiment in rural Finland.* Helsinki: Finnish Foundation for Alcohol Studies.

Landis, B.Y. (1945). Some economic costs of inebriety. In *Alcohol, science and society.* New Haven, CT: *Journal of Studies on Alcohol.*

Lanu, K.E. (1956). *Control of deviating drinking behavior.* Helsinki: Finnish Foundation for Alcohol Studies.

Lee, A. (1944). Techniques of social reform: An analysis of the new prohibition drive. *American Sociological Review, 9,* 65–77.

Lemert, E. (1954a). Alcohol and the Northwest Coast Indians. *University of California Publications in Society and Culture, 2,* 303–406.

Lemert, E. (1954b). The life and death of an Indian state. *Human Organization, 13,* 23–27.

Lemert, E. (1958a). An interpretation of society's efforts to control the use of alcohol. In *Alcoholism: Society's responsibility.* Berkeley, CA: California State Department of Health.

Lemert, E. (1958b). The use of alcohol in three Salish Indian tribes. *Quarterly Journal of Studies on Alcohol, 19,* 90–107.

Lemert, E. (1962). Alcohol use in Polynesia. *Tropical & Geographical Medicine, 14,* 183–191.

Levy, H. (1951). *Drink: An Economic and Social Study* (pp. 136–141). London: Routledge and Kegan Paul.

Lolli, G., Serianni, E., Golder, G.M., & Luzzatto-Fegiz, P. (1958). *Alcohol in Italian culture.* Glencoe, IL: Free Press.

MacLeod, W.C. (1928). *The American Indian frontier* (chap. III). London: Kegan Paul, French, Trubner.

Mead, M. (1937). Public opinion mechanisms among primitive people. *Public Opinion Quarterly, 1,* 5–16.

Mignot, A. (1955). *L'Alcoolisme: Suicide collectif de la nation* (pp. 83f). Paris: Cahiers des Amis de la Liberté.

McIlwraith, T.F. (1948). *The Bella Coola Indians* (Vol. II). Toronto: University of Toronto Press.

Moore, M. (1942). The alcohol problem in military service. *Quarterly Journal of Studies on Alcohol, 3,* 244–256.

Myerson, A. (1940). Alcohol: A study in social ambivalence. *Quarterly Journal of Studies on Alcohol, 1,* 13–20.

Nordhoff, C.B., & Hall, J.N. (1934). *Pitcairn's Island.* Boston: Little, Brown.

Poirier, J. (1956). Les sources de l'alcool. *Alcool en Oceanie.* Paris, Mission des Iles, No. 66.

Salone, E. (1904). Les sauvages du Canada et les malades importées de France au XVIIIe siècle: La picote de l'alcoolisme. *Journal de la Societé des Américanistes, 4,* 1–17.

Samuelson, J. (1878). *The history of drink.* London: Trubner.

Sariola, S. (1956). *Drinking patterns in Finnish Lapland.* Helsinki: Finnish Foundation for Alcohol Studies.

Smith, A. (1950). *The wealth of nations* (pp. 456–457). London: Methuen.

Snyder, C.R. (1958). *Alcohol and the Jews.* Glencoe, IL: Free Press.

Straus, R., & Bacon, S. (1953). *Drinking in college* (chaps. 6, 9). New Haven, CT: Yale University Press.

Thursten, L.A. (no date). The liquor question in Hawaii. University of Hawaii Library, Manuscript Collection.

Translation of the laws of the Hawaiian Islands established in the reign of Kamehameha III. (1842). Unpublished manuscript, Hawaiian Missionary Children's Society Library, Honolulu.

White, L. (1942). The Pueblo of Santa Ana. *American Anthropologist,* n.s. *44,* 69.

Wraxall, P. (1915). *An abridgement of the Indian affairs, contained in four folio volumes, transacted in the Colony of New York, from the year 1678 to the year 1751* (C.M. McIlwaine, Ed.). Cambridge: Harvard University Press.

CHAPTER 36

The AA Group

Norman K. Denzin

In this chapter I will examine the structures of interpretation the recovering alcoholic finds in Alcoholics Anonymous. With Maxwell (1984, pp. 38–39) I assume that the alcoholic who enters AA finds a new social world. This world is structured around AA's Twelve Steps, the Twelve Traditions, and around a fellowship in which "being an alcoholic" is the primary identity that is shared (Denzin, 1987; Rudy, 1986).

My intentions are to bring the structures of AA before the reader, as these structures are given to the new member. By offering a combination of historical, structural, textual, and interactional analyses of the "AA experience," I hope to reveal how the recovering alcoholic can in fact become a part of a structure of experience that is larger than the member is. I will show how AA requires the member's presence and that of other newcomers for its continued existence. In so doing I will examine how the recovering alcoholic realizes his or her "universal singularity" in the company of fellow alcoholics who are also universal singulars in their recovery experiences. By drawing upon this universal singularity of each of its members, AA solves its own organizational problematic of membership replacement (and recruitment), while giving each of its members a context of interaction wherein recovery from alcoholism can occur. How this oneness of experience and purpose generates a structure of common experience that is mutually beneficial to all parties—AA and alcoholic alike—is my topic. Central to this process is the cultural and personal history AA rests upon—most important, the history of the first AA group. I briefly turn to this history as it is made available to the new member.

Norman K. Denzin is with the Department of Sociology, University of Illinois at Urbana-Champaign, Urbana, Ill. This chapter was written especially for this book. Portions of this chapter are drawn from *The Recovering Alcoholic,* by N.K. Denzin (Beverly Hills: Sage, 1987), chap. 4.

The First AA Group and Lived History

AA tradition states: "Two or more alcoholics meeting together for the purpose of sobriety may consider themselves as an AA group." Consider the following conversation between two AA members. M, 48 years old, has been sober for over three years. He has arrived early at the meeting site of an AA group for its usual Thursday night (8:00 o'clock) meeting. D, a 21-year old college student, has recently received a medallion for 6 months of continuous sobriety. It is now 8:05 and they are the only two AA members present. M speaks first:

> Do you want a meeting? You know two alcoholics can have a meeting if they want to. Remember Bill W. and Dr. Bob? That first meeting was just the two of them.

D nods his head and replies:

> Yes, I do. I remember that story. I heard it in Chicago. I need a meeting. I've only had two this week. I'll chair, if you'll read "How It Works" and "The Thought for the Day." [field conversations, as reported, October 2, 1983]

The meeting begins. A third AA member joins the group at 8:10, and the meeting started between M and D continues until 9:00, with each of the members speaking in turn.

This account instances the reenactment of AA's original two-member group. M and D drew upon their knowledge of the first meeting between Bill Wilson and Dr. Robert Smith (Bill W. and Dr. Bob), using that pivotal moment in AA history as justification for their two-person AA meeting. That D had heard this story in a meeting in Chicago reveals how this key moment in AA history is kept alive and passed on to new members.

AA dates its inception from the first meeting that occurred between Bill Wilson and Dr. Robert Smith, starting at 5:00 p.m. and ending at 11:15 p.m. on Sunday, May 12, 1935, in the home of Henrietta Seiberling and her husband in Akron, Ohio. The first meeting between Smith and Wilson has subsequently been redefined, within AA folklore, as the first AA group.

The importance of the foregoing is elaborated in the following statements made by AA members in meetings. The first speaker has just returned from a Young People's Convention in Chicago. Sober slightly more than one year, he states:

> It was great. The keynote speaker was sponsored by Bill W. He talked of the first groups in Cleveland and Akron. He told us to work the steps and follow

> the traditions. He talked about the fights they had in the early days in Detroit over the membership requirement of six months sobriety before you could come to a meeting. It was great! I'm really glad I went. [field conversation, September 20, 1984]

This speaker conveys the importance of having heard an AA member sponsored by Bill Wilson. By listening to this man he is taken back into the early days of AA. He is given the essential message of Wilson and Smith in 1935. He was a witness to an instance of lived history within AA.

The next speaker is 76 years old. He has been in AA for 35 years. He knew the Reverend Sam Shoemaker, the minister who worked with Wilson in 1934 when Wilson was trying to sober up men in Shoemaker's church in New York:

> Old Sam Shoemaker used to say, "Keep it simple. Listen to what is inside you, practice the spiritual principles but don't come down on this God thing too hard for newcomers." [field conversation, July 4, 1983]

This AA member keeps AA history alive by mention of Sam Shoemaker, one of the early advisers of Wilson and a person listed by Wilson as having been central to the early beginnings of AA.

The third speaker is a female, sober seven years. She states, in regard to self and ego:

> Her majesty the baby. That's me. I'm a spoiled brat. I want things my way, and I want it now. I keep getting in my own way. The Big Book [the semiformal name for *Alcoholics Anonymous,* the basic text of AA] and the Twelve and Twelve [Twelve Steps and Twelve Traditions, another important AA book] talk about me—I prefer her majesty the baby! [field conversation, November 25, 1982]

The phrase *his majesty the baby* is attributed to Freud by Tiebout (1954, p. 612), who used it to describe the narcissism of alcoholics. Tiebout also employed the phrase in a letter to Bill Wilson, suggesting that Wilson was trying to live out infantile grandiose demands (Kurtz, 1979, p. 127). Alcoholics often apply this phrase to themselves. The speaker, perhaps not knowing the source of his phrase, was, as she spoke, connecting herself to a moment in AA history when Tiebout was chastising Wilson for his own self-centeredness.

The above interactions reveal how AA's history, its key figures, its mythology, its folklore, and its key phrases are kept alive in meetings and in AA's oral tradition.

Four Types of Alcoholic Understanding

Understanding refers to "the process of interpreting, knowing and comprehending the meaning intended, felt and expressed by another" (Denzin, 1984a, p. 284). *Alcoholic understanding* refers to the process whereby each in a pair of alcoholics interprets, knows, and comprehends the meanings intended by the other in terms of previously experienced interactions with the active and recovering phases of alcoholism. Four forms of alcoholic understanding may be distinguished. The first is the understanding conveyed between a recovering alcoholic and an alcoholic who is still drinking but wishes to stop or is seeking help. This form may be termed *authentic alcoholic understanding.* The second form of alcoholic understanding is that conveyed between two or more alcoholics who are recovering and are in, or have been in, Alcoholics Anonymous. The three statements given by alcoholics in the discussion of "lived history" displays this form of alcoholic understanding, which may be termed *AA understanding.* Each member who listened to those accounts knew who Bill Wilson was, had perhaps heard of Sam Shoemaker, and could relate to the phrase *his majesty the baby.*

The third form of alcoholic understanding transpires between drinkers who are active alcoholics and who may or may not have had experience with Alcoholics Anonymous. I term this *insincere alcoholic understanding.* In this mode the alcoholics justify their active alcoholism through recourse to the term *alcoholic,* or its equivalents—*lush, drunk, alkie.*

The fourth form of alcoholic understanding is, as the foregoing might suggest, *spurious alcoholic understanding.* Alcoholics who wish to recover, or who have recovered, have experienced this form of understanding many times in the past. Their network of significant others—including enablers and victims—have bombarded them with the spurious understandings nonalcoholics bring to bear upon the experiences and the relationships they have with a practicing and a recovering alcoholic. Although these others may feel sympathy and compassion for the alcoholic who is still drinking, they do not have the experiences of the active alcoholic. They have not felt the pains of withdrawal. They have never hidden a supply. They have never felt the fear of a police car stopping them for drunk driving. They have not been hospitalized for alcoholic-related illnesses. They have not felt the embarrassment of seeing the word *alcoholic* on a medical record. They do not know what it means to try to stop drinking and not be able to. They have never attempted to work while under the addictive influence of alcohol or drugs.

In the absence of these and related experiences, the alcoholic's other stands as an "outsider" to the inner, lived experiences of the alcoholic who is attempting to control and manage his or her alcoholism. All of the good intentions of the other produce misplaced, spurious alcoholic understandings. The following words by an alcoholic married for 45 years to the same woman address this point:

> I've lived with that woman for 45 years. For 30 of those years I was an alcoholic, drunk every day. And after those 45 years she'll never come within a hair on a gnat's ass in understanding me. She can't. She's not an alcoholic. I don't care how many of those Al-Anon meetings she goes to. She'll never understand me! Never. [field observation, February 24, 1984]

This fourth form of alcoholic understanding draws the recovering alcoholic back to Alcoholics Anonymous, for in AA he or she finds a community of others who are understanding in ways that are not spurious. In the fields of experience that AA offers, the recovering alcoholic experiences, over and over again, authentic emotional understanding.

Every alcoholic who recovers has experienced this process of having "received" the message of recovery from another alcoholic. If the alcoholic has gone through treatment, this process of coming to understand alcoholic understanding will have been repeatedly experienced. As the alcoholic learns to attend AA meetings and becomes a regular AA member of the AA social world in his or her community, the word *understanding* will be one of the most frequently repeated terms he or she hears. Alcoholic understanding, in all its forms, and the meanings and terms that flow from that phrase, constitute the frame of reference or universe of discourse from which all other AA discourse flows.

I turn next to the topics of the AA group and the AA meetings. It is in the context of these two phenomena that recovering alcoholics learn to achieve and experience the kind of self-understanding of alcoholism that was first promised when "the message" was carried to them. My intentions are to offer a view of the AA group that is consistent with Sartre's (1960/1976) dialectical theory of groups. It will be necessary in this discussion to define such terms as *group, fused group, pledge,* and *"third party."*

The AA Group

An AA group is a historical structure of dyadic and triadic relationships that coheres within a shared universe of discourse. That universe of meanings (Mead, 1934) turns on the omnipresence of five unifying forces: alcohol, alcoholics, recovery (or not drinking), AA history, and AA's conception of a "higher power." These processes give the AA group its reasons for being. The AA group is pledged, in Sartre's sense (1960/1976, pp. 419–420), to the primary purpose of "staying sober and helping other alcoholics to achieve sobriety." This purpose externalizes the group in the form of a pledge. That pledge fuses, or joins, the members in the pursuit of this common goal, which is both individual and collective. The group, then, mediates between the individual and active and recovering alcoholism. It provides an arena of interaction wherein recovery from alcoholism on a daily basis may be accomplished. Individual action is realized only through group action. Similarly, the group is individual action; that is, the group inserts itself into each individual. Each AA member embodies the AA group. The group extends itself through and into each group member. Conversely, the group embodies each member as a part of itself. The group, then, has a reality that is both individual and collective.

This interrelationship between group and individual action and purpose is succinctly stated in AA's First and Fifth Traditions, which respectively state: "Our common welfare should come first; personal recovery depends upon AA unity"; and "Each group has but one primary purpose—to carry its message to the alcoholic who still suffers." These two traditions are mediated by the Second Tradition, which states: "For our group purpose there is but one ultimate authority—a loving God as He may express Himself in our group conscience. Our leaders are but trusted servants; they do not govern."

The Second Tradition thus positions God, or the higher power, above and within every group. Group conscience, or the will of the group membership, is seen as embodying the ultimate authority of God. Because personal recovery depends on AA unity (Tradition One), each individual becomes an embodiment of the group. Each member's conception of God is thus mediated through the common welfare of the group, which in turn becomes a collective sense of God and group purpose, which is also individual purpose.

Groups exist in and through their meetings; hence, to study an AA meeting is to study the AA group. However, groups exist over and above the meetings that they hold, for group members have interactions that occur outside the boundaries of meetings. Groups are registered in the annual AA directory of groups (for AA members only). This recording of the group's existence lists its name and the names and telephone numbers of two AA members who may be contacted concerning the meetings of the group. The two AA members whose name are listed are typically the group's service representative (GSR) and its secretary-treasurer, or alternative GSR. These two representatives of the group are nominated and elected through the group conscience. They typically are members who have displayed a commitment to the group and its meetings. (See Alcoholics Anonymous, 1983–1984, pp. 34–37, for a discussion of the functions of the GSR.)

Group names may take any form, drawing, for example, on AA slogans such as the Day at a Time Group, the Serenity Group, or the Goodwill Open Discussion Group. On the other hand, the group may take its name from the address where it meets, calling itself, for example, the Oak Street Group, the Downtown Group, or the Campus Group. The group may use the name of the day that it meets as its name, for example, the Tuesday Night Group, or connect the name of a group with its meeting place, for example, the Tuesday Night Group at the Treatment Center. Special-interest groups (women, gays, nonsmokers, Latinos) may take their name from this special interest (such as the Tuesday Night Women's Group).

By assuming a special name, the group identifies itself and its membership and locates itself within the universe of AA meetings that exist within the local AA community. Each group is, then, an autonomous structure within the AA social structure. Start with the address of a church; add the presence of coffee and a coffee pot, along with a multitude of ashtrays for cigarette smokers, a table with chairs, and posters with AA slogans and any AA member will know that he or she has found the site of an AA meeting. These are the universal signs of an AA meeting.

If the First Step is not the topic of discussion, the chair may suggest the thought for the day. Or a member may come forth with a topic, such as depression, loneliness, resentment, anger, fear, the desire to drink, overconfidence, family, work, anonymity, gratitude, or the holidays and staying sober. Once a topic is selected, the chair may say: "O.K., let's talk about resentment. Who would like to start?" The chair then either calls on someone, or a person volunteers to begin speaking. Each member at the meeting speaks in turn, usually for 1 to 3 minutes, the length depending on the size of the meeting and the speaker's inclination.

A Group in Action

The following is an account of a closed meeting in which the topic of discussion was the Twelfth Step ("carrying the message to others"). I offer the interaction of that meeting as an instance of an AA group in action.

Chair: Does anyone have any problems or topics for discussion?

Tom: I do. Twelve Stepping. I have a problem. I got a neighbor who is fighting this thing. I've talked to him several times. Yesterday I went over and I gave it to him straight. I said he couldn't fight this by himself. He needed help. Either a treatment center, a psychiatrist, a minister and religion, or AA. I got overinvolved. I know in my guts that I let my emotional self get too far involved. I'd like to hear your thoughts on this.

Les: I'm Les. I'm an alcoholic. I wasn't going to speak today. Eight years ago today they took me to the fifth floor. Not the second, not the third, but straight to the fifth. The floor for the crazies. I was Twelve Stepped while I was there. You had a successful Twelve Step, Tom. You came back sober. You carried the message.

Cathy: I'm Cathy. I'm an alcoholic. I don't know what got me sober. It was many different things. It had to be a sum that was greater than its parts. It was 10% this part, 20% this part. It was things I did and didn't do. I know that it finally worked and it worked when I was ready. I don't know how it worked. I don't think it is any specific thing we do.

Carl: I'm Carl. I'm an alcoholic. Two-and-a-half months ago my father killed himself. The day of the night before he killed himself I felt that something was wrong. I was at the club in [city] and I called the man who got me into AA. I said, "Do you need a meeting tonight?" He said that he did. He came over, and he told me, "I don't know if I should tell you this but 10 years ago I took your father to some AA meetings." This made me feel good because I knew there was nothing I could do for my Dad.

Dave: I'm Dave. I'm an alcoholic. I don't know. I'm working with someone right now. He's sober, sober, then drunk, drunk, then sober, then drunk. I don't know what works. I know I was successfully Twelve Stepped. I hope to be able to do that someday. Thank you, I'm glad to be here today.

The Prose of the Group

From this record of a group meeting I wish to extract the following point. First, it will be noted that each member speaks "on topic," addressing in every instance AA's Twelfth Step. However, only a specific part of the Twelfth Step is discussed—carrying the message. Second, the discourse that is produced reflects back upon itself. Member after member turns to the original topic announced by Tom, and then adds his or her own experience of either being Twelve Stepped or having done a Twelfth Step. Third, as each member personalizes the topic, a part of his or her biography is shared with the other group members, much as the life stories of other members are shared in the *Big Book*. Fourth, the "mystery" of sobriety and of AA is spoken to. No member claims to know how he or she got sober. Fifth, reports on alcoholics who still drinking are brought into the meeting. The "presence" of these absent others serves to remind all of the members of how it was when they were still drinking. Sixth, humor emerges as problems in speaking to alcoholics who are still drinking are discussed. Seventh, each member evidences gratitude for being sober, and conveys a sense of thanks to the others for being present at the meeting. Eighth, a successful Twelve Step is defined for the group. If you carry the message and return sober, you have been successful. This point is critical, for when the AA member enters the "alcoholic situation" of a person who asks for help, alcohol is often present. If the member is uncertain of his or her sobriety, the desire to drink may return and the member may in fact drink on the Twelve Step call, in which case the group may lose a member to active alcoholism, where the intent of the Twelve Step call was to help another alcoholic.

These eight points must be positioned within what I call the "prose of the AA group." This prose is an individual and collective production that speaks the language of ordinary people woven through the understandings of AA and recovery. It is proselike in structure, coming forth in full sentences, and in logical and illogical sequences. Each utterance is framed within the shared interpretive structures of Alcoholics Anonymous. In each member's talk thoughtful, meaningful, biographically specific information is produced and shared. Meanings taken for granted (Garfinkel, 1967) (*Twelve Stepping, crazies, message, treatment, alcoholic)* receive (and need) no explanation, as each member speaks to the topic at hand. Any speaker may (as many do) draw on personal experiences in discussing the topic.

These are not the utterances of the workplace, the home, the telephone, the letter, the hospital emergency room, the psychiatrists's office, or group therapy (see Grimshaw, 1981, pp. 222–226, for a review of these other forms of talk, and Goffman, 1981). These utterances are embedded in the sequential talk of AA members having an AA meeting. The sequentiality, the biographical detail, the hovering presence of alcohol as an organizing "third party," and the shared constraints of AA traditions and ritual serve to produce a structure of "understanding" discourse that is perhaps unique to AA meetings. The talk and the prose of AA is sober talk. It is poetic, poignant, nuanced by the dialects, accents, and speaking idiosyncrasies of each member. It is a prose that is at once personal and collective; as members speak for themselves, they speak to the group as a collectivity. But because no member speaks for AA, the talk that is shared is "self-talk" (Goffman, 1981). By offering his or her prose to the group, each member thus contributes to a group discourse that is greater than the sum of its spoken parts. The totality of these parts gives a poetic and narrative unity to the group's meeting. Thus, a historical continuity in the group's life is produced. This continuity derives from the contributions of each member.

Face-Work and Emotion

Face-work and impression management (Goffman, 1967) are not integral to the members' talk. AA talk places a value on, and incorporates into its discourse, talk that in ordinary conversation would be defined as displaying a "loss of face." Crying, the revelation of deviance while under the influence of alcohol, discussions of bouts of insanity, mentions of crippling fears or depressions, and talk of failures in marriages and social relationships are all sanctioned and accepted within the talk of AA meetings. Positive face-work and the maintenance of face, or self, through the usual means of interactional social control (Grimshaw, 1981, p. 225) are not problematics in an AA meeting, as they apparently are in other areas of everyday social life (Goffman, 1974). Indeed, shows of emotionality and the apparent loss of face are valued and treated with compassion and care within the AA meeting.

These features are most evident when a member reveals a "slip" to the group, for at one level a "slip" could be regarded as a loss of face. But within AA, members are praised for speaking of their "slips," for to do so indicates that the member reaffirms his or her desire to be a member of AA. Consider the following account. The speaker is 37 years old. He had been sober for three months and then drank for four days. As he speaks, his hands shake, his face is unshaven, and alcohol can still be smelled on his breath. He is crying as he talks:

> I'm back here because this is where I need to be. Always before I would wait until I was healthy again. I didn't want anyone to know that I had been drinking. My false pride would keep me away, even after I drank. I let things pile up in my head. Resentments, fears, anxieties, little things and big things. Then I decided that a drink would be all right. I started with one, then I got a bottle and then another bottle, and then another. I was afraid to stop. I couldn't bear the DTs and the dreams. But I made it through last night. I want to get better [breaks down crying again]. I don't know what's wrong with me. I need help, God I need help. [field conversation, November 30, 1985]

Each member who spoke after this individual thanked him for coming back. A box of tissues was passed to him as he cried. Members offered him rides to other meetings. His show of emotion was not taken, then, as a sign of the loss of face.

The following account speaks to another instance when face is not lost when a member shows emotion. The speaker is a valued member of the group. He has been sober over 10 years. He, too, cries as he begins speaking:

> I don't know what to do. Decided I had to bring it to you people. My 15-year old daughter told me and mom that she's pregnant. Told us last night. I ordered her to her room and didn't speak to her 'til this morning. I could kill her and him. Christ, I don't know what to do. Never expected this to happen. I sure as hell ain't goin to drink 'bout it. That's for damned sure. But what in the fuck are we s'posed to do? [field conversation, December 2, 1985]

Here the member breaks down over a family crisis. Each of the 16 AA members at the group thanked him for sharing his problem and each offered advice, based on personal experience. Two female members of the group, who were social workers, shared their telephone numbers with him and offered family-planning assistance.

The AA group exists, then, in and through a shared oral tradition that is structured by the rituals of Alcoholics Anonymous. This oral tradition is, as just indicated, personal and collective, but always given meaning within the overall interpretive system of Alcoholics Anonymous. In order for groups to have meetings, they must have members. But there cannot be members without groups for members to belong to. Hence, a dialectic of the personal and the group is woven through every structure of AA. The meeting, then, is the interactional site for AA in action.

It is in the meeting, as the above transcripts reveal, that the alcoholic subject is transformed into an AA member who is recovering from alcoholism. By announcing himself or herself as an alcoholic, each member makes his or her history of recovery available for others to draw upon. By sharing this history of recovery, each member becomes part of a group that is recovering together. This collective recovery can only be accomplished in and through the talk of each member. Talk, then, is the means to recovery. That talk, proselike in structure, autobiographical in nature, anchors the personal history of each member in the collective and shared history of the AA group. In this way the member's personal life becomes a part of the shared, group consciousness. The AA group becomes a public structure of private lives.

Ritual and Oral History

Interactionally, rituals symbolize the problematics of a group. At the heart of any organized group exists a set of rituals that, when communicated to the newcomer, serve to draw him or her into the inner fabric of the group. Rituals, then, stand at the intersection of individuals, societies, and groups (see Denzin, 1984b, pp. 246–247).

AA ritual, enacted through the meeting, yet permanently recorded in the texts and readings of AA, may be analyzed in terms of these dimensions. I shall take up the key AA rituals in turn, beginning with the reading that structure an AA meeting.

The reading of the AA "Preamble," "How It Works," the Twelve Traditions, and the "Thought for the Day," brings into every AA meeting the collective history of Alcoholics Anonymous. Although many groups do not read "The Thought for the Day" or "The Traditions" or all of "How It Works," the AA "Preamble" is always read. These AA readings are endowed with solemn ceremony. Members listen quietly as they are read. Each reading references a problematic in the AA program and a problematic in the member's own program. When the meeting is called to order by the chair the ritual self of being an AA member at the meeting is brought into existence. From this moment until the last words of the Lord's Prayer, the members are within the "frame" of an AA meeting. This becomes a sacred moment in their day, as attending Mass might be for another individual. The AA readings permit few variations. They are read in order, exactly as they were read at the last meeting. Wherever an AA member goes these readings will be presented in the same sequence, with the same words. In this sense AA ritual admits of no variations.

These readings embody the AA worldview. By listening to them being read the member becomes a part of AA's oral history. Furthermore, by hearing these readings read out loud the member obtains some measure of protection against taking a drink that day. In this way the AA rituals serve as mechanisms for ensuring sobriety in a world that is regarded as uncontrollable and unmanageable. Because they focus explicitly on the problematics of AA—sobriety and its maintenance—these ritual readings keep AA's primary purpose constantly in front of the member. These rituals are positive, joining rituals. They bring members into one another's presence, providing a bridge between the loneliness of alcoholism and the community of AA recovery. By allowing the newcomer to read, AA invites that member to move more deeply into the inner structures of the fellowship. In this way newcomers are incorporated into group life.

Interpretations

Beginning with the original AA meeting between Wilson and Smith, I have presented the historical, textual, and interactional structures that AA rests upon. The several levels of AA, which range from its historical texts to the network of meetings that exist in a single community and extend to the international community of AA and to the New York offices, find their immediate meaning in the biography of the individual AA member who comes to find himself or herself as a recovering alcoholic within the AA meeting. Each AA meeting, like the members who make it up, is an universal singular, epitomizing in its structures, its talk, and its rituals all AA meetings that have occurred in the past or will occur in the future. The single-mindedness of purpose that underlies all of the structures of AA permits this historical continuity that joins the past with the present. AA, at all its levels and in all its forms—personal, group, in texts, in oral histories, in its bureaucratic structures, in its rules, as a social movement, in its rituals, and in its traditions—is a unique social structure. Although none of the units or forms that make up the AA structure are themselves unique, the overall structure is.

The key to this uniqueness is alcoholic understanding and the shared experiences that AA draws upon. These experiences permit AA members to form social groups among strangers, yet to do so within a historical structure of understanding that makes no alcoholic a stranger to a recovering alcoholic. How this task is accomplished has been the topic of this chapter.

The AA group stands in stark contrast to the primary groups studied by Cooley (1902/1956) and Schutz (1964, pp. 106–119). The AA group is a group of strangers who create a fellowship of interaction that is based on shared, common experiences that the broader culture and alcoholism have produced for each individual. Yet this community of strangers is often interpreted by the member as being family:

> I lost everything. Family, home, wives, kids, job, everything. Even my parents turned against me. I fought like hell to get it all back and it didn't work. They went off to be who they were and left me to find myself. I think they hated me. Lot of self-pity on my part. I finally found you people. Now you're my family. Wherever I go, you're there. Wherever I go. But I got me a "home group" back in LA and that's where my new permanent family is. Gives me everything I ever wanted and ever looked for. I feel like I'm needed again. [field observation, as reported September 2, 1982; 47-year old male, salesman, over three years sobriety]

This family that is found within AA envelopes the member within a noncompetitive collective structure that is larger than the member (Bateson, 1972). Largely male based, this AA culture, and the "families" that it spawns, permits the creation of shared pasts, feelings of solidarity, and the sharing of common futures (Couch, Saxtin, and Katovich,1986).[1] These processes are, of course, at the center of all long-standing social groups, including families. In this sense AA universalizes the desire for "groupness" that lies within our culture (see Bellah et al., 1985).

Note

1. See Denzin (1991) for a discussion of how AA (and its groups) have been treated in American film over the past half-century.

References

Alcoholics Anonymous, (1983–1984). *The AA service manual combined with twelve concepts for world service by Bill W.* New York: Alcoholics Anonymous World Services.

Bateson, G. (1972) The cybernetics of 'self': A theory of alcoholism. In G. Bateson (Ed.), *Steps to an ecology of mind.* (pp. 309–337). New York: Ballantine.

Bellah, R., et al. (1985). *Habits of the heart: Individualism and commitment in American life.* Berkeley: University of California Press.

Cooley, C.H. (1956) *The two major works of C.H. Cooley.* New York: Free Press (original work published in 1902).

Couch, C.J., Saxton, S.L., & Katovich, M.A. (Eds.). (1986). *Studies in symbolic interaction: The Iowa School.* Greenwich, CT: JAI.

Denzin, N.K. (1984a). *On understanding emotion.* San Francisco: Jossey-Bass.

Denzin, N.K. (1984b). Ritual behavior. In R.J. Corsini (Ed.) *Encyclopedia of psychology* (Vol. 3, pp. 246–247). New York: Wiley.

Denzin, N.K. (1987). *The alcoholic self.* Beverly Hills, CA: Sage.

Denzin, N.K. (1991). *Hollywood shot by shot: Alcoholism in American cinema.* New York: Aldine de Gruyter.

Garfinkel, H. (1967). *Studies in ethnomethodology.* Englewood Cliffs, NJ: Prentice-Hall.

Goffman, E. (1967). *Interaction ritual.* New York: Doubleday.

Goffman, E. (1974). *Frame analysis.* New York: Basic Books.

Goffman, E. (1981). *Forms of talk.* Philadelphia: University of Pennsylvania Press.

Grimshaw, A.D. (1981). Talk and social control. In M. Rosenberg & R.H. Turner (Eds.), *Social psychology: Sociological perspectives* (pp. 200–234). New York: Basic Books.

Kurtz, E. (1979). *Not-God: A history of Alcoholics Anonymous.* Center City, MN: Hazelden Educational Materials.

Maxwell, M.A. (1984). *The Alcoholics Anonymous experience: A close-up view for professionals.* New York: McGraw-Hill.

Mead, G.H. (1934). *Mind, self and society.* Chicago: University of Chicago Press.

Rudy, D. (1986)*Becoming alcoholic: Alcoholics Anonymous and the reality of alcoholism.* Carbondale: Southern Illinois University Press.

Sartre, J.P. (1976). *Critique of dialectical reason.* London: NLP (original work published 1960).

Schutz, A. (1964). *Collected papers: Vol. II. Studies in social theory.* A. Brodersen, Ed. The Hague: Martinus Nijhoff.

Tiebout, H.M. (1954). The ego factors in surrender in alcoholism. *Quarterly Journal of Studies on Alcohol, 15,* 610–621.

CHAPTER 37

The Adult Children of Alcoholics Movement: A Social Constructionist Perspective

DAVID R. RUDY

Social constructionist views on social problems repeatedly have demonstrated the ways in which social problem definitions and perceptions shape the ways that we respond to problems. Recent work in the sociology of alcohol illustrates on both the macro- and microlevels that definitions of alcoholism are social constructions. It also shows the ways in which various definitions have shaped alcohol policy (Fingarette, 1988; Peele, 1985; Reinarman, 1988; Room, 1978; Rudy, 1986; Schneider, 1978; Tournier, 1979). Gusfield's classic work, *Symbolic Crusade* (1963) highlights historical definitions of alcoholism as "sin," "moral weakness," "crime," and "disease." Of particular importance in much of this work is the view that definitions of problems and responses to problems frequently have more to do with a social and political reality than with the actual characteristics of the problem as reflected in the research literature. In Schneider's (1978) sense alcoholism is more a social and political accomplishment than a scientific one. This chapter examines the most rapidly expanding arena in the alcoholism enterprise, the adult children of alcoholics movement. My utilization of the term *ACOA movement* includes the related issues of children of alcoholics (COAs) and co-dependents.

In the past few years there has been an explosion of interest, media coverage, and publishing in the area of alcohol and the family, particularly in the specific topics of children of alcoholics, adult children of alcoholics, and co-dependents. Health Communications Inc. listed approximately 20 new volumes along with 80 backlisted volumes, audio tapes, and the announcement of 14 conferences in its Spring 1989 catalogue highlighting "self-discovery and recovery." The 1990 preliminary conference schedule

David R. Rudy is with the Department of Sociology, Social Work, and Corrections, Morehead State University, Morehead, Ky. This chapter was written especially for this book.

highlights 21 regional, national, and international conferences. Fully 58 journal articles written in the past 10 years are referenced in the *Social Science Index* under *children of alcoholics*. The majority (49) of these articles have appeared in the past 5 years. In the area of popular literature, *Readers' Guide to Periodical Literature* lists 34 articles over 10 years, 21 of these having appeared within the past 3 years. *Newsweek*'s cover story on January 18, 1988, presents a shattered family portrait with a spilled booze bottle and the caption "Growing Up with Alcoholic Parents Can Leave Scars for Life." Treatment programs and national conferences have proliferated, a foundation and a national association for COAs have been created, and Janet Woititz's *Adult Children of Alcoholics* (1983) has made the *New York Times* best-seller list and has sold over 1 million copies. In bookstores the ACOA movement's publications warrant separate sections where readers can find personal accounts, workbooks, recovery novels, poetry anthologies, audio cassettes, calendars, and video tapes. Two magazines, *Changes: For and About Adult Children*, and *Focus on Chemically Dependent Families*, regional and national conferences, and training programs and materials are also mainstays of this burgeoning movement.

What factors, themes, and conditions in the contemporary United States explain the beginnings as well as the tremendous growth in the ACOA movement? Are there millions of ACOAs in need who are seeking services? Has the research literature recently documented a wide range of pathologies that affect children of alcoholics requiring treatment? What organizations, interest groups, and treatment providers are involved in ACOA issues? This chapter explores some of the inputs of the adult children of alcoholics movement; it raises some problematic issues and questions and suggests some possible cautions.

Theoretical and Methodological Issues

The general theoretical stance that I utilize in examining the ACOA movement is symbolic interactionist-constructionist. My approach is similar to the one I developed in *Becoming Alcoholic* (1986), where I examined the world as constructed by "alcoholics," as well as by "alcohologists." In that work I developed a relativist-interactionist definition of alcoholism as I regarded alcoholism as a social, cultural, and historical accomplishment and construction. Although the characteristics of ethyl alcohol and human beings show little variance from culture to culture and throughout history, what comes to be regarded as "alcoholism" varies and will continue to vary along complicated social, historical, scientific, and political continua. Of the millions of people in the United States who drink heavily, including those who drink with considerable social consequence, relatively few are regarded by themselves or by others as "alcoholic." My view of alcoholism remains problematic; accordingly, my view of COAs, ACOAs, and co-dependents is likewise problematic. This, then, is an analysis of views, beliefs, claims, and facts about the ACOA movement and about persons who are participants in it. The work explores new definitions and constructions as well as changes in definitions regarding relatives of "alcoholics." In addition, it undertakes a general comparison between the beliefs of persons leading the movement, primarily clinicians, coupled with the beliefs of researchers who have studied the consequences of "alcoholism" in the family. Finally, it postulates some speculative connections and implications of the movement.

Data sources will include the scholarly and the popular literatures on various aspects of the ACOA movement. In addition to these sources, newsletters, foundation reports, trade journals, and agency brochures will be examined. Finally, my earlier work in AA exposed me to many spouses and children of members. These contacts have expanded through my being regularly approached by students, colleagues, and friends who hear or read about my interest in the ACOA movement and in "alcoholism" and who regard themselves as children of "alcoholics."

Organizational Influences

Although the scientific literature on the effects of alcoholism on the family may not be overwhelming in terms of its frequency and scope, it has, nevertheless, been consistent for several decades. For example, Joan Jackson's work (1954) highlighted the crisis nature of alcoholism for the family; Fox (1962), drawing from her psychiatric practice, articulated the deleterious consequences of alcoholism for children of alcoholics; Nyelander (1960) conducted one of the first systematic studies specifying a degree of symptomatology that differentiated children of alcoholics from children of nonalcoholics; and Cork (1969) presented anecdotal and clinical data showing the distrust, anger, and low self-esteem of children of alcoholic parents. These studies and others certainly have influenced and generated some of the contemporary research studies on COAs and ACOAs, but in my view they have had little direct impact on the ACOA movement. In fact, Cork's work, *The Forgotten Children* (1969), although frequently cited in the ACOA and COA literature, is actually a retrospective validation rather than a generator of the movement. Its title is synonymous with one of the dominant themes of the movement: that spouses and family members have been relatively ignored or forgotten in the alcoholism rehabilitation process. If the scientific research literature didn't initiate the adult children of alcoholics movement, what did?

One of the influences behind the movement is Alcoholics Anonymous and its satellite organizations, Al-Anon and Alateen. Al-Anon developed in the early 1950s, and Alateen began shortly after, around 1957. In fact, the AA convention in 1955 in St. Louis sponsored a session called "Children of Alcoholics." From the perspective of these organizations:

> Many of the symptoms of alcoholism are in the behavior of the alcoholic. The people who are involved with the alcoholic react to his behavior. They try to control it, make up for it or hide it. They often blame themselves for it and are hurt by it. Eventually they become emotionally disturbed themselves. (Al-Anon, 1987, p. 6)

Alcoholism is not only a physical, spiritual, and emotional affliction but also a "disease of relationships." AA ideology, as well as that of its satellite organizations, is officially portrayed through their publishing activities. However, it is also portrayed by clinicians, researchers, and writers who have become familiar with AA and by AA members who have become a vocal and vital part of the alcoholism treatment industry. Approximately 72 % of counselors working in the more than 10,000 substance abuse centers are "recovering" from substance abuse (Sobell & Sobell, 1987). Given the dominance of AA as a treatment program, the majority of these counselors are likely to be AA members. AA ideology also comes to the ACOA movement more directly through members who write as adult children of alcoholics and members who are selected in studies on children of alcoholics. Numerous research efforts have drawn their samples of adult children and children of alcoholics from Al-Anon and Alateen groups.

A second dominant influence on the ACOA movement has been the efforts of state and federal agencies in promoting attention and services relative to children of alcoholics. With the publication of *An Assessment of the Needs of and Resources for Children of Alcoholic Parents* (1974), the National Institute on Alcohol Abuse and Alcoholism (NIAAA) brought national attention along with a commitment of funds for demonstration and research projects on children of alcoholics. In 1979 NIAAA sponsored a national conference, "Services for Children of Alcoholics," in Silver Spring, Maryland. The conference addressed the issues of "identification, intervention, treatment and prevention" and produced a monograph in the NIAAA Research Monograph Series, *Services for Children of Alcoholics (1979).* Among the conference participants were clinicians, program directors, and others, including Janet Woititz. The following year NIAAA awarded a contract that eventually led to another publication, *A Growing Concern: How to Provide Services for Children from Alcoholic Families* (1983). Although these publications and the conference represent a very small portion of NIAAA activity they were important in initiating, validating, and encouraging interest, research, and, most important, services for COAs.

Even more influential than these activities are the series of conferences and publications initiated in 1982 by the State of New York, Division of Alcoholism and Alcohol Abuse. The initial report, *Children of Alcoholics* (Balis, 1989), was commissioned by the former secretary of Health, Education and Welfare, Joseph A. Califano, in his role as special counselor to Governor Hugh L. Carey. Using a combination of factors including national estimates of problem drinkers and demographic data, the report estimated that over 9 % of New York State children, (age 19 or younger) live in alcoholic families. The report also reviewed some of the research literature and discussed the need for additional programs and public awareness efforts to reach the estimated 555,000 children of alcoholics in the state. Finally, out of this initial conference came the nucleus of the Children of Alcoholics Foundation. The foundation has organized other conferences, published literature reviews and awareness brochures, and has estimated the number of children of alcoholics at over 28 million nationally. In addition to a board of directors, the Children of Alcoholics Foundation lists over 40 honorary advisers, including many political, business, and entertainment elites and a scientific advisory board including Henri Begleiter, Marc Schuckit, Robert Straus, and George Vaillant.

Another influential organization in the movement is the National Association for Children of Alcoholics (NACoA), a nonprofit organization formed in 1983. NACoA has 21 state chapters, publishes a quarterly newsletter, and provides a range of services for children of alcoholics and for professionals who work with children of alcoholics. NACoA leadership consists primarily of clinicians and counselors. Some are adult children and some, including Claudia Black, Timmen Cermak, Robert Ackerman, Robert Subby, and Sharon Wegscheider-Cruse, are frequent contributors to the semipopular literature on children of alcoholics and are featured conference speakers around the nation. Wegscheider-Cruse, Woititz, Black, and Ackerman were featured speakers at the Fifth National Convention on Children of Alcoholics in February 1989, which enjoyed a 500% increase in participation over the 1985 conference. NACoA is much more involved than the Children of Alcoholics Foundation in establishing ACOA groups, sponsoring conferences, and, through its member-authors, providing literature to its members and ACOAs in general.

The leadership, honorary leadership, and members of both organizations write extensively, appear on television and radio talk shows, and are frequent speakers on the rapidly expanding national and regional conference circuit. Recently, Migs Woodside and Joseph Califano from the Children of Alcoholics Foundation and Cathleen Brooks, Phil Diaz, and Ellen Morehouse from NACoA testified at the Congressional Subcommittee on Children, Family, Drugs, and Alcoholism (1987). Such activity is extremely significant in mobilizing a movement. Many persons share common characteristics and experiences but their crystallization into an organized interest group requires perception by others, increased interaction, symbolic expressions of collectivity, and other features that highlight commonalities of norms and values (Gamson, 1968). Leadership, visibility, and publicity regarding ACOAs have been very effective in attracting members and in attracting a sympathetic audience.

Literature Themes and Problems

Recently, a number of writers (Ackerman, 1987; Balis, 1989; Black, 1981; Castine, 1989; Wegscheider, 1981; Wegscheider-Cruse, 1989; Woititz, 1983; Wood, 1987) have produced monographs that have highlighted the problems and recoveries of ACOAs and co-dependents. Woititz's work made the *New York Times* best-seller list and has sold over 1 million copies. Because these studies draw on clinical and anecdotal data, they tend to overemphasize and stereotype the characteristics of adult children of alcoholics. That is to say that the adult children of alcoholics who present themselves to clinicians and self-help groups for treatment are, by self-selection, more seriously affected by the alcoholism of their parent (or parents) than those who do not present themselves. In addition, because most studies draw children of alcoholics from Al-Anon, Alateen, ACOA groups, and delinquency and mental health programs, their reflections are likely to be shaped by these organizations as well as by their status of being a child of an alcoholic. Likewise, other studies examine children of alcoholics whose parents are in treatment. It is likely that these persons are in the most seriously affected alcoholic families and consequently may experience higher levels of deleterious effects. Another of the many complicating factors biasing much of the research is that the children of alcoholics in some studies also are alcoholic. For example, Woititz's (1983) research is based upon the members of her therapy groups in which half of the group members are recovering alcoholics. When Beardslee, Son, and Vaillant (1986), removed alcoholic subjects from their analysis, the differences between ACOAs and non-ACOAs disappeared. A balanced view of the consequences of parental alcoholism upon offspring must include persons outside of treatment programs and self-help groups as well as persons inside these settings.

In *Adult Children of Alcoholics,* Woititz (1983, p. 4) highlights 13 characteristic behaviors of ACOAs that are common in her treatment groups, including: guessing at what constitutes normal behavior, having difficulty completing projects, lying when telling the truth is just as easy, judging themselves without mercy, finding difficulty in having fun, taking themselves very seriously, constantly seeking approval and affirmation, and so on. Woititz cautions that her generalizations are not scientific, nor are they necessarily shared by every ACOA. She tells us nothing about the therapy group members on whom she bases her generalizations, except that half are recovering and half are women. Also, some are married, some have children, the youngest member is 23, and all are committed to self-growth (Woititz, 1983, p.4). The obvious methodological and inferential weaknesses of the most influential book in the ACOA movement unfortunately are also rampant in most of the movement's research as well as in the published journal literature. Substantive literature reviews (Heller, Sher, & Benson, 1982; Jacob, Favorini, Meisel, & Anderson, 1978; Watters & Theimer, 1978; West & Prinz, 1987) conclude that methodological weaknesses "overpredict" the vulnerability of children of alcoholics and that many qualifying, rival, causal factors remain unexplored. Although it seems safe to say that something is happening out there, one cannot be sure of the extent and the way in which parental alcoholism produces problems in offspring (West & Prinz, 1987).

The methodological limitations of past ACOA studies are not significantly resolved in Ackerman's "National Adult Children of Alcoholics Research Study." Ackerman, a sociologist and founding board member of NACA, has presented findings (Ackerman, 1987; Ackerman & Gondolf, 1989) from a "purposive cluster sample." The "sample" was drawn from attendees at 62 human service conferences in 38 states. However, little detail beyond background characteristics of the subjects is presented; and the data analysis rarely controls for other spurious variables that could easily relate to the higher ACOA scores for alcoholic offspring. One interesting finding of note is that ACOAs who sought treatment had higher pathology scores than those who had not sought treatment. The explanation of this "counterintuitive" finding is the "heightened awareness" brought about during the treatment process (Ackerman & Gondolf, 1989, p. 9). Goglia (1985) provides a similar explanation of negative findings in hypothesizing that ACOAs not in treatment deny their symptoms and hold back other information because they do not trust interviewers. After all, denial and lack of trust are "symptoms" of ACOAs. Another possible interpretation of these findings is that ACOAs are socialized into the values and ideology of the movement. The movement may produce the "symptoms" as much as the "symptoms" produce the movement (Miller, 1987).

Other clinician-writers within the movement have similar views on the consistency of ACOA symptoms and the necessity of treatment. Claudia Black (1979) argues that all children of alcoholics should be treated whether or not they present pathology. Many clinicians argue that when COAs and ACOAs do not resolve their problems through treatment the problems automatically carry over to the next generation. Even when COAs seem to be doing well, they are typically viewed as "supercopers" and "overachievers," and as people still in need in treatment. This line of reasoning is further developed in Robinson's *Work Addiction* (1989), where it is argued that a large number of ACOAs "use work as their drug of choice.... Switching addictions into work is the most acceptable yet dangerous form of addiction for which little help is available" (p. viii).

Cermak (1986), Subby (1987), Wegscheider-Cruse (1984), Wood (1987), and Young (1987) highlight the consequences of alcoholism for children and spouses in the family. In their view many family members suffer from the disease of "co-dependency." "Co-dependency is a primary disease and a disease within every member of an alcoholic family. It is what happens to family members when they try to adapt to a sick family system that seeks to protect and enable the alcoholic" (Wegscheider-Cruse, 1984, p. 1). Cermak (1986, 1989) has consistently called for a view of co-dependency as a type of personality disorder and a "disease" of relationships. Like ACOAs, wives of "alcoholics" as co-dependents show significantly more diversity and less pathology than clinical data suggest (Corder, Hendricks, & Corder, 1964; Edwards, Harvey, & Whitehead, 1973). Additionally, the research on family rituals shows that family adaptations and responses to problem drinking are variables, in their own right, that can mediate the consequences of the drinking upon family members (Brower, 1987; Wolin, Bennett, Noonan, & Teitelbaum, 1980).

We have been describing the popular view of children of alcoholics and have suggested that it is, in part, a stereotype. That is not to say that children of alcoholics are not without pathology. Some children have been hurt immeasurably by the drinking of their parent (or parents). For some the damage may be long-term; for others it may be short-lived. Despite the views of most clinicians, there is research (Moos & Billings, 1982) that shows that children of alcoholics improve when their parents improve and that many are resilient and show minimal or no significant pathology differences from comparison groups (Berkowitz & Perkins, 1988; Chafetz, Blane, & Hill, 1971; Collins, 1982; Goodwin, 1987; Herjanic, Herjanic, Perick, Tomelleri, & Ambruster, 1977; Kammeier, 1971; Werner, 1986). On the other hand, therapists argue that

> the effects of alcoholism on the family occur even when the active drinking is not present. Second, the alcoholic system will recreate itself generation-after-generation if the family is not treated. A great number of ACOA's come from families where there is no alcoholic drinking taking place. If alcoholism is in their family history, they will have ACOA issues. (Kritsberg, 1985, p. 8)

Views such as these are carried to further extremes in the recent work by Ann Smith, *Grandchilden of Alcoholics* (1988). On the rear cover flap the publisher states:

> We first recognize certain personality characteristics that predominated in Adult Children of Alcoholics. Now sociological studies are discovering similar patterns in the *children* of the Adult Children of Alcoholics. The characteristics can occur in seemingly alcohol-free families where it is found later that a grandparent who died at an early age "sometimes drank too much," but it was never talked about.

Like similar studies on children of alcoholics, all of Smith's respondents are in treatment, with over a third having problems of substance abuse. The major problems that grandchildren of alcoholics report include: difficulty with relationships, being out of touch with feelings, poor self-worth, and feeling angry a great deal. (A. Smith, 1988, pp. 59–63).

The dominant theme in the children of alcoholics literature is that most, if not all, children of alcoholics could benefit from treatment. Furthermore, screening techniques need to be refined and implemented in medical settings and in school systems so that we can identify most easily and quickly the approximately 28 million children of alcoholics. Treatment programs in hospitals and freestanding alcoholism treatment centers range from individual and group therapy to workshops, dramatic presentations, and lengthy inpatient programs up to 21 days. There are programs for co-dependents, ACOAs, COAs, and members of dysfunctional families. Some programs promise to give a "greater sense of well being," and a "satisfying, happy, life." Other programs promise "a lifetime of protection from substance abuse" for their clients—children from 2 to 8 years of age who participate in a treatment program in which the sponsors use puppets. Some programs are inexpensive and brief whereas others advertise 2 or 3 years of weekly sessions.

Structural and Ideological Concomitants of the ACOA Movement

In *The Sociological Imagination* (1959), C. Wright Mills extols the value of connecting personal troubles with public issues and of connecting personal biography with history in order to understand better and to develop a humanistic sociology. Recently, Norman Denzin (1989) in *Interpretive Interactionism* developed a similar argument in which detailed, "thick description" of participants' experiences and epiphanies in their struggle with problems can give understanding to contemporary conditions of everyday life. In Denzin's view a sociological perspective of any arena must emphasize the participants' interpretations and meanings. It must also examine the connection between individuals and their society and social conditions in analyzing the construction of meaning and in deconstructing the interpretations as well. Here I would like to explain a number of themes that are related to the development of the ACOA movement. Some are more obvious and less speculative than others; however, they represent my sociological interpretation of the movement from reading its literature and listening to some of its participants, and through my participation in the "alcoholism" literature.

The ACOA movement represents a merging of therapeutic currents and ideologies. The expansion of AA and Al-Anon represents only a small facet of what is more broadly called the 12-Step movement. Mutual-help groups for common and uncommon maladies abound in the contemporary United States and elsewhere. There is belief in the utility, simplicity, superiority, and effectiveness of 12-Step programs particularly on the popular, personal, and clinical level, if not on the research level. Likewise, there is an ever-growing interest among clinicians, particularly social workers and psychologists, in microsystems and family therapies. Since many "alcoholism" counselors are "recovered alcoholics," and since many persons in treatment have "alcoholic" families, it is easy to imagine how these individuals could become strong supporters and proselytizers of ACOA treatment and the movement. Like other 12-Step programs, ACOA programs are cast at a general level and provide meaning, interpretation, and assistance in accessible form. The clinical view of ACOA characteristics and issues is so general that most people in the United States probably qualify. It really doesn't matter whether one's problems are related to familial "alcoholism" or to some other "dysfunction." Current conferences and scores of therapy groups treat drugs, food, sex, money, work, and relationships as addictions that are implicated in one way or another as ACOA issues. Through ACOA meetings, conferences, and publications members learn how to make sense of their past lives and current problems primarily from the perspective of family drinking or some other family dysfunction. Like AA (Greil & Rudy, 1984; Rudy & Greil, 1989), ACOA groups are profitably viewed as "social cocoons" and "quasi religions." Members come to see their existing lives and selves as "unmanageable" and as tied to a faulty view of the world (Denzin, 1987; Bateson, 1971). ACOA groups provide an ideology and social support network that helps members to reconstruct their lives and reinterpret them in a new way.

Changes in the organization and ideology of social control responses to alcohol problems are significant in generating and promoting both ACOA treatment and, correspondingly, the ACOA movement. As the medicalization and the medical treatment of alcohol problems have expanded and become commonplace, so, too, has the alcoholism treatment industry. Problem maximization (Room, 1978) and amplification (Room, 1984), public health approaches to alcohol-related problems, professionalization in "alcoholism" programs, profit, and third-party payments have each had an impact on definitions of "alcoholism" and alcohol-related problems. It is understandable and not malicious that "alcoholism" treatment interests, recovered "alcoholic" clinicians, and others should warmly respond to a market of 28 million COAs and to additional millions of co-dependents, particularly in an era when more and more behaviors are discovered or maximized as alcohol-related. Nearly 1,000 of the 1,500 or more treatment providers who advertised in the 1988–89 treatment directory published by the *U.S. Journal* provided services for COAs and ACOAs. Although most services remain on an outpatient basis, some residential programs with average stays of 20 or more days are also available.

Changes in organizational social control are reciprocally linked with changes in personal social control. As Room (1988) has suggested, there is a movement toward increasing personal temperance. Decreases in consumption, movement away from distilled spirits, and personal concern with health and fitness are indicators of an increasingly sober consciousness. For many members ACOA ideology provides a beacon around which their existing attitudes and concerns about alcohol and alcohol-related problems can be organized and focused.

Relationship problems and role responses to family drinking as well as affiliation with ACOA and co-dependent groups are tied to sociological and interactional variables as well as to personal variables. The overwhelming majority of participants in the ACOA movement are women. Published studies, "samples" in treatment settings, and comments from other alcohologists suggest that 70-75 % is a reasonable estimate. Although this figure may be partially explained by the higher relative frequency of male "alcoholics," other factors are probably operative. In her dissertation, Asher (1988) points to "social power" and "cultural resources" as factors that shape the perception as well as the adaptation of wives to their husbands' drinking. Martin (1988) sees co-dependency as a label that may mask some of the broader issues associated with inequality between men and women in U.S. society. Room (1988) interprets aspects of the movement as constituting "a cloaked feminist critique of men's drinking and its impact on women" (p. 824).

Ridlon (1988) uses the concept of "status insularity" to explain why fewer women are viewed as "alcoholics" and why they are subject to greater stigmatization than male drinkers. It may be that dependency needs and status vulnerability make women more likely to define their circumstances in ways that lead them to ACOA groups. A flier advertising an ACOA group for women in late 1989 demonstrates a general approach that, no doubt, appeals to many. Questions asked on the flier include: Do you have problems expressing your feelings? Do you have problems asking for what you need? Do you experience difficulties with intimacy in relationships? Other questions deal with depression, anxiety, and trust. ACOA conference advertisements are increasingly cast to include "those who identify." It may be that the positive aspects of treatment outweigh any potential deleterious aspects for all who participate. However, to the extent that persons' problems may not be due to the consequences of alcohol in the family, the resolution of their problems may be hindered and not enhanced by ACOA participation.

The ACOA movement fits well with the renewed interest in genetic theories of alcoholism. ACOA literature frequently cites the increased vulnerability of children of "alcoholics" for "alcoholism." Movement participants believe that co-dependency and ACOA issues, like "alcoholism," are generational. Some clinicians (Kritsberg, 1985; A. Smith, 1988) strongly maintain that ACOA problems are carried from generation to generation if intervention does not take place. Although the differences between non-ACOAs and ACOAs are seen in behaviors, traits, and personality characteristics and not genetics, they are considered as real as physical differences. The insistence of the generational nature and similarity of symptoms in ACOAs is noteworthy in light of what one might call weak research evidence, at best. This tendency is strikingly similar to what recent writers (Allen, 1986; Rafter, 1986; J.D. Smith, 1985) describe in their analysis of the American Eugenics movement of the late 1800s and early 1900s and the studies of "degenerate clans." These writers illustrate how beliefs and good intentions in bettering society led to a confusion between beliefs and science. J.D. Smith (1985) documents how aspects of biographies were overlooked and photographs touched-up so that the beliefs of the eugenicists could be more advantageously portrayed. "Feebleminded" persons and "degenerate clans" were believed to be responsible for a wide range of social ills, including crime, prostitution, poverty, and alcoholism; and their identification, institutionalization, and, in some cases, sterilization were likewise believed to be necessary to protect and improve society.

Although today's clinicians are not calling for sterilization, they are calling for earlier and increased identification of all of the alcohol-related syndromes. Alcohol's presence in any problem situation is usually viewed as direct causal evidence that alcohol caused the accident, abuse, violence, and so on. In our desire to remedy problems we over-simplify and gloss over the complexities of alcohol-related behavior as well as the consequences of "alcoholism" on family life. We mix belief with science when we fail to see that "alcoholics" and ACOAs are much more like us than unlike us. In today's eugenics movement we believe that the discovery of the physical (genetic) causes of alcoholism, crime, and depression are just around the corner. As in the earlier movement, we emphasize individual and familial explanations of alcoholism and co-dependency and not social or structural explanations. Similarly, we push with religious zeal for increased support and funding to allow for earlier identification and increased treatment of COAs and others so that we may repair the pathological families and improve our society.

Issues and Implications

The bodies of literature on social movements and the social construction of knowledge suggest that public recognition of social problems is frequently as much a consequence of societal definitions and perceptions as it is a response to actual objective conditions. This argument has been developed and empirically substantiated by numerous researchers (Becker, 1963; Conrad & Schneider, 1980; Gusfield, 1963, 1981; Spector & Kitsuse, 1987). How social problems come to be publicly defined and represented has specific consequences for how these problems are treated, who does the treatment, who receives the treatment, who pays for it, and so forth. Disease definitions of alcoholism allow for hospitalization, insurance payments, disability payments, and the like. Behavioral definitions of alcoholism allow for an entirely different set of responses, including personal responsibility, confrontation, and imprisonment. A similar case can easily be made for adult children of alcoholics. For example, one of the larger growth industries in the human service sector is alcohol rehabilitation. In fact, drug and alcohol programs and expanding mental health services are two principal areas that contemporary U.S. hospitals are emphasizing to compensate for empty hospital beds. The point is that hospitals, private drug programs, and other clinical settings have a potential market of 28 million children of alcoholics, should this group become defined as suffering from an identifiable disorder. Again, it must be emphasized that adult children of alcoholics do suffer from their parent's alcoholism. However, the magnitude, patterning, characteristics, and complexity of this suffering have not been empirically verified. Equally important, the play of economic and political interests have become a source of widespread misinformation. Misinformation or stereotypical information is easily proselytized in the contemporary era, with its emphasis on wars on drugs, its problem amplification (Room, 1984), and its moral panics (Ben Yehuda, 1986). Heavy alcohol use is both a health problem and a public health problem. However, blaming alcohol as the culprit in every malady in which it is found frequently has the effect of diminishing other possible explanations. So, too, for children of alcoholics. Some have problems because of their parent's (or parents') alcoholism and more so because of the dynamics in some alcoholic families. Some have problems because of other circumstances and situations, and some do not have problems at all. To suggest that a condition as complex and varied as alcoholism produces a list of specific consequences in the family members that share the alcoholic's life is too simple an explanation, given what we know about the imagination, adaptability, and resiliency of human actors. Although such a view may identify some children of alcoholics who need treatment, it will also identify children who do not need it. To the extent that our approach to alcohol problems continues to emphasize laws, regulations, and professional control, and to the extent that our programs are based upon stereotypical conceptions of persons affected, we will continue to blow a lot of steam and spend a lot of money while guaranteeing a constant if not an increasing pool of problems.

References

Ackerman, R.J. (1987). *Let go and grow: Recovery for adult children*. Deerfield Beach, FL: Health Communications.

Ackerman, R.J., & Gondolf, E.W. (1989). Differentiating adult children of alcoholics: The effects of background and treatment on ACOA symptoms. Paper presented at the Annual Meeting of the American Sociological Association, San Francisco, CA.

Al-Anon Family Group Headquarters. (1987). *Alateen: Hope for children of alcoholics*. New York: Author.

Allen, Garland E. (1986). The Eugenics Record Office at Cold Spring Harbor, 1910–1940: An Essay in institution history. *Osiris*, 2, 225–264.

Asher, R.M. (1988). *Ambivalence, moral career and ideology: A sociological analysis of the lives of women married to alcoholics*. Ph. D. diss., University of Minnesota.

Balis, S. (1989). *Children of alcoholics*. Deerfield Beach, FL: Health Communications.

Bateson, G. (1971). The cybernetics of "self": A theory of alcoholism. *Psychiatry, 34*, 1–18.

Beardslee, W.R., Son, L., & Vaillant, G.E. (1986). Exposure to parental alcoholism during childhood and outcome in adulthood: A prospective longitudinal study. *British Journal of Psychiatry, 149*, 584–591.

Becker, H.S. (1963). *Outsiders: Studies in the sociology of deviance*. New York: Free Press.

Ben-Yehuda, N. (1986). The sociology of moral panics: Toward a new synthesis. *Sociological Quarterly, 27*, 495–513.

Berkowitz, A., & Perkins, H.W. (1988). Personality characteristics of children of alcoholics. *Journal of Counseling and Clinical Psychology, 56*, 206–209.

Black, C. (1979). Children of alcoholics. *Alcohol Health and Research World, 4*, 23–27.

Black, C. (1981). *It will never happen to me*. Denver: M.A.C. Printing.

Brower, S. (1987). *Effect of the family environment on the social adjustment of adult children of alcoholics*. Unpublished doctoral dissertation, North Carolina State University, Raleigh.

Castine. J. (1989). *Recovery from rescuing*. Deerfield Beach, FL: Health Communications.

Cermak, T. (1986, November/December). Diagnosing and treating co-dependence. *Alcoholism and Addiction Magazine*, p. 57.

Cermak, T. (1989). *A primer on adult children of alcoholics*. Deerfield Beach, FL: Health Communications.

Chafetz, M.E., Blane, H.T., & Hill, M.J. (1971). Children of alcoholics: Observations in a child guidance clinic. *Quarterly Journal of Studies on Alcohol, 32*, 687–698.

Collins, T. (1982). *An Analysis of the degree of internalization of selected attitudes and their relationship with locus of control and sex of children of alcoholics and a comparison group.* Unpublished doctoral dissertation, University of Maine, Orono.

Conrad, P., & Schneider, J. (1980). *Deviance and medication.* St. Louis: C.V. Mosby.

Corder, B.F., Hendricks, A., & Corder, R.F. (1964). An MMPI study of a group of wives of alcoholics. *Quarterly Journal of Studies on Alcohol, 25*, 551–554.

Cork, R.M. (1969). *The forgotten children: A study of children with alcoholic parents.* Toronto: Addiction Research Foundation of Ontario.

Denzin, N. (1987). *The alcoholic self.* Beverly Hills: Sage Publications.

Denzin, N. (1989). *Interpretive interactionism.* Beverly Hills: Sage Publications.

Edwards, P., Harvey, C., & Whitehead, P.C. (1973). Wives of alcoholics: A critical review and analysis. *Quarterly Journal of Studies on Alcohol, 34*, 112–132.

Fingarette, H. (1988). *Heavy drinking: The myth of alcoholism as a disease.* Berkeley: University of California Press.

Fox, R. (1962). Children in the alcoholic family. In W. Bier (Ed.), *Problems in addiction: Alcoholism.* New York: Fordham University Press.

Gamson, W. (1968). *Power and discontent.* Homewood, IL: Dorsey Press.

Goglia, L. (1985). *Personality characteristics of adult children of alcoholics.* Unpublished doctoral diss., Georgia State University.

Goodwin, R. (1987). Adult children of alcoholics. *Journal of Counseling and Development, 66*, 162–163.

Greil, A.L., & Rudy, D. (1984). Sociological cocoons: Encapsulation and identity change organizations. *Sociological Inquiry, 54*, 260–278.

Gusfield, J. (1963). *Symbolic-crusade: Status passage and the American temperance movement.* Urbana, IL: University of Illinois Press.

Gusfield, J. (1981). *The culture of public problems: Drinking, driving and the symbolic order.* Chicago: University of Chicago Press.

Heller, K., Sher, K., & Benson, C. (1982). Problems associated with risk overprediction in studies of offspring of alcoholics: Implications for prevention. *Clinical Psychology Review, 2,* 183–200.

Herjanic, B., Herjanic, M., Penick, E., Tomelleri, C., & Ambruster, R. (1977). Children of alcoholics. In F.A. Seixas (Ed.), *Currents in Alcoholism* (Vol. 3, pp. 445–455). New York: Grune & Stratton.

Jackson, J.K., (1954). The adjustment of the family to the crisis of alcoholism. *Quarterly Journal of Studies on Alcohol, 15*, 562–586.

Jacob, T., Favorini, A., Meisel, S., & Anderson, C. (1978). The alcoholic's spouse, children and family interactions, *Journal of Studies on Alcohol, 39*, 1231–1251.

Kammeier, M.L. (1971). Adolescents from families with and without alcohol problems. *Quarterly Journal of Studies on Alcohol, 32*, 364–371.

Kritsberg, W. (1985). *The adult children of alcoholics syndrome: From discovery to recovery.* Pompano Beach, FL: Health Communications.

Martin, D. (1988). A review of the popular literature on co-dependency. *Contemporary Drug Problems, 15,* 383–398.

Miller, W.R. (1987). Adult cousins of alcoholics. *Psychology of Addictive Behaviors*, 1(1), 74–76.

Mills, C.W. (1959). *The sociological imagination.* New York, NY: Oxford University Press.

Moos, R., & Billings, A. (1982). Children of alcoholics during the recovery process: Alcoholic and matched control families. *Addictive Behaviors,* 7, 155–163.

NIAAA (1974). *An assessment of the needs of and resources for children of alcoholic parents.* Rockville, MD: U.S. Department of Health and Human Services.

NIAAA (1979). *Services for children of alcoholics.* Rockville, MD; U.S. Department of Health and Human Services.

NIAAA (1983). *A growing concern: How to provide services for children from alcoholic families.* Rockville, MD: U.S. Dept. of Health and Human Services.

Nyelander, I. (1960). Children of alcoholic fathers. *Acta Paediatrika, 49* (Suppl. 121), 1–134.

Peele, S. (1985). *The meaning of addiction.* Lexington, MA: Lexington Books.

Rafter, N.H. (1988). *White trash: The eugenic family studies, 1877–1919.* Boston: Northeastern University Press.

Reinarman C. (1988). The social construction of an alcohol problem: The case of mothers against drunk drivers and social control in the 1980's. *Theory and Society, 17,* 91–120.

Ridlon, F.V. (1988). *A fallen angel: The status insularity of the female alcoholic.* Lewisburg: Bucknell University Press.

Robinson, B.E. (1989). *Work addiction: Hidden legacies of adult children.* Deerfield Beach, FL: Health Communications.

Room, R. (1978). *Governing images of alcohol and drug problems: The structure, sources, and sequels of conceptualizations of intractable problems.* Unpublished doctoral dissertation, University of California, Berkeley.

Room, R. (1984). Alcohol and ethnography: A case of problem deflation? *Current Anthropology 25,* 161–191.

Room, R. (1988). Cultural changes in drinking and trends in alcohol problems indicators: Recent U.S. experience. In R.B. Waahlberg (Ed.), *Prevention and control/realities and aspirations: Proceedings of the 35th International Congress on Alcoholism and Drug Dependence* (pp. 820–831). Oslo: National Directorate for Alcohol and Drug Problems.

Rudy, D. (1986). *Becoming alcoholic: Alcoholics Anonymous and the reality of alcoholism.* Carbondale: Southern Illinois University Press.

Rudy, D.R., & Greil, A.L. (1989). Is Alcoholics Anonymous a religious organization? Meditations on marginality. *Sociological Analysis, 50* (1), 41–51.

Schneider, J. (1978). Deviant drinking as a disease: Alcoholism as a social accomplishment. *Social Problems, 25,* 361–372.

Smith, A. (1988). *Grandchildren of alcoholics.* Pompano Beach, FL: Health Communications.

Smith, J.D. (1985). *Minds made feeble: The myth and legacy of the Kallikaks.* Rockville, MD: Aspen Systems Corportation.

Sobell, M., & Sobell, L. (1987). Conceptual issues regarding goals in the treatment of alcohol problems. *Drugs and Alcohol, 2/3,* 1–37.

Spector, M., & Kitsuse, J. (1987). *Constructing social problems.* New York: Aldine De Gruyter.

Subby, R. (1987). *Lost in the shuffle: The codependent reality.* Pompano Beach, FL: Health Communications.

Tournier, R. (1979). Alcoholics Anonymous as treatment and as ideology. *Journal of Studies on Alcohol, 40,* 230–238.

Watters, T., & Theimer, W. (1978). Children of alcoholics: A critical review of some literature. *Contemporary Drug Problems,* 7, 195–201.

Wegscheider, S. (1981). *Another chance: Hope and health for the alcoholic family.* Palo Alto, CA: Science and Behavior Books.

Wegscheider-Cruse, S. (1989). *The miracle of recovery.* Deerfield Beach, FL: Health Communications.

Werner, E.E. (1986). Resilient offspring of alcoholics: A longitudinal study from birth to age 18. *Journal of Studies on Alcohol, 47*, 34–40.

West, M., & Prinz, R. (1987). Parental alcoholism and childhood psychopathology. *Psychopathology Bulletin, 102,* 204–218.

Wolin, S., Bennett, L., Noonan, D., & Teitelbaum, M. (1980). Disrupted family rituals: A factor in the intergenerational transmission of alcoholism. *Journal of Studies on Alcohol 41,* 199–214.

Woititz, J. (1983). *Adult children of alcoholics.* Hollywood, FL: Health Communications.

Wood, B. (1987). *Children of alcoholism.* New York: New York University Press.

Young, E. (1987). Co-alcoholism as a disease: Implications for psychotherapy. *Journal of Psychoactive Drugs, 19,* 257–268.

CHAPTER 38

The Chronic Drunkenness Offender: Before and After Decriminalization

EARL RUBINGTON

Before 1970, the jail was the key agency in the social control of chronic drunkenness offenders (defined as persons arrested for public drunkenness three or more times in a given year). After 1970, the alcoholic detoxification center (hereafter referred to as *detox*), became the major control agency, particularly in those states that had both decriminalized public drunkenness and established civil detoxification facilities. Accordingly, this chapter, after a brief sketch of offenders' characteristics and their way of life, compares jail and detox influences on chronic drunkenness offenders.

Experience with social control agencies teaches offenders how to perform agency roles. Through repeated social interaction with keepers, helpers, and fellow-inmates, they learn how to act as prisoners or patients. Thus, chronic drunkenness offenders are made, not born. Whether learning jail culture or detox culture, they adopt ways of acting that not only comply with agency rules but also make it possible for them to achieve personal objectives. And although compliance may not necessarily foster achievement of agency goals, it does help to sustain the agency as a going concern.

Offenders and Their Ways

Offenders, older and less educated than the general population, come out of the lower social classes. They have worked for most of their lives, first regularly, then casually, at semiskilled or unskilled labor. Most offenders are not "family men." Half never married; the other half either deserted their wives and family or were divorced, separated, or widowed. Their health as measured by rates of accidents, injuries, and illnesses, both

Earl Rubington is with the Department of Sociology, Northeastern University, Boston, Mass. This chapter was written especially for this book.

physical and mental, is much poorer than that of the general population. Similarly, their death rates are much higher. And offenders, as heavy drinkers, are in various stages of alcohol dependence or abuse.

Unattached men obtain benefits of companionship at the cost of heavy drinking. Heavy drinking connects offenders with people who drink the way they do and, at the same time, disconnects them from people who are not heavy drinkers. Offenders, whose ties to conventional groups such as family, work, social, and religious groups are either weak or nonexistent, are correspondingly freed from restraints group membership imposes. In time, they withdraw from or are expelled from whatever family group in which they held membership. They seek out or are thrust into the company of men and groups who drink more or less the way they do. Such groups are more likely to be found wherever unattached men live, work, and play.

Unattached men on Skid Row share in its deviant subculture. It consists of a set of beliefs, values, and norms designed by residents for solving problems of Skid Row life. Elements of this subculture include skills, self-image, norms, and ideology. The skills comprise techniques for "getting by"; the self-image consists of alternate views Skid Row men fashion for themselves as a defense against the hostile rejection of the conventional social world; the norms refer to rules on how to act and include a considerable tolerance of both dependency and deviance and a taboo on asking any personal questions; and the ideology justifies independence, nonalignment with the conventional social world, and an excess of the appearance, if not the actual possession, of the masculine virtues. Prescribed heavy drinking becomes the vehicle for both achieving and expressing the values of this deviant way of life.

Rank among heavy drinkers on Skid Row depends on how well persons manage the dual problem of the heavy-drinking life: gaining the rewards of drinking while avoiding or reducing its social punishments. Where men drink, how they act while drinking, what they drink, how they obtain alcoholic beverages, how they share them, and how they weather the crisis that heavy drinking brings become the bases for assigning rank and respect. Older pensioners are more often *room drinkers*, whether the room is a furnished one in a rooming house, an apartment, or a cubicle in a cage hotel. They will drink alone as often as they will share a bottle with a trusted friend (Cohen & Sokolovsky, 1989). *Bar drinkers* get money for drink from working at casual labor and, in time, become the regulars in Skid Row taverns. The elite among Skid Row habitues have more money, are in better physical shape, can buy more alcohol of better quality, can play the role of the "live one" in bars, and can more likely abide by barroom etiquette (Jackson & Connor, 1953). If bar drinkers participate in tavern society, then *bottle drinkers* are more often street people. Bottle drinkers more often have to panhandle, to join bottle gangs in order to drink, and to be concerned about "flop money," finding places to sleep, and so forth. (Peterson & Maxwell, 1958; Pittman & Gordon, 1958; Rooney, 1961; Rubington, 1968; & Spradley, 1970). Bottle drinkers have less control over the drinking situation and are much more exposed to the risks of social punishment. But the risk they run more often than other kinds of heavy drinkers is that of getting arrested (Rubington, 1962).

Extended drinking bouts produce numerous alcoholic crises. The ways men manage these crises change the pattern of their dependency on alcohol, people, and social agencies. Crises break down the barriers to contact between the several ranks of drinkers; in the process people find out how people on the next level cope. Patterns of social control along with heavier drinking disrupt the system of drinking relations in the homeless men's quarter and generate cross-rank contact. For example, unemployed white-collar men moved to Chicago shelters during the Great Depression. Shelters provided inexpensive food and lodging while the men went out each day seeking employment. But shelters afforded contact with other men who had lost the incentive to look for work and who were now living day-to-day, forgetting their troubles in group drinking. A process of shelterization occurred (Sutherland & Locke, 1936). As morale sagged, some white-collar men began to look for drinking companions instead of continuing the almost hopeless task of finding a job. More often, however, it is the jail that fosters association among heavy drinkers and makes for a different set of social arrangements after release.

The Jail and the Chronic Drunkenness Offender

In the homeless men's quarters heavy drinkers are at risk of arrest. Street drinkers get arrested more often than bar drinkers who, in turn, get arrested more often than room drinkers. Arrests vary according to police department policy, protests from downtown merchants, seasons of the year, the condition of the intoxicated persons, their attitude towards police, police officers' feelings about drunks, and so on (Bittner, 1967; Nimmer, 1971; Wiseman, 1970). But the consequences of jail contact, particularly during the early stages of offender careers, turn out to be uniformly the same.

The jail serves as the principal agency controlling people found drunk in public. Its stated goals are confinement, punishment, and deterrence. For persons going to jail, it acts as the entry point into the criminal justice system (Irwin, 1985). Intended to deter, the jail actually plays a major part in the making of chronic drunkenness offenders. But inmates do not become chronic drunkenness offenders overnight. There are at least two phases in the development of the chronic drunkenness offender: the breaking and the remaking of social ties.

Initial incarceration, regardless of its length, disrupts preexisting relationships the arrested offender has to people on the outside. Being locked up is only the first of a series of shocks, among which are being housed in cramped quarters, being thrust into intimate company with strangers, being deprived of one's personal property and freedom, being moved around as cattle, being pushed and shoved and manhandled by one's keepers, and being subjected to verbal and physical abuse by one's captors and fellow inmates alike. All of these events serve only to lower one's good opinion of self. A sense of deep injustice arouses embitterment. Simply for getting drunk in public, drunkenness offenders find jail personnel and many inmates treating them as if they were the worst criminals in the world. Retaliation comes after release when many offenders, to purge themselves of the entire jail experience, find that the only way to blot it out of memory is to get drunk as soon as possible (Pittman & Gordon, 1958; Spradley, 1970).

After the initial sequence of arrests and incarceration for public intoxication, however, subsequent arrests lead to the stage of redefinition. Learning how to "do time," how to "jail," occurs during this stage. Making inside ties that will lead to outside connections now becomes possible. Drunkenness offenders seek out or are sought by other drunkenness offenders. Companionship, understanding, and social support in the immediate situation become available. Out of this association offenders learn ways of coping with the hardships of imprisonment, how to adapt, and, in some instances, how to "work the system." Involuntary association with people in the same boat also paves the way for voluntary association after release. The circle of one's acquaintances has expanded greatly. After discharge, the offender now knows a larger number of people on whom to count for drinking companionship, obtaining funds, managing alcoholic crises, and so on.

Also, during the redefinition stage, it sometimes becomes possible for offenders to develop ties with keepers and even to find a place for themselves while serving time. As they begin to return to jail more frequently, offenders change their definition of the situation. Many offenders are often assigned a "good" job in the jail whenever they are resentenced; they frequently find a degree of self-respect and status that is completely absent whenever they go back out on the streets (Giffen, 1966). In some county workhouses offenders sometimes receive treatment for alcohol withdrawal from guards and often develop friendly relations with them (Andrews, 1986).

Guards appreciate the fact that chronic alcohol offenders, generally speaking, are most tractable inmates (Amir, 1966; Kantor & Blacker, 1958). Offenders rarely give guards any discipline problems and are much more likely than other inmates to defer to the guard's authority. A consequence of the high frequency of interaction between jail guards and drunkenness offenders is personal relationships. Over time, inmates and guards come to know each other as individuals. Predictability results from these extended contacts, and each comes to count on the other not to make trouble. Herein lies the basis for the symbiotic relationship between jail guards and offenders. Offenders perform work while in jail and assist guards; in exchange, they receive bed and board and time to recover from their drinking bouts.

The repeated cycle of arrest-incarceration-release of individuals labels them as chronic drunkenness offenders. It assigns them a status and sets forth behaviors expected of them while in jail and back on the streets. It defines them as outcasts and assigns them the function of negative role models when released (Lemert, 1951; Pittman & Gillespie, 1967). In the development of this strange and enduring relationship is the fact that the jail does not deter offenders from drinking but rather sustains them in the heavy-drinking life the jail was designed ostensibly to reduce and punish, if not eliminate completely. The failure of jails to deter chronic drunkenness offenders from drinking led to the movement to decriminalize public drunkenness.

Decriminalization of Public Drunkenness

James I, King of England, made public drunkenness a crime in 1606. Settlers in colonial America, bringing English law with them, made drunkenness a crime as well as a sin. Through the years, however, people have questioned the costs, morality, and effectiveness of jailing people found drunk in public view (Pittman & Gordon, 1958).

In the 1940s, the current alcoholism movement emerged. It was based on the assumption that alcoholics were sick people who needed help and should be a public health responsibility. Variations on these themes heard more and more in the post–World War II world were: that jail was "no place for the alcoholic" and that chronic drunkenness offenders were actually "serving a life sentence on the installment plan." Pittman and Gordon in 1958 documented these charges in copious detail in their aptly titled book, *Revolving Door.* They described offender characteristics, detailed the jail's failure to deter subsequent drunkenness, and called for treatment and rehabilitation to replace punishment and custodial care.

A decade later three influential national commissions made similar recommendations and, in addition, called for the decriminalization of public drunkenness (Plaut, 1976; *Report of the President's Commission on Crime in the District of Columbia*, 1966; President's Commission on Law Enforcement and Administration of Justice, 1967). In 1970, with the passage of the Hughes Act, there emerged for the first time a federal policy on alcoholism and alcohol-related problems that declared alcoholism a public health problem. Included in the ideology of the alcoholism treatment establishment was the notion that people intoxicated in public should go to a medical facility rather than to jail.

By 1980, some 30 states had adopted their version of the model Uniform Alcoholism and Intoxication Treatment Act (hereafter *Uniform Act*), first drafted by the National Conference of Comissioners on Uniform State Laws in 1971 (Alcohol and Health, 1973). The various versions of the Uniform Act as adopted by the several states changed the system of social control of public drunkenness. If public drunkenness is a crime, then the criminal justice system is in charge. But when public drunkenness is decriminalized, detox replaces the drunk tank. The planners who drafted the Uniform Act intended that detox become the entry point into the health care system. In jail people began their criminal careers as "public drunks"; in detox people are supposed to terminate their criminal careers and begin their treatment careers as "alcoholics." Section 1 of the Uniform Act read:

> It is the policy of this State that alcoholics and intoxicated persons may not be subjected to criminal prosecution because of their consumption of alcoholic beverages but rather should be afforded a continuum of treatment in order that they may lead normal lives as productive members of society. (Alcohol and Health, 1973, p. 269)

The Uniform Act emphasized that treatment should be both voluntary and continuous; because alcoholism is a chronic illness, neither relapses nor dropping out of treatment should be penalized.

States in their implementing the Uniform Act did so in one of three ways: they decriminalized public drunkenness without making any provision for civil detoxification facilities, or they established detoxes that were either medical or nonmedical. Medical detoxification treats acute intoxication or alcohol withdrawal as a medical emergency and employs tranquilizing drugs such as Librium or Valium to treat the condition. In nonmedical detoxification (usually called "social detox") staff members generally "talk" the intoxicated person "down" without using any medication (Sparadeo et al., 1982).

The Detox and the Chronic Drunkenness Offender

Detox Outcomes: Research Findings

After the passage of the various state versions of the Uniform Act, detox outcome studies began to appear. Most took place in California, the District of Columbia, Illinois, Massachusetts, Minnesota, Missouri, New York, Rhode Island, and Ontario, Canada. They all address three components of the Uniform Act: diversion, treatment, and rehabilitation. The diversion (referral to detox) question of these studies examines the role of police as gatekeepers. The questions on length of stay and post-detox referral examine the individualized treatment plan and the continuity-of-care principle; and the question on readmission deals with the broad topic of rehabilitation. Outcome studies quantified these questions by using agency records (from detoxes, halfway houses, etc.). The questions were: How do people come into detoxes? How long do they stay? Where do they go after detox? How often do they return for detoxification?

The diversion question produced three kinds of answers. All detox studies show sharp declines in police referrals over time. Police stop taking intoxicated persons to the detox when there are no beds, when they have to wait a long time for the detox personnel to decide to admit the individual, and when they begin to see the same persons drunk on the street whom they have recently driven to the detox. In general, the responses show that the diversion goals of the Uniform Act have been "imperfectly achieved" (Finn, 1985, p. 14).

The length-of-stay question produced similar answers regardless of the detox type or its desired length of stay. Most detoxes prefer a 5-day stay; a few prefer 10 days. Regardless of staff preferences, however, the average stay is 3.5 days. Although there is some disagreement, studies of successful treatment outcomes indicate that the longer alcoholics stay in treatment, the more likely the treatment is to succeed (Ogborne, 1978; Baekeland & Lundwall, 1975).

The after-detox question produced answers showing that on the average, for the detoxes studied, at most a third of all those admitted accept referral to an aftercare facility. And, similarly, 35% of all those admitted have been in detox two or more times; some have returned to detox frequently.

Thus, on all three measures of treatment and rehabilitation effectiveness, the detoxes score extremely low in relation to the aspirations and expectations of alcohol reformers, planners, detox directors, their staff members, and their clients.

Most observers say detox doors revolve just as jail doors do, but with two important differences: Detox doors, now "padded," spin much faster than jail doors (Fagan & Mauss, 1978). A "spinning door" makes little or no contribution to treatment and rehabilitation goals. A "padded door" more likely "coddles" drunks, fosters dependence, fails to "straighten out" chronic drunkenness offenders, and according to some observers, wastes taxpayers' money.

Detox Outcomes: Explanations

Most studies report the facts, content to describe the outcomes. Others offer one of two kinds of explanation for detox failures to achieve their mandate: client qualities or treatment-client mismatch (Regier, 1979). The first explanation perhaps the most accepted, is based on research generalizations on successful treatment outcomes. Alcoholics who are socially stable (i.e., have homes, wives, families, jobs, etc.) and who have experienced less severe impairment from heavy drinking have a better chance of recovering than socially unstable alcoholics who have been impaired more severely from drinking (Ogborne, 1978). Thus, the argument runs, most chronic drunkenness offenders belong to the socially unstable category. And so the detox fails with them because they have weaker or nonexistent motivation for sobriety, a lessened capacity to obtain it, and fewer opportunities to maintain it. The mismatch explanation holds that offenders differ from middle-class alcoholics in many ways and have a greater variety of medical, social, and economic problems that must be solved in addition to, if not in advance of, their drinking problems. Middle-class detox staff members couch their clients' problems in a medical-moral context; their clients see them as immediate practical-material problems. Therefore, the staff members' chances of reaching, let alone influencing, clients decrease markedly (Regier, 1979).

Both arguments have considerable merit. But a more general explanation is that detox outcomes are consequences of acculturation, which itself is the product of frequent staff-client interaction as well as client-client interaction. In jail, inmate roles develop through inmate-guard and inmate-inmate relations. Detox acculturation differs in that jails house, process, and maintain inmate-guard and inmate-inmate contacts over a longer time period. Similarly, whereas jails have a greater range of coercive sanctions to bring to bear on their inmates, detox sanctions differ by virtue of the centers' health care mandate. Whereas jails can only expand or contract punishments, detoxes can only contract but cannot deny treatment. Thus, detoxes operate at a disadvantage when attempting to control clients.

Detox staffs devise ways of coping with the situation. In time, staffs sort out types of clients and develop ways of managing the problems these different types of individuals present. And, similarly, clients work out ways of making the best of their situation while in detox. The process again is very much the same as the one in jail. The significant differences, of course, lie in the kinds of outcomes. Although the vast majority of jail inmates are chronic drunkenness offenders, only a minority of detox clients are. However, a significant minority of chronic drunkenness offenders in detoxes become an apparent majority in the eyes of the detox staff members. And this perception of rampant recidivism has consequences for how both staffs and clients define the day-to-day situation in detox.

The Metro City Detox Study

The acculturation explanation suggested above stems from one year's field work in six Metro City detoxes (Rubington, 1980; Rubington, 1986).[1] Research methods included observation, interviews with staff members, and analysis of detox records. Baseline data on length of stay, type of discharge, referrals, and readmissions come from the same 2-month period for all six detoxes. This presentation lumps together all six detoxes, averages outcomes, and advances a sociological interpretation of the observed regularities.

All six detoxes were established from 1970 to 1975. The state in which they are located passed its version of the Uniform Act in 1971, with decriminalization to take effect in 1973. North Detox, the first to be established, started with federal funding in 1970 and with 56 beds; the other five are all 20-bed detoxes. North Detox's civilian rescue team of recovering alcoholics patrolled the streets of the North End in search of voluntary clients for detoxification. Planners placed the detox in this area because it had the highest rates of arrest for public drunkenness, the most liquor outlets, the most rooming houses, and a large, unattached male population, some portion of which were heavy drinkers, if not already chronic alcoholics.

After a few years, a standard referral procedure has evolved. Acutely intoxicated persons or those approaching or already in a state of alcohol withdrawal go or are taken to the Metro City Alcoholism Clinic. The clerk calls one of the six detoxes and announces that the clinic has a candidate. After a brief exchange of information about the person, if the detox has a bed, a staff member tells the clerk it will be held for two hours.

At the detox the candidate is taken through a set procedure. A nurse or paramedic screens the candidate to decide whether he (or, on occasions, she) needs detoxification. If the candidate is admitted, an aide stores the clothing and valuables while the client showers and then is escorted to a bed on the receiving ward. A nurse or paramedic completes a physical examination, takes a drinking history, paying special attention to the drinking bout that led to the admission (beverages, amounts, duration of bout, etc.). After the nurse has injected Thiamine and the first of a series of doses of the tranquilizing drug (usually Librium) to treat medically the symptoms of acute alcohol withdrawal distress, the aide puts the client to bed. This is the first phase of a process that takes 3 days, after which the client has been withdrawn from both alcohol and the tranquilizing drug.[2] Thus, whenever staff discuss clients in case conferences, they always precede their account of status and progress by saying that he is "Day 1", "Day 2", and so forth, as the case may be.

Medical aides take the vital signs on the newly admitted patients every four hours (pulse, blood pressure, temperature). At admission a counselor is assigned to each client. The counselor introduces himself to the patient, usually on the second day in detox, and makes arrangements for the two of them to meet as soon as the patient feels better. They discuss the patient's drinking history, whether he thinks he has a drinking problem, and what he would like to do after he leaves detox. The counselor lists a variety of the post-detox treatment options such as a halfway house, a mental health clinic, an outpatient alcoholism clinic, a hospital, custodial care, and so on. Once ambulatory, patients are supposed to attend a number of activities. These include resource meetings where halfway house personnel describe their program, slide shows, alcoholism and alcohol education films, talks, group discussions (usually called "rap sessions"), AA meetings, and the like. All the detoxes prefer that clients stay a minimum of 5 days.

Clients leave detox in one of four ways: they take placement (TP), leave with their own plan (OP), leave against medical advice (AMA), or they are released, in a procedure called administrative discharge (AD). The percentages for each type of discharge for the 2-month baseline period were as follows: TP, 15% ; OP, 45% ; AMA, 35% ; and AD, 5% . It may prove helpful to add that the designation *TP* means that the patient agreed to continue treatment after his discharge, usually in a halfway house; *OP*, that the patient stayed five or more days in detox, completed medical detoxification, but declined any post-detox referral; *AMA*, that the person left before completion of the medical detoxification with alcohol, drugs, or both in the bloodstream; and *AD*, that the person was required to leave, generally for major violations of detox rules (verbal or physical abuse of staff, bringing in or consuming alcohol or drugs on the premises, etc.).

All alcohol treatment organizations face the problem of relapse. As organizations age, more patients return for treatment while the cohort for potential relapses increases yearly. Detoxes, because of their definition as emergency alcoholism treatment services, confront this problem within a relatively short time. Their problem becomes more acute because by law they are not supposed to deny treatment to anyone. Since "open admission" is the policy, detox staffs would seem to have little control over the situation. But, as readmissions begin to mount, the detox, its procedures, its staff, and its clientele come under scrutiny (Regier, 1979). This is the point at which detox staffs begin to redefine their situation. Relapse poses a serious problem for them, and they begin to evolve some solutions, which they share, transmit, and revise.

Staff adaptations evoke their reciprocals, that is, client adaptations. In turn, daily work contingencies sometimes hamper staff members and prevent them from acting on their redefinition of the situation. In time, however, detoxes attain a relative degree of organizational harmony and become going concerns to the extent that they begin to draw on a set of stable and reliable repeaters. Just as the jail shapes and fits drunkenness offenders to its needs for stable and dependable roles (and probably dependent roles as well), so does life in detox help an important segment of its clientele to assume and perform stable and dependable (albeit dependent also) roles.

Staff Conceptions of Detox Clients

Shortly after being hired, aides, nurses, and counselors form a set of conceptions about the kinds of people who enter detox. As their experience broadens, they begin to develop specific types of ideas about detox clients. Two axes of interest form the basis of their client typology. One has to do with the amount, kind, and frequency of troubles clients make for the staff. The other turns on the typical way patients are likely to depart from detox (TP, OP, AMA, or AD). Beliefs, values, and norms emerge and become important elements in staff culture, and all of these influence the answers to questions on what to feel, think, say, and do with particular types of clients. The kinds of trouble clients make varies with where they are in the detox process and with which staff members they come into contact. Thus, for example, small aides are more apt to have difficulty undressing and showering large, uncooperative, intoxicated clients. Patients who try to "wheedle medications" are most apt to do so with nurses. And those who are "playing games," who can't make up their mind whether they have a drinking problem or to which halfway house they should go after detox, are more apt to present these kinds of problems to their counselors.

The second axis of interest concerns the typical ways clients leave detox. The amount, kind, and frequency of troubles staff members seem willing to accept are all relative to the discharge disposition. As a rough rule of thumb, the amount of time staff members are willing to give to clients and the amount, kind, and frequency of trouble they are willing to accept from clients are roughly proportional to the probable discharge disposition. That is, for example, the more likely a patient's TP discharge, the more time and trouble he is worth. The alcoholism movement enunciates the dictum that the alcoholic is a sick person and worthy of help. Treatment personnel, however, in daily contact with alcoholics decide that some alcoholics are more worthy of help than others (Wiseman, 1970). The time-trouble/disposition ratio from which the staff members derive their typology of detox clients makes it possible for staff members to decide on the relative worth of their clients.

A major part of staff problems stems from their lack of control over admissions and readmissions. Staffs cannot keep out all of the difficult clients referred to them or those who occasionally seek admission as self-referrals. Only two options are open to them: restrict readmissions whenever possible, and limit contact with readmitted clients while they are in detox. If staff members feel a former client has not made proper use of detox, they can place him on a 30-day, 60-day, or 90-day restriction as the case may warrant. For example, a person can be placed on the restricted list because he "always leaves AMA." A number of former clients are permanently banned and placed on the "do not admit" (DNA) list because they are considered to be "inappropriate referrals" (for example, drug addicts, psychotics, persons suffering from a chronic medical condition that the detox is not equipped to treat, etc.) or because they have a history of violation of detox rules. Detox staffs generally refer to the DNA list as the "blacklist." Given the great demand for detox beds, however, these restrictions cover only a small number of possible candidates for detoxification. And so reduction of staff-client contact in detox becomes the more frequent staff option.

Within limits, staff members can reduce the frequency, duration, and intensity of client contacts. However, their ability to limit contact varies with their kind of work. Aides, for instance, can do little about the frequency of contact with clients, whether the latter are first-timers or repeaters. Aides must perform a number of regularly scheduled routine activities, such as storing clothes and valuables, showering patients, and taking vital signs. On the other hand, nurses spend much less time with repeaters. Both physical examinations of repeaters and paperwork on them take nurses much less time than that spent with repeaters by aides. In addition, as nurses come to know more about repeaters, they can distinguish those who have "easy" withdrawals from those who have "hard" withdrawals and can choose accordingly. Of all the staff, counselors are in the best position to manage client contacts. In their first brief meetings, clients state directly or give signs that they have no interest in seeing a counselor, talking about alcoholism, or making plans for aftercare; and counselors are permitted to respect clients' wishes. By virtue of their diverse locations in detox work flow, then, staff members differ in their ability to manage the time-trouble ratio; nonetheless all do what they can.

Alcohol treatment organizations have the problem "of deciding what patients will benefit from treatment" (Cahalan, 1987, p. 127). Given the perception of a rising tide of readmissions, staff place a high premium on the persons coming into detox for the very first time.[3] One nurse indicated the very high value of a "new face" when she said: "I will go out of my way to get a new one in. It's a terrific morale-booster for the staff."

Generally speaking, staff members classify clients into three types: (1) those who seek shelter, (2) those who just want to "dry out," and (3) those who want detoxification and will accept placement. And those types often match their discharge dispositions: (1) AMA, those seeking shelter just for the night, (2) OP, those who just want to "dry out", and (3) TP, those who want both detoxification and a post-detox referral to some kind of continuing alcoholism treatment. But of all the types who come into detox, the ones who leave AMA arouse the most moral indignation among the staff.

At any given time, around a third of the clients of the six detoxes are repeaters. The staff knows them personally and has a fairly good idea of how they withdraw from alcohol, what their desire for additional medications is, whether they are interested in taking placement after they have completed detoxification, and so on. Thus, the staff has ideas on how repeaters come in, how they act while in detox, and how they will leave. This working knowledge helps them solve detox work problems.

Learning to be a Detox Client

Just as chronic drunkenness offenders learn in jail how to "jail," how to "do time," and, in the process, how to become and continue as chronic drunkenness offenders, a number of repeaters learn how to be detox clients. How does this transformation occur? Clients learn how to "do detox" through frequent contact with the staff and with other clients. Primarily, they learn what the staff expects of them as they move through the sequences of their particular detoxifications. Their acculturation comes to them through observation, experience, and, in many instances, through instruction by staff members, and fellow-clients alike.

Clients acquire a body of information on how to act while in detox. In the main, it consists of a body of rules that the detox staff has promulgated. Clients can infer these rules from observation, from what staff members say about other clients, from some of the questions they ask, and from what other clients tell them. The detox rules are as follows: (1) Don't come into detox unless you're hurting; (2) say you'll stay five days; (3) don't be a PITA;[4] (4) talk about your drinking problem; (5) admit you're an alcoholic; (6) attend meetings; (7) say you'll take placement; (8) stay five days; (9) take placement; and (10) don't come back right away. These rules state how clients ought to act. However, in the teaching and learning of rules, these are variations dependent upon the kind of drinking and detox life. Participation in drinking life varies with whether one is a street (bottle), bar, or room drinker, and detox life varies with whether one is a *revolver* or a *rounder* and whether one is a continuous or an intermittent client of detox (Wiseman, 1970, p. 318).

A revolver is a person who goes to only one of the six detoxes. A rounder is a person who becomes a frequent client of two or more of the six detoxes. Continuous patronage, whether as revolver or as rounder, occurs in a regular, patterned sequence (every other day, every week, etc.). Intermittent patronage follows no particular pattern at all. A revolver who appears continuously at a given detox becomes well known and will be typed rather quickly as a chronic repeater. A revolver who appears intermittently is less apt to "wear out his welcome"; it will take the passage of considerable time before he is recognized as a familiar face around the detox.

The same applies to the rounder, but in each instance more time has to elapse before his identity as a chronic repeater becomes well established. An intermittent rounder is apt to be labeled much later in his detox career as a chronic repeater than a continuous rounder. The sequence, then, of being assigned the identity of repeater goes the following way, quickest to slowest: continuous revolver, intermittent revolver, continuous rounder, and intermittent rounder.

Types of Detox Careers

As the frequency of readmissions begins to mount, detox clients, like staff members, devise ways of coping with detox. Clients acquire these abilities through experience, observation, and interaction with the staff and with other clients. From our research we have constructed the types of roles that street (bottle) drinkers, bar drinkers, and room drinkers are most likely to adopt in the course of their careers as detox clients. Important sources of variation in these careers are the kind of life the drinker has been living outside detox and his experiences with the detox staff.

Street drinkers. Street drinkers have the highest frequency of readmissions. Whether revolvers or rounders, they are more often continuous than intermittent detox users. Their degree of alcohol withdrawal severity varies inversely with the number of times they have been detoxified. Thus, in the early stages of their detox careers, their admissions are marked by relatively severe withdrawal. In the middle stages of their career, the admitting withdrawal state is moderate. And, in the later stages of their career, it is mild. On the whole, street drinkers are less apt to be typed as PITAs by the detox staff. They are more likely to "suffer in silence." Nurses sometimes note that many of these patients show an interesting pain-to-complaint ratio: the more they hurt, the less they complain. Once detoxified, however, they avoid talk about drinking problems, deny alcoholism, skip meetings, and so on. Most of the time they leave AMA, occasionally OP, but rarely TP. Because of their immersion in the drinking subculture, they have most of their contacts outside of detox with other street drinkers. Given their extended circle of drinking acquaintances, they are more likely than other types of drinkers to know other clients who are being detoxified about the same time as they are. As a consequence, sometimes two acquaintances will suddenly leave AMA in order to resume drinking together. Mostly set in their street ways and definitions of the drinking situation, they show the strongest resistance to the relatively narrow definition of alcoholism, and the ways to recover from it, that the staff presents. As a result of their strong resistance to redefinition, coupled with their conformity to detox rules, their short stays, and their frequent readmissions, the staff begins to relax its pressure to convert them to detox definitions.

Bar drinkers. Bar drinkers, on the other hand, are at risk of becoming street drinkers. They are more apt to retain some conventional views on alcohol and alcoholism. Consequently, staff members find them less resistant to detox ideology on alcoholism than street drinkers. The staff presses them hard to adopt detox ideology. In comparison to street drinkers, they appear to be better risks for successful treatment because of the pattern of their contacts with detoxes. Although they may have a moderate rate of readmission, whether revolvers or rounders, they are more often intermittent detox users. Consequently, they are much more likely to be mistaken for first-timers. Generally they are admitted in severe or moderate withdrawal distress. Unlike street drinkers, in time they adhere less to client etiquette. With more experience in detox, they are more likely to become known as PITAs. As they learn detox ideology, they will announce their intentions of staying 5 days and of taking placement. They become skillful at manipulating staff members and at finding fault with halfway houses they had agreed to go to on discharge. Because they do not look or act like the detox stereotype of the chronic repeater, staff members invest much time with them, are more tolerant of their complaints, and are more sympathetic of their plight because of their greater willingness to talk about their drinking problem and their seemingly ready acceptance of self-definition as an alcoholic. Unlike street drinkers, bar drinkers retain some contacts with significant others outside of detox who share the same sentiments about alcoholism as the detox staff. Thus, bar drinkers have contact with definitions both favorable and unfavorable to the idea of stopping drinking. This duality contributes, in part, to their seemingly strange change in mind in the course of any given detox stay. They are clearly of two minds about their situation. Their vacillation could just as easily be defined as manipulation if they were more frequent or more continuous detox users. the net result is that the staff spends a good deal of time with them relative to the increasing amount of troubles they bring. Because these troubles are spread out over different readmissions, often at different detoxes, staff members continue to give them the benefit of the doubt about their sincerity about recovery from alcoholism. Most frequently then, bar drinkers depart from detox OP, frequently TP (particularly in wintertime), and occasionally AMA.

Room drinkers. Room drinkers, unlike street drinkers and bar drinkers, are more often social isolates who have minimal contacts with other drinkers and minimal or nonexistent ties to conventional significant others. They are the least frequently readmitted. Their pattern is intermittent, and considerable time elapses between each admission. They return to detox usually in severe withdrawal distress, much of it emotional. On the whole, they are well-behaved clients who rarely are categorized as PITAs (their infrequent appearance in detox qualifies them as first-timers). On admission they are quiet, serious, and concerned, as well as confused, about their drinking. They talk to a moderate degree with the staff; with other clients they do more listening than talking. They attend meetings although they are not quite sure whether they are alcoholics or not. Their career pattern is both progressive and, at the same time, characterized by sudden shifts. After the first series of admissions, they leave AMA. The next couple of times, they go out OP. But in the last series of admissions, they take placement. Over time, their doubts about being alcoholic have been resolved because they have come to regard the detox staff members as their significant others. In time, they come to view themselves and what they have been doing through the eyes of the detox staff. Outside detox, they have few close associates who might offer an alternative definition of their drinking.

Recovery from alcoholism requires two conditions: accepting the definition of oneself as alcoholic that significant others present and then applying their rules for making a recovery. Bar drinkers, room drinkers, and street drinkers all vary in their degree of acceptance of the social definition of the word *alcoholic* and the extent to which they abide by detox rules. And the ways in which staff members interact with them influence the detox rules. Room drinkers, for example, at first reject the definition and the rules, then almost as quickly both accept the definition and abide by the rules. From this research we offer an explanation for the high rates of unsuccessful detox outcomes. The great bulk of detox chronic repeaters are street drinkers. Once they are identified as repeaters, the staff members give them as little of their time as possible. In turn, repeaters remain the kinds of people they were on each admission, make little or no trouble for the staff, usually leave AMA and return as regularly as they can, going through numerous detoxifications. Room drinkers, the least frequent of repeater types, get little staff time while giving little trouble in the beginning sequence of admissions; they alter this pattern in subsequent sequences, getting more time while giving less trouble. In the acculturative process, they accept definition as alcoholics, they adopt staff perspectives of their situation and what they ought to do about it, and they take post-detox referral, thereby having successful detox outcomes. Bar drinkers, typecast as good treatment risks, get the most time from staff members while making increasing trouble for them. In relation to the effort expended on them, they account for a relatively small number of the successful detox outcomes.

Typecasting yields relatively few successful outcomes. Street drinkers fill the beds and play by the detox rules. When staff members define street drinkers as chronic repeaters, they also reduce contact with them in exchange for the repeaters giving them no trouble and making their expected early departure. When staff members define bar drinkers as good prospects, they in turn spend considerable resources on them only to find their troubles mounting proportionately to the time they give and their efforts being rewarded by relatively unsuccessful outcomes for these clients. Only room drinkers, the least frequent of repeaters, because of their zigzag detox career pattern provide staffs with successful detox outcomes in which there have been moderate investments in time and relatively little trouble.

Summary and Conclusions

The jail's failure to deter or change chronic drunkenness offenders paved the way for decriminalization. In many states, detoxes replaced drunk tanks. Despite their promise, detoxes have not fulfilled inflated expectations. Research shows that detox doors revolve much faster than jail doors. Most analysts attribute detox failures to client quality; some explain them by treatment-client mismatch. This chapter subsumes both explanations under the rubric of acculturation. Just as jail and chronic drunkenness offenders worked out a symbiotic relationship, so have detoxes and their band of chronic repeaters. However, because of variations in drinking life, three different types of detox careers have emerged. Whereas street drinkers help fill the beds (thus meeting agency goals), bar drinkers give staff "good" treatment risks to work on, thus satisfying the needs of service workers who want to help. Only room drinkers, who evade typecasting, seem to make a somewhat larger contribution to successful detox outcomes.

What are the future prospects of detoxes, given these modest results? As Skid Row areas contract and as homeless populations grow in diversity, conjecture suggests that the older subculture, which afforded considerable support to typical chronic drunkenness offenders, will lose numerous adherents (Hoch & Slayton, 1989). As the varieties of people unattached change, as new styles of homelessness emerge, and as multiple substance abuse spreads, there will be a need for more rather than fewer detoxes. If the number of street drinkers diminishes and if the number of room drinkers increases, rates of successful detox outcomes should increase correspondingly.

Three conditions are likely to increase successful detox outcomes: (1) development of broader views about people who have drinking problems, (2) diversification of programs for detox clients, both before and after detox, and (3) more experience with detoxes and more systematic research on these experiences (McCarthy, Mulligan, & Angeriou, 1987).

The great growth of the alcoholism treatment system will no doubt make a substantial difference in alcoholic careers. For one thing, the increased number and variety of treatment agencies have simply expanded the circuit that all varieties of alcoholics can now make. This expansion may help to remove relapse from the categories of sin, crime, or moral failure, as many treatment personnel view it, and cause it to be seen as an event that needs to be understood. A reexamination of the detox informed by a new, broader understanding of the nature of relapse may help to show where detox fits in the servicing and treatment of chronic drunkenness offenders. Jails have existed for more than 300 years. The detox, as an alternative to jailing, is only 25 years old.

Acknowledgments

Research reported in this chapter was supported in part by National Institute of Alcohol Abuse and Alcoholism grant AA-02900 and by a grant from the Northeastern University Research and Scholarship Development Fund.

Notes

1. All names of persons and places are fictitious.
2. North Detox medical staff members referred to the people they administered as patients. The counseling staff called them clients. This chapter uses both terms.
3. Because of the lack of information, staff members often do not know that a first-timer in their detox has already become a "chronic repeater" at one or more of the other five detoxes.

4. PITA stands for "pain in the ass." Clients who are "chronic complainers" or "whiners," who make a number of petty requests repeatedly, who try to "wheedle medications" from nurses, who, in general, make a great deal of what staff members conceive of as unnecessary trouble are generally awarded the appellation of PITA. The term often appears in nurses' notes, occasionally even on the DNA list.

References

Alcohol and Health (Report from the Secretary of Health, Education, and Welfare). (1973). New York: Charles Scribner's Sons.

Amir, M. (1966). Sociological study of the house of correction. *American Journal of Corrections, 28,* 20–25.

Andrews, F.K. (1986). Experiences of incarcerated alcoholics: Chronic drunkenness offenders in a county workhouse. In D. Strug, S. Priyadarsini, & M.H. Hyman (Eds.), *Alcohol interventions: Historical and sociocultural approaches* (pp. 129–141). New York: Haworth Press.

Baekeland, F., & Lundwall, L. (1975). Dropping out of treatment: A critical review. *Psychological Bulletin, 82,* 738–783.

Bittner, E. (1967). The police on Skid Row: A study of peace-keeping. *American Sociological Review, 32,* 701–706.

Cahalan, D. (1987). *Understanding America's drinking problem.* San Francisco: Jossey-Bass.

Cohen, C.T., & Sokolovsky, J. (1989). *Old men of the bowery: Strategies for survival among the homeless.* New York: Guilford.

Fagan, R.W., Jr., & Mauss, A.L. (1978). Padding the revolving door: An initial assessment of the Uniform Alcoholism and Intoxication Act in practice. *Social Problems, 26,* 232–247.

Finn, P. (1985). Decriminalization of public drunkenness: Response of the health care system. *Journal of Studies on Alcohol, 46,* 7–23.

Giffen, P.J. (1966). The revolving door: A functional interpretation. *Canadian Review of Sociology and Anthropology, 3,* 154–166.

Hoch, D., & Slayton, R.A. (1989). *New homeless and old: Community and the skidrow hotel.* Philadelphia: Temple University Press.

Irwin, J. (1985). *The jail.* Berkeley: University of California Press.

Jackson, J., & Connor, R. (1953). The Skid Road alcoholic. *Quarterly Journal of Studies on Alcohol, 14,* 468–486.

Kantor, D., & Blacker, E. (1958). *A survey of Bridgewater: An institution for the chronic drunkenness offender.* Commonwealth of Massachusetts: Office of the Commissioner on Alcoholism.

Lemert, E.M. (1951). *Social pathology.* New York: McGraw Hill.

McCarthy, D., Mulligan, D.H., & Argeriou, M. (1987). Admission and referral patterns among alcohol detoxification patients. *Alcoholism Treatment Quarterly, 4,* 79–90.

Nimmer, R.T. (1971). *Two million unnecessary arrests: Removing a social service concern from the criminal justice system.* Chicago: American Bar Foundation.

Ogborne, A.C. (1978). Patient characteristics as predictors of treatment outcomes for alcohol and drug abusers. In Y. Israel, F.B. Glaser, H. Kalant, R.E. Popham, V. Schmidt, & R.G. Smart (Eds.), *Research advances in alcohol and drug problems* (pp. 177–224). New York: Plenum Press.

Peterson, W.J., & Maxwell, M.A. (1958). The skid road 'wino.' *Social Problems, 5,* 308–316.

Pittman, D.J., & Gordon, C.W. (1958). *Revolving door.* New Haven, CT: Yale Center of Alcohol Studies.

Pittman, D.J., & Gillespie, D.G. (1967). Social policy as deviancy reinforcement: The case of the public intoxication offender. In D.J. Pittman (Ed.), *Alcoholism* (pp. 106–124). New York: Harper & Row.

Plaut, T.F.A. (1967). *Alcohol problems: A report to the nation by the cooperative commission on the study of alcoholism.* New York: Oxford.

President's Commission on Law Enforcement and the Administration of Justice (1967). *Task force report: Drunkenness.* Washington, DC: U.S. Government Printing Office.

Regier, M.G. (1979). *Social policy in action: Perspectives on the implementation of alcoholism reforms.* Lexington, MA: Lexington Books.

Report of the President's Commission on Crime in the District of Columbia. (1966). Washington, DC: U.S. Government Printing Office.

Rooney, J.F. (1961). Group processes among Skid Row winos: A reevaluation of the undersocialization hypothesis. *Quarterly Journal of Studies on Alcohol, 22,* 444–460.

Rubington, E. (1962). "Failure" as a heavy drinker: The case of the chronic-drunkenness offender on Skid Row. In D.J. Pittman & C.R. Snyder (Eds.), *Society, culture, and drinking patterns* (pp. 146–153). New York: Wiley.

Rubington, E. (1968). The bottle gang. *Quarterly Journal of Studies on Alcohol, 29,* 943–955.

Rubington, E. (1980). The social organization of relapse. *Sociological Abstracts, 28,* 107.

Rubington, E. (1986). Staff culture and public detoxes. In D. Strug, S. Priyadarsini, & M.E. Hyman (Eds.), *Alcohol interventions: Historical and sociocultural approaches* (pp. 97–128). New York: Haworth Press.

Sparadeo, F.R., Zwick, W.R., Ruggiero, S.D., Meek, D.A., Carloni, J.A., & Simone, S.S. (1982). Evaluation of a social-setting detoxication program. *Journal of Studies on Alcohol, 43,* 1124–1136.

Spradley, J.P. (1970). *You owe yourself a drunk: An ethnography of urban nomads.* Boston: Little, Brown.

Sutherland, E.H., & Locke, H.J. (1936). *Twenty thousand homeless men.* Philadelphia: Lippincott.

Wiseman, J.P. (1970). *Stations of the lost.* Englewood Cliffs, NJ: Prentice-Hall.

CHAPTER 39

The Medicalized Conception of Alcohol-Related Problems: Some Social Sources and Some Social Consequences of Murkiness and Confusion

PAUL M. ROMAN AND TERRY C. BLUM

In the realm of the intangible, little has received more scholarly attention than the disease concept of alcoholism and its accompanying intellectual baggage. Although there is no known claim of a bibliography of every work that represents an attempt to describe, analyze, attack, or defend this concept, surely the list would be massive; even partial lists are overwhelming (e.g., Fingarette, 1988, pp. 147–162). Likely even more massive would be the confusion, distress, and despair that would attend an endeavor directed toward absorbing and integrating all that has been written about the disease concept of alcoholism.

However, this scope of attention does not diminish the continuing importance of the disease concept and its influence on both research and practice. If there is any single theme that has haunted the sociocultural analysis of alcohol and alcohol problems since its earliest days, it is surely this conceptual ball of yarn, which includes the notions of illness, progression, control by the self, control by others, addiction, habits, desire, defeat, submission, and will, as well as many others.

In an attempt to capture much of this "package," the referent used in this chapter is the "medicalized concept of alcohol-related problems." Here the analytic approach is neither to attack nor to defend this concept. Instead we take a different tack by examining sociocultural sources of ambiguity and confusion that may be undermining cultural and political support for the general view of alcohol problems that has been dominant in North America for the past 20 or 30 years. To a considerable extent we

Paul M. Roman is with the Department of Sociology, University of Georgia, Athens. Terry C. Blum is with the School of Management, Georgia Institute of Technology, Atlanta. This chapter was written especially for this book.

view actions and directions within the alcoholism movement itself as contributing to that ambiguity and confusion. What is presented here may be seen as an organizational analysis framed within concepts of culture.

While much commentary on the medicalized concept of alcoholism is centered on its growth or decline, we are not operating within that linear model. We instead project the concept's dominance as more or less a given for the foreseeable future. We do not discern a distinctive "recriminalization" of alcohol abuse and alcoholism (Roman, 1978, March), despite forces that might be interpreted as movement in such a direction. What we are attempting to describe and explain are ambiguity and confusion, with the accompanying expectation of a decline in the readiness of the public in the United States to support intervention and referral to treatment of alcohol-troubled individuals. At the very least we expect a decline in the growth of such readiness as has been apparent over the past decade.

The Murkiness of the Medicalized Conception

We have selected the term *murkiness* to summarize the current status of the cultural meaning of the medicalized conception of alcohol problems. What is this murkiness? Essentially it can be described through the apparent problems that would pervade hypothetical answers to questions such as these:

Is alcohol a drug? Are there significant commonalities between alcoholics and those addicted to other drugs? Are these commonalities so significant that these people really belong in a single category of illness or deviance? Is repeated excessive alcohol consumption the result of individual volition, or of involuntary "biopsychosocial" processes? Is alcoholism the major and central "alcohol-related problem" or should the definition of the "alcohol-related problem" include the multiple social, economic, and physical costs of drinking? Is alcoholism primarily a drinking-related disorder, or does it embody or reflect broader psychiatric disorders? Is the availability of alcohol significantly linked to alcohol-related problems? Murkiness can be found and elaborated in the ambiguous and contradictory answers offered by the students of alcohol-related phenomena to these and other closely related questions that are prominent in both the public mind and scientific investigation.

What is the evidence that this murkiness is reflected in this so-called public mind? Partial answers to this question are contained in much of the material making up the remainder of this chapter. There is, however, a critical set of data-based observations that describe a very important dimension of murkiness.

As is well known, research on public attitudes toward alcohol problems has tended to demonstrate over time a growing public acceptance of the medicalized concept of alcoholism. Without doubt, these public attitudes are a major and dramatic "social accomplishment" (Schneider, 1978). Yet beginning with an analysis carried out by Mulford and Miller (1961), there has been persistent evidence of murkiness, namely, that the medicalized concept and its supposed opposite, the "moral weakness" conception, are concomitantly held by a substantial proportion of the population—a finding evident in both the 1970s (Orcutt, Cairl, & Miller, 1979) and the 1980s (Blum, Roman, & Bennett, 1989; Caetano, 1987).

The original Mulford and Miller finding substantially predated the intense federally funded campaigns of the 1970s, which were geared toward a sociocultural and organizational transformation of alcohol problems. These efforts encompassed public education, research, demonstration projects, service delivery, and legislative influence, which not only attempted attitudinal change but provided much of the foundation for a new system of alcoholism treatment and intervention, which is now in place. Thus, one might expect that data collected in the 1980s would reflect a diminution of this ambivalence, but this does not appear to be the case. Therefore, primary evidence of the "impact" of murkiness is found in the persistent "confusion" among a substantial proportion of the public over the meaning of alcoholism, a condition they simultaneously characterize as the responsibility of individuals yet something out of their control.

What are the consequences of murky conceptions of alcoholism? Do such conceptions make a difference? Again, much of what follows is directed toward answering these questions. But to address the issue of consequences of murkiness, we turn to the observations of Don Cahalan, a social epidemiologist who has been in many ways a central figure in the social scientific study of alcohol-related problems over the past 30 years, especially noteworthy for his advocacy of the concept of alcohol-related problems as a partial alternative to that of alcoholism.

Recently Cahalan (1987) has written that despite all of the energies that have been directed toward dealing with alcohol-related problems in the United States since the early 1970s, it is nearly impossible to marshal any societal-level evidence that the social and economic impact of alcohol abuse and alcoholism on U.S. society has been affected in any way. At best it might be said that a "holding operation" has been successfully achieved, that the impacts of alcohol problems would be considerably more intense had not these efforts occurred. Elsewhere, Cahalan (1988) attributes this lack of progress to the minimal emphasis on the primary prevention of alcohol problems, an emphasis that would have been expected as a natural part of an overall strategy stemming from the designation of alcoholism as an illness. He argues that there are numerous implicit organizational conflicts between treatment interests and advocates of primary prevention, with the much greater sociopolitical strength of the former group effectively resisting a prevention emphasis.

Alternatively, it is clear that a popular explanation of the apparently less-than-desirable impact on the U.S. alcohol problem can be centered on the relative ineffectiveness of alcoholism treatment (Saxe, Dougherty, Esty, & Fine, 1983). Here issues become quickly muddled and confused. What type of treatment is being assessed, what definition of validity is implicit in the research design, and are any measures comparing treatment with no treatment? Indeed, this classic requirement of experimental design might be viewed as the Achilles' heel of treatment evaluation. In a world where alcoholism treatment consists primarily of social interaction and where social support and interaction can surround alcoholics wherever they are, what can the notion of "no treatment" possibly mean? None of these concerns addresses the most basic research problem of the sources, conditions at entry, and external contingencies characterizing clients admitted for alcoholism treatment.

In any event, the use of treatment effectiveness as a major "account" for scanty evidence of impact on the U.S. alcohol problem presumably has the convenience of being "inside the tent" of the field of alcohol-related problems. Though of much interest to those who pay the bills for alcoholism treatment, the treatment effectiveness issue has much less of the potential for generating major conflict with "outsiders" that is evident with preventive strategies that seriously address the availability of alcohol. Treatment evaluation also offers itself to the ready solution of more scientific research (Gordis, 1987).

The Central Problem: Use of Interventions

The core "action" of social processes that lead to referral to treatment of alcohol-related problems is where the central consequences of murkiness of the medicalized concept of alcohol-related problems are found. Two consistent findings (Institute of Medicine, 1989), repeated across studies of the treatment of alcohol-related problems, are (1) the low proportion of cases that come to treatment in relation to their presumed presence in the population and (2) the chronicity or "lateness" of the alcohol-related problems observed among those who enter treatment. Both of these observations can be brought together as an outline of the central consequences of murkiness.

If it is accepted that intervention in alcohol-related problems is culturally, socially, and economically desirable in the United States, then this intervention should be maximized in relation to the presence of alcohol-related problems in the population. In order for intervention techniques to be used, action by members of the general public is necessary, in the form of either self-referral or referral of significant others to those expert in intervention and treatment.

These actions are motivated by cognitions and beliefs about the nature of the perceived problem and the nature of the appropriate solution. These actions are also affected by the cognitions and beliefs manifest in the formal and informal organizations and networks to which persons turn for social support. These actions (or the relative absence of them) directed toward intervention in alcohol-related problems hinge upon the diffusion and acceptance of a consistent message about what these problems are and what ought to be done about them.

To summarize this logic: It is evident that treatment has become defined as the culturally appropriate means of instrumentally dealing with people who are defined to have alcohol problems, that is, whose behavior has exceeded the tolerance of their social networks. It is evident that the proportion of the population that holds mixed or ambivalent attitudes toward alcohol-related problems as health problems is substantial, is somewhat stable as a proportion of the population, and perhaps can grow in size with continuing ambiguity of informational messages about alcohol-related problems. Finally, there is accumulated evidence that the overall referral and treatment intervention system already does not work very effectively in terms of reducing U.S. problems with alcohol (Institute of Medicine, 1989).

The key point is that regardless of the efficacy of treatment, it is of little consequence if it is not utilized. Although the use of treatment for alcohol-related problems has seen an increase, represented in part by the very rapid growth in the numbers of inpatient and outpatient treatment centers, most centers remain underutilized. Further, a total census of treated cases does not come even close to the estimated prevalence of alcohol dependence and other alcohol-related problems in the U.S. population. Finally, as mentioned, much entry into treatment occurs at the "later" stages of problems. All of these bits of evidence highlight the importance of the infrastructure of the culturally dominant definitions of what alcohol-related problems are and what ought to be done about them. The importance of this attitudinal infrastructure is its effects on the use of alcoholism counseling and treatment.

Thus, this society's system of intervention in alcohol-related problems can be expected to remain inefficient or to deteriorate, or both, if the social messages regarding the nature of alcohol-related problems and their solutions continue to be confused or become increasingly muddled. This chapter is a beginning effort to bring together observations of some of these organizational processes. It is not intended to be exhaustive, however, and it considers only two sources of definitional confusion: (1) transformation of the clinical nature of alcohol-related problems and their treatment toward a psychiatric model in such a way that the traditional disease model is increasingly "irrelevant," and (2) introduction of mandatory health warning labels on alcoholic beverage containers as a universal cultural statement about the nature of alcohol and the potential consequences of drinking.

Transformation of Alcoholism Treatment

Regardless of its actual derivation from "science," the medicalized concept of alcoholism is deeply rooted in Alcoholics Anonymous (AA), which from its founding in 1935 has demonstrated effectiveness in producing "recovering alcoholics." Posing alcoholism as a biologically based condition beyond the etiological responsibility of the alcoholic is vital to AA's regimen of recovery. That a disease could be effectively managed solely by lay intervention created what was for many years a tension, albeit one hardly visible. Effective lay treatment was, however, both a barrier to acceptance of alcoholism as a "disease like any other," for which sick-role assignment could be expected, and a direct contradiction of this imagery. The primacy of AA, coupled with the absence of societal resources directed toward dealing with alcohol-related problems, minimized professionals' design of alternative interventions and likely were barriers to research interest as well.

If it can be reasonably argued that AA has been primary custodian of the medicalized concept of alcoholism for the past 50 years, then it is important to examine aspects of the intermingling of the medicalized concept, AA, and alcoholism treatment. It is clear that within all three of these there is an emphasis upon the uniqueness of alcoholism. This uniqueness is underlined by AA's apparent hesitancy to transform the disease model of alcoholism to a disease model of addiction. It is further indicated by the extent to which AA uses biology either directly or indirectly to attribute basic "difference" to the alcoholic population, creating the circumstance in which, according to AA beliefs, this biological individuality precludes the alcoholic from being able to drink "normally." Finally, the uniqueness of alcoholism is demonstrated by AA's strong resistance to any notion that alcoholic behavior is secondary to a psychiatric disorder. This notion in turn contributes to a more subtle AA perspective that psychiatric symptoms apparent among recovering individuals are acute problems that will be principally resolved through conscientiously "working the program."

These attributions of alcoholic uniqueness are reflected in the strategies of intervention dominated by AA principles that have characterized the recent past, namely, a combination of abstinence with an adopted lifestyle of sobriety, which in turn represents lifestyle changes in the personal, interpersonal, and spiritual realms. This outcome may be achieved through a variety of means, but almost always includes some form of medical backup to assure that unanticipated physical problems do not threaten the client's health or otherwise hamper the course of recovery.

Heavy reliance upon AA for the treatment of alcoholism in the U.S. was typical until the early 1970s. Establishment of the National Institute on Alcohol Abuse and Alcoholism (NIAAA) in 1970 was the stage for a professionalized alcoholism movement, with large-scale resource investments made in three interdependent strategies to "mainstream" alcoholism treatment into the U.S. health care system: (1) mandated health insurance coverage of inpatient alcoholism treatment, (2) growth of inpatient treatment organizations, and (3) workplace-based intervention programs, which generated clients with insurance coverage for entry into alcoholism treatment (Roman & Blum, 1987).

NIAAA's three-pronged support for the medicalization of alcoholism led to the assignment of responsibility for treatment to "experts" embedded in bureaucratic systems that parallel other strategies of rational health care. In contrast to lay-based self-help groups, bureaucratic interventions can be accountable and can create an interface with work organizations and health insurance reimbursements. It was clear that despite the perceived effectiveness of AA, its organizational nature and principles precluded the accountability that is essential to fee-for-service health care, which is in turn the defining characteristic of the mainstream of the health care delivery system. Furthermore, if work organizations were to become major consumers of these services, it was essential for them to be able to form interorganizational linkages with parallel organizational structures, a requirement that obviously could not be met by the exclusive or heavy use of AA.

At the same time, psychotherapists began to move in large numbers into alcoholism treatment. This movement was spurred by the perception of significant new opportunities for professional therapists, created by the infusion of resources through the NIAAA supports. The entry of formally trained therapists was welcomed by those desiring alcoholism to be dealt with as "a disease like any other." While psychotherapy and its vaguely defined surrogate, "counseling," constitute the operative technology of many of the occupations engaged in direct human services, they had not been prominent in alcoholism treatment. From its inception, formal alcoholism treatment had been driven by the principles of recovery embodied in AA. These principles have historically included an implicit repudiation of psychotherapy as an effective intervention to deal with alcoholism, based upon the following "beliefs," which may or may not be accurate: (1) the belief that psychotherapists are generally ignorant of the "nature" of alcoholism, (2) the belief that psychotherapists commonly encourage the client to address problems believed by them to "underlie" alcoholism rather than deal directly with the client's drinking, (3) the belief that psychotherapists may encourage attempts at controlled drinking, and (4) the belief that individual psychotherapy does not offer the supportive group context believed to be vital for recovery to occur. Thus the initial wave of inpatient treatment centers did not emphasize psychotherapy but were heavily based around the principles of AA.

As more members of therapeutic and counseling occupations have been attracted to work in these centers, the use of therapy and counseling has increased. The success of the treatments has seen the costs of the standard 4 weeks of inpatient care rise to $10,000 and more, necessitating justification of the medical legitimacy of the regimen.

In the typical alcoholism treatment center established during the 1970s, the treatment setting is decidedly "low tech" in comparison with drug-based psychiatric treatment and especially with other medical care for which third-party reimbursements are made. Furthermore, the technology used is often so heavily based on AA that treatment centers, in order to be viewed as medically legitimate, must counter the image that their programs constitute little more than "residential AA."

Thus a claims-making problem for alcoholism treatment lies in its low degree of "mystification," which is vital to accepted medicalized interventions for which substantial payments for professional services are expected. This image problem may be compounded by the "spiritual" emphases of the AA recovery program, which find their way into most professionalized alcoholism treatment; it is indeed remarkable that at least to date "rationally" based third-party payments are forthcoming for regimens that are, in part, openly and explicitly "spiritual" in their design and content.

The institutional transformation of responsibility for alcoholism rehabilitation from AA to professional treatment is well described in a widely respected volume written by an alcoholism treatment researcher and therapist (Brown, 1986). Although this book appears intended as a comprehensive guide for the therapist desiring to learn more about treating alcoholism, for the social scientist it offers fascinating documentation of claims-making processes.

The anonymity of AA generally precludes well-controlled evaluation of its efficacy compared with that of other interventions, but AA bears enough anecdotal and face validity for most professionals to support its model. Baldly, many therapeutic professionals through the 1950s and 1960s supported AA as the primary modality for dealing with alcoholism because the professional community had little interest in becoming involved in the treatment of alcoholism. It was not a particularly reputable disorder, alcoholics tended to be stereotyped as low-status and socially disaffiliated, and alcoholism treatment was rarely supported by third-party reimbursement.

At the same time the use of the AA strategy within the AA structure kept alcoholism treatment within the realm of "folk medicine" and prevented the "mainstreaming" essential for the status enhancement of the alcoholism movement. One means of achieving this enhanced status would have been for the scientific and professional community to repudiate AA and replace it with a rationalized strategy. But, as mentioned, there was widespread acceptance of AA in the professional community, together with genuine belief that it "works." Further, as a "folk" strategy, AA has not followed the typical pattern of developing an image of social marginality; but through the shedding of anonymity on the part of many of its members, it has instead successfully "mainstreamed" itself into societal structures of class and stratification.

It should also be pointed out that in contrast to much folk medicine, AA is remarkably bureaucratic through its 12 Steps, 12 Traditions, and other organizational guidelines, together with a relatively complete and readily accessible printed literature. In many respects it is transferable technology par excellence. Although the "spirituality" dimension of the AA recovery program is indeed central, it appears a remarkably demystified form of spirituality, in many ways consistent with the other organized aspects of the Fellowship.

Therefore the challenge to the psychotherapeutic community is effective "colonization" of AA with a minimum of conflict. Brown's (1986) work is important to social scientists through its illustration of how psychotherapists are not only achieving this task but also securing broader claims on alcoholism treatment. The author first presents a detailed model of alcoholic recovery and then justifies the value of psychotherapy in the process. An effective political stage is set through the first chapters, which are in many respects paeans to AA. The author avoids challenges to the AA ideology and, indeed, regards its use as essential for alcoholic recovery.

She then, however, moves carefully through a rather slender data base that confirms for her the contention that psychotherapy is crucial in alcoholism treatment, especially after abstinence has begun. Again no challenge to AA ideology or principles is found. In a very subtle but crucial translation, AA's formal commitment to "cooperation" with the professional community becomes couched in the language of a "partnership" between AA and psychotherapists.

Thus, although not unique in its claims for the importance of psychotherapy in alcoholism treatment, Brown's (1986) work is a classic example of diplomacy in claims-making, effectively transmitting three messages simultaneously: (1) praising and accepting the value of the AA program, (2) supporting the "colonization" of AA within the alcoholism treatment enterprise and (3) avowing the critical role of psychotherapists in this process. By assigning AA's importance to the initiation and early stages of recovery but describing psychotherapy's importance in maintaining the healthy accompaniments of abstinence that AA calls sobriety, Brown accomplishes an anticipatory disarming of critics by distinctively (and perhaps dramatically) placing psychotherapy in the very front lines against relapses, the events that some in alcoholism treatment call their "number one enemy."

This enlisting of psychotherapy against relapses calls attention to a notable discontinuity between the predominant use of AA and the development of formalized alcoholism treatment. It is evident from numerous studies that (1) many treatment centers do not provide systematic or adequate follow-up with clients after they have completed treatment, and (2) systematic follow-up contributes very substantially to the treatment's long-term success. Discontinuity with earlier reliance upon AA is evident in the fact that the essence of AA is follow-up, with the individual member's ultimate achievement of the "sponsor" role as a long-term means of sustaining sobriety. By contrast, treatment regimens, lodged as they are within a medical model, have a determinate period of care for which billings are made, and around which organizational planning and routines are developed within a mode of "people processing."

It is possible and indeed an ideal for the treatment center to place the exiting patient successfully within an AA group or network, but AA does not function as a bureaucracy to which referrals can be made and from which accountability can be expected. Thus, treatment centers are faced with a long-term and indeterminate need for following clients to increase long-term effectiveness. Such an open-ended commitment does not fit with a typical organizational design, with the consequence that the quantity and quality of treatment centers' follow-up activities continues to be poor.

Brown's (1986) work may be seen as recasting follow-up into psychotherapy, offering a very different set of assumptions. If the alcoholic's psychiatric disorder requires care, then such care is not follow-up but another phase of the treatment process. The postalcoholism treatment phase thus becomes a new regimen rather than a continuing "booster" for the input that occurred during alcoholism treatment. The actual content of follow-up and psychotherapy may be identical, but a reconceptualization and relabeling offer a very different set of implications for organizational processes and for the assessment of effectiveness.

The transformation of alcoholism treatment is also supported by other research, which has been focused upon the psychiatric characteristics of alcoholics, with the ultimate goal of finding clues for improving treatment effectiveness. The focus of this research is *comorbidity,* a distinctively medical term describing the simultaneous presence of several disorders. Depression has been found to be very frequently present in recovering alcoholics and appears to be a precursor of severe alcohol problems as well. Co-morbidity creates clear requirements for psychotherapeutic expertise in alcoholism treatment and is a further contributor to the transformation of patterns of treatment delivery.

The infusion of psychotherapy together with emphases on co-morbidity may help solve organizational problems for alcoholism treatment centers by increasing mystification and enhancing medicalization. These aspects of psychiatric practice introduce new core technologies that do not have the "face simplicity" of AA and thus cannot be quickly or easily comprehended by outsiders. This distancing may aid the centers in their relationships with third-party payers who in the present day are looking in every corner for means to reduce costs and essentially minimize service use to bare essentials.

To date, it appears that alcoholism treatment centers have been especially vulnerable to external critiques of their effectiveness because of the simplicity of their technology. Adding psychiatric technologies that have well-established track records of mystification can reduce this vulnerability. Reluctance on the part of third-party payers can also be reduced by the introduction of familiar treatment modalities for psychiatric conditions that at the very least have a longer history of eligibility for health insurance coverage.

At the same time, co-morbidity can also be very helpful to centers' relationships with critical outsiders concerned with health care cost containment. Co-morbidity provides an "account" for the ineffectiveness of alcoholism treatment, showing that the patient's failure to recover fully was because of an additional, more insidious problem or because of the emergence of a new problem following the achievement of abstinence. This pattern of outcome and explanation takes the traditional measure of posttreatment abstinence as the primary criterion of alcoholism treatment effectiveness and throws it "into a cocked hat."

Although there may be considerable organizational advantages embedded in these changes, they are not without some distinctive costs. These outlined trends, which can be expected to continue as the psychotherapeutic and counseling occupations and professions continue to grow, clearly undermine the unique medicalized conception of alcohol-related problems and many of the implications of this conception. Such a trend may be seen as a parody of the early campaigns of the National Council on Alcoholism: "Alcoholism is a psychiatric disorder like any other." Indications that successful treatment of alcoholism is simply an extension of typical psychiatric care encourage a "cultural reclassification" of alcoholism. At the same time, the emphasis on the crucial importance of psychotherapy and the co-morbidity with psychiatric disorders undermines the apparent simplicity of the AA regimen of abstinence and sobriety as the keys to recovery. It must be recognized that this simplicity not only may facilitate entry into treatment but also supports social acceptance of the recovering alcoholic (Trice & Roman, 1970). Both of these processes may be substantially undermined by the additional "disablement" of the alcoholic produced by attributions of substantial co-morbidity.

A movement in response to these possibilities may be found in the attempt to establish and promote "addictionology" as a medical specialty to which claims over alcoholism and other addiction treatment would be ceded, and which would supersede psychiatry or psychotherapists in any claims they are attempting to make. Such a movement appears to be somewhat "after the fact" of psychotherapists' claims-making in alcoholism treatment. Furthermore, if addictionologists can "know" psychotherapy, why not the reverse?

Thus, ironically, some of the basic tenets of the alcoholism field, as well as some of its most cherished totems, may be lost in the successful "mainstreaming" of alcoholism into the health care system. To the extent that this mainstreaming was a planned goal of NIAAA and much of the rest of the alcoholism field in the early 1970s, it appears that there was a miscalculation of the acceptance of alcoholism's "uniquenesses" with the organizational complex of medical care. The initial growth of alcoholism treatment based in freestanding centers or in distinctively segregated units within larger medical care organizations supported the assumptions of uniqueness. But now it appears that organizational processes are moving alcohol problems into the psychiatric arena, a context where at earlier points in history effective attention to such problems was not impressive.

Complicating these trends are the previously discussed increases in polydrug abuse and the common treatment of alcohol and other drug addiction in the same settings. These seemingly more complex problems are clearly less responsive to traditional treatment regimens than primary alcoholism. The foreseeable future will likely hold substantial increases in societal resources directed toward treating substance abuse problems other than those primarily involving alcohol. As treatment centers expand their scope of coverage of substance abuse problems and reduce their rates of effectiveness, cultural optimism about effective treatment for one addiction may be affected by the rates of effectiveness in dealing with parallel or polyaddictions. This development may contribute further to the murkiness of the medicalized conception of alcohol-related problems.

The Introduction of Health Warning Labels

For the past 100 years U.S. public policy has demonstrated a tension between locating the source of alcohol-related problems in the substance, alcohol, versus locating the source in the drinker (Room, 1984). This tension is part of the overall theme of this chapter and is demonstrated in the contrasts between Prohibition, which may be viewed as conceptually locating the alcohol-related problem in alcohol, and medicalized public policy strategy, which locates the problem in the drinker.

We contend that the medicalized public policy is the major governing image of alcohol problems (Room, 1978) in U.S. culture in 1990. The introduction of federally mandated alcohol warning labels in late 1989 contrasts with this governing image in terms of both its implicit ideology and its distinctive character as a "technological shortcut" to social and behavioral change (Etzioni & Remp, 1973).

The warning label strategy assumes literacy on the part of the drinker and assumes as well that an important segment of drinkers purchase or consume legally produced alcohol directly from containers rather than use illegal alcohol or have alcohol served to them by others. The technology of attaching a warning label to beverage containers is, however, remarkably simple in comparison with the techniques embedded in the prohibition or medicalization strategies. Furthermore, one cannot simply classify the warning label as a technique that locates the alcohol-related problem in the substance alcohol, for the content of the label is a specification of three "risky" drinking situations (pregnancy, operation of a motor vehicle, operation of machinery) followed by a nonspecific (and last on the list) suggestion that "consumption of alcoholic beverages . . . may cause health problems." Nevertheless, the importance of these messages can easily be underestimated, especially given the universal presence of them on beverage containers as contrasted to the typical circumstances where individuals may or may not choose exposure to "data" about alcohol and alcohol-related problems diffused through other types of media.

Indeed, the presence of health warning labels on alcohol containers may be a necessary step in a transformation of cultural attitudes toward alcohol and alcohol-related problems that is viewed as desirable by some policymakers and legislators. It is tempting to suggest that cigarette warning labels have constituted such a necessary step in the eventual transformation of cultural attitudes toward smoking. At bottom, however, the significance of warning labels as a source of information about alcohol and alcohol-related problems lies in the remarkable extent to which the labels will be diffused and in the extent to which information about the labels will be diffused through the mass media and incorporated into education campaigns.

There is no doubt that the past 50 years have been dominated by public education efforts to diffuse the concept of alcoholism as an illness and to encourage problem drinkers and their significant others to seek expert help in resolving these problems. As described previously, data on the public acceptance of these concepts indicate widespread agreement with the medicalized concept of alcoholism, but these survey data also indicate a significant proportion of the population holding a mixture of attitudes wherein alcoholism is an illness in which the individual's "moral weakness" has played an etiological role (Blum et al., 1989; Caetano, 1987). Thus, key empirical questions are (1) how the introduction of the additional "data" contained in the alcohol warning label will affect public attitudes toward alcohol, alcohol-related problems, and alcoholism; (2) whether the presence of labels will increase the proportion of the population that regards alcohol as the primary source of alcohol-related problems instead of ascribing etiology to differences in the psychobiology of individual drinkers, and (3) whether the presence of labels will increase the proportion of the population that holds ambivalent attitudes toward the nature and etiology of alcohol problems. Following from these issues is how attitudinal changes will be associated with changes in behavioral predispositions, not only in regard to personal drinking-related behaviors but also in possible courses of action for dealing with deviant drinking among significant others.

In order for alcohol warning labels to exert their impact on the outcomes discussed above, individuals must gain some information from the labels. Although there is evidence that persons often are not aware of or do not read warning labels and that they frequently cannot comprehend or recall the information presented in them, an extensive research review commissioned by the federal government (Department of Health and Human Services, 1987) concluded that warning labels can exert a significant influence on consumer knowledge, attitudes, and behavior if designed properly.

Although the consequences of alcohol warning labels will only become known with the passage of time, a general hypothesis that may be offered is that warning labels will increase the perceived risk associated with alcohol consumption and that the consequences of consumption will be perceived negatively. If so, then persons who indulge in the consumption, or have problems related to the consumption, of this substance also may come to be viewed more negatively than they have been viewed in a social environment where warning labels were not present.

Further, social disruptions and impaired role performances resulting from consumption of alcohol may engender more adverse social reactions. If it can be assumed that the warning labels will increase perceptions of alcohol as a dangerous substance, then it is likely that there will be increased assignment of responsibility to persons for their drinking behavior and for the consequences of alcohol consumption. Both of these attitudinal changes may be associated with decreased tolerance of alcohol-related problems and of persons who experience these problems.

All of these factors may operate separately or additively both to lessen the readiness of persons perceived as problem drinkers to seek access to treatment opportunities and to increase the stigmatization of these persons. Other research has indicated that when victims of stigmatizing conditions are perceived as responsible for their condition, they are more likely to meet with social rejection than victims who are not considered responsible (Farina, Holland, & Ring, 1966; Weiner, 1981). Certainly, research has demonstrated that those who view alcoholics as responsible for their problems are less tolerant and compassionate than are those who do not ascribe responsibility to the alcoholic (Orcutt et al., 1979; Reis, 1977). Thus, if responsibility for alcohol-related problems becomes centered on individual choices, an effect of the introduction of alcohol warning labels may be an increase in the stigma associated with problem drinking and with alcoholism.

The change in attributions of responsibility for problems related to alcohol may undermine and confuse the substantial popular acceptance of the medicalized conception of such problems. This change may encourage moral conceptions of alcohol-related problems or the entry of a greater proportion of the public into the fairly common category of holding both medical and moral conceptions of these problems (Blum et al., 1989; Caetano, 1987).

Independent of those who are directly afflicted with alcohol-related problems are decision makers who function as gatekeepers to alcoholism treatment, that is, corporate human resources managers who design and recommend benefits packages. Labels may come to influence their thinking as well, creating financial barriers to the use of treatment that are extremely difficult to surmount. Thus, although its effect is easily underestimated, the diffuse presence of alcohol warning labels may come to be a major contributor to the murkiness of the medicalized conception of alcohol-related problems.

Probably most social and behavioral scientists expect few if any effects from the introduction of warning labels on alcoholic beverage containers. We agree about the short run but believe that exposure to these labels from generation to generation will not be benign. As persons view these labels on every alcoholic beverage container that they see, there will be a slow but steady growth in cultural support structures for the attitude that alcohol-related problems and alcoholism are the consequence of imprudence and irresponsibility.

Support for this prediction lies in part in the obvious "persistent presence" the labels will enjoy. Of equal importance is the fact that the labels' message refers to potential damage not only to the drinker but to others in the social environment around the drinker. Thus, for example, one's ideology might support the attitude that the creation of increased risk for chronic illness and premature death is the cigarette smoker's "own business." By contrast, the emphasis in the alcohol labels on impacts beyond the individual provide additional "points of entry" through the ideological and experiential barriers for making changes in individual attitudes toward the etiology of and attribution of responsibility for alcohol-related problems.

Support for the Medicalized Approach to Alcohol-Related Problems

The practical question of primary interest lies in what the future may hold. Three alternatives are possible:

One, the medicalized conception has reached its zenith. In an environment of cultural beliefs and organizational behaviors that are increasingly in conflict, the medicalized conception is now being transformed as a basic cultural conception toward definitions that increase the imputation of individual responsibility to those with alcohol-related problems while reducing societal compassion for them.

Two, the medicalized conception will continue to flourish as the dominant cultural conception. Its organizational bases and constituency strength will render impotent alternative conceptions or challenges to its internal consistency.

Three, the medicalized conception will remain a somewhat confused and confusing conception from an intellectual point of view but retain enough vigor to remain the center for organizing social responses to alcohol issues.

It is important that our observations be tempered by our own distinct projection that the medicalized conception of alcohol problems is here to stay, at least for the foreseeable future. But although we are not contending that this dominant approach to dealing with alcohol-related problems in the United States is about to disappear, the future will likely be considerably more contentious and the environment less supportive of intervention in such problems than it has been over the past two decades. Indeed, it is because of the persistence and tenacity of the medicalized approach to alcohol-related problems that we project continuing and growing conflict.

Although early in his career the first author contributed to sociological critique of the disease model of alcoholism (Roman & Trice, 1968), more recently he has written about the supports that are found for the medicalized concept of alcoholism within the context of U.S. culture (Roman, 1988a). There is still no doubt of the importance of those cultural supports and the "fit" that is found between the medicalized concept and basic aspects of U.S. culture.

Beyond the cultural supports is the sheer magnitude of organizational support structures and vested interests (Cahalan, 1988). As has also been described previously (Roman, 1988b; Roman & Blum, 1987), the organizational structures that have grown up around a generic medicalized approach to alcohol-related problems constitute a formidable system of support for the maintenance of this approach. These structures include not only the obvious example of for-profit alcoholism treatment centers, but also the organizational components that generate the flow of clients into these systems, such as employee assistance programs (EAPs) in workplaces and student assistance programs in educational institutions.

There are numerous other components of the organizational complex that supports the medicalized approach to alcoholism and alcohol-related problems. These include: local councils on alcoholism and their AA constituents who intensely support a disease model; the bureaucratic practices of third-party reimbursement for counseling and treatment which are fully couched in disease-model language and assumptions; the commitments and mandates of state and local substance abuse agencies that are typically divisions of larger health or mental health departments; and the organized interests of therapists, counselors, and EAP specialists who benefit strongly from an occupational identification with medical and health-related systems. The medicalized approach is the principal dimension of the "core technology" of the interventions around which all of these organizational processes revolve (Thompson, 1967). There can be little doubt that these constituent systems and processes will act to protect against environmental intrusions that suggest alcoholism or alcohol-related problems should be defined in other than a medicalized framework.

Beyond the organizational supports that have grown over the past 20 years, the continued centrality of the medicalized approach can also be discerned from another perspective, namely, consideration of suggested alternatives (Roman, 1980). It is evident that most critics of the medicalized approach do not clearly suggest alternative strategies or, if they do, present alternatives that are of dubious practical value (such as the legalization of all substances, with users held directly responsible for the consequences of their substance-use-related behaviors [Szasz, 1974]), or indeed offer strategies that can be barely distinguished from what is being criticized, such as Szasz's (1970) conception of "contractual psychiatry" as an alternative to the evils he perceived in "institutional psychiatry."

One illustration of such confusion is found in the current struggle to deal effectively with alcohol abuse and alcoholism in the Soviet Union, a cultural setting in which there are equally intense forces for the adoption and for the rejection of a disease model of alcoholism (Roman & Gebert, 1979). The Soviet interest has led most recently to initiation of cooperative ventures with U.S. alcoholism researchers and practitioners. A Soviet official is quoted as criticizing U.S. reliance on the disease model instead of recognizing that alcoholism is "primarily social in nature" (Holden, 1989). Subsequently the same official indicates that the greatest promise for dealing with alcoholism in the Soviet Union lies in the development of AA groups. Although one might be able to envision an adaptation of AA without the spirituality emphasis, it is difficult to conceive of the program without its basic disease model assumptions.

An example of an apparent but misleading alternative to the medicalized approach is the "addiction model" of alcohol-related problems discussed by Cahalan (1988) and others. This approach uses principles of behavioral modification as an alternative to counseling and is the model often associated with the goal of controlled drinking. Even with an abstinence goal, however, it is hard for the following reasons to distinguish this approach at its basic level from the medicalized conception:

1. Most of the contemporary strategies used to "motivate" the drinker into treatment are a form of behavioral modification, particularly when loss of job or family relationships is a primary lever.
2. The AA-type strategies prominent in both AA groups and formalized treatment are loaded with negative reinforcements through, for example, the content of "stories," the use of "chips" to reward and socially recognize sobriety, and the manner in which "slips" can be "setups" to "teach" recovering alcoholics that they cannot control their drinking.
3. Despite its offer as an alternative to medicalized conception, the addiction model is typically implemented under some form of quasi-medicalized conditions, albeit settings dominated by clinical psychologists. These professionals are rarely found dissociating themselves from organizational arrangements conducive to third-party reimbursements for health care services.

The distinctively clear alternative to the medicalized conception of alcohol-related problems is some form of criminalization, or recriminalization. Although portions of the public appear to desire retribution for some alcohol-associated events of injury, there is practically no evidence of a ground swell of public opinion for the abandonment of "humane," "enlightened," and "modern" strategies of intervention. Perhaps we are still much too close in history to Prohibition and the caricatures that are the basis for cultural memories of Prohibition.

Furthermore, if recriminalization were to be effective, receptivity on the part of the criminal justice system would be essential. Yet it is evident that the overcrowding of prisons and the perceived ineffectiveness of the criminal justice system in dealing with many different behavior patterns have been major forces supporting the medicalization of deviance in the first place (Conrad & Schneider, 1980). Thus there is little support for the possibility that the criminal justice system is either ready or eager to "take back" the problems of alcoholism and alcohol abuse.

Conclusion

There are additional sources of murkiness in the medicalized conception of alcohol-related problems that are not considered in detail here. These include:

1. An increasingly organized group of researchers and policy activists who identify themselves somewhat loosely with "public health" and who are primarily concerned with issues related to the advertising of, availability of, and distribution policies associated with alcoholic beverages (Mosher & Jernigan, 1989).
2. Social and legal changes that focus upon criminological definitions of drinking behavior associated with destructive outcomes, especially the definition of drunk driving (Jacobs, 1989), together with a low-key but persistent societal interest in the association between psychoactive substance use and the commission of crime.
3. A new self-help movement, Adult Children of Alcoholics, which has shown magnetic popularity reflected in extremely rapid growth. The ideology of this group affords members explanations of problems in adulthood on the basis of their having been raised in an alcoholic home. The group's principles characterize alcoholics as interpersonally destructive in the extreme, regardless of any recovery or attempts at repentance they might make or have made, the apparent assumption being that maintenance of this "blame" will sustain the mental health of the alcoholic's adult child.
4. Ongoing battles among individuals and groups working in the alcohol problem field, one example being repeated critiques that are in most instances sincere and well intended but that to the outsider may support the impression that treatment and other interventions in alcohol-related problems are minimally effective from both substantive and economic perspectives (Cahalan, 1987; Fingarette, 1988; Gordis, 1987; Peele, 1989).

These other forces are intermingled in many ways with those that are considered here. Collectively, all of these changes and forces represent the interplay between culture, "policy" in both the public and private sectors, and organizational dynamics, which deserve much more attention by social scientists.

What is the locus of the impact of these changes? From this discussion, it appears likely that individuals who develop alcohol-related problems may be the ones who suffer the most from the growing murkiness of the medicalized conception of alcohol-related problems. Despite our projection of continued dominance of the medicalized conception, we also predict that stigma will be increasingly associated with alcohol-related problems in both their active and recovery phases and that there will be a decline in the readiness of both those with alcohol-related problems and their significant others to utilize counseling and treatment.

With increasing confusion over both the cause of alcohol-related problems and individual responsibility for these problems—confusion created by a multiplicity of mixed messages—there may be increased ambivalence (in comparison with the present) among significant others, which will further slow taking action aimed at helping the deviant drinker. This ambivalence may be compounded by the imagery associated with the treatment enterprise as it broadens and becomes both more psychotherapeutic and more oriented toward drug treatment. These and other factors may interact to increase the stigma that is associated with alcohol-related problems—especially stigma that is placed upon individuals who have been through treatment, as that treatment comes to be less often viewed as effective and increasingly viewed as tinged with psychiatric labels.

Does the perspective of this chapter offer anything new? We hope so. When social scientists write about the medicalized concept of alcoholism, a critique of the logic of the medicalized definition of apparently voluntary choice behavior is typically expected (e.g., Fingarette, 1988). Such critiques are typically bound in a historical context that describes the dramatic success of claims making by those who have espoused this approach and subsequently constructed various forms of empires around these assumptions (Beauchamp, 1979; Wiener, 1981). From a teaching perspective, much of this work is often social science at its very best, especially when the discussion is broadly based in considering historical and organizational forces.

But, by the same token, this corpus of work is remarkably redundant and frustrating in the manner in which it more or less poses hegemony and even conspiracy, downplaying an appreciation of the flow of large-scale social and cultural forces that has occurred independently of individual or organizational plans within the alcoholism movement. We hope that in this chapter we have offered an alternative approach—one that avoids both direct and indirect attribution of "blame," instead describing the complexities that arise through the interplay between cultural definitions of problems and solutions and the organizational forms within which these definitions are implemented.

Acknowledgments

The authors acknowledge partial support during the preparation of this chapter from National Institute on Alcohol Abuse and Alcoholism grants RO1-AA-07218 and No. T32-AA-07473 to the University of Georgia, and grants RO1-AA-07250 and RO1-AA-07192 to Georgia Institute of Technology.

References

Beauchamp, D. (1979). *Beyond alcoholism.* Philadelphia: Temple University Press.

Blum, T.C., Roman, P.M., & Bennett, N. (1989). Public images of alcoholism: Data from a Georgia survey. *Journal of Studies on Alcohol, 50,* 5–14.

Brown, S. (1986). *The treatment of alcoholism.* New York: Wiley.

Cahalan, D. (1987). *Understanding America's drinking problem: How to combat the hazards of alcohol.* San Francisco: Jossey-Bass.

Cahalan, D. (1988). Implications of the disease concept of alcoholism. *Drugs and Society, 2,* 49–68.

Caetano, R. (1987). Public opinions about alcoholism and its treatment. *Journal of Studies on Alcohol, 48,* 153–160.

Conrad, P., & Schneider, J. (1980). *Deviance and medicalization: From badness to sickness.* St. Louis: C.S. Mosby.

Department of Health and Human Services. (1987). Review of the research literature on the effects of health warning labels: A report to the U.S. Congress (Contract No. ADM 281-86-0003). Washington, DC: National Institute on Alcohol Abuse and Alcoholism.

Etzioni, A., & Remp, R. (1973). *Technological shortcuts to social change.* New York: Russell Sage Foundation.

Farina, A., Holland, C.H., & Ring, K. (1966). The role of stigma and set in interpersonal interaction. *Journal of Abnormal Psychology, 71,* 471–478.

Fingarette, H. (1988). *Heavy drinking: The myth of alcoholism as a disease.* Berkeley: University of California Press.

Gordis, E. (1987). Accessible and affordable health care for alcoholism and related problems: Strategy for cost containment. *Journal of Studies on Alcohol, 48,* 579–585.

Holden, C. (1989). Soviets seek U.S. help in combating alcoholism. *Science, 246,* 878–879.

Institute of Medicine. (1989). *Prevention and treatment of alcohol problems: Research opportunities.* Washington, DC: National Academy Press.

Jacobs, J. (1989). *Drunk driving.* Chicago: University of Chicago Press.

Mosher, J.F., & Jernigan, D.H. (1989). New directions in alcohol policy. *Annual Review of Public Health, 10,* 245–279.

Mulford, H.A., & Miller, D.E. (1961). Public definitions of the alcoholic. *Quarterly Journal of Studies on Alcohol, 22,* 312–320.

Orcutt, J.D., Cairl, R.E., & Miller, E.T. (1980). Professional and public conceptions of alcoholism. *Journal of Studies on Alcohol, 41,* 652–660.

Peele, S. (1989). The diseasing of America. Lexington, MA: D.C. Heath.

Reis, J.K. (1977). Public acceptance of the disease concept of alcoholism. *Journal of Health and Social Behavior, 18,* 338–344.

Roman, P.M. (1978, March). *The recriminalization of alcoholism?* Paper presented at the meeting of the Southern Sociological Society, Knoxville, TN.

Roman, P.M. (1980). Alternatives to the medicalization of deviant behavior. *Psychiatry, 43,* 168–174.

Roman, P. (1988a). The disease concept of alcoholism: Sociocultural and organizational bases of support. *Drugs and Society, 2,* 5-32.

Roman, P. (1988b). Growth and transformation in workplace alcoholism programing. In M. Galanter (Ed.), *Recent developments in alcoholism (vol. 6, pp. 131–158). New York: Plenum Press.*

Roman, P., & Blum, T. (1987). Notes on the new epidemiology of alcoholism in the USA. *Journal of Drug Issues, 17,* 321–332.

Roman, P., & Gebert, P.J. (1979). Alcohol abuse in the U.S. and the U.S.S.R.: Divergence and convergence in policy and ideology. *Social Psychiatry, 14,* 207–216.

Roman, P., & Trice, H. (1968). The sick role, labeling theory and the deviant drinker. *International Journal Social Psychiatry, 14,* 114–136.

Room, R. (1978). *The governing images of alcohol problems.* Unpublished doctoral dissertation, University of California at Berkeley.

Room, R. (1984). Alcohol control and public health. *Annual Review of Public Health, 5,* 293–317.

Saxe, L., Dougherty, D., Esty, K., & Fine, M. (1983). *Health technology case study 22: The effectiveness and costs of alcoholism treatment* (Report No. OTA-HCS-22). Washington, DC: Office of Technology Assessment.

Schneider, J. (1978). Deviant drinking as disease: Alcoholism as a social accomplishment. *Social Problems, 25,* 361–372.

Szasz, T.S. (1970). *The manufacture of madness.* New York: Harper & Row.

Szasz, T.S. (1974). *Ceremonial chemistry.* Garden City, NY: Doubleday.

Thompson, J. (1967). *Organizations in action.* New York: McGraw-Hill.

Trice, H.M., & Roman, P.M. (1970). Delabeling, relabeling and Alcoholics Anonymous. *Social Problems, 17,* 538–546.

Wiener, C. (1981). *The Politics of Alcoholism.* New Brunswick, NJ: Transaction Books.

CHAPTER 40

The New Temperance Movement

David J. Pittman

For the third time in the 20th century the United States is in the midst of a war on drugs. The first occurred in the period bounded by the passage of the Harrison Act in 1914 and the repeal of Prohibition in 1933. The second is here defined to include the period from the appointment by President Johnson of his Commission on Law Enforcement and the Administration of Justice in 1965 to the withdrawal of U.S. troops from Vietnam in the early 1970s. Currently we are in a third war on drugs, marked by a new temperance movement and federal drug legislation, "The Anti-Drug Abuse Act of 1986," and "The Omnibus Drug Act of 1988."

To better understand the development of the more restrictive social and regulatory climate toward alcoholic beverages in U.S. society (which this chapter terms a new temperance movement), we propose to use the social constructionist theoretical model for the development of social problems. Kitsuse and Spector (1973, p. 415) have defined social problems as, "the activities of groups making assertions of grievances and claims with respect to some putative condition." The term *putative* is "intentionally, even ostentatiously, careful talk, allowing one to speak of something without commitment to its actuality" (Rains, 1975, p. 3). In this chapter, this term refers to alleged (it must be emphasized) alcohol-related deviance. We choose to use the term *putative conditions* (1) to avoid the implication that there is a normative consensus, (2) to avoid the value judgment that deviance is always bad and conformity is always good, and (3) to avoid certifying the truth or falsity of institutional agents' beliefs that rationalize their social control actions (Pittman & Staudenmeier, 1988).

Therefore within this framework of social problems construction the theory must account for "the emergence and maintenance of claim-

David J. Pittman is with the Department of Psychology, Washington University, St. Louis, Mo. This chapter was written especially for this book. An earlier version was presented at the American Sociological Association meetings (Issues in Social Movements: II), San Francisco, August 1989.

making and responding activities" (Kitsuse & Spector, 1973, p. 415)—in this case, the new temperance movement. Thus, the chapter will discuss (1) some major social factors related to the development of the new temperance movement and (2) the major goals of this new U.S. temperance movement.

The Sociocultural Context of the New Temperance Movement

In 1933, the United States repealed the constitutional amendment that prohibited the manufacture, distribution, and sale of alcoholic beverages, except for medicinal and religious purposes. But in the last decade U.S. society has witnessed an emergence of a new temperance movement, which has as its goal the reduction of the per capita consumption of alcoholic beverages and the incidence of alcohol-related damage in the population, as well as the prohibition of drinking in specific sex, age, and status groups, as well as in allegedly biologically vulnerable groups. Claim makers in the movement have asserted that a reduction in the apparent per capita consumption of alcoholic beverages in the United States will result in a reduced rate of alcohol-related problems in the society.

The sociological question is, what convergence of factors has been responsible for the reemergence of a new temperance movement in the latter part of the 20th century? First, it is crucial to remember that the United States, along with Canada, the United Kingdom, Iceland, Norway, and Sweden, are temperance cultures. These countries, basically Protestant in religious beliefs and historically characterized by drinking distilled spirits, were swept by strong temperance movements in the 19th and 20th centuries. In these countries alcoholic beverages were used as powerful symbols of the form that evil took and as an explanation for all of society's major social problems, from poverty to family discord. As Harry Levine (1988) has noted, these countries stand in strong contrast to those of the non-temperance cultures in the Mediterranean basin, where alcoholic beverages, viewed as food, are not used to explain social and family problems. Thus, U.S. society is one in which there is a strongly ambivalent attitude toward alcoholic beverages; abstinent sentiments as reflected in the beliefs of religious groups, such as Mormons, Christian Scientists, and many fundamentalist Protestants, coexist with the norms of heavy drinking and drunkenness in some macho-oriented youth groups. This U.S. ambivalence to drinking limits the development of stable attitudes to drinking that characterizes nontemperance cultures and insulates the drinking practices of some subgroups from social controls.

Ironically the new temperance movement has developed partially as a response to the disease concept of alcoholism. The scientific anchors for it resided originally in the Yale University Center of Alcohol Studies and in the dissemination of empirical studies through its *Quarterly Journal of Studies on Alcohol* (both founded in 1940). The disease concept of alcoholism, originally promulgated in the United States by Dr. Benjamin Rush in 1785 and others too numerous to mention, reemerged in the late 1930s. Instrumental in the public diffusion of this medical orientation to alcoholism were: (1) the founders of Alcoholics Anonymous, in 1935, one of whom was a physician, who conceptualized alcoholism partially as being an "allergy to alcohol"; (2) those, especially Marty Mann, credited with founding the National Council on Alcoholism, in 1945, which emphasized that "alcoholism was a treatable and beatable" disease; and (3) scientists such as E.M. Jellinek, whose research was published in 1960 in his book *The Disease Concept of Alcoholism,* which outlined the empirical basis for considering alcoholism a disease.

Thus, the social construction of alcohol-related problems through the period of 1930 until the mid-1970s was based on the disease model. Leaders of the various social movements to change the status of the alcoholic were cognizant that in Western society the victims of a disease are the primary responsibility of medical and social helping professions—not of police, courts, and correctional institutions.

However, by medicalizing the condition of alcoholism and emphasizing the control of alcoholism the leaders of the social movements did not address the question of alcohol-related problems leading to inappropriate behaviors in the course of drinking, such as drunken driving. The disease concept of alcoholism could not be stretched to include all these behaviors in its net. Furthermore, the disease concept seemed to divide the drinking population into two groups of people—those who were addicted to alcohol and therefore needed medical treatment and the others, who were nonproblematic drinkers. It was obvious that such a dichotomy was not only fallacious but simplistic in its formulation. Garland Allen (1987, p. 18), a biologist, has written:

> The same can be said of alcoholism: there is no way to define rigorously or independently. It can be given a clinical definition, but this is arbitrary and always dependent on social context: for example, setting, prevalence of drinking in the society at large, stress, etc. Alcoholism is clearly not defined by quantity of alcohol consumed, frequency of drinking or of getting drunk. It encompasses some notion of addiction or dependency, but that is obviously influenced strongly by environmental conditions such as stress or fatigue. Alcoholism *per se* cannot even be determined by measuring levels of alcohol dehydrogenase or other alcohol-metabolizing enzymes in the body, since such levels appear to bear no correlation to the behavior we loosely call alcoholism. It is a social phenotype whose very definition is meaningful only in a particular social context.

His position is one with which many scholars agree in that the facets of alcoholism are determined by the cultural context in which they occur.

Given the numerous critiques of the disease model of alcoholism and its inability to explain such disparate phenomenon as drunken driving, drunkenness and rowdy behavior among some young people, and so on, the competing explanation, the distribution-of-consumption model, developed by the French scholar S. Ledermann (1956), has gained many adherents throughout the last three decades. [See chapters 7 and 29, by Skog, in this volume.] Alcohol-related damage in a society was, according to this model, directly related to its per capita consumption of alcohol; this per capita consumption could be decreased by measures that reduced the availability of alcohol and increased its cost. Thus, a key strategy in the new temperance movement was to reemphasize alcoholic beverages as the major source of alcohol-related problems; and in this way the attitudes, beliefs, and customs of individuals and groups toward them were reduced to a secondary role. Numerous critiques of the scientific basis of the distribution-of-consumption model for explaining alcohol-related problems have appeared in the literature (Parker & Harman, 1978). Despite the model's shortcomings, it has focused the attention of social policymakers on such questions as the availability of the beverage, its price, and the marketing practices, including advertising by the producers of the product. In short, the new temperance movement has been enhanced by the disease model of alcoholism's inability to explain alcohol-related problems.

Since the end of World War II, the U.S. consumer movement has made significant strides in being a force to be reckoned with in the discussion of major policy issues. Simply stated, this movement's goal is to provide the population with a maximum degree of information concerning the products they purchase either for consumption or for use; this information includes, among other things, price, potential defects, ingredients, and the risks to the individual and society of using the product. Federal action to implement these goals is carried out through such agencies as the Federal Trade Commission, the Bureau of Alcohol, Tobacco and Firearms, the Environmental Protection Agency, the Food and Drug Agency, and so on, and various laws such as the "Truth in Lending Law," which requires the lender to disclose interest rates to the consumer, and the Pure Food and Drug Act of 1906 and subsequent amendments.

Patricia O'Gorman, former director of the Prevention Branch of the National Institute on Alcohol Abuse and Alcoholism (NIAAA), was among the first to realize the potential of the consumer movement for achieving certain goals of the temperance movement, such as health warning labels on containers of alcohol beverages. In a 1980 speech to the annual meeting of the National Council on Alcoholism she expressed disappointment with the alcoholism field's inability to achieve better diffusion of information about alcoholic beverages and their effect on health, especially for pregnant women.

O'Gorman's call for consumer group involvement in the field of alcohol policy was answered by the Center for Science in the Public Interest (CSPI), an offshoot of Ralph Nader's public interest group. In 1981 this group advocated tax increases on beer, wine, and liquor to discourage consumption of alcoholic beverages; in 1982, CSPI formed a new organization called Citizens Coalition on Alcohol Advertising; and in 1983 CSPI petitioned the Federal Trade Commission to review and tighten regulation of alcohol advertising. Of relevance was the group's pseudoscientific publication *The Booze Merchants, The Inebriating of America* (Jacobson, Atkins, & Hacker, 1983), which was a political tract. It was, however, a major factor in CSPI's launching its project "Stop Marketing Alcohol on Radio and Television" (Project SMART). Although this public action group was successful in its drive to have congressional hearings in 1985 on the role of alcohol advertising in creating and maintaining alcohol-related problems, no federal legislation has been enacted to ban the advertising of malt beverages (beer) and wine from the electronic media. Distilled spirits are not advertised, except in print media, by action of their industry originally and later by the code of the National Association of Broadcasters.

Moreover, the 1980s in the United States witnessed the increased involvement of other consumer and public interest groups in efforts to determine alcohol policy; the Consumers' Union became active in the California battle to label alcoholic beverages as reproductive toxins, the National Federation of Parents for Drug-Free Youth gained increased visibility by 1984, and the American Public Health Association became increasingly concerned with alcohol-related issues. Many other groups, too numerous to mention here, were mobilized by new temperance advocates to make alcoholic beverages less accessible and more expensive.

Another major impetus to the growth of the new temperance movement has been the emphasis upon physical fitness by vast segments, especially in the middle and upper classes, of the U.S. population. The health promotion movement, reinforced by constant coverage by the mass media, emphasizes that the key to longevity is found in a moderate lifestyle. This lifestyle involves either stopping smoking or never starting, reducing weight to appropriate limits, following a low-cholesterol diet, exercising regularly in a way that invigorates the cardiovascular system, and consuming alcohol in moderation, if at all. Despite the rash of studies that show a negative association between moderate drinking of alcoholic beverages and heart attack (Baum-Baicker, 1985), this relationship should be researched more thoroughly. Some advocates of the health promotion movement stress the avoidance of drugs, for they believe that these agents contaminate the "holy temple" of the body and that one therefore should not use nicotine, caffeine, alcohol, or illicit drugs, such as marijuana, cocaine, and the like. Thus, today there is a lifestyle, adopted by some Americans, that is alcohol-free; admittedly an even more numerous group advocates moderation in drinking. Both groups have minimal tolerance for either public or private displays of drunkenness. This negative sanctioning of inappropriate drinking and behavior while drinking is a positive change in U.S. values. Given its emphasis upon physical fitness, U.S. culture can be characterized as health-centered.

A major impetus for claim-making activity by groups in the new temperance movement focusing on the negative consequences of alcohol consumption was the concern about drunken driving. It was not that alcohol-related crashes were a new phenomenon in the United States. Earlier attention to this issue had resulted in the passage of the National Highway Safety Act in 1966, when the death rate was 5.7 people for every 100 million miles of motor vehicle travel. By 1982, when claim making in this area was at its peak, the death rate had dropped to 2.95, a reduction of 48% (Marshall, 1983, p. 5). In claim-making activity, the objective fact is most often of little importance to those making claims. Charismatic leadership, the social and political climate, and media responsiveness, to note only three factors, will frequently determine the fate of a claim. In the case of drunken driving, Candy Lightner, a charismatic personality whose daughter was killed by an intoxicated driver, was able to mobilize not only individuals whose children and other relatives had been killed or injured by drunken drivers, but also the mass media, including such popular publications as *Reader's Digest,* and political leaders in a crusade against drunken driving. She was the major figure in organizing individuals into the group that would later be called Mothers Against Drunk Drivers (MADD) (Reinarman, 1988). One of the major domestic issues that both federal and state legislators in the 1980s have confronted is what penalties to assign to those convicted for driving under the influence of alcohol. Literally hundreds of new laws have been enacted to make penalties more severe for intoxicated drivers, including mandatory jail sentences. Attitudinal shifts throughout the 1980s have moved drunken driving from a "tolerated evil" to a behavior that must be severely sanctioned.

This discussion has addressed some of the major sociocultural factors responsible for the reemergence of alcohol-related damage as a major social policy issue in the latter part of the 20th century. The following section addresses a number of the major goals of the new temperance movement.

Some Major Goals of the New Temperance Movement

In a previous work (Pittman, 1980) the author outlined the major goals of the new temperance movement. Briefly, the following action items were listed together with some additions (8–10):

1. Placing restrictions on the advertising of beverage alcohol, including the banning of advertisements in both print and electronic media.
2. Disallowing alcoholic beverage advertising as a business tax deduction.
3. Placing mandatory health warning labels on alcoholic beverage containers.
4. Raising the minimum age for the purchase of alcoholic beverages to 21.
5. Increasing taxes on beverage alcohol.
6. Levying earmarked taxes on beverage alcohol to help defray the cost of alcoholic rehabilitation programs.
7. Placing restrictions on the availability of alcoholic beverages:
   a. Restricting the number of commercial outlets (both on- and off-premise sales locations).
   b. Limiting the hours and days of sales.
   c. Limiting the number of sales outlets in lower-income neighborhoods.
8. Legislating mandatory increases in the cost of alcoholic beverages to match the increase in the Consumer Price Index.
9. Legislating mandatory public information messages by the alcohol beverage industry to point out the health and social damage of their products.
10. Requiring ingredient labeling, including alcohol content.

It is interesting to note a decade later that the agenda items with a physical health focus have been the ones to receive the greatest public support and to be enacted into law, that is, the policy that one must be 21 years old to purchase alcohol and the placement of health warning labels on containers of alcoholic beverages. Thus, the strategy of the new temperance movement in linking avoidance of alcohol to the physical and mental health of specific population groups has had major consequences; these are segmented prohibition for youth under age 21, women in childbearing years, and children of alcoholics, which are referred to in this chapter as age, gender, and status prohibition.

Age Prohibition

Before 1970, the minimum purchase age for alcoholic beverages was 21 in almost all U.S. states, despite the fact that surveys at that time and today, indicated that almost 70% of the high school seniors (age 17–19) had consumed alcoholic beverages in the last 30 days (Johnston, O'Malley & Bachman, 1988, p. 40). However, the Vietnam War, from 1965 to 1973, was a major catalyst for social change in U.S. society, especially attitudes toward youth by adults and attitudes of youth toward themselves and adults. Given the fact that the average age of the U.S. enlisted personnel serving in Vietnam and sometimes dying in combat was 19, many viewed an alcohol purchase age of 21 years as moral hypocrisy. Therefore, approximately half of the states lowered the purchase age to 18 or 19 to conform to the belief that if one were old enough to fight and die then one was old enough to purchase alcohol. With the end of the Vietnam War in 1973, the movement to lower the drinking age in the remaining states ground to a halt.

In the late 1970s a new social movement began in the United States to raise the age for legal purchasing back to 21 in those states that had lowered it to 18 or 19. The rationale was constructed on the following assumptions: (1) the use of alcoholic beverages by teenagers had reached alarming proportions and the nation was in danger of having a teenage alcoholism epidemic (Chauncey, 1980); (2) lowering the purchase age to 18, an age at which a number of teenagers are still in high school, created a situation in which these youth served as conduits to making alcoholic beverages available to even younger individuals; (3) raising the purchase age to 21 would aid the goal of reducing per capita consumption of alcohol; and (4) lowering the purchase age would result in significant decreases in highway accidents and fatalities and drunken driving arrests of those aged 16 to 21 years. It was, however, the last factor that had the greatest effect on the population's acceptance of a higher purchase age. New temperance movement advocates pointed out that such a policy would save thousands of teenagers from premature death in car crashes involving alcohol (Wagenaar, 1981). The enactment by the United States Congress in 1984 of a minimum purchase age of 21 under the threat of partial loss by the various states of their federal highway traffic safety funds was facilitated by the country's obsessive concern about drunken driving during this period. In 1988, the last state, Wyoming, to resist increasing its minimum purchase age enacted the 21-years-of-age law.

However, the age restriction on the purchase of alcoholic beverages and the message that any drinking by those under 21 years old is unhealthy and dangerous are part of 20th-century U.S. society's protective tendency towards its youth, as evidenced by such legislation as child labor, marriage, and compulsory school attendance laws. These restrictions, although aimed at protecting the health and well-being of our youth, have a concomitant effect of delaying adulthood.

Gender Prohibition

In the 1970s, U.S. researchers, the mass media, and the public in general rediscovered the fetal alcohol syndrome. That the consumption of alcoholic beverages could have an effect on conception and the health of the fetus was well known to the residents of the Greek city states in the ancient world. This knowledge appears to have remained part of the folk wisdom of many societies. Scientists have established that heavy consumption of alcohol during pregnancy places the fetus at severe risk of damage. However, the major scientific and social policy questions revolve around the issues of (1) whether there is a safe level of drinking during pregnancy, and (2) how women in the child-bearing years of 15–44 should be informed of the risk of fetal damage by excessive drinking. The answer to the first question is still the subject of scientific debate in that certain researchers maintain that even small doses of alcohol may have a deleterious effect on the fetus, whereas other reputable scientists maintain that there is no definitive scientific study which demonstrates that low doses of alcohol have a negative impact on fetal development (Rosett & Weiner, 1984). However, the major federal agencies and the medical profession have adopted policies advising women to abstain from alcohol during pregnancy. In 1981, the United States surgeon general as well as the NIAAA advised women not to drink during pregnancy; in 1982, the House of Delegates of the American Medical Association endorsed a similar position. Thus, the judgment of these official bodies has been that since a safe limit of alcohol ingestion is as yet undetermined, it is best to err on the conservative side and to recommend abstinence.

The second question, how women in the child-bearing years should be informed of the risk placed on the fetus by their drinking alcoholic beverages has been a contentious area of social policy debate for over a decade in U.S. society. Many have advocated that health warning labels should be placed on containers of alcoholic beverages to warn women of the risk. This is not a new idea in the United States. In 1945 the Legislative Special Commission to Study the Problems of Drunkenness in Massachusetts proposed the following cautionary label for alcoholic beverages:

> Directions for use: Use moderately and not on successive days. Eat well while drinking, and if necessary, supplement food by vitamin tablets while drinking. Warning: If this beverage is indulged in consistently and immoderately, it may cause intoxication (drunkenness), later neuralgia and paralysis (neuritis) and serious mental derangement such as delirium tremens and other curable and incurable mental diseases, as well as kidney and liver damage. (Haggard, 1945, p. 1)

This warning label was rejected by the legislature at that time. Thus, the rediscovery of the fetal alcohol syndrome first by scientists and later in 1975 by the mass media triggered the demand for health warning labels by various members of the U.S. congress (especially Senator Strom Thurmond [R-S.C.]), some alcoholism constituency groups, and members of the new temperance movement. In opposition to health warning labels were members of the alcoholic beverage industry and some medical and social science professionals who felt that the major problem with warning labels was that they were a cosmetic and simplistic approach to the problem of alcohol-related damage in general and, more specifically, the damage that excessive drinking may cause to the fetus. In 1980, the U.S. retailers, wholesalers, and producers of beverage alcohol, in response to a report by the Departments of Health and Human Services and the Treasury, mounted a major education campaign to inform not only women in the child-bearing years but their physicians of the dangers of excessive consumption of alcohol during pregnancy. But this educational campaign, not only by the alcoholic beverage industry but by both private and public organizations unaffiliated with them, did not satisfy those who advocated warning labels. Various cities, led by New York in 1983 and followed by others such as Philadelphia, Los Angeles, and Columbus, required retail establishments dispensing alcohol to display health warning posters. In 1987, California authorities, acting on the law required by their Proposition 65 initiative, labeled alcohol as a reproductive toxin; in 1988 warning posters were required throughout that state where alcoholic beverages were sold.

The health warning advocates moved to include more than fetal damage in their labels; they pressed the idea that health warnings should include references to alcohol's being a drug and to other health problems that could result from the ingestion of alcoholic beverages. More specifically, in 1988, the proposed Thurmond-Conyers law would have required producers of alcoholic beverages to use five rotating warning labels. These were:

Warning: The Surgeon General has determined that the consumption of this product, which contains alcohol, during pregnancy can cause mental retardation and other birth defects.

Warning: Drinking this product, which contains alcohol, impairs your ability to drive a car or operate heavy machinery.

Warning: This product contains alcohol and is particularly hazardous in combination with some drugs.

Warning: The consumption of this product, which contains alcohol, can increase the risk of developing hypertension, liver disease and cancer.

Warning: Alcohol is a drug which may be addictive.

The new temperance movement advocates strongly supported the proposal that warning labels should be rotated on containers of alcoholic beverages.

In the mid-1980s the alcoholic beverage industry began to be a target of product liability suits for its failure to inform consumers about the risks of addiction and health damage. Legal action was begun against Brown-Forman and G. Heileman Brewing Company in Illinois in 1986 for failure to warn that their products were addictive. Further suits involving alleged fetal damage by alcohol use were filed in Seattle, Washington. Because of its obsessive concern with the potential product liability awards, the alcohol beverage industry acquiesced to warning labels. Therefore, the Omnibus Drug Act, enacted by the U.S. Congress in 1988, requires that all containers of alcoholic beverages produced for domestic consumption carry the following warning labels as of November 19, 1989:

1. According to the Surgeon General, women should not drink alcoholic beverages during pregnancy because of the risk of birth defects.
2. Consumption of alcoholic beverages impairs your ability to drive a car or operate machinery, and may cause health problems.

The controversy over both warning labels and the best way to avoid fetal damage from excessive consumption of alcohol can be linked to broader concerns in the nation. In 1986, the U.S. Congress enacted the Emergency Planning and Community Right to Know Act, which related to reporting requirements involving the emission of certain noxious chemicals into the atmosphere and the stockpiling of these dangerous chemicals (*Insight,* 1988). The "right to know" not only the ingredients but the risks of various products is a major aspect of the consumer movement discussed earlier. Moreover, the fact that U.S. society is a health-centered one and that medical science has been unable to determine a safe level of drinking during pregnancy leads inevitably to the recommendation by some in the new temperance movement of gender prohibition, the idea being that women in the child-bearing years with the ability to conceive should abstain from alcoholic beverages because damage may occur to the fetus before the woman is aware she is pregnant.

Status Prohibition

Status prohibition as defined here is based on the empirical fact that certain groups and individuals are at higher risk for developing alcohol-related problems than others. Epidemiological studies have already identified groups in U.S. society whose rates of alcoholism and alcohol-related damage are higher than those expected by chance. Specifically, problems developing from alcohol misuse are higher in minority groups such as Native Americans (Indians, Eskimos and Aleuts), lower-income blacks, and gay men. Members of these groups are not only stigmatized but discriminated against in U.S. society, or their historic cultural systems, which could have partially protected them against alcohol misuse, were destroyed by the majority culture (as in the cases of American Indians and blacks) or never existed (as in the case of gay men). Conversely, cultural and religious groups that have abstinence as a primary tenet of their belief systems, for example, Mormons, Seventh-day Adventists, Moslems, many Pentecostal and fundamentalistic Christian religious groups, have few adherents with alcohol problems. Also, cultural and religious groups, such as the Chinese in their home country as well as abroad, the Orthodox Jews, the Italians, and the Greeks, living in societies in which drinking is well integrated into the religious, dietary, and related cultural patterns of the group, have minimal difficulties with alcohol misuse. Thus, the fundamental fact of whether alcohol misuse will occur in the group is centered not on whether drinking is proscribed but on how drinking behavior is integrated into the individual's own belief system about drinking, which is communicated to him or her through the family unit (the basic means by which cultural values are transmitted). From this perspective, we can better understand that specific subgroups in any complex society have a higher chance than expected to develop alcoholism and to have alcohol-related problems.

On the individual level, numerous studies have demonstrated that children of alcoholics are at greater risk for developing alcohol dependency problems; the NIAAA posits that "genetic reasons" are the major factor in this relationship rather than environmental ones (*Sixth Special Report,* 1987). Thus, the major question is why currently there is so much attention being paid to genetic factors in alcoholism. Surely such attention cannot be based completely on scientific findings. For example, the concordance rate for alcoholism derived with the use of standardized diagnostic criteria for monozygotic (identical) twins is not 100%. Kaij, (1960) in Sweden, found only a 70% concordance for alcohol abuse in his identical twins. There are critiques of the twins studies in Sweden and elsewhere by Murray, Clifford, and Gurling (1983). Or, as Goodwin (1985) has noted, approximately half of those hospitalized for alcoholism deny a family history of alcoholism. Therefore other factors, namely social, psychological, and cultural, must be relevant to the development of alcohol abuse. [See also chapter 30, by Goodwin, in this volume.]

None of the above is to deny that there may be biological differences in how individuals metabolize alcohol, including the reaction of various enzymes to the ethanol. In view of the fact that there are humans who cannot tolerate various substances, for example, penicillin, it is to be expected that some will be allergic to ethanol.

Biological and genetic constructions of alcoholism serve both the purpose of allowing individuals who become abusive users of alcohol to assign the responsibility for their situation to an inborn defect that is specific to alcohol and also allow these same individuals and others to mount vigorous campaigns to prevent this "noxious poison" from overwhelming the vulnerable populations, such as the children of alcoholics. Furthermore, it sets the stage for the biological scientists to hold out the promise that a diagnostic test may be developed to screen the population for potential alcoholics. For example, Ernest Noble (1984, p. 146), former director of the NIAAA, has written:

> Given that hereditary factors play a role in certain forms of alcoholism, it is not unduly optimistic to predict that accurate biological markers for some types of alcoholism and alcohol-related problems will become available in the near future.

And furthermore the nonprofit National Foundation for the Prevention of Chemical Dependency Disease has as one of its major goals the development of a biochemical test to determine whether the individual is predisposed to be chemically dependent.

Thus, the search for a biochemical test and for a genetic marker for alcoholism has assumed a major priority in research in this country. In the absence of such a discovery, then, the children of alcoholics, a high-risk group for developing alcohol-related problems, are admonished by new temperance advocates in the clinical profession to be cautious about their ingestion of alcohol. That these children have been, on the whole, scarred by their parent's (or parents') alcoholism is an undeniable fact. One of the major social movements in the last decade has been the organization in the United States of hundreds of support groups for adult children of alcoholics, in which they can learn to cope with dysfunctional traits that their family's alcoholism left on their personality; these same individuals become more aware of the effect of excessive alcohol consumption by their parent (or parents) on their own lives and in some cases choose to become abstainers themselves. [See also chapter 37, by Rudy, in this volume.]

Health Prohibition

Given the fact that U.S. society's concern about health is a primary attribute, the effect of alcohol on the physical well-being of individuals easily becomes a major focus of debates on social policy. Numerous studies have documented the relationship between excessive drinking and damage to the major organ systems of the body, ranging from the central nervous to the cardiovascular system. Therefore the question of what a safe level of drinking for the ordinary man or woman is becomes a crucial one to answer. The guidelines for individual drinking developed in Great Britain by the Health Education Council, the London College of Physicians, and the College of Psychiatrists suggest that 21 units for men and 14 for women a week are safe levels of consumption and that over 50 units for men and 35 for women a week are definitely hazardous levels of drinking (Kendall, 1987). Although these guidelines are approximations, at least they do recognize that there are sex differences and levels above which consumption becomes dangerous to the individual and society. Of course clinicians among others will object that these "safe levels of consumption" are inappropriate for recovering alcoholics or drug-dependent individuals, pregnant females, individuals who are taking medications with which alcohol is contraindicated, and those in certain occupations such as the fields of medicine, transportation, and athletics.

Of more recent importance to health issues is the 1987 Working Group's report of the International Agency for Research on Cancer (IARC), which examined and reviewed many studies that looked at the relationship between drinking alcohol and various types of human cancer. They concluded that "the occurrence of malignant tumors of the oral cavity, pharynx, larynx, oesophagus and liver is causally related to the consumption of alcoholic beverages" (World Health Organization, 1988, p. 259).

This interpretation of the scientific literature of a causal connection between drinking and various carcinomas will have a tremendous impact, given the U.S. population's fear of this disease. The scientific panel that advised former California governor, George Deukmejian on implementing that state's new law requiring the public to be warned of toxins and carcinogens in the environment declared that alcoholic beverages cause cancer. The remaining question the panel must address is, at what dose level is cancer triggered by alcohol in humans (*San Francisco Chronicle,* 1988)? Alcohol as a carcinogen becomes a powerful new weapon in the arsenal of the new temperance movement's emphasis upon health.

Potentially of major significance is the relationship of alcohol to the immune system and to the acquired immune deficiency syndrome (AIDS). The impact of various doses of alcohol on the immune system is currently a major area of investigation, but already questions have been posed concerning alcohol's role as a co-factor both in a person's becoming infected by the HIV virus and in the course of the illness. Of course there is no question that unsafe levels of drinking may impair an individual's judgment in a sexual encounter and thus lead to unsafe practices.

However, one should be skeptical about studies reporting that a sizable proportion of people with AIDS have used drugs, including alcohol, without there being a control group (Siegal, 1986). Given the hysterical response of the U.S. population, with notable exceptions, to AIDS, any implication that alcohol use plays any significant role in the transmission of the virus and subsequent infection of the individual will have a major impact in increasing the number of abstainers.

Conclusion

This chapter has postulated a number of social and cultural trends in U.S. society that have facilitated the rebirth of a new temperance movement in the latter part of the 20th century. The major goal of the movement to reduce the per capita consumption of alcoholic beverages by the drinking age population has been constructed on health risks associated with drinking alcohol. Therefore this temperance movement has been most successful in mobilizing health and consumer groups to its cause. In contrast to their support of the temperance movement of the first part of the 20th century, major religious groups have been notably silent on this issue.

References

Allen, G. (1987). The gene fix: The social origins of genetic determinism (Unpublished paper). St. Louis: Washington University, Department of Biology.

Baum-Baicker, C. (1985). The health benefits of moderate alcohol consumption: A review of the literature. *Drug and Alcohol Dependence, 15,* 207–227.

Chauncey, R. (1980). New careers for moral entrepreneurs. *Journal of Public Health Policy, 2,* 206–225.

Goodwin, D. (1985). Alcoholism and genetics: The sins of the fathers. *Archives of General Psychiatry, 42,* 171–174.

Haggard, H.W. (1945). The proposed Massachusetts "label" and its place in education against inebriety. *Quarterly Journal of Studies on Alcohol, 6,* 1–3.

Insight. (1988, May 23), p. 9.

Jacobson, M., Atkins, R., & Hacker, G. (1983). *The booze merchants: The inebriating of America.* Washington: CSPI Books.

Jellinek, E.M. (1960). *The disease concept of alcoholism.* New Brunswick, NJ: Rutgers Center of Alcohol Studies.

Johnston, L.D., O'Malley, O., & Bachman, J.G. (1988). *Illicit drug use, smoking, and drinking by America's high school students, college students, and young adults, 1975–1981* (p. 40). Rockville, MD: National Institute on Drug Abuse.

Kaij, L. (1960). *Alcoholism in twins.* Stockholm: Almquist and Wiksel.

Kendall, R.E. (1987). Drinking sensibly. *British Journal of Addiction, 82,* 1279–1288.

Kitsuse, J., & Spector, M. (1973). Toward a sociology of social problems: Social conditions, value judgments, and social problems. *Social Problems, 20,* 415.

Ledermann, S. (1956). *Alcool, alcoolisme, alcoolisation: Données scientifiques de caractère physiologique, économique et social.* Institut Nationale d'Études Démographiques, Travaux et Documents (Cah. No. 29). Paris: Presses Universitaires de France.

Levine, H. (1988, May). Overview of worksite prevention and intervention issues from a historical perspective. Paper presented at Research Conference on Alcohol and the Workplace, Jekyll Island, Ga.

Marshall, R.L. (1983). The systems approach to DWI reduction. In *Beverage alcohol in American society* (pp. 5–14). National Conference of State Liquor Administrators.

Murray, R.M., Clifford, C.A., & Gurling, H.M.A. (1983). Twin and adoption studies: How good is the evidence for a genetic role? In M. Galanter (Ed.), *Recent developments in alcoholism* (Vol. 1). New York: Plenum Press.

NIAAA. (1987). *Sixth special report to the U.S. Congress.* Rockville, MD: U.S. Department of Health and Human Services.

Noble, E. (1984). Prevention of alcohol abuse and alcoholism. In L. West (Ed.), *Alcoholism and Related Problems.* Englewood Cliffs, New Jersey: Prentice-Hall.

Parker, D.A., & Harman, M.S. (1978). The distribution of consumption model of prevention of alcohol problems: A critical assessment. *Journal of Studies on Alcohol, 39,* 377-399.

Pittman, D.J. (1980). *Primary prevention of alcohol abuse and alcoholism: An evaluation of the control of consumption policy.* St. Louis: Washington University, Social Science Institute.

Pittman, D.J., & Staudenmeier, W.J. (1988, August). Types and sources of social control: The case of alcohol and other drugs. Paper presented at the 35th International Congress on Alcoholism and Drug Dependence, Oslo, Norway.

Rains, P. (1975). Imputations of deviance: A retrospective essay on the labeling perspective. *Social Problems, 23,* 1–11.

Reinarman, C. (1988). The social construction of an alcohol problem. The case of Mothers Against Drunk Driving and social control in the 1980's. *Theory and Society, 17,* 91–120.

Rosett, H.L., & Weiner, L. (1984). *Alcohol and the fetus: A clinical perspective.* New York: Oxford University Press.

San Francisco Chronicle (1988, April 23), p. 1.

Siegal, L. (1986). AIDS: Relationship to alcohol and other drugs. *Journal of Substance Abuse Treatment, 3,* 271–274.

Wagenaar, A.C. (1981). Effects of an increase in the legal minimum drinking age. *Journal of Public Health Policy, 2,* 206–225.

World Health Organization. (1988). *Alcohol drinking* (Vol. 44). (IARC monographs on the evaluation of carcinogenic risks to humans). Geneva: World Health Organization.

Index

INDEX